# PEGylated Nanocarriers in Medicine and Pharmacy

Rakesh Kumar Tekade
Narendra Kumar Jain
Editors

# PEGylated Nanocarriers in Medicine and Pharmacy

Springer

*Editors*
Rakesh Kumar Tekade
National Institute of Pharmaceutical Education and Research (NIPER) Ahmedabad, An Institute of National Importance, Government of India, Department of Pharmaceuticals, Ministry of Chemicals and Fertilizers, Palaj, Opp. Air force station, Gandhinagar, Gujarat, India

Narendra Kumar Jain
Department of Pharmaceutical Sciences
Dr. Hari Singh Gour University
Sagar, Madhya Pradesh, India

ISBN 978-981-97-7315-2 ISBN 978-981-97-7316-9 (eBook)
https://doi.org/10.1007/978-981-97-7316-9



This Springer imprint is published by the registered company Springer Nature Singapore Pte Ltd.
The registered company address is: 152 Beach Road, #21-01/04 Gateway East, Singapore 189721, Singapore

# Preface

The editors welcome the readers into the fascinating world of PEGylation through this unique book, which embarks on a voyage via an inclusive landscape of historical perspectives, groundbreaking discoveries, and technologies in medicine and the healthcare sector. Notably, since the advent of this technology in 1970, it seemingly offers remarkable properties to therapeutic molecules such as proteins, peptides, drugs, drug carriers, etc., through the properties bestowed upon them. PEGylation technology undeniably emerged as the show's topper; the rest is history. This book answers all basic concepts and scientific queries, beginning with fundamental to contemporary advancements in the field.

With this book, readers will be able to understand how PEGylation technology can enhance safety as well as therapeutic potential of drugs. It will shed light on how this technology can transform difficult-to-deliver therapeutic moiety into an extraordinary one. PEGylation enhances the solubility of a drug and makes it more water-friendly. PEG layering grants immunity from biological degradation and lowers immune recognition. It precisely extends the biological half-life of an agent within the body, as if it offers a VIP tag to it by allowing it to stay longer in the physiological compartment. It is projected that the readers will be exceptionally amazed by the versatility of PEGylation technology as they progress to read the book page-by-page. From solubilization, drug delivery, and targeting avenues to diagnosis and theragnosis, it has an impactful resonance and echoes across all the fields of healthcare and medicine. The editors of this book invite readers to embark on an exciting journey to explore the infinite potential of PEGylation, a simple yet immensely versatile technology.

Ahmedabad, Gujarat, India — Rakesh Kumar Tekade
Sagar, Madhya Pradesh, India — Narendra Kumar Jain

# Contents

# Editors and Contributors

## About the Editors

**Rakesh Kumar Tekade,** currently working as an Associate Professor at NIPER-Ahmedabad, is an academic researcher with more than 15 years of teaching and research experience. Before joining NIPER-Ahmedabad, he served as a Faculty at the Indian Institute of Technology (IIT), Jammu and at International Medical University, Malaysia, Kuala Lumpur. His current research involves developing fit-to-purpose, translational, and novel targeted specific nano-biomaterials for biomedical applications. He has co-authored more than 350 publications (>16500 citations; h-index: 65). He has edited 10 reference books with Elsevier (as editor-in-chief), 05 invited editorial articles, and 11 patent applications to his credit. For the past four consecutive years, he has been listed as one of the top 2% of researchers in Pharmacy and Pharmacology by Stanford University, USA. He was a recipient of several acclaimed awards, including Commonwealth Fellowships (British Council; University of Central Lancashire, UK), National Doctoral Fellowship (AICTE-NDF), Irish Government Postdoctoral Award (IRISH-PDF), and CSIR Senior Research Fellowship (CSIR-SRF), India and Junior Research Fellowship, India (AICTE-JRF), Best Associate Professor Award 2023, etc. to name the few.

Dr. Tekade underwent two postdoctoral trainings from The University of Hawaii, USA, and The University of Texas Southwestern Medical Center, USA. So far, he has trained 05 postdocs in his lab, guided 16 Ph.D. (08 completed and 8 in progress), and supervised 75 master's and biomedical science dissertations. He served/serving as a Principal Investigator for various extramurally funded projects (worth more than 2 Crore) funded by DST, ICMR, MOSTI, etc. He is a recipient of the Pharma Innova Award by Troika Pharmaceuticals Ltd. for Best Research Mentor and the Best Thesis Award. Dr. Tekade is a grant reviewer for the FONDECYT funds (Chile), the National Science Centre (Poland), the French National Research Agency (ANR) (France), the Khwarizmi International Award (KIA) (Iran), and a referee for more than 30 reputed international journals. Dr. Tekade is an editor-in-chief of a book series entitled *Advances in Pharmaceutical Product Development and Research Series*. https://www.elsevier.com/catalog/pharma/pharmaceutical-science/drug-delivery/advances-in-pharmaceutical-product-development-and-research.

**Narendra Kumar Jain** is an internationally renowned academician and researcher and a senior professor in India who superannuated in June 2014 after rendering more than 40 years of dedicated and distinguished service as a teacher and 25 years as full professor. Starting his academic career in 1973, he worked as Lecturer in Pharmacy, Holkar College, Indore; Reader in Pharmaceutical Technology; M.S. University, Baroda; Professor and Principal, College of Pharmacy, Nasik; Professor and Head, Department of Pharmaceutical Sciences, Dr. H. S. Gour University, Sagar; Principal, College of Pharmacy, New Delhi; Visiting Professor, Dubai Pharmacy College, Dubai (UAE); and Head, Pharmacy Department, Delhi University. Subsequent to his superannuation, Prof. Jain rendered his services as Director, ISF College of Pharmacy, Moga, Punjab, India, and as Emeritus Fellow (U.G.C.) at Rajiv Gandhi Technological University, Bhopal, M.P., India.

Prof. Jain is the author of over two dozen celebrated books in pharmaceutical sciences including a series on controlled and novel drug delivery systems and has contributed over 60 chapters in national and international books. He has to his credit over 500 publications in reputed pharmaceutical journals and has supervised 55 Ph.D. and 142 M.Pharm. candidates. He has been a reviewer for several international and national research journals. Professor Jain has been the recipient of IDMA

Research Award in Industrial Pharmacy in 1990, 1997, and 2004; Motan Devi Dandiya Award for Best Publication in Pharmacy in 1999; M.L. Khorana Prize for Best Research Paper in Pharmaceutics and Biopharmaceutics published in IJPS in 2003; Prof. M.L. Khorana Memorial Prize for Best Paper in the field of Pharmaceutics and Biopharmaceutics published in IJPS during the year 2006; "Teacher of the Year 2003" award from the Association of Pharmaceutical Teachers of India; "Dr. (Mrs.) Manjushree Pal Memorial Award of the Association of Pharmaceutical Teachers of India for Best Pharmaceutical Scientist" in 2008; Schroff Memorial National Award for his outstanding contributions as an author, academician, and scientist by the Indian Hospital Pharmacists' Association in 2008; and Motan Devi Dandiya Biennial Prize in Pharmacy for Best Publication in 2010.

## Contributors

**Samah Hamed Almurisi** Department of Pharmaceutical Technology, School of Pharmacy, International Medical University, Bukit Jalil, Kuala Lumpur, Malaysia

**Sweta Acharya** National Institute of Pharmaceutical Education and Research (NIPER) Ahmedabad, An Institute of National Importance, Government of India, Department of Pharmaceuticals, Ministry of Chemicals and Fertilizers, Palaj, Opp. Air force station, Gandhinagar, Gujarat, India

**Tejas Girish Agnihotri** Department of Pharmaceutics, National Institute of Pharmaceutical Education and Research (NIPER)-Ahmedabad, Gandhinagar, Gujarat, India

**Natasha Akojwar** Department of Pharmaceutics, University Department of Pharmaceutical Sciences, Nagpur University Campus, Nagpur, MS, India

**Ala' Adnan Ali** Faculty of Pharmacy, Philadelphia University, Amman, Jordan

**Girish Meravanige Basavarajappa** Department of Biomedical Sciences, College of Medicine, King Faisal University, Al-Ahsa, Saudi Arabia

**Sahebrao Boraste** Department of Pharmaceutics, GES's Sir Dr. M. S. Gosavi College of Pharmaceutical Education and Research, Nashik, Maharashtra, India

**Kanak Chahar** Department of Pharmaceutics, ISF College of Pharmacy, Moga, Punjab, India

**Salome A. Chime** Department of Pharmaceutical Technology and Industrial Pharmacy, University of Nigeria, Nsukka, Nigeria

**Vaishnavi Chinkure** National Institute of Pharmaceutical Education and Research (NIPER) Ahmedabad, An Institute of National Importance, Government of India, Department of Pharmaceuticals, Ministry of Chemicals and Fertilizers, Palaj, Opp. Air force station, Gandhinagar, Gujarat, India

**Dwiptesha Dahake** Department of Pharmaceuticals, National Institute of Pharmaceutical Education and Research (NIPER), Ahmedabad, An Institute of National Importance, Government of India, Ministry of Chemicals and Fertilizers, Palaj, Opposite Air Force Station, Gandhinagar, Gujarat, India

**Pran Kishore Deb** Department of Pharmaceutical Sciences and Technology, Birla Institute of Technology (BIT), Ranchi, Jharkhand, India

**Vinod Gaikwad** Department of Pharmaceutics, National Institute of Pharmaceutical Education and Research (NIPER), Hajipur, Bihar, India

**Sanyam Gandhi** Takeda Pharmaceuticals, Boston, MA, USA

**Shyam Sudhakar Gomte** Department of Pharmaceutics, National Institute of Pharmaceutical Education and Research (NIPER)-Ahmedabad, Gandhinagar, Gujarat, India

**Sourajyoti Goswami** Department of Pharmacy, Indira Gandhi National Tribal University, Amarkantak, Anuppur, Madhya Pradesh, India

**Naveen Gupta** Patel Institute of Pharmacy, Madhyanchal Professional University, Bhopal, Madhya Pradesh, India

**Sajidul Hoque** Department of Pharmaceutics, National Institute of Pharmaceutical Education and Research (NIPER), Hajipur, Bihar, India

**Aakanchha Jain** Department of Pharmaceutics, National Institute of Pharmaceutical Education and Research (NIPER)-Ahmedabad, Gandhinagar, Gujarat, India

**Heet Jani** National Institute of Pharmaceutical Education and Research (NIPER) Ahmedabad, An Institute of National Importance, Government of India, Department of Pharmaceuticals, Ministry of Chemicals and Fertilizers, Palaj, Opp. Air force station, Gandhinagar, Gujarat, India

**Megha Joshi** PHC, Amarkantak, Amarkantak, Madhya Pradesh, India

**Ceyda Oksel Karakus** Department of Bioengineering, Izmir Institute of Technology, Izmir, Turkey

**Gagandeep Kaur** National Institute of Pharmaceutical Education and Research (NIPER) Ahmedabad, An Institute of National Importance, Government of India, Department of Pharmaceuticals, Ministry of Chemicals and Fertilizers, Palaj, Opp. Air force station, Gandhinagar, Gujarat, India

**Simranjit Kaur** National Institute of Pharmaceutical Education and Research (NIPER) Ahmedabad, An Institute of National Importance, Government of India, Department of Pharmaceuticals, Ministry of Chemicals and Fertilizers, Palaj, Opp. Air force station, Gandhinagar, Gujarat, India

**Satyam Khare** Department of Pharmaceutical Quality Assurance, ISF College of Pharmacy, Moga, Punjab, India

**Lakshmi Kumari** Department of Pharmaceutical Quality Assurance, ISF College of Pharmacy, Moga, Punjab, India

**Balak Das Kurmi** Department of Pharmaceutical Quality Assurance, ISF College of Pharmacy, Moga, Punjab, India

Department of Pharmaceutics, ISF College of Pharmacy, Moga, Punjab, India

**Niyati Lad** National Institute of Pharmaceutical Education and Research (NIPER) Ahmedabad, An Institute of National Importance, Government of India, Department of Pharmaceuticals, Ministry of Chemicals and Fertilizers, Palaj, Opp. Air force station, Gandhinagar, Gujarat, India

**Thiagarajan Madheswaran** Department of Pharmaceutical Technology, School of Pharmacy, International Medical University, Bukit Jalil, Kuala Lumpur, Malaysia

**Randa S. H. Mansour** Faculty of Pharmacy, Philadelphia University, Amman, Jordan

**Seyedeh Zohreh Mirjalili** Faculty of Pharmacy, Drug and Food Control Department, Tehran University of Medical Sciences, Tehran, Iran

**Ankit Mishra** Faculty of Pharmacy, VNS Group of Institutions, Neelbad, Bhopal, MP, India

VNS Group of Institutions, Bhopal, Madhya Pradesh, India

**Lopamudra Mishra** Department of Pharmaceutical Quality Assurance, ISF College of Pharmacy, Moga, Punjab, India

**Mahima Mishra** National Institute of Pharmaceutical Education and Research (NIPER) Ahmedabad, An Institute of National Importance, Government of India, Department of Pharmaceuticals, Ministry of Chemicals and Fertilizers, Palaj, Opp. Air force station, Gandhinagar, Gujarat, India

**Pranali Mishra** Faculty of Pharmacy, VNS Group of Institutions, Neelbad, Bhopal, MP, India

**Sehasree Mohanta** School of Biological Sciences, Indian Association for the Cultivation of Science, Kolkata, West Bengal, India

**Mumuni A. Momoh** Department of Pharmaceutics, University of Nigeria, Nsukka, Nigeria

**Shubham Ramdas Mule** National Institute of Pharmaceutical Education and Research (NIPER) Ahmedabad, An Institute of National Importance, Government of India, Department of Pharmaceuticals, Ministry of Chemicals and Fertilizers, Palaj, Opp. Air force station, Gandhinagar, Gujarat, India

**Anuja Muley** Department of Pharmaceuticals, National Institute of Pharmaceutical Education and Research (NIPER), Ahmedabad, An Institute of National Importance, Government of India, Ministry of Chemicals and Fertilizers, Palaj, Opposite Air Force Station, Gandhinagar, Gujarat, India

**Sreeharsha Nagaraja** Department of Pharmaceutics, Vidya Siri College of Pharmacy, Bengaluru, Karnataka, India

Department of Pharmaceutical Sciences, College of Clinical Pharmacy, King Faisal University, Al-Ahsa, Saudi Arabia

**Prasanthi Sri Nagindera Rao** Department of Pharmaceutical Technology, School of Pharmacy, International Medical University, Bukit Jalil, Kuala Lumpur, Malaysia

**Aniket Navale** National Institute of Pharmaceutical Education and Research (NIPER) Ahmedabad, An Institute of National Importance, Government of India, Department of Pharmaceuticals, Ministry of Chemicals and Fertilizers, Palaj, Opp. Air force station, Gandhinagar, Gujarat, India

**Piyush Neware** Department of Pharmaceutics, National Institute of Pharmaceutical Education and Research (NIPER), Hajipur, Bihar, India

**Shivam Otavi** National Institute of Pharmaceutical Education and Research (NIPER) Ahmedabad, An Institute of National Importance, Government of India, Department of Pharmaceuticals, Ministry of Chemicals and Fertilizers, Gandhinagar, Gujarat, India

**Preeti Patel** Department of Pharmaceutical Chemistry, ISF College of Pharmacy, Moga, Punjab, India

**Vasu Peddinti** Department of Pharmaceutics, National Institute of Pharmaceutical Education and Research (NIPER)-Ahmedabad, Gandhinagar, Gujarat, India

**Prashant Pingale** Department of Pharmaceutics, GES's Sir Dr. M.S. Gosavi College of Pharmaceutical Education and Research, Nashik, Maharashtra, India

**Aprameya Prasad** Drug Product Development and Delivery, Therapeutics Development and Supply, Janssen Research and Development, Malvern, PA, USA

**Aprameya Ganesh Prasad** Department of Chemical and Biomolecular Engineering, Johns Hopkins University, Baltimore, MD, USA

**Amarjitsing Rajput** Department of Pharmaceutics, Bharati Vidyapeeth Deemed to be University, Poona College of Pharmacy, Pune, Maharashtra, India

**V. Ravichandiran** National Institute of Pharmaceutical Education and Research (NIPER), Hajipur, Bihar, India

**Biswajit Rout** Department of Pharmaceutics, National Institute of Pharmaceutical Education and Research (NIPER)-Ahmedabad, Gandhinagar, Gujarat, India

**Maliheh Safavi** Department of Biotechnology, Iranian Research Organization for Science and Technology, Tehran, Iran

**Sombir Saharan** Department of Pharmaceutics, National Institute of Pharmaceutical Education and Research (NIPER), Hajipur, Bihar, India

**Rakesh Sahu** Sanjivani Institute of Pharmacy, Ganiyari, Bilaspur, Chhattisgarh, India

**Sweety Shah** National Institute of Pharmaceutical Education and Research (NIPER) Ahmedabad, An Institute of National Importance, Government of India, Department of Pharmaceuticals, Ministry of Chemicals and Fertilizers, Palaj, Opp. Air force station, Gandhinagar, Gujarat, India

**Raheleh Shakeri** Faculty of Science, Department of Biological Science, University of Kurdistan, Sanandaj, Iran

**Yash Sharma** Department of Pharmaceutics, ISF College of Pharmacy, Moga, Punjab, India

**Dilpreet Singh** University Institute of Pharma Sciences, Chandigarh University, Gharuan, Mohali, India

**Pranay Soni** Department of Pharmacy, Indira Gandhi National Tribal University, Amarkantak, Anuppur, Madhya Pradesh, India

**Muktika Tekade** School of Pharmacy, Devi Ahilya Vishwavidyalaya, Takshila Campus, Indore, Madhya Pradesh, India

School of Pharmaceutical Sciences, Lovely Professional University, Phagwara, Punjab, India

**Rakesh Kumar Tekade** National Institute of Pharmaceutical Education and Research (NIPER) Ahmedabad, An Institute of National Importance, Government of India, Department of Pharmaceuticals, Ministry of Chemicals and Fertilizers, Palaj, Opp. Air force station, Gandhinagar, Gujarat, India

**Akhilesh Tiwari** Department of Pharmacy, Indira Gandhi National Tribal University, Amarkantak, Anuppur, Madhya Pradesh, India

**Mansi Upadhyay** Department of Pharmaceuticals, National Institute of Pharmaceutical Education and Research (NIPER), Ahmedabad, An Institute of National Importance, Government of India, Ministry of Chemicals and Fertilizers, Palaj, Opposite Air Force Station, Gandhinagar, Gujarat, India

**Sakshi Wani** Department of Pharmaceutics, GES's Sir Dr. M.S. Gosavi College of Pharmaceutical Education and Research, Nashik, Maharashtra, India

# PEGylated Pharmaceutical Nanocarriers

1

Prashant Pingale, Sakshi Wani, Sahebrao Boraste, and Amarjitsing Rajput

**Abstract**

Bioactive compounds benefit significantly from PEGylation in both pharmacological and biological applications. The most popular technique for giving drug nanocarriers stealth characteristics is now "PEGylation." Target-site drug delivery systems are becoming more popular in the pharmaceutical industry because of their many benefits, including greater bioavailability and increased drug dosage capacity. However, specific current issues must be resolved. The interplay between drug delivery systems and blood proteins serves as one example. Based on alterations to the polyethylene glycol surface, this problem has a strong candidate. The most well-known and complex technique to create drug delivery systems with an extended blood circulation period is the surface coating of different medicinal nanocarriers with polyethylene glycol. For drug-loaded nanocarriers to effectively accumulate in target organs or tissues, prolonged circulation is often necessary. Targeted distribution is made possible by polyethylene glycol's protection against potential exterior interactions with other chemicals. These innovative methods can help cure diseases like cancer, diabetes, hemophilia, and pain. This research examines key polyethylene glycol and other polymer characteristics that can be leveraged to create long-circulating nanocarriers. This review also includes descriptions of PEGylated liposomes, metal nanoparticles, poly-

P. Pingale · S. Wani · S. Boraste
Department of Pharmaceutics, GES's Sir Dr. M. S. Gosavi College of Pharmaceutical Education and Research, Nashik, Maharashtra, India

A. Rajput (✉)
Department of Pharmaceutics, Bharati Vidyapeeth Deemed to be University, Poona College of Pharmacy, Pune, Maharashtra, India
e-mail: amarjit.rajput@bharatividyapeeth.edu


R. K. Tekade, N. K. Jain (eds.), *PEGylated Nanocarriers in Medicine and Pharmacy*, https://doi.org/10.1007/978-981-97-7316-9_1

meric nanoparticles, and other drug delivery systems. This strategy is expected to offer more viable and valuable applications soon.

**Keywords**

PEGylation · PEGylated nanocarriers · Drug delivery · PEG products · PEG density · Targeting · Toxicity

## Abbreviations

| | |
|---|---|
| BCS | Biopharmaceutical classification system |
| CAC | Critical aggregation concentration |
| CuNPs | Copper nanoparticles |
| DA | Diabetes |
| DEX | Dexamethasone |
| DNR | Daunorubicin |
| DOX | Doxorubicin |
| DSPE | 1,2-distearoyl-sn-glycero-3-phospho-ethanolamine |
| EPR | Enhanced permeation and retention |
| GA | Gambogic acid |
| GDH | Glutamate dehydrogenase |
| LSPR | Localized surface plasmon resonance |
| mPEG | Mono-methoxylated PEG |
| MPS | Mononuclear phagocyte system |
| NLCs | Nanostructured lipid carriers |
| NLS | Localization signal peptide |
| NMR | Nuclear magnetic resonance |
| PEG | Polyethylene glycol |
| PEI | Polyethyleneimine |
| PLA | Polylactic lactic acid |
| PLGA | Poly (lactic-co-glycolic acid) |
| RGD | Arginine-glycine-aspartic acid peptide |
| SARs | Structural activity relationships |
| SCID | Severe combined immunodeficiency disease |
| SLNs | Solid lipid nanoparticles |
| SMEDDS | Self-micro-emulsifying drug delivery system |
| SPIO | Superparamagnetic iron oxide |
| $T_f$ | Transferrin |
| TfR | Transferrin receptor |
| TGA | Thermogravimetric analysis |
| $TiO_2$ | Titanium dioxide |
| TMZ | Demozolomide |
| VEGF | Pegylated anti-vascular endothelial growth factor |
| XPS | X-ray photoelectron spectroscopy |

## 1.1 PEGylation

A peptide, protein, or nonpeptide monomer is modified when one or more polyethylene glycol (PEG) chains are joined. Davies and Abuchowsky initially discussed PEGylation in two critical articles on albumin and catalase modification in the 1970s. This was a significant turning point since, at the time, it was not thought possible to significantly alter an enzyme while maintaining its activity. Since proteins were considered susceptible, only a few mild changes with low-molecular-weight products were made, mainly to study structural activity relationships (SARs). Since then, the initially stated PEGylation process has dramatically grown and improved, and various chemical and enzymatic conjugation methods are now available. The range of modification techniques provides the possibility of addressing the needs of various proteins. Only the desired amino acids in the sequence can be modified using a more reactive PEG. The original PEG targets were amino groups via acylation or alkylation reactions. Still, PEG can also be conjugated to thiol, hydroxyl, or amide groups utilizing a variety of specialized chemical or enzymatic techniques. This polymer is safe, immunogenic, and antigenic and has FDA approval. It is also highly soluble in water. The PEG-drug conjugates have many benefits, including an extended residence in the body, a reduced rate of metabolic enzyme breakdown, and a decreased or eliminated level of protein immunogenicity (Thakur et al. 2015).

Monomeric ethylene oxide undergoes ring-opening polymerization and produces PEG. Water may begin the polymerization reaction, producing chains of bifunctional polymers, or methanol can produce monofunctional PEG. When connecting two proteins, monofunctional PEG is usually used to avoid utilizing a similar polymer chain again. Mono-methoxylated PEG (mPEG), generated through methanol initiation, is widely used as the starting material for PEGylating proteins. However, some dihydroxyl PEG (a bifunctional polymer) is usually present, which results in undesirable crosslinked products. To avoid the cross-linking reaction, a method was created by Roberts et al. to change diols into completely methylated inert PEG (Zhao et al. 2019). For protein conjugation, the free hydroxyl end group on mPEG is functionalized (Rondon et al. 2021).

### 1.1.1 PEG Characteristics

The hydrophilic oligomers or polymers known as polyethylene glycols are created from ethylene oxide and include the repeating unit (O-CH2-CH2). PEGs are made with many different molecular weights. They can be "polydispersed" polymers, which are more common, or "monodispersed" polymers, which have a Gaussian distribution of chain lengths and molecular weights. The advantages of PEG polymers have been substantially increased by the capability to add a range of reactive functional groups to their terminal locations. Bifunctional PEGs are best suited for cross-linking agents or spacers between two chemical entities. In contrast, hetero- and homo-bifunctional PEG derivatives are particularly suited for single-point attachment to targets due to their sterically bulky structure (Nascimento et al. 2021).

PEG is well known for having a very flexible structure, biocompatible, amphiphilic, no steric hindrance, and high hydration capacity. PEGs with MW >2000 are hard crystalline solids with melting temperatures around 63 °C, while those with MW between 100 and 700 are liquids at ambient temperature. PEGs with a molecular mass of 1000–2000 are soft solids. PEG stands out among structural polymers with a similar structure due to its high polarity, which encourages hydrophilicity and enhances water solubility. PEG is highly soluble in both organic and inorganic solvents. The contact angle is a significant determinant of hydrophilicity. PEG-coated glass and gold plates can both lower the contact angle by 34° and 50°, respectively. According to research by Moore and colleagues, PEG is hydrophilic and includes functional terminals that are electrically neutral and active at all pH levels, making them physiologically inert (Walkowiak-Kulikowska et al. 2020).

This imparts an in vivo anti-biofouling effect on biomedical and bone scaffolds, whereas, in nanoparticles, it enhances medication targeting and bioavailability in addition to anti-biofouling. The glass transition temperature and melting point can be altered by copolymerizing or mixing with the PEG of various MWs. Its mechanical qualities are based on its plasticizing process. PEGs with low melting points are essential in hot melt extruded admixtures. Different molecular engineering approaches are feasible because PEGs have limited chemical and radiochemical stability. PEGs are the ideal excipient for liquid dosage forms since they prevent fungus growth and do not develop bad odors. For cutaneous applications, ingredients with good occlusive qualities are chosen (Sánchez-Cid et al. 2022).

### 1.1.2 PEG Metabolism in the Body

Early research on PEG's aerobic degradation suggested that the process must involve oxidation. After PEG is first oxidized to an aldehyde and a monocarboxylic acid, the ether bond is broken, shortening the PEG molecule by one glycol unit. It is also feasible for the molecule's two terminal alcohol groups to oxidize simultaneously. A similar reaction with the monocarboxylic acid might be used to depolymerize. When this process is repeated, depolymerized PEG is the result. An alcohol dehydrogenase/oxidase is an enzyme that changes terminal alcohol groups into carboxylic acid groups and splits ether bonds to produce glyoxylate. PEG-glycolic acid is the best molecule to employ because it is a metabolite of PEG. Because monoalkyl PEG was used but not dialkyl PEG, it is possible that a terminal alcohol group is a component that causes PEG to break down exogenously. PEG degradation of a polymer molecule produces depolymerized chemicals, while exogenous breakdown cannot. PPG and PTMG are both harmful to developing cultures. As a result, exogenous biodegradation begins with a terminal group and solely depends on the chemical makeup of the monomer units (Rondon et al. 2021).

Pearce and Heydeman suggested a nonoxidative elimination of PEG units as acetaldehyde utilizing the membrane-bound, new oxygen-sensitive enzyme diPEGlyase. In contrast, Haines and Alexander suggested that an enzyme may digest PEG 20,000 to create oligomers as metabolic products. Only cyanocobalamin and adenosylcobalamin accelerated the process among the cofactors examined, albeit the effect differed between preparations. These authors evaluated the incubation mixture's enzyme activity using vapor-phase chromatography. However, measuring chemicals when unknown metabolizable materials are present is dangerous. Schöberl was the first to propose that a C1 step catabolizes PEG, releasing formate, which is metabolized by a serine pathway. He then changed his mind about the C1 theory, noting that dimer–tetramer interactions produced dicarboxy products, showing that an ether bond was broken to release glycolic acid (Kawai 2005; Simone 2008).

### 1.1.3 Pharmaceutical Uses of PEG

Pharmaceutical formulations for parenteral, topical, ocular, oral, and rectal delivery use PEGs, which have also been explored in controlled-release devices with biodegradable polymeric matrices. Polyethylene glycol mixtures have various advantages over lipids when employed as suppository bases. For instance, the suppository's melting point can be increased to endure exposure to warmer regions. Furthermore, the physical stability of the suppository during storage is improved, the drug's release is not dependent on the suppository's melting point, and suppositories are readily miscible with rectal fluids. It was hypothesized and reported that the dimer tetramer yielded dicarboxy products, indicating that an ether bond was broken to release glycolic acid. The amount of release of drugs that are water soluble decreases as the molecular weight of polyethylene glycol increases. Mucous membranes are often more irritated by polyethylene glycols than by lipids (Ciolacu et al. 2020).

Polyethylene glycols can function as emulsion stabilizers when combined with other emulsifiers. Soft gelatin capsules' contents are dissolved in liquid polyethylene glycols that are water soluble. They could, however, cause the gelatin in the capsule shell to harden by absorbing moisture from it. PEG 300 and PEG 400 facilitate parenteral dosage forms in up to 30% v/v concentrations. If agitates the material while it cools, it turns paste-like and crystallizes into granules. The preparation of dosage forms like lozenges can be done using this technique when delayed disintegration is necessary (D'souza and Shegokar 2016). Polyethylene glycols having a molecular weight of 6000 can be used to improve permeability in enteric-coated tablets. Polyethylene glycols are used as plasticizers to prevent from rupturing when compressed into tablets. PEGs can improve the water solubility or dissolving properties of poorly soluble substances by creating solid dispersions with the right polyethylene glycol. As steroid osmotic pump solvents, polyethylene glycols have also been used in animal investigations (Saikh 2021).

#### 1.1.3.1 Passive Targeting Agent

As a result of the medication's high concentration in the tumor area, the passive targeted drug delivery strategy mainly relies on the gradient in concentration between the intracellular and extracellular spaces. Through leaky vasculature and inadequate lymphatic drainage, PEG conjugates benefit from the enhanced permeation and retention (EPR) effect carried out by tumors and accumulate in the pathophysiological milieu of tumor arteries. The size-dependent effect cannot be explored with low molecular weight medicines that freely extravasate and cause systemic toxicity. PEGylation enhances the medications' solubility, molecular mass, size, and serum stability. PEGylation is regarded as one of the most excellent techniques for passive targeting of anti-cancer treatments for all these reasons (Yadav and Dewangan 2021).

#### 1.1.3.2 Solubility Enhancers

Due to their low water solubility, biopharmaceutical classification system (BCS) classes 2 and 4 have minimal bioavailability. Physical and chemical alterations, particle size reduction, the creation of nanoemulsions, cyclodextrin inclusion complexes, self-micro-emulsifying drug delivery system (SMEDDS), and micellar solubilization are all methods for enhancing solubilization and penetration. These technologies need the appropriate concentrations of raw materials, financial resources, and expertise in technology. Another problem is that these formulations are unstable at large dilutions. It exhibits a level of solubility in both organic and inorganic solvents. Because water's strong lattice structure prevents hydrophobic pharmaceuticals from doing so, co-solvents like PEG make it easier for medications to dissolve by reducing the solvent system's polarity (Bhalani et al. 2022).

PEG 300 and 400 are allowed in parenteral up to 30% v/v. High MW PEGs are frequently utilized to microencapsulate active pharmaceutical ingredients. PEGs reduce the use of strong solvents throughout encapsulation. PEG is a lubricant that is used in eye drops. PEG is a perfect excipient for liquid dosage forms since it resists rancidity and mold growth. PEGs in liquid form (MW up to 1000) are typically utilized as emulsion stabilizers and suspending agents in conjunction with other stabilizers. Additionally, it serves as a solubilizer for liquid-filled soft gelatin capsules or drops (for ophthalmic or otic treatments) (Pham Le Khanh et al. 2022).

The use of micellar solubilization frequently achieves hydrophobic medication solubilization. The surfactant increases the wetting of solids by first lowering surface tension. Micelles then develop above the CMC, encasing the medication and improving its solubilization. PEG and hydrophobic polymers like PLGA, poly amino acids, and PCL can easily self-assemble to generate nanoscale core-shell structures that resemble micellar structures. These nanoscale structures make hydrophobic moieties, such as those found in medicines, proteins, peptides, and genes, more soluble. It is generally known that macromolecular co-polymers have a critical aggregation concentration (CAC), below which the co-polymers exist as unimers. As the overall concentration increases to a level equivalent to the CAC value, aggregates of micellar-like structures begin to grow. Over CAC, they generate micellar structures with a typical core shell. These thermodynamically stable molecules can facilitate the solubilization of hydrophobic medicines. The PEG shell's presence guarantees stabilization in an aqueous medium. It gives it an advantage over surfactant micelles, mainly when diluted in blood, the CMC falls, and the drug precipitates (Luan et al. 2019).

### 1.1.4 PEG Toxicity in PEGlyted Products

Many articles, most recently that of Fruijtier-Pölloth (2005), have thoroughly studied and evaluated the toxicity of PEG. However, oral administration has been used in most PEG toxicity research. Pharmacokinetic results unequivocally show that oral absorption reduces as PEG molecular weight rises. Therefore, it is impossible to distinguish between toxicity and absorption changes using the data. PEGs delivered intravenously, subcutaneously, or intraperitoneally enable direct comparison of the toxicological exposure relationship. Most PEG toxicological investigations have used 'regular' PEG, which has both hydroxyl groups present. When observed, toxicity is typically linked to the kidneys and can cause extreme structural vacuolation of the proximal renal tubules (Ciolacu et al. 2020). In general, these compounds have little to no toxicity. However, it is thought that data from the rat and rabbit are likely comparable, particularly as renal excretion of PEG is thought to be the primary clearance method in both situations. It is unclear what causes PEG to be toxic to the kidneys (Karabasz et al. 2019).

Prior investigations on toxicity utilizing polymeric micelles made of PEG-P(Asp (Bzl) block copolymers of poly(ethylene glycol) and poly(aspartate) block were conducted using rats of the Donryu strain. All-trans retinoic acid and camptothecin were synthetic retinoids transported via a block copolymer. Researchers used rats of the Donryu strain to investigate the associated toxicities. They administered intravenously delivered injections five times (low dose) or 200 mg/kg (high dose) daily. The brain, thymus, lymph node, heart, lungs, liver, spleen, and kidneys underwent histological examinations to determine the weights of the principal organs and the change in body weight at the time of sacrifice.

However, knowing a few potential issues with PEG as a medication carrier is vital. The faster blood clearance caused by PEGylation carriers is the first issue. Numerous investigations have discovered that while a PEGylated carrier's initial dosage resulted in extended circulation in vivo, the subsequent treatment led to fast clearance. This was explained by the development of anti-PEG antibodies, which might have sped up the removal of the nanocarriers. Hypersensitivity, caused by PEG injection inducing an immune response in vivo, would be a second issue. The immune system's nonspecific and specific recognition of the PEG (by anti-PEG antibodies) has been blamed for the reaction. The nonbiodegradability of PEG presents a third issue; additional research is still required to determine whether PEG is eliminated from the body and whether any negative consequences may result from potential PEG accumulation (Padín-González et al. 2022). Pharmaceutical uses of PEG is shown in (Fig. 1.1) and clinically approved PEG products are listed in (Table 1.1).

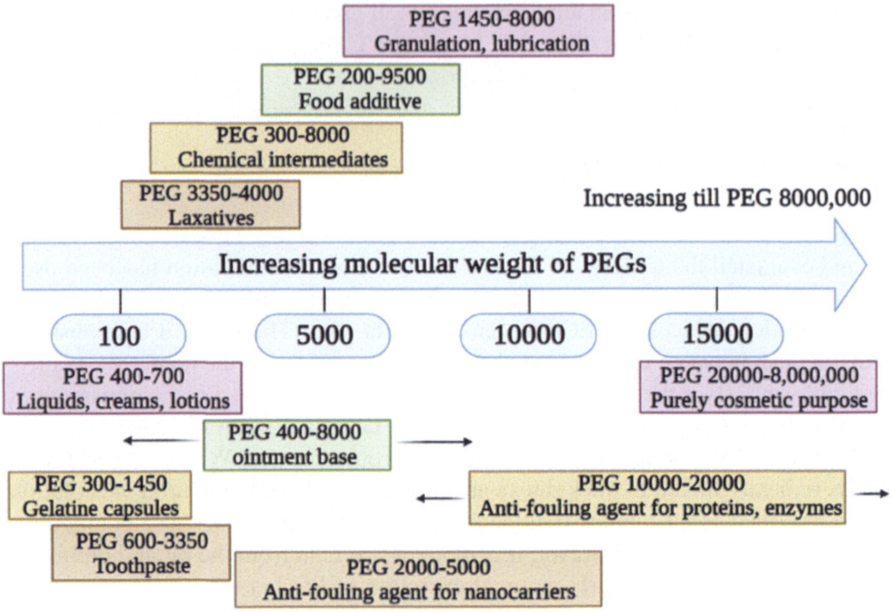


**Fig. 1.1** Pharmaceutical uses of PEG

**Table 1.1** Clinically approved PEG products

| Brand name | Company | Indication | Comment |
|---|---|---|---|
| Adagen | Enzon | Severe combined immunodeficiency disease (SCID) | PECylated adenosine deaminase |
| Cimzia | Nektar UCB Pharma | Rheumatoid arthritis and Crohn's disease | PEGylated Fab' fragment of a humanized TNF inhibitor monoclonal antibody |
| Daxil/ calyx | Ortho Biotech/ Schering-Plough | Cancer | The PEGylated liposome of doxorubicin |
| Macugen | Pfizer | Neovascular age-related macular degeneration | Pegylated anti-vascular endothelial growth factor (VEGF) aptamer |
| Mircera | Roche | Anemia associated with chronic kidney disease | PEGylated erythropoietin |
| Neulasta | Amgen | Chemotherapy-induced neutropenia | PEGylated recombinant methionyl human granulocyte colony-stimulating factor |
| Omontys | Affymax/Takeda Pharmaceuticals | Anemia associated with chronic kidney disease | PEGylated synthetic peptide analog of erythropoietin |
| Oncaspar | Enzon | Acute lymphoblastic leukemia | PEGylated L-asparaginase |

(continued)

**Table 1.1** (continued)

| Brand name | Company | Indication | Comment |
|---|---|---|---|
| PEGASYS | Hoffmann-La Roche | Hepatitis B and hepatitis C | PEGylated interferon alpha |
| Pegintron | Schering-Plough/ Enzon | Hepatitis B and hepatitis C | PEGylated interferon alpha |
| Pegloticase | Savient | Gout | PEGylated uricase |
| Somavert | Pfizer | Acromegaly | PEGylated human growth hormone mutein antagonist |

## 1.2 PEGylation Determination

The method for quantitatively determining the degree of PEGylation of protein bioconjugates utilizing $^{1}H$ NMR spectroscopy and technical safety precautions is presented in the chapter "Mechanisms of Activity Loss for a Multi-PEGylated Protein by Experiment and Simulation" by Zaghmi et al. As an example of a bioconjugate system, glutamate dehydrogenase (GDH), modified with several copies of mPEG (0.5–20 kDa), is employed. This quantitative approach can be applied to additional proteins and is adaptable enough to function with other polymers. An essential element for defining a bioconjugate is the quantity of methoxy poly (ethylene glycol) (mPEG) chains attached to a protein (Zaghmi et al. 2019).

### 1.2.1 Confirmation of PEG Chain

PEG can be used in various lengths and densities to create PEG chains with mushroom or brush conformations. PEG shape affects protein corona patterns and the relative abundance of different proteins in the corona. Clusterin is more likely to bind to PEG chains with a brush shape, and serum albumin is deposited in the corona at a lower rate. With restricted phagocytic uptake, the PEG brush shape in the low protein absorbance regime and a surface enriched in clusterin results in the stealth behavior of nanocarriers. The molecular structure of the grafted PEG chains was determined by observing the $^{1}H$ NMR relaxation time of the chains, which offers details on the dynamics of macromolecular chains close to a solid surface. Low grafting densities severely limit local segment mobility because polymer chains spread out in a flat manner over the surface. Polymer chains resist one another and assume the shape of a longer "brush" when the grafting ratio rises. The correlation time of local segmental motions is crucial for understanding grafted polymer chains' dynamics. The spin-lattice relaxation time T1, which decreases as the mobility of the polymer chains increases, is used to estimate this parameter (Li et al. 2021).

#### 1.2.1.1 Impact of Molecular Weight of PEG on Molecular Confirmation Changes

The molecular weight of PEG impacts PEGylation of nanoparticles (Zhang et al. 2017a, b). Lu et al. reported that the diffusion rate of free heparin sulfate-PEG in water decayed exponentially with an increase of molecular weight. It causes enhancement in PEG density on the surface of nanoparticles as its molecular weight decreases (Lu et al. 2019). Rabanel, Chan, and Teramura also noticed that a same phenomenon occurred by the excluded volume impacts (Teramura et al. 2016; Rabanel et al. 2019; Chan et al. 2020).

#### 1.2.1.2 Impact of Solvent on Conformation Changes

The impact of solvent on the PEG confirmation is very substantial. Selli et al. evaluated the PEG chain's confirmation on the surface of tin oxide nanoparticles using various solvents, such as dichloromethane and water, based on atomistic molecular dynamics simulations. In the presence of water, changes from a mushroom conformation to a brush confirmation begin only at a high density (2.25 chains per $nm^2$). It changes to brush conformation in dichloromethane because of more $O_{PEG}$–Ti bonds among the PEG (terminal group) and Ti atoms (2.25 chains per $nm^2$) (Selli et al. 2019).

### 1.2.2 Quantitative Determination of Surface PEG Density

For more than 50 years ago, experimental methods for measuring charge densities on dielectric surfaces have been investigated. Surface charge densities (SCDs) were initially determined through the examination of electrets. SCDs are now a hot topic of study in fields like semiconductors, liquid crystals, and hybrid materials. Quantitative information from the SCD can offer material-specific characteristics, including details on lattice disruptions, an estimation of the degree of alignment, or further information on interactions in compositions at the microscale. SCD measurements additionally provide the quantitative analysis of screening mechanisms. Electrical characterization methods that do not involve touch are typically used to measure the SCD. One can use the surface photovoltage effect to ascertain the SCD when dealing with semiconductors. This method cannot be used with all materials since it depends on producing electron–hole pairs by light. The vibrating probe approach, invented in the 1960s and eventually evolved into a scanning technique, provides a more versatile way to quantify SCD (Pannuzzo et al. 2020).

#### 1.2.2.1 Thermal Gravimetric Analysis (TGA)

A sample is heated and continuously weighed in thermogravimetric analysis (TGA) using an inert gas atmosphere. The most often used types are as follows:

1. Dynamic TGA: The temperature rises over time as the mass is measured. This makes it possible to simultaneously determine how much gas is evacuated and what temperature it occurs.

2. Static TGA: The temperature is constant during the mass measurement. This can be used to investigate a material's resistance to a particular temperature or discover more about a decomposition that occurs at a particular temperature.
3. Semi-static TGA: The sample is repeatedly heated to various temperatures and kept there for a long time until the mass stabilizes. This is ideal for looking at substances that decompose differently at various temperatures and better characterizing how they decompose (De Blasio 2019).

They used to evaluate the thermal decomposition of PEG at temperatures ranging from room temperature to 600 °C in an $N_2$ environment at a heating rate of 10 °C/min to distinguish between photo-dissociation of PEG and thermal dissociation. PEG began to decompose at about 340 °C and completed around 415 thermally. The backbone chain's -C-O and -C-C- bonds are considered, where PEG thermally breaks down. Thermal dissociation only happens in a small temperature range because PEG has a basic linear-chain-bond structure of 66% -C-O- and 33% -C-C- in the backbone chain with identical bond energies in the 82–83 kcal/mol range. By employing UV/O GPC, the substrate temperature was limited to 200 °C, at which thermal dissociation of PEG did not significantly occur (Burkeyev et al. 2022).

#### 1.2.2.2 Nuclear Magnetic Resonance (NMR)

When specific atomic nuclei are exposed to a strong enough fixed magnetic field, they selectively absorb very high-frequency radio waves. This is known as NMR. The initial observations of these occurrences were made independently in 1946 by the physicists Edward M. Purcell and Felix Bloch (Rinck 2019).

Because the signals from the repeating units overlap to produce prominent peaks, polymer spectra frequently include carbon satellite peaks. A doublet evenly spaced and cantered on either side of the principal peak makes up the signal for 1.1% of $^{1}H$ nuclei near a $^{13}C$. This splitting results in two smaller peaks for a sp3-hybridized C, separated by 115–140 Hz and representing the gyromagnetic ratio of $^{13}C$. The integration of each secondary peak is 0.55% of the primary peak. When present, this splitting is typically overlooked because it occurs within the baseline for the majority of tiny molecules. For 2000 and 5000 g/mol PEG, the ratio of the functionalized terminal group to the repeating polymer unit is generally 2 to 0.9%; the $^{1}H$-$^{13}C$ coupling can, nevertheless, become important when analyzing the functionalization of polymers like PEG with a single functionalized terminal group. The integration of a CH2 signal from the terminal (functionalized) group ($e = 3.55$ ppm) is therefore close to that of the sidebands because of the $^{1}H$-$^{13}C$ link of the repeating CH2-CH2-O unit (= 3.46 ppm) and the 5000 g/mol PEG (Pasek-Allen et al. 2023).

#### 1.2.2.3 Detection of Coordination Complex

Neutral molecules or anions, often known as ligands, form coordinate covalent bonds with a main metal atom (or ion) in a Lewis acid-base process. The final product is a coordination complex. Coordination is the term used to describe the dipolar "coordinate covalent bonds" between the ligands and the central atom. Initially, a complex was believed to consist of reversible chemical bonds joining molecules, atoms, or ions. In terms of coordination chemistry, the significance has altered. Relatively strong bonds join many metal complexes, creating specific metal complexes irrevocably (Li and Zuo 2020).

To make conjugates of monomethoxy-poly (ethylene glycol) and cisplatin (MPEG-DA/CDDP), a poly (ethylene glycol) (PEG) derivative with a terminal dicarboxylic acid group was coupled with cisplatin (CDDP) through chelate-type coordination bonds. These conjugates have excellent anti-cancer efficacy, high water solubility, and low adverse effects. The cytotoxic activity of the MPEG-DA/CDDP compound was somewhat less than that of free CDDP. The MPEG-DA/CDDP conjugation maintained a higher level of cytotoxicity even while treatment in medium with serum lowered the cytotoxic property of free CDDP and the control conjugate (Zhang et al. 2022).

#### 1.2.2.4 X-Ray Photoelectron Spectroscopy

X-ray photoelectron spectroscopy (XPS), a quantitative technique for analyzing a material's surface's chemical composition, also determines the constituent elements' binding states. The normal XPS probe depth is 10 nm. The material must first be evacuated because XPS needs a high vacuum before analysis begins. Photons of specific energy are absorbed by electrons in a sample and released once the photon has left the material. XPS stands out for its popularity, adaptability, and utility compared to many other techniques. Several applications will be used to demonstrate the adaptability and effectiveness of XPS. This reasonably practical technology might also be more exciting by looking at spectroscopic imaging and prospects (Gengenbach et al. 2021).

The PEG lubricant's spectra contained the high-intensity peaks C 1s at 285.09 eV and O 1s at 531.7 eV. This demonstrated that most PEG lubricants comprised the C and O components. Using XPS peak software, additional analysis of the two peaks showed that the element C was present primarily as C-C, C-O, C-H, and C-N. Both PEG monomers had significant amounts of C-C, C-O, and C-H, although C-N predominated. The most prevalent form of the element O was C-OH. These outcomes thus verified the inclusion of both PEGs in the manufactured lubricant. At 23.3 °C and 26.1 °C, PEG showed two comparatively high, narrow, and strong diffraction peaks, demonstrating a higher level of crystallinity. It can be seen that the composite sample, which also included two reasonably high and narrow diffraction peaks at 23.3 °C and 26.3 °C, increased the strength of the peak at 23.3 °C (Gao et al. 2022).

## 1.3 PEGylation of Nanocarriers

### 1.3.1 Nanoparticulate System

Nanoparticulate systems may be essential in imaging, surgery, and medication delivery. Since nanoparticulate systems are more significant than most "small molecule" type drugs but smaller than cells, they may have a longer retention time (Dang and Guan 2020).

#### 1.3.1.1 Solid Lipid Nanoparticles

Solid lipid nanoparticles (SLNs) provide benefits by minimizing some of the individual drawbacks of emulsions, polymeric nanoparticles, and liposomes. Typically, they have a solid hydrophobic matrix core covered in phospholipids. Therefore, it is envisaged that solid lipid nanoparticles will more effectively trap hydrophobic medicines in their core compared to traditional liposomes (Borges et al. 2020).

#### 1.3.1.2 Nanostructured Lipid Carriers (NLCs)

When solid and liquid lipids are combined to create NLCs, the resulting structures are less organized, allowing for a firmer inclusion of the drug molecules inside the matrix over the course of the shelf life. The structural parity of two lipids in NLCs, which has been consistently reported, causes defects in their structure and increases space for drug accommodation during solidification, and is the cause of the improved entrapment efficiency in NLCs. Additionally, the more excellent solubility of medicines in liquid lipids compared to solid lipids is the cause of the increased entrapment efficiency. NLCs have a higher payload capacity and lengthy shelf storage stability than traditional lipid-based systems. Additionally, NLCs can include both hydrophilic and lipophilic medications. They may also deliver medications to the site of action and offer a prolonged release of the substances (Elmowafy and Al-Sanea 2021). Numerous studies have demonstrated how these nanoplatforms improve oral medication bioavailability by encouraging intestinal absorption. Due to its ability to be sustained for extended periods and modulated by therapeutic efficacy, this system has also given hope for treating chronic disorders.

Solid lipid nanoparticles (SLNs) were used to create the improved generation of lipid nanoparticles known as NLCs. They minimize several SLN-related issues, such as drug leakage during storage and a finite amount of drug that can be loaded into them, while retaining many of the benefits of SLNs, including regulated drug release, biocompatibility, and the potential for large-scale industrial production. According to reports, PEGylated nanoparticles have an increased half-life in circulation and are less likely to be absorbed by the RES (Garg et al. 2022). Additionally, PEG can lengthen the time cancer cells are exposed to nanoparticles carrying an anti-tumor medicine and can improve the enhanced permeability and retention. In the current study, PEGylated NLCs (PEG-NLCs) were developed as carriers to stabilize the lactone form of HCPT, extending its duration in circulation and boosting its anti-tumor efficacy against lung cancer. In this study, researchers discovered a brand-new PEGNLC lung-targeting effect following i.v. injection in mice (Dhiman et al. 2021).

#### 1.3.1.3 Polymeric Nanoparticles

Polymeric nanoparticles have been used to treat various illnesses, including cancer, diabetes, malaria, and tuberculosis. In comparison to other carriers, they offer several benefits. The method of synthesis and choice of ingredients for the polymeric nanoparticles depend on some factors, including the desired physicochemical qualities, the makeup of the medicine, and the expected course of therapy (Zielińska et al. 2020).

### 1.3.2 Metal Nanoparticles

The use of metallic nanoparticles in several biomedical applications like bioimaging, biosensors, target/sustained drug delivery, hyperthermia, and photoablation treatment, has gained popularity in recent years. Additionally, these nanoparticles have been altered and functionalized with particular functional groups that enable them to bind to medicines, antibodies, and other ligands, making these systems more attractive for use in biomedical applications (Jamkhande et al. 2019; Saravanan et al. 2021).

#### 1.3.2.1 Silver Nanoparticles

Silver nanoparticles are extremely small silver atoms with sizes between 1 and 100 nm. Ionic silver has a lengthy history and was first employed in yellow-stain glass, like gold nanoparticles. Work is currently being done to include silver nanoparticles in various medical products, such as surgical masks, bone cement, and other items. Additionally, it has been demonstrated that ionic silver can cure wounds when used in the proper dosages (Yaqoob et al. 2020; Yin et al. 2020). Silver nanoparticles have taken over from silver sulfadiazine as the preferred method for treating wounds. On the surfaces of home appliances, Samsung has also developed and sold a product called Silver Nano that contains silver nanoparticles.

Additionally, these nanomaterials have drawn much interest in biological imaging employing SERS because of their appealing physiochemical features. Individual silver nanoparticles are excellent candidates for molecular tagging due to their surface plasmon resonance (Abdelfattah et al. 2022). As a result, numerous targeted silver oxide nanoprobes are being created right now. They are commonly produced by reducing a silver salt in the presence of a colloidal stabilizer, such as sodium borohydride. The most widely used colloidal stabilizers are polyvinyl alcohol, poly (vinylpyrrolidone), bovine serum albumin (BSA), citrate, and cellulose. (Park et al. 2020). One of the more recent novel methods is ion implantation, which produces silver nanoparticles using starch as a stabilizer and D-glucose as a reducing sugar (Abdel-Hameed et al. 2022). It is vital to remember that not all nanoparticles produced are made equally. It has been demonstrated that its effectiveness is influenced by its size and form. Elechiguerra et al. showed that silver nanoparticles interact with HIV-1 in a size-dependent manner, with particles in the 1–10 nm range adhering to the virus (Elechiguerra et al. 2005).

#### 1.3.2.2 Gold Nanoparticles

A suspension (or colloid) of gold nanoparticles is called colloidal gold, also called gold nanoparticles. These colloidal solutions have a long history, from the Roman era when they were employed to decorate glass with stains. These gold nanoparticles' intriguing optical characteristics result from their particular interaction with light. In the presence of the light's pulsating electromagnetic field, the free electrons of the metal nanoparticles move in relation to the metal lattice (Zhang et al. 2021). This mechanism, known as localized surface plasmon resonance (LSPR), is resonant at a specific light frequency. Following absorption, the surface plasmon degenerates either nonradiatively by turning the light absorbed into heat or radiatively, leading to light dispersion (Sadalage et al. 2021). By precisely delivering AuNPs to the nucleus of cancer cells, El Sayed et al. used gold nanoparticles for cancer imaging. To specifically deliver the AuNPs to the cancer cell nucleus, they coupled an arginine-glycine-aspartic acid peptide (RGD) and a nuclear localization signal peptide (NLS) to a 30-nm AuNPs via PEG (Kang et al. 2010).

#### 1.3.2.3 Titanium Dioxide Nanoparticles

There are two crystalline forms of titanium dioxide: rutile and anatase, the latter chemically more active. Titanium dioxide fine particles ($TiO_2$ FPs) are another name for the rutile form of $TiO_2$ nanoparticles. Compared to the rutile phase, the anatase phase is significantly more hazardous to healthy cells due to the increased crystalline structure of the anatase form, which increases the formation of reactive oxygen species (Selli et al. 2019).

Additionally, the chemical modifications made to nanoparticle surfaces vary in how active $TiO_2$ NPs are. In human SMMC-7721 hepatocarcinoma cells, for instance, one-dimensional $TiO_2$ Ws amplified the lethal effects of DNR by raising its dosage. Similarly, combining TMZ with $TiO_2$ nanostructures may enhance its anti-tumor activity in treating brain gliomas (Shah et al. 2019; Siani et al. 2022). Additionally, the release of valproic acid in several disorders was caused by its gradual but persistent encapsulation in $TiO_2$ matrices. Compared to their medications, GA and DNR from GA-$TiO_2$ and DNR-$TiO_2$ nanocomposites showed greater potential anti-tumor efficacy in human leukemia K562 cells (Cin 2021).

#### 1.3.2.4 Copper Nanoparticles

Copper nanoparticles (CuNPs) have become a more popular due to their accessibility, affordability, and potent anti-bacterial effects. CuNPs' primary benefit over gold and silver nanoparticles is that they are less expensive and more readily available, leading to sample synthesis and various uses for CuNPs. Copper nanoclusters (Tf-CuNCs) with increased luminescence were formed due to experimental investigations on creating targeted medication delivery and bioimaging molecules (Okyere et al. 2022). After being tested in vivo on mice bearing Dalton's lymphoma ascites and the transferrin receptor (TfR), the newly created nanomaterials showed improved tumor growth suppression and increased animal survival.

Curcumin-capped CuNPs have been identified by Kamble et al. as potential inhibitors of human breast cancer cells and angiogenesis when compared to natural curcumin (Kamble et al. 2016). Mupirocin-linked copper nanoparticles were created by Vikram et al. to overcome *Staphylococcus aureus*'s antibiotic resistance, which causes cutaneous skin infections. CuNPs exhibit 96.5% drug release in an in vitro release study and have significant antibacterial action against *Staphylococcus aureus* (Verma and Kaushik 2020).

### 1.3.3 Vesicular Systems

Vesicular drug delivery systems have certain benefits, including extending the duration of the medication's presence in systemic circulation and possibly reducing toxicity if selective uptake is possible due to the drug being delivered directly to the infection site. It enhances bioavailability, particularly for medicines that are difficult to dissolve (Carter et al. 2019).

#### 1.3.3.1 Liposomes

The Greek terms "lipo" and "soma," which mean "fat" and "body," respectively, are the source of the phrase "liposome." A hydrophilic "head" and a hydrophobic "tail" (hydrocarbon chain) make up each phospholipid molecule. The self-association of phospholipid layers during the spontaneous synthesis of liposomes takes place (Mohamed et al. 2019). The liposome's hydrophilic center is located between two hydrophobic layers. With the help of this unique characteristic, we can insert hydrophilic or hydrophobic medicines between the two phospholipid layers (Singh et al. 2020).

Due to their specific individuality, liposomes, vesicular nanocarriers, are widely used as a successful delivery mechanism for brain targeting. A lipid bilayer surrounds an aqueous core in liposomes, simulating the structure of a cell membrane. This distinction facilitates the fusing of liposomes with cell membranes and subsequent cell uptake. Because liposomes are amphipathic, they can be used to encapsulate hydrophilic and hydrophobic medications. Conventional liposomes have drawbacks such as quick blood clearance and slow RES absorption because of the accumulating plasma proteins (Ghaferi et al. 2020). Liposomes are quickly cleared from the bloodstream and accumulate in the liver and spleen due to RES absorption. Biocompatible PEG polymers have been used to stabilize conventional liposomes sterically. In addition to passive accumulation in solid tumors, typical liposomes are PEGylated to significantly extend their half-life in circulation (Roces et al. 2020). It was also attempted to directly couple ligands to the surface of the liposome using PEGylated liposomes. However, the PEG chains showed a potent shielding effect that prevented the attached ligand from interacting with its receptor. Targeting ligands are linked to the ends of PEG chains to improve the vector's flexibility and accessibility to the brain, where PEG is primarily used as a spacer (McSweeney et al. 2019).

Transferrin ($T_f$)-anchored PEGylated liposomes loaded with horseradish peroxidase (HRP) and tested for their ability to target BCEC's Tf receptor in vitro. The binding results of Tf-anchored liposomes were four times greater than those of conventional liposomes. Additionally, PEGylated liposomes were used to reduce RES absorption and lengthen the time that polyethyleneimine/DNA (PEI/DNA) polyplexes remained in the body (Ko et al. 2009).

#### 1.3.3.2 Niosomes

Due to the strict conditions needed to handle liposomes in cryogenic environments, the nonionic surfactant is now used in vesicular drug delivery systems instead of phospholipids (Haroun et al. 2022). Niosomes can be altered or modified by adding additional excipients to the membrane, such as cholesterol, and they can have one or more lipid bilayers surrounding an aqueous core (Haroun et al. 2022). Niosomes have an internal structure that primarily combines hydrophobic and hydrophilic molecules, allowing them to accommodate medicinal molecules with various solubilities (Khodabakhsh et al. 2022). Cholesterol addition causes an ordered liquid phase, giving the bilayer stiffness and causing less leaky niosomes. Diacetyl phosphate is well recognized for expanding vesicles and giving them a charge, boosting entrapment effectiveness. Stearyl amine and diacylglycerol are additional charge-inducers that aid in the electrostatic stabilization of the vesicles. Over liposomes, niosomes have particular advantages (Shahbazi et al. 2023). Even in their emulsified state, niosomes are remarkably stable structures. Cholesterol improves the hydrodynamic diameter and trapping effectiveness of niosomes. They are chemically stable and require no specific storage or protection conditions, like an inert atmosphere or low temperature. Because the cost of the raw materials is relatively inexpensive, it can be manufactured industrially. Niosomes have been created using various substances, including sucrose ester surfactants and polyoxymethylene alkyl ether surfactants. Like liposomes, niosomes can entrap solutes (Huang et al. 2008). By delaying clearance from circulation, increasing accessibility to a specific region, simply shielding the drug from its biological environment, and delivering the drug under controlled conditions to a specific location, niosomes can enhance the performance of drug molecules (Davarpanah et al. 2018).

#### 1.3.3.3 Ethosomes

A cutting-edge method of delivering medication, ethosomes predominantly penetrate biological membranes through the skin. Ethosomes are lipid vesicles utilized primarily for transdermal medication administration, including phospholipids, alcohol (ethanol and isopropyl alcohol) in relatively high concentrations, and water (Pandey et al. 2021). Since ethosomes penetrate the skin more quickly than liposomes, they can often be utilized instead of liposomes. However, the precise method by which ethosomes can better penetrate deeper epidermal layers is still unclear (Singhvi et al. 2020). The various PEGylated nanocarriers are shown in (Fig. 1.2) and PEGylated nanocarriers marketed formulations are listed in (Table 1.2).

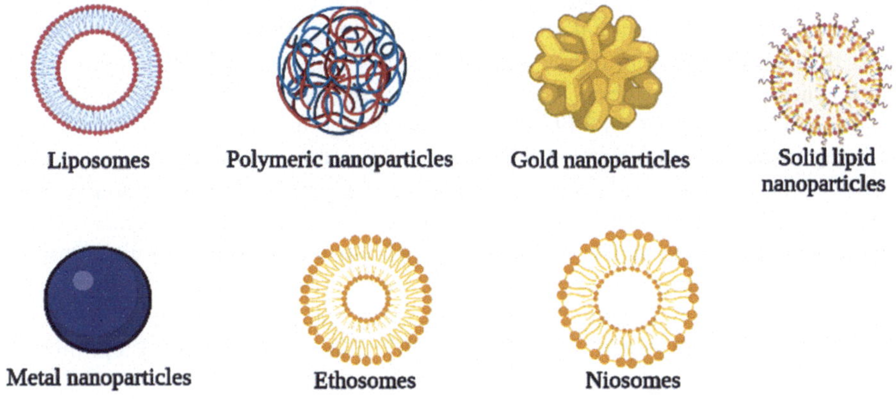


**Fig. 1.2** Representation of various PEGylated nanocarriers

**Table 1.2** PEGylated nanocarriers marketed formulations

| Company | Brand name | Indication |
|---|---|---|
| Hoffimann-La Roche | PEGASYS | Hepatitis B and hepatitis C |
| Nektar pharma | Cimzia | Rheumatoid arthritis |
| Pfizer | Macugen | Neovascular age-related macular degeneration |
| Amgen | Neulasta | Chemotherapy-induced neutropenia |
| Ortho Biotech | Doxil | Cancer |
| Pfizer | Somavert | Acromegaly |

## 1.4 Application of PEGylated Nanocarriers in Various Diseases

### 1.4.1 Cancer

Existing chemotherapeutic drugs along with immune-regulatory and anti-tumor active chemicals, increase the likelihood of curing cancer. Although it is a targeted therapy, the fundamental drawback of the methods employed to treat this widespread disease is that they cannot distinguish between cancer and healthy cells. As a result, much study has been done on cancer treatment. By employing PEG as a carrier for active compounds like doxorubicin, camptothecin, and paclitaxel, targeted therapy for malignant cells can be obtained (Sanchez Armengol et al. 2022). Interleukin (IL) 2 has been altered to create aldesleukin, a recombinant protein. Some of the cell types it stimulates in terms of proliferation and differentiation are B-cells, T-helper cells, and natural killer cells. As an immunotherapy, it is applied to treating metastatic kidney cancer. However, if the dose of this active ingredient is increased due to adverse effects, the number of individuals eligible for this therapy declines. This results in a higher level of T-helper cell activation in the polymer-drug combination NKTR-214, a prodrug of this molecule that maintains the amino acid sequence of aldesleukin. Additionally, because PEG is utilized as a carrier, the medicine is released more gradually, preserving a continuous dose (Akkın et al. 2021).

The existence of functional groups on the exterior of PEGylated dendritic structures that permit particular interactions with cancer cells is a different hypothesis. It is influenced by the number of functional groups present on the surface of the dendrimer and the length of the PEG chain. The capacity of cancer cells to circulate is growing in cases with advanced malignancy. Since recovery under these circumstances is very challenging, focused therapy is a viable tactic. Polyamidoamine dendrimers (PAMAM dendrimers), in particular, have intriguing features. These ethylenediamine-based dendrimers exhibit a pH-dependent surface affinity. Despite this, PAMAM dendrimers exhibit hemolytic and poisonous characteristics. The PAMAM dendrimer's cationic surface most likely causes this. Research based on this emphasizes both boosting biocompatibility and, on the one hand, optimizing surface structure to lessen cationic characteristics (Li et al. 2022).

According to Daniels et al., the hypervascularity of solid tumors encourages PEG-protein conjugates' permeability. But human erythroleukemia (K562), human epidermoid carcinoma (KB), normal human hepatocyte LO2 cells, murine sarcoma 180 (S-180), and human erythroleukemia 180 (S-180) all responded more favorably to transferrin (Tf)-conjugated systems. When Tf-PEG-protein conjugates were compared to PEG-protein conjugates alone, the half-life was extended by 9.83 h. Additionally, Tf conjugates' selectivity improved active site targeting and delayed blood clearance (Sanchez Armengol et al. 2022).

### 1.4.2 Gene Delivery

The inability of nanoparticles to target specific sick areas after delivery is another obstacle to clinical gene therapy. The most crucial idea is that a ligand receptor-mediated active targeting approach is needed to improve cellular uptake of nanoparticles by interacting with the receptors of targeted cells, such as antibodies, proteins, peptides, and aptamers. Strong specific interactions between ligand and receptor pairs and overexpression of the target receptors on target cells as compared to normal cells are necessary for effective nanoparticle distribution. Passive targeting is used, particularly for tumors with weak vascular architecture and insufficient lymphatic drainage, because these characteristics permit nanoparticles to extravasate into tumor tissues and promote their retention in the interstitial space. PEGylation of nanoparticles is a method to reduce serum aggregation by extending circulation periods and reducing potential serum aggregation. After being exposed to sick tissues or outside stimuli like pH, temperature, light, or magnetic fields, stimulus-responsive nanoparticles may undergo physical or chemical alterations. Response to internal or external stimuli allows one to precisely customize the timing and location of gene therapy, considerably increasing the likelihood that these nanoparticles will be used in clinical settings (Mollé et al. 2022).

By improving the expression of the cytokine interleukin-12, polyethylene glycol polyethyleneimine-cholesterol (PEG-PEI-cholesterol) has been formulated as gene delivery for ovarian cancer. The vast majority of clinical tests involving the transfer of genes via nanoparticles focus on the lipidation of nucleic acids. Alnylam Pharmaceuticals' ALN-VSP uses systemic therapy to target vascular endothelial growth factor (VEGF siRNAs) and reduce liver metastases, potentially sensitizing cancer cells to chemotherapy. Patients with TTR amyloidosis received ALN-TTR02, a lipid-based siRNA formulation, intravenously. The safety, tolerability, pharmacokinetics, and pharmacodynamics of multiple dosages of ALN-TTR02 were assessed (Nguyen et al. 2020).

### 1.4.3 Diagnostics Imaging

The new discipline of nanomedical imaging and therapy relies heavily on nanoparticles. PEG often shields these compounds from the immune system when used in vivo. Significant developments in PEG size, shape, density, loading level, molecular weight, charge, and purification have been established using a wide range of nanoparticle coating and characterization techniques. Superparamagnetic iron oxide (SPIO), one of the first NP applications in imaging, was first described in publications from the 1960s and was widely used by the 1990s. SPIO and ultrasmall SPIO (USPIO) can have their t12 extended by up to 200 min and have their dextran or PEG coating applied. Targeted SPIO makes it possible to perform molecular imaging using MRI without the need for exogenous contrast chemicals. In the 1970s, gold nanoparticles (gold NPs) were first utilized as immunogold to contrast transmission electron microscopy, but they are now used to contrast computed tomography and radiographs. Since then, numerous uses for PEGylated gold NPs and nanorods have been documented (Chen et al. 2021).

### 1.4.4 Vaccines

The polymer utilized in these vaccinations to encapsulate the SLNPs, polyethylene glycol (PEG), has been identified as the main trigger of these allergic responses. Meals, cosmetics, and prescription medication formulas are among the additional products that contain PEG. It is essential to keep nanoparticles colloidally stable in biological fluids and prevent their absorption by filter organs, increasing their effectiveness and safety after vaccination (Garvey and Nasser 2021). Pre-incubating Doxil, a PEGylated liposome used for cancer therapy, activates the complement of anti-PEG antibodies linked to nanoparticles, according to an in vitro investigation using mouse plasma. This was accomplished by increasing the level of C3a.

Nevertheless, the study's use of commercial monoclonal antibodies did not fairly reflect the in vivo setting, where the average level of anti-PEG antibodies in participants who tested positive was 52 ng/mL. The scientists found pre-existing anti-PEG antibodies in plasma samples; however, the same investigation found no correlation between this finding and the complement activation during Doxil incubation. This was accurate when human volunteer plasma was directly evaluated. Doxil, from Pfizer/BioNTech and Moderna vaccines, has a PEGylated surface that can interact with circulating antibodies and serum components, resulting in anaphylactic episodes in females (Bavli et al. 2020).

### 1.4.5 Rheumatoid Arthritis

To create 1,2-distearoyl-sn-glycero-3-phospho-ethanolamine-poly (DSPE-PEG2000), MMP-responsive PEGylated lipid nanoparticles, the ester bond of triglycerol monostearate (TGMS), and the PEG chain (PEG) are being used to create a new treatment strategy for RA. After being administered intravenously to arthritic rats, it has been demonstrated that this method reduces joint swelling and lowers the production of TNF- and IL-1 in joint tissues. The released dexamethasone (Dex) was found to be an effective medication delivery strategy for the treatment of RA (He et al. 2020).

### 1.4.6 Hemophilia

PEGylation improves a protein, peptide, or small-molecule drug's pharmacokinetic, pharmacodynamic, and immunological properties by covalently adding PEG. Today, a proven technique known as PEGylation of proteins is used to address various therapeutic issues. To increase the factor VIII (FVIII) or factor IX circulation half-life, many PEGylated coagulation proteins are created for hemophilia A and B. The hemophiliac individuals' quality of life and adherence to therapy can be significantly enhanced by the lengthening of the half-life, which leads to fewer injections. According to the currently available safety data for PEGylated proteins with high molecular weight PEG, there are no known safety concerns with long-term (chronic) use in human or animal models. The chronic use of PEGylated products that are now on the market has been proven safe, opening the door for those with hemophilia to do so (Chowdary 2020).

### 1.4.7 Pain Therapy

Although there is a growing need for effective, safe pain medications, few active components are now being studied in clinical trials. This provides potential for future research and development. An opioid receptor antagonist with PEG compatibility is naloxegol. Despite being utilized as an antagonist to an agonist, research has found that the mechanism also applies to a chemical with a comparable composition. According to a 2019 study by Ekladious et al., using PEG on oxycodone considerably lowers the side effects. The two medications are polymer-drug conjugates. Opioid-based pain management is no longer the exclusive option; PEG coupling and research into nonsteroidal, anti-inflammatory pharmaceuticals have been made possible. A well-known anti-inflammatory cytokine called interleukine-10 (IL-10) has demonstrated promising results in managing neuropathic pain. Soderquist et al. modified IL-10 using two distinct PEGylation methods to assess the feasibility of the process. The findings indicated that PEGylation of IL-10 would be an effective method for treating pain. To assess cellular proliferation in more depth, in vitro tests were carried out, and total ATP concentrations and EC50 values were established. When compared to unmodified IL-10, the EC50 of PEG-5000-IL-acylation, PEG-5000-IL-amination, and PEG-20000-IL-amination increased by 34.4, 3.09, and 3.15 times, respectively (Sánchez-Cid et al. 2022).

### 1.4.8 Diabetes

Diabetes (DA) is a metabolic disorder brought on by insulin resistance or insufficiency (types 1 and 2). DA symptoms can include polyuria, glucosuria, vision disturbances, or issues with wound healing, among others. But type 2 diabetes is increasingly becoming a problem for the general public. Polymer–protein conjugates have decreased glucose levels through exogenous insulin delivery (Kumar et al. 2020) successfully.

Calceti et al. revealed their findings in 2004 for the oral administration of insulin utilizing a PEG-based nanocarrier. The outcomes demonstrated that PEGylated insulin was progressively released from the polymeric carrier. In this region, PEG-based modified insulin does not show more permeability than unmodified, contrary to past studies that claimed that the better paracellular transport that follows from conjugating low molecular weight PEGs to pharmaceuticals could boost drug permeability. All insulin types supplied had comparable biological and pharmacological activity, according to the findings of in vivo experiments conducted on diabetic mice. Insulin that has been changed using PEG does not have better permeability in this area than unmodified insulin. All insulin types supplied had comparable biological and pharmacological activity, according to the findings of in vivo experiments conducted on diabetic mice (Sanchez Armengol et al. 2022).

PEX168, a different pharmaceutical substance, is currently in the third stage of clinical trials and is most likely to be approved. It is thought to be used for type 2 diabetes and renal failure and is referred to as a long-acting GLP-1 receptor agonist. Patients with mild or moderate renal impairment had $AUC_0$ increases of 13 and 100.7% higher than expected and had longer half-lives (Javia et al. 2022).

### 1.4.9 Others

PEGylation, the covalent attachment of a PEG derivative to molecules, enhances a molecule's water solubility and biocompatibility, particularly helpful for medication development. To achieve regulatory compliance for medical uses, PEGylated compounds need to undergo rigorous characterization using sophisticated analytical techniques. PEGylation of peptides, proteins, and small compounds, including folate, mannose, prodrugs, oligonucleotides, cells, nanoparticles, virus particles, and surfaces, commonly uses bifunctional PEG derivatives. Multi-arm PEG derivatives are primarily used to create hydrogels for controlled release of medicines, medical devices, regenerative medicine, and various other applications, such as cell culture, wound sealing, and wound healing (Ibrahim et al. 2022).

PEGs have been used in a wide range of applications for many years, according to reports in the scientific literature. Their use is growing in various research and development fields as regulatory authorities worldwide gain more expertise and comfort using these materials in drug and medical device applications. PEGs are employed in creating medical devices, drug research, and diagnostics in various ways, including medication administration, wound healing, cell culture models, and tissue regeneration (Janrao et al. 2022).

## 1.5 Limitations and Challenges Associated with PEGylation of Nanocarriers

PEGylation negatively impacts the physicochemical properties of NPs, increasing particle size—especially with high molar weight PEG, which could increase RES absorption—and cause a fast release of the therapeutic contents from NPs within a short period. For instance, it has been shown that doxorubicin-loaded PEGylated liposomes release drugs more quickly than their non-PEGylated counterparts; 90% of the loaded drug was released 3 h after IV administration. Therefore, assuming that the loaded drug cargo is released after entering the target cells was incorrect. Compared to polylactic lactic acid (PLA) polymer particles, research has revealed that morphine escapes from poly (L-lactide) or PLA-PEG block copolymer particles much faster. Additionally, it has been demonstrated that the amount of PEG in the copolymer affects the release rate. The initial burst release is more noticeable in particles with a PEG concentration of 5% instead of 3%.

Instead of a burst release, 0–1% PEG-containing particles showed a continuous, persistent release throughout time. The T80% of particles without PEG was around 15 and 30 times longer than that of particles with 3 and 5% PEG, respectively. T80% denotes the moment when morphine's cumulative release percentage surpassed 80%. This shows that laden components may be freed from PEG-containing particles before the polymer degrades. In other words, particles may become empty of their loaded cargo while moving before they degrade. Furthermore, a comparable propensity was shown by PLA and PEGylated poly (lactic-co-glycolic acid) (PLGA) NPs (Sebak 2018).

PEGylated nanocarriers have better physicochemical characteristics and pharmacokinetic profiles than those that have not. However, they may not advantage from the enhanced permeability and retention of the tumor's leaky vasculature or be unable to escape the reticulo endothelial system (RES) sequestration and clearance. Additionally, PEGylation has the potential to produce undesirable particles and post-in vivo administration restrictions on them, such as restricted RES uptake evasion, allergic reactions, reduced intracellular accumulation, and interference with subcellular processing (Nunes et al. 2022). Figure 1.3 shows the major limitations of PEGylation and PEGylated products in a clinical trial are listed in (Table 1.3).

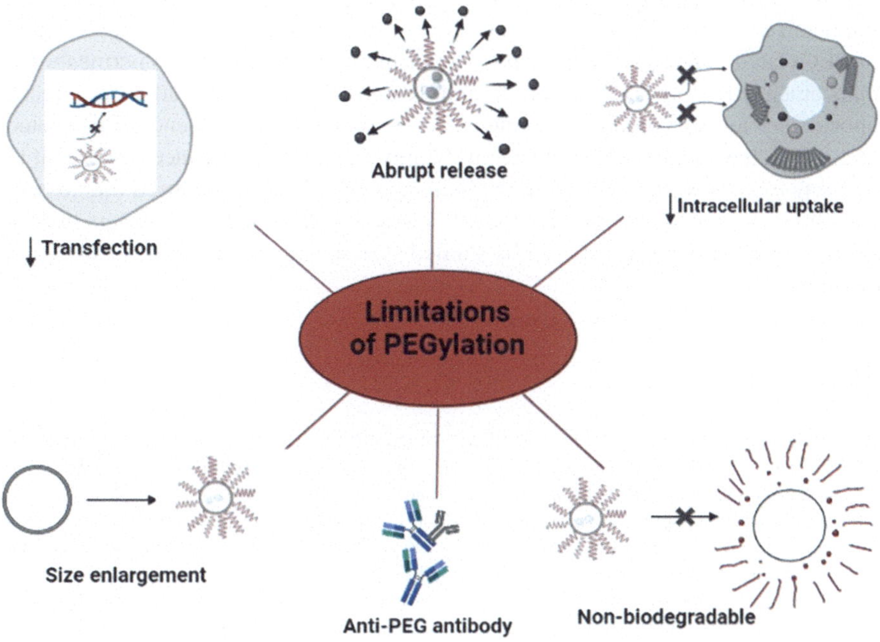


**Fig. 1.3** Major limitations of PEGylation

**Table 1.3** PEGylated products in a clinical trial

| Products | Active agent | Indication | Company | Phase |
|---|---|---|---|---|
| ThermoDox | Doxorubicin | Hepatocellular carcinoma and also recurring chest wall breast cancer | Celsion | Phase III |
| MM-302 | Doxorubicin | Breast cancer | Merrimack | Phase I/II (Terminated) |
| S-CKD602 | Potent topoisomerase I inhibitor | Cancer | Alza Corporation | Phase I |
| 2B3-102 (ENX-201) | Methylprednisolone | Acute relapses of multiple sclerosis | 2-BBB Therapeutic | Phase I |
| SPI-077 | Cisplatin | Lung, head and neck cancer | Alza Corporation | Phase II |
| 2B3-101 (2X-111) | Doxorubicin | Brain metastases and recurrent malignant glioma | 2-BBB Therapeutic | Phase I/IIa |

## 1.6 Conclusion

This chapter has discussed all the aspects related to PEGylation like PEGylation, determination of PEGylation, PEGylation of nanocarriers, applications of PEGylation, and challenges associated with PEGylation. It also explains the characteristics of PEG that play an important role in drug delivery. PEG serves as a useful component as a solubility enhancer and for passive targeting. It is also highlighted that PEGylated nanocarriers show prolong drug circulation, increase permeability, enhancement in drug accumulation and retention, and avoid adverse effects. Nanocarriers act as potential treatments for various diseases. Hence, PEGylated nanocarriers have an excellent chance to be more efficient anti-cancer treatment in upcoming era. They can be used in the management of diseases caused due to bacteria, fungi, and viruses. Moreover, the efficiency of PEGylated nanocarriers in drug delivery for specific targets should be studied systemically to attain the desired and safe output.

## References

Abdelfattah A, Aboutaleb AE, Abdel-Aal ABM, Abdellatif AA, Tawfeek HM, Abdel-Rahman SI (2022) Design and optimization of PEGylated silver nanoparticles for efficient delivery of doxorubicin to cancer cells. J Drug Deliv Sci Technol 71:103347

Abdel-Hameed M, Farrag NS, Aglan H, Amin AM, Mahdy M (2022) A new modality in targeted delivery of epirubicin for tumor theranosis based on PEGylated silver nanoparticles: design, radiolabeling and bioevaluation. Int J Pharm 629:122358

Akkın S, Varan G, Bilensoy E (2021) A review on cancer immunotherapy and applications of nanotechnology to chemoimmunotherapy of different cancers. Molecules 26(11):3382

Bavli Y, Chen B-M, Roffler SR, Dobrovolskaia MA, Elnekave E, Ash S, Barenholz Y, Turjeman K (2020) PEGylated liposomal methyl prednisolone succinate does not induce infusion reac-

tions in patients: a correlation between in vitro immunological and in vivo clinical studies. Molecules 25(3):558
Bhalani DV, Nutan B, Kumar A, Singh Chandel AK (2022) Bioavailability enhancement techniques for poorly aqueous soluble drugs and therapeutics. Biomedicine 10(9):2055
Borges A, de Freitas V, Mateus N, Fernandes I, Oliveira J (2020) Solid lipid nanoparticles as carriers of natural phenolic compounds. Antioxidants (Basel) 9(10):998
Burkeyev M, Tazhbayev Y, Bolatbay A, Minayeva Y, Davrenbekov S, Nassikhatuly Y, Kovaleva A, Kutzhanova K, Kazhmuratova A (2022) Study of thermal decomposition of the copolymer based on polyethylene glycol fumarate with acrylic acid. J Chem 2022(7):1–8
Carter P, Narasimhan B, Wang Q (2019) Biocompatible nanoparticles and vesicular systems in transdermal drug delivery for various skin diseases. Int J Pharm 555:49–62
Chan C-H, Poignant F, Beuve M, Dumont E, Loffreda D (2020) Effect of the ligand binding strength on the morphology of functionalized gold nanoparticles. J Phys Chem Lett 11(7):2717–2723
Chen B-M, Cheng T-L, Roffler SR (2021) Polyethylene glycol immunogenicity: theoretical, clinical, and practical aspects of anti-polyethylene glycol antibodies. ACS Nano 15(9):14022–14048
Chowdary P (2020) Extended half-life recombinant products in haemophilia clinical practice–expectations, opportunities and challenges. Thromb Res 196:609–617
Cin AN (2021) Identifying characteristic properties and antioxidant effect of quercetin loaded titanium dioxide nanoparticles
Ciolacu DE, Nicu R, Ciolacu F (2020) Cellulose-based hydrogels as sustained drug-delivery systems. Materials (Basel) 13(22):5270
D'souza AA, Shegokar R (2016) Polyethylene glycol (PEG): a versatile polymer for pharmaceutical applications. Expert Opin Drug Deliv 13(9):1257–1275
Dang Y, Guan J (2020) Nanoparticle-based drug delivery systems for cancer therapy. Smart Mater Med 1:10–19
Davarpanah F, Khalili Yazdi A, Barani M, Mirzaei M, Torkzadeh-Mahani M (2018) Magnetic delivery of antitumor carboplatin by using PEGylated-Niosomes. Daru 26:57–64
De Blasio C (2019) De Blasio and technology. In: Thermogravimetric analysis (TGA), pp 91–102
Dhiman N, Awasthi R, Sharma B, Kharkwal H, Kulkarni GT (2021) Lipid nanoparticles as carriers for bioactive delivery. Front Chem 9:580118
Elechiguerra JL, Burt JL, Morones JR, Camacho-Bragado A, Gao X, Lara HH, Yacaman MJ (2005) Interaction of silver nanoparticles with HIV-1. J Nanobiotechnol 3:6
Elmowafy M, Al-Sanea MM (2021) Nanostructured lipid carriers (NLCs) as drug delivery platform: advances in formulation and delivery strategies. Saudi Pharm J 29(9):999–1012
Fruijtier-Pölloth C (2005) Safety assessment on polyethylene glycols (PEGs) and their derivatives as used in cosmetic products. Toxicology 214(1–2):1–38
Gao L, Chen L, Zhang T, Li P, Lu H, An R (2022) Preparation and characterization of polyethylene glycol/chitosan composite water-based wound healing lubricant. Front Bioeng Biotechnol 10:990249
Garg J, Pathania K, Sah SP, Pawar S (2022) Nanostructured lipid carriers: a promising drug carrier for targeting brain tumours. Fut J Pharmaceut Sci 8(1):25
Garvey LH, Nasser S (2021) Anaphylaxis to the first COVID-19 vaccine: is polyethylene glycol (PEG) the culprit? Br J Anaesth 126(3):e106–e108
Gengenbach TR, Major GH, Linford MR, Easton CD (2021) Practical guides for x-ray photoelectron spectroscopy (XPS): interpreting the carbon 1s spectrum. J Vacuum Sci Technol A Vacuum Surf Films 39(1):013204
Ghaferi M, Asadollahzadeh MJ, Akbarzadeh A, Ebrahimi Shahmabadi H, Alavi SE (2020) Enhanced efficacy of PEGylated liposomal cisplatin: in vitro and in vivo evaluation. Int J Mol Sci 21(2):559
Haroun M, Elsewedy HS, Shehata TM, Tratrat C, Al Dhubiab BE, Venugopala KN, Almostafa MM, Kochkar H, Elnahas HM (2022) Significant of injectable brucine PEGylated niosomes in treatment of MDA cancer cells. J Drug Deliv Sci Technol 71:103322

He L, Fan D, Liang W, Wang Q, Fang J (2020) Matrix metalloproteinase-responsive PEGylated lipid nanoparticles for controlled drug delivery in the treatment of rheumatoid arthritis. ACS Appl Bio Mater 3(5):3276–3284

Huang Y, Chen J, Chen X, Gao J, Liang W (2008) PEGylated synthetic surfactant vesicles (Niosomes): novel carriers for oligonucleotides. J Mater Sci Mater Med 19:607–614

Ibrahim M, Ramadan E, Elsadek NE, Emam SE, Shimizu T, Ando H, Ishima Y, Elgarhy OH, Sarhan HA, Hussein AK (2022) Polyethylene glycol (PEG): the nature, immunogenicity, and role in the hypersensitivity of PEGylated products. J Control Release 351:215–230

Jamkhande PG, Ghule NW, Bamer AH, Kalaskar MG (2019) Metal nanoparticles synthesis: an overview on methods of preparation, advantages and disadvantages, and applications. J Drug Deliv Sci Technol 53:101174

Janrao C, Khopade S, Bavaskar A, Sudhakar Gomte S, Agnihotri TG, Jain A (2022) Recent advances of polymer based nanosystems in cancer management. J Biomater Sci Polym Ed 34:1274–1335

Javia A, Vanza J, Bardoliwala D, Ghosh S, Misra A, Patel M, Thakkar H (2022) Polymer-drug conjugates: design principles, emerging synthetic strategies and clinical overview. Int J Pharm 623:121863

Kamble S, Utage B, Mogle P, Kamble R, Hese S, Dawane B, Gacche R (2016) Evaluation of curcumin capped copper nanoparticles as possible inhibitors of human breast cancer cells and angiogenesis: a comparative study with native curcumin. AAPS PharmSciTech 17:1030–1041

Kang B, Mackey MA, El-Sayed MA (2010) Nuclear targeting of gold nanoparticles in cancer cells induces DNA damage, causing cytokinesis arrest and apoptosis. J Am Chem Soc 132(5):1517–1519

Karabasz A, Szczepanowicz K, Cierniak A, Mezyk-Kopec R, Dyduch G, Szczęch M, Bereta J, Bzowska M (2019) In vivo studies on pharmacokinetics, toxicity and immunogenicity of polyelectrolyte nanocapsules functionalized with two different polymers: poly-L-glutamic acid or PEG. Int J Nanomedicine 14:9587–9602

Kawai F (2005) Biodegradation of polyethers (polyethylene glycol, polypropylene glycol, polytetramethylene glycol, and others). In: Biopolymers, p 9

Khodabakhsh F, Bourbour M, Yaraki MT, Bazzazan S, Bakhshandeh H, Ahangari Cohan R, Tan YN (2022) pH-responsive PEGylated niosomal nanoparticles as an active-targeting cyclophosphamide delivery system for gastric cancer therapy. Molecules 27(17):5418

Kumar A, Arora A, Sharma P, Anikhindi SA, Bansal N, Singla V, Khare S, Srivastava A. Is diabetes mellitus associated with mortality and severity of COVID-19? A meta-analysis. Diabetes & Metabolic Syndrome: Clinical Research & Reviews. 2020;14(4):535–45.

Ko YT, Bhattacharya R, Bickel U (2009) Liposome encapsulated polyethylenimine/ODN polyplexes for brain targeting. J Control Release 133(3):230–237

Li CH, Zuo JL (2020) Self-healing polymers based on coordination bonds. Adv Mater 32(27):1903762

Li M, Jiang S, Simon J, Paßlick D, Frey M-L, Wagner M, Mailänder V, Crespy D, Landfester K (2021) Brush conformation of polyethylene glycol determines the stealth effect of nanocarriers in the low protein adsorption regime. Nano Lett 21(4):1591–1598

Li H, Zha S, Li H, Liu H, Wong KL, All AH (2022) Polymeric dendrimers as nanocarrier vectors for neurotheranostics. Small 18(45):2203629

Lu J, Xue Y, Shi R, Kang J, Zhao C-Y, Zhang N-N, Wang C-Y, Lu Z-Y, Liu K (2019) A non-sacrificial method for the quantification of poly(ethylene glycol) grafting density on gold nanoparticles for applications in nanomedicine. Chem Sci 10(7):2067–2074

Luan H, Gong L, Yue X, Nie X, Chen Q, Guan D, Que T, Liao G, Su X, Feng Y (2019) Micellar aggregation behavior of alkylaryl sulfonate surfactants for enhanced oil recovery. Molecules 24(23):4325

McSweeney MD, Price LS, Wessler T, Ciociola EC, Herity LB, Piscitelli JA, DeWalle AC, Harris TN, Chan AK, Saw RS (2019) Overcoming anti-PEG antibody mediated accelerated blood clearance of PEGylated liposomes by pre-infusion with high molecular weight free PEG. J Control Release 311:138–146

Mohamed M, Abu Lila AS, Shimizu T, Alaaeldin E, Hussein A, Sarhan HA, Szebeni J, Ishida T (2019) PEGylated liposomes: immunological responses. Sci Technol Adv Mater 20(1):710–724
Mollé LM, Smyth CH, Yuen D, Johnston APR (2022) Nanoparticles for vaccine and gene therapy: overcoming the barriers to nucleic acid delivery. Wiley Interdiscip Rev Nanomed Nanobiotechnol 14(6):e1809
Nascimento ÍF, Guimarães ATB, Ribeiro F, de Lima Rodrigues AS, Estrela FN, da Luz TM, Malafaia G (2021) Polyethylene glycol acute and sub-lethal toxicity in neotropical Physalaemus cuvieri tadpoles (Anura, Leptodactylidae). Environ Pollut 283:117054
Nguyen KG, Vrabel MR, Mantooth SM, Hopkins JJ, Wagner ES, Gabaldon TA, Zaharoff DA (2020) Localized Interleukin-12 for cancer immunotherapy. Front Immunol 11:575597
Nunes SS, de Oliveira Silva J, Fernandes RS, Miranda SEM, Leite EA, de Farias MA, Portugal RV, Cassali GD, Townsend DM, Oliveira MC (2022) PEGylated versus non-PEGylated pH-sensitive liposomes: new insights from a comparative antitumor activity study. Pharmaceutics 14(2):272
Okyere D, Manso RH, Tong X, Chen J (2022) Stability of polyethylene glycol-coated copper nanoparticles and their optical properties. Coatings 12(6):776
Padín-González E, Lancaster P, Bottini M, Gasco P, Tran L, Fadeel B, Wilkins T, Monopoli MP (2022) Understanding the role and impact of poly (ethylene glycol) (PEG) on nanoparticle formulation: implications for COVID-19 vaccines. Front Bioeng Biotechnol 10:882363
Pandey M, Choudhury H, Gorain B, Tiong SQ, Wong GYS, Chan KX, They X, Chieu WS (2021) Site-specific vesicular drug delivery system for skin cancer: a novel approach for targeting. Gels 7(4):218
Pannuzzo M, Esposito S, Wu L-P, Key J, Aryal S, Celia C, Di Marzio L, Moghimi SM, Decuzzi P (2020) Overcoming nanoparticle-mediated complement activation by surface PEG pairing. Nano Lett 20(6):4312–4321
Park T, Lee S, Amatya R, Cheong H, Moon C, Kwak HD, Min KA, Shin MC (2020) ICG-loaded pegylated BSA-silver nanoparticles for effective photothermal cancer therapy. Int J Nanomedicine 15:5459–5471
Pasek-Allen JL, Wilharm RK, Bischof JC, Pierre VC (2023) NMR characterization of polyethylene glycol conjugates for nanoparticle functionalization. ACS Omega 8:4331–4336
Pham Le Khanh H, Nemes D, Rusznyák Á, Ujhelyi Z, Fehér P, Fenyvesi F, Váradi J, Vecsernyés M, Bácskay I (2022) Comparative investigation of cellular effects of polyethylene glycol (PEG) derivatives. Polymers (Basel) 14(2):279
Rabanel J-M, Adibnia V, Tehrani SF, Sanche S, Hildgen P, Banquy X, Ramassamy C (2019) Nanoparticle heterogeneity: an emerging structural parameter influencing particle fate in biological media? Nanoscale 11(2):383–406
Rinck PA (2019) Magnetic resonance in medicine: a critical introduction. BoD–Books on Demand, Norderstedt
Roces CB, Port EC, Daskalakis NN, Watts JA, Aylott JW, Halbert GW, Perrie Y (2020) Rapid scale-up and production of active-loaded PEGylated liposomes. Int J Pharm 586:119566
Rondon A, Mahri S, Morales F, Dumoulin M, Vanbever R (2021) Protein engineering strategies for improved pharmacokinetics. Adv Funct Mater 31:2101633
Sadalage PS, Patil RV, Havaldar DV, Gavade SS, Santos AC, Pawar KD (2021) Optimally biosynthesized, PEGylated gold nanoparticles functionalized with quercetin and camptothecin enhance potential anti-inflammatory, anti-cancer and anti-angiogenic activities. J Nanobiotechnol 19(1):1–17
Saikh MAA (2021) Aqueous film coating the current trend. J Drug Deliv Therapeut 11(4-S):212–224
Sanchez Armengol E, Unterweger A, Laffleur F (2022) PEGylated drug delivery systems in the pharmaceutical field: past, present and future perspective. Drug Dev Ind Pharm 48(4):129–139
Sánchez-Cid P, Jiménez-Rosado M, Romero A, Pérez-Puyana V (2022) Novel trends in hydrogel development for biomedical applications: a review. Polymers (Basel) 14(15):3023
Saravanan A, Kumar PS, Karishma S, Vo D-VN, Jeevanantham S, Yaashikaa P, George CS (2021) A review on biosynthesis of metal nanoparticles and its environmental applications. Chemosphere 264:128580

Sebak AA (2018) Limitations of PEGylated nanocarriers: unfavourable physicochemical properties, biodistribution patterns and cellular and subcellular fates. Int J Appl Pharmaceut 10(5):6–12

Selli D, Motta S, Di Valentin C (2019) Impact of surface curvature, grafting density and solvent type on the PEGylation of titanium dioxide nanoparticles. J Colloid Interface Sci 555:519–531

Shah Z, Nazir S, Mazhar K, Abbasi R, Samokhvalov IM (2019) PEGylated doped-and undoped-TiO2 nanoparticles for photodynamic therapy of cancers. Photodiagnosis Photodyn Ther 27:173–183

Shahbazi R, Jafari-Gharabaghlou D, Mirjafary Z, Saeidian H, Zarghami N (2023) Design and optimization various formulations of PEGylated niosomal nanoparticles loaded with phytochemical agents: potential anti-cancer effects against human lung cancer cells. Pharmacol Rep 75:442–455

Siani P, Frigerio G, Donadoni E, Di Valentin C (2022) Molecular dynamics simulations of cRGD-conjugated PEGylated TiO2 nanoparticles for targeted photodynamic therapy. J Colloid Interface Sci 627:126–141

Simone EA (2008) Filamentous and spherical polymer nanocarriers for enzyme therapeutics. University of Pennsylvania, Philadelphia, PA

Singh A, Neupane YR, Shafi S, Mangla B, Kohli K (2020) PEGylated liposomes as an emerging therapeutic platform for oral nanomedicine in cancer therapy: in vitro and in vivo assessment. J Mol Liq 303:112649

Singhvi G, Rapalli VK, Nagpal S, Dubey SK, Saha RN (2020) Nanocarriers as potential targeted drug delivery for cancer therapy. Springer, Cham, pp 51–88

Teramura Y, Kuroyama K, Takai M (2016) Influence of molecular weight of PEG chain on interaction between streptavidin and biotin–PEG-conjugated phospholipids studied with QCM-D. Acta Biomater 30:135–143

Thakur S, Kesharwani P, Tekade RK, Jain NK (2015) Impact of pegylation on biopharmaceutical properties of dendrimers. Polymer 59:67–92

Verma V, Kaushik D (2020) Mupirocin mounted copper nanoparticle offered augmented drug delivery against resistant bacteria. Indian J Pharmaceut Educ Res 54:637–646

Walkowiak-Kulikowska J, Wolska J, Koroniak H (2020) Biopolymer membranes in fuel cell applications. In: Biopolymer membranes and films. Elsevier, Amsterdam, pp 423–476

Yadav D, Dewangan HK (2021) Pegylation: an important approach for novel drug delivery system. J Biomater Sci Polym Ed 32(2):266–280

Yaqoob AA, Umar K, Ibrahim MNM (2020) Silver nanoparticles: various methods of synthesis, size affecting factors and their potential applications—a review. Appl Nanosci 10:1369–1378

Yin IX, Zhang J, Zhao IS, Mei ML, Li Q, Chu CH (2020) The antibacterial mechanism of silver nanoparticles and its application in dentistry. Int J Nanomedicine 15:2555–2562

Zaghmi A, Greschner AA, Mendez-Villuendas E, Liu JY, de Haan HW, Gauthier MA (2019) Determination of the degree of PEGylation of protein bioconjugates using data from proton nuclear magnetic resonance spectroscopy. Data Brief 25:104037

Zhang H, Wang W, Akinc M, Mallapragada S, Travesset A, Vaknin D (2017a) Assembling and ordering polymer-grafted nanoparticles in three dimensions. Nanoscale 9(25):8710–8715

Zhang H, Wang W, Mallapragada S, Travesset A, Vaknin D (2017b) Macroscopic and tunable nanoparticle superlattices. Nanoscale 9(1):164–171

Zhang Y, Wang P, Mao H, Zhang Y, Zheng L, Yu P, Guo Z, Li L, Jiang Q (2021) PEGylated gold nanoparticles promote osteogenic differentiation in in vitro and in vivo systems. Mater Des 197:109231

Zhang C, Xu C, Gao X, Yao Q (2022) Platinum-based drugs for cancer therapy and anti-tumor strategies. Theranostics 12(5):2115

Zhao C, Ma Z, Zhu XX (2019) Rational design of thermoresponsive polymers in aqueous solutions: a thermodynamics map. Prog Polym Sci 90:269–291

Zielińska A, Carreiró F, Oliveira AM, Neves A, Pires B, Venkatesh DN, Durazzo A, Lucarini M, Eder P, Silva AM (2020) Polymeric nanoparticles: production, characterization, toxicology and ecotoxicology. Molecules 25(16):3731

# Methods and Protocols for the Synthesis of PEGylated Pharmaceutical Nanocarriers

# 2

Ala' Adnan Ali, Rakesh Kumar Tekade,
and Randa S. H. Mansour

**Abstract**

This chapter focuses on topics related to synthesizing PEGylated nanocarriers and their characterization. The types of poly(ethylene glycol) (PEG) derivatives and conformation in addition to PEGylation techniques are detailed. Moreover, the methods employed for synthesizing PEGylated nanocarriers are categorized and discussed in detail. The methods explained are classified into self-assembly from PEG-containing molecules by either nanoprecipitation or emulsification and surface modification of nanocarriers with PEG chain covalently and noncovalently. Indirect assessments of surface PEGylation efficiency using particle size, surface hydrophilicity, zeta potential, etc. are emphasized, in addition to other methods used for direct assessments, including colorimetric methods, chromatographic quantification methods, UV and fluorescence spectroscopy, and more. Finally, alternative polymers to PEG are briefly illustrated, as well as the challenges of nanocarriers PEGylation.

**Keywords**

PEGylated nanocarriers · Nanoprecipitation · Zeta potential · Polymers · Solvent evaporation · Solvent diffusion · Chemical conjugation · Physical adsorption · Mushroom conformation · Brush conformation

A. A. Ali (✉)
Faculty of Pharmacy, University of Jordan, Amman, Jordan

R. K. Tekade
National Institute of Pharmaceutical Education and Research (NIPER) Ahmedabad, An Institute of National Importance, Government of India, Department of Pharmaceuticals, Ministry of Chemicals and Fertilizers, Palaj, Opp. Air force station, Gandhinagar, Gujarat, India

R. S. H. Mansour
Faculty of Pharmacy, Zarqa University, Zarqa, Jordan



R. K. Tekade, N. K. Jain (eds.), *PEGylated Nanocarriers in Medicine and Pharmacy*, https://doi.org/10.1007/978-981-97-7316-9_2

## Abbreviations

| | |
|---|---|
| APTES | Amino propyl triethoxy silane |
| APTMS | Amino propyl trimethoxy silane |
| $CDCl_3$ | Deuterated chloroform |
| Chln | Cholane |
| Chol | Cholesterol |
| CMC | Critical micelle concentration |
| $D_2O$ | Deuterated water |
| DLS | Dynamic light scattering |
| DMAP | 4-Dimethylaminopyridin |
| DMSO | Dimethyl sulfoxide |
| DSPE | Distearoyl-sn-glycero-3-phosphoethanolamine |
| EDC | N′-ethylcarbodiimide |
| ELSD | Evaporative light scattering detector |
| FT-IR | Fourier transform-infrared spectroscopy |
| G4-PAMAM | Generation 4 poly(amidoamine) |
| HIC | Hydrophobic interaction chromatography |
| HPLC | High-pressure liquid chromatography |
| LC | Liquid chromatography |
| mPEG | Methoxy-poly(ethylene glycol) |
| MS | Mass spectrometry |
| NA | Nucleic acid |
| NHS | N-hydroxysuccinimide |
| NMR | Nuclear magnetic resonance |
| PCL | Poly(ε-caprolactone) |
| PCS | Photon correlation spectroscopy |
| PEG | Poly(ethylene glycol) |
| PEI | Poly(ethylene imine) |
| PEO-PPO-PEO | Polyethylene oxide-b-polypropylene oxide-b-polyethylene oxide |
| PGA | Poly(glutamic acid) |
| PGs | Poly(glycerols) |
| PHDCA | Poly(hexadecyl cyanoacrylate) |
| PHEMA | Poly(2-hydroxyethyl methacrylate) |
| PHPMA | Poly(hydroxypropyl methacrylate) |
| PLA | Poly(lactic acid) |
| PLGA | Poly(lactide-co-glycolide) |
| PLL | Poly(L-lysine) |
| POX | Poly(oxazolines) |
| PPG | Poly(propylene glycol) |
| PSA | Polysebacic acid |
| pSLN | PEGylatec solid lipid nanoparticles |
| PVL | Poly(δ-valerolactone) |
| PVP | Poly(vinylpyrrolidone) |
| RH | Hydrodynamic radius" |

| | |
|---|---|
| RI | Refractive index |
| SEM | Scanning electronic microscopy |
| SLN | Solid lipid nanoparticles |
| SM | Sphingomyelin |
| TEA | Triethanolamine |
| TEM | Transmission electron microscopy |
| THF | Tetrahydrofuran |
| UV | Ultra-violet |
| XPS | X-ray photoelectron spectroscopy |

## 2.1 Introduction

Several hydrophilic polymers have been utilized to coat the surface of nanocarriers, such as polysaccharides, polyacrylamide, poly(glycerols), poly(vinyl alcohol), poly(oxazolines), poly(ethylene glycol) (PEG), and PEG-containing co-polymers. Among all the polymers, PEG and PEG-copolymer are the most commonly used and recognized as the gold standard. PEGylated nanocarriers have several advantages, such as avoidance of the nonspecific recognition by serum proteins, cells, tissues, and organs, reducing toxicity, and extending circulation time (Moghimi and Szebeni 2003; Vonarbourg et al. 2006).Therefore, PEGylated nanocarriers can be considered valid new pharmaceutical entities compared with their non-PEGylated counterparts.

PEGylation refers to attaching or coating the nanocarrier surface with PEG via adsorption, grafting, or entrapment methods (Howard et al. 2008). In recent few years, a considerable amount of research has been directed toward PEGylation chemistry, allowing the researcher to improve the PEGylation strategies from a "random" to a "site-specific" PEGylation. In addition, PEGylation is currently focused on using bifunctional PEG chains, which react from one side with the particle surface and on the other side with a targeting agent.

The objectives of this chapter are to provide the readers with the methods and protocols of PEGylation, the tools to assess PEG surface quantitatively and qualitatively, and the main alternative polymers to PEG.

### 2.1.1 Physical and Chemical Properties of PEG

PEG is a synthetic polymer of ethylene glycol ($HO\text{-}CH_2\text{-}CH_2\text{-}OH$) that is available in different molecular weights ranging from hundreds to several thousands of Daltons (Chen et al. 2005). PEGs have good water solubility that decreases with increases in chain lengths(Sanchez Armengol et al. 2022). PEGs with low molecular weight (less than 800 Daltons) are highly soluble in water and are liquids at room temperature (Chen et al. 2005).

Different useful features make PEG one of the most popular polymers for modifying nanocarriers. PEG is soluble in both water and many organic solvents due to the presence of both hydrophilic and hydrophobic groups in its structure (Pasut and Veronese 2012). Moreover, the PEG molecule has high flexibility due to the absence of a bulky substituent group on the PEG backbone (Mohapatra et al. 2019). Another feature of PEG is its distinctive stability against decomposition by acids, bases, moderately high temperatures, and hydrogen peroxide, oxidation, and reduction (Chen et al. 2005). In addition, PEG can be functionalized by selectively oxidizing the terminal OH group with various terminal end groups or by attaching large ligands, such as biomolecules (Karakoti et al. 2011).

The synthesis of PEG (Fig. 2.1) involves a condensation polymerization where ethylene is reacted with aluminum oxide to obtain oxirane. Then, the polymerization of diethylene glycol as the starting molecule with oxirane takes place at 130 °C in the presence of alkaline catalysts, as it eliminates water, subsequently the reaction is terminated by adding an acid (Li and Kao 2003).

## 2.1.2 Uses of PEG

PEG has been widely used in biomedical applications, especially in pharmaceutics, due to its high solubility in aqueous media, good tolerance, biocompatibility, FDA-approved, and negligible interference with the drug release (Knop et al. 2010).

A breakthrough in PEGylation was the modification of albumin and catalase by Abuchowski in 1977, which decreased the immunogenicity of the proteins and increased their circulating time in blood (Abuchowski et al. 1977). The first PEGylated nanocarrier introduced in 1995 in the US market was Doxil®, a formulation of PEGylated liposomes of doxorubicin (Barenholz 2012). Later, various FDA-approved PEGylated products have emerged like cytokines (interferon-α2a and interferon-α2b), enzymes (bovine adenosine deaminase, Lasparginase, and urate oxidase), hormones (epoetin-β), antibodies and their fragments, other organic molecules (pegvisomant and pegatinib), and others are in clinical trials (Pasut and Veronese 2009). In addition to enzymes and proteins, PEGylation has been widely utilized in the modification of nanocarriers delivery systems, including polymeric nanocarriers (Ebrahimnejad et al. 2011), liposomes (Allen et al. 1995), solid lipid nanoparticles (SLN) (Yuan et al. 2013), and micelles (Lee et al. 2011). This results in improving their in vivo stability and solubility, increasing their accumulation at the target site, and reducing their clearance rate from circulation (Torchilin 1998; Maruyama 2002).

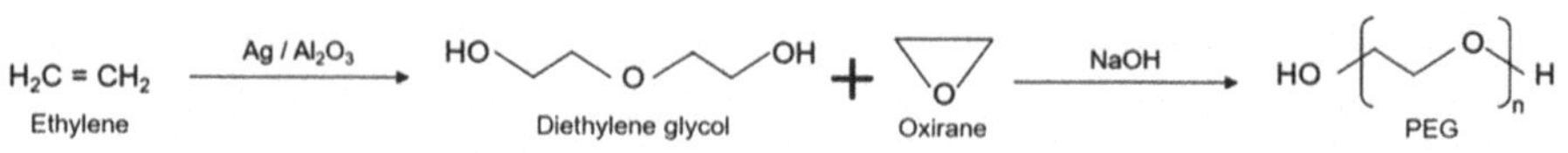


**Fig. 2.1** Synthesis of PEG. ((Sanchez Armengol et al. 2022) with permission) . ©Copyright 2022. Published by Informa UK Limited, trading as Taylor & Francis)

### 2.1.3 Types of PEG Derivatives

Diverse types of PEG derivatives with differences in shapes, functionality, and molecular weights affect the option of a suitable PEG moiety for therapeutic applications (Mohapatra et al. 2019).

#### 2.1.3.1 According to the Shape of the Polymer

According to the shape of the polymer, PEGs are classified into (Fig. 2.2):

(a) *Linear PEG derivatives*: they are the simplest agents, having one reactive functional group that aids in conjugation, and the other end is a methoxy group or a targeting molecule to make the PEG monofunctional. Also, bifunctional PEGs that have two functional groups (refers to Sect. 2.1.3.2) (Mohapatra et al. 2019).
(b) *Y-shaped PEG derivatives*: two linear PEG derivatives are linked to active groups (Santos et al. 2018), they are characterized by higher in vivo stability against physiological conditions like pH, temperature, and enzymatic environment (Veronese et al. 1997).
(c) *Forked PEG derivatives*: they pose multi-proximal reactive groups at one or both ends of a linear PEG chain (Veronese et al. 1997).

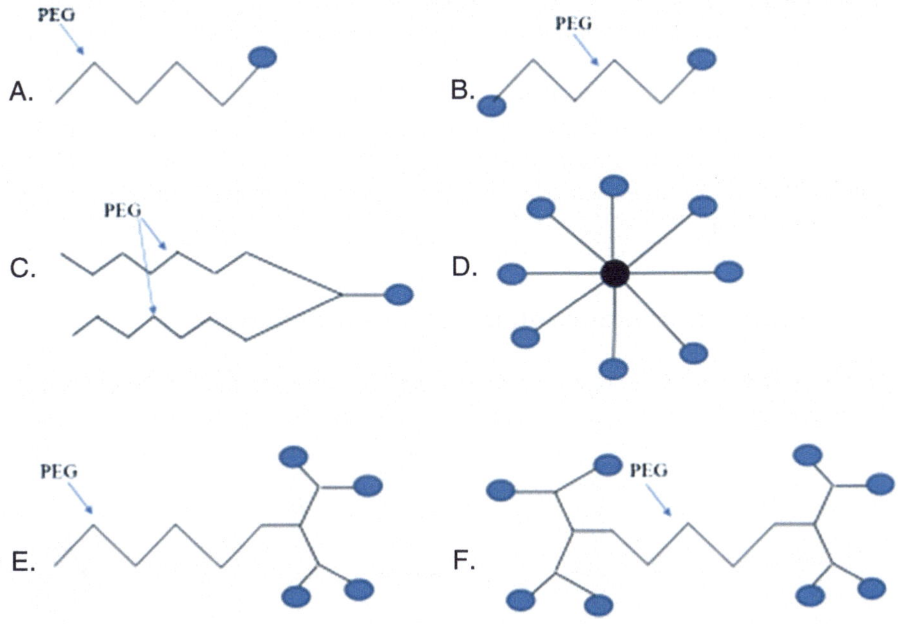


**Fig. 2.2** Different-shaped PEGs. (**a**) Linear PEG. (**b**) Bifunctional PEG. (**c**) Y-shaped PEG. (**d**) Multi-arm PEG. (**e**) Fork-shaped PEG; providing at one PEG chain end multi-proximal reactive groups. (**f**) Fork-shaped PEG; providing at both PEG chains end multi-proximal reactive groups. (●) represents functional groups of PEG derivatives

(d) *Multi-arm PEG derivatives*: they are star-like structures with multi-hydroxyl or functional groups. Thus, the number of active sites and molecular weight are higher (Kim et al. 2016).

#### 2.1.3.2 According to Functionality

PEG functionalization of nanocarriers can enhance the targeting, controlling the release of drugs from colloidal nanostructures (Howard et al. 2008), and increasing the in vivo circulation half-life (Steenpaß et al. 2006). The structure of PEG possesses two equivalent hydroxyl groups in which a variety of reactive functional groups or bioactive species can be attached by covalent coupling (Gajbhiye et al. 2007).

According to functionality, PEGs are classified into monofunctional PEGs and bifunctional PEGs. In monofunctional PEGs, one of the hydroxyl groups is replaced with a reactive functional group (Turecek et al. 2016). Examples of those functional groups are bromo, amino, carboxymethyl, succinimido succinate, tosylate, mesylate, aldehyde, octadesylamine, monopalmitate, stearyl oxy, or methoxy (Harris 1985).

To illustrate, Steenpa et al. prepared monofunctional soy sterol-PEG1300 ethers by attaching a tresyl group to the end of the PEG chain. Then the PEG chain was attached to bovine serum albumin, and the sterol–PEG–protein formed was inserted into the outer liposome monolayer using the post-insertion method. This resulted in an increasing in vivo circulation half-life of those particles (Steenpaß et al. 2006).

On the other hand, bifunctional PEGs have two reactive functional groups. They are usually used for cross-linking and conjugation of drugs, targeting ligands, dyes, and proteins.

Bifunctional PEGs could be homo-biofunctionalized or heterobifunctional. The homo-functionalized PEGs contain two similar functional groups, and the heterobifunctional PEGs include two different groups. An example of a linear heterobifunctional PEG is thiol-PEG-coumarin for functionalizing gold nanoparticle surfaces (Fig. 2.3) (Fu et al. 2004).

### 2.1.4 PEG Conformations on the Nanocarrier Surface

PEG forms a flexible layer on the surface of nanocarriers (Heald et al. 2002), blocking the adsorption of opsonins (proteins) by steric hindrance (Drobek et al. 2005) and their consequent uptake by phagocytic cells (Storm et al. 1995). The PEG-chain

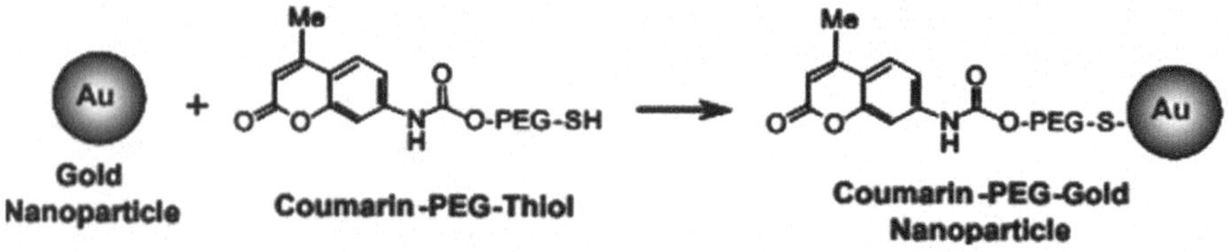


**Fig. 2.3** Schematic representation of the functionalization of gold nanocarrier with coumarin PEG-thiol. ((Fu et al. 2004) with permission)

conformation on the particle surface is often described as a mushroom or a brush model (Amoozgar and Yeo 2012).

When the surface PEG density is low, it results in the mushroom state configuration. As the PEG density increases, it results in the brush state configuration (Tirosh et al. 1998) as in Fig. 2.4.

The "mushroom" configuration has meager surface coverage, leading to large areas for protein binding. On the contrary, the "brush" configuration is characterized by a very high surface coverage, which provides a higher anti-opsonization effect resulting in more effective binding, prolonging the circulation time of nanocarriers compared to the former (Photos et al. 2003).

The optimal PEG density varies by system. For example, for poly(lactic acid) (PLA) nanocarriers (Sheng et al. 2009) or poly(lactide-co-glycolide) (PLGA) nanocarriers 10 wt.% PEG density was reported as optimal regarding particle dispersibility and stealth effect (Beletsi et al. 2005). However, Gref et al. considered that the optimal PEG surface density for PLA, PLGA, and polycaprolactone nanocarriers was 5 wt.% and that any further increase in the levels of PEG content will not result in further reduction in protein adsorption (Gref et al. 2000).

In liposomes, a mushroom-like shape is obtained using <5% PEG, mushroom- or brush-like shape is obtained using 5–15% PEG, while >15% PEG gives a brush-like shape. Hence, the higher the molar PEG–lipid/lipid composition ratio, the higher the surface density (Nosova et al. 2019).

### 2.1.5 Types of PEGylation Techniques

The key step of PEGylation is a conjugation of the PEG chains by different spacers or linkers with nanocarriers like polymeric nanocarriers, micelles, liposomes, or inorganic nanocarriers. Numerous derivatives of PEG (i.e., linear PEG and Y-shaped PEG) are used in the PEGylation techniques (Zacchigna et al. 2011; Chen et al. 2011). There are three types of PEGylation techniques: first, second, and third generations.

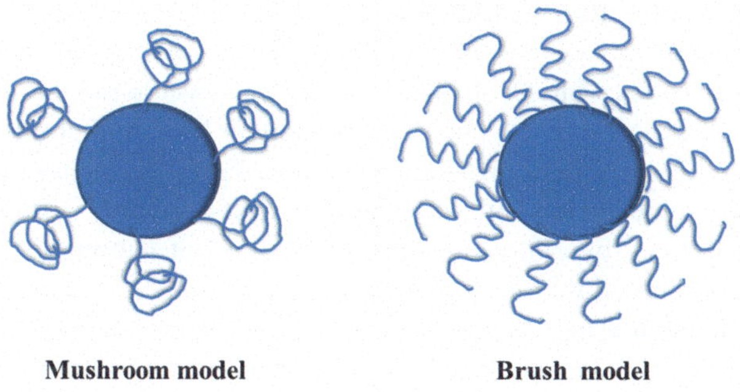


**Fig. 2.4** PEG-chain conformations on the nanocarrier surface, () represent PEG chains

In first-generation PEGylation, linear and low molecular weight PEG is conjugated with a reactive group with hydroxyl groups like carbonates, chlorides, and anhydrides (Damodaran and Fee 2010). This type of PEGylation results in non-specific random conjugations and is inefficient for protein conjugation. In second-generation PEGylation, branched structured PEG derivatives are used, which are more site-specific because more functional groups like esters, aldehydes, and amides are available (Damodaran and Fee 2010). In addition, branched PEGylation is more effective in diminishing immunogenicity and increasing the circulation time than linear PEG. On the other hand, it might affect the biomolecules' activities; therefore, the third generation is developed to conserve drug bioactivity (Monfardini et al. 1995). In third-generation PEGylation branched, Y-shaped structures revealed a reduction in viscosity and lack of organ accumulation (Singh 2015).

## 2.2 Methods and Protocols for the Synthesis of Pegylated Nanocarriers

Due to variations in the chemical makeup, the methods and protocols used for the PEGylation of nanocarriers can be classified under polymeric nanocarriers, liposomes, micelles, and inorganic nanocarriers.

### 2.2.1 PEGylation of Polymeric Nanocarriers

The chemistry of PEGylation of polymeric nanocarriers cannot be defined under a single heading, as that of proteins, due to high variance in the availability of functional groups and mode of preparation. PEGylation of a nanocarrier surface can be achieved in different ways (Fig. 2.5), including the preparation of nanocarriers by self-assembly with block co-polymers of PEG and hydrophobic polymers (Gref et al. 1994). Otherwise, it can achieved by the addition of PEGs on preformed nanocarriers by physical adsorption or chemical conjugation of functional PEGs to the available surface reactive groups (Owens and Peppas 2006).

The main drawback of physical adsorption methods is the risk of desorption, the removal of the coating layer in vivo, and consequent loss in pharmacokinetic outcomes that the polymer should provide (Neal et al. 1998). To overcome the desorption, PEGylated nanocarriers are prepared by self-assembly using a modified derivative of polymers by choosing a proper method of nanoparticle preparation (emulsification and precipitation) in which the hydrophobic portions of PEG chains embedded inside the matrix of nanocarriers and hydrophilic portions orient toward the aqueous phase (Yoncheva et al. 2005). However, this does not ensure that PEG chains are localized only on the surface of nanocarriers, as PEGs may retain within the particle core (Vila et al. 2004). Therefore, the chemical conjugation method for PEGylation of nanocarriers' surface is preferred.

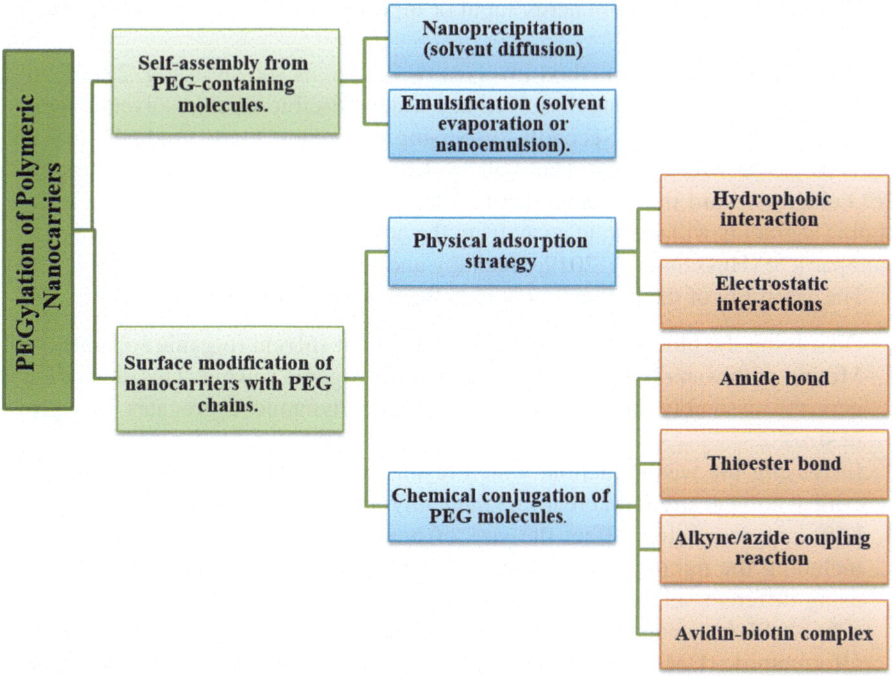


**Fig. 2.5** Methods of PEGylation of polymeric nanocarriers

#### 2.2.1.1 Self-Assembly from PEG-Containing Molecules

To prepare PEGylated nanocarriers using the self-assembly method, PEG is linked/bound with hydrophobic molecules such as lipids or hydrophobic polymers to create amphiphilic moieties. After that, it proceeds by two basic approaches: nanoprecipitation (also called solvent diffusion) or emulsification (also called solvent evaporation or nanoemulsion) (Huckaby and Lai 2018; Hussain et al. 2019).

It is anticipated that PEG segments will segregate completely toward the water while the hydrophobic part of the polymer will be retained in the particle core. However, some of the PEG segments can also be retained in the particle core, thus diminishing PEG concentration at the nanocarrier surface (Vila et al. 2004). This phenomenon can be attributed to several reasons, such as entanglements, loss of small polymeric chains in the water phase, physical processes related to nanocarrier preparation (for example, diffusion of solvent and water, viscosity differences, and polymer–polymer interactions.), solidification processes, or the existence of aqueous cavities within the particle (Rabanel et al. 2014).

##### Nanoprecipitation (Solvent Diffusion)

Many PEGylated nanocarriers have been prepared by the nanoprecipitation method (Kolishetti et al. 2010; Gu et al. 2008). The nanoprecipitation method has several advantages, including it is a one-step process, an environmentally friendly approach, and it does not require high-energy input or high-shear forces, so it is a low-energy process.

The nanoprecipitation method involves two steps, as shown in Fig. 2.6. Firstly, PEG-containing moieties are dissolved in a water-miscible organic solvent (such as acetonitrile) with drug(s). Secondly, dispersing the organic phase gradually into the aqueous phase with mixing. This will lead to the partitioning of the hydrophilic PEG chains toward the aqueous phase at the nanoparticle surface, and the hydrophobic segments will separate away from the aqueous phase to form the core of the nanoparticle (Hussain et al. 2019; Huckaby and Lai 2018).

The assembly of the polymer occurs in three steps (Bovone et al. 2022):

1. Dissolving the block co-polymers in an excellent solvent (organic solvent).
2. Mixing with water to lower the strength of the organic solvent. This change triggers the assembly of block co-polymer into dynamic aggregates that grow in size.
3. Increasing the water fraction, which results in growth arrest. This occurs at a critical solvent strength, creating kinetically trapped core–shell nanocarriers.
4. Different parameters affect the size of nanocarriers prepared by this method including the mixing time and the solvent type. A study reported the effect of using different solvents on the size of poly(ethylene glycol)-block-polylactide (PEG-b-PLA) nanocarriers. The smallest nanocarriers were formed with dimethylformamide (DMF), followed by acetone, acetonitrile, tetrahydrofuran (THF), and dimethyl sulfoxide (DMSO) (Bovone et al. 2022).

#### Emulsification (Solvent Evaporation or Nanoemulsion)

The emulsification method involves several steps, which require emulsifier/s and/or a high-pressure homogenizer or sonicator. On the other hand, drug loading capacity is higher using the emulsification method compared to the nanoprecipitation method.

The PEG-linked moieties/polymers are dissolved with drugs in a water-immiscible organic solvent. This organic solvent is dispersed gently into an aqueous

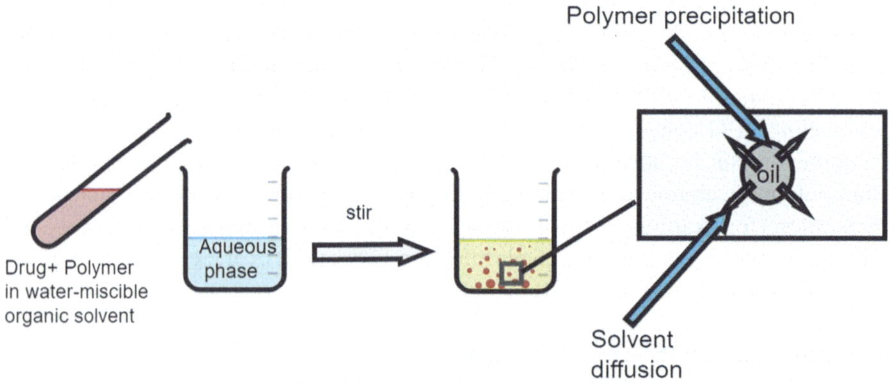


**Fig. 2.6** Schematic representation of preparing PEGylated nanocarriers by the nanoprecipitation method

solvent with the assistance of emulsifiers and/or by high-energy homogenization or sonication. Then the organic solvent is removed by evaporation, letting the hydrophilic PEG chains partition toward the oil/water interface of the nanocarrier surface (Fig. 2.7) (Xu et al. 2015; Huckaby and Lai 2018).

For example, Prawatborisut et al. prepared shellac-PEG (SHPEG) nanocarriers by the miniemulsion solvent evaporation method. Firstly, at room temperature, SHPEG was dissolved in a variety of tetrahydrofuran (THF) and they used a chloroform:THF mixture (2.3:1 w:w). The solution was mixed then with water with stirring at room temperature, followed by ultrasonication in an ice bath. Next, a rotary evaporator was used to evaporate chloroform and THF (Prawatborisut et al. 2019).

Nanocarriers composed of a biodegradable diblock co-polymer of polysebacic acid and polyethylene glycol (PSA-PEG) have been prepared by self-assembly using both nanoprecipitation and emulsification methods (Tang et al. 2009). The surface PEG coating density for the PLGA-PEG nanocarriers was higher in nanocarriers prepared by emulsification (~100% of incorporated PEG found on nanocarrier surface compared to ~89% using nanoprecipitation) (Xu et al. 2015) in addition to a higher level of drug-loading capacity (Hrkach et al. 2012). Despite those advantages, drawbacks of this method include high shear and high energy requirements, which can be impractical for sensitive drugs and biologics (Karnik et al. 2008).

#### 2.2.1.2 Surface Modification of Nanocarriers with PEG Chains

PEGs can be attached to the surface of preformed nanocarriers. By using this technique, the PEG chains will present totally at the surface instead of being embedded within the particle core. However, the main limitation of this method is the low density of PEG on the surface of the nanocarrier which is inadequate to reach the brush conformation mainly due to the steric hindrance of adjacent bulky PEG chains. Moreover, a purification step is required to remove the excess of nonadsorbed or unreacted PEG chains. This is a challenging step, especially in the case of physical adsorption due to the weak interaction and the risk of desorption (Rabanel et al. 2014).

As shown in Fig. 2.8, PEGylation of nanocarriers via surface modification can be done by (1) direct PEGylation, where the PEG molecules are physically adsorbed

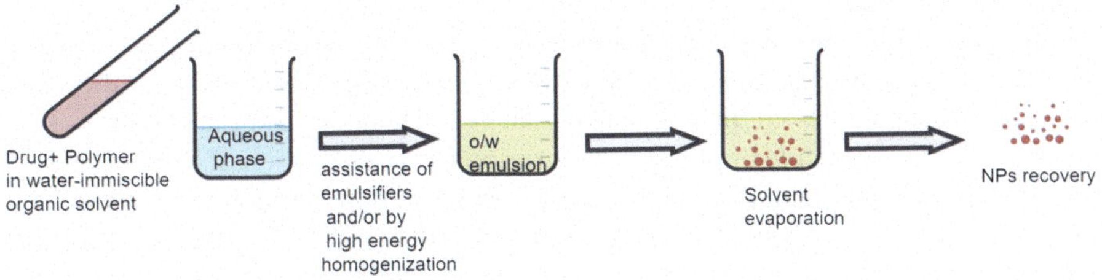


**Fig. 2.7** Schematic representation of preparing PEGylated nanocarriers by the emulsification method

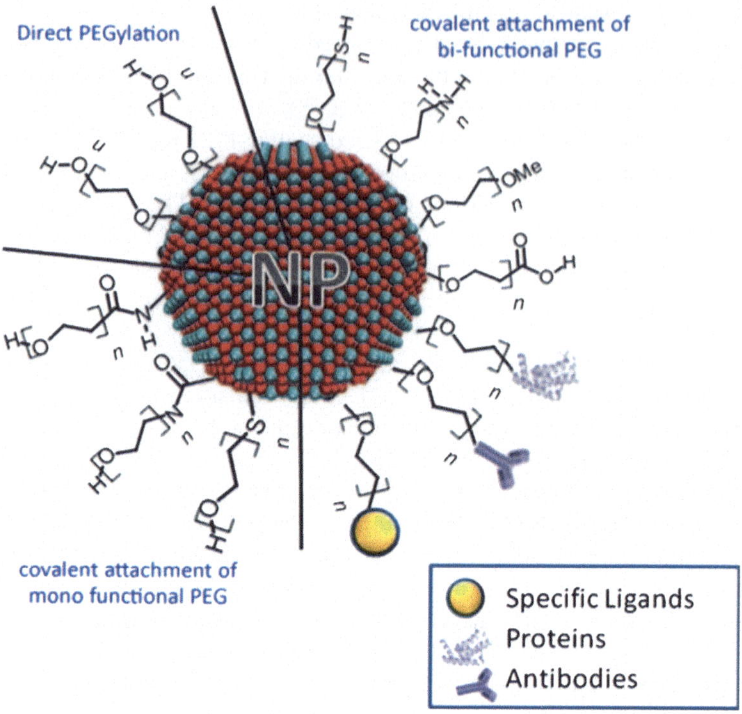


**Fig. 2.8** Different strategies for PEGylation of nanocarriers via surface modification. ((Karakoti et al. 2011) with permission)

on the nanocarrier surface or (2) chemical conjugation via covalent bonds of either monofunctional PEG or bifunctional PEG molecules (Huckaby and Lai 2018).

**Physical Adsorption Strategy**

Physical adsorption is a simple, direct technique for PEG coating based on noncovalent association either by electrostatic interactions with charged particle surfaces (Vandevondele et al. 2003) or by hydrophobic interactions (Redhead et al. 2001; Huckaby and Lai 2018); nevertheless, those interactions are feeble which results in the desorption of PEGs from the surfaces of nanocarriers in vivo thus limiting the efficacy of this technique (Pulkkinen et al. 2008; Cu and Saltzman 2009).

PEG coating by physical adsorption is done by simply incubating pre-formed nanocarriers with an aqueous solution of PEG-containing molecules and allowing the PEGs to associate with the nanocarrier surface (Huckaby and Lai 2018).

An illustrative example of physical adsorption is the adsorption of Pluronics onto the nanocarrier's surfaces. Pluronics is a triblock co-polymer of polyethylene oxide-b-polypropylene oxide-b-polyethylene oxide (PEO-PPO-PEO) that can be adsorbed onto the hydrophobic nanocarrier's surfaces via hydrophobic interaction with the PPO segment (Yang et al. 2011, 2014). Another example is the hydrophobic interaction of the poly(propylene glycol) (PPG) chain of poloxamers (nonionic block co-polymers of (PPG) surrounded by two hydrophilic chains of PEG) with the nanocarrier surface (Storm et al. 1995). In addition, the PEG coating can also be generated based on electrostatic interactions such as the coating of negatively charged polylactic-co-glycolic acid (PLGA) nanocarriers with positively charged PEG block polymers such as PEG-poly(ethylene imine) PEI and PEG- poly(L-lysine) PLL (Wang et al. 2010).

**Chemical Conjugation of PEG Molecules**

A preferred method for PEGylation of nanocarriers is by chemical conjugation due to the formation of a stable PEG coating (Hussain et al. 2019). Several types of bonds can be formed between PEG and the nanocarrier surface by chemical conjugation, such as:

1. *Amide bond*: involves a carbodiimide coupling reaction between a carboxyl-modified nanocarrier with amine-modified PEG (Meng et al. 2004; Suh et al. 2007; Lai et al. 2007, 2010; Wang et al. 2008; Nance et al. 2012) or between a carboxyl-modified PEG with amine-modified nanocarriers. In addition, stable amide bonds can be formed by the reaction between N-hydroxysuccinimide ester (NHS ester)-modified PEG with primary amines (Kim et al. 2013; Moser et al. 2015).
2. *Thioester bond*: formed by reacting thiols with modified PEG yielding thioester bonds (Matsumoto et al. 2014).
3. *Alkyne/azide coupling reaction:* performed with or without the presence of copper catalyst (Cavalli et al. 2006; Kumar et al. 2010; O'Mahony et al. 2012; Breed et al. 2009).
4. *Avidin–biotin complex*: represents the strongest noncovalent, high affinity, protein–ligand interactions. Biotinylated PEG is reacted with avidin-coated nanocarriers to form stable coatings on nanocarrier surfaces (Cu and Saltzman 2009; Park et al. 2009; Jazayeri et al. 2016; Friedman et al. 2013).

For example, generation 4 poly(amidoamine) dendrimers (G4 PAMAM) are nanoscale macromolecules, highly branched, with several amine groups on the surface. PEGylated G4 PAMAM dendrimers were prepared by conjugating PEG on the surface of G4-NH2 dendrimers via an amide bond. For this, hydroxy-terminated PEG was modified into succinyl PEG, having a free carboxylic group using succinic anhydride. Then using a carbodiimide reaction, the modified PEG was conjugated into G4-NH2 dendrimers. The synthesis procedure for preparing PEGylated G4 dendrimers is represented in Fig. 2.9 (Jangid et al. 2022; Yang et al. 2004; Singh et al. 2008).

### 2.2.2 PEGylation of Liposomes

The FDA has approved numerous drugs based on PEGylated liposomes that are available in the market. Such as PEGylated liposomes of doxorubicin (Hensley et al. 2001; Wibroe et al. 2016), vincristine (Sarris et al. 2000), and daunorubicin (Allen and Cullis 2013).

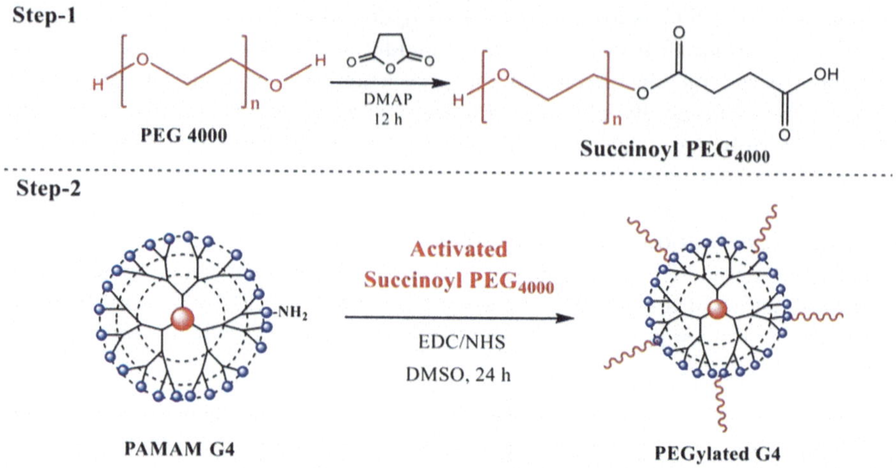


**Fig. 2.9** Scheme representing the synthesis of PEGylated G4 dendrimer. In step one, the hydroxyl (–OH) group of PEG4K was modified into a carboxylic group (–COOH) by reacting it with succinic anhydride and 4-dimethylaminopyridin (DMAP). In step two, the –COOH group was activated by dissolving mPEG4K-COOH, N′-ethyl carbodiimide (EDC), and N-hydroxysuccinimide (NHS) in anhydrous dimethyl sulfoxide (DMSO). Next, adding amine-terminated PAMAM G4 dendrimers and triethanolamine (TEA) to the above reaction mixture. ((Jangid et al. 2022) with permission)

Liposome PEGylation is performed either by mixing PEG–lipid conjugates with a mixture of lipids at a fixed ratio before liposome formation (pre-insertion method) or by mixing PEG–lipids with preformed liposomes (post-insertion method) (Nosova et al. 2019; Amoozgar and Yeo 2012).

#### 2.2.2.1 Pre-Insertion PEGylation

Usually, PEGylated liposomes are prepared by the pre-insertion method. Liposomal particles are formed by mixing cationic, anionic, or neutral lipids with PEG lipids, followed by the drug-loading step (Kapoor and Burgess 2012; Bouxsein et al. 2007). Figure 2.10 represents the formation of PEGylated lipoplexes using this method, where nucleic acid (NA) is added after forming PEGylated liposomes (Halder et al. 2006; Whitehead et al. 2014).

The traditional anchoring PEG–lipids used for liposomal PEGylation (Fig. 2.11) are monomethoxy-poly(ethylene glycol) (mPEG) conjugated to phospholipids such as 1,2-distearoyl-sn-glycero-3-phosphoethanolamine (DSPE) [DSPE-mPEG], where mPEG-OH is chemically conjugated by urethane linkage (−HN−COO−) to the terminal amino group of the lipid head (Lavan et al. 2002). Others are mPEG-ceramide and mPEG-cholesterol. Those PEG–lipids vary in their lipophilic domain, fatty acid type, and PEG chain length (Fu et al. 2004).

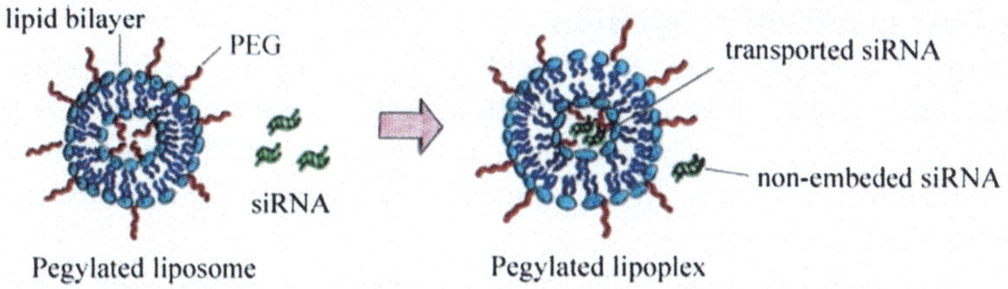


**Fig. 2.10** Scheme represents the preparation of the PEGylated lipoplex by the pre-insertion method. ((Nosova et al. 2019) with permission)

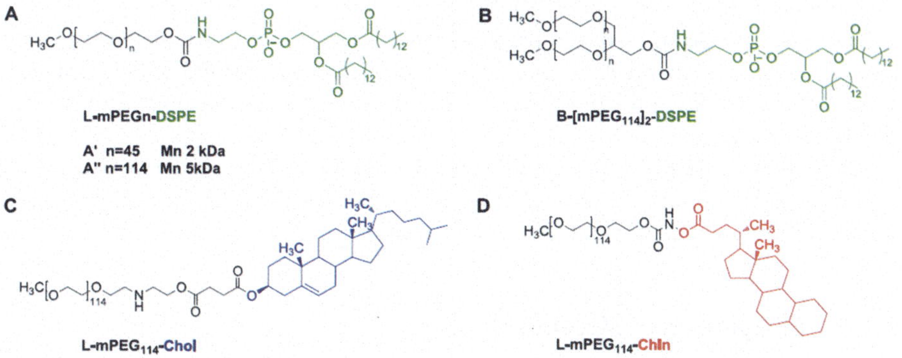


**Fig. 2.11** Chemical structure of mPEG–lipid derivatives used for liposome coating. (**a′**) Linear mPEG45-DSPE (L-mPEG45-DSPE); (**a″**) linear mPEG114-DSPE (L-mPEG114-DSPE); (**b**) branched mPEG114-DSPE [B-(mPEG114)2-DSPE]; (**c**) linear mPEG114-cholesterol (L-mPEG114-Chol); (**d**) linear mPEG114-cholane (L-mPEG114-Chln). ((Mastrotto et al. 2020) with permission)

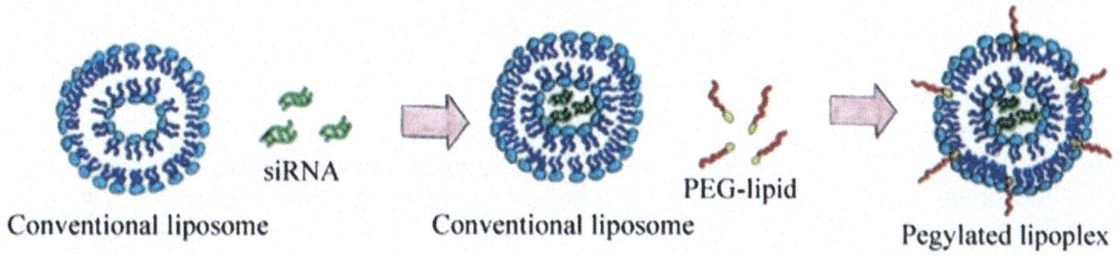


**Fig. 2.12** Scheme representing the preparation of the PEGylated lipoplex by the post-insertion method. ((Nosova et al. 2019) with permission)

Using the pre-insertion method, a low PEG density on the periphery can be obtained (approximately 1%) (Fehring et al. 2014). Besides that, PEG chains will exist on both the external and internal parts of the liposome. As a result, drug loading can be affected due to the PEG moieties present inside the liposome. However, using the post-insertion method, PEG can exist only at the outer liposome surface (Nosova et al. 2019).

#### 2.2.2.2 Post-Insertion PEGylation

The post-insertion method adds PEG–lipid solution or PEG micelles to preformed liposomes after the drug loading step. Subsequently, a purification step is done to remove unreacted PEG derivatives via centrifugation and filtration. Figure 2.12 shows an example of the post-insertion method preparation for a PEGylated lipoplex (Nosova et al. 2019).

Using this technique, stable PEGylated liposomes are formed with a high level of PEGylation (more than 5% PEG) (Uster et al. 1996). Moreover, PEG presents only at the outer surface of the liposomal membrane; this will result in higher encapsulation rates than PEGylated liposomes prepared using the pre-insertion method (Awasthi et al. 2004; Visser et al. 2005).

The type of bonding between the PEG derivative and the liposome surface can vary. For example, chemical bonding can be formed between PEG with a terminal azide group and a hydrophobic alkyne-ended anchor built into the lipid bilayer by a "click" reaction.

As illustrated in Fig. 2.13, a high linear correlation is reported between the amounts of PEG–lipid solutions added to liposomes for PEGylation and the actual amount of modified PEG–lipid (Nakamura et al. 2012).

### 2.2.3 PEGylation of Micelles

Polymeric micelles are formed by self-assembling diblock and triblock amphiphilic co-polymers, comprising both hydrophobic and hydrophilic segments at concentrations above the critical micelle concentration (CMC). PEG is employed as the hydrophilic segment of those co-polymers. When co-polymers are dispersed in aqueous media with hydrophobic drugs, they will assemble into nano-sized spherical micelles (Fig. 2.14), where the core involves the hydrophobic segments

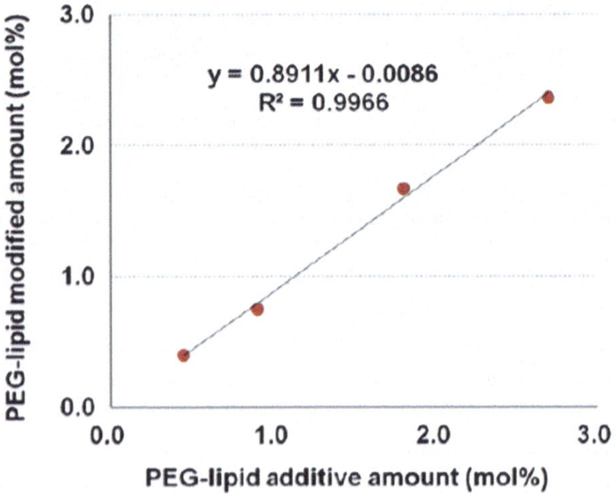


**Fig. 2.13** The modified PEG–lipid amount as a function of applied PEG–lipid amount. For PEGylation of the liposomes, different amounts of PEG–lipid solution were added, and after PEGylation, the unmodified PEG–lipids were removed using gel permeation chromatography. ((Nakamura et al. 2012) with permission)

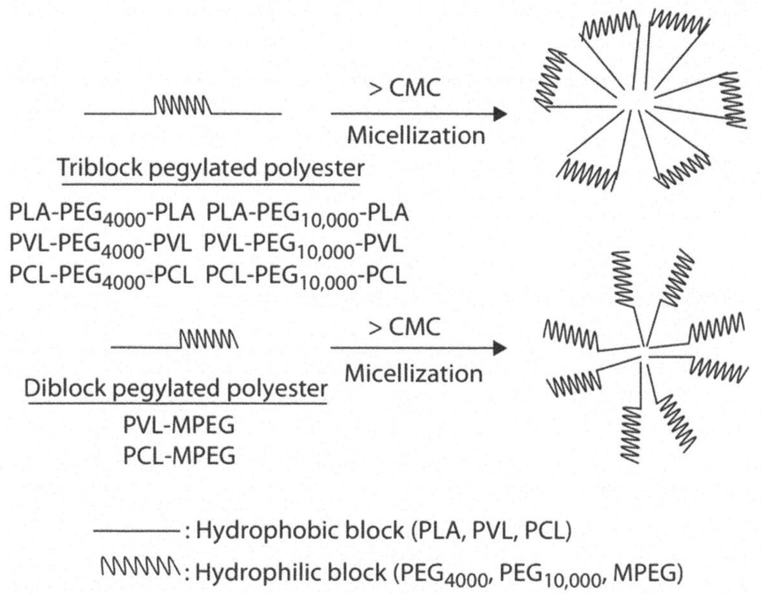


**Fig. 2.14** Polymeric micelles conformation made by triblock and diblock amphiphilic pegylated polyesters. *PLA* poly(L-lactide), *PVL* poly(δ-valerolactone), *PCL* poly(ε-caprolactone), *PEG4000* poly(ethylene glycol) with molecular weight ~4000 Da, *PEG10,000* poly(ethylene glycol) with molecular weight ~10,000 Da, *MPEG* methoxy poly(ethylene glycol). ((Lin et al. 2010) with permission)

and the drug. In contrast, the hydrophilic segments form the shell (Aliabadi et al. 2008).

Polymeric micelles are more stable than surfactant micelles because they have a lower CMC. In other words, polymeric micelles are more able to maintain their core–shell conformation after dilution with a bulk volume of blood in circulation (Lin et al. 2010).

Several methods are available to prepare polymeric micelles, mainly depending on the degree of hydrophilicity of the copolymer. The direct dissolution method is used for the hydrophilic copolymer, in which the micelles are formed directly after being dissolved in water above its critical micelle concentration (Zhang et al. 2009). Another method is the film rehydration method, wherein a volatile solvent is used to dissolve the co-polymers and drugs, followed by solvent evaporation to form a membrane, next adding water or buffer solution with stirring to dissolve the co-polymer membranes to form the micelles (Zhan et al. 2010).

An example of the copolymer used to form polymeric micelles is the PEG-PLGA block copolymer. PEG serves as a hydrophilic corona and PLGA acts as a hydrophobic core. In which hydrophobic drugs such as paclitaxel can be incorporated into the particle core (Bockstaller et al. 2005; Grubbs 2005; Vaia and Maguire 2007; Pozzo and Walker 2008; Ohshima 2016).

Nishiyama et al. prepared micelles by self-assembling block co-polymers composed of PEG and polyaspartate loaded with cisplatin (Nishiyama et al. 2003). In another approach, Bae et al. developed a novel technique for the conjugation of doxorubicin with PEG-b-poly(aspartic acid) diblock co-polymers, in which doxorubicin was linked to co-polymer by a pH-responsive hydrazine linkage, which allowed its release in the acidic tumor microenvironment (Bae and Kataoka 2009).

### 2.2.4 PEGylation of Inorganic Nanocarriers

Inorganic nanocarriers made of calcium phosphate, gold, silica, and iron oxide can be prepared, resulting in nanocarriers with uniform size with easy surface functionalization (Karakoti et al. 2011). Inorganic nanocarriers have low stability and could be toxic in biological systems. However, surface PEGylation can improve their biological stability and biocompatibility.

PEGylation methods for gold nanocarriers have been frequently done using thiol (SH) terminated PEGs since the strong binding of thiol groups to gold (SAu bond energy = 47 kcal/mol) (Fig. 2.3). Accordingly, the stability of gold colloids increases against aggregation in different buffers/mediums and at high ionic concentrations. Short-chain low molecular weight PEG (below 5000 Da) is often used to offer enough surface coverage to cover the surface of nanocarriers completely. Monofunctional PEG-SH can be used to render the surface of gold nanocarrier passive when no other surface ligands are required (Karakoti et al. 2011). On the other hand, Brown et al. use heterobifunctional PEG with a thiol end coated on a naked

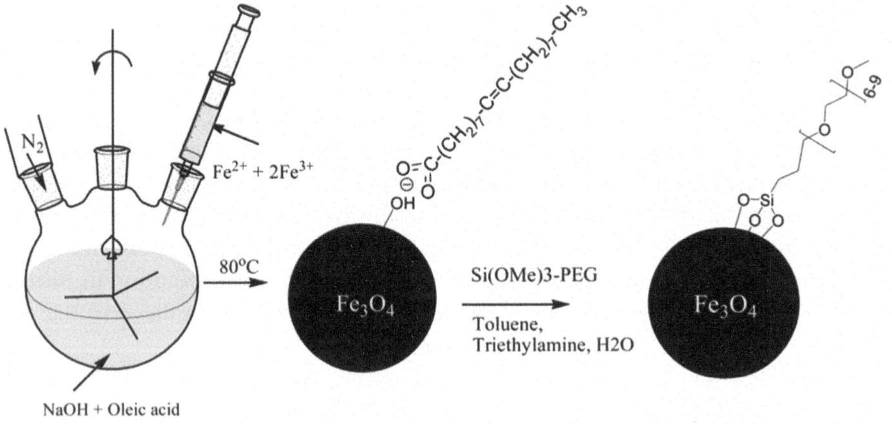


**Fig. 2.15** Preparation of PEGylated magnetic iron oxide nanocarriers. In the first step, iron ions are precipitated under basic conditions in the presence of oleic acid, producing oleic-acid-coated nanocarriers. In the second step, performed in toluene, oleic acid is replaced with a silane group using ligand exchange with silane-PEG, resulting in PEGylated magnetic iron oxide nanocarriers MNPs. ((Larsen et al. 2009) with permission)

nanoparticle surface and carboxylate functionalities to conjugate oxaliplatin to the gold nanocarriers (Brown et al. 2010).

PEGylation of magnetic iron oxide nanocarriers is frequently done via a combination of PEG–silane due to the very high binding affinity of the silane group ($SiH_4$) with the surface of oxide nanocarriers, Fig. 2.15 shows an example of preparing PEGylated iron oxide nanocarriers. Two ways are mainly used for preparing PEGylated magnetic iron oxide nanocarriers:

1. Coating the nanocarriers first with a silane group via APTMS (aminopropyl trimethoxy silane) or APTES (aminopropyl tri ethoxy silane) followed by functionalizing the terminal amine group with a carboxy terminated PEG.
2. Producing PEG–silane first and then reacting with magnetic iron oxide nanocarriers to get PEGylated magnetic iron oxide nanocarriers in one step with silane as the primary shell and PEG forming the external shell (Larsen et al. 2009).

PEGylation of silicon nanocarriers can be performed by hydrosilylation of the Si_H terminated silicon nanocarriers with alkene-terminated PEG using chloroplatinic acid as a catalyst (Mangeney et al. 2002).

## 2.3 Assessments of Surface PEGylation Efficiency

### 2.3.1 Indirect Assessment (Qualitative Assessment)

To evaluate the existence of PEG on a particle surface, many indirect, qualitative measurements are used based on the changes in the physical or chemical properties upon PEGylation. Indirect assessment includes monitoring the particle size, measuring zeta potential ($\zeta$), the surface measurements of hydrophilicity on the PEGylated nanocarrier, and microscopic techniques (Rabanel et al. 2014).

#### 2.3.1.1 Particle Size

Several methods are available to measure particle diameters. The most used method is dynamic light scattering (DLS), also called photon correlation spectroscopy (PCS). The measured particle size depends on the core size of the nanocarrier and any attached or loosely bound polymer or solvent/ion molecules to the particle surface (Fig. 2.16). The measured particle size by DLS is expressed as a "hydrodynamic radius" (RH). Additionally, it offers values comparable to size values obtained by techniques such as transmission electron microscopy (TEM) (Rabanel et al. 2014).

Typically, attaching PEG on a nanocarrier surface increases the nanocarrier diameter. Consequently, measuring the hydrodynamic diameters of nanocarrier before and after PEG grafting has been used to evaluate the degree of surface PEGylation, wherein the diameter increases as the PEG layer thickness increases, particularly in the brush regime (Fig. 2.16). Both the PEG molecular weight and coverage-density determine the PEG layer thickness, and it usually takes a value between 1 and 10 nm (Thierry and Griesser 2012; Gaumet et al. 2008).

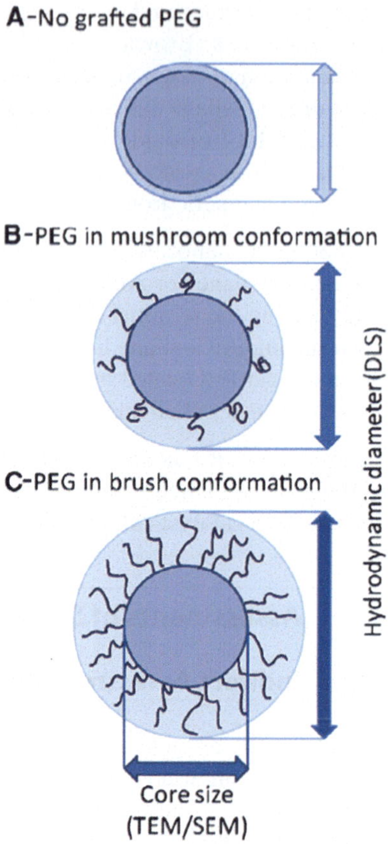


**Fig. 2.16** Core diameter of the nongrafted nanocarrier vs. hydrodynamic diameter PEGylated nanocarrier. (**a**) Nongrafted nanocarrier; (**b**) PEGylated nanocarrier with mushroom conformation; and (**c**) PEGylated nanocarrier with brush conformation. ((Rabanel et al. 2014) with permission)

For example, the particle size of the silica nanocarrier before and after attaching PEG 5 kD revealed differences in diameter consistent with the radius of gyration of PEG (Thierry and Griesser 2012). Similarly, Redhead et al. attach PEG by physical adsorption to the surface of PLGA nanocarrier by incubating them with Poloxamine 908®. Which causes an increase in mean particle size after physisorption but with low values (4–6 nm layer thickness) (Redhead et al. 2001).

The DLS technique has the following limitations:

1. Only spherical nanocarriers, monodisperse (narrow width of distribution) particles with sizes below 1 μm can be analyzed (Gaumet et al. 2008). Other methods, such as microscopy, should be used for nanocarriers with different shapes (Gaumet et al. 2008; Thierry and Griesser 2012).
2. Several factors limit the sensitivity of DLS. Minor differences in the layer thickness may not be detected when measuring particle sizes larger than this range (>100–200 nm). Conversely, little differences can be seen for smaller particles (<20 nm), with a constant polydispersity index (PDI) before and after grafting (Thierry and Griesser 2012; Gaumet et al. 2008).

3. Several factors can significantly affect the measurements, such as viscosity, pH, temperature, the release of surfactants from the particle surface, salt concentration changes, and nanocarrier concentration, which should be controlled to preserve similar conditions (Hackey and Clogston 2010).
4. Only recommended if the PEG layer is added after the formation of the nanocarrier (Rabanel et al. 2014).

#### 2.3.1.2 Zeta Potential

Measuring the zeta potential to assess the PEGylation efficiency is a simple, widely used indirect technique. It has been applied to various types of nanocarriers, including liposomes (Webb et al. 1998; Dadashzadeh et al. 2008), polymeric nanocarriers (Gref et al. 2000; Dunn et al. 1994; Shi et al. 2006; Li et al. 2001; Craparo et al. 2006; Peracchia et al. 1998), solid lipid nanoparticles (SLNs) (Bocca et al. 1998), nanocapsules (Mosqueira et al. 1999; Loch-Neckel et al. 2007), gold nanocarriers (Owens et al. 2007), nano complexes (Neu et al. 2007), magnetic nanocarriers (Jain et al. 2005), and core–shell nanocarriers (Zahr et al. 2006).

An alteration in the $\zeta$ potential upon the adsorption of PEGylated macromolecules on a different type of particles has been reported (Redhead et al. 2001; Stolnik et al. 1994; Suma et al. 2012; Poon et al. 2011). A reduction in the $\zeta$ potential occurs by increasing the surface coverage density of PEG chains due to an increase in the PEG layer thickness (Meng et al. 2004). However, by increasing PEG chain lengths, comparable results could be obtained. Craparo et al. reported a reduction in the surface charge of poly(hydroxy ethyl aspartamide methacrylate) (PHM) nanocarriers as the amount of PEG2000 used was increased, as shown in Table 2.1 (Craparo et al. 2006). In another study, the absolute value of the zeta potential for PEGylated solid lipid nanoparticles (pSLN) was reduced from about 20 to 15 mV by increasing the amount of PEG (Yuan et al. 2013).

However, the zeta potential method has several limitations including:

1. It is challenging to have quantitative correlations between PEG coverage density and $\zeta$. In most studies, a plateau of the zeta potential value is reached quickly at a low PEG content (Gref et al. 2000). Rahme et al. showed a non-linear relationship between PEG molecular weight and the $\zeta$ potential of gold particles, with a plateau (a zeta potential minimum) at around 20 kD PEG (Rahme et al. 2013).

**Table 2.1** Mean diameter (nm) and the zeta potential for PEGylated poly(hydroxyethylaspartamide methacrylated) nanocarrier

| Extent of PEGylation[a] | Mean diameter (nm) | Zeta potential (mV) |
|---|---|---|
| 0 | 147.0 | −49.10 ± 6.05 |
| 15 | 184.2 | −19.63 ± 5.72 |
| 30 | 189.6 | −16.45 ± 4.32 |
| 50 | 237.5 | −15.86 ± 4.65 |

A decrease in the zeta potential of nanocarriers in water is noticed as the extent of PEGylation increases (Craparo et al. 2006)

[a]Expressed as % of PHM-PEG2000 co-polymer weight on the total co-polymer amount

On the other hand, Meng et al. reported a linear relationship between PEG surface concentration and ζ (Meng et al. 2004).

2. The zeta potential value depends on other factors, not only the PEG layer, such as pH, buffer type, and salt concentration, so it is impractical to compare results obtained from different environments (Ebbesen et al. 2013).
3. No correlations between the zeta potential and PEG grafting are evident in some cases, such as when PEG is being utilized to coat an already neutral surface. For example, Webb et al. noted no difference between the surface charges of liposomes of egg sphingomyelin/cholesterol (SM/chol) and PEGylated ones, as near zero potential was measured for both (Webb et al. 1998). In addition, PEGylation of PLA or PLGA nanocarriers has revealed a weak effect on the highly negative zeta potential despite adding a significant amount of PEG (Duncanson et al. 2007; Tobio et al. 1998).

#### 2.3.1.3 Surface Hydrophilicity

Several methods can be used to assess the surface hydrophilicity of nanocarriers, where PEGylated nanocarriers show higher surface hydrophilicity compared to non-PEGylated nanocarriers.

One of those methods is the Rose Bengal method (Doktorovova et al. 2012). Figure 2.17 represents the surface hydrophobicity of PEGylated zein nanocarriers using the Rose Bengal test as a function of the PEG-to-zein ratio. As the amount of

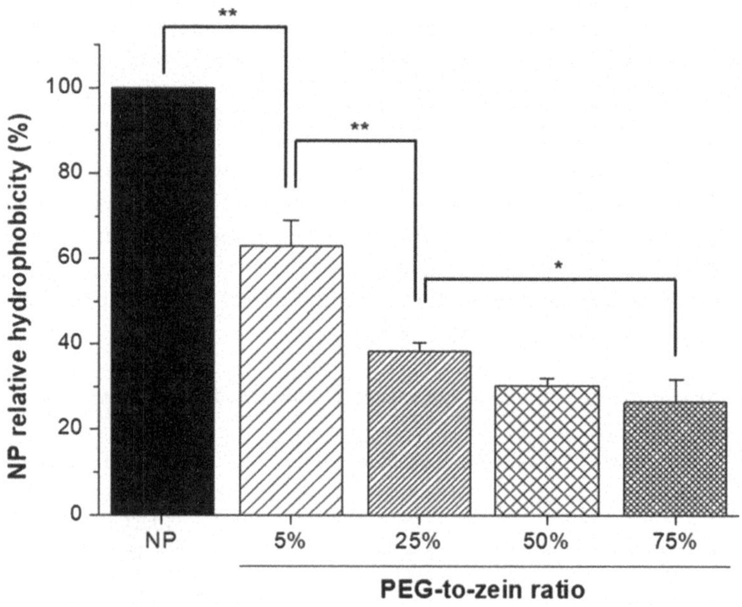


**Fig. 2.17** Surface hydrophobicity of zein nanocarriers as a function of the PEG-to-zein ratio. ((Reboredo et al. 2021) with permission)

PEG used during the coating process increases, the surface hydrophobicity decreases (Reboredo et al. 2021).

Another method used is comparing the HIC (hydrophobic interaction chromatography) ratios for PEGylated nanocarriers with non-PEGylated ones. This method is based on the measurement of the surface hydrophobicity/hydrophilicity of a particle by washing the particles present in a column twice. The more hydrophilic particles will be eluted in the first wash, while the more hydrophobic particles will be eluted in the second wash. Then HIC ratio is calculated using Eq. (2.1).

$$\text{HIC ratio} = \frac{\text{AUC elution peak}}{\text{AUC wash peak}} \times 100 \tag{2.1}$$

More hydrophilic particles exhibit a larger HIC ratio. So the higher the degree of PEGylation, the higher the HIC ratio. For illustration, Peracchia et al. utilized this method to confirm the PEGylation of poly(hexadecyl cyanoacrylate) (PHDCA) nanocarriers (Peracchia et al. 1998). Nevertheless, this method has some limitations. It is not quantitative, it only applies to particles with a highly hydrophobic surface, and it also cannot be used alone to expect in vivo behavior of a nanocarrier (Howard et al. 2008).

### 2.3.1.4 Microscopic Techniques

A possible way to examine PEGylated nanocarrier is by using microscopic techniques, such as scanning electron microscopy (SEM) and electronic transmission microscopy (TEM). They have been used quite often to characterize the morphology of nanocarriers (Howard et al. 2008) and measure the size of the core of the particles (Bonevich and Haller 2010; Sitterberg et al. 2010; Roe et al. 2004) (Fig. 2.16). However, they are not commonly applied as a characterization technique for PEGylation (Howard et al. 2008) since most polymeric layers on

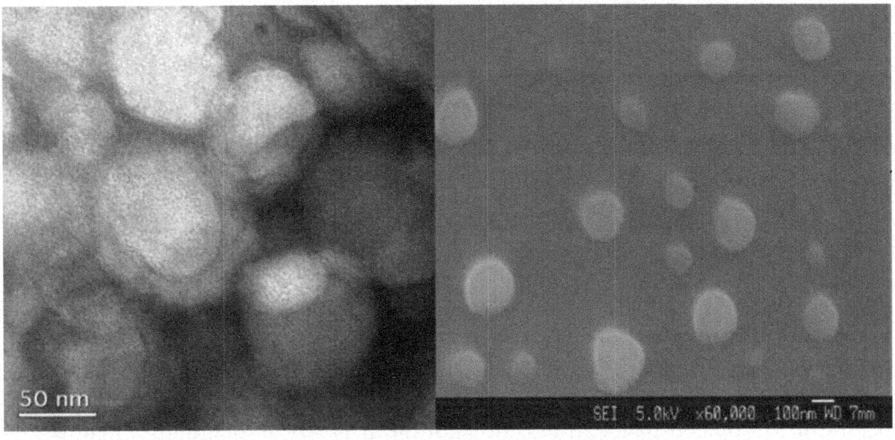


**Fig. 2.18** TEM and SEM micrographs of diblock and star-shaped copolymer micelle-like nanocarriers. ((Jie et al. 2005) with permission)

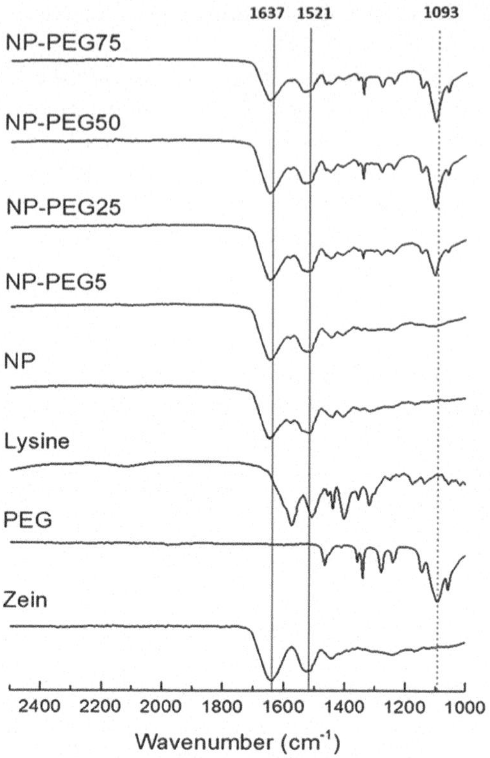


**Fig. 2.19** FTIR spectra of zein, PEG 35,000, lysine, bare zein nanocarriers, and PEG-coated zein nanocarriers. Straight lines correspond to 1637 and 1521/cm stretching vibration bands. The dashed line corresponds to the 1093/cm band ((Reboredo et al. 2021) with permission)

nanocarrier are not easily visible as they collapse in vacuum and have low electronic density (Sitterberg et al. 2010; Roe et al. 2004). Generally, staining procedures can be used to characterize the PEGylation, such as negative staining of PEGylated micelle-like nanocarriers with phosphotungstic acid that results in a bright hydrophobic core surrounded by a gray hydrophilic shell (Fig. 2.18) (Jie et al. 2005).

#### 2.3.1.5 Fourier Transform-Infrared Spectroscopy (FT-IR)

Fourier transform-infrared spectroscopy (FT-IR) is a qualitative method used to assess the PEGylation of nanocarriers (Howard et al. 2008). PEGylation is generally confirmed by the following absorption bands:

- The stretching vibration of C–O–C (1143/cm).
- The vibration of $CH_2$ groups (1465/cm).
- The C–O vibration of the OH end group of PEG (1093/cm).

Figure 2.19 demonstrates the FTIR spectra of the different nanocarriers and the raw materials utilized in their preparation. The PEGylation of the nanocarrier is confirmed by the presence of PEG vibration bands (1465, 1143, and 1058/cm) in the spectra of PEGylated nanocarriers. In addition, it is worth mentioning that as the PEG-to-zein ratio increases, the size of these signals increases (Reboredo et al. 2021).

### 2.3.2 Direct Assessment (Quantitative Assessment)

#### 2.3.2.1 Colorimetric Methods

Colorimetric assays have been utilized to quantify and localize PEG chains (Shi et al. 2006; Bazile et al. 1995). In the iodine/potassium iodide colorimetric assay or Baleux's assay, a reaction between iodine and PEG produces a blue coordination complex which is quantified via spectrophotometry (Baleux 1972). By comparing the concentration of PEG quantified before and after alkaline hydrolysis of nanocarriers, it is possible to determine if PEG is located on the particle's surface (Howard et al. 2008). Based on the measured size of the nanocarriers, an estimation of PEG coverage density and distance between adjacent chains can be done assuming a homogeneous surface (Howard et al. 2008).

An additional calorimetric method is the aqueous ferro-thiocyanate assay. It detects the PEG conjugate or nonionic PEGylated surfactant by creating a colored complex when PEG partitions in an organic phase (Cheng et al. 2012; Nag et al. 1996; Al-Hanbali et al. 2007). In both methods degrading of PEG from the nanocarrier surface is required since only PEG in the solution can be measured. Moreover, in these methods, fresh solutions and precise measurement time points are required to obtain reliable data due to the limited sensitivity and a slow decline of absorbance over time (Budijono et al. 2010).

#### 2.3.2.2 Chromatographic Methods

Chromatographic methods have been developed to detect free and un-grafted PEG chains using HPLC coupled with refractive index (RI) or viscosimetric detectors, evaporative light scattering detector (ELSD), mass spectrometry (MS), rather than UV, as PEGs lack a chromophore (Auriola et al. 1993).

An illustrative example is the quantification of free PEGs (nonattached to the gelatin nanocarrier) that separate from nanocarriers via asymmetric flow field-flow fractionation (AF4) (Fraunhofer and Winter 2004). Then HPLC coupled to a refractive index (RI) detector is used to detect and quantify free PEG in the effluent (Zillies et al. 2007). The amount of PEG grafted on nanocarriers is calculated as the difference between the initial PEG quantity and the free PEG quantified by HPLC (Zillies et al. 2007). RI detectors have low sensitivity, unstable baseline, long equilibration time, and high limit of detection due to the sensitivity of RI to impurities, temperature, and flow. On the other hand, quantification methods using ELSD and LC–MS detectors have higher sensitivity than RI and with less interference (Nair et al. 2006).

#### 2.3.2.3 UV and Fluorescence Spectroscopy

The amount of PEG can be quantified via UV and fluorescence spectroscopy using different strategies and via coupling of PEG with a chromophore or fluorophore (Rabanel et al. 2014). Two strategies of PEG quantification by fluorescence or UV spectroscopy are (Fig. 2.20) the grafting of fluorescent PEG conjugates and the labeling of grafted PEG chains.

In the first strategy, the nanocarriers are PEGylated using fluorescent PEG conjugates such as fluorescein-PEG 5 kD (Perry et al. 2012). Then the bound PEG

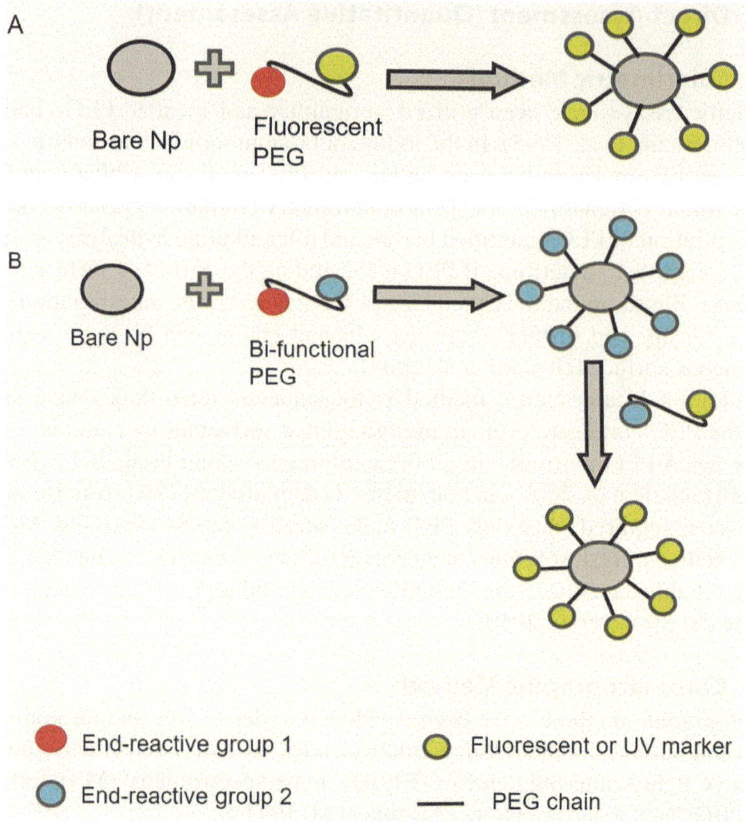


**Fig. 2.20** Methods of PEG quantification by fluorescence or UV spectroscopy. (**a**) Grafting of fluorescently labeled PEG and (**b**) labeling of grafted PEG chains with a fluorescent or UV marker

molecules are quantified by fluorescence spectroscopy. However, the problem with this approach is that conjugating the fluorescent tag to PEG could affect its physical and chemical properties, affecting chain conformation, grafting yield, and eventually surface coverage density (Rabanel et al. 2014).

The second strategy (labeling of grafted PEG chains) requires the presence of a reactive group at the end of the PEG chain to attach a fluorescent dye or specific ligand. The quantitative results are related to the yield and the rate of coupling. An example is the Kaiser test (primary amine test), which is based on the reaction of the amine group ($NH_2$) of bound PEG with ninhydrin to produce a deep blue color (Kaiser et al. 1970).

The main drawback of the UV/fluorescence approach is the need to prepare labeled particles. Hence, the quantitative measurements rely on the yield of the labeling reaction, so they may not be representative of the real particles. On the other hand, NMR and XPS analyses do not need those modifications (Rabanel et al. 2014).

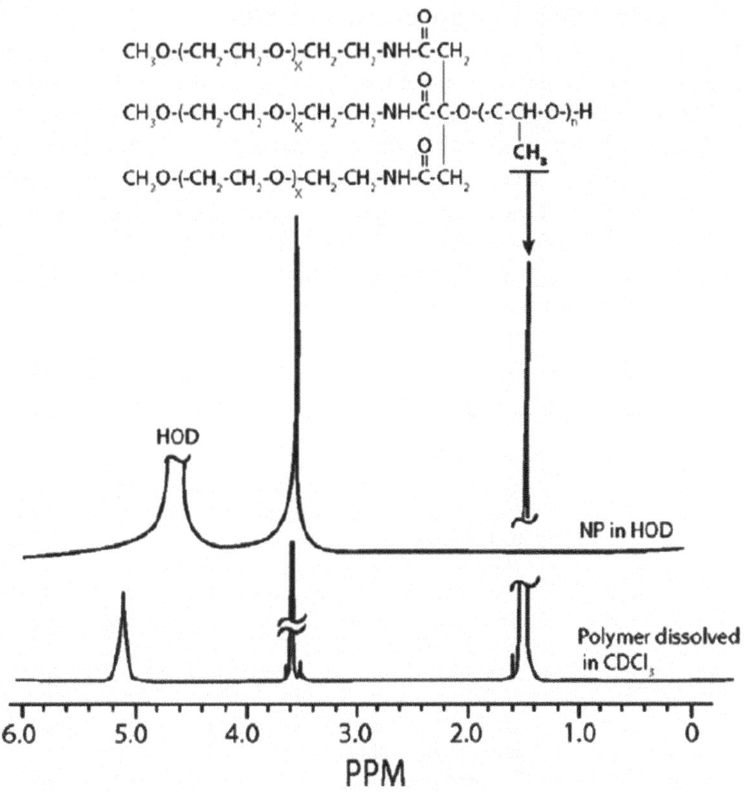


**Fig. 2.21** $^1$H NMR spectra of PEG-PLA branched multiblock co-polymer (PLA attached to three blocks of PEG) suspended in $CDCl_3$, and $^1$H NMR spectra of nanocarriers prepared with the same polymer by emulsification-solvent evaporation and suspended in $D_2O$. ((Rabanel et al. 2014) with permission)

### 2.3.2.4 Nuclear Magnetic Resonance (NMR)

NMR can assess the degree of PEGylation without any post-modification. It simply depends on the presence or absence of specific chemical moieties. HrKach et al. characterize qualitatively the PEGylation of nanocarriers made of the PEG-PLA diblock. The PEGylation of nanocarrier was confirmed by the presence of a signal in $^1$H NMR compared to free PEG in deuterated water ($D_2O$). At the same time, the polymeric solid inner core made of hydrophobic PLA segments gave no NMR signals (Fig. 2.21) (Hrkach et al. 1997). In addition, PEG coating efficacy (% of total PEG found on the surface) of PLA-PEG nanocarriers can be calculated as the ratio of surface PEG acquired in $D_2O$ analysis to the total PEG obtained from nanocarriers dissolved in deuterated chloroform ($CDCl_3$) (Sheng et al. 2009).

#### 2.3.2.5 X-Ray Photoelectron Spectroscopy (XPS)

X-ray photoelectron spectroscopy (XPS) is a technique used to characterize the chemical composition of surfaces and allows for some quantification (Briggs 1990; Briggs and Grant 2003). It is mainly used to identify the relative percentage of the different chemical states of an atom; therefore, the chemical functionalities that exist at the nanocarrier surface. The presence of PEG on the surface can be confirmed by comparing spectra before and after PEGylation (Howard et al. 2008). For example, suppose the particle has no ether compounds. In that case, the appearance of an ether bond specific to PEG (C–O–C peak) and the signal integration are related to PEG concentration (Brindley et al. 1995).

**Table 2.2** A list of potential alternatives of PEG

| Polymers | Comments |
|---|---|
| Poly(glycerols) (PGs) | • Could be hyperbranched or linear, with a high degree of functionalization due to multiple hydroxyl groups (Siegers et al. 2004)<br>• Have low immunogenicity and good biocompatibility<br>• Hyperbranched PGs are more highly resistant to oxidation or thermal stress than PEG and demonstrate long plasma half-lives (Siegers et al. 2004)<br>• Application to liposomes was demonstrated (Hofmann et al. 2010)<br>• It tends to accumulate in the kidneys and liver, limiting their use (Abbina and Parambath 2018) |
| Poly(oxazolines) (POX) | • Soluble in both hydrophilic and hydrophobic solvents (Abbina and Parambath 2018)<br>• Less prone to oxidation reactions, unlike PEG with no bioaccumulation (Khutoryanskiy 2018)<br>• Thermo-sensitiveness (Abbina and Parambath 2018)<br>• Difficult, costly synthesis (Khutoryanskiy 2018; Hadjesfandiari and Parambath 2018)<br>• Studies are still to be done to prove their effectiveness as an alternative to PEG (Pytela et al. 1989) |
| Poly(acrylamide) and poly(methacrylamide) | • Such as poly(hydroxypropyl methacrylate) (PHPMA) and poly(2-hydroxyethyl methacrylate) (PHEMA) are nonionic polymers that are widely used in biomedical applications (Nunvářová et al. 2019)<br>• Have low immunogenicity, biocompatibility, and a prolonged circulation time. However, the highly toxic monomers limit their use as PEG alternatives (Abbina and Parambath 2018) (Hadjesfandiari and Parambath 2018)<br>• PHPMA shows excellent efficacy in preclinical studies as a carrier for chemotherapeutic drugs and has already entered clinical trials (Abbina and Parambath 2018) |

(continued)

**Table 2.2** (continued)

| Polymers | Comments |
|---|---|
| Poly(vinyl pyrrolidone) (PVP) | • Highly hydrated structure, biocompatible, and nonimmunogenic, so considered an excellent alternative to PEG (Abbina and Parambath 2018; Zhang et al. 2016; Hadjesfandiari and Parambath 2018)<br>• Limited use due to unclear immunological behavior, the nonbiodegradability tendency for bioaccumulation, and the carcinogenicity of their free monomers (Abbina and Parambath 2018) (Hadjesfandiari and Parambath 2018; Zhang et al. 2016) |
| Poly(amino acids) | • Have low toxicity and do not undergo bioaccumulation due to their degradation by protease enzymes (Romberg et al. 2007)<br>• Comparable to PEG in increasing the blood circulation of nanocarriers (Romberg et al. 2007; Metselaar et al. 2003)<br>• An example of poly(amino acids) is poly(glutamic acid) (PGA) which has already entered clinical trial (phase III) in the form of PGA–paclitaxel conjugate (Brocchini et al. 2008) |
| Polybetaines | • Such as sulfobetaine and carboxy betaine are zwitterionic polymers with a significant reduction in nonspecific protein adsorption (Zhang et al. 2008b)<br>• Poly(carboxy betaine) has multiple functional groups, which can be further functionalized to produce multifunctional nanocarriers (Jiang and Cao 2010)<br>• Poly(carboxy betaine) has been conjugated to a variety of nanocarriers such as iron oxide (Zhang et al. 2010), silica (Jia et al. 2009), PLGA (Cao et al. 2010), gold (Yang et al. 2009), and hydrogel nanocarriers (Zhang et al. 2011; Cheng et al. 2010) |
| Polysaccharides | • Include derivatives of chitosan (Kim et al. 2010; Fan et al. 2010), dextran (Mehvar 2000; Li et al. 2009), hyaluronic acid (Choi et al. 2010), and heparin (Park et al. 2007; Hou et al. 2011)<br>• Biodegradable with little toxicity and immunogenicity (Park et al. 2007; Kean and Thanou 2010; Li et al. 2011)<br>• Polysaccharide-based nanocarriers show long circulation times and improvement in their accumulation in tumors (Amoozgar and Yeo 2012) |

Limitations of this method include the difficulty in quantifying PEG on polymeric nanocarriers due to the similar composition (carbon and oxygen) for both the PEG and the polymer. In addition, the interference by carbon contamination from environmental exposure (Rabanel et al. 2014; Howard et al. 2008). Moreover, the quantification using XPS is not restricted to the surface as the probing depth is 1–10 nm.

## 2.4 Alternative Polymers for Protecting Nanocarrier Surface

Despite the numerous beneficial properties of PEG, there are several rare side effects of PEG, including hypersensitivity reaction, which can proceed to anaphylactic shock (Abuchowski et al. 1977), and the tendency to form blood clotting, which causes embolism. In addition, the limited stabilities of PEG due to chemical changes triggered by oxygen, water, and energy (Porter and Casale 1985).

Many studies focused on developing alternative polymers that can replace PEG and mimic its physicochemical properties with advantages over PEG in biodegradability, ability to functionalize, or avoiding immune responses. Table 2.2 elucidates the main alternatives to PEG briefly.

## 2.5 Conclusion and Future Perspective

PEGylation is a vital technique to improve the performance of nanocarriers in different ways, such as increasing their biocompatibility and circulation time. In addition, it increases their accumulation in the brain/tumor. Many PEGylated nanocarriers are approved for therapeutic use or undergoing clinical trials (Zhang et al. 2008a). PEGylation of nanocarriers can be achieved in different ways, either by adding PEG during the nanocarrier formation or by attaching PEG to the formed nanocarrier physically or chemically. On the other hand, it is necessary to continue improving or modifying the methods of preparing PEGylated nanocarriers and the methods to evaluate the existence of PEG on the nanocarrier surface. In addition, to find alternative polymers to PEG with superior properties.

First, more efficient methods of PEGylation must be developed rather than changing the chain conformation and length of PEG or modifying the methods of attaching PEG to nanocarriers. Recently, more focus has been on increasing the flexibility of PEG molecules which creates denser coverage and more uniform water cloud to prevent penetration by opsonins (Torchilin et al. 1994). Secondly, it may be necessary to improve the methods of characterization. There are several qualitative and quantitative assessment methods, but each has its limitations. More efforts should be made to develop a new rapid, inexpensive strategy whose results would describe the homogeneity of PEG on the nanocarrier surface and correlate with in vivo results (Howard et al. 2008).DisclosuresThere is no conflict of interest and disclosures associated with the manuscript.

## References

Abbina S, Parambath A (2018) PEGylation and its alternatives: a summary. In: Engineering of biomaterials for drug delivery systems. Elsevier, Amsterdam

Abuchowski A, Van Es T, Palczuk N, Davis F (1977) Alteration of immunological properties of bovine serum albumin by covalent attachment of polyethylene glycol. J Biol Chem 252:3578–3581

Al-Hanbali O, Onwuzo NM, Rutt KJ, Dadswell CM, Moghimi SM, Hunter AC (2007) Modification of the Stewart biphasic colorimetric assay for stable and accurate quantitatitive determination of Pluronic and Tetronic block copolymers for application in biological systems. Anal Biochem 361:287–293

Aliabadi HM, Shahin M, Brocks DR, Lavasanifar A (2008) Disposition of drugs in block copolymer micelle delivery systems: from discovery to recovery. Clin Pharmacokinet 47:619–634

Allen TM, Cullis PR (2013) Liposomal drug delivery systems: from concept to clinical applications. Adv Drug Deliv Rev 65:36–48
Allen TM, Hansen CB, de Menezes DEL (1995) Pharmacokinetics of long-circulating liposomes. Adv Drug Deliv Rev 16:267–284
Amoozgar Z, Yeo Y (2012) Recent advances in stealth coating of nanoparticle drug delivery systems. Wiley Interdiscip Rev Nanomed Nanobiotechnol 4:219–233
Auriola SO, Rönkkö KM, Urtti A (1993) Determination of polyethylene glycols by high-performance liquid chromatography—thermospray mass spectrometry. J Pharm Biomed Anal 11:1027–1032
Awasthi V, Garcia D, Klipper R, Goins BA, Phillips WT (2004) Neutral and anionic liposome-encapsulated hemoglobin: effect of postinserted poly (ethylene glycol)-distearoylphosphatidylethanolamine on distribution and circulation kinetics. J Pharmacol Exp Ther 309:241–248
Bae Y, Kataoka K (2009) Intelligent polymeric micelles from functional poly (ethylene glycol)-poly (amino acid) block copolymers. Adv Drug Deliv Rev 61:768–784
Baleux B (1972) Colorimetric determination of nonionic polyethylene oxide surfactants using an iodine-iodide solution. CR Acad Sci Ser C 279:1617–1620
Barenholz YC (2012) Doxil®—the first FDA-approved nano-drug: lessons learned. J Control Release 160:117–134
Bazile D, Prud'homme C, Bassoullet MT, Marlard M, Spenlehauer G, Veillard M (1995) Stealth Me. PEG-PLA nanoparticles avoid uptake by the mononuclear phagocytes system. J Pharm Sci 84:493–498
Beletsi A, Panagi Z, Avgoustakis K (2005) Biodistribution properties of nanoparticles based on mixtures of PLGA with PLGA–PEG diblock copolymers. Int J Pharm 298:233–241
Bocca C, Caputo O, Cavalli R, Gabriel L, Miglietta A, Gasco MR (1998) Phagocytic uptake of fluorescent stealth and non-stealth solid lipid nanoparticles. Int J Pharm 175:185–193
Bockstaller MR, Mickiewicz RA, Thomas EL (2005) Block copolymer nanocomposites: perspectives for tailored functional materials. Adv Mater 17:1331–1349
Bonevich JE, Haller WK (2010) Measuring the size of nanoparticles using transmission electron microscopy (TEM). NSIT-NCL Joint Assay Protocol, PCC-7
Bouxsein NF, Mcallister CS, Ewert KK, Samuel CE, Safinya CR (2007) Structure and gene silencing activities of monovalent and pentavalent cationic lipid vectors complexed with siRNA. Biochemistry 46:4785–4792
Bovone G, Cousin L, Steiner F, Tibbitt MW (2022) Solvent controls nanoparticle size during nano-precipitation by limiting block copolymer assembly. Macromolecules 55:8040–8048
Breed DR, Thibault R, Xie F, Wang Q, Hawker CJ, Pine DJ (2009) Functionalization of polymer microspheres using click chemistry. Langmuir 25:4370–4376
Briggs D (1990) Practical surface analysis. In: Auger and X-ray photoelecton spectroscopy, vol 1. Wiley, Chichester, pp 151–152
Briggs D, Grant JT (2003) Surface analysis by auger and X-ray photoelectron spectroscopy. IM Publications LLP, Chichester
Brindley A, Davis S, Davies M, Watts J (1995) Polystyrene colloids with surface-grafted polyethylene oxide as model systems for site-specific drug delivery: I. Preparation and surface chemical characterization using SIMS and XPS. J Colloid Interface Sci 171:150–161
Brocchini S, Godwin A, Balan S, Choi J-W, Zloh M, Shaunak S (2008) Disulfide bridge based PEGylation of proteins. Adv Drug Deliv Rev 60:3–12
Brown SD, Nativo P, Smith J-A, Stirling D, Edwards PR, Venugopal B, Flint DJ, Plumb JA, Graham D, Wheate NJ (2010) Gold nanoparticles for the improved anticancer drug delivery of the active component of oxaliplatin. J Am Chem Soc 132:4678–4684
Budijono SJ, Russ B, Saad W, Adamson DH, Prud'Homme RK (2010) Block copolymer surface coverage on nanoparticles. Colloids Surf A Physicochem Eng Asp 360:105–110
Cao Z, Yu Q, Xue H, Cheng G, Jiang S (2010) Nanoparticles for drug delivery prepared from amphiphilic PLGA zwitterionic block copolymers with sharp contrast in polarity between two blocks. Angew Chem Int Ed 49:3771–3776

Cavalli S, Tipton AR, Overhand M, Kros A (2006) The chemical modification of liposome surfaces via a copper-mediated [3+ 2] azide–alkyne cycloaddition monitored by a colorimetric assay. Chem Commun:3193–3195

Chen J, Spear SK, Huddleston JG, Rogers RD (2005) Polyethylene glycol and solutions of polyethylene glycol as green reaction media. Green Chem 7:64–82

Chen B, Van Der Poll DG, Jerger K, Floyd WC, Fréchet JM, Szoka FC (2011) Synthesis and properties of star-comb polymers and their doxorubicin conjugates. Bioconjug Chem 22:617–624

Cheng G, Mi L, Cao Z, Xue H, Yu Q, Carr L, Jiang S (2010) Functionalizable and ultrastable zwitterionic nanogels. Langmuir 26:6883–6886

Cheng T-L, Chuang K-H, Chen B-M, Roffler SR (2012) Analytical measurement of PEGylated molecules. Bioconjug Chem 23:881–899

Choi KY, Chung H, Min KH, Yoon HY, Kim K, Park JH, Kwon IC, Jeong SY (2010) Self-assembled hyaluronic acid nanoparticles for active tumor targeting. Biomaterials 31:106–114

Craparo EF, Cavallaro G, Bondì ML, Mandracchia D, Giammona G (2006) PEGylated nanoparticles based on a polyaspartamide. Preparation, physico-chemical characterization, and intracellular uptake. Biomacromolecules 7:3083–3092

Cu Y, Saltzman WM (2009) Controlled surface modification with poly (ethylene) glycol enhances diffusion of PLGA nanoparticles in human cervical mucus. Mol Pharm 6:173–181

Dadashzadeh S, Vali A, Rezaie M (2008) The effect of PEG coating on in vitro cytotoxicity and in vivo disposition of topotecan loaded liposomes in rats. Int J Pharm 353:251–259

Damodaran VB, Fee C (2010) Protein PEGylation: an overview of chemistry and process considerations. Eur Pharm Rev 15:18–26

Doktorovova S, Shegokar R, Martins-Lopes P, Silva A, Lopes CM, Müller R, Souto EB (2012) Modified Rose Bengal assay for surface hydrophobicity evaluation of cationic solid lipid nanoparticles (cSLN). Eur J Pharm Sci 45:606–612

Drobek T, Spencer ND, Heuberger M (2005) Compressing PEG brushes. Macromolecules 38:5254–5259

Duncanson WJ, Figa MA, Hallock K, Zalipsky S, Hamilton JA, Wong JY (2007) Targeted binding of PLA microparticles with lipid-PEG-tethered ligands. Biomaterials 28:4991–4999

Dunn SE, Brindley A, Davis SS, Davies MC, Illum L (1994) Polystyrene-poly (ethylene glycol)(PS-PEG2000) particles as model systems for site specific drug delivery. 2. The effect of PEG surface density on the in vitro cell interaction and in vivo biodistribution. Pharm Res 11:1016–1022

Ebbesen M, Whitehead B, Ballarin-Gonzalez B, Kingshott P, Howard K (2013) Surface analysis of PEGylated nano-shields on nanoparticles installed by hydrophobic anchors. Pharm Res 30:1758–1767

Ebrahimnejad P, Dinarvand R, Jafari MR, Tabasi SAS, Atyabi F (2011) Characterization, blood profile and biodistribution properties of surface modified PLGA nanoparticles of SN-38. Int J Pharm 406:122–127

Fan L, Li F, Zhang H, Wang Y, Cheng C, Li X, Gu C-H, Yang Q, Wu H, Zhang S (2010) Co-delivery of PDTC and doxorubicin by multifunctional micellar nanoparticles to achieve active targeted drug delivery and overcome multidrug resistance. Biomaterials 31:5634–5642

Fehring V, Schaeper U, Ahrens K, Santel A, Keil O, Eisermann M, Giese K, Kaufmann J (2014) Delivery of therapeutic siRNA to the lung endothelium via novel lipoplex formulation DACC. Mol Ther 22:811–820

Fraunhofer W, Winter G (2004) The use of asymmetrical flow field-flow fractionation in pharmaceutics and biopharmaceutics. Eur J Pharm Biopharm 58:369–383

Friedman AD, Claypool SE, Liu R (2013) The smart targeting of nanoparticles. Curr Pharm Des 19:6315–6329

Fu W, Shenoy D, Li J, Crasto C, Jones G, Dimarzio C, Sridhar S, Amiji M (2004) Biomedical applications of gold nanoparticles functionalized using hetero-bifunctional poly (ethylene glycol) spacer. MRS Online Proc Libr 845(AA5):4

Gajbhiye V, Vijayaraj Kumar P, Kumar Tekade R, Jain N (2007) Pharmaceutical and biomedical potential of PEGylated dendrimers. Curr Pharm Des 13:415–429

Gaumet M, Vargas A, Gurny R, Delie F (2008) Nanoparticles for drug delivery: the need for precision in reporting particle size parameters. Eur J Pharm Biopharm 69:1–9
Gref R, Minamitake Y, Peracchia MT, Trubetskoy V, Torchilin V, Langer R (1994) Biodegradable long-circulating polymeric nanospheres. Science 263:1600–1603
Gref R, Lück M, Quellec P, Marchand M, Dellacherie E, Harnisch S, Blunk T, Müller R (2000) 'Stealth' corona-core nanoparticles surface modified by polyethylene glycol (PEG): influences of the corona (PEG chain length and surface density) and of the core composition on phagocytic uptake and plasma protein adsorption. Colloids Surf B Biointerfaces 18:301–313
Grubbs RB (2005) Hybrid metal–polymer composites from functional block copolymers. J Polym Sci A Polym Chem 43:4323–4336
Gu F, Zhang L, Teply BA, Mann N, Wang A, Radovic-Moreno AF, Langer R, Farokhzad OC (2008) Precise engineering of targeted nanoparticles by using self-assembled biointegrated block copolymers. Proc Natl Acad Sci 105:2586–2591
Hackey VA, Clogston JD (2010) Measuring the size of nanoparticles in aqueous media using batch-mode dynamic light scattering. NIST-NCL Joint Assay Protocol
Hadjesfandiari N, Parambath A (2018) Stealth coatings for nanoparticles: polyethylene glycol alternatives. In: Engineering of biomaterials for drug delivery systems. Elsevier, Amsterdam
Halder J, Kamat AA, Landen CN Jr, Han LY, Lutgendorf SK, Lin YG, Merritt WM, Jennings NB, Chavez-Reyes A, Coleman RL (2006) Focal adhesion kinase targeting using in vivo short interfering RNA delivery in neutral liposomes for ovarian carcinoma therapy. Clin Cancer Res 12:4916–4924
Harris JM (1985) Laboratory synthesis of polyethylene glycol derivatives. J Macromol Sci 25:325–373
Heald CR, Stolnik S, Kujawinski K, De Matteis C, Garnett MC, Illum L, Davis SS, Purkiss S, Barlow RJ, Gellert P (2002) Poly (lactic acid)−poly (ethylene oxide)(PLA−PEG) nanoparticles: NMR studies of the central solidlike PLA core and the liquid PEG corona. Langmuir 18:3669–3675
Hensley ML, Hoppe B, Leon L, Sabbatini P, Aghajanian C, Chi D, Spriggs DR (2001) The costs and efficacy of liposomal doxorubicin in platinum-refractory ovarian cancer in heavily pretreated patients. Gynecol Oncol 82:464–469
Hofmann AM, Wurm F, Hühn E, Nawroth T, Langguth P, Frey H (2010) Hyperbranched polyglycerol-based lipids via oxyanionic polymerization: toward multifunctional stealth liposomes. Biomacromolecules 11:568–574
Hou L, Fan Y, Yao J, Zhou J, Li C, Fang Z, Zhang Q (2011) Low molecular weight heparin-all-trans-retinoid acid conjugate as a drug carrier for combination cancer chemotherapy of paclitaxel and all-trans-retinoid acid. Carbohydr Polym 86:1157–1166
Howard MD, Jay M, Dziubla TD, Lu X (2008) PEGylation of nanocarrier drug delivery systems: state of the art. J Biomed Nanotechnol 4:133–148
Hrkach JS, Peracchia MT, Bomb A, Langer R (1997) Nanotechnology for biomaterials engineering: structural characterization of amphiphilic polymeric nanoparticles by 1H NMR spectroscopy. Biomaterials 18:27–30
Hrkach J, Von Hoff D, Ali MM, Andrianova E, Auer J, Campbell T, De Witt D, Figa M, Figueiredo M, Horhota A (2012) Preclinical development and clinical translation of a PSMA-targeted docetaxel nanoparticle with a differentiated pharmacological profile. Sci Transl Med 4:128ra39
Huckaby JT, Lai SK (2018) PEGylation for enhancing nanoparticle diffusion in mucus. Adv Drug Deliv Rev 124:125–139
Hussain Z, Khan S, Imran M, Sohail M, Shah SWA, De Matas M (2019) PEGylation: a promising strategy to overcome challenges to cancer-targeted nanomedicines: a review of challenges to clinical transition and promising resolution. Drug Deliv Transl Res 9:721–734
Jain TK, Morales MA, Sahoo SK, Leslie-Pelecky DL, Labhasetwar V (2005) Iron oxide nanoparticles for sustained delivery of anticancer agents. Mol Pharm 2:194–205
Jangid AK, Patel K, Joshi U, Patel S, Singh A, Pooja D, Saharan VA, Kulhari H (2022) PEGylated G4 dendrimers as a promising nanocarrier for piperlongumine delivery: synthesis, characterization, and anticancer activity. Eur Polym J 179:111547

Jazayeri MH, Amani H, Pourfatollah AA, Pazoki-Toroudi H, Sedighimoghaddam B (2016) Various methods of gold nanoparticles (GNPs) conjugation to antibodies. Sensing Biosensing Res 9:17–22

Jia G, Cao Z, Xue H, Xu Y, Jiang S (2009) Novel zwitterionic-polymer-coated silica nanoparticles. Langmuir 25:3196–3199

Jiang S, Cao Z (2010) Ultralow-fouling, functionalizable, and hydrolyzable zwitterionic materials and their derivatives for biological applications. Adv Mater 22:920–932

Jie P, Venkatraman SS, Min F, Freddy BYC, Huat GL (2005) Micelle-like nanoparticles of star-branched PEO–PLA copolymers as chemotherapeutic carrier. J Control Release 110:20–33

Kaiser E, Colescott RL, Bossinger CD, Cook P (1970) Color test for detection of free terminal amino groups in the solid-phase synthesis of peptides. Anal Biochem 34:595–598

Kapoor M, Burgess DJ (2012) Efficient and safe delivery of siRNA using anionic lipids: formulation optimization studies. Int J Pharm 432:80–90

Karakoti AS, Das S, Thevuthasan S, Seal S (2011) PEGylated inorganic nanoparticles. Angew Chem Int Ed 50:1980–1994

Karnik R, Gu F, Basto P, Cannizzaro C, Dean L, Kyei-Manu W, Langer R, Farokhzad OC (2008) Microfluidic platform for controlled synthesis of polymeric nanoparticles. Nano Lett 8:2906–2912

Kean T, Thanou M (2010) Biodegradation, biodistribution and toxicity of chitosan. Adv Drug Deliv Rev 62:3–11

Khutoryanskiy VV (2018) Beyond PEGylation: alternative surface-modification of nanoparticles with mucus-inert biomaterials. Adv Drug Deliv Rev 124:140–149

Kim K, Kim JH, Park H, Kim Y-S, Park K, Nam H, Lee S, Park JH, Park R-W, Kim I-S (2010) Tumor-homing multifunctional nanoparticles for cancer theragnosis: simultaneous diagnosis, drug delivery, and therapeutic monitoring. J Control Release 146:219–227

Kim AJ, Boylan NJ, Suk JS, Hwangbo M, Yu T, Schuster BS, Cebotaru L, Lesniak WG, Oh JS, Adstamongkonkul P (2013) Use of single-site-functionalized PEG Dendrons to prepare gene vectors that penetrate human mucus barriers. Angew Chem 125:4077–4080

Kim J, Kong YP, Niedzielski SM, Singh RK, Putnam AJ, Shikanov A (2016) Characterization of the crosslinking kinetics of multi-arm poly (ethylene glycol) hydrogels formed via Michael-type addition. Soft Matter 12:2076–2085

Knop K, Hoogenboom R, Fischer D, Schubert US (2010) Poly (ethylene glycol) in drug delivery: pros and cons as well as potential alternatives. Angew Chem Int Ed 49:6288–6308

Kolishetti N, Dhar S, Valencia PM, Lin LQ, Karnik R, Lippard SJ, Langer R, Farokhzad OC (2010) Engineering of self-assembled nanoparticle platform for precisely controlled combination drug therapy. Proc Natl Acad Sci 107:17939–17944

Kumar A, Erasquin UJ, Qin G, Li K, Cai C (2010) "Clickable", polymerized liposomes as a versatile and stable platform for rapid optimization of their peripheral compositions. Chem Commun 46:5746–5748

Lai SK, O'Hanlon DE, Harrold S, Man ST, Wang Y-Y, Cone R, Hanes J (2007) Rapid transport of large polymeric nanoparticles in fresh undiluted human mucus. Proc Natl Acad Sci 104:1482–1487

Lai SK, Wang Y-Y, Hida K, Cone R, Hanes J (2010) Nanoparticles reveal that human cervicovaginal mucus is riddled with pores larger than viruses. Proc Natl Acad Sci 107:598–603

Larsen EK, Nielsen T, Wittenborn T, Birkedal H, Vorup-Jensen T, Jakobsen MH, Østergaard L, Horsman MR, Besenbacher F, Howard KA (2009) Size-dependent accumulation of PEGylated silane-coated magnetic iron oxide nanoparticles in murine tumors. ACS Nano 3:1947–1951

Lavan DA, Lynn DM, Langer R (2002) Moving smaller in drug discovery and delivery. Nat Rev Drug Discov 1:77–84

Lee S-W, Yun M-H, Jeong SW, In C-H, Kim J-Y, Seo M-H, Pai C-M, Kim S-O (2011) Development of docetaxel-loaded intravenous formulation, Nanoxel-PM™ using polymer-based delivery system. J Control Release 155:262–271

Li J, Kao WJ (2003) Synthesis of polyethylene glycol (PEG) derivatives and PEGylated–peptide biopolymer conjugates. Biomacromolecules 4:1055–1067

Li Y, Zhou Z, Pei Y, Zhang X, Gu Z, Yuan W (2001) PEGylated polycyanoacrylate nanoparticles as salvicine carriers: synthesis, preparation, and in vitro characterization. Acta Pharmacol Sin 22:645–650

Li YL, Zhu L, Liu Z, Cheng R, Meng F, Cui JH, Ji SJ, Zhong Z (2009) Reversibly stabilized multifunctional dextran nanoparticles efficiently deliver doxorubicin into the nuclei of cancer cells. Angew Chem 121:10098–10102

Li H, Niu R, Yang J, Nie J, Yang D (2011) Photocrosslinkable tissue adhesive based on dextran. Carbohydr Polym 86:1578–1585

Lin W-J, Juang L-W, Wang C-L, Chen Y-C, Lin C-C, Chang K-L (2010) Pegylated polyester polymeric micelles as a nano-carrier: synthesis, characterization, degradation, and biodistribution. J Exp Clin Med 2:4–10

Loch-Neckel G, Nemen D, Puhl AC, Fernandes D, Stimamiglio MA, Silva MA, Hangai M, Silva MCS, Lemos-Senna E (2007) Stealth and non-stealth nanocapsules containing camptothecin: in-vitro and in-vivo activity on B16-F10 melanoma. J Pharm Pharmacol 59:1359–1364

Mangeney C, Ferrage F, Aujard I, Marchi-Artzner V, Jullien L, Ouari O, Rékaï ED, Laschewsky A, Vikholm I, Sadowski JW (2002) Synthesis and properties of water-soluble gold colloids covalently derivatized with neutral polymer monolayers. J Am Chem Soc 124:5811–5821

Maruyama K (2002) PEG-immunoliposome. Biosci Rep 22:251–266

Mastrotto F, Brazzale C, Bellato F, De Martin S, Grange G, Mahmoudzadeh M, Magarkar A, Bunker A, Salmaso S, Caliceti P (2020) In vitro and in vivo behavior of liposomes decorated with PEGs with different chemical features. Mol Pharm 17:472–487

Matsumoto M, Matsusaki M, Akashi M (2014) Preparation of biodegradable peptide nanospheres with hetero PEG brush surfaces. Macromol Biosci 14:142–150

Mehvar R (2000) Dextrans for targeted and sustained delivery of therapeutic and imaging agents. J Control Release 69:1–25

Meng F, Engbers GH, Feijen J (2004) Polyethylene glycol–grafted polystyrene particles. J Biomed Mater Res Pt A 70:49–58

Metselaar JM, Bruin P, De Boer LW, De Vringer T, Snel C, Oussoren C, Wauben MH, Crommelin DJ, Storm G, Hennink WE (2003) A novel family of L-amino acid-based biodegradable polymer-lipid conjugates for the development of long-circulating liposomes with effective drug-targeting capacity. Bioconjug Chem 14:1156–1164

Moghimi SM, Szebeni J (2003) Stealth liposomes and long circulating nanoparticles: critical issues in pharmacokinetics, opsonization and protein-binding properties. Prog Lipid Res 42:463–478

Mohapatra A, Uthaman S, Park I-K (2019) Polyethylene glycol nanoparticles as promising tools for anticancer therapeutics. In: Polymeric nanoparticles as a promising tool for anti-cancer therapeutics, pp 205–231

Monfardini C, Schiavon O, Caliceti P, Morpurgo M, Harris JM, Veronese FM (1995) A branched monomethoxypoly (ethylene glycol) for protein modification. Bioconjug Chem 6:62–69

Moser M, Behnke T, Hamers-Allin C, Klein-Hartwig K, Falkenhagen J, Resch-Genger U (2015) Quantification of PEG-maleimide ligands and coupling efficiencies on nanoparticles with Ellman's reagent. Anal Chem 87:9376–9383

Mosqueira VCF, Legrand P, Gref R, Heurtault B, Appel M, Barratt G (1999) Interactions between a macrophage cell line (J774A1) and surface-modified poly (D, L-lactide) nanocapsules bearing poly (ethylene glycol). J Drug Target 7:65–78

Nag A, Mitra G, Ghosh PC (1996) A colorimetric assay for estimation of polyethylene glycol and polyethylene glycolated protein using ammonium ferrothiocyanate. Anal Biochem 237:224–231

Nair LM, Konkel J, Thomas M, Koberda M (2006) Comparison of electrospray ionization mass spectrometry and evaporative light scattering detections for the determination of Poloxamer 188 in itraconazole injectable formulation. J Pharm Biomed Anal 41:725–730

Nakamura K, Yamashita K, Itoh Y, Yoshino K, Nozawa S, Kasukawa H (2012) Comparative studies of polyethylene glycol-modified liposomes prepared using different PEG-modification methods. Biochim Biophys Acta 1818:2801–2807

Nance EA, Woodworth GF, Sailor KA, Shih T-Y, Xu Q, Swaminathan G, Xiang D, Eberhart C, Hanes J (2012) A dense poly (ethylene glycol) coating improves penetration of large polymeric nanoparticles within brain tissue. Sci Transl Med 4:149ra119

Neal JC, Stolnik S, Schacht E, Kenawy ER, Garnett MC, Davis SS, Illum L (1998) In vitro displacement by rat serum of adsorbed radiolabeled poloxamer and poloxamine copolymers from model and biodegradable nanospheres. J Pharm Sci 87:1242–1248

Neu M, Germershaus O, Behe M, Kissel T (2007) Bioreversibly crosslinked polyplexes of PEI and high molecular weight PEG show extended circulation times in vivo. J Control Release 124:69–80

Nishiyama N, Okazaki S, Cabral H, Miyamoto M, Kato Y, Sugiyama Y, Nishio K, Matsumura Y, Kataoka K (2003) Novel cisplatin-incorporated polymeric micelles can eradicate solid tumors in mice. Cancer Res 63:8977–8983

Nosova A, Koloskova O, Nikonova A, Simonova V, Smirnov V, Kudlay D, Khaitov M (2019) Diversity of PEGylation methods of liposomes and their influence on RNA delivery. Med Chem Commun 10:369–377

Nunvářová K, Charvátová B, Šlouf M, Hermanová S, Merna J (2019) Synthesis of amphiphilic copolymers based on dendritic polyethylene grafted by polyhydroxyethylmethacrylate and polyhydroxypropylmethacrylate and their use for construction of nanoparticles. Eur Polym J 115:193–200

Ohshima H (2016) Encyclopedia of biocolloid and biointerface science, 2 volume set. Wiley, Hoboken, NJ

O'Mahony AM, Ogier J, Desgranges S, Cryan JF, Darcy R, O'Driscoll CM (2012) A click chemistry route to 2-functionalised PEGylated and cationic β-cyclodextrins: co-formulation opportunities for siRNA delivery. Org Biomol Chem 10:4954–4960

Owens DE III, Peppas NA (2006) Opsonization, biodistribution, and pharmacokinetics of polymeric nanoparticles. Int J Pharm 307:93–102

Owens DE III, Eby JK, Jian Y, Peppas NA (2007) Temperature-responsive polymer–gold nanocomposites as intelligent therapeutic systems. J Biomed Mater Res Pt A 83:692–695

Park K, Lee SK, Park SA, Kim K, Chang HW, Jeong E-J, Park R-W, Kim I-S, Kwon IC, Byun Y (2007) The attenuation of experimental lung metastasis by a bile acid acylated-heparin derivative. Biomaterials 28:2667–2676

Park J, Fong PM, Lu J, Russell KS, Booth CJ, Saltzman WM, Fahmy TM (2009) PEGylated PLGA nanoparticles for the improved delivery of doxorubicin. Nanomedicine 5:410–418

Pasut G, Veronese FM (2009) PEG conjugates in clinical development or use as anticancer agents: an overview. Adv Drug Deliv Rev 61:1177–1188

Pasut G, Veronese FM (2012) State of the art in PEGylation: the great versatility achieved after forty years of research. J Control Release 161:461–472

Peracchia MT, Vauthier C, Desmaële D, Gulik A, Dedieu J-C, Demoy M, D'Angelo J, Couvreur P (1998) Pegylated nanoparticles from a novel methoxypolyethylene glycol cyanoacrylate-hexadecyl cyanoacrylate amphiphilic copolymer. Pharm Res 15:550–556

Perry JL, Reuter KG, Kai MP, Herlihy KP, Jones SW, Luft JC, Napier M, Bear JE, Desimone JM (2012) PEGylated PRINT nanoparticles: the impact of PEG density on protein binding, macrophage association, biodistribution, and pharmacokinetics. Nano Lett 12:5304–5310

Photos PJ, Bacakova L, Discher B, Bates FS, Discher DE (2003) Polymer vesicles in vivo: correlations with PEG molecular weight. J Control Release 90:323–334

Poon Z, Chang D, Zhao X, Hammond PT (2011) Layer-by-layer nanoparticles with a pH-sheddable layer for in vivo targeting of tumor hypoxia. ACS Nano 5:4284–4292

Porter RS, Casale A (1985) Recent studies of polymer reactions caused by stress. Polym Eng Sci 25:129–156

Pozzo D, Walker L (2008) Macroscopic alignment of nanoparticle arrays in soft crystals of cubic and cylindrical polymer micelles. Eur Phys J E 26:183–189

Prawatborisut M, Seidi F, Yiamsawas D, Crespy D (2019) PEGylation of shellac-based nanocarriers for enhanced colloidal stability. Colloids Surf B Biointerfaces 183:110434

Pulkkinen M, Pikkarainen J, Wirth T, Tarvainen T, Haapa-Aho V, Korhonen H, Seppälä J, Järvinen K (2008) Three-step tumor targeting of paclitaxel using biotinylated PLA-PEG nanoparticles and avidin–biotin technology: formulation development and in vitro anticancer activity. Eur J Pharm Biopharm 70:66–74

Pytela J, Saudek V, Drobnik J, Rypáček F (1989) Poly (N5-hydroxyalkylglutamines). IV. Enzymatic degradation of N5-(2-hydroxyethyl)-L-glutamine homopolymers and copolymers. J Control Release 10:17–25

Rabanel J-M, Hildgen P, Banquy X (2014) Assessment of PEG on polymeric particles surface, a key step in drug carrier translation. J Control Release 185:71–87

Rahme K, Chen L, Hobbs RG, Morris MA, O'Driscoll C, Holmes JD (2013) PEGylated gold nanoparticles: polymer quantification as a function of PEG lengths and nanoparticle dimensions. RSC Adv 3:6085–6094

Reboredo C, González-Navarro C, Martínez-Oharriz C, Martínez-López A, Irache J (2021) Preparation and evaluation of PEG-coated zein nanoparticles for oral drug delivery purposes. Int J Pharm 597:120287

Redhead H, Davis S, Illum L (2001) Drug delivery in poly (lactide-co-glycolide) nanoparticles surface modified with poloxamer 407 and poloxamine 908: in vitro characterisation and in vivo evaluation. J Control Release 70:353–363

Roe G, McDonnell L, Ghanem A (2004) A method for measuring the size distribution of latex particles by scanning force microscopy. Ultramicroscopy 100:319–329

Romberg B, Metselaar JM, Baranyi L, Snel CJ, Bünger R, Hennink WE, Szebeni J, Storm G (2007) Poly (amino acid) s: promising enzymatically degradable stealth coatings for liposomes. Int J Pharm 331:186–189

Sanchez Armengol E, Unterweger A, Laffleur F (2022) PEGylated drug delivery systems in the pharmaceutical field: past, present and future perspective. Drug Dev Ind Pharm 48:129–139

Santos JHPM, Torres-Obreque KM, Meneguetti GP, Amaro BP, Rangel-Yagui CO (2018) Protein PEGylation for the design of biobetters: from reaction to purification processes. Braz J Pharm Sci 54:e01009

Sarris A, Hagemeister F, Romaguera J, Rodriguez M, McLaughlin P, Tsimberidou A, Medeiros L, Samuels B, Pate O, Oholendt M (2000) Liposomal vincristine in relapsed non-Hodgkin's lymphomas: early results of an ongoing phase II trial. Ann Oncol 11:69–72

Sheng Y, Yuan Y, Liu C, Tao X, Shan X, Xu F (2009) In vitro macrophage uptake and in vivo biodistribution of PLA–PEG nanoparticles loaded with hemoglobin as blood substitutes: effect of PEG content. J Mater Sci Mater Med 20:1881–1891

Shi B, Fang C, Pei Y (2006) Stealth PEG-PHDCA niosomes: effects of chain length of PEG and particle size on niosomes surface properties, in vitro drug release, phagocytic uptake, in vivo pharmacokinetics and antitumor activity. J Pharm Sci 95:1873–1887

Siegers C, Biesalski M, Haag R (2004) Self-assembled monolayers of dendritic polyglycerol derivatives on gold that resist the adsorption of proteins. Chemistry 10:2831–2838

Singh AK (2015) Engineered nanoparticles: structure, properties and mechanisms of toxicity. Academic Press, San Diego, CA

Singh P, Gupta U, Asthana A, Jain NK (2008) Folate and folate– PEG– PAMAM dendrimers: synthesis, characterization, and targeted anticancer drug delivery potential in tumor bearing mice. Bioconjug Chem 19:2239–2252

Sitterberg J, Özcetin A, Ehrhardt C, Bakowsky U (2010) Utilising atomic force microscopy for the characterisation of nanoscale drug delivery systems. Eur J Pharm Biopharm 74:2–13

Steenpaß T, Lung A, Schubert R (2006) Tresylated PEG-sterols for coupling of proteins to preformed plain or PEGylated liposomes. Biochim Biophys Acta 1758:20–28

Stolnik S, Dunn SE, Garnett MC, Davies MC, Coombes AG, Taylor D, Irving M, Purkiss S, Tadros TF, Davis SS (1994) Surface modification of poly (lactide-co-glycolide) nanospheres by biodegradable poly (lactide)-poly (ethylene glycol) copolymers. Pharm Res 11:1800–1808

Storm G, Belliot SO, Daemen T, Lasic DD (1995) Surface modification of nanoparticles to oppose uptake by the mononuclear phagocyte system. Adv Drug Deliv Rev 17:31–48

Suh J, Choy K-L, Lai SK, Suk JS, Tang BC, Prabhu S, Hanes J (2007) PEGylation of nanoparticles improves their cytoplasmic transport. Int J Nanomedicine 2:735–741

Suma T, Miyata K, Anraku Y, Watanabe S, Christie RJ, Takemoto H, Shioyama M, Gouda N, Ishii T, Nishiyama N (2012) Smart multilayered assembly for biocompatible siRNA delivery featuring dissolvable silica, endosome-disrupting polycation, and detachable PEG. ACS Nano 6:6693–6705

Tang BC, Dawson M, Lai SK, Wang Y-Y, Suk JS, Yang M, Zeitlin P, Boyle MP, Fu J, Hanes J (2009) Biodegradable polymer nanoparticles that rapidly penetrate the human mucus barrier. Proc Natl Acad Sci 106:19268–19273

Thierry B, Griesser HJ (2012) Dense PEG layers for efficient immunotargeting of nanoparticles to cancer cells. J Mater Chem 22:8810–8819

Tirosh O, Barenholz Y, Katzhendler J, Priev A (1998) Hydration of polyethylene glycol-grafted liposomes. Biophys J 74:1371–1379

Tobio M, Gref R, Sanchez A, Langer R, Alonso M (1998) Stealth PLA-PEG nanoparticles as protein carriers for nasal administration. Pharm Res 15:270–275

Torchilin V (1998) Polymer-coated long-circulating microparticulate pharmaceuticals. J Microencapsul 15:1–19

Torchilin VP, Omelyanenko VG, Papisov MI, Bogdanov AA Jr, Trubetskoy VS, Herron JN, Gentry CA (1994) Poly (ethylene glycol) on the liposome surface: on the mechanism of polymer-coated liposome longevity. Biochim Biophys Acta 1195:11–20

Turecek PL, Bossard MJ, Schoetens F, Ivens IA (2016) PEGylation of biopharmaceuticals: a review of chemistry and nonclinical safety information of approved drugs. J Pharm Sci 105:460–475

Uster PS, Allen TM, Daniel BE, Mendez CJ, Newman MS, Zhu GZ (1996) Insertion of poly (ethylene glycol) derivatized phospholipid into pre-formed liposomes results in prolonged in vivo circulation time. FEBS Lett 386:243–246

Vaia RA, Maguire JF (2007) Polymer nanocomposites with prescribed morphology: going beyond nanoparticle-filled polymers. Chem Mater 19:2736–2751

Vandevondele S, Vörös J, Hubbell JA (2003) RGD-grafted poly-l-lysine-graft-(polyethylene glycol) copolymers block non-specific protein adsorption while promoting cell adhesion. Biotechnol Bioeng 82:784–790

Veronese FM, Caliceti P, Schiavon O (1997) Branched and linear poly (ethylene glycol): influence of the polymer structure on enzymological, pharmacokinetic, and immunological properties of protein conjugates. J Bioact Compat Polym 12:196–207

Vila A, Gill H, McCallion O, Alonso MAJ (2004) Transport of PLA-PEG particles across the nasal mucosa: effect of particle size and PEG coating density. J Control Release 98:231–244

Visser CC, Stevanović S, Voorwinden LH, Van Bloois L, Gaillard PJ, Danhof M, Crommelin DJ, De Boer AG (2005) Targeting liposomes with protein drugs to the blood–brain barrier in vitro. Eur J Pharm Sci 25:299–305

Vonarbourg A, Passirani C, Saulnier P, Benoit J-P (2006) Parameters influencing the stealthiness of colloidal drug delivery systems. Biomaterials 27:4356–4373

Wang YY, Lai SK, Suk JS, Pace A, Cone R, Hanes J (2008) Addressing the PEG mucoadhesivity paradox to engineer nanoparticles that "slip" through the human mucus barrier. Angew Chem 120:9872–9875

Wang H, Zhao P, Su W, Wang S, Liao Z, Niu R, Chang J (2010) PLGA/polymeric liposome for targeted drug and gene co-delivery. Biomaterials 31:8741–8748

Webb MS, Saxon D, Wong FM, Lim HJ, Wang Z, Bally MB, Choi LS, Cullis PR, Mayer LD (1998) Comparison of different hydrophobic anchors conjugated to poly (ethylene glycol): effects on the pharmacokinetics of liposomal vincristine. Biochim Biophys Acta 1372:272–282

Whitehead KA, Dorkin JR, Vegas AJ, Chang PH, Veiseh O, Matthews J, Fenton OS, Zhang Y, Olejnik KT, Yesilyurt V (2014) Degradable lipid nanoparticles with predictable in vivo siRNA delivery activity. Nat Commun 5:4277

Wibroe PP, Ahmadvand D, Oghabian MA, Yaghmur A, Moghimi SM (2016) An integrated assessment of morphology, size, and complement activation of the PEGylated liposomal doxorubicin products Doxil®, Caelyx®, DOXOrubicin, and SinaDoxosome. J Control Release 221:1–8

Xu Q, Ensign LM, Boylan NJ, Schon A, Gong X, Yang J-C, Lamb NW, Cai S, Yu T, Freire E (2015) Impact of surface polyethylene glycol (PEG) density on biodegradable nanoparticle transport in mucus ex vivo and distribution in vivo. ACS Nano 9:9217–9227

Yang H, Morris JJ, Lopina ST (2004) Polyethylene glycol–polyamidoamine dendritic micelle as solubility enhancer and the effect of the length of polyethylene glycol arms on the solubility of pyrene in water. J Colloid Interface Sci 273:148–154

Yang W, Zhang L, Wang S, White AD, Jiang S (2009) Functionalizable and ultra stable nanoparticles coated with zwitterionic poly (carboxybetaine) in undiluted blood serum. Biomaterials 30:5617–5621

Yang M, Lai SK, Wang YY, Zhong W, Happe C, Zhang M, Fu J, Hanes J (2011) Biodegradable nanoparticles composed entirely of safe materials that rapidly penetrate human mucus. Angew Chem 123:2645–2648

Yang M, Yu T, Wang YY, Lai SK, Zeng Q, Miao B, Tang BC, Simons BW, Ensign LM, Liu G (2014) Vaginal delivery of paclitaxel via nanoparticles with non-mucoadhesive surfaces suppresses cervical tumor growth. Adv Healthc Mater 3:1044–1052

Yoncheva K, Lizarraga E, Irache JM (2005) Pegylated nanoparticles based on poly (methyl vinyl ether-co-maleic anhydride): preparation and evaluation of their bioadhesive properties. Eur J Pharm Sci 24:411–419

Yuan H, Chen C-Y, Chai G-H, Du Y-Z, Hu F-Q (2013) Improved transport and absorption through gastrointestinal tract by PEGylated solid lipid nanoparticles. Mol Pharm 10:1865–1873

Zacchigna M, Cateni F, Drioli S, Bonora GM (2011) Multimeric, multifunctional derivatives of poly (ethylene glycol). Polymers 3:1076–1090

Zahr AS, Davis CA, Pishko MV (2006) Macrophage uptake of core-shell nanoparticles surface modified with poly (ethylene glycol). Langmuir 22:8178–8185

Zhan C, Gu B, Xie C, Li J, Liu Y, Lu W (2010) Cyclic RGD conjugated poly (ethylene glycol)-co-poly (lactic acid) micelle enhances paclitaxel anti-glioblastoma effect. J Control Release 143:136–142

Zhang L, Gu F, Chan J, Wang A, Langer R, Farokhzad O (2008a) Nanoparticles in medicine: therapeutic applications and developments. Clin Pharmacol Therapeut 83:761–769

Zhang Z, Vaisocherová H, Cheng G, Yang W, Xue H, Jiang S (2008b) Nonfouling behavior of polycarboxybetaine-grafted surfaces: structural and environmental effects. Biomacromolecules 9:2686–2692

Zhang H, Xia H, Wang J, Li Y (2009) High intensity focused ultrasound-responsive release behavior of PLA-b-PEG copolymer micelles. J Control Release 139:31–39

Zhang L, Xue H, Gao C, Carr L, Wang J, Chu B, Jiang S (2010) Imaging and cell targeting characteristics of magnetic nanoparticles modified by a functionalizable zwitterionic polymer with adhesive 3, 4-dihydroxyphenyl-l-alanine linkages. Biomaterials 31:6582–6588

Zhang L, Xue H, Cao Z, Keefe A, Wang J, Jiang S (2011) Multifunctional and degradable zwitterionic nanogels for targeted delivery, enhanced MR imaging, reduction-sensitive drug release, and renal clearance. Biomaterials 32:4604–4608

Zhang P, Sun F, Liu S, Jiang S (2016) Anti-PEG antibodies in the clinic: current issues and beyond PEGylation. J Control Release 244:184–193

Zillies JC, Zwiorek K, Winter G, Coester C (2007) Method for quantifying the PEGylation of gelatin nanoparticle drug carrier systems using asymmetrical flow field-flow fractionation and refractive index detection. Anal Chem 79:4574–4580

# Topographic Properties of PEGylated Nanocarriers

3

Aprameya Ganesh Prasad, Shivam Otavi, Mahima Mishra,
and Rakesh Kumar Tekade

**Abstract**

Nanoparticles have been widely used as vectors for bioactive cargo due to their high therapeutic index. Despite their attractiveness for safe and efficacious drug delivery, nanoparticles are known to suffer from rapid systemic clearance. PEGylation, conjugation of the nanocarrier with polyethylene glycol (PEG) chains, is one of the strategies that has been developed to combat this challenge. PEGylation confers steric stability to the nanoparticle through a "stealth" coating which masks the core of the nanoparticle thereby evading a response from the immune and reticuloendothelial systems. PEG chains are also critical to ensure the stability of the nanoparticles as they prevent aggregation in colloidal suspensions. Due to the crucial role they play in developing an optimal formulation, it is imperative to establish robust analytical tools to characterize the surface topography and morphology of PEGylated nanocarriers. Surface analytical techniques are frequently modified to enable characterization at a nanoscale. This chapter describes the major characterization tools we have at our disposal to effectively analyze critical surface attributes. Examples of each technique are also provided to demonstrate their application, benefits, and drawbacks.

A. G. Prasad (✉)
Department of Chemical and Biomolecular Engineering, Johns Hopkins University, Baltimore, MD, USA

S. Otavi · M. Mishra · R. K. Tekade
National Institute of Pharmaceutical Education and Research (NIPER) Ahmedabad, An Institute of National Importance, Government of India, Department of Pharmaceuticals, Ministry of Chemicals and Fertilizers, Palaj, Opp. Air force station, Gandhinagar, Gujarat, India


R. K. Tekade, N. K. Jain (eds.), *PEGylated Nanocarriers in Medicine and Pharmacy*, https://doi.org/10.1007/978-981-97-7316-9_3

**Keywords**

PEGylation · PEG chain · Steric stability · Characterization · Topography

## 3.1 Introduction

Nanoparticles are widely used as carriers for a variety of payloads such as small molecule drugs, oligonucleotides, and diagnostic agents. Their attractiveness in the realm of drug delivery systems can be attributed to their superior physicochemical properties including high drug loading efficiencies, enhanced transport across vascular membranes, and tunable uptake and internalization in the case of targeted delivery (Mitchell et al. 2021; Shi et al. 2021). Nanocarriers have thus been exploited to treat multiple disease indications through a variety of dosage forms and drug product presentations. A primary challenge that is associated with nanoparticles in blood circulation is their poor pharmacokinetic profile (Zhu et al. 2022; Zelepukin et al. 2020). Factors such as size, charge distribution, and surface functionalization can trigger an immune response that results in rapid clearance from circulation (Alexis et al. 2008; Zolnik et al. 2010; Hooftman and O'Neill 2022).

Polyethylene glycol (PEG) is a nontoxic, hydrophilic polymer that has been extensively used in the field of nanomedicine. It was approved by the Food and Drug Administration (FDA) in 1990 and has since been used in several commercially available pharmaceutical formulations (Suk et al. 2016; Swierczewska et al. 2015). When coated on the surface of nanoparticles, PEG prevents the adsorption of serum proteins and masks charges, if any, thereby modulating the biodistribution and pharmacokinetics of the drug. Therefore, PEGylation is crucial to prevent clearance by the reticuloendothelial system (RES) (Jokerst et al. 2011; Mohamed et al. 2019). In addition to reducing the immunogenicity of nanocarriers, PEG can sometimes also be used for controlled or sustained-release applications (Stras et al. 2020; Peleg-Shulman et al. 2004; Prasad et al. 2021). PEG also prevents aggregation of nanoparticles and is therefore used as a stabilizer in nanomedicine (Larson et al. 2012). A lot of research has focused on modifying nanocarriers with PEG in different ways—through covalent and noncovalent grafting, chemical conjugation, and sometimes as a constituent of a copolymer matrix (Rabanel et al. 2014). It is therefore of critical importance to be able to characterize the efficiency of the preparation methods, confirm successful PEG conjugation, and monitor the stability of the conjugated nanocarriers over time. After systemic administration, the interaction of PEG chains with the physiological environment may also need to be assessed to predict the in vivo efficacy of such nanocarriers.

Though nanoparticles are routinely characterized using several analytical tools, there are limited methods that can investigate the surface topography of PEGylated nanocarriers (Mourdikoudis et al. 2018; Baer et al. 2010; Joudeh and Linke 2022). Standard characterization tools may not provide sufficient resolution to probe the surface at the nanoscale level (Takechi-Haraya et al. 2022). Modifications to conventional methods and a combination of techniques may sometimes be necessary to determine the topographical parameters precisely and accurately. Here, we discuss multiple characterization techniques spanning microscopic, spectroscopic, calorimetric, and gravimetric analyses. Principles of each technique are discussed along with the topographical information that can be obtained from each. Surface structure, composition, and morphological characterization are also considered due to their interdependence on topography. Sample preparation methods and limitations of each technique are also discussed to guide the selection of appropriate methods. Discussion of each technique is also accompanied by examples of their application in the characterization of PEGylated nanocarriers.

## 3.2 Characterization Techniques

### 3.2.1 Topographical Characterization Methods for PEGylated Nanocarriers

| Technique | Information available | Limitations | References |
|---|---|---|---|
| Infrared spectroscopy | Surface composition<br>Ligand-binding | Nonquantitative<br>Low sensitivity | Coelho et al. (2015), Yoncheva et al. (2005) and Retout et al. (2019) |
| Raman spectroscopy | Chemical structure<br>Crystallographic orientation<br>Conformation | Nonquantitative<br>Low sensitivity | Katuwavila et al. (2020), León et al. (2017) and Levin et al. (2006) |
| XPS/ESCA | Surface composition<br>Electronic structure<br>Chemical state<br>Ligand-binding<br>Particle size (<~10 nm) | Dry state<br>Specialized equipment<br>Highly trained personnel required<br>Poor lateral resolution | Ebbesen et al. (2013), Rabanel et al. (2015), Sant et al. (2008), Essa et al. (2010), Gref et al. (2001), Damodaran et al. (2010) and Gref et al. (1994) |
| NMR | Molecular structure<br>Phase information<br>Ligand density and arrangement<br>Atomic composition | Need for particle resuspension in $D_2O$ or lyophilization | Coelho et al. (2015), Yoncheva et al. (2005), Retout et al. (2019), Ebbesen et al. (2013), Rabanel et al. (2015), Essa et al. (2010), Vila et al. (2004), Conte et al. (2019), Lu et al. (2019), Xu et al. (2015), Chen et al. (2016), Essa et al. (2011), Garcia-Fuentes et al. (2004) and Vernooij et al. (1999) |
| EDS/EDX | Elemental composition | Complicated data interpretation due to surface contamination | Ebadi et al. (2019), Karim and Chowdhury (2022), Iranpour et al. (2021) and Maghsoudi et al. (2023) |

| Technique | Information available | Limitations | References |
| --- | --- | --- | --- |
| Mass spectroscopy | Elemental composition<br>Molecular orientation and conformation<br>Size and size distribution | Highly trained personnel required<br>Dry state | Hinterwirth et al. (2013), Sebby and Mansfield (2015), Ara et al. (2014) |
| TGA | Mass<br>Composition | Destructive analysis<br>Large sample requirement | Sebby and Mansfield (2015), Butterworth et al. (2001), Mansfield et al. (2014) and Maccarini et al. (2010) |
| DSC | Phase transition<br>Molecular interactions | Destructive analysis | Rabanel et al. (2015) and Essa et al. (2010) |
| AFM | Particle shape and size<br>Morphology<br>Texture and roughness | High instrumentation cost<br>Long scan time<br>Potential damage to the sample from probe | Sant et al. (2008), Essa et al. (2010), Essa et al. (2011), Sacchetti et al. (2013) and Backmann et al. (2010) |
| SEM | Morphology | High instrumentation cost<br>Tedious sample preparation<br>Limited spatial resolution | Yoncheva et al. (2005), Ebbesen et al. (2013) and Gref et al. (2001) |
| TEM | Morphology<br>Structure<br>Size and size distribution | High instrumentation cost<br>Tedious sample preparation<br>Dry state | Coelho et al. (2015), Rabanel et al. (2015), Gref et al. (2001), Vila et al. (2004), Xu et al. (2015), Chen et al. (2016), Maccarini et al. (2010), Pelaz et al. (2015), Walkey et al. (2012) and Allard-Vannier et al. (2017) |

## 3.3 Infrared Spectroscopy

In infrared (IR) spectroscopy, the molecules on the surface are exposed to an infrared light source, and absorption in the near-infrared region (0.7–2.5 μm) creates changes in dipole moments. A polychromatic source is useful to generate a spectrum over the full infrared range, whereas a monochromatic laser source is more suitable to study vibrational resonance at a specific wavelength. The absorption or transmission spectra are measured by a dispersive spectrometer where individual spectral bands correspond to specific covalent bonds and functional groups. Through the information about bonds and intermolecular interactions, the spectra can establish the chemical identity and molecular structure of the surface.

Fourier transform infrared (FTIR) spectroscopy is more suitable for the mid-infrared (2.5–25 μm) and far-infrared (25–1000 μm) regions. In this technique, different wavelengths are scanned using an interferometer. Total light energy detected at each position of the mounted mirror is converted to an energy spectrum by applying Fourier transformation across the wavelengths. FTIR can be used to deduce not only the surface composition but also to confirm and investigate ligand-binding density and structural arrangements as in the case of PEGylated and other functionalized nanocarriers.

When it is not possible to obtain the spectrum by transmission, attenuated total reflectance (ATR)-FTIR provides an improved sensitivity at the interface compared to the FTIR technique. Here, a crystal is placed next to the sample in the direction of incident light. Using the phenomena of total internal reflection, the molecules in the sample absorb energy from the standing evanescent wave thereby resulting in an attenuated reflected beam. Atomic force microscopy (AFM) can be coupled with IR spectroscopy to include imaging capabilities. This technique is also called photothermal-induced resonance (PTIR) and provides a better resolution at the nanoscale (Dazzi et al. 2005). Upon illumination by a laser IR source, the heat energy in the molecules results in thermal expansion. Such local spatial changes on the surface are detected by the AFM cantilever at high sensitivity. A lateral resolution of as low as ~10 nm is possible using this technique (Mathurin et al. 2018).

Sample preparation methods depend on the nature of the sample. Neat liquid samples can be directly measured by placing a drop between sandwiched salt plates. Solutions of liquids may also be used by accounting for IR absorption of the solvent. Aqueous samples are not suitable because water dissolves the potassium bromide (KBr) and sodium chloride (NaCl) plates that are typically used for dissolved solid and liquid samples, respectively. Water and alcohols, like methanol and ethanol, also strongly absorb in the infrared region thereby masking critical bands in the sample. Thin-film liquid cells can also be used for solution-phase IR spectroscopy. Solid samples can be dissolved in a solvent as described above. If using a solvent, it is important to avoid volatile liquids as they can evaporate while the spectrum is being acquired. However, it is typical to use a "Nujol mill" or "KBr press" for solid samples. Nujol is a viscous hydrocarbon liquid in which fine solid particles are dispersed. Alternatively, the sample can be mixed with solid KBr and this mixture is ground and pressed in a die. For ATR analysis, liquid and solid samples are placed in a trough crystal and flat crystal, respectively.

Even though IR spectroscopy is convenient, it only provides for a qualitative analysis. The technique is not as sensitive as other spectroscopic techniques such as mass spectroscopy. There is a limited choice of solid substrates that can be used to enhance the sensitivity. In the majority of cases, the sample needs to be in a dried state which is achieved by dehydration or the use of deuterium oxide. This restricts the ability to discern the surface elemental analysis of dispersed nanocarriers in their native aqueous state. AFM-IR suffers from some of the same limitations associated with AFM including the need to dilute and immobilize samples.

### 3.3.1 Application in PEGylated Nanocarriers

Silvia Coelho et al. used FTIR spectroscopy in conjunction with NMR spectroscopy to study PEGylated gold nanoparticles designed for drug delivery to tumors. The nanoparticles were encapsulated with Bortezomib and FTIR transmission spectra were analyzed to investigate the extent of inclusion and interaction of the drug with the PEGylated nanocarrier. Figure 3.1 shows the spectra of PEGylated nanoparticles with (2) and without (1) the drug. The common PEG functional group resulted in vibrations at 1091 and 1117/cm (C–O–C asymmetric group in the PEG chain), and 1710/cm (C=O group in the PEG layer) for both variants. However, the spectra for drug-loaded nanoparticles showed a shift in the stretching of the C=O group (1628/cm) and missing 1250/cm resonance. Based on this observation, the authors deduced electrostatic attraction between the positively charged Bortezomib and negatively charged PEGylated gold nanoparticle surface. This hypothesis was corroborated by a substantial change in the zeta potential of the nanoparticles loaded with the drug (Coelho et al. 2015). This demonstrates the benefit of utilizing this technique to investigate interactions between the drug and PEG, thereby shedding light on the localization of encapsulated cargo at the surface of nanocarriers.

Retout and co-authors developed gold nanoparticles coated with PEG utilizing different chemistries. The PEG ligands were functionalized with a thiol group at one end, and a methoxy, carboxylate or alkyne group at the other end. The PEG layers formed by coating the nanoparticles with varying proportions of the two PEG ligands were systematically analyzed using IR spectroscopy. As shown in Fig. 3.2, all variants showed the characteristic C–O–C asymmetric stretching band at 1100/cm. All samples with the carboxylate group in Fig. 3.2a gave a similar spectrum independent of the composition except at 1580/cm where a significant absorption

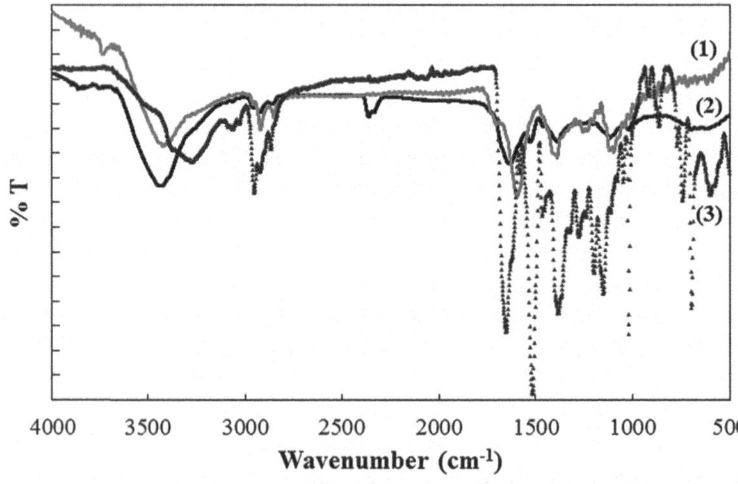


**Fig. 3.1** FTIR transmission spectra of (2) Bortezomib-loaded and (1) empty PEGylated gold nanoparticles. (Figure adapted with permission from Silvia Coelho et al. (2015) Copyright © 2015, Royal Chemical Society)

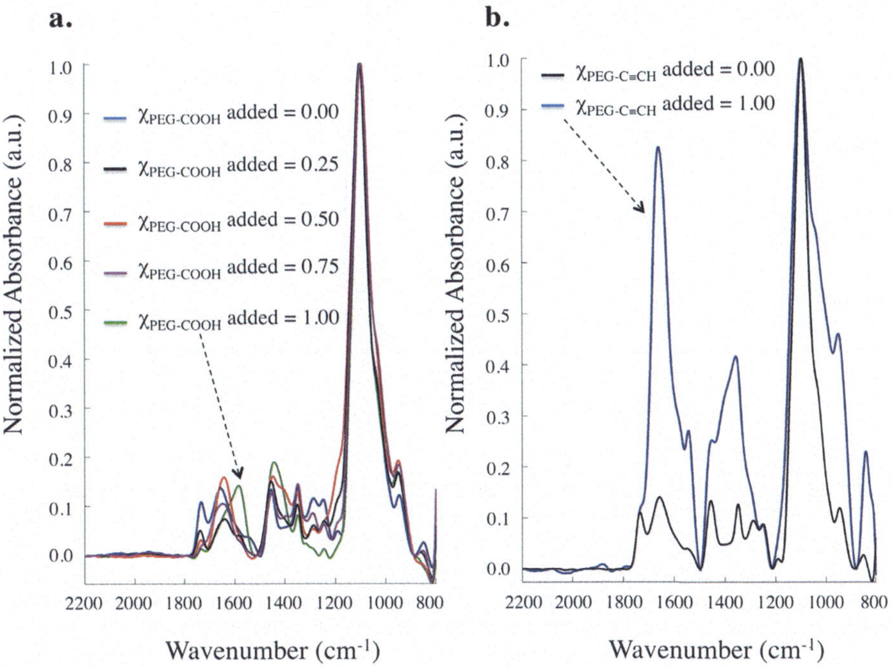


**Fig. 3.2** Normalized IR absorbance spectra of gold nanoparticles functionalized with (**a**) HS-PEG-CH3/HS-PEG-COOH and (**b**) HS-PEG-CH3/HS-PEG-C ≡ CH. (Figure adapted with permission from Retout et al. (2019) Copyright ©2019 Elsevier)

band was seen only for pure carboxylate-PEG functionalized nanoparticles. For the variants functionalized with the alkyne group shown in Fig. 3.2b, there was no characteristic signal seen in any of the samples. However, the intensity of the absorption band at 1680/cm was seen to be significantly higher for alkyne functionalized nanoparticles compared to pure thiol-PEG functionalized nanoparticles. This difference was attributed to the presence of citrate ions on the surface of nanoparticles pointing to a low grafting density of this functional group (Retout et al. 2019). This study illustrates how IR spectroscopy can be effectively used as a tool to discern differences in the degree of grafting density of various PEG groups.

Krassimiri Yoncheva et al. designed PEGylated nanoparticles made of poly(methyl vinyl ether-co-maleic anhydride) (PVM/MA). To characterize the way PEG was attached to the nanoparticles, IR spectroscopy was used. Figure 3.3 depicts the IR spectra obtained for PVM/MA nanoparticles before and after PEGylation. In addition to the peaks common to PVM/MA, the PEGylated nanoparticles also showed a peak at 1730–1740/cm which corresponds to the carbonyl group. The authors attributed this to potential ester bonds that may have formed as a result of the PEG-coupling reaction (Yoncheva et al. 2005). This report underscores the importance of infrared spectroscopy in confirming the identity of PEGylated nanocarriers.

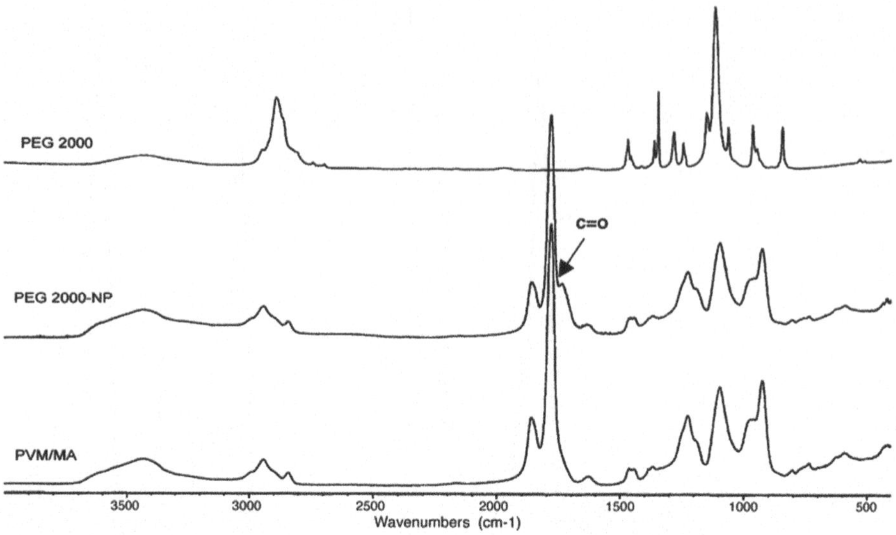


**Fig. 3.3** IR spectra of PEG 2000, PEG 2000-NP, and PVM/MA. The arrow shows the new peak appeared in the spectrum of PEGylated nanoparticles missing in the spectrum of PVM/MA. (Figure adapted with permission from Yoncheva et al. (2005) Copyright © 2005 Elsevier)

## 3.4 Raman Spectroscopy

Raman spectroscopy is a vibrational spectroscopic technique that works based on the principle of inelastic scattering of light upon irradiation of a monochromatic laser source. When a material is irradiated, part of the incident energy is elastically scattered at the same wavelength and energy. There is also a small fraction of incident energy that gets scattered at a different wavelength (inelastic scattering). This difference in energy arising due to the shift in wavelength is known as the "Raman shift" and the magnitude of this shift is unique to a molecule and its constituent bonds. The shift in wavelength can also be regarded as the vibrational energy of the irradiated molecule. As the shift tends to be typically very small, there is a need to use a monochromatic laser light source.

The irradiation causes perturbations in the electric dipole of the molecule thereby changing its electrical field. The resulting molecular vibrations known as photons and phonons can be recorded and the Raman intensity is plotted versus the frequency of the shift to obtain the spectrum. Sharp bands in the spectrum correspond to unique functional groups and this can be used to ascertain the identity, chemical structure, and crystallographic orientation of the irradiated material. Raman spectroscopy not only provides insights into the covalent bonds, intra and intermolecular interactions but also valuable information about interfacial interactions polymer chain conformation. The main advantage of using this technique is that the presence of water in the sample does not affect the spectra in any way.

Only moderately strong signals can be obtained due to the low cross section available in conventional Raman scattering. To overcome this challenge and improve the sensitivity and specificity of the technique, surface-enhanced Raman Spectroscopy (SERS) uses electromagnetic enhancement and a chemical effect that amplifies the polarizability of the molecule due to charge-transfer. This is particularly useful in the case of metallic nanoparticles where enhancement can be exploited to deduce information about surface binding and modifications in functional groups of molecules.

Tip-enhanced Raman Scattering (TERS) is analogous to SERS but also includes a scanning probe microscopy such as AFM or scanning tunneling microscopy (STM). The molecules on a metal nanoparticle directly under the sharp metal tip show enhanced localized surface plasmon resonance (LSPR) thereby allowing for single molecule detection (Bailo and Deckert 2008). The enhancement in the Raman signal is due to the creation of a "hot spot" between the tip and the substrate. This overcomes the limitations of spatial resolution and substrates in SERS. As the TERS signal comes from the small gap between the apex of the tip and the substrate, a very high resolution of 1–10 nm can be obtained (Zhang et al. 2016). The major benefit of TERS is its applicability to characterize samples both in air and liquid. This is particularly useful to study surfaces of modified nanocarriers in solution.

Even though Raman spectroscopy requires minimal to no sample preparation, it is important to select the right substrate based on the sample and application. The samples can be analyzed either as powders or resuspended in PBS to simulate physiologic conditions. Dried or colloidal suspensions of nanoparticles can be deposited onto glass slides that are typically coated with poly(4-vinlypyridine) (PVP). To stably immobilize the nanoparticles on the glass support, the slides can also be functionalized with (3-aminopropyl) triethoxysilane.

As with IR spectroscopy, Raman spectroscopy may not be ideal for quantitative analysis. The choice of substrate for particle immobilization may be limited to a transparent substrate like glass. Though SERS and TERS have been successfully used to study surfaces of metallic nanoparticles, whether they can also be applied to nonmetallic nanoparticles is still under investigation.

### 3.4.1 Application in PEGylated Nanocarriers

Nuwanthi Katuwavila and co-workers described a sustained-release drug delivery system for cephalexin (CEF) using PEGylated graphene oxide. Raman spectroscopy was used to confirm the successful conjugation of PEG. Raman spectra of graphene oxide (GO) and PEGylated graphene oxide (GO-PEG) are shown in Fig. 3.4. The characteristic peaks corresponding to the D and G bands were located for both nanocarriers at 1380/cm and 1580/cm. Intensity ratios of the G band to D band were found to be very similar to GO and GO-PEG, at 0.840 and 0.844, respectively. Similarity in this ratio was used to confirm that conjugation with PEG did not affect the aromatic structure of GO nanoparticles (Katuwavila et al. 2020). Therefore,

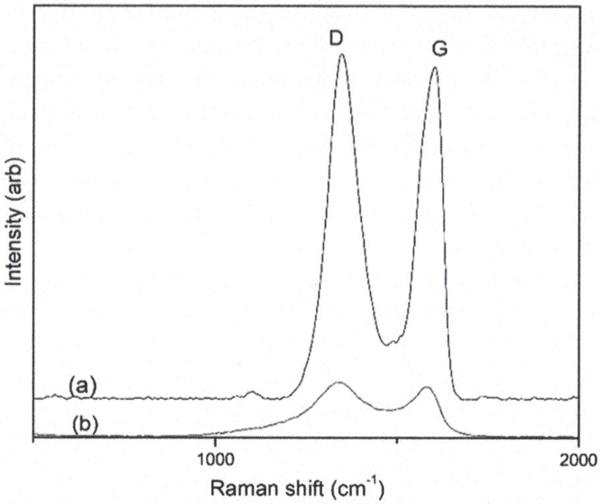


**Fig. 3.4** Raman spectra of (**a**) GO and (**b**) GO-PEG where Go was functionalized with PEG by a simple esterification. adapted with permission from Katuwavila et al. (2020) Copyright © 2020 Published by Elsevier)

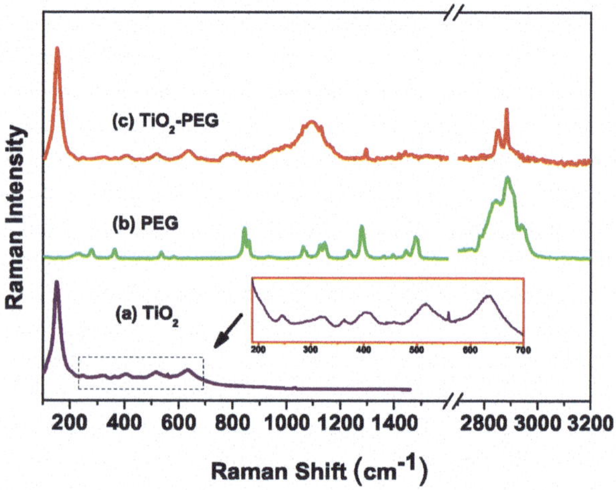


**Fig. 3.5** Raman spectra for (**a**) $TiO_2$, (**b**) PEG, and (**c**) $TiO_2$–PEG, recorded in the zone 150–3200/cm. (Figure adapted with permission from León et al. (2017) under Creative Common CC BY license)

Raman spectroscopy can be used to assure the integrity of the core structure of nanocarriers.

León and colleagues developed PEGylated $TiO_2$ nanocarriers for targeted delivery of 2-methoxyestradiol (2ME). Raman spectroscopy was used to follow the PEG conjugation process. In Fig. 3.5, we see the Raman spectra for $TiO_2$ (a), PEG (b), and composite $TiO_2$–PEG (c). The major characteristic peaks for PEG were observed at 2938/cm, 2886/cm, and 2843/cm. As there is no appreciable shift in the bands in $TiO_2$–PEG compared to $TiO_2$, it was concluded that modification with PEG did not affect the crystalline structure of the $TiO_2$ nanocarriers. Moreover, functionalization with PEG was confirmed by locating broad peaks around 1000–1200/cm that correspond to C–C and C–O stretching vibrations. Sharp characteristic peaks around

2900/cm also prove that the nanocarriers were successfully PEGylated (León et al. 2017).

Carly Levin et al. used surface-enhanced Raman spectroscopy to determine the packing density of thiolated poly(ethylene glycol) adsorbates on gold nanoshells. They developed a nondestructive assay to calculate the number of 2000 MW and 5000 MW PEG molecules bound to gold nanoshell surfaces. *p*-Mercaptoaniline (pMA) was used as the linker molecule and its Langmuir isotherm was interpolated to deduce the packing density. Figure 3.6 shows the SERS spectra for gold nanoshells coated with (a) 5000 MW pMA-PEG-Fl conjugate, (b) 2000 MW pMA-PEG-Fl conjugate, and (c) pMA only. The predominant bending and stretching modes arising due to the benzene moiety of pMA were observed on all spectra. This result confirmed successful thiol-mediated conjugation of the fluorescent PEG moiety. Furthermore, the spectra for PEGylated nanoshells showed an additional peak at 1330/cm corresponding to the xanthene ring of the fluorescent moiety. Interestingly, the spectrum for the conjugate with 2000 MW PEG showed a more defined Stokes feature for fluorescein. This was attributed to a higher number of PEG molecules on the surface of the nanoshell compared to the 5000 MW PEG conjugates that had a larger molecular footprint. By comparing the estimated and theoretical PEG densities for PEGylated nanoshells, the authors were also able to determine the conformation of PEG chains. Their analysis on packing densities showed that both PEG conjugates compactly bound to the interfaces covering surface areas that are consistent with the "brush" conformation, rather than the extended "mushroom" configuration (Levin et al. 2006). This report presented a quantitative analysis that could be utilized to assess not only the number of PEG molecules bound to the surface but also deduce the conformation of PEG chains.

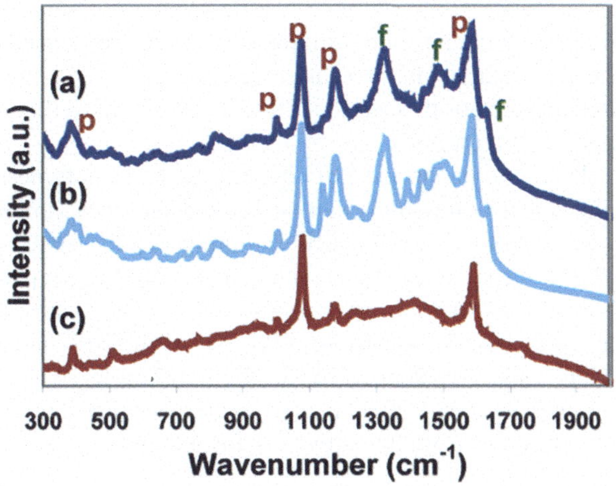


**Fig. 3.6** SERS spectra for Au-silica nanoshells coated with (**a**) pMA-PEG5000-Fl, (**b**) pMA-PEG2000-Fl, and (**c**) pMA (offset for clarity). (Figure adapted with permission from Levin et al. (2006) Copyright © 2006 American Chemical Society)

## 3.5 X-Ray Photoelectron Spectroscopy

X-ray photoelectron spectroscopy (XPS) is a powerful surface analysis technique that can be used to characterize polymeric nanomaterials. As it provides valuable chemical information at the surface, it is also known as electron spectroscopy for chemical analysis (ESCA). This technique is based on the principle of photoelectric effect wherein irradiation by a monochromatic X-ray beam results in the emission of photoelectrons from the inner shell of atoms. The photoelectrons are collected by detectors that identify and quantify their kinetic energy. The kinetic energy is equal to the binding energy of the electron and this information is used to identify the element. With this method, the surface composition can be determined which can be used to further evaluate the coverage density of the polymer.

The maximum sampling depth that corresponds to the upper layers of the sample is typically less than 10 nm. Unlike microscopic techniques, XPS provides elemental composition along the direction of the electron beam. Except hydrogen, XPS can detect all other atoms on the surface. The kinetic energy of the photoelectrons can be related to the electronic structure and oxidation states of the elements. Other surface characteristics and properties relevant to ligand binding, attachment of functional moieties can be deduced through a high-resolution scan that can identify and quantify chemical bonds. As XPS lacks spatial (lateral) resolution, individual nanoparticles cannot be analyzed. However, collections of particles that form a single layer can be surveyed to obtain elemental compositions. In comparison to other techniques like Auger electron spectroscopy (AES), XPS is less damaging to the sample due to relatively lower energy of X-rays.

Most of the limitations of this technique arise due to the complexity of sample handling. Operation of the XPS equipment also requires frequent maintenance and highly trained users. XPS samples need to be prepared and handled carefully as minute impurities can affect the analysis. To prevent contamination, the sample surface is thoroughly washed with volatile organic solvents and dried. The analysis is performed on dehydrated samples in an ultra-high vacuum environment. For colloidal samples, the sample must be dried and immobilized on a flat substrate only after removing all salts from the solvent. As mentioned earlier, a major drawback of this technique is its poor lateral resolution which complicates quantitative analysis and interpretation of results. This can be aggravated by the presence of environmental carbon impurities.

Precision in surface PEG quantification in polymeric nanocarriers composed of other ether compounds can be limited due to a similar proportion of carbon and oxygen atoms. Additionally, sampling depths for C1s and O1s are ~10 nm and ~8 nm, respectively which can cause further discrepancies. This challenge can be overcome by including heteroatoms such as nitrogen or sulfur in the PEG functional group. Another drawback of XPS analysis for PEG measurements has to do with the deleterious consequences of drying the samples. The sample in vacuum has a collapsed dehydrated PEG layer whose surface distribution and thickness can be markedly different from the native conformation in an aqueous solution.

### 3.5.1 Application in PEGylated Nanocarriers

Damodaran and coworkers used XPS to calculate PEG layer thickness and grafting densities in amine functionalized poly(ethylene glycols) (PEGs). The amine conjugation method was used to graft 2000 and 4000 Da PEGs on hydrophilic and hydrophobic derivatives. Elemental surface composition was evaluated by survey scans and high-resolution C1s scans were used for detailed surface chemical analysis. High-resolution C1s scans (Fig. 3.7) show C–O peaks with increased intensity at 286.5 eV. Increasing concentrations of PEG resulted in increasing intensities and this confirmed successful PEG coupling. The authors also calculated the thickness of the PEG layer followed by surface conjugation and density. The grafting densities were correlated with brush and mushroom conformations for the polystyrene and Sephadex matrices, respectively (Damodaran et al. 2010).

Rabanel et al. synthesized PEGylated PLA polymeric nanocarriers with varying PEG surface densities in the "brush" conformation. The orientation of PEG chains at the surface was investigated by both XPS and nuclear magnetic resonance (NMR). The two techniques showed exceptionally similar results confirming the segregation of PEG chains at the surface. Minor discrepancies between XPS and NMR were attributed to the depth resolution of XPS (~10 nm) which may not be clearly distinguishable between PEG chains at and underneath the nanoparticle surface. Figure 3.8 shows high-resolution C1s (**a**) and O1s (**b**) scans for PEGylated nanoparticles. The authors used these scans to ascertain the constituent chemical bonds that enable the identification of ether groups. As ether groups are specific to PEG, they were able to successfully distinguish between PEG and PLA (Rabanel et al. 2015).

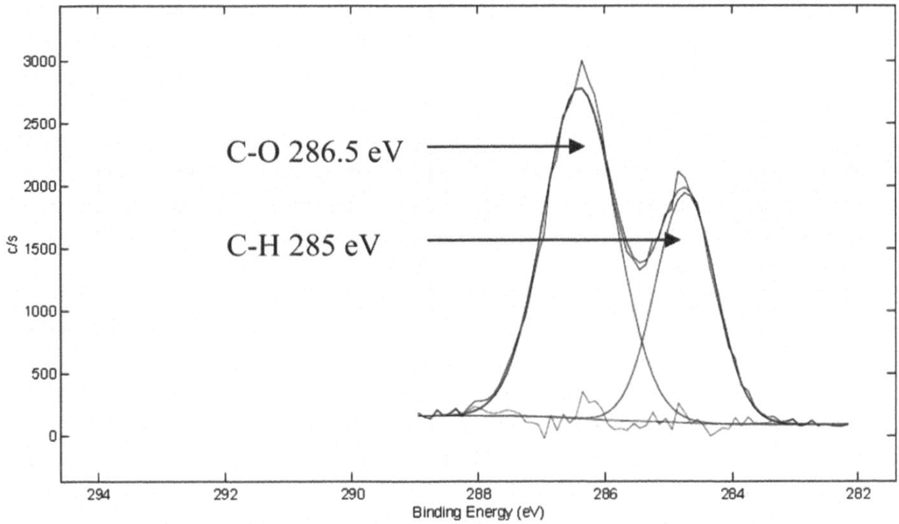


**Fig. 3.7** XPS high-resolution C1s scans. (Figure adapted with permission from Damodaran et al. (2010) Copyright © 2010 American Chemical Society)

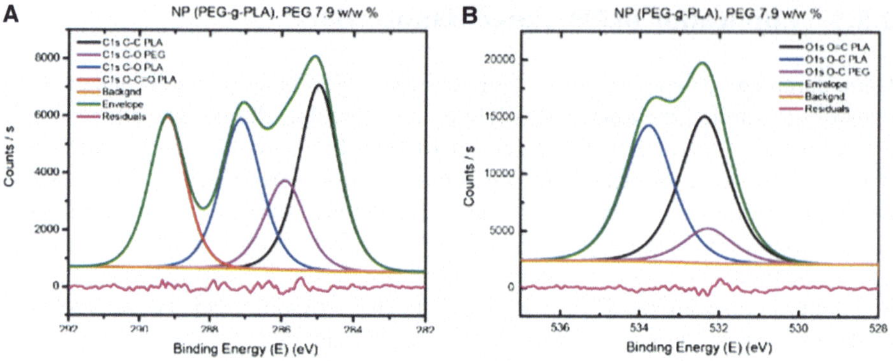


**Fig. 3.8** XPS high-resolution spectra with deconvoluted C1s (**a**) and O1s peaks (**b**) for PEG-g-PLA NP with a 7.9% PEG content. (Figure adapted with permission from Rabanel et al. (2015) Copyright © 2015 American Chemical Society)

## 3.6 Nuclear Magnetic Resonance

Nuclear magnetic resonance (NMR) is a quantitative analytical tool that exploits the magnetic nature of nuclei of atoms of isotopes like $^{1}H$, $^{13}C$, $^{19}F$, and $^{31}P$ to characterize the structure of materials at nanoscale. When a strong magnetic field is applied, the nuclei spin about the direction of the field and at a frequency proportional to the magnetic field strength.

The "spin-up" and "spin-down" states of the nuclei correspond to their alignment in the direction of the applied field and its opposite direction, respectively. The energy difference due to the transitions between these states can be investigated by radiofrequency electromagnetic waves. The electromagnetic waves are irradiated at right angles to the applied magnetic field at a frequency close to the Larmor frequencies of the nuclei. When the frequency of nuclear precession and electromagnetic waves matches, magnetic resonance occurs which is the basis of this technique. The resonance frequency depends both on the sample and the solvent. The energy differences are measured, and the absorption energy is plotted versus chemical shift, which is typically expressed in ppm. Chemical shift is the difference in absorption frequencies for the sample and a standard normalized by the absorption frequency of the standard. The integrated peak surface in the resulting plot is proportional to the number of protons detected.

NMR enables the sensitive measurement of protein displacement and is also versatile to characterize particles in suspensions as well as gaseous environments. NMR relies on the detection of spin coherences and can be used to investigate the topographical properties of nanoparticles. Chemical shifts detected by NMR can be used to obtain critical information about chemical bonds including bond lengths, electron polarization, and dynamics. These properties help ascertain the chemical identity of atoms at the surface and NMR; therefore, enabling monitoring and controlling the local structure. Though NMR is not the primary technique for topographical characterization of nanoparticles, chemical functionalization and the enhanced surface area of nanoparticles make it amenable to investigate the chemical structure at the surface. In PEGylated nanocarriers, NMR can be used to study ligand binding interactions at the surface. In addition to the density and orientation of ligands at the surface, it can also provide information on specific chemical bonds, atomic composition, and molecular structure. PEG density at the surface can also be estimated as exemplified later in this section.

Nanoparticles for analysis can be directly prepared in 0.5% $D_2O$ solution with a surfactant like cholic acid sodium salt (CHA). Internal standards like 3-(trimethylsilyl)-1-propanesulfonic acid, sodium salt (DSS), and tetramethylsilane (TMS) are required for $^1$H NMR analysis. Serial dilutions of PEG in $D_2O$ with the internal standard are generated to obtain a calibration curve for the PEG signal. For lyophilized samples, dissolution in deuterated chloroform ($CDCl_3$) enables the determination of the total PEG content.

A major limitation associated with NMR is that the nanoparticles must be dissolved in a deuterated solvent. However, with this resuspension step, the PEG chains at the surface and the core can no longer be distinguished in the NMR spectrum. If it can be guaranteed that the PEG is grafted exclusively on the surface, the NMR signal can be considered to represent the surface-bound PEG. If lyophilized nanocarriers need to be characterized without dispersion in $D_2O$, another challenge arises. Particles typically aggregate upon freeze-drying and the concomitant decrease in surface area can result in erroneous analysis that underestimates the PEG surface area. Lyoprotectants may be added to prevent this, but this could alter NMR signals due to interference with the internal standard of PEG. Though NMR can be an effective tool to characterize diamagnetic and antiferromagnetic substances, it is not suitable for ferromagnetic and ferrimagnetic materials. Due to their high saturation magnetization, local magnetic fields are affected, and signal frequencies are shifted. This results in broad signal peaks that cannot be interpreted for accurate measurements.

### 3.6.1 Application in PEGylated Nanocarriers

Xu and colleagues synthesized poly(lactic-co-glycolic acid) (PLGA) based nanocarriers with varying surface densities of PEG to study the impact of PEG density on interactions between nanoparticles and mucus. They used $^1$H NMR to determine the PEG content phase separated at the surface and compared it to the total PEG content, obtained by dissolving lyophilized nanoparticles in $CDCl_3$. Figure 3.9 shows the $^1$H NMR spectra for (a) PLGA nanoparticles without PEG with DSS as the internal standard, (b) PLGA nanoparticles with 10% PEG, dispersed in $D_2O$ with DSS as the internal standard, and (c) lyophilized PLGA nanoparticles with 10% PEG, dissolved in $CDCl_3$ with TMS as the internal standard. DSS shows its characteristic peaks at 2.91, 1.76, 0.65, and 0 ppm. Figure 3.9b shows an additional peak at 3.65 ppm confirming PEG functionalization at the nanoparticle surface. This also allows for the quantitation of PEG density on the nanocarrier surface. Figure 3.9c shows the same peak at 3.65 ppm demonstrating that all the PEG content is at the surface (Xu et al. 2015). In this study, the authors used $^1$H NMR

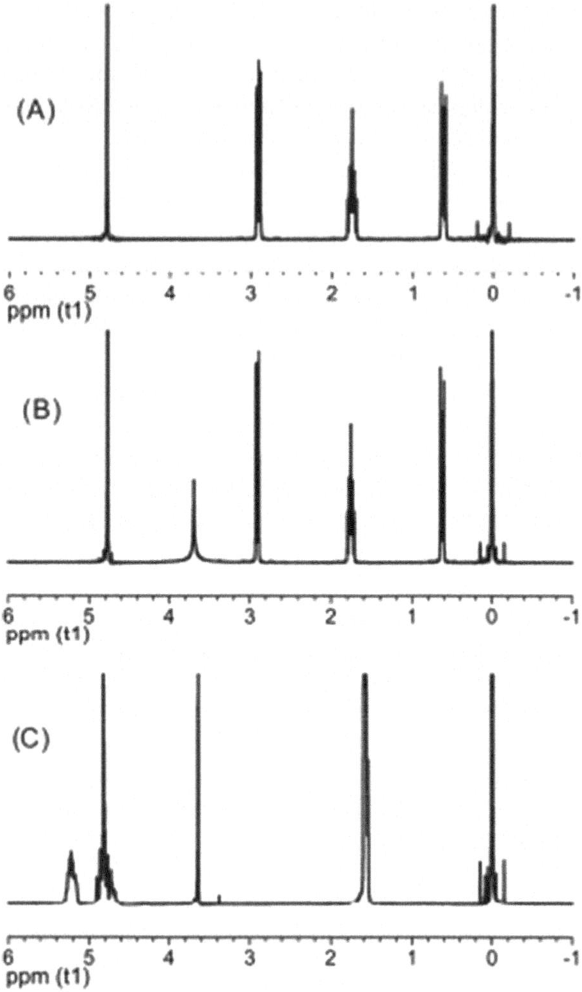


**Fig. 3.9** Representative $^{1}$H NMR spectra of (**a**) PLGA (0% PEG) nanoparticles with 1 wt% DSS as the internal standard, (**b**) PLGA-PEG10% nanoparticles suspended in $D_2O$ with 1 wt% DSS as the internal standard, and (**c**) lyophilized PLGA-PEG10% nanoparticles dissolved in $CDCl_3$ with TMS as the internal standard. (Figure adapted with permission from Xu et al. (2015) Copyright © 2015 American Chemical Society)

spectroscopy to not only confirm the conjugation process but also quantitatively prove that the method of synthesis enabled PEG to completely phase separate on the surface of the nanocarriers.

Garcia-Fuentes et al. developed a novel drug nanocarrier made of tripalmitin, lecithin, and PEG-stearate. They used NMR methods to characterize the structure, composition, and architectural organization of the PEGylated nanocarriers. Specifically, PEG drafting on the surface of the nanoparticles was confirmed by studying its interactions with the external aqueous phase. The authors generated a nuclear Overhauser effect (NOE) spectrum (Vogeli 2014) (Fig. 3.10) which demonstrates molecular interactions between the PEG chains and water. The 3.6 ppm PEG signal confirms the localization of PEG on the surface of nanocarriers (Garcia-Fuentes et al. 2004).

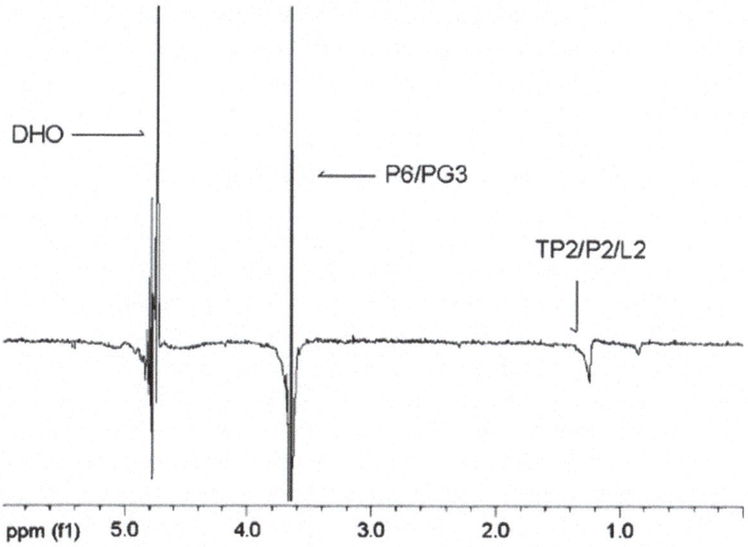


**Fig. 3.10** NOE spectrum between $H_2O$ and the nanoparticle functional groups. (Figure adapted with permission from Garcia-Fuentes et al. (2004) Copyright © 2004 American Chemical Society)

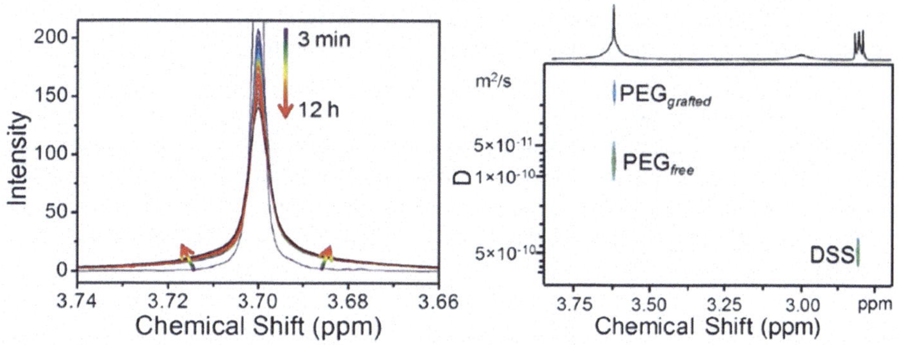


**Fig. 3.11** Temporal evolution of the $^{1}$H NMR spectra and DOSY spectrum of the PEG signal upon grafting to the surface of gold nanoparticles. (Figure adapted with permission from Lu et al. (2019) under a Creative Commons Attribution-NonCommercial 3.0 Unported Licence)

Lu et al. developed a nondestructive $^{1}$H NMR method to determine the grafting density of PEG on gold nanoparticles. NMR signals from free and grafted PEG were distinguished using a multi-Lorentzian-splitting algorithm. As seen in Fig. 3.11, the PEG grafting process was monitored with time, and the intensity of the PEG peak at 3.70 ppm decreased gradually. The authors inferred the broadening of the bottom parts of the peak to correlate to the existence of two types of PEG chains. This conclusion was also corroborated by diffusion-ordered spectroscopy (DOSY) (Groves et al. 2004) which showed two peaks at 3.70 ppm with diffusion coefficients corresponding to free and grafted PEG chains. The authors were therefore able to investigate the temporal evolution of the grafting process and analyze the grafting rates and surface densities of PEG by a unique NMR approach (Lu et al. 2019).

## 3.7 Energy-Dispersive X-Ray Spectroscopy

Energy-dispersive X-ray spectroscopy (EDS/EDX) is primarily used to quantify surface elemental compositions. It is sometimes combined with scanning electron microscopy (SEM) wherein an electron beam is incident on the sample. As this excites the surface atoms, electrons from the inner shells are ejected and valence electrons occupy their place. This transition results in the emission of X-rays whose wavelengths can be analyzed by a detector. Each element emits X-rays at a unique wavelength, thereby enabling the determination of the surface compositions. EDX is also compatible with other electron microscopic techniques like transmission electron microscopy (TEM) and scanning transmission electron microscopy (STEM). The spatial resolution of the data acquired is ~10 nm and the enhanced resolution of the morphology depends on the associated microscopic technique.

EDS samples used for TEM/STEM are required to be electron transparent. Hence, samples are prepared on thin supporting foils, typically made from carbon. The X-rays are emitted isotropically and the detector, usually placed at an elevation angle of ~35°, can collect a very small fraction of the total X-rays emitted. It is also important to note that X-ray photon energy may be significantly attenuated due to scattering and absorption before reaching the detector. The detector should therefore be sensitive and have a larger angular area to receive the photons. Extending measurement times is another strategy to compensate for low signal-to-noise ratios and overly attenuated photon energies. Although EDX is useful for surface compositional analysis, it does not provide accurate quantitative data. However, peak intensities in the spectrogram can be used to estimate the abundance of each surface element.

### 3.7.1 Application in PEGylated Nanocarriers

Karim et al. fabricated pH-responsive nanoparticles composed of strontium sulfite (SSNs) that were modified by biotinylated PEG to prevent rapid clearance from systemic circulation. They used EDX to analyze the elemental compositions of the nanocarriers. Figure 3.12 shows a comparative analysis of nanoparticles before and after PEGylation. Common elements such as C, O, Sr, and S were detected in both the constructs although PEG-SSNs contained a significantly higher percentage of C and O. This demonstrated the presence of biotin-PEG in the PEGylated nanocarriers. Interestingly, Si and Pt were also detected, and the authors suggested this could be due to the use of a glass holder and Pt sputtering during sample preparation (Karim and Chowdhury 2022). This report underscores the utility of EDX to confirm PEGylation as well as its limitations in data interpretation that arise due to surface contamination.

A)

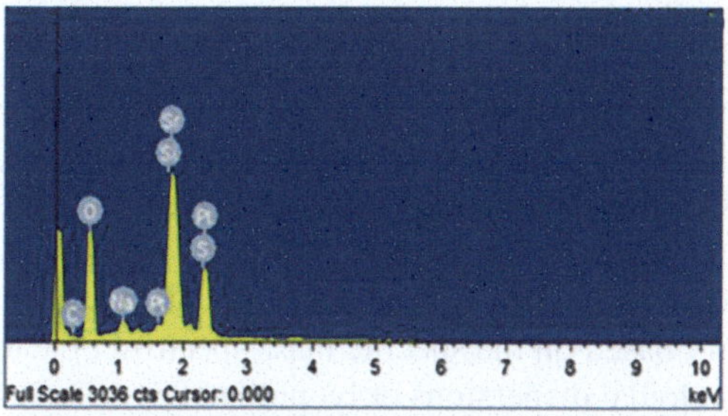


B)

| Element | Weight % | Atomic % |
|---|---|---|
| C | 2.87 | 7.28 |
| O | 30.07 | 57.35 |
| Na | 1.36 | 1.81 |
| Si | 4.43 | 4.81 |
| S | 14.31 | 13.61 |
| Sr | 40.61 | 14.14 |
| Pt | 6.36 | 0.99 |

C)

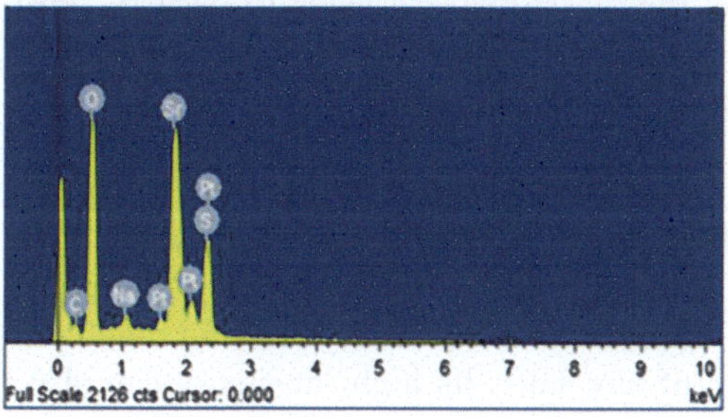


D)

| Element | Weight % | Atomic % |
|---|---|---|
| C | 5.54 | 12.80 |
| O | 36.16 | 62.73 |
| Na | 0.90 | 1.09 |
| S | 12.76 | 11.05 |
| Sr | 34.31 | 10.87 |
| Pt | 10.34 | 1.47 |

**Fig. 3.12** Elemental analysis of SSNs and PEG-SSNs by EDX. (**a**) EDX spectrum of SSNs, (**b**) weight and atomic percentage of SSNs, (**c**) EDX spectrum of PEG-SSNs, and (**d**) weight and atomic percentage of PEG-SSNs. (Figure adapted with permission from Karim and Chowdhury (2022) under the terms and conditions of the Creative Commons Attribution (CC BY) license (https://creativecommons.org/licenses/by/4.0/))

## 3.8 Mass Spectroscopy (MS)

Mass spectroscopy (MS) can be used for the quantitative analysis of the structure of nanocarriers. When a sample is ionized, it degrades into fragments which can be analyzed based on the differences in atomic masses. As discussed below, there are several common mass spectroscopic techniques that can be used to probe critical topographic information of nanocarriers.

Secondary ion mass spectroscopy (SIMS) is a technique that can provide information about the polymer surface structure and composition. Due to its smaller sampling depth, SIMS provides high spectral resolution. The sample is bombarded by an ion beam at a low current density, and secondary ions are sputtered from the surface in response. Low beam density is important to ensure the topography of the carriers is not damaged or altered. The sputtered ions are analyzed by a mass analyzer and the distribution of the molecules on the surface can be ascertained.

In time-of-flight secondary ion mass spectroscopy (TOF–SIMS), the mass of the ions can be determined using time-of-flight. TOF–SIMS is a highly sensitive chemical analysis technique that can be used to investigate the topmost layer of the polymeric surface (~5 nm). As the resulting mass spectrum is of a high resolution, detailed information at the molecular level can be acquired at detection limits as low as ppb–ppm. These benefits make TOF–SIMS a more attractive technique compared to other spectroscopic methods like XPS and AES. Matrix-assisted laser desorption/ionization (MALDI) is a method employed in mass spectrometry wherein a high-energy laser is used for sample desorption and ionization. MALDI-TOF MS provides relatively better estimates of dispersity. In addition to topographical properties, MALDI can also be used to determine the size of the nanocarrier.

Sample preparation methods vary based on the specific mass spectroscopic technique and application. Typically, colloidal dispersions or lyophilized powder can be used with minimal processing. As the mass analyzer tends to be very sensitive, samples should be devoid of all contaminants. To achieve this, high-purity solvents are required, and wearing silicone-free gloves is recommended while handling the samples and operating the instrument.

Though mass spectroscopy enables topographic characterization at a high resolution, it suffers from a few limitations. The main drawback of SIMS is that it is not amenable to quantitative analysis. TOF–SIMS requires the sample to be dry and the analysis can only be performed in high vacuum. Before drying, care should also be taken to ensure the solvent is free of any salts. Its high sensitivity can also be a shortcoming if the sample surface has contaminants. Samples should therefore be handled and processed very carefully. As with SIMS, quantitative analysis may be challenging due to the matrix effect (dependence of secondary ion emission on the sample matrix). Even though high-depth resolution is possible, the lateral resolution is relatively low (~50 nm). If the ion beam is at a high current density, the sample surface will be damaged resulting in poor data interpretation. Due to these reasons, advanced mass spectroscopic techniques like TOF–SIMS and MALDI need highly trained users.

### 3.8.1 Application in PEGylated Nanocarriers

Hinterwirth and coworkers synthesized PEGylated gold nanoparticles with different PEG chain lengths. PEGylation was achieved through thiol-based conjugation and surface coverage by PEG was determined using inductively coupled plasma mass spectrometry (ICP-MS). The results were also compared with a lipophilic thiol ligand. The authors reported a novel gold-to-sulfur ratio measurement that could not only estimate the surface PEG density but also predicted the impact of chain length and nanoparticle size on the stability induced by Van der Waals interactions (Hinterwirth et al. 2013).

In a similar report, Sebby et al. modified 10 nm and 30 nm gold nanoparticles with fluorescently labeled PEG. The authors determined PEG surface coverage for nanoparticles conjugated through thiolate as well as lipoic acid-based linkers. While thermogravimetric analysis was the primary technique used for analysis, MALDI-TOF-MS was utilized for the determination of molecular weights of the PEG derivatives. Figure 3.13 shows the Gaussian curve fits for lipoic acid-PEG-amine (LA-PEG-$NH_3$) and Thiol-PEG-FITC (HS-PEG-FITC). The estimates obtained

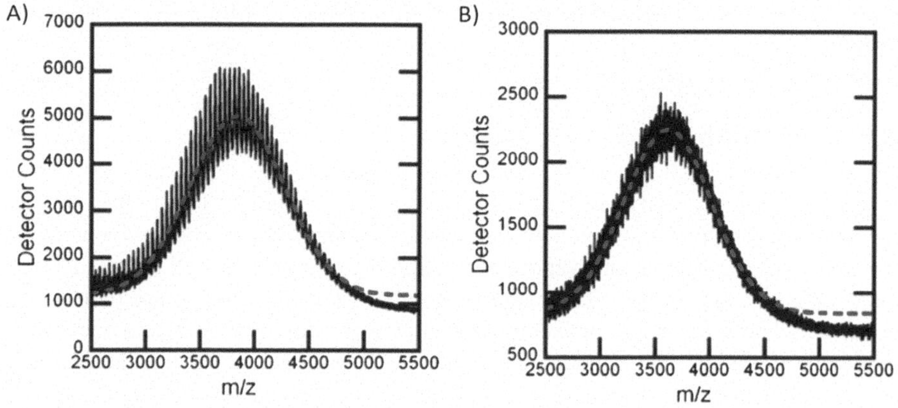


**Fig. 3.13** MALDI-TOF-MS data (solid, black lines) with Gaussian fits (red, dashed lines) for (**a**) HS-PEG-FITC and (**b**) LA-PEG-$NH_3$. (Figure adapted with permission from Sebby and Mansfield (2015) Copyright © 2015, Springer Nature)

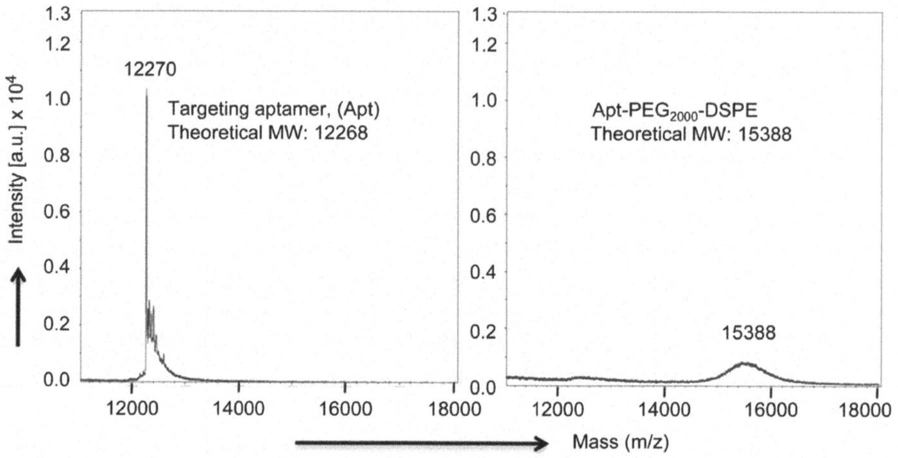


**Fig. 3.14** MALDI-TOF-MS data for the naked aptamer and PEGylated aptamer. Figure adapted with permission from Ara et al. (2014) Copyright © 2014 Elsevier Ltd.)

from the Gaussian fits were 3900 g/mol and 3850 g/mol, respectively, and were in close agreement with the manufacturer's specifications of 3400 g/mol (Sebby and Mansfield 2015). This work illustrates how mass spectroscopy can complement other surface analysis techniques to deduce critical topographical information about nanocarriers.

MALDI-TOF-MS can also be used to monitor the PEG conjugation process. Ara and colleagues developed PEGylated liposomes that were also modified with an aptamer to endow the nanocarrier with a targeting functionality. As shown in Fig. 3.14, MALDI-TOF-MS was employed to confirm the PEGylation of nanocarriers (Ara et al. 2014). Hence, mass spectroscopy can detect small changes in molecular weights enabling the detection and monitoring of the PEG conjugation process.

## 3.9 Thermogravimetric Analysis

Thermogravimetric analysis (TGA) is based on the principle of loss in the mass of the sample upon heating it to controlled elevated temperatures. The components of the nanocarrier degrade and vaporize at different temperatures. A thermally isolated, sensitive weighing balance and temperature-tunable furnace are part of the apparatus. The relationship between mass and temperature is plotted as a decomposition curve. Oxidation temperature and loss of mass are detected and the residual mass, which includes inorganic components of the nanocarriers, residual metals, and other sample impurities, can be obtained. An IR spectrometer can be combined with TGA to identify and analyze gases produced during thermal decomposition.

The main advantages of TGA include the precise quantification in a wide detection range, fast, and reliable analysis. In the context of PEGylated nanocarriers, TGA can be used to characterize the surface and determine the purity. TGA can also be used to determine the PEG surface density of nanoparticles by calculating the degradation of PEG in the temperature range of 300–450 °C. The mass fraction of the organic components can be obtained at a precision as high as $10^{-4}$ (Corbierre et al. 2004).

TGA analysis does not involve cumbersome sample preparation except for drying. However, the sample requirement can be relatively high, in the order of several milligrams. This poses a serious limitation when the sample quantity is limited. It is therefore recommended to combine TGA with other analytical tools to characterize the topography. To improve the sensitivity of thermogravimetric analysis for smaller sample amounts, an improved microscale method has been developed that provides comparable results at sample amounts three orders of magnitude less than the requirement for the conventional TGA method (Mansfield et al. 2014).

### 3.9.1 Application in PEGylated Nanocarriers

Mansfield et al. used a micro thermogravimetric analysis (μ-TGA) method to characterize PEGylated silica nanoparticles. The authors were able to confirm the presence of PEG and quantify surface-bound PEG content. The effectiveness of the developed microscale technique was verified by comparing it with the conventional TGA method. Figure 3.15 shows the thermogram for $SiO_2$–PEG nanoparticles. The decline in mass above 100 °C is due to the decomposition of surface PEG chains. Above 500 °C, mass increases due to the oxidation of silica nanoparticles. This method quantified a mass change of 31.79% (*blue line*) compared to 34.1% by conventional TGA measurement (*red line*) (Mansfield et al. 2014).

Maccarini et al. used several analytical techniques including TGA to characterize gold nanoparticles derivatized with thiolated PEG. The authors were able to detect

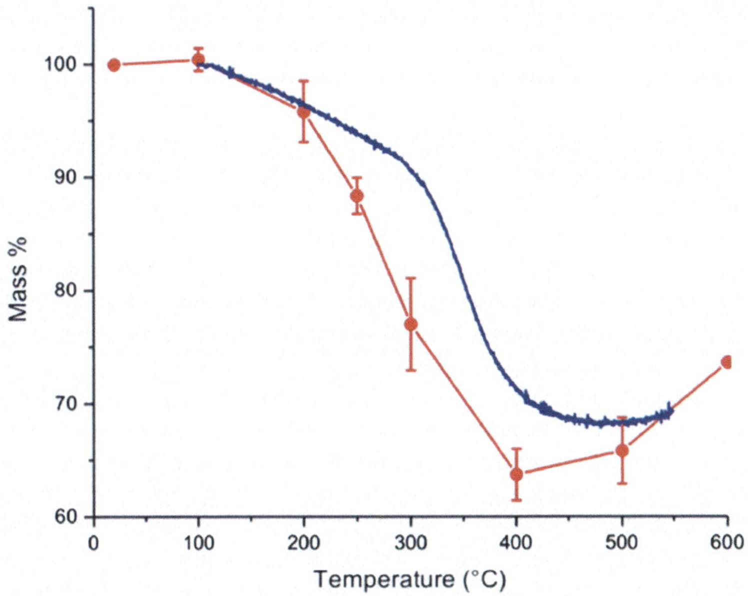


**Fig. 3.15** Mass % versus temperature thermogram of $SiO_2$ – PEG ($n = 25$) nanoparticle samples evaluated using μ-TGA (red) as compared to a conventional TGA measurement (blue). (Figure adapted with permission from Mansfield et al. (2014) Copyright © 2014, American Chemical Society)

PEG chain terminations at the water interface and obtain information about the hydrated structure of PEGylated nanocarriers. TGA could precisely determine the gold to polymer ratio in desiccated samples (Maccarini et al. 2010).

## 3.10 Differential Scanning Calorimetry

Differential scanning calorimetry (DSC) is a thermal analysis method that measures heat flows associated with transitions in materials. The transitions can include physical or chemical changes that result in a change in the heat capacity of the material. Sample temperature may change due to several events like solvent loss, phase transition, or thermal decomposition. To monitor such processes, DSC allows for a very wide temperature range from −60 °C to greater than 1500 °C.

DSC can be run on two modes—power compensation and heat flux. In the power compensation mode, the sample and an inert reference material are maintained at the same temperature. The instrument records the amount of heat needed to increase the temperature of each as a function of temperature. The heat capacity of the sample relative to the reference is equivalent to the energy required to maintain both at an equal temperature. Here, enthalpy change is expressed as energy per unit time and is plotted against time. In the heat flux mode, the analyte and reference are assembled in a single furnace system. Changes in enthalpy of the sample result in a temperature difference between the sample and reference. The subsequent flow of heat that brings the system to a thermal equilibrium can be correlated with temperature changes. This relationship is established through a calibration curve prior to sample analysis.

DSC can be used to determine the phase transition points for melting and crystallization. Latent heat capacity and the heat of fusion may also be ascertained. For polymeric nanoparticles, DSC may be used to investigate the molecular interactions between the drug and PEG chains upon PEGylation. Differences in the glass transition temperature ($T_g$) can point to changes in the topography of the nanocarriers as well as the crystalline state of the encapsulated drug. The main advantage of DSC is its very low sample requirement (a few milligrams), and the determination of weak phase transitions can be fairly accurate.

Solid and liquid samples can be analyzed by DSC. To improve the peak sharpness and resolution of the thermogram, the contact surface between the sample and the crucible should be maximized. This ensures low resistance to heat flow through the system. For solid samples, a thin film or fine powder spread at the bottom of the crucible is ideal. It is important to note that there are no interactions between the sample and the crucible. Even though the measurement requires a very small amount of sample, DSC is a destructive analysis technique. There are also limitations with respect to volatile samples which pose difficulties due to evaporation during sample preparation. Increased resolution may be achieved at the cost of sensitivity and vice versa. As a topographical characterization tool, DSC does not provide direct structural information. However, thermal transitions may be indicative of structural changes.

### 3.10.1 Application in PEGylated Nanocarriers

Essa et al. developed poly(D, L-lactide) (PLA) based polymeric nanocarriers that were PEGylated in two ways—grafted on the surface as 7%PEG-g-PLA and as a multiblock copolymer (PLA–PEG–PLA)n. Both formulations had a similar PEG insertion ratio and chain lengths. The effect of PEGylation on the thermal properties of both types of nanocarriers was evaluated using DSC. Pure PLA was measured to have a $T_g$ of 46.4 °C. As seen in Fig. 3.16, PEG-grafted formulation showed an increase in $T_g$ of ~7 °C (a) that was attributed to increased chain rigidity. On the other hand, the multiblock copolymer formulation showed a lower $T_g$ value of 39 °C (b). This could be due to the higher PEG content of 8.9 mol% and comparatively lower molecular weight. The authors inferred this result to also be indicative of a fraction of PEG chains being present inside the PLA core of the nanoparticles. Furthermore, the interaction between the loaded drug and polymeric matrix was also investigated from the DSC thermograms. As shown in Fig. 3.16a, pure ibuprofen has an endothermic peak at 78 °C, whereas the melting endotherm of the drug as a physical mixture with PEG-grafted polymer dropped to 74 °C. This shift in peak could be due to weak interactions between ibuprofen and the polymers. The

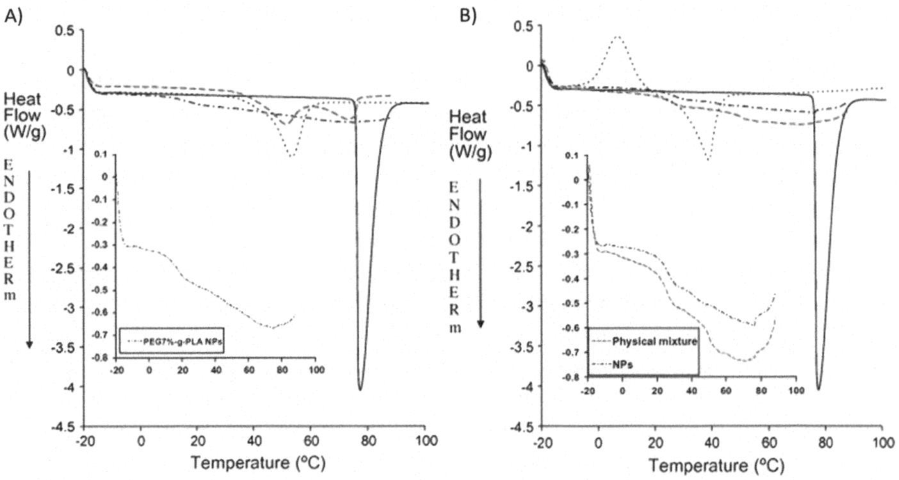


**Fig. 3.16** DSC curves of ibuprofen (solid line), (**a**) physical mixture of PEG7%-g-PLA with ibuprofen (dashed pink line), PEG7%-g-PLA polymer (dotted line), and ibuprofen-loaded NPs (dash-dotted line); (**b**) physical mixture of (PLA–PEG–PLA)n with ibuprofen (dashed pink line), (PLA–PEG–PLA)n polymer (dotted line), and ibuprofen-loaded NPs (dash-dotted line). (Figure adapted with permission from Essa et al. (2010) Copyright © 2010 Elsevier)

authors also used DSC to demonstrate that the drug exists in a crystalline state in both formulations (Essa et al. 2010).

As discussed previously, Rabanel et al. synthesized PEGylated polymeric nanoparticles in the "brush" conformation. The authors found that the $T_g$ of PEG-g-PLA polymers had a strong dependence on the concentration of PEG. Compared to the bulk polymer, the lyophilized nanocarriers showed a very distinct thermogram. In PEG-g-PLA polymers, the presence of PEG was confirmed by a PEG fusion endotherm and a reduction in $T_g$. However, with an increase in the PEG content, there was no further change in the glass transition temperature. These results were also corroborated by NMR and XPS analyses (Rabanel et al. 2015).

## 3.11 Atomic Force Microscopy

Atomic force microscopy (AFM) is a robust scanning probe microscopy technique that is based on the determination of interaction forces between the sample and a probe. The fine probe of the AFM is coupled to a cantilever that can retrace the topography of the sample. As the probe traverses the sample surface, there is a repulsion between the tip and the sample. A constant force is maintained between the sample and the tip which causes deflection of the cantilever. The movements of the cantilever are supported by a piezoelectric scanner. A laser beam captures the degree of bending of the cantilever which correlates with the variation in the reflected beam. This variation is detected by a photodiode, and along with the rigidity of the cantilever, the interaction force and height of the surface can be quantified.

AFM can be operated on three modes based on contact between sample and probe: contact, noncontact, and tapping modes. Contact mode is accompanied by strong lateral forces and this can potentially distort the sample surface and may also damage the probe. The intermittent contact (tapping) mode, most commonly used for nanoparticle characterization, is sensitive to the amplitude of the oscillating tip and is hence also referred to as the oscillating mode. In the tapping mode, oscillation amplitudes can be typically close to 200 nm compared to ~10 nm in the noncontact mode. For sensitive samples that are susceptible to contact by the pointed probe, noncontact mode is ideal.

AFM can be used to generate magnified 3-D images of the surface topography at a very high resolution (~nm). High resolution is possible due to its wide surface scanning range, the sensitivity of the piezoelectric crystal, and the nanometric size of the probe. The resolution is not only similar to that obtained with SEM and TEM, but AFM is also a less-expensive technique that can be easily installed while requiring a low footprint in the laboratory. Surface morphology, including the shape and size of nanoparticles, can be determined precisely. AFM is frequently combined with other spectroscopic techniques like FTIR and TOF–SIMS to improve spatial resolution and enable the detection of elemental composition, respectively. AFM is also one of a few available techniques that can be used to determine the surface texture and roughness of nanomaterials. AFM can also be used to investigate thermal and magnetic interactions if the probe is functionalized based on the application.

Samples can be imaged directly without any surface modification, and measurement is possible in any environment—gas, liquid, or vacuum. However, the sample must be stably immobilized on an inert substrate to enable accurate determination of the morphology. The concentration of colloidal dispersions should also be optimized to prevent the formation of aggregates which adversely affect the topographical analysis. Factors like tip curvature radius and surface properties of the material affect the measured topological properties. Particle characterization results may be difficult to interpret if tip artifacts are not dealt with appropriately.

Even though AFM is a powerful topographical characterization tool, it suffers from several limitations. Its high-resolution images require significantly longer scan times compared to electron microscopic techniques. If multiple samples need to be tested, the low throughput of this method causes further delays in measurement. Properties of the probe tip like shape and curvature can complicate measurements and interpretation of results.

### 3.11.1 Application in PEGylated Nanocarriers

Backmann and coworkers reported a microcantilever-based sensor that was used to characterize the conformation of PEG chains grafted on a gold surface. The authors investigated two solvent conditions—PBS and 20% 2-propanol, by AFM imaging and force–distance curve measurements. Figure 3.17 shows that the PEG chains are

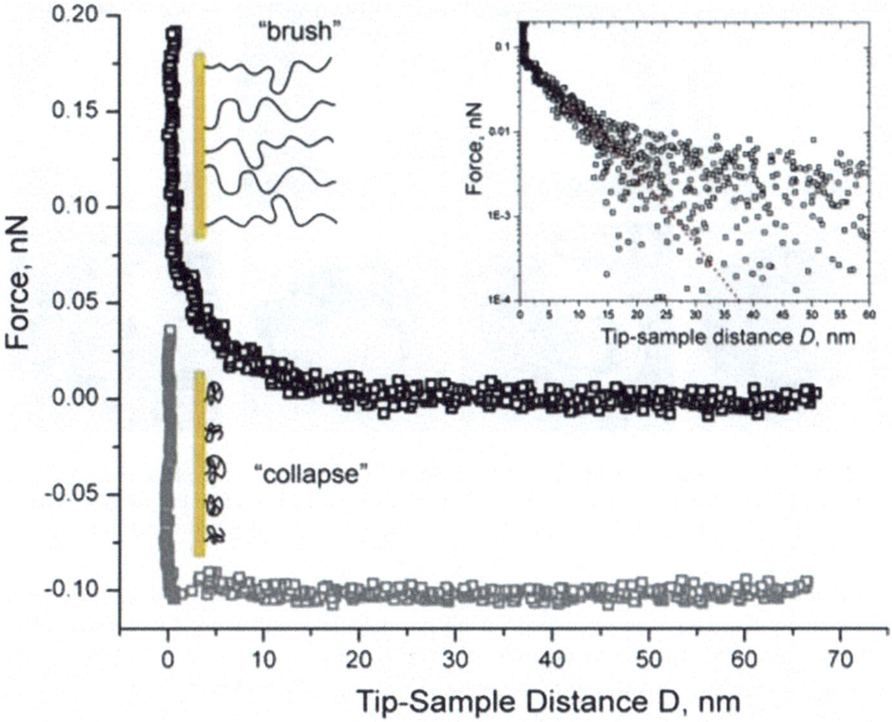


**Fig. 3.17** Representative force curves obtained by approaching the AFM tip to the Au surface grafted with 20 kDa mPEG–SH immersed in PBS (black squares) and 20% 2-propanol (gray squares). (Figure adapted with permission from Backmann et al. (2010) under the terms of the Creative Commons Attribution License (http://creativecommons.org/licenses/by/2.0))

in a "brush" configuration when immersed in PBS (black squares). When the tip–sample distance ($D$) = 0, we are in the region of "hardwall repulsion." As $D$ increases to ~25 nm, the long-range repulsive force decays exponentially indicating the characteristic repulsive forces of a polymer brush. However, when the solvent was switched to 20% 2-propanol (gray squares), the long-range repulsion disappeared and a weak repulsion was observed at ~5 nm. Here, the PEG monolayer was in a "collapsed" state. The authors further investigated the solvent effect by acquiring images at varying contact imaging forces. Their observations corroborated the results of the force measurements and the authors concluded that grafted PEG monolayers reversibly change their conformation when exposed to "good" and "bad" solvents (Backmann et al. 2010).

Shilpa Sant and colleagues developed multiple types of PEGylated nanoparticles with PEG grafted on the surface or as a constituent of a copolymer matrix. All nanocarriers were characterized for their surface morphology using AFM. The authors found that the PEG content was comparatively lower for the multiblock copolymer. This was attributed to some chains that may have interpenetrated inside the nanoparticle core. Figure 3.18 shows AFM phase images (lower panel) and topography

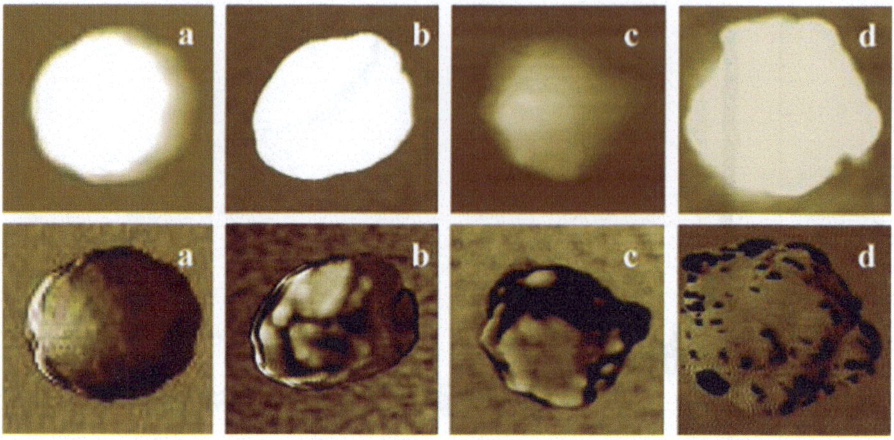


**Fig. 3.18** Tapping mode AFM images of NPs: topography (upper panel) and phase images (lower panel). PLA (**a**), PEG1%-g-PLA (**b**), PEG5%-g-PLA (**c**), and PLA–PEG–PLA)n (**d**). (Figure adapted with permission from Sant et al. (2008) Copyright © 2008 Wiley Periodicals, Inc.)

(upper panel) for both types of PEGylated nanocarriers. As seen in the lower panel, sharp phase contrast was observed on all nanocarriers that were PEGylated. The authors suggested this could be due to PVA coating on PLA on non-PEGylated particles. However, PVA did not adsorb on hydrophilic PEGylated particles thereby leading to a sharper contrast (Sant et al. 2008).

Essa et al. studied the effect of PEG surface PEG density of PEG-g-PLA nanoparticles on their physicochemical and biological properties including protein binding and macrophage uptake. AFM was used to study the morphology of nanocarriers pre- and post-lyophilization. Figure 3.19 shows AFM images (surface morphology on the left and phase image on the right) of these nanocarriers encapsulated with rhodamine B. Before lyophilization, nanoparticles were spherical in shape, whereas aggregates were observed after lyophilization. The authors also determined the surface chemistry of the particles and confirmed the presence of PEG monolayers at the surface of nanoparticles (Essa et al. 2011).

## 3.12 Scanning Electron Microscopy

Scanning electron microscopy (SEM) is an electron microscopy technique wherein the sample surface is scanned by a focused electron beam to produce a 3-D topographic image. The primary electron beam, collimated through a probe, hits the sample to generate X-rays and secondary electrons which are collected by detectors. Although it is run at low voltages of 500 V to 30 kV, SEM can produce images of 10–20 nm particles at a high resolution. SEM is a very useful tool to characterize the morphology and structure of nanoparticles. Combining it with spectroscopic techniques like EDX can provide the elemental composition of the sample surface. In

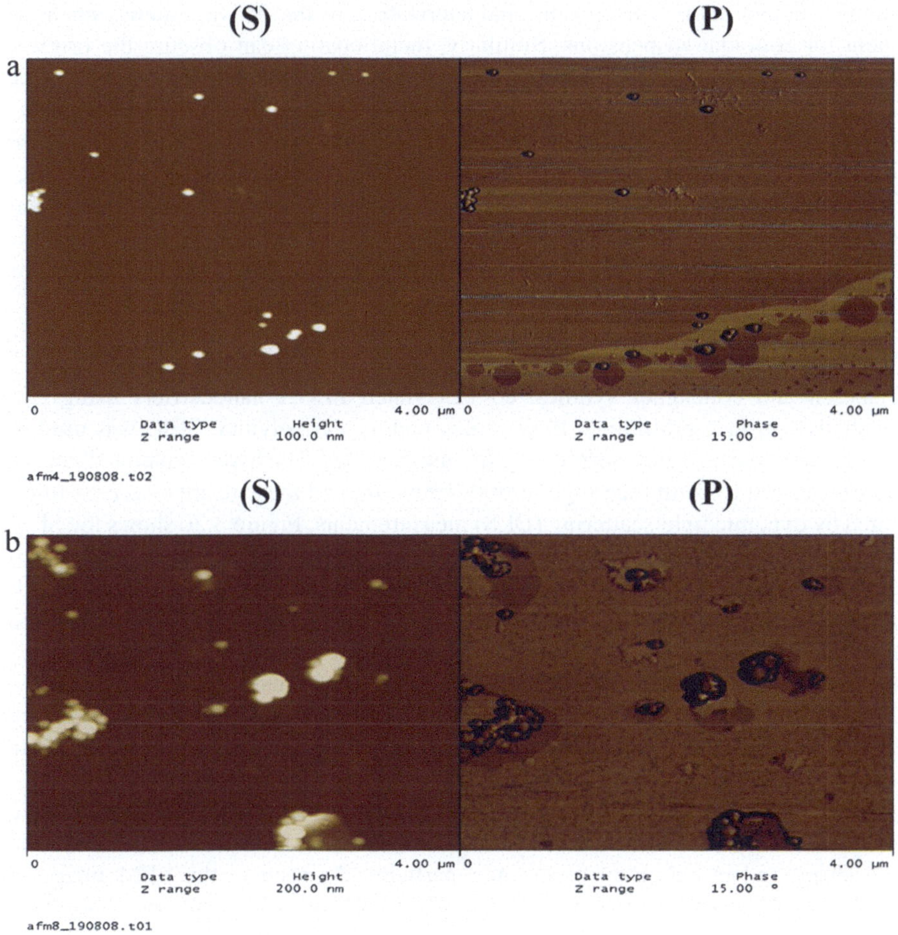


**Fig. 3.19** AFM images of PEG7%-g-PLA rhodamine B-loaded nanocarriers before (**a**) and after (**b**) lyophilization; surface morphology (left, S) and phase image (right, P). (Figure adapted with permission from Essa et al. (2011) Copyright © 2011 Elsevier)

addition to particle size distribution, SEM can also be used to investigate the aggregation and dispersion of colloidal nanocarriers.

SEM requires specific sample preparation protocols based on the application. For topographic and morphological investigation, nonconductive samples need to be coated with a metal like platinum or gold. This not only improves imaging quality by enhancing secondary electron signal but also protects samples from radiation damage. For any type of characterization, the samples should also be dried, and a completely dry environment must be ensured in the vacuum chamber.

Most limitations of SEM arise due to its tedious sample preparation requirements. Dehydration of the sample may cause damage in the form of shrinking or aggregation of particles. Even if solvent removal is not detrimental to a given sample, dehydration will prevent characterization in the native aqueous environment for colloidal suspensions. Similarly, metal coating can obscure the original topographic features of the specimen. In the context of polymeric nanoparticles like PEGylated nanocarriers, vacuum conditions can collapse the grafted PEG chains thereby distorting the topography. Though it is a very valuable tool to characterize the surface of the nanocarrier, SEM is not suitable for studying the internal structure. In addition to these technical challenges, SEM is also an expensive instrument to install and maintain.

### 3.12.1 Application in PEGylated Nanocarriers

Ebbesen and colleagues synthesized PEGylated PLGA nanocarriers using two methods—surface grafting of PEG chains and as a copolymer. SEM was used to characterize the size and morphology of nanoparticles. The hydrodynamic diameter was measured to be in the range of 500–1500 nm, and these results were corroborated by dynamic light scattering (DLS) measurements. Figure 3.20 shows the SEM micrographs for (a) non-PEGylated nanocarriers; PEGylated nanocarriers with (b) 30% nonanchoring PEG, (c) 6% PLGA-b-PEG, and (d) 30% PLGA-b-PEG. All nanocarriers were observed to be polydisperse with a spherical morphology and smooth topography. Interestingly, PLGA-b-PEG nanocarriers with 6% PEG showed significantly larger particles, whereas other PEGylated nanocarriers did not show any apparent difference in comparison to the blank (0% PEG) nanocarriers (Ebbesen et al. 2013). In this report, the authors demonstrated the utility of SEM to investigate the shape, size distribution, and morphology of polymeric nanoparticles.

In another report, Yoncheva et al. developed poly(methyl vinyl ether-co-maleic anhydride) (PVM/MA) based PEGylated nanocarriers. SEM was used to determine the shape and morphology of the nanoparticles. As seen in the SEM image in Fig. 3.21, nanoparticles displayed a size of 280–300 nm with a spherical morphology. These results agreed with the photon correlation spectroscopy analysis. Based on this, the authors concluded that grafting the surface with PEG chains had no significant influence on the size of the PEGylated nanocarriers (Yoncheva et al. 2005).

## 3.13 Transmission Electron Microscopy

Transmission electron microscopy (TEM) works on a similar principle to that of SEM. The main difference between the two techniques is the higher voltages of 80–300 keV used in TEM. A tungsten or lanthanum hexaboride electric gun emits high-energy electrons. The electrons are focused through an electromagnetic lens and the fine beam passes through the specimen. Scattered and transmitted electrons are detected and the difference in electron densities produces 2-D images with contrast on a phosphorescent screen. The brightness of the image is directly related to the fraction of transmitted electrons.

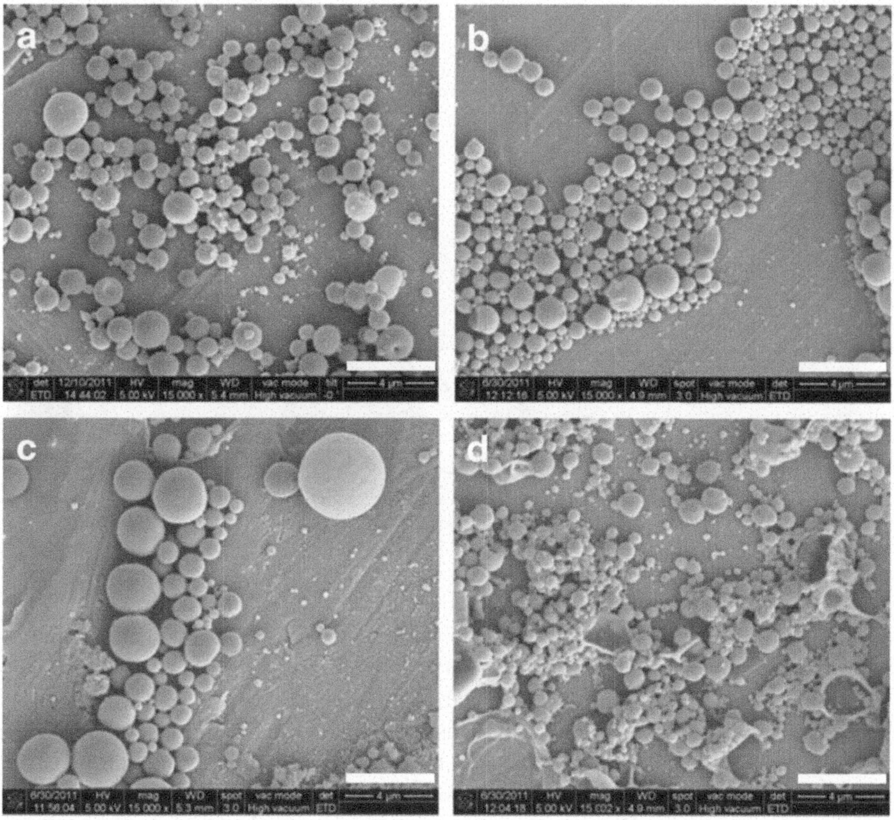


**Fig. 3.20** SEM images of nanocarriers: (**a**) 0% (blank), (**b**) 30% nonanchoring PEG, (**c**) 6% PLGA-b-PEG, and (**d**) 30% PLGA-b-PEG. (Figure adapted with permission from Ebbesen et al. (2013) Copyright © 2013, Springer Nature)

TEM can be used to determine the size distribution, morphology, and atomic structure of the sample at ultra-high resolution. Elemental analysis is also possible if emitted X-rays are detected or when combined with spectroscopic techniques. In addition to the topography of the sample, TEM is also routinely used to characterize interactions of nanoparticles with biological milieu shedding light on aspects such as uptake and localization in cells. Cryo-TEM is particularly apt for such applications wherein aqueous dispersion of particles can be observed to analyze them in their native environment.

As is the case with SEM, sample preparation for TEM is cumbersome. Polymeric and other nanoparticles are drop-casted before immobilizing the dried specimen onto a copper grid coated with a carbon film. This is followed by the negative staining with phosphotungstic acid, uranyl acetate, or ammonium molybdate solution. Negative staining is important to ensure an opaque background that enables the translucent specimen to be viewed easily. TEM requires the sample to be thin for the

**Fig. 3.21** SEM images of PVM/MA-based PEGylated nanocarriers (Figure adapted with permission from Yoncheva et al. (2005) Copyright © 2004 Elsevier).

beam to get transmitted through it. If the sample is larger than ~500 nm, ultramicrotomy or a focused ion beam is necessary.

Though it is the most preferred electron microscopy technique, TEM suffers from several limitations as follows. The instrumentation and maintenance of equipment are not only expensive but it also requires highly trained users. All drawbacks of SEM apply to TEM as well. As it is operated in vacuum conditions, the samples must be completely dry. Drying protocols can result in sample damage or distort the topography and produce artifacts at best. This is a major challenge, particularly for the characterization of polymeric nanocarriers. This can be addressed by spiking the staining solution with a small concentration of glucose. Interaction with high-energy scattered electrons may also damage the sample. This can be prevented by preparing samples that are thinner than 500 nm.

### 3.13.1 Application in PEGylated Nanocarriers

Ana Vila et al. fabricated PEGylated poly (lactic acid)-poly (ethylene glycol) (PLA–PEG) nanoparticles and studied the impact of PEG coating density on transport across the mucosal membrane. PLA–PEG copolymers of different molecular weights were used to form the nanocarriers through nanoprecipitation and multiple emulsion techniques. The authors used TEM to investigate the topography and morphology of nanocarriers. TEM images (Fig. 3.22) showed that particles synthesized with different molecular weights of PEG displayed no significant differences in appearance for a given preparation technique. However, different preparation techniques influenced changes in the size and morphology of nanocarriers. As seen in

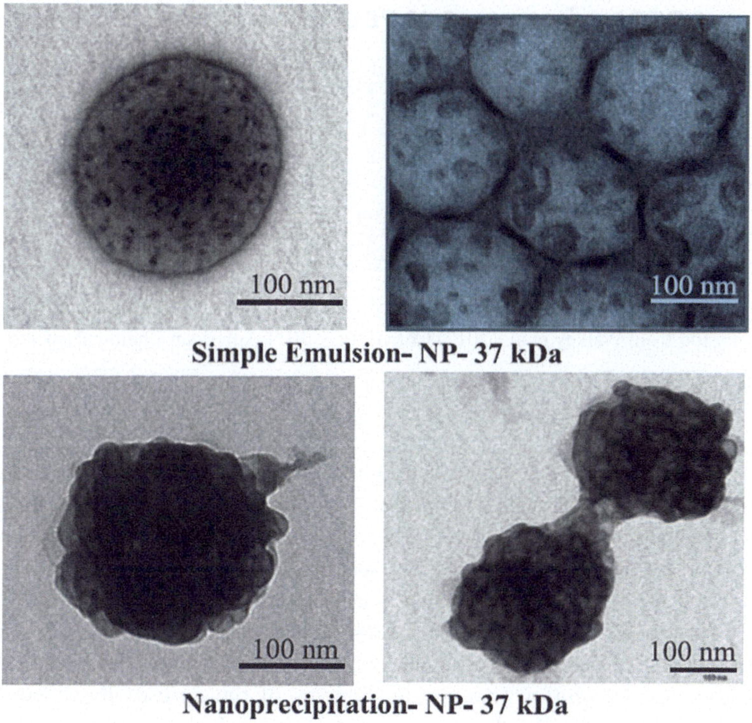


**Fig. 3.22** TEM images of nanocarriers prepared with PLA–PEG copolymers of 37 kDa weight, using different preparation techniques. (Figure adapted with permission from Vila et al. (2004) Copyright © 2004 Published by Elsevier)

the image, the emulsion technique formed particles that were more spherical than those prepared using the nanoprecipitation technique (Vila et al. 2004). This report demonstrates how TEM can be used to not only study the morphology of PEGylated nanocarriers but also compare different preparation techniques.

Carl Walkey and colleagues prepared gold nanoparticles of different sizes and studied the impact of size and PEG coating on serum protein adsorption and macrophage uptake. The authors used TEM to prove that PEGylated nanocarriers localized in the endosomes are independent of PEG grafting density. Figure 3.23 shows that non-PEGylated carriers formed dense aggregates, whereas PEGylated carriers were well dispersed (Walkey et al. 2012). TEM can therefore be used to analyze biological samples and study interactions of nanocarriers with cells and serum proteins.

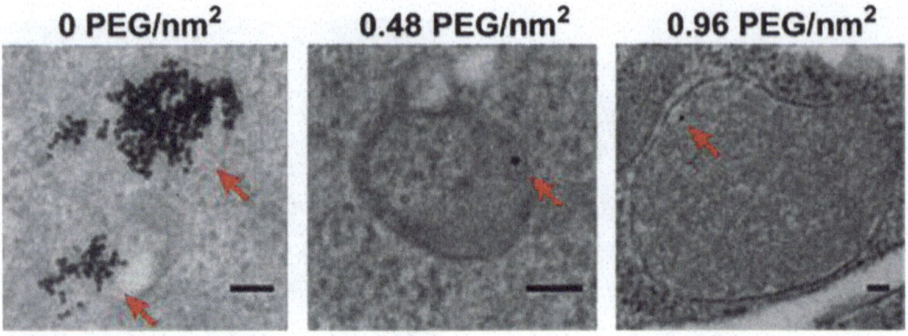


**Fig. 3.23** TEM images of nanocarriers with PEG at 0, 0.48, and 0.96 PEG/$nm^2$. (Figure adapted with permission from Walkey et al. (2012) Copyright © 2011 American Chemical Society)

## 3.14 Conclusion

Surface characterization is a very essential tool to probe the surface topography and morphology of nanomaterials. As interfacial parameters dictate the stability and properties of PEGylated nanocarriers, there is a need to determine the critical attributes accurately and precisely. Several spectroscopic techniques have been routinely applied to investigate the surface topography of nanocarriers. Modifications to the existing methods can provide useful information about the conformation and arrangement of PEG chains. Advanced electron microscopic techniques and AFM can achieve high-resolution topographic images of the core of the nanocarriers, whereas other methods like thermogravimetric and calorimetric analysis provide vital complementary information about the tethered PEG chains. As all techniques suffer from specific limitations with respect to sample preparation and method applicability, combining two or three tools would be the most effective approach to characterize the topographic features of PEGylated nanocarriers.

## References

Alexis F et al (2008) Factors affecting the clearance and biodistribution of polymeric nanoparticles. Mol Pharm 5(4):505–515

Allard-Vannier E et al (2017) Folic acid-capped PEGylated magnetic nanoparticles enter cancer cells mostly via clathrin-dependent endocytosis. Biochim Biophys Acta Gen Subj 1861(6):1578–1586

Ara MN et al (2014) An aptamer ligand based liposomal nanocarrier system that targets tumor endothelial cells. Biomaterials 35(25):7110–7120

Backmann N et al (2010) Sensing surface PEGylation with microcantilevers. Beilstein J Nanotechnol 1:3–13

Baer DR et al (2010) Application of surface chemical analysis tools for characterization of nanoparticles. Anal Bioanal Chem 396(3):983–1002

Bailo E, Deckert V (2008) Tip-enhanced Raman scattering. Chem Soc Rev 37(5):921–930

Butterworth MD, Illum L, Davis SS (2001) Preparation of ultrafine silica- and PEG-coated magnetite particles. Colloids Surf A Physicochem Eng Asp 179(1):93–102

Chen H et al (2016) DUP1 peptide modified micelle efficiently targeted delivery paclitaxel and enhance mitochondrial apoptosis on PSMA-negative prostate cancer cells. SpringerPlus 5:362

Coelho SC et al (2015) Structural characterization of functionalized gold nanoparticles for drug delivery in cancer therapy: a NMR based approach. Phys Chem Chem Phys 17(29):18971–18979

Conte C et al (2019) Surface exposure of PEG and amines on biodegradable nanoparticles as a strategy to tune their interaction with protein-rich biological media. Nanomaterials (Basel) 9(10):1354

Corbierre MK, Cameron NS, Lennox RB (2004) Polymer-stabilized gold nanoparticles with high grafting densities. Langmuir 20(7):2867–2873

Damodaran VB et al (2010) Conformational studies of covalently grafted poly(ethylene glycol) on modified solid matrices using X-ray photoelectron spectroscopy. Langmuir 26(10):7299–7306

Dazzi A et al (2005) Local infrared microspectroscopy with subwavelength spatial resolution with an atomic force microscope tip used as a photothermal sensor. Opt Lett 30(18):2388–2390

Ebadi M et al (2019) Synthesis and properties of magnetic nanotheranostics coated with polyethylene glycol/5-fluorouracil/layered double hydroxide. Int J Nanomedicine 14:6661–6678

Ebbesen MF et al (2013) Surface analysis of PEGylated nano-shields on nanoparticles installed by hydrophobic anchors. Pharm Res 30(7):1758–1767

Essa S, Rabanel JM, Hildgen P (2010) Effect of polyethylene glycol (PEG) chain organization on the physicochemical properties of poly(D, L-lactide) (PLA) based nanoparticles. Eur J Pharm Biopharm 75(2):96–106

Essa S, Rabanel JM, Hildgen P (2011) Characterization of rhodamine loaded PEG-g-PLA nanoparticles (NPs): effect of poly(ethylene glycol) grafting density. Int J Pharm 411(1–2):178–187

Garcia-Fuentes M et al (2004) Application of NMR spectroscopy to the characterization of PEG-stabilized lipid nanoparticles. Langmuir 20(20):8839–8845

Gref R et al (1994) Biodegradable long-circulating polymeric nanospheres. Science 263(5153):1600–1603

Gref R et al (2001) Development and characterization of CyA-loaded poly(lactic acid)-poly(ethylene glycol)PEG micro- and nanoparticles. Comparison with conventional PLA particulate carriers. Eur J Pharm Biopharm 51(2):111–118

Groves P et al (2004) Diffusion ordered spectroscopy as a complement to size exclusion chromatography in oligosaccharide analysis. Glycobiology 14(5):451–456

Hinterwirth H et al (2013) Quantifying thiol ligand density of self-assembled monolayers on gold nanoparticles by inductively coupled plasma-mass spectrometry. ACS Nano 7(2):1129–1136

Hooftman A, O'Neill LAJ (2022) Nanoparticle asymmetry shapes an immune response. Nature 601(7893):323–325

Iranpour S et al (2021) Improving anti-cancer drug delivery performance of magnetic mesoporous silica nanocarriers for more efficient colorectal cancer therapy. J Nanobiotechnol 19(1):314

Jokerst JV et al (2011) Nanoparticle PEGylation for imaging and therapy. Nanomedicine (Lond) 6(4):715–728

Joudeh N, Linke D (2022) Nanoparticle classification, physicochemical properties, characterization, and applications: a comprehensive review for biologists. J Nanobiotechnol 20(1):262

Karim ME, Chowdhury EH (2022) PEGylated strontium sulfite nanoparticles with spontaneously formed surface-embedded protein Corona restrict off-target distribution and accelerate breast tumour-selective delivery of siRNA. J Funct Biomater 13(4)

Katuwavila NP et al (2020) Graphene oxide-based nanocomposite for sustained release of cephalexin. J Pharm Sci 109(2):1130–1135

Larson TA, Joshi PP, Sokolov K (2012) Preventing protein adsorption and macrophage uptake of gold nanoparticles via a hydrophobic shield. ACS Nano 6(10):9182–9190

León A et al (2017) FTIR and Raman characterization of TiO2 nanoparticles coated with polyethylene glycol as carrier for 2-Methoxyestradiol. Appl Sci 7(1):49

Levin CS et al (2006) Determining the conformation of thiolated poly(ethylene glycol) on au nanoshells by surface-enhanced Raman scattering spectroscopic assay. Anal Chem 78(10):3277–3281

Lu J et al (2019) A non-sacrificial method for the quantification of poly(ethylene glycol) grafting density on gold nanoparticles for applications in nanomedicine. Chem Sci 10(7):2067–2074

Maccarini M et al (2010) Characterization of poly(ethylene oxide)-capped gold nanoparticles in water by means of transmission electron microscopy, thermogravimetric analysis, mass density, and small angle scattering. J Phys Chem C 114(15):6937

Maghsoudi S et al (2023) Development of doxorubicin-encapsulated magnetic liposome@PEG for treatment of breast cancer in BALB/c mice. Drug Deliv Transl Res 13:2589

Mansfield E et al (2014) Determination of nanoparticle surface coatings and nanoparticle purity using microscale thermogravimetric analysis. Anal Chem 86(3):1478–1484

Mathurin J et al (2018) How to unravel the chemical structure and component localization of individual drug-loaded polymeric nanoparticles by using tapping AFM-IR. Analyst 143(24):5940–5949

Mitchell MJ et al (2021) Engineering precision nanoparticles for drug delivery. Nat Rev Drug Discov 20(2):101–124

Mohamed M et al (2019) PEGylated liposomes: immunological responses. Sci Technol Adv Mater 20(1):710–724

Mourdikoudis S, Pallares RM, Thanh NTK (2018) Characterization techniques for nanoparticles: comparison and complementarity upon studying nanoparticle properties. Nanoscale 10(27):12871–12934

Pelaz B et al (2015) Surface functionalization of nanoparticles with polyethylene glycol: effects on protein adsorption and cellular uptake. ACS Nano 9(7):6996–7008

Peleg-Shulman T et al (2004) Reversible PEGylation: a novel technology to release native interferon alpha2 over a prolonged time period. J Med Chem 47(20):4897–4904

Prasad A et al (2021) Transport-driven engineering of liposomes for delivery of α-particle radiotherapy to solid tumors: effect on inhibition of tumor progression and onset delay of spontaneous metastases. Eur J Nucl Med Mol Imaging 48(13):4246–4258

Rabanel JM, Hildgen P, Banquy X (2014) Assessment of PEG on polymeric particles surface, a key step in drug carrier translation. J Control Release 185:71–87

Rabanel JM et al (2015) Effect of the polymer architecture on the structural and biophysical properties of PEG-PLA nanoparticles. ACS Appl Mater Interfaces 7(19):10374–10385

Retout M et al (2019) Limits of thiol chemistry revealed by quantitative analysis of mixed layers of thiolated-PEG ligands grafted onto gold nanoparticles. J Colloid Interface Sci 557:807–815

Sacchetti C et al (2013) Surface polyethylene glycol conformation influences the protein corona of polyethylene glycol-modified single-walled carbon nanotubes: potential implications on biological performance. ACS Nano 7(3):1974–1989

Sant S, Poulin S, Hildgen P (2008) Effect of polymer architecture on surface properties, plasma protein adsorption, and cellular interactions of pegylated nanoparticles. J Biomed Mater Res A 87(4):885–895

Sebby KB, Mansfield E (2015) Determination of the surface density of polyethylene glycol on gold nanoparticles by use of microscale thermogravimetric analysis. Anal Bioanal Chem 407(10):2913–2922

Shi L et al (2021) Effects of polyethylene glycol on the surface of nanoparticles for targeted drug delivery. Nanoscale 13(24):10748–10764

Stras S et al (2020) Growth of metastatic triple-negative breast cancer is inhibited by deep tumor-penetrating and slow tumor-clearing chemotherapy: the case of tumor-adhering liposomes with interstitial drug release. Mol Pharm 17(1):118–131

Suk JS et al (2016) PEGylation as a strategy for improving nanoparticle-based drug and gene delivery. Adv Drug Deliv Rev 99:28–51

Swierczewska M, Lee KC, Lee S (2015) What is the future of PEGylated therapies? Expert Opin Emerg Drugs 20(4):531–536

Takechi-Haraya Y et al (2022) Current status and challenges of analytical methods for evaluation of size and surface modification of nanoparticle-based drug formulations. AAPS PharmSciTech 23(5):150

Vernooij EA et al (1999) 1H NMR quantification of poly(ethylene glycol)-phosphatidylethanolamine in phospholipid mixtures. Pharm Res 16(10):1658–1661

Vila A et al (2004) Transport of PLA-PEG particles across the nasal mucosa: effect of particle size and PEG coating density. J Control Release 98(2):231–244
Vogeli B (2014) The nuclear Overhauser effect from a quantitative perspective. Prog Nucl Magn Reson Spectrosc 78:1–46
Walkey CD et al (2012) Nanoparticle size and surface chemistry determine serum protein adsorption and macrophage uptake. J Am Chem Soc 134(4):2139–2147
Xu Q et al (2015) Impact of surface polyethylene glycol (PEG) density on biodegradable nanoparticle transport in mucus ex vivo and distribution in vivo. ACS Nano 9(9):9217–9227
Yoncheva K, Lizarraga E, Irache JM (2005) Pegylated nanoparticles based on poly(methyl vinyl ether-co-maleic anhydride): preparation and evaluation of their bioadhesive properties. Eur J Pharm Sci 24(5):411–419
Zelepukin IV et al (2020) Fast processes of nanoparticle blood clearance: comprehensive study. J Control Release 326:181–191
Zhang Z et al (2016) Tip-enhanced Raman spectroscopy. Anal Chem 88(19):9328–9346
Zhu GH, Gray ABC, Patra HK (2022) Nanomedicine: controlling nanoparticle clearance for translational success. Trends Pharmacol Sci 43(9):709–711
Zolnik BS et al (2010) Nanoparticles and the immune system. Endocrinology 151(2):458–465

# PEGylated Nanocarriers for Drug Delivery Applications

# 4

Salome A. Chime and Mumuni A. Momoh

**Abstract**

Polyethylene glycol (PEG), also called macrogol, is a polyether made up of ethoxy units often obtained by polymerizing the ring opening of ethylene oxide. PEG polymers are generally linear and have chemically active hydroxyl groups, facilitating conjugate formation with other functional groups. Thus, biomolecules and nanocarriers can form conjugates with PEG in a process called PEGylation. It is an approach employed by others to enhance the efficiency of drugs, vaccines, and gene delivery, and it is often utilized for cell and tissue targeting. It is a process of modification of the surfaces of particles, which usually occurs by adsorption, entrapment, or grafting of known polymer chain lengths of PEG. Incorporating PEG polymers into various nanocarriers improves the pharmacokinetic profile of drugs in the nanocarriers. PEGylation also enhances the diffusion of nanocarriers drug delivery systems across different biological systems, including the mucosal systems, extracellular spaces, and the brain. It also improves the physicochemical profiles of varying drug delivery systems and improves their biocompatibility by facilitating drug solubility and reducing their toxicity. PEGylated formulations have elongated residence in the body, enhanced drug stability, reduced metabolic enzymes-induced degradation, and minimized protein elimination. In this work, the process of PEGylation of different nanocarriers including lipid-based and polymeric nanoparticles, liposomes, and dendrimers will be explored. The applications of PEGylated nanocarriers in drug delivery will also be exhaustively discussed.

S. A. Chime (✉)
Department of Pharmaceutical Technology and Industrial Pharmacy, University of Nigeria, Nsukka, Nigeria
e-mail: salome.chime@unn.edu.ng

M. A. Momoh
Department of Pharmaceutics, University of Nigeria, Nsukka, Nigeria



R. K. Tekade, N. K. Jain (eds.), *PEGylated Nanocarriers in Medicine and Pharmacy*, https://doi.org/10.1007/978-981-97-7316-9_4

**Keywords**
Drug delivery · Generally regarded as safe · Methods of PEGylation

## 4.1 Introduction

A lot of active pharmaceutical agents require specialized formulation technologies and strategies to target them to the particular organ of the body where the drug's effect is mainly needed. There are also significant challenges in controlling the drug release, biodistribution, pharmacokinetic profiles, and circulation time of drugs. Hence, different drug delivery systems (DDS) and strategies have been developed to solve this problem, including polymeric and lipid-based nanoparticles, hydrogels, and surface-functionalized nanoparticles (Liu et al. 2016). Surface modification of drug carrier systems with polyethylene glycol (PEG) has been used to control non-specific and specific reactions of drugs and some significant components of blood and avoid opsonization (Vllasaliu et al. 2014; Abuchowski et al. 1977).

PEG is a "stealth" polymer generally employed in drug delivery (DD) due to its record of nontoxicity in humans. They were classified by the FDA as Generally Regarded as Safe (GRAS) (Jain and Manoj 2010); hence, they are safe and have received approval from the FDA for systemic applications (Haris 1992). Polyethylene glycol (PEG), often called macrogol, is a polyether consisting of ethoxy units obtained by polymerizing the ring opening of ethylene oxide. It is a hydrophilic polyether diol that is nonionic linear with outstanding biocompatibility and excellent biological and physiochemical characteristics such as nontoxicity, good solubility in aqueous as well as organic solvent, the absence of immunogenicity, and antigenicity. The neutral nature of PEG polymers emanates from the weak hydrogen bond acids of the hydroxyl end groups and ether linkages that are mainly weakly basic, forming their backbone. The low-grade PEGs are viscous colorless liquids, while the high molecular weight grades (2000–5000) are waxy. Polymers of PEG are traditionally often linear and have chemically active hydroxyl groups facilitating conjugate formation with various functional groups. Hence, nanocarriers and biomolecules can conjugate with PEG in a process called PEGylation process (Jain and Manoj 2010). PEG polymers enhance the delivery of therapeutics agents and have been employed to deliver drug-entrapped micro and nanocarriers (Vllasaliu et al. 2014).

PEGylation enhances the efficiency of drugs, vaccines, and gene delivery and is often employed to target cells and tissues (Vllasaliu et al. 2014). It is a process of modification of the surfaces of particles, which often occurs through the adsorption, entrapment, or grafting of known polymer chain lengths of PEG. Incorporating PEG polymers into various nanocarriers is a functional method to enhance the pharmacokinetic profile of drugs in the nanocarriers. PEGylation also enhances the diffusion of nanocarriers drug delivery systems across different biological systems, including the mucosal systems, extracellular spaces, and the brain. It also enhances various drug delivery systems' biocompatibility profiles and physicochemical properties by facilitating drug solubility and reducing their toxicity. PEGylated formulations have improved drug stability, prolonged in vivo body residence, minimized toxicity, reduced degeneration by metabolic enzymes, and reduced protein elimination. PEGylation has been used to maximally improve the pharmacokinetic profile of different novel drug delivery nanoformulations. It also enhances the nanoparticles' stability giving it improved superficial energy with tendencies of aggregation and improved diffusion via biological systems, viz. mucosal cells, brain, and extracellular spaces (Vllasaliu et al. 2014). In this work, the process of PEGylation of different nanocarriers, including lipid-based and polymeric nanoparticles, liposomes, and dendrimers, will be explored. Also, the applications of PEGylated nanocarriers in drug delivery will be exhaustively discussed.

## 4.2 PEGylated Nanocarriers and Drug Delivery Benefits

Nanotechnology has advanced tremendously over the years and has made huge successes in drug delivery and nanomedicines. Nanocarriers include both polymeric and lipid nanoparticles, viz. solid lipid nanoparticles (SLNs), nanostructured lipid carriers (NLC), liposomes, micelles, and dendrimers, amon others. These nanocarriers have numerous advantages over conventional delivery systems due to their small size and targeting capabilities, which provide site-specific delivery and increased local concentrations, decreasing systemic toxicity (Howard et al. 2008; Owens and Peppas 2006; Sahoo and Labhasetwar 2003; Mumper et al. 2003). Nanocarriers can influence the drug release profile of most drugs and are used for specific targeting of active pharmaceutical ingredients (APIs) to significant organs of the body, for example, brain (Howard et al. 2008). Drug targeting is vital in cancer chemotherapy, to circumvent several toxic effects of drugs, and in the treatment of disorders of neurons, which are limited by the inability of many drugs to cross the blood–brain barrier (Howard et al. 2008; Torchilin 2007; Francis et al. 1996; Calv et al. 2001; Sinha et al. 2006). Nanocarriers also protect drugs from enzymatic degradation, have low levels of toxicity, and have high stability. They also result in improved drug solubility with good potential to be functionalized to yield controlled release systems (Chimé et al. 2014, 2022; Yildirimer et al. 2011; Ou et al. 2018; Yang et al. 2018; Pinheiro et al. 2021). However, nanocarriers suffer from many disadvantages despite the huge successes recorded in the field of drug delivery, including issues of stability and drug leakage, reticuloendothelial system blood clearance (RES uptake), hemolytic toxicity, immunogenicity, and hydrophobicity. These disadvantages could be improved upon by nanocarriers PEGylation (Gajbhiye et al. 2020).

Nanoparticles (NPs) have the potential to deliver therapeutic payloads to various tissues and organs of the body, prolonging the blood circulation time and partitioning into target organs (Suk et al. 2016). However, the administration of these NPs by systemic drug delivery has been limited by the influence of the mononuclear phagocyte system (MPS). The MPS includes the granulocytes, blood monocytes, dendritic cells, and tissue-resident macrophages in the liver, spleen, and lymph nodes in charge of the clearing process and the decomposition of the exogenous materials in the bloodstream (Moghimi et al. 2001). Endothelia cells in MPS-associated organs mostly contain fenestration, which helps to screen circulating bodies based on particle size. Hence, NPs of about 100 nm penetrate the endothelial fenestrae in the spleen, liver, and lymph nodes (Alexis et al. 2008; Braet et al. 2007). MPS serves as a defense mechanism protecting the body against pathogens and also eliminates drug-loaded NPs rapidly from the blood. Therefore, NPs present in the bloodstream are detected and entrapped by opsonins, which are serum proteins and complement compounds such as fibronectin, immunoglobulins, and apolipoproteins (Vonarbourg et al. 2006), a process called opsonization. NPs are more responsive to phagocytosis via MPS cells, hence, the need for PEGylation. Opsonization of NPs occurs mainly by hydrogen bond interactions and hydrophobic and electrostatic interactions (Yoon et al. 1998; Roser et al. 1998; Walkey et al. 2012; Gessner et al. 2000). However, NPs may directly be captured by opsonin-independent scavenger receptors macrophages, which recognize mostly repeating patterns (Monopoli et al. 2012; Liu and Liu 1996). Also, numerous other types of serum proteins readily associate with NPs circulating within the bloodstream, yielding protein corona on the NPs' surface that improves NPs uptake by endothelial cells (Liu and Liu 1996; Tenzer et al. 2013). It is worth noting that protein absorption minimizes the time of circulation of functionalized NPs and also weakens their targeting abilities (Salvati et al. 2013). Generally, nonfunctionalized NPs are removed within 10 min after systemic administration from the bloodstream, no matter their composition (Suk et al. 2016; Moghimi et al. 2001; Salvati et al. 2013).

PEGylation technology is one of the most successful approaches used clinically to improve the pharmacokinetic properties of drug-loaded NPs and minimize their immunogenicity (Marwa et al. 2020; Veronese and Mero 2008). PEGylation of nanocarriers improved the pharmacokinetics of drugs as they help to prevent the opsonins of nanocarriers, leading to an increase in molecular weight (MW) of nanocarriers and particle size, leading to changes in the properties of nanocarriers. The density and MW of PEG chains can alter the circulation time of nanocarriers as well as affect the biocompatibility, size, charge, and interaction with macrophages (Gajbhiye et al. 2020). Complement proteins, immunoglobulin, and fibronectin are the major opsonin proteins. However, dominant opsonin depends on the specific properties of the nanocarriers, with the majority exhibiting improved adsorption of many types of proteins (Buyens et al. 2012; Andriyanov et al. 2017). Hence, for these nanocarriers to function as improved alternatives to the existing DDS, they should not be readily recognized by the opsonins. PEGylation is one of the simplest and most reliable methods of decreasing protein adsorption and imparting "stealthiness" to nanocarriers (Howard et al. 2008). The mechanisms by which PEG improves the stealthiness of nanocarriers include the reduction in hydrophobicity and surface charges of nanocarriers, thereby decreasing the attractive forces between opsonin proteins and nanocarriers. Therefore, PEGylation increases the plasma half-life of different drugs, viz. enzymes, proteins, and small molecular drugs, achieved mainly by circumventing their clearance from circulation by opsonins, leading to an enhanced therapeutic index (Marwa et al. 2020; Veronese and Mero 2008).

Apart from improving the pharmacokinetics of drugs, PEGylation is also important in the mucosal delivery of drugs. Mucosal delivery of drug-entrapped nanocarriers is based on their ability to target the cell internalization pathways of epithelial cells (Vllasaliu et al. 2014). The epithelial cells and mucosa present barriers in the mucosal surface that affect or impede the passage of nanocarriers (Vllasaliu et al. 2014; Vllasaliu et al. 2011). Also, proteases at the mucosal surface function as barriers preventing the availability of nanocarrier-delivered protein drugs, influencing their absorption (Lee 1990). Epithelial tight junctions also limit the passage of macromolecules greater than 1000 Da, therefore preventing most nanocarriers from traversing the paracellular space (Illum 2000). PEGylated nanocarriers in mucosal drug delivery have been reported to have a major breakthrough in nanocarrier diffusion across the mucosal surfaces (Ensign et al. 2013; Cu and Saltzman 2009; Wang et al. 2008). NPs diffusion is influenced and may be limited by the PEG MW and the density used in the coating; also, coating some PEGs with short lengths at high density improves the diffusion of NPs through the mucosa in combination with shape, size, and the kind of NPs (Vllasaliu et al. 2014; Cu and Saltzman 2009; Wang et al. 2008; Shi et al. 2021).

## 4.3 Methods of PEGylation of Nanocarriers

PEGylation significantly alters the surface properties of nanocarriers; most nanocarriers present hydrophobic surfaces, which may be difficult to bind PEG, leading to low conjugation efficiency (Tobio et al. 2000). Several methods are adopted in achieving PEGylation, including the physical or chemical adsorption, grafting, and the incorporation of PEG chain (Jain and Manoj 2010; Li et al. 2001; Calvo et al. 2001; Gref et al. 1995, 2000). The major limitation in simple adsorption is the displacement of the PEG layers of coating in vivo; therefore, the PEGylation of nanocarriers by the covalent method is preferred. Derivatization and activation of PEG is a vital step in synthesizing the PEGylated system. Chemical derivatization of the end groups of PEG, which is often an essential first step in the preparation of bioconjugates, yields stable products (Jain and Manoj 2010). The methods involved in nanocarriers PEGylation are discussed below.

### 4.3.1 Nanoparticles PEGylation

There are three methods of applying PEG coating onto the surface of nanoparticles: physical adsorption, chemical conjugation, and molecular self-assembly.

**Physical Adsorption** PEG adheres physically to nanoparticles by physical adsorption, including hydrophobic and electrostatic interaction. Coating of the surface of nanoparticles with PEG is physical adsorption and is the most common method of PEGylation. This process is simple and takes place under easily controlled conditions. This method is primarily suitable when low-density PEG and PEG derivatives and substratum have good adsorption potentials (Tobio et al. 2000). In this method, the charged or hydrophobic groups of the nanomaterial adsorb to PEG based on electrostatic adsorption or hydrophobic ability. Physical adsorption, however, suffers significant limitations due to low adsorption strength, sometimes causing detachment of PEG chains from the surface of NPs (Tobio et al. 2000; Kaur et al. 2008; Xiao et al. 2013).

**Chemical Conjugation** In chemical conjugation, stable chemical bonds are formed when PEG is securely grafted on the NPs' surface. PEGylated nanoparticles can be formed by covalent coupling to prevent PEG separation or desorption from the surface of nanocarriers. This involves a chemical reaction between the exposed NPs surface and end groups of PEG having minimal chemical interactions, yielding firm bonds between PEG and NPs (Tami et al. 2004). In this method, PEG is present only at the NPs' surface, resulting in the presence of PEG in the core of the nanoparticles (Tobio et al. 2000). Hence, this could ensure that the shelf-life of PEGylated products is prolonged longer than non-PEGylated products. However, surface graft density could be a problem while trying to attain application requirements due to issues of active surface for PEG grafting, and there may be batch-to-batch variations in the graft ratio due to differences in reaction rate and space hindrances (Rabanel et al. 2014; Liu et al. 2018).

**Molecular Self-Assembly** The self-assembly method utilizes the hydrophilic and amphiphilic nature of PEG; hence, when PEG is combined with hydrophobic polymers and lipids molecules, they self-assemble with their molecules. Two basic methods employed in formulating PEGylated nanoparticles by self-assembly are the nanoprecipitation method, also known as solvent diffusion, and the emulsification method (also called nanoemulsion or solvent evaporation). The molecular self-assembly method of PEGylation of nanoparticles occurs mainly by nanoprecipitation, also called solvent diffusion. Self-assembly could also be formed by emulsification/solvent evaporation. Spherical nanoparticles form by self-assembly by amphiphilic polymers in the presence of water without other compounds. In this method, hydrophilic PEGs bind to hydrophobic lipids or polymers (Serra et al. 2006). The nanoparticles form a hydrophobic core, while PEG coatings make the layers outside it (Tobio et al. 2000).

Nanoprecipitation often yields long-circulating PEGylated nanoparticles that could be given via intravenous route (Huang et al. 2000; Gu et al. 2008). It is a step procedure during formulation that does not require high shear. Here, drugs and PEG are dissolved in an organic solvent that mixes well with water, such as acetonitrile, tetrahydrofuran, and acetone. This organic phase gradually transfers into the aqueous phase with or without that may contain surfactants (Suk et al. 2016). Changing the organic phase may be used to control the PEG density and the particle size of NPs (Karnik et al. 2008). The limitation of this method is that the PEG and drugs must dissolve in the same solvent mixture. Hence, formulation challenges exist with poorly soluble and unstable drugs or bioactive (Suk et al. 2016). Also, the nanoprecipitation method yields low drug loading. It has the limitation of poor mixing, which may lead to incomplete self-assembly, causing the PEG chains to remain in the core of the NP buried intact instead of partitioning to the surface of the NPs (Suk et al. 2016; Xu et al. 2015).

The emulsification method of NPs PEGylation involves the application of adequate emulsifiers into the aqueous phase and mixing the molecules containing PEG dissolved in organic phase water-immiscible solvent into this aqueous phase with proper mixing and sonication. After that, the organic solvent is evaporated gradually, causing the hydrophilic PEG chains to partition toward the oil/water interface in the emulsion droplets (Suk et al. 2016). Partitioning of the chains of PEG is allowed to the NP surface fully due to the relatively slow solidification of the emulsion droplets (Hrkach et al. 2012). The emulsification method also yields higher drug loading than the nanoprecipitation method (Torchilin and Weissig 2003) and can easily be scaled up. Also, different types of emulsion structures of NP could be formulated (Xu et al. 2015). However, this method's limitations include the uses of high energy and shear, which could denature sensitive APIs. There may be a burst release of drugs due to the accumulation of APIs at the water–particle interface, causing burst release due to the evaporation of the organic solvent from the droplets of emulsions (Xu et al. 2015). Different types of NPs could be PEGylated using self-assembly and adsorption methods, including polyplexes, PEGylated liposomes or polymersomes, solid lipid nanoparticles, and nanostructured lipid carriers, among others (Suk et al. 2016; Petersen et al. 2002a).

### 4.3.2 Polyplexes (PP) PEGylation

Polyplexes (PP) are core-shell formulations/structures formed by the condensation of polycation and PEG or PEG and lipid or with plasmid DNA. Here, the driving force for self-assembly is the electrostatic reaction in the cationic groups of the PEG-containing molecules and phosphate groups of the DNA. PEG could interact covalently with a region of the cationic block co-polymers or could be grafted, forming comb-like structures after interactions with the cationic regions (Suk et al. 2014, 2016; Nakamura et al. 2012).

### 4.3.3 PEGylation of Liposomes (Polymersomes)

Liposomes could be easily PEGylated by preparing PEG-conjugated lipid and mixing the PEGylated lipid along with other lipids during the formulation processes (Jain and Manoj 2010). Conjugates of PEG-lipid could penetrate into pre-formed liposomes. Here, the lipid tail (hydrophobic) is attracted and penetrates the lipid bilayer. This pre-insertion needs minimal PEG-lipid conjugate to achieve the same surface PEGylation as the pre-insertion and have a more prolonged half-life in the bloodstream (Suk et al. 2016; Uster et al. 1996). Also, the amount of PEG-lipids added should be less than the critical micellar concentration. PEG-lipid solutions are slowly incorporated in the lipids, temperatures should be close to the lipid melting temperature to circumvent the formation of micelles (Torchilin 2005). In the post-insertion method, the bilayer of the liposomes' outer surface could be altered rather than in a self-assembly technique that allows the insertion of PEG randomly into the core of liposomes. Post-conjugation could also be used to prepare PEGylated liposomes, usually for targeting ligands (Gajbhiye et al. 2007).

Self-assembly process (pre-insertion) is another technique for preparing polymersomes. Here, liposomes are prepared by thin film hydration with an aqueous medium (Petersen et al. 2002a), and PEGylated lipids could be added during the self-assembly process, which will be inserted into the bilayer. At the same time, the hydrophilic PEG chains will extend toward the aqueous phase. Partitioning of the PEG molecules in the inner and outer layers of liposomes lipid bilayer could happen. To PEG in the core, which could affect drug loading, the PEG can be added by post-conjugation or post-insertion methods (Suk et al. 2016). Monitored PEGylation of liposome surface could circumvent liposome aggregation, while a large quantity of PEG deposit on the surface has a limitation of compromising liposome formation and stability (Suk et al. 2016).

### 4.3.4 PEGylation of Dendrimers

PEGylation of dendrimers can be attained based on the orientation of the chains of PEG needed in the dendrimers. Here, PEGylated dendrimers having PEG chains conjugated to their periphery could simply be developed (Jain and Manoj 2010). Also, dendrimers with inner PEG core could be developed. This creates more distance between branching units of the dendrimers, preventing the dumping of the drug in the dendritic inner core. PEGylated dendrimers with branching monomer units of PEG may also be formulated (Jain and Manoj 2010). Drug-conjugated PEGylated dendrimers are another strategy for developing this formulation. Here, PEG chains come between the dendrimers and the drug, forming steric hindrance over the dendrimer, resulting in controlled drug release (Dunn et al. 1994).

## 4.4 Characterization of PEGylation

### 4.4.1 Qualitative Assessments of PEGylation

The qualitative assessment of PEGylated NPs is done using the methods discussed below:

**Zeta Potential** This method is the simplest method of assessing the quality of the PEGylated nanoformulations used to determine the surface charge of NPs (Howard et al. 2008). The surface coating of PEG and the PEGylating agent can influence the zeta potential of the NPs. The zeta potential of NPs may also affect the liver uptake. Liver uptake was found to decrease when the zeta potential neared zero (Craparo et al. 2006). Also, the surface charge of PEGylated NP may decrease with an increase in PEG as reported by some researchers (Churae and Nikologorskaja 1991). However, there is difficulty in determining the quantitative amount of PEG on the surface of NP (Howard et al. 2008). The thickness of the hydrodynamic layer around the NP could be calculated from the zeta potential (Suzawa and Shirahama 1991; Webb et al. 1998). There is also difficulty in measuring the zeta potential in NPs with neutral surfaces and low surface charges (Peracchia et al. 1998).

The comparison of hydrophobic interaction chromatography (HIC) ratios: This method involves the measurement of the surface hydrophobicity/hydrophilicity of NPs (Howard et al. 2008). Here, NPs are transferred into a column and washed twice. The first wash will elute the hydrophilic particles more, while the second is to elute more hydrophobic particles. Calculation of the HIC ratio is carried out from the peaks of their area under the curve (AUC) thus:

$$\text{HIC ratio} = \frac{\text{AUC elution peak}}{\text{AUC wash peak}} \times 100 \tag{4.1}$$

Here, the more hydrophilic particles show a greater HIC ratio, hence giving information on the PEG localization on the NPs' surface and the amount of PEGylation (Howard et al. 2008; Brigger et al. 2000). However, this method is inadequate for the quantitative evaluation of PEGylated NPs and not for use in particles without hydrophobic surfaces (Howard et al. 2008).

**Near-Infrared (NIR) Spectroscopy** Principal component analysis (PCA) in combination with NIR can also be used to evaluate the hydrophilic and hydrophobic balance of PEGylated nanoparticles to confirm PEGylation (Hu et al. 2006).

**Fourier Transform-Infrared Spectroscopy (FTIR), Microscopy, and $^{13}$C-NMR** The presence of the absorption bands of C–O–C and $CH_2$ in PEGylated NPs is qualitatively determined by FTIR, and it is one of the qualitative confirmations of the presence of PEG in NPs (Howard et al. 2008; Arima et al. 2008; Jie et al. 2005). NMR also reveals that some specific chemical groups confirm PEGylation. The decrease in CH and $CH_2$ peaks from a 13C-NMR spectrum has been used to verify PEGylation (Arima et al. 2008).

Microscopic techniques have been used to characterize PEGylation. Transmission electron microscopy (TEM) is applied in studying the morphological properties of nanocarriers, but it has not commonly been applied as a characterization technique for PEGylation (Howard et al. 2008). However, upon staining PEGylated micelle-like nanoparticles with phosphotungstic acid, a bright hydrophobic core surrounded by a gray hydrophilic shell was obtained (Khosravi-Darani et al. 2007). Atomic force microscopy (AFM), scanning tunneling microscopy, and scanning probe microscopy are all applied for surface imaging of NPs (Xia et al. 2012).

### 4.4.2 Quantification of PEG Surface Density

PEGylation usually affects the properties of nanocarriers, including the hydrodynamic diameter, protein-binding capability, hydrophilicity, and surface charge (Howard et al. 2008; Suk et al. 2016). Hence, the quantitative and qualitative measurement of PEGylated is essential. The methods that could be used for quantitative determination of PEGylated NPs surface include:

**TGA (Thermogravimetric Analysis)** TGA measures the content of PEG and the difference in content pre- and post-PEGylation as a result of thermal denaturation. However, it has limitations because of inorganic materials and the need for a significant amount of samples (Suk et al. 2016). The interaction between functionalized PEG and terminal groups, viz. -SH,-$NH_2$, etc. labeled with detectable dyes for absorbance readings can be employed in measuring the PEG that is unreacted in the supernatant. Hence, content PEG grafted on the surface of NPs can be indirectly calculated (Suk et al. 2016; Walkey et al. 2012; Perry et al. 2012; Valencia et al. 2011). The PEG and ligand densities, respectively, could be quantified by measuring the amount of ligand grafted to PEG on the NP surface (Suk et al. 2016; Garcia-Fuentes et al. 2004).

**NMR (Nuclear Magnetic Resonance)** NMR is utilized in the study of the structural conformations of nanoparticle components (Zabaleta et al. 2007). The new NMR pulse sequences have been used to increase diffusion-ordered NMR spectroscopy sensitivity to identify slow diffusion molecules with typical NMR probes (Howard et al. 2008). The quantity of PEG on the surface of the nanocarrier could be resolved here by comparison between the integrals of the $^1$H NMR ($\delta = 3.6$ ppm) PEG peak and the peak of an internal standard (benzylsulfonic sodium salt). The density of PEG coatings is determined with reference to the SA (surface area) and the content of surface PEG. Hence, new NMR techniques could be used for the quantification and characterization of PEGylated NPs (Howard et al. 2008). Therefore, the NMR technique could be employed in quantitatively and qualitatively assessing the PEG density on NPs surfaces (PEG peak typically observed at ~3.65 ppm) (Howard et al. 2008; Suk et al. 2016).

**HPLC Analysis and Anti-PEG Antibody Binding** The amount of PEG deposited on NPs' surface can be quantified using the HPLC method (Howard et al. 2008; Nair et al. 2006). Chromatography can be employed to quantify the PEG amount and does not require conjugation or dye modification (Suk et al. 2016). However, this method cannot assess PEG localization within the NPs, *and* they also have low sensitivity to contaminants, unstable baselines, and high sensitivity limits when the refractive index (RI) detectors are used (Suk et al. 2016; Chuang et al. 2010). Anti-

PEG antibodies, which are highly sensitive in picomolar ranges, can specifically bind PEG molecules. In some biological processes, this phenomenon can be utilized to quantify PEG density (Cheng et al. 2005; Su et al. 2010; Zillies et al. 2007). However, this binding could be influenced by steric interactions in densely packed neighboring PEG chains. Determination of the unreacted PEG molecules could be done with a refractive index (RI) detector or evaporative light scattering detector; hence, the amount of free PEG could be quantified after chromatographic separation, such as HPLC (Nair et al. 2006; Liu et al. 2004). Therefore, HPLC is unable to determine PEG localization within the NPs but is used when coupled with other techniques such as mass spectrometry, evaporative light scattering detection (ELSD), and refractive index detection (RID) due to the lack of a chromophore in PEG for UV detection (Nair et al. 2006; Gauchel et al. 2003; Auriola et al. 1993; Zhang et al. 2007). Comparison of the thickness and particle size of non-PEGylated NPs is employed during the determination of the grafting thickness (Howard et al. 2008).

**XPS (X-Ray Photoelectron Spectroscopy)** This parameter provides useful information on the elements present on the surface of PEGylated nanocarrier (from the NPs surface, ~ 8–10 nm depth with a margin of error of ~10–20%) and is ideal for quantitative determination (Howard et al. 2008; Suk et al. 2016). Here, spectra are obtained before and after PEGylation (Suk et al. 2016). This technique could validate the presence of PEG on NPs. Many researchers have utilized XPS to evaluate PEGylated nanoparticles successfully and confirmed that the values obtained corresponded very well to the amount of PEG added (Churae and Nikologorskaja 1991; Yang and Lai 2015). However, the limitations of XPS here include the difficulty in quantifying PEG when the NPs core has a similar composition of elements as the PEG within a probing depth of 1–10 nm, and there is the possibility of contaminants interfering with results (Howard et al. 2008).

**Colorimetric Assays** This method is utilized in the determination of the localization and concentration of PEG molecules (Howard et al. 2008; Hu et al. 2006). The amount of PEG measured before and after alkaline hydrolysis of NPs is compared. Here, the potassium iodide complex method is utilized, and this gives information on the positions of PEG on the surface of the nanocarrier. For particles that have a homogenous surface, the surface density PEG and distance between neighboring molecule chains could be determined with reference to the measured particle size of NPs (Howard et al. 2008).

## 4.5 Drug Delivery Applications of PEGylated Nanocarriers

PEGylation of nanocarriers is applied in various routes of drug delivery, including systemic and nonsystemic routes, passive and active targeting, and vaccine delivery, as discussed below.

### 4.5.1 Systemic Drug Delivery

The systemic drug delivery of NPs has many advantages compared to the traditional delivery system. NPs have the potential to deliver therapeutic payloads to various tissues and organs of the body, prolonging the blood circulation time and partitioning into target organs. However, as earlier stated, conventional NPs are generally cleared from the blood circulation within 10 min after systemic administration, notwithstanding the composition of the NPs (Suk et al. 2016; Moghimi et al. 2001; Yoo et al. 2010), hence the need to PEGylate. Therefore, PEG molecules are hydrophilic and could be grafted on NPs to create a hydrated environment around the NP that sterically hinders their interaction with other NPs or blood components (Klibanov et al. 1990). The PEG chains are flexible, imparting them with high conformational freedom, which opposes the penetration of foreign bodies into the corona of PEG (Vonarbourg et al. 2006). PEGylation improves the half-life of systemically administered drug-loaded NPs and minimizes the liver uptake (Suk et al. 2016; Matsumura and Maeda 1986). Following the systemic administration of PEGylated NPs, it provides enhanced permeability and retention (EPR), resulting in the prolonged circulation of PEGylated NPs, which can be beneficial in drug targeting of various tumors, for example, Human studies showed that Doxil® provided long circulation half-life of 3–4 days and passive accumulation in tumors (Dawidczyk et al. 2014; Nance et al. 2014a). Also, the systemic administration of PEGylated NPs could be used to target other diseases caused by abnormal neovascularization, such as multiple sclerosis, ocular disorders, asthma, and diabetes (Alexis et al. 2008). PEGylated NPs could be decorated with specific ligands for specific targeting to different body organs such as the brain (Suk et al. 2016; van der Meel et al. 2013; Dobrovolskaia et al. 2008).

PEGylation of NPs confers on the NPs the ability to minimize the hemotoxicity of some NPs. The systemic toxicity of NPs is not to be overlooked because NPs administered via systemic delivery route would reach most vascularized tissues in the body. Hence, NPs may directly associate with erythrocytes in the bloodstream, resulting in aggregation of erythrocytes and/or hemolysis followed by the release of hemoglobin (Suk et al. 2016; Fischer et al. 2003). Erythrocytes' disruption is often seen in cationic NPs because of their interaction with cells that are negatively charged by electrostatic interactions (Ziemba et al. 2012; Eliyahu et al. 2002). PEGylation has been shown to circumvent the hemotoxic properties of nanocarriers. PEGylation reduces aggregation and hemolysis of erythrocytes induced by polyethyleneimine (PEI-based) nanocarriers (Petersen et al. 2002b; Rukmani et al. 2009). Hemolysis reduction could be another mechanism of improving the blood circulation of NPs by PEGylation (Suk et al. 2016).

### 4.5.2 Nonsystemic Drug Delivery

PEGylation of NPs is not only advantageous in systemic drug delivery, but it could also be employed in the delivery of drugs through nonparenteral routes and controls drug release after oral administration. The nonsystemic applications of PEGylation are summarized below.

#### 4.5.2.1 Oral Administration of PEGylated Nanocarriers

Oral drug delivery remains one of the most convenient, efficacious, and safe routes of drug delivery to the gastro intestinal tract (GIT). The GIT, being complex in nature, requires specialized formulation techniques like PEGylation to target drugs to various locations of the GIT, as well as protect some drugs from denaturation by the acidic upper GIT or design the drug to release at a location of utmost absorption. PEGylated oral insulin NPs protect this drug from stomach-induced acidic degradation (Yuan et al. 2013). Nonsteroidal anti-inflammatory drugs, for example, diclofenac, may be PEGylated, preventing its release in the stomach and avoiding ulceration. Chime et al. prepared an oral PEGylated nanostructured lipid carrier and found out that PEGylation resulted in enhanced physical stability on the NLC, which resulted in significantly higher controlled drug release over time compared to non-PEGylated NLC. The NLC also showed potential for preventing gastric ulceration often seen in patients on diclofenac sodium (Chime et al. 2022). PEGylation, therefore, improves the oral bioavailability of most orally administered NPs, enhances the hydrophilicity of the SLN, and reduces mucoadhesion (Suk et al. 2009, 2016).

#### 4.5.2.2 PEGylated Nanocarriers for Improved Nasal and Pulmonary Administration

NPs are often administered through the pulmonary route for local and systemic actions in some disease conditions. The mucus layer covers the pulmonary system, which rapidly regenerates and clears via mucociliary clearance (MCC) mechanisms (Suk et al. 2016). In cystic fibrosis (CF) and chronic obstructive pulmonary disease (COPD), the airway mucus creates a complex penetrating barrier due to the viscoelasticity of the mucus barrier. However, PEG coatings on NP enhance its penetration through the mucus membrane of the airways, leading to reduced toxicity and uniform distribution of NPs payload (Lai et al. 2011).

PEGylated NPs are also an effective DDS through the mucosa of the nostrils lined by the mucus. This mucus serves as a barrier to NP and drug absorption. In patients with chronic rhinosinusitis (CRS), highly viscoelastic mucus accumulates in the sinuses, leading to immobilization of non-PEGylated NPs in CRS mucus. PEGylated NPs, at about 200 nm, could penetrate CRS mucus rapidly (Brooking et al. 2001). Some researchers found that PEGylation reduced intranasal NP uptake attributed to reduced interactions between the PEG-coated nanocarrier and the nasal-associated lymphoid tissue (NALT), some other researchers found that PEG coatings enhance nanocarrier absorption across the nasal mucosa (Suk et al. 2016; Tobio et al. 1998; Kenechukwu et al. 2018).

#### 4.5.2.3 PEGylated Intravaginal Nanocarriers

PEGylation techniques offer enhanced drug efficacy by enhancing dwelling time and increasing the drug concentration at the site of action (Dawidczyk et al. 2014;

Lee et al. 2005; Ensign et al. 2012). PEGylation could be employed to develop local delivery to the vaginal tract for controlled vaginal delivery of drugs for effective treatment of infections such as vulvovaginal candidiasis and other diseases that affect the female reproductive tract, for example, bacterial vaginosis, cervical cancer, and sexually transmitted diseases (Suk et al. 2016; Lee et al. 2005).

The primary barrier to NP-based drugs in the vaginal tract is the rapidly cleared mucus layers coating the cervicovaginal area and the viscoelasticity of the vaginal mucosa (Ensign et al. 2012; Cone 2009; Lai et al. 2009). Hence, non-PEGylated NPs are trapped in mucus by adhesive interactions (Lai et al. 2007). Due to their mucoadhesiveness, PEGylated NPs interpenetrate the vaginal mucosa (Suk et al. 2016; Huang et al. 2000; Gu et al. 2008; Sahlin and Peppas 1997). When NPs are coated with PEG of low MW (2 or 5 kDa) and with a particle size of 200 nm, PEG rapidly diffuse through the cervical mucous, but when coated with PEG of about 10 kDa, PEG becomes trapped in the mucosa (Suk et al. 2016). Also, the dense coating of NP with low MW PEG leads to the shielding of the core of the NP, preventing its interactions with mucin and mucus (Wang et al. 2008).

#### 4.5.2.4 Brain Delivery of PEGylated Nanocarriers

PEGylated NPs could be used to increase drug-brain delivery. The extracellular matrix (ECM) in the brain presents an extra barrier that prevents drugs entrapped in NPs from reaching target cells (Thorne and Nicholson 2006). However, the density of PEGylated NPs is critical in delivering drugs to the brain. Researchers observed that exceptionally densely coated PEGylated NPs of about 114 nm could rapidly diffuse more across the brain tissues of rats (Nance et al. 2012; Mastorakos et al. 2015). In contrast, non-PEGylated NPs of any size were adhesively immobilized in the various types of brain tissues. Densely PEG-coated NPs also would enhance drug delivery and efficacy in malignant glioma due to increased nanoparticle spread and distribution (Nance et al. 2014b). PEG coatings improve the stability of the NPs in cerebrospinal fluid and reduce cell toxicity without significantly compromising cellular uptake (Suk et al. 2016).

#### 4.5.2.5 Ocular Delivery of PEGylated Nanocarriers

A good number of barriers prevent the eyes from effectively utilizing various drug delivery systems/dosage forms designed to treat different ocular diseases that may predispose one to blindness. The traditional dosage forms in eye drops undergo poor absorption and rapid clearance; hence, less than 5% of the applied drops reach the intraocular tissues (Jarvinen et al. 1995). Therefore, other more efficient delivery systems, such as NPs and PEGylated NPs, have been explored to enhance drug residence time (DRT) and absorption via the ocular route. PEG enhances mucoadhesion via hydrogen bonding to the mucus and interpenetration (Suk et al. 2016; Wang et al. 2008). Hence, PEGylated NPs could serve as a mucoadhesive to promote DRT in the mucosal area of the eye (Suk et al. 2016; Giannavola et al. 2003). PEGylated NPs are primarily stable with enhanced mobility. Hence, PEGylation increases the mobility of NPs in the vitreous, while non-PEGylated NPs form large, immobile aggregates, whereas most are in the vitreous gel (Martens et al. 2013).

#### 4.5.2.6 Vaccines Entrapped PEGylated Nanocarriers

Vaccine-entrapped PEGylated nanocarriers increase the circulation time and, hence, enhance the properties of the entrapped vaccines than the non-PEGylated nanocarriers. PEGylation is often employed in reducing reactions between nanocarriers and proteins, cells, etc. During circulation to reduce clearance, (Suk et al. 2016) PEGylation will also minimize desirable cell uptake, for example, NP-based vaccine delivery to antigen-presenting cells (APCs). Coating NPs with PEG protects vectors from existing neutralizing antibodies against the virus (Weaver and Barry 2008). PEGylation of NPs was found to increase the relative ratio of dendritic cells to B cells. Also, there was an increase in the amount of NPs internalized by the dendritic cell, suggesting a relative increase in cellular uptake in vivo (Zhan et al. 2012). Vaccine-entrapped PEGylated liposome, PEGylated lipoplexes, PEG-polyplexes, and PEGylated nanoparticles could be formulated and administered via nonsystemic routes, including dermal route of administration for improved delivery of vaccines (van den Berg et al. 2010). Also, the intravaginal delivery of vaccines has been recently explored (Xie et al. 2014).

### 4.5.3 Roles of PEGylation in Passive and Active Targeting of Drugs

PEGylated NPs cause a significant improvement in the penetration of drugs into various tumors and cancer cells. PEGylation enhances drug solubility and serum stability of drugs; hence, they have been utilized for drug-targeting cancer cells, especially (Deepa and Hitesh 2020). Targeting of cancer cells via PEGylatyed nanocarriers can be active or passive targeting (Sutradhar and Amin 2014).

**Active Targeting** In active targeting, PEGylated NPs containing therapeutic agent(s) may be designed to interact directly with the cancer cells. Active targeting is initiated by molecular recognition, which involves surface modification of the PEGylated NPs through antibody–antigen recognition or ligand–receptor interaction (Cho et al. 2008; Yezhelyev et al. 2006). Overexpressed receptors on cancerous cell surfaces abound and are different from normal cells at the molecular level. When complementary ligands attach to PEGylated NPs surfaces, it allows them to specifically target cancerous cells. Once the ligand PEG NPs bind to these receptors, they undergo phagocytosis, or receptor-mediated endocytosis by cells, resulting in cell internalization entrapped drug in the NPs (Sutradhar and Amin 2014). Some of these receptors and their targeting ligands used for cancerous cells specific targeting include ligand–receptor interaction, viz. Folate Receptor, Transferrin Receptor, Asialoglycoprotein (ASGPRs) receptors and Luteinizing Hormone-Releasing Hormone Receptor (LHRHR). Active targeting can also occur by antibody-mediated targeting. Immunoglobulin (Ig) antibodies have gained popularity due to their high specificity and unique in vivo properties (Sutradhar and Amin 2014). These classes of Ig could be used as ligands to target different cancer cells, viz. IgG, IgA, IgM, IgD, and IgE (Sutradhar and Amin 2014), for example,

*Trastuzumab* (*Herceptin*, Genentech, Inc., South San Francisco, California) is a highly purified recombinant DNA-derived humanized monoclonal *immunoglobulin* G1 used for the design of NPs for (MDA-MB-468 BT-474, NCI-H520, PC9, SK-BR-3) breast and (PC3) prostate cancers (Sutradhar and Amin 2014). Angiogenesis is one of the hallmarks of cancer, as it allows the tumor to gain oxygen in large amounts and allows nutrients to thrive (Sudarshan et al. 2005). PEGylated NPs encapsulating anticancer drugs can be conjugated to antibodies that bind to receptors overexpressed in angiogenic endothelial cells, thus majoring in both antiangiogenic and cytotoxic effects to improve therapeutic efficiency. The angiogenic targets of antibody-functionalized NPs for cancer include the vascular endothelial growth factor (VEGF) and its receptors and matrix metalloproteinases (Ferrara 2005).

**Passive Targeting** Nanoparticles and indeed PEGylated nanocarriers can also target cancer through passive targeting. The size of the pores in leaky endothelial cancerous cells ranges from 100 to 780 nm, depending on the type of cancer. Hence, PEGylated nanocarriers below that size can easily penetrate these pores (Shubik 1982; Baban and Seymour 1998). PEGylated NPs can be targeted to specific areas of capillary endothelium to increase drug concentration within a particular organ, perforating the tumor cells by convection or passive diffusion (Sutradhar and Amin 2014).

## 4.6 Factors That Affect the Circulation Time of PEGylated Nanocarriers

Several factors affect the circulation time of PEGylated nanocarriers. These factors include the molecular weight of PEG used during the formulation, surface density of PEG, physicochemical parameters of NPs, and other physicochemical properties that influence the systemic circulation of nanocarriers (Suk et al. 2016; Alexis et al. 2008; Vonarbourg et al. 2006; Gref et al. 1995; Mozar and Chowdhury 2018).

### 4.6.1 Physicochemical Properties of Nanoparticles

The physicochemical parameters of NPs, viz. size and surface charge are important parameters (Albanese et al. 2012). They influence the interactions between the biological environment and the NPs, affecting internalization, distribution, and clearance. The binding of NPs to drug components like the opsonins and particle aggregation in circulation is influenced by the hydrophobicity of NPs (Mozar and Chowdhury 2018; Karra and Benita 2012; Brigger et al. 2002). In aqueous systems, uncharged and hydrophobic NPs are easily aggregated via hydrophobic and Van der Waals forces, while highly charged hydrophilic nanocarriers maintain their colloidal stability due to repulsive forces in them (Tenzer et al. 2013; Mozar and

Chowdhury 2018). Therefore, PEG coating of NPs is highly desirable to prevent opsonization by serum factors (Bhadra et al. 2002). PEG coating on nanoparticles neutralizes the surface charges, yielding "stealth" properties, minimizing RES opsonization, and improving blood retention time. Hydrophilicity and steric hindrance offered by surface-anchored PEG prevent nonspecific interactions between NPs and proteins; hence, reducing phagocytosis mediated by opsonin, causing prolonged residence time of NPs in circulation (Mozar and Chowdhury 2018; Pai et al. 2006; Harrington et al. 2000). Drug encapsulation and loading into NPs may be affected by factors such as matrix composition and the physicochemical properties of drugs (Mozar and Chowdhury 2018; Govender et al. 1999, 2000).

The physicochemical properties of the NPs core influence the adsorption of proteins and their circulation time. Smaller size NPs interact less with the end groups within cell surfaces of PEG chains. In contrast, NPs of larger sizes at a given PEG surface density may attach themselves to MPS cells more firmly than smaller NPs due to weak, multivalent interactions between the PEG chains and/or terminal groups with cell surfaces (Unsworth et al. 2008). Hence, smaller NPs show more resistance against macrophage uptake than larger NPs and have increased surface curvature, which may need larger PEG density to shield them effectively (Suk et al. 2016; Choi et al. 2011; Perrault et al. 2009; Fang et al. 2006).

### 4.6.2 PEG Molecular Weight (MW)

PEG MW is crucial for effective shielding of the surface of NPs in order to prevent interactions with MPS cells and serum proteins (Suk et al. 2016). Protein adsorption on PEGylated NPs is affected by the MW. Hence, the association between the ligand and PEGylated NPs is affected by PEG surface coating density on NPs. PEG with MW of 10 kDa is more effective in reducing protein absorption than PEG with MW of 2 kDa and 5 kDa, respectively. Therefore, high MW PEGs prevent phagocytic uptake protein adsorption and increase circulation time (Suk et al. 2016; Fang et al. 2006). Generally, the MW of grafted PEG chains is proportional to the polymer chain length. Increasing the MW of PEG from 2 kDa to 20 kDa prevents the adsorption and aggregation of NP to blood components, thereby increasing its circulation time (Suk et al. 2016; Mozar and Chowdhury 2018). PEG with MW of 2 kDa or higher could shield particle surfaces, circumventing recognition by the MPS and protein adsorption (Mozar and Chowdhury 2018; Pai et al. 2006).

### 4.6.3 Surface Density, PEG Content and Conformation

The surface density of the grafted PEG layer is a crucial parameter that influences the PEGylated NPs' ability to resist protein adsorption and clearance after systemic administration (Suk et al. 2016; Vonarbourg et al. 2006). PEGylated liposomes prepared with 10 mol % of PEGylated lipids prevented liposome aggregation in whole blood, while formulations with 5 and 3 mol % of PEG agglomerated over time

(Braeckmans et al. 2010). Also, increasing the content of PEG on nanocarriers may not ensure more excellent surface coverage of PEG. There is a threshold for maximum achievable PEG surface density, which depends on the type of NPs and formulation methods. Hence, an accurate determination of the surface PEG density during the interpretation of the effect of PEG density on NP circulation is essential (Suk et al. 2016).

The structural conformation of PEG molecules on the surface and the effectiveness of the PEG in shielding the NP surface is calculated by the average distance between neighboring PEG chains on an NP surface ($D$). If $D$ is greater than the Flory radius of the PEG chain,

$$\text{Flory radius, RF} = aN^{3/5} \tag{4.2}$$

In the above equation, $N$ = the degree of polymerization, which increases with an increase in PEG MW, and $a$ = the effective monomer length = 0.35 nm; then, the neighboring PEG chains are said to be in a "mushroom" conformational regime and will not overlap ($RF/D \leq 1$) (Suk et al. 2016). Increasing the surface PEG density such that adjacent PEG chains overlap ($RF/D > 1$) will cause the PEG chain to stretch away from the nanocarrier surface, resulting in a "brush" layer. When the surfaces of PEG densities change from mushroom-to-brush transitions, it is beneficial to circumvent the adsorption of serum proteins, avoiding NPs uptake by MPS cells (Owens and Peppas 2006; Mozar and Chowdhury 2018; Jokerst et al. 2011). Also, higher RF/D values could result in longer PEG chains ($\geq$ 10 kDa) because of entanglement in the neighboring chains (Suk et al. 2016; Yang et al. 2014a).

Mushroom state conformation results from low-density PEGylation, whereas brush state configuration results from high-density PEGylation and is more effective in prolonging the circulation time of PEGylated NPs by higher anti-opsonization effect (Mozar and Chowdhury 2018). However, having an opsonization effect that is too strong could prevent the interactions between NPs and cells, reducing some properties like their tumor uptake, an effect called "PEG Dilemma." (Mozar and Chowdhury 2018)

## 4.7 PEGylated Nanocarriers Products

There are currently so many PEGylated nanocarriers in the market, some of which are shown in Table 4.1 (Mozar and Chowdhury 2018; Abdellatif and Alsowinea 2021; Patra et al. 2018). It is also worth noting that an uncountable number of different nanocarriers are undergoing various stages of clinical trials (Mozar and Chowdhury 2018).

Adagen® (Pegademase bovine) is a PEGylated polymeric nanocarrier approved in 1990 for the treatment of severe combined immunodeficiency disease (SCID) (Abdellatif and Alsowinea 2021; Patra et al. 2018). Doxil® is a PEGylated liposomal doxorubicin approved by FDA and European Medicines Evaluation Agency (EMA) in 1999 and 2000, respectively, (Mozar and Chowdhury 2018; Abdellatif and Alsowinea 2021) for the treatment of advanced ovarian cancer patients with platinum-based therapy resistance. Doxil® has reduced toxicity with five to tenfold higher tumor levels and improved circulation time, especially cardiotoxicity. Also, various marketed PEGylated protein products are available, viz. PEGylated

**Table 4.1** PEGylated nanocarriers products

| Loaded drug | Nanocarrier | Brand (Company) | Use | Date of approval |
|---|---|---|---|---|
| Pegademase bovine (Abdellatif and Alsowinea 2021; Patra et al. 2018) | Polymeric nanoparticles | Adagen® (Sigma-Tau Pharmaceuticals) | Severe combined immunodeficiency disease (SCID) | 1990 |
| Glatopa | Polymeric nanoparticles | Copaxone® (Teva) | Multiple sclerosis (MS) | 1996 |
| Pegfilgrastim | PEGylated GCSF protein | Neulasta (Amgen) | Leukopenia by chemotherapy | 2002 |
| Peginterferon alfa-2A | PEGylated IFN alpha-2a protein | Pegasys (Hoffman-La Roche) | Hepatitis B and C | 2002 |
| Peginterferon alfa-2B | PEGylated IFN alpha-2b protein | PegIntron (Schering) | Hepatitis C | 2001 |
| Pegvisomant | PEGylated HGH receptor antagonist | Somavert (Pharmacia) | Acromegaly | 2003 |
| Pegaspargase | Polymer–protein conjugate PEGylated L-asparaginase | Oncaspar (Sigma Tau) | Acute lymphocytic blood clot | 1994 |
| Pegloticase | Polymer–protein conjugate (PEGylated porcine-likeuricase) | Krystexxa (Horizon) | Chronic gout | 2010 |
| Peginterferon beta-1A | Polymer–protein conjugate (PEGylated IFNbeta-1a) | Plegridy (Biogen) | Multiple sclerosis | 2014 |
| PEGylated factor VIII | Polymer–protein conjugate (PEGylated factor VIII) | Adynovate (Baxalta) | Hemophilia | 2015 |
| Paclitaxel | Polymeric micelle | Genexol-PM (Samyang) | Metastatic breast and lung cancer | 2007 |
| Doxorubicin | PEGylated liposomal | DOXIL (Ben Venue Laboratories) | Ovarian, Kaposi's sarcoma, and Multiple myeloma | 1999 and 2000 |

interferon, PEGylated arginine deaminase, and PEGylated L-asparaginase (Mozar and Chowdhury 2018; Abdellatif and Alsowinea 2021; Patra et al. 2018).

## 4.8 Limitations of Nanocarriers PEGylation

Despite an array of the advantages of PEGylation of NPs in drug delivery, it suffers from some major limitations such as particle size enlargement, batch-to-batch variability, high variability, and diversity of results obtained with PEGylated NPs and nonbiodegradability of PEG polymers among others (Howard et al. 2008; Suk et al. 2016).

### 4.8.1 High Variability and Diversity of Results Obtained with PEGylated NPs

PEGylated nanocarriers may show significant variability in results outcomes depending on the type of NPs PEGylated due to the individual properties of the excipient or composition of the NPs (Sebak 2018). Variability in outcomes of PEGylation may also arise due to the desorption of PEG, nonhomogeneity of PEG on particle surfaces, and the attraction of a variety of plasma proteins (Suk et al. 2016).

### 4.8.2 Nonbiodegradability of PEG Polymers

PEG polymers are nonbiodegradable despite their advantages and their biocompatibility. Hence, there could be some potential side effects occurring from their nonbiodegradability (Sebak 2018). There may be situations that can lead to lysosome bioaccumulation in healthy tissue (Verhoef and Anchordoquy 2013; Lowe et al. 2015).

### 4.8.3 Disadvantageous Physicochemical Properties

The PEGylation of NPs affects their physicochemical properties and could lead to particle size enlargement, predisposing NPs to enhance RES uptake. This could be worsened by the use of considerable molecular weight PEG, resulting in the sudden release of loaded drugs from NPs. The loaded drugs could be released from PEGylated NPs before degradation of the polymer (Sebak 2018). Hence, particles could be circulating without the loaded drugs before their degradation. There have been several reports of rapid release of hydrophilic drugs from PEGylated NPs, which could be due to the hydrophilic nature of PEG chains, which attract water molecules, causing wetting of PEGylated NPs and faster drug release (Sebak 2018; Chen et al. 2013; Yang et al. 2014b).

### 4.8.4 Unfavorable Cellular and Subcellular Fate

PEGylated NPs sometimes exhibit limited intracellular uptake significantly lower than non-PEGylated NPs, resulting from steric hindrance from polymer chains of PEG and the hydrophilic nature of the surfaces of NPs. Hence, hydrophilic and neutral NPs form of PEGylated NPs exhibit slower internalization in target cells than Nps with charges (Oh and Park 2014; Mao et al. 2013). Therefore, despite being able to prevent the isolation of drug-loaded NP by RES cells and increasing extravasation of NPs through the unstable tumor vasculature, PEGylation reduces target tissue interaction (Verhoef and Anchordoquy 2013)· and inhibits endosomal escape, leading to the loss of activity of some delivery systems like nucleic acid and gene delivery systems in the treatment of cancer where the NPs degrade in the acidic digestive medium of the endolysosomal system (Sebak 2018; Hatakeyama et al. 2013).

### 4.8.5 Limited RES Evasion Capacity

Improper selection and optimization of PEG molecular weight and density, coatings of PEG, and inadequate nanocarrier core properties could affect RES evasion capacity. Hence, despite the ability of PEG to provide steric hindrance and mask the surface charge of NPs, preventing opsonization, their ability to avoid this opsonization has been shown to be limited, leading to several reports of opsonization and RES clearance even after PEGylation (Sebak 2018; Verhoef and Anchordoquy 2013; Wan et al. 2017; Zhang et al. 2016).

Reduced RES evasion capacity by PEGylated NPs could be caused majorly by the production of anti-PEG antibodies, size enlargement of nanocarriers, and development of an immune response that leads to faster clearance of PEGylated NPs from the bloodstream and removal of PEG from surfaces of NPs (Sebak 2018).

**Particle Size Enlargement** PEGylation of NPs leads to the particle size increase, which could affect some pharmacokinetics of NPs and may also lead to an abnormal reduction in the MRT of nanocarriers in blood, affecting RES avoidance by PEG (Rafiei and Haddadi 2017; Xu et al. 2013).

**Insufficient PEG Coating or Desorption of PEG Layer** Improper surface coating of PEG or polymer desorption from the surface of NPs could present surfaces or holes where opsonins can bind (Sebak 2018; Kettiger et al. 2013).

**Immunogenicity of PEG** PEG polymers were once known to be nonantigenic. However, recent reports showed that they could develop an immune response, causing an effect known as the ABC phenomenon, because of the generation of anti-PEG IgM antibodies produced from the spleen in response to a previous dose

(Kettiger et al. 2013; Chen et al. 2016). These antibodies form a complex with other serum components, resulting in more sequestration by RES (Verhoef and Anchordoquy 2013; Zhang et al. 2014). It has been discovered that a second dose of PEGylated injected liposomes exhibited poor circulation time and high accumulation in the spleen and liver, more than the first dose (Zhang et al. 2014). Also, in 250 healthy blood donors screened, isotypes of anti-PEG antibodies of both IgG and IgM were discovered in 25%. This could be due to increased exposure of the masses to PEG-containing pharmaceutical and cosmetics products for a prolonged period of time (Sebak 2018; Hatakeyama et al. 2013). This explains the loss of activity that could occur with repeated dosing of PEGylated products.

## 4.9 Conclusion

Nanocarriers consist of polymeric and lipid-based NPs arrays with different constituents, formulation methods, and functionalities. Adequate selection of the particular type of NP to coat with PEG for effective and enhanced NP properties is essential to achieve better results with PEGylation. Also, proper preformulation studies should be carried out to choose an optimized coating composition in terms of PEG MW, density, and layer of coating to avoid desorption that could lead to the failure of PEGylated NPs and abnormal functioning in the body. PEGylation of NPs remains an area for scientists to explore further so that all the advantages embodied in these nanocarriers and PEG polymers can be utilized for better treatment of diseases of man and better treatment outcomes.

PEGylation has become an essential aspect of drug delivery utilized effectively in the formulation of challenging delivery systems used to treat many chronic and life-threatening illnesses like different forms of cancers. Its use in the formulation of herbal formulations and nutraceuticals is yet to be fully explored; hence, it should be an area for future development. Also, methods and processes that will prevent the problems encountered in the coating of NPs to avoid RES reversal and developing antiPEG antibodies are all areas under study.

## References

Abdellatif AAH, Alsowinea AF (2021) Approved and marketed nanoparticles for disease targeting and applications in COVID-19. Nanotechnol Rev 10:1941–1977

Abuchowski A, Van Es T, Palczuk NC, Davis FF (1977) Alteration of immunological properties of bovine serum albumin by covalent attachment of polyethylene glycol. J Biol Chem 252:3578–3581

Albanese A, Tang PS, Chan WC (2012) The effect of nanoparticle size, shape, and surface chemistry on biological systems. Annu Rev Biomed Eng 14:1–16

Alexis F, Pridgen E, Molnar LK, Farokhzad OC (2008) Factors affecting the clearance and biodistribution of polymeric nanoparticles. Mol Pharm 5:505–515

Andriyanov AV et al (2017) Therapeutic efficacy of combined PEGylated liposomal doxorubicin and radiofrequency ablation: comparing single and combined therapy in young and old mice. J Control Release 257:2–9

Arima Y, Toda M, Iwata H (2008) Complement activation on surfaces modified with ethylene glycol units. Biomaterials 29:551
Auriola SO, Ronkko KM, Urtti A (1993) Determination of polyethylene glycols by high-performance liquid chromatographythermospray mass spectrometry. J Pharm Biomed Anal 11:1027
Baban DF, Seymour LW (1998) Control of tumour vascular permeability. Adv Drug Deliv Rev 34:109–119
Bhadra D, Bhadra S, Jain P, Jain NK (2002) Pegnology: a review of PEGylated systems. Pharmazie 57:5–29
Braeckmans K, Buyens K, Bouquet W, Vervaet C, Joye P, De Vos F, Plawinski L, Doeuvre L, Angles-Cano E, Sanders NN, Demeester J, De Smedt SC (2010) Sizing nanomatter in biological fluids by fluorescence single particle tracking. Nano Lett 10:4435–4442
Braet F, Wisse E, Bomans P, Frederik P, Geerts W, Koster A, Soon L, Ringer S (2007) Contribution of high-resolution correlative imaging techniques in the study of the liver sieve in three-dimensions. Microsc Res Tech 70:230–242
Brigger I, Chaminade P, Desmaële D, Peracchia MT, d'Angelo J, Gurny R, Renoir M, Couvreur P (2000) Near infrared with principal component analysis as a novel analytical approach for nanoparticle technology. Pharm Res 17:1124
Brigger I, Dubernet C, Couvreur P (2002) Nanoparticles in cancer therapy and diagnosis. Adv Drug Deliv Rev 54(5):631–651
Brooking J, Davis SS, Illum L (2001) Transport of nanoparticles across the rat nasal mucosa. J Drug Target 9:267–279
Buyens K et al (2012) Liposome based systems for systemic siRNA delivery: stability in blood sets the requirements for optimal carrier design. J Control Release 158(3):362–370
Calv OP, Gouritin B, Chacun H, Desmaële D, D'Angelo J, Noel JP, Georgin D, Fattal E, Andreux JP, Couvreur P (2001) Long circulating PEGylated polycyanoacrylate nanoparticles as new drug carrier for brain delivery. Pharm Res 8(8):1157–1166
Calvo P, Gouritin B, Chacun H, Desmaele D, D'Angelo J, Noel JP, Georgin D, Fattal E, Andreux JP, Couvreur P (2001) Long circulating PEGylated polycyanoacrylate nanoparticles as new drug carrier for brain delivery. Pharm Res 18:1157–1166
Chen F, Yin G, Liao X, Yang Y, Huang Z, Gu J (2013) Preparation, characterization and *in vitro* release properties of morphine-loaded PLLA-PEG-PLLA microparticles via solution enhanced dispersion by supercritical fluids. J Mater Sci Mater Med 24:1693–1705
Chen BM, Su YC, Chang CJ, Burnouf PA, Chuang KH, Chen CH et al (2016) Measurement of pre-existing IgG and IgM antibodies against polyethylene glycol in healthy individuals. Anal Chem 88:10661–10666
Cheng T-L, Cheng C-M, Chen B-M, Tsao D-A, Chuang K-H, Hsiao S-W, Lin Y-H, Roffler SR (2005) Monoclonal antibody-based quantitation of poly(ethylene glycol)-derivatized proteins, liposomes, and nanoparticles. Bioconjug Chem 16:1225–1231
Chime SA, Ugwu CE, Ejiofor EU, Onunkwo GC (2022) Application of *Cyperus esculentus* oil in the development of sustained release diclofenac sodium-loaded nanostructured lipid carrier. J Curr Biomed Res 2(3):145–159
Chime SA, Kenechukwu FC, Attama AA (2014) Nanoemulsions—advances in formulation, characterization and applications in drug delivery. In: Application of nanotechnology in drug delivery. InTech, London, pp 77–126. https://doi.org/10.5772/15371
Cho K, Wang X, Nie S, Chen Z, Shin DM (2008) Therapeutic nanoparticles for drug delivery in cancer. Clin Cancer Res 14(5):1310–1316
Choi CH, Zuckerman JE, Webster P, Davis ME (2011) Targeting kidney mesangium by nanoparticles of defined size. Proc Natl Acad Sci U S A 108:6656–6661
Chuang KH, Tzou SC, Cheng TC, Kao CH, Tseng WL, Shiea J, Liao KW, Wang YM, Chang YC, Huango BJ, Wu CJ, Chu PY, Roffler SR, Cheng TL (2010) Measurement of poly(ethylene glycol) by cell-based anti-poly(ethylene glycol) ELISA. Anal Chem 82:2355–2362
Churae NV, Nikologorskaja EA (1991) Brownian and electrophoretical mobility of silica particles coated with absorbed layers of polyethylenoxides. Colloids Surf 59:71

Cone RA (2009) Barrier properties of mucus. Adv Drug Deliv Rev 61:75–85
Craparo EF, Cavallaro G, Bondi ML, Mandracchia D, Giammona G (2006) PEGylated nanoparticles based on a polyaspartamide. Preparation, physico-chemical characterization, and intracellular uptake. Biomacromolecules 7:3083
Cu Y, Saltzman WM (2009) Controlled surface modification with poly(ethylene)glycol enhances diffusion of PLGA nanoparticles in human cervical mucus. Mol Pharm 6:173–181
Dawidczyk CM, Kim C, Park JH, Russell LM, Lee KH, Pomper MG, Searson PC (2014) State-of-the-art in design rules for drug delivery platforms: lessons learned from FDA-approved nanomedicines. J Control Release 187:133–144
Deepa Y, Hitesh KD (2020) PEGylation: an important approach for novel drug delivery system. J Biomater Sci Polym Ed 32:266. https://doi.org/10.1080/09205063.2020.1825304
Dobrovolskaia MA, Aggarwal P, Hall JB, McNeil SE (2008) Preclinical studies to understand nanoparticle interaction with the immune system and its potential effects on nanoparticle biodistribution. Mol Pharm 5:487–495
Dunn SE, Brindle A, Davis SS, Davies MC, Illum L (1994) Polystyrene-poly(ethylene glycol) (PS-PEG2000) particles as model systems for site specific drug delivery. 2. The effect of PEG surface density on the in vitro cell interaction and in vivo biodistribution. Pharm Res 11:1016
Eliyahu H, Servel N, Domb AJ, Barenholz Y (2002) Lipoplex-induced hemagglutination: potential involvement in intravenous gene delivery. Gene Ther 9:850–858
Ensign LM, Schneider C, Suk JS, Cone R, Hanes J (2012) Mucus penetrating nanoparticles: biophysical tool and method of drug and gene delivery. Adv Mater 24:3887–3894
Ensign LM, Henning A, Schneider CS et al (2013) Ex vivo characterization of particle transport in mucus secretions coating freshly excised mucosal tissues. Mol Pharm 10:2176–2182
Fang C, Shi B, Pei YY, Hong MH, Wu J, Chen HZ (2006) In vivo tumor targeting of tumor necrosis factor-alpha-loaded stealth nanoparticles: effect of MePEG molecular weight and particle size. Eur J Pharm Sci 27:27–36
Ferrara N (2005) VEGF as a therapeutic target in cancer. Oncology 69(3):11–16
Fischer D, Li Y, Ahlemeyer B, Krieglstein J, Kissel T (2003) In vitro cytotoxicity testing of polycations: influence of polymer structure on cell viability and hemolysis. Biomaterials 24:1121–1131
Francis GE, Delgado C, Fisher D, Malik F, Agrawal AK (1996) Polyethylene glycol modification: relevance of improved methodology to tumour targeting. J Drug Target 3(5):321–340
Gajbhiye V, Vijayarajkumar P, Rakesh T, Jain NK (2007) Pharmaceutical and biomedical potential of PEGylated dendrimers. Curr Pharm Design 13:415–429
Gajbhiye KR, Pawar A, Mahadik KR, Gajbhiye V (2020 Mar) PEGylated nanocarriers: a promising tool for targeted delivery to the brain. Colloids Surf B Biointerfaces 187:110770. https://doi.org/10.1016/j.colsurfb.2019.110770
Garcia-Fuentes M, Torres D, Martín-Pastor M, Alonso MJ (2004) Application of nmr spectroscopy to the characterization of pegstabilized lipid nanoparticles. Langmuir 20:8839
Gauchel G, Huppertz B, Feiertag H, Keller R (2003) Clinical use of polyethylene glycols as marker substances and determination in urine by liquid chromatography. J Chromatogr B Analyt Technol Biomed Life Sci 787:271
Gessner A, Waicz R, Lieske A, Paulke B, Mader K, Muller RH (2000) Nanoparticles with decreasing surface hydrophobicities: influence on plasma protein adsorption. Int J Pharm 196:245–249
Giannavola C, Bucolo C, Maltese A, Paolino D, Vandelli MA, Puglisi G, Lee VH, Fresta M (2003) Influence of preparation conditions on acyclovir-loaded poly-d,l-lactic acid nanospheres and effect of PEG coating on ocular drug bioavailability. Pharm Res 20:584–590
Govender T, Stolnik S, Garnett MC, Illum L, Davis SS (1999) PLGA nanoparticles prepared by nanoprecipitation: drug loading and release studies of a water soluble drug. J Control Release 57:171–185
Govender T, Riley T, Ehtezazi T et al (2000) Defining the drug incorporation properties of PLA-PEG nanoparticles. Int J Pharm 199:95–110

Gref R, Domp A, Quellec P, Blunk T, Muller RH, Verbavatz JM, Langer R (1995) The controlled intravenous delivery of drugs using PEG coated sterically stabilized nanospheres. Adv Drug Deliv Rev 16:215–233

Gref R, Luck M, Quellec P, Marchand M, Dellacherie E, Harnishch S, Blunk T, Muller RH (2000) 'Stealth' corona-core nanoparticles surface modified by polyethylene glycol (PEG): influences of the corona (PEG chain length and surface density) and of the core composition on phagocytic uptake and plasma protein adsorption. Colloids Surf B Biointerfaces 18(3–4):301–313

Gu F, Zhang L, Teply BA, Mann N, Wang A, Radovic-Moreno AF, Langer R, Farokhzad OC (2008) Precise engineering of targeted nanoparticles by using self-assembled biointegrated block copolymers. Proc Natl Acad Sci U S A 105:2586–2591

Haris JM (1992) Poly (ethylene glycol) chemistry: biotechnical and biomedical applications. Plenum Press, New York, p 228

Harrington KJ, Rowlinson-Busza G, Uster PS, Stewart JS (2000) Pegylated liposome-encapsulated doxorubicin and cisplatin in the treatment of head and neck xenograft tumours. Cancer Chemother Pharmacol 46(1):10–18

Hatakeyama H, Akita H, Harashima H (2013) The polyethyleneglycol dilemma: advantage and disadvantage of PEGylation of liposomes for systemic genes and nucleic acids delivery to tumors. Biol Pharm Bull 36:892–899

Howard MD, Jay M, Dziubla TD, Lu X (2008) PEGylation of nanocarrier drug delivery systems: state of the art. J Biomed Nanotechnol 4(2):133–148

Hrkach J, Von Hoff D, Ali MM, Andrianova E, Auer J, Campbell T, De Witt D, Figa M, Figueiredo M, Horhota A, Low S, McDonnell K, Peeke E, Retnarajan B, Sabnis A, Schnipper E, Song JJ, Song YH, Summa J, Tompsett D, Troiano G, Hoven TVG, Wright J, LoRusso P, Kantoff PW, Bander NH, Sweeney C, Farokhzad OC, Langer R, Zale S (2012) Preclinical development and clinical translation of a PSMA-targeted docetaxel nanoparticle with a differentiated pharmacological profile. Sci Transl Med 4:4

Hu FX, Neoh KG, Cen L, Kang ET (2006) Cellular response to magnetic nanoparticles "PEGylated" via surface-initiated atom transfer radical polymerization. Biomacromolecules 7:809

Huang YB, Leobandung W, Foss A, Peppas NA (2000) Molecular aspects of muco- and bioadhesion: tethered structures and site-specific surfaces. J Control Release 65:63–71

Illum L (2000) Transport of drugs from the nasal cavity to the central nervous system. Eur J Pharm Sci 11:1–18

Jain NK, Manoj N (2010) PEGylated nanocarriers for systemic delivery. In: Grobmyer SR, Moudgil BM (eds) Cancer nanotechnology. Methods in molecular biology, vol 624. Springer Science+Business Media, LLC, Berlin, pp 221–234

Jarvinen K, Jarvinen T, Urtti A (1995) Ocular absorption following topical delivery. Adv Drug Deliv Rev 16:3–19

Jie P, Venkatraman SS, Min F, Freddy BY, Huat GL (2005) Micelle-like nanoparticles of star-branched PEO-PLA copolymers as chemotherapeutic carrier. J Control Release 110:20

Jokerst JV, Lobovkina T, Zare RN, Gambhir SS (2011) Nanoparticle PEGylation for imaging and therapy. Nanomedicine (Lond) 6:715–728

Karnik R, Gu F, Basto P, Cannizzaro C, Dean L, Kyei-Manu W, Langer R, Farokhzad OC (2008) Microfluidic platform for controlled synthesis of polymeric nanoparticles. Nano Lett 8:2906–2912

Karra N, Benita S (2012) The ligand nanoparticle conjugation approach for targeted cancer therapy. Curr Drug Metab 13(1):22–41

Kaur H, Das T, Kumar R, Ajore R, Bharadwaj LM (2008) Biosystems 92:69–75

Kenechukwu FC, Attama AA, Ibezim EC, Nnamani PO et al (2018) Novel intravaginal drug delivery system based on molecularly pegylated lipid matrices for improved antifungal activity of miconazole nitrate. Biomed Res Int 2018:1–18. https://doi.org/10.1155/2018/3714329

Kettiger H, Schipanski A, Wick P, Huwyler J (2013) Engineered nanomaterial uptake and tissue distribution: from cell to organism. Int J Nanomedicine 8:3255–3269

Khosravi-Darani K, Pardakhty A, Honarpisheh H, Rao VS, Mozafari MR (2007) The role of high-resolution imaging in the evaluation of nanosystems for bioactive encapsulation and targeted nanotherapy. Micron 38:804

Klibanov AL, Maruyama K, Torchilin VP, Huang L (1990) Amphipathic polyethyleneglycols effectively prolong the circulation time of liposomes. FEBS Lett 268:235–237

Lai SK, O'Hanlon DE, Harrold S, Man ST, Wang YY, Cone R, Hanes J (2007) Rapid transport of large polymeric nanoparticles in fresh undiluted human mucus. Proc Natl Acad Sci U S A 104:1482–1487

Lai SK, Wang YY, Hanes J (2009) Mucus-penetrating nanoparticles for drug and gene delivery to mucosal tissues. Adv Drug Deliv Rev 61:158–171

Lai SK, Suk JS, Pace A, Wang YY, Yang M, Mert O, Chen J, Kim J, Hanes J (2011) Drug carrier nanoparticles that penetrate human chronic rhinosinusitis mucus. Biomaterials 32:6285–6290

Lee VH (1990) Protease inhibitors and penetration enhancers as approaches to modify peptide absorption. J Control Release 13:213–223

Lee GK, Maheshri N, Kaspar B, Schaffer DV (2005) PEG conjugation moderately protects adeno-associated viral vectors against antibody neutralization. Biotechnol Bioeng 92:24–34

Li YP, Pei YY, Zhang XY, Gu ZH, Zhou ZH, Yuan WF, Zhou JJ, Zhu JH, Gao XJ (2001) PEGylated PLGA nanoparticles as protein carrier; synthesis, preparation and bio distribution in rats. J Control Release 71:203–211

Liu F, Liu DX (1996) Serum independent liposome uptake by mouse liver. BBA-Biomembranes 1278:5–11. [PubMed: 8611606]

Liu M, Xie C, Xu W, Lu W (2004) Separation of polyethylene glycols and their amino-substituted derivatives by high-performance gel filtration chromatography at low ionic strength with refractive index detection. J Chromatogr A 1046:121

Liu F, Sun Y, Kang C, Zhu H (2016) Pegylated drug delivery systems: from design to biomedical applications. Nano LIFE 6(3 and 4):1642002

Liu X, Shen X, Sun X, Peng Y, Li R, Yun P, Li C, Liu L, Su F, Li S (2018) Polym Adv Technol 29:205–215

Lowe S, O'Brien-Simpson NM, Connal LA (2015) Antibiofouling polymer interfaces: poly(ethylene glycol) and other promising candidates. Polym Chem 6:198–212

Mao Z, Zhou X, Gao C (2013) Influence of structure and properties of colloidal biomaterials on cellular uptake and cell functions. Biomater Sci 1:896

Martens TF, Vercauteren D, Forier K, Deschout H, Remaut K, Paesen R, Ameloot M, Engbersen JF, Demeester J, De Smedt SC, Braeckmans K (2013) Measuring the intravitreal mobility of nanomedicines with single-particle tracking microscopy. Nanomedicine (Lond) 8:1955–1968

Marwa M, Eman A, Amal H, Hatem AS (2020) Liposomes and PEGylated liposomes as drug delivery systems. J Adv Biomed Pharm Sci 3:80–88

Mastorakos P, Zhang C, Berry S, Oh Y, Lee S, Eberhart CG, Woodworth GF, Suk JS, Hanes J (2015) Highly PEGylated DNA nanoparticles provide uniform and widespread gene transfer in the brain. Adv Healthc Mater 4:1023

Matsumura Y, Maeda H (1986) A new concept for macromolecular therapeutics in cancer chemotherapy: mechanism of tumoritropic accumulation of proteins and the antitumor agent smancs. Cancer Res 46:6387–6392

Moghimi SM, Hunter AC, Murray JC (2001) Long-circulating and target-specific nanoparticles: theory to practice. Pharmacol Rev 53:283–318

Monopoli MP, Aberg C, Salvati A, Dawson KA (2012) Biomolecular coronas provide the biological identity of nanosized materials. Nat Nanotechnol 7:779–786

Mozar FS, Chowdhury EH (2018) Impact of PEGylated nanoparticles on tumor targeted drug delivery. Curr Pharm Des 24:3283–3296

Mumper RJ, Cui Z, Oyewumi MO (2003) Nanotemplate engineering of cell specific nanoparticles. J Dispers Sci Technol 24(3–4):569–588

Nair LM, Konkel J, Thomas M, Koberda M (2006) Comparison of electrospray ionization mass spectrometry and evaporative light scattering detections for the determination of Poloxamer 188 in itraconazole injectable formulation. J Pharm Biomed Anal 41:725–730

Nakamura K, Yamashita K, Itoh Y, Yoshino K, Nozawa S, Kasukawa H (2012) Comparative studies of polyethylene glycol-modified liposomes prepared using different PEG-modification methods. Biochim Biophys Acta 1818:2801–2807

Nance EA, Woodworth GF, Sailor KA, Shih TY, Xu Q, Swaminathan G, Xiang D, Eberhart C, Hanes J (2012) A dense poly(ethylene glycol) coating improves penetration of large polymeric nanoparticles within brain tissue. Sci Transl Med 4:4 149ra119

Nance E, Timbie K, Miller GW, Song J, Louttit C, Klibanov AL, Shih TY, Swaminathan G, Tamargo RJ, Woodworth GF, Hanes J, Price RJ (2014a) Non-invasive delivery of stealth, brain-penetrating nanoparticles across the blood-brain barrier using MRI-guided focused ultrasound. J Control Release 189:123–132

Nance E, Zhang C, Shih TY, Xu Q, Schuster BS, Hanes J (2014b) Brain-penetrating nanoparticles improve paclitaxel efficacy in malignant glioma following local administration. ACS Nano 8:10655–10664

Oh N, Park JH (2014) Endocytosis and exocytosis of nanoparticles in mammalian cells. Int J Nanomedicine 9:51–63

Ou W, Thapa RK, Jiang L, Soe ZC, Gautam M, Chang JH et al (2018) Regulatory T cell-targeted hybrid nanoparticles combined with immuno-checkpoint blockage for cancer immunotherapy. J Control Release 281:84–96

Owens DE III, Peppas NA (2006) Opsonization, biodistribution, and pharmacokinetics of polymeric nanoparticles. Int J Pharm 307(1):93–102

Pai AS, Rubinstein I, Onyuksel H (2006) PEGylated phospholipid nanomicelles interact with beta-amyloid(1-42) and mitigate its beta beta-sheet formation, aggregation and neurotoxicity *in vitro*. Peptides 27(11):2858–2866

Patra JK, Das G, Fraceto LF et al (2018) Nano based drug delivery systems: recent developments and future prospects. J Nanobiotechnol 16(71):1–33

Peracchia MT, Authier CV, Desmaële D, Gulik A, Dedieu J-C, Demoy M, d'Angelo J, Couvreur P (1998) PEGylated nanoparticles from a novel methoxypolyethylene glycol cyanoacrylate-hexadecyl cyanoacrylate amphiphilic copolymer. Pharm Res 15:550

Perrault SD, Walkey C, Jennings T, Fischer HC, Chan WC (2009) Mediating tumor targeting efficiency of nanoparticles through design. Nano Lett 9:1909–1915

Perry JL, Reuter KG, Kai MP, Herlihy KP, Jones SW, Luft JC, Napier M, Bear JE, DeSimone JM (2012) PEGylated PRINT nanoparticles: the impact of PEG density on protein binding, macrophage association, biodistribution, and pharmacokinetics. Nano Lett 12:5304–5310

Petersen H, Fechner PM, Martin AL, Kunath K, Stolnik S, Roberts CJ, Fischer D, Davies MC, Kissel T (2002a) Polyethylenimine-graft-poly(ethylene glycol) copolymers: influence of copolymer block structure on DNA complexation and biological activities as gene delivery system. Bioconjug Chem 13:845–854

Petersen H, Fechner PM, Fischer D, Kissel T (2002b) Synthesis, characterization, and biocompatibility of polyethylenimine-graft-poly(ethylene glycol) block copolymers. Macromolecules 35:6867–6874

Pinheiro RGR, Coutinho AJ, Pinheiro M, Neves AR (2021) Nanoparticles for targeted brain drug delivery: what do we know? Int J Mol Sci 22:11654. https://doi.org/10.3390/ijms222111654

Rabanel JM, Hildgen P, Banquy X (2014) J Control Release 185:71–87

Rafiei P, Haddadi A (2017) Docetaxel-loaded PLGA and PLGA-PEG nanoparticles for intravenous application: pharmacokinetics and biodistribution profile. Int J Nanomedicine 12:935–947

Roser M, Fischer D, Kissel T (1998) Surface-modified biodegradable albumin nano- and microspheres. II: effect of surface charges on in vitro phagocytosis and biodistribution in rats. Eur J Pharm Biopharm 46:255–263

Rukmani R, Willi P, Chandra PS (2009) Synthesis and characterization of PEGylated calcium phosphate nanoparticles for oral insulin delivery. J Biomed Mater Res B Appl Biomater 88(1):41–48

Sahlin JJ, Peppas NA (1997) Enhanced hydrogel adhesion by polymer interdiffusion: use of linear poly(ethylene glycol) as an adhesion promoter. J Biomater Sci Polym Ed 8:421–436

Sahoo SK, Labhasetwar V (2003) Nanotech approaches to drug delivery and imaging. Drug Discov Today 8(24):1112–1120
Salvati A, Pitek AS, Monopoli MP, Prapainop K, Bombelli FB, Hristov DR, Kelly PM, Aberg C, Mahon E, Dawson KA (2013) Transferrin-functionalized nanoparticles lose their targeting capabilities when a biomolecule corona adsorbs on the surface. Nat Nanotechnol 8:137–143
Sebak AA (2018) Limitations of PEGylated nanocarriers: unfavourable physicochemical properties, biodistribution patterns and cellular and subcellular fates. Int J App Pharm 10(5):6–12
Serra L, Domenech J, Peppas NA (2006) Design of poly(ethylene glycol)-tethered copolymers as novel mucoadhesive drug delivery systems. Eur J Pharm Biopharm 63:11–18
Shi L, Zhang J, Zhao M, Tang S, Cheng X, Zhang W, Li W et al (2021) Effects of polyethylene glycol on the surface of nanoparticles for targeted drug delivery. Nanoscale 13:10748–10764
Shubik P (1982) Vascularization of tumors: a review. J Cancer Res Clin Oncol 103(3):211–226
Sinha R, Kim GJ, Nie S, Shin DM (2006) Nanotechnology in cancer therapeutics: bioconjugated nanoparticles for drug delivery. Mol Cancer Ther 5:1909–1917
Su Y-C, Chen B-M, Chuang K-H, Cheng T-L, Roffler SR (2010) Sensitive quantification of PEGylated compounds by second-generation anti-poly(ethylene glycol) monoclonal antibodies. Bioconjug Chem 21:1264–1270
Sudarshan S, Holman DH, Hyer ML et al (2005) *In vitro* efficacy of Fas ligand gene therapy for the treatment of bladder cancer. Cancer Gene Ther 12(1):12–18
Suk JS, Lai SK, Wang YY, Ensign LM, Zeitlin PL, Boyle MP, Hanes J (2009) The penetration of fresh undiluted sputum expectorated by cystic fibrosis patients by non-adhesive polymer nanoparticles. Biomaterials 30:2591–2597
Suk JS, Kim AJ, Trehan K, Schneider CS, Cebotaru L, Woodward OM, Boylan NJ, Boyle MP, Lai SK, Guggino WB, Hanes J (2014) Lung gene therapy with highly compacted DNA nanoparticles that overcome the mucus barrier. J Control Release 178:8–17
Suk JS, Xu Q, Kim N, Hanes J, Ensign ML (2016) PEGylation as a strategy for improving nanoparticle-based drug and gene delivery. Adv Drug Deliv Rev 99(Pt A):28–51
Sutradhar KB, Amin M (2014) Nanotechnology in cancer drug delivery and selective targeting. Nanotechnology 2014:1–12. https://doi.org/10.1155/2014/939378
Suzawa T, Shirahama H (1991) Adsorption of plasma proteins onto polymer latices. Adv Colloid Interface Sci 35:139
Tami BHC, Lasseter L, Abbott NL, Hamers RJ (2004) J Am Chem Soc 126:10220–10221
Tenzer S, Docter D, Kuharev J, Musyanovych A, Fetz V, Hecht R, Schlenk F, Fischer D, Kiouptsi K, Reinhardt C, Landfester K, Schild H, Maskos M, Knauer SK, Stauber RH (2013) Rapid formation of plasma protein corona critically affects nanoparticle pathophysiology. Nat Nanotechnol 8:772–781
Thorne RG, Nicholson C (2006) In vivo diffusion analysis with quantum dots and dextrans predicts the width of brain extracellular space. Proc Natl Acad Sci U S A 103:5567–5572
Tobio M, Gref R, Sanchez A, Langer R, Alonso MJ (1998) Stealth PLA-PEG nanoparticles as protein carriers for nasal administration. Pharm Res 15:270–275
Tobio M, Sanchez A, Vila A, Soriano I, Evora C, Vila-Jato JL, Alonso MJ (2000) The role of PEG on the stability in digestive fluids and in vivo fate of PEG– PLA nanoparticles following oral administration. Colloids Surf B Biointerfaces 18(3–4):315–323
Torchilin VP (2005) Recent advances with liposomes as pharmaceutical carriers. Nat Rev Drug Discov 4:145–160
Torchilin VP (2007) Targeted pharmaceutical nanocarriers for cancer therapy and imaging. AAPS J 9(2):E128–E147
Torchilin V, Weissig V (2003) Liposomes: a practical approach, 2nd edn. Oxford University Press, Oxford
Unsworth LD, Sheardown H, Brash JL (2008) Protein-resistant poly(ethylene oxide)-grafted surfaces: chain density-dependent multiple mechanisms of action. Langmuir 24:1924–1929
Uster PS, Allen TM, Daniel BE, Mendez CJ, Newman MS, Zhu GZ (1996) Insertion of poly(ethylene glycol) derivatized phospholipid into pre-formed liposomes results in prolonged in vivo circulation time. FEBS Lett 386:243–246

Valencia PM, Hanewich-Hollatz MH, Gao W, Karim F, Langer R, Karnik R, Farokhzad OC (2011) Effects of ligands with different water solubilities on self-assembly and properties of targeted nanoparticles. Biomaterials 32:6226–6233

van den Berg JH, Oosterhuis K, Hennink WE, Storm G, van der Aa LJ, Engbersen JF, Haanen JB, Beijnen JH, Schumacher TN, Nuijen B (2010) Shielding the cationic charge of nanoparticle-formulated dermal DNA vaccines is essential for antigen expression and immunogenicity. J Control Release 141:234–240

van der Meel R, Vehmeijer LJ, Kok RJ, Storm G, van Gaal EV (2013) Ligand-targeted particulate nanomedicines undergoing clinical evaluation: current status. Adv Drug Deliv Rev 65:1284–1298

Verhoef JJF, Anchordoquy TJ (2013) Questioning the use of PEGylation for drug delivery. Drug Deliv Transl Res 3:499–503

Veronese FM, Mero A (2008) The impact of PEGylation on biological therapies. BioDrugs 22(5):315–329

Vllasaliu D, Fowler R, Garnett M et al (2011) Barrier characteristics of epithelial cultures modelling the airway and intestinal mucosa: a comparison. Biochem Biophys Res Commun 415:579–585

Vllasaliu D, Fowler R, Snow S (2014) PEGylated nanomedicines: recent progress and remaining concerns. Expert Opin Drug Deliv 11(1):139–154

Vonarbourg A, Passirani C, Saulnier P, Benoit JP (2006) Parameters influencing the stealthiness of colloidal drug delivery systems. Biomaterials 27:4356–4373

Walkey CD, Olsen JB, Guo H, Emili A, Chan WC (2012) Nanoparticle size and surface chemistry determine serum protein adsorption and macrophage uptake. J Am Chem Soc 134:2139–2147. [PubMed: 22191645]

Wan X, Zhang J, Yu W, Shen L, Ji S, Hu T (2017) Effect of protein immunogenicity and PEG size and branching on the anti-PEG immune response to PEGylated proteins. Process Biochem 52:183–191

Wang YY, Lai SK, Suk JS, Pace A, Cone R, Hanes J (2008) Addressing the PEG mucoadhesivity paradox to engineer nanoparticles that "slip" through the human mucus barrier. Angew Chem Int Ed 47:9726–9729

Weaver EA, Barry MA (2008) Effects of shielding adenoviral vectors with polyethylene glycol on vector-specific and vaccine-mediated immune responses. Hum Gene Ther 19:1369–1382

Webb MS, Saxon D, Wong FM, Lim HJ, Wang Z, Bally MB, Choi LS, Cullis PR, Mayer LD (1998) Comparison of different hydrophobic anchors conjugated to poly(ethylene glycol): effects on the pharmacokinetics of liposomal vincristine. Biochim Biophys Acta 1372:272

Xia X, Yang M, Wang Y, Zheng Y, Li Q, Chen J, Xia Y (2012) Quantifying the coverage density of poly(ethylene glycol) chains on the surface of gold nanostructures. ACS Nano 6:512–522

Xiao X-F, Jiang X-Q, Zhou L-J (2013) Chin J Anal Chem 41:445–453

Xie Z, Ji Z, Zhang Z, Gong T, Sun X (2014) Adenoviral vectors coated with cationic PEG derivatives for intravaginal vaccination against HIV-1. Biomaterials 35:7896–7908

Xu J, Gattacceca F, Amiji M (2013) Biodistribution and pharmacokinetics of EGFR-targeted thiolated gelatin nanoparticles following systemic administration in pancreatic tumour-bearing mice. Mol Pharm 10:2031–2044

Xu Q, Ensign LM, Boylan NJ, Schon A, Gong X, Yang J-C, Cai S, Yu T, Freire E, Hanes J (2015) Impact of surface polyethylene glycol (PEG) density on biodegradable nanoparticle transport in mucus ex vivo and distribution in vivo. ACS Nano 9:9217–9227

Yang Q, Lai SK (2015) Anti-PEG immunity: emergence, characteristics, and unaddressed questions. Wiley Interdiscip Rev Nanomed Nanobiotechnol 7:655

Yang Q, Jones SW, Parker CL, Zamboni WC, Bear JE, Lai SK (2014a) Evading immune cell uptake and clearance requires PEG grafting at densities substantially exceeding the minimum for brush conformation. Mol Pharm 11:1250–1258

Yang X, Wu S, Wang Y, Li Y, Chang D, Luo Y et al (2014b) Evaluation of self-assembled HCPT-loaded PEG-b-PLA nanoparticles by comparing with HCPT-loaded PLA nanoparticles. Nanoscale Res Lett 9:1–8

Yang YS, Moynihan KD, Bekdemir A, Dichwalkar TM, Noh MM, Watson N et al (2018) Targeting small molecule drugs to T cells with antibody-directed cellpenetrating gold nanoparticles. Biomater Sci 7(1):113–124

Yezhelyev MV, Gao X, Xing Y, Al-Hajj A, Nie S, O'Regan RM (2006) Emerging use of nanoparticles in diagnosis and treatment of breast cancer. Lancet Oncol 7(8):657–667

Yildirimer L, Thanh NT, Loizidou M, Seifalian AM (2011) Toxicology and clinical potential of nanoparticles. Nano Today 6:585–607

Yoo JW, Chambers E, Mitragotri S (2010) Factors that control the circulation time of nanoparticles in blood: challenges, solutions and future prospects. Curr Pharm Des 16:2298–2307

Yoon JY, Kim JH, Kim WS (1998) Interpretation of protein adsorption phenomena onto functional microspheres. Colloid Surf B 12:15–22

Yuan H, Chen CY, Chai GH, Du YZ, Hu FQ (2013) Improved transport and absorption through gastrointestinal tract by PEGylated solid lipid nanoparticles. Mol Pharm 10:1865–1873

Zabaleta V, Campanero MA, Irache JM (2007) An HPLC with evaporative light scattering detection method for the quantification of PEGs and Gantrez in PEGylated nanoparticles. J Pharm Biomed Anal 44:1072–1078

Zhan X, Tran KK, Shen H (2012) Effect of the poly(ethylene glycol) (PEG) density on the access and uptake of particles by antigen-presenting cells (APCs) after subcutaneous administration. Mol Pharm 9:3442–3451

Zhang L, Gu FX, Chan JM, Wang AZ, Langer RS, Farokhzad OC (2007) Nanoparticles in medicine: therapeutic applications and developments. Clin Pharmacol Ther 83:761

Zhang F, Liu M, Wan H (2014) Discussion about several potential drawbacks of PEGylated therapeutic proteins. Biol Pharm Bull 37:335–339

Zhang P, Sun F, Liu S, Jiang S (2016) Anti-PEG antibodies in the clinic: current issues and beyond PEGylation. J Control Release 244:184–191

Ziemba B, Matuszko G, Bryszewska M, Klajnert B (2012) Influence of dendrimers on red blood cells. Cell Mol Biol Lett 17:21–35

Zillies JC, Zwiorek K, Winter G, Coester C (2007) Method for quantifying the PEGylation of gelatin nanoparticle drug carrier systems using asymmetrical flow field-flow fractionation and refractive index detection. Anal Chem 79:4574–4580

# PEGylated Nanocarriers for Solubilization

# 5

Randa S. H. Mansour, Rakesh Kumar Tekade, and Ala' Adnan Ali

**Abstract**

This chapter is dedicated to explaining the principle of PEGylation in the solubilization of nanocarriers for pharmaceutical and biomedical applications, in addition to the enhancement of the propensity of these carriers to solubilize drugs. The chapter introduces the general rationale for nanocarrier PEGylation with a specific focus on the topic of solubilization in addition to the solubility characteristics of PEG. PEGylated small molecular weight drugs and proteins are briefly introduced to emphasize the stages of pharmaceutical solubilization by PEGylation. Furthermore, protein PEGylation as a means to solubilize enzymes and proteins for facilitated nanoparticle production is highlighted. Proposed mechanisms and factors affecting the drug solubilization propensity of PEGylated nanocarriers are also explored. Illustrative examples and reviewed applications are included wherever appropriate.

**Keywords**

PEGylated nanocarriers · Aqueous solubility · Organic solvents · Nanocarrier solubilization · Drug solubilization · PEGylation extent · PEG chain length

R. S. H. Mansour (✉)
Faculty of Pharmacy, Zarqa University-Jordan, Zarqa, Jordan
e-mail: rmansour@zu.edu.jo

R. K. Tekade
National Institute of Pharmaceutical Education and Research (NIPER) Ahmedabad, An Institute of National Importance, Government of India, Department of Pharmaceuticals, Ministry of Chemicals and Fertilizers, Palaj, Opp. Air force station, Ahmedabad, Gujarat, India

A. A. Ali
Faculty of Pharmacy, Zarqa University-Jordan, Zarqa, Jordan

Faculty of Pharmacy, University of Jordan, Amman, Jordan





R. K. Tekade, N. K. Jain (eds.), *PEGylated Nanocarriers in Medicine and Pharmacy*, https://doi.org/10.1007/978-981-97-7316-9_5

## Abbreviations

| | |
|---|---|
| C | Carbon |
| CAC | Critical aggregation concentration |
| FDA | Food and Drug Administration |
| mPEG | Methyl PEG |
| PAMAM | Poly-amidoamine |
| PEG | Polyethylene glycol |
| PEG-b-PAMA | Poly(ethylene glycol)-block-poly((2-*N*,*N*-dimethylamino)ethyl methacrylate) |
| PEG-PLGA | Poly(ethylene glycol)-poly(lactic-*co*-glycolic acid) |
| Si | Silicon |

## 5.1 Introduction

Polyethylene glycol (PEG), commercially known as Macrogol, is a polyether composed of repeated ethylene glycol units [$-(CH_2CH_2O)_n$] (D'Souza and Shegokar 2016). It is synthesized via anionic polymerization of ethylene oxide and a hydroxyl initiator such as water, ethylene glycol, or any diols. Alternatively, it can be obtained by ring-opening polymerization of epoxyethane. Commercially, PEG is available with different degrees of polymerization and activated functional groups yielding variable derivatives that are characterized by their molecular weights ranging from hundreds to several thousands of Daltons, typically from 200 to 6000 (Abuchowski et al. 1977; Thomas et al. 2014; Chen et al. 2005). The numerical designation of PEGs generally indicates their average molecular weight.

PEG is inert, nonimmunogenic, nonantigenic, and biocompatible with a well-established human safety profile; additionally, it has excellent hydration capacity and water solubility. It is approved by the US-FDA for versatile pharmaceutical applications and classified as generally regarded as a safe (GRAS) polymer (D'Souza and Shegokar 2016; Bhadra et al. 2002; Mohapatra et al. 2019; Luong et al. 2016; Suk et al. 2016; Knop et al. 2010). Moreover, PEG is an uncharged, amphiphilic, flexible polymer (Sanchez Armengol et al. 2022; Kulhari et al. 2011). These properties make it suitable for an extensive range of drug-delivery applications and technologies. For example, it has been used in dermatological products, tablets, soft gelatin capsules, suppositories, and parenterals (D'Souza and Shegokar 2016). The use of PEG was later extended to conjugate (PEGylate) drugs and proteins (Abuchowski et al. 1977), and this concept was then applied to different types of nanocarriers, as PEG imparts steric stabilization of colloids with the added advantage of escaping the immune defense mechanisms of the host as a result of low immunogenicity (Sanchez Armengol et al. 2022).

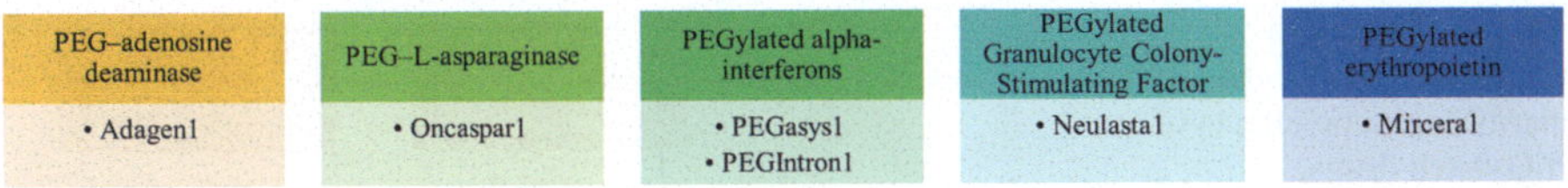


**Fig. 5.1** Examples of marketed PEGylated bioactives

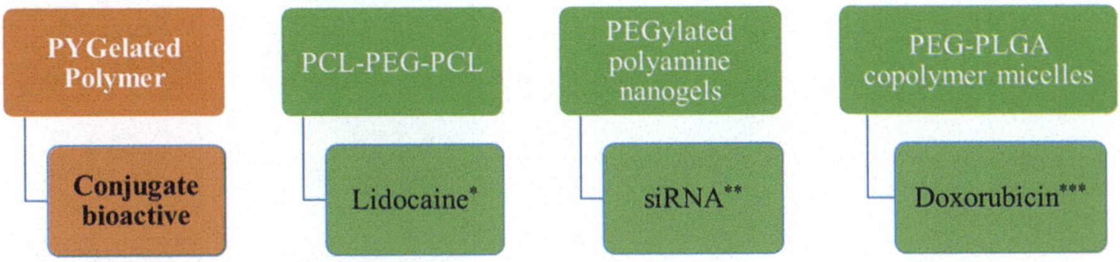


**Fig. 5.2** Some PEGelated polymers for drug delivery. * Yin et al. (2009), ** Tamura et al. (2009), *** Yang et al. (2007a)

### 5.1.1 Rational for PEGylation of Drugs/Drugs Nanocarriers

PEG can be used to PEGylate a wide range of bioactive and their carriers, including small molecular weight drug molecules, peptides and proteins, micro- and nanoparticles (Otsuka et al. 2003; Wattendorf and Merkle 2008; Karakoti et al. 2011; Jokerst et al. 2011), polymers (Joralemon et al. 2010; Freichels et al. 2011), oligonucleotides (Ravelli et al. 2012), in addition to nanocarriers (Howard et al. 2008). Some PEGylated bioactives currently marketed (González et al. 2011) are illustrated in Fig. 5.1, whereas Fig. 5.2 shows some PEGylated biodegradable polymers that have been investigated as encapsulating materials for several drugs (Elbert and Hubbell 1996). Concerning nanocarriers, PEGylation proves to be the most successful method of surface modification (Vllasaliu et al. 2014) to provide enhanced properties.

PEGylation imparts desirable pharmacokinetic and pharmacodynamics properties to drugs and drug carriers (Sanchez Armengol et al. 2022; Ravelli et al. 2012; Mirkin and Taton 2000; Hussain et al. 2019; Owens and Peppas 2006; Suk et al. 2016; Karra and Benita 2012; Elbert and Hubbell 1996; Awasthi et al. 2004; Ho and Gibaldi 2013; D'Souza and Shegokar 2016; Roberts et al. 2002; Vllasaliu et al. 2014; Sweet et al. 2009; Fung et al. 1997; Mozar and Chowdhury 2018), eventually enhancing pharmaceutical and therapeutic performance, particularly in cancer diagnosis and therapy as shown in Fig. 5.3. While this chapter focuses on the solubilization effect exerted by PEG on the PEGylated species, other chapters are devoted to discussing other results.

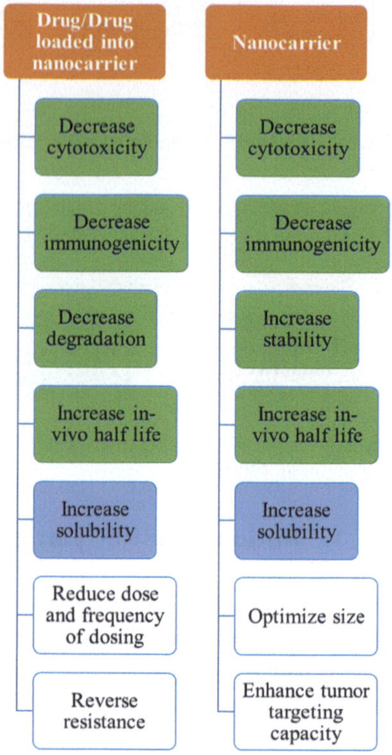


**Fig. 5.3** Enhanced pharmacodynamic and pharmacokinetic effects of PEGylated drugs, nanocarriers, and drugs loaded into nanocarriers

### 5.1.2 PEG Solubility Characteristics

PEG is readily soluble in aqueous media. Due to its hydrophilicity, it has high water binding capacity as each ethylene oxide unit can bind two to three water molecules, thereby showing a high hydrodynamic volume (Roberts et al. 2002). This pattern of interaction is unique between PEG and the aqueous environment, unlike the other closely related polymers such as poly(propylene oxide), poly(methylene oxide) (Israelachvili 1997), and isomeric polyacetaldehyde (Harris 1992) as these polymers are considered hydrophobic and water-insoluble (Harris 1992). The Hildebrand solubility parameter of PEG is 10.3 (cal cm$^{-3}$)$^{0.5}$ (Bailey 1990).

While all grades of PEG are soluble in water (handbook of ph ex), the extent of solubility is actually dependent on the chain length (Table 5.1). It is observed that the solubility decreases with increasing molecular weight of PEG (Sanchez Armengol et al. 2022). PEGs with molecular weight lower than 800 Da are liquid at room temperature and are highly soluble in water, whereas PEGs with molecular weights 1000–2000 are soft waxy substances. PEGs with molecular weight higher than 2000 occur as hard crystalline solids with melting points around 63 °C (Otsuka et al. 2003; Chen et al. 2005). Aqueous solutions of high molecular weight grades may form gels (Aranda-Lara et al. 2021).

**Table 5.1** Aqueous solubility and state of variable PEG grades

| PEG grade | State at room temperature[a] | Aqueous solubility (g/100 g)[b] |
|---|---|---|
| PEG 200 | Clear liquid | Highly soluble |
| PEG 300 | Clear liquid | Highly soluble |
| PEG 400 | Clear liquid | Highly soluble |
| PEG 600 | Clear liquid | Highly soluble |
| PEG 1000 | Semi-cystalline solid | 75 |
| PEG 4000 | Semi-cystalline solid | 55 |
| PEG 10000 | Semi-cystalline solid | 53 |
| PEG 35000 | Semi-cystalline solid | 50 |

[a] Gullapalli and Mazzitelli (2015)
[b] Wiley-VCH (2012), with permission

On the other hand, PEG has high solubility in most organic and inorganic solvents (Bhadra et al. 2005; Jain et al. 1996). Liquid PEGs are soluble in acetone, alcohols, benzene, glycerin, and glycols, whereas the solid ones are soluble in acetone, dichloromethane, ethanol (95%), and methanol, and slightly soluble in aliphatic hydrocarbons and ether, but insoluble in fats, fixed oils, and mineral oil (Aranda-Lara et al. 2021).

### 5.1.3 PEG as Solubilizer

PEG is known to act as a solubilizer as it enhances the water solubility of hydrophobic moieties. This effect is mediated through different mechanisms depending on the employed solubilization system.

The high solubility of PEG in both aqueous and organic media gives it the ability to form a monolayer at interfaces, a typical property of amphiphilic molecules, and this property is affected by the molecular weight of PEG rather than its concentration (Bhadra et al. 2003; Pasut and Veronese 2012). PEG can decrease the contact angle of substances. Coating of the gold plate with PEG resulted in a 50° decrease in its contact angle. Similarly, the PEG-coated glass plate exhibited a 34° reduction. Thus, PEG can significantly decrease the contact angle of hydrophobic materials (Bhadra et al. 2003). These characteristics of PEG make it able to increase the hydrophilicity of hydrophobic drugs (Pasut and Veronese 2012) and decrease the polarity of the aqueous solvent, i.e., act as a co-solvent (Nayak and Panigrahi 2012), thus permitting the aqueous solubilization of hydrophobic drugs.

Upon the use of PEG in preparation for solid dispersions, the solubilization of the drug is aided by permitting the drug to be trapped in the interstitial spaces of PEG when the molten PEG is solidified (Chiou and Riegelman 1971) or the formation of metastable crystalline polymorphs of the drug (Pasut and Veronese 2012; Martinez-Oharriz et al. 1999).

Solubilization by the formation of polymeric or copolymeric PEG micelles also occurs when macromolecular PEG polymers or copolymers are added above their corresponding critical aggregation concentration (CAC), as they readily self-assemble to form micellar-like, nanosized, thermodynamically stable systems with core–shell structures (Hagan et al. 1995; Kazunori et al. 1993; Otsuka et al. 2003). The subsequent effect will facilitate the solubilization of hydrophobic moieties such as drugs, proteins, peptides, and genes (Muralidharan et al. 2014). These micellar systems can increase the solubilization of hydrophobic drugs (Otsuka et al. 2003). In the case of PEG copolymers, the PEG shell ensures stabilization in an aqueous medium (Savić et al. 2006). Unlike these systems, surfactant micelles tend to precipitate the drug upon dilution by the blood, as the surfactant concentration will drop below its critical micelle concentration.

### 5.1.4 PEGylation of Drugs/Drugs Nanocarriers for Solubility Enhancement

Good water solubility is a prerequisite for the therapeutic utility of bioactiveness. PEG is the most commonly employed polymer to conjugate moieties targeting their aqueous solubility enhancement within this context. PEGylation is considered a cornerstone in increasing the aqueous solubility of small molecular weight drugs, proteins, and nanocarriers.

## 5.2 Water-Soluble PEGylated Small Molecule Drugs

In this section, some examples of small molecule drug PEGylation for the sake of drug solubility enhancement are illustrated.

PEGylation of taxol at its 7-OH functionality via a urethane or carbonate linkage produced highly water-soluble taxol derivatives. The solubility of PEG 350 and PEG 750 derivatives was $1.87 \times 10^{-3}$ mmol/mL, which is much higher than the native taxol with a solubility value of $3 \times 10^{2}$ nM. The higher molecular weight PEG derivatives (2000 and 5000) resulted in a superior solubility of (0.1 mmol/mL), which is approximately 30,000 times that of native taxol (Greenwald et al. 1995).

The aqueous solubility of the poorly water-soluble paclitaxel was also enhanced by PEGylation. The estimated solubility of it is 10–20 μM. In one study, esterification with PEG 5000 increased the solubility to 120 mg/mL (Lee et al. 2005). Another study showed that the oral bioavailability of PEGylated paclitaxel in rats was 3.94-fold higher than that of non-PEGylated paclitaxel (Choi and Jo 2004).

PEGylation of curcumin was attempted to increase its dissolution rate and solubility. The poor water solubility of curcumin limits its use for treating corneal bacterial infections. For increasing solubility, curcumin was conjugated to PEG 6000. The solubility of PEGylated curcumin was increased up to $93 \pm 3.2\%$ compared to curcumin (Hanif et al. 2022).

## 5.3 Soluble PEGylated Proteins/Enzymes

It is well known that the attachment of PEG can be utilized to improve water solubility, but it could also enhance the solubility of moieties in organic solvents. Both effects are helpful for the enhancement of the solubility of proteins/enzymes in both types of solvents. The latter effect is of particular importance in the enhancement of enzyme activity in dry organic solvents (Harris 1992). In 1985, Veronese and co-workers were the first to introduce a surface modification of proteins by PEG (Veronese et al. 1985). PEGylation of enzymes used in organic solvents was demonstrated in 1986 (Inada et al. 1986).

### 5.3.1 PEGylated Proteins Soluble Under Physiological Conditions

Many proteins can be used as therapeutic agents; nevertheless, the utility of some of these is limited by their poor water solubility. A classic example of PEGylated water-soluble proteins is the uricase enzyme. Humans lack the enzyme uricase, or urate oxidase, which can be used to treat gout. Nevertheless, there are some challenges for the practical use of this enzyme, including its insolubility in water. The first U.S. patent claiming "water-soluble nonimmunogenic polypeptide compositions" was issued in 1979. It described a PEGylated uricase (Davis et al. 1979). A water-soluble PEGylated recombinant mammalian uricase possessing sufficient solubility, in addition to other desirable properties, was developed. The enzyme PEGylated with PEG-10000 was soluble under physiological conditions to enable good bioavailability, unlike the native enzyme, which was soluble only at high pH values (Williams et al. 2003; Sherman et al. 2004).

### 5.3.2 Organic Solvent–Soluble PEGylated Proteins/Enzymes

In industrial biotechnology, organic solvents are frequently employed in enzymatic reactions as an alternative to aqueous media. This approach offers valuable advantages such as enhanced solubilization of hydrophobic substrates, shifting the thermodynamic equilibrium to favor synthesis over hydrolysis, and suppressing water-dependent side reactions (Carrea and Riva 2000). The main obstacles to the employment of this approach are the low stability, solubility, and catalytic activity of the enzymes in organic media as they tend to denature and precipitate (Stepankova et al. 2013).

Modified enzymes were demonstrated to dissolve in organic solvents (Inada et al. 1986); consequently, PEGylated enzymes were prepared and exhibited solubility, stability, and catalytic activity in organic media due to both the hydrophilic and hydrophobic properties of PEGs (Stepankova et al. 2013; Castillo et al. 2006, 2008; Kwon et al. 1999; Hernaiz et al. 1997; Castellanos et al. 2005).

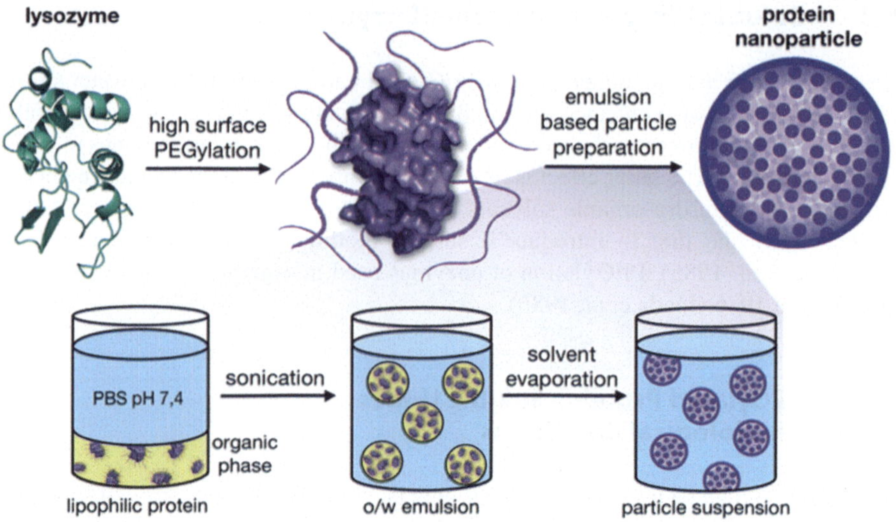


**Fig. 5.4** Formation of the protein nanoparticles mediated by enhanced solubility of the protein in organic solvents through PEGylation [(Radi et al. 2016) with permission]

A study demonstrated the solubilization of lysozyme in organic solvents by PEGylation with methyl PEG (mPEG) derivatives. This modification leads to an increase in the solubility of lysozyme in organic solvents such as dichloromethane while preserving its native structure. Subsequently, the enhanced solubility in organic solvents permitted the employment of an emulsion-based solvent evaporation method to form enzyme nanoparticles (Fig. 5.4) (Radi et al. 2016).

## 5.4 Water-Soluble PEGylated Drug Nanocarriers

Due to many favorable characteristics (Md et al. 2019; Ndlovu et al. 2019; Gorain et al. 2018; Hussain et al. 2017a, b; Tran et al. 2017; Albanese et al. 2012), nanocarriers offer various benefits (Gao et al. 2018; Khan et al. 2018; Choudhury et al. 2018; Hussain et al. 2018; Howard et al. 2008; Xia et al. 2013; Shi et al. 2017) in cancer diagnosis and therapy. Despite their paramount pharmaceutical and therapeutic applications, the significance of nanocarriers is limited due to their short plasma half-lives, low solubility, and low biocompatibility, among other reasons. Surface modification of nanocarriers is usually required to enhance their properties, and PEG has been proven to have a tremendous beneficial influence on the characteristics of nanocarriers (Vllasaliu et al. 2014; Karra and Benita 2012), including the enhancement of their solubility and that of the drug they carry (Fig. 5.3). Indeed, among all the polymers tested for surface modification of nanocarriers, PEG and PEG-copolymers are currently the most ideal, widely recognized, popular, and effective (Ravelli et al. 2012; Joralemon et al. 2010; Howard et al. 2008; Bhadra et al. 2002).

### 5.4.1 Water-Soluble PEGylated Silicon Nanocarriers

The fluorescent properties of silicon (Si) nanoparticles are interesting for sensing and tagging applications (Erogbogbo et al. 2008). They have potential use in cellular and assay labeling techniques, as well as in deep-tissue imaging (Alivisatos 2004; Bruchez et al. 1998; Michalet et al. 2005; Pinaud et al. 2006), in addition to a diverse range of materials and cell culture applications (Sudeep et al. 2008). The solubility of Si nanoparticles in water is crucial for utilizing them in biological applications. Surface modification of Si nanoparticles with water-soluble functionalities has been attempted with amines (Alivisatos 2004; Bruchez et al. 1998; Michalet et al. 2005; Pinaud et al. 2006; Rosso-Vasic et al. 2008; Tilley et al. 2005) carboxylic acids (He et al. 2009), and poly(acrylic acid) (Li and Ruckenstein 2004).

PEGylation of Si nanoparticles has also been described to render them soluble in aqueous media as well as organic solvents. Si particles PEGylated by PEG 1100 were successfully prepared and yielded optically clear solutions in many solvents, including water, methanol, and chloroform. Moreover, these nanoparticles could be stored as solids and then redispersed readily in the same solvents (Sudeep et al. 2008). In another study, water-soluble Si nanoparticles were obtained through PEGylation with PEG 2000 (Xu et al. 2015).

### 5.4.2 Water-Soluble PEGylated Carbon Nanotubes

Like Si nanoparticles, PEGylation of carbon nanotubes (C nanotubes) is necessary to render them more water-soluble (Ravelli et al. 2012; Zhao et al. 2005; Balasubramanian and Burghard 2010), permitting their biological and medicinal applications. C nanotubes are potentially used in targeted drug therapy, chemical and biological imaging, and sensing (Lacerda et al. 2006; Madani et al. 2011; Elhissi et al. 2012; Prato et al. 2008; Liu et al. 2007). The low solubility and susceptibility of C nanotubes in biological media and other obstacles limit their utility (Bottini et al. 2011; Hong et al. 2006; Liu et al. 2008). Novel water-soluble PEGylated single-walled C nanotubes were prepared and provided a means of attaching drugs. The study showed that a specific extent of coverage of PEG chains on nanotubes was both necessary and sufficient to impart aqueous solubility of the nanotubes without aggregation, and PEG chains extending into the water were responsible for the enhanced solubility (Liu et al. 2007). In another study, the suspendability in water of single-walled C nanotubes PEGylated with PEG 400 was superior to that of the native nanotubes in water and in water-PEG 400 media. Figure 5.5 demonstrates that the nonfunctionalized nanotubes immediately precipitated from the water solution and formed a dispersion in the water–PEG 400 medium that was stable for only a few minutes. In contrast, the PEGylated nanotubes produced a more stable dispersion. The study also confirmed that the higher the PEGylation degree, the higher the dispersion stability (Ravelli et al. 2013).

Figure 5.6 shows another single-walled C nanotube that was noncovalently functionalized with fluorescein-PEG 5000 and exhibited excellent water solubility at

| C nanotubes | Native | Native | PEGylated |
|---|---|---|---|
| Dispersion Medium | Water | Water-PEG 400 | Water |
| After sonication |  | | |
| After four days standing | 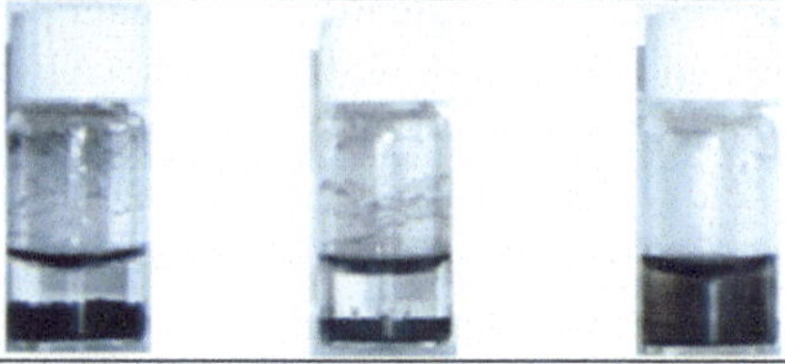 | | |

**Fig. 5.5** Improved suspendability of single-walled C nanotubes by PEGylation [(Ravelli et al. 2013), reproduced with permission]

different pH values attributed to PEG chains (Nakayama-Ratchford et al. 2007). In another attempt, single-walled C nanotubes that were noncovalently functionalized with phospholipid-PEG 2000 or 5000 were prepared to enhance the aqueous solubility of the nanotubes, and aqueous solubility of 0.84 mg/mL was achieved for PEG 5000 conjugates (Hadidi et al. 2011). C nanotubes covalently functionalized with PEG 600 were also prepared, resulting in an aqueous solubility of 5.9 mg/mL (Zhao et al. 2005). PEG 800 or PEG 750 monomethyl ether functionalized double-walled carbon nanotubes exhibited good aqueous solubility and formed stable suspensions in water, whereas the unfunctionalized nanotubes immediately settled in water, PEG 800, and PEG 750 monomethyl ether. The saturated aqueous concentrations of PEG 800 and PEG 750 monomethyl ether PEGylated nanotubes were 0.86 mg/mL (equivalent nanotubes concentration of 0.36 mg/mL) and 0.93 mg/mL (equivalent nanotubes concentration of 0.37 mg/mL), respectively (Nie et al. 2010).

### 5.4.3 Water-Soluble PEGylated Metal/Metal Oxides Nanoparticles

Many factors, including their water instability, hinder the diagnostic and therapeutic applications of metal nanoparticles. Accordingly, surface modification of these inorganic nanoparticles is crucial to impart the desired properties, including the enhancement of solubility in physiological media (Tilley et al. 2005). PEGylation of inorganic nanoparticles can modify their properties and overcome the limitations of their applicability (Otsuka et al. 2003).

PEGylation of gold nanoparticles to provide water solubility has been described in the literature, such as gold nanoparticles containing a noncovalently bound silicon phthalocyanine (Cheng et al. 2008, 2011a, b; Meyers et al. 2015) and gold

Day 1
No precipitation is observed.

Day 4
After >48hs, we can start to see black precipitates from the pH 12 solution.

Day 5
SWNT precipitates have settled at the bottom from the pH 12 solution, but more are still precipitating out. No precipitates were observed for the lower pHs.

**Fig. 5.6** Solubility of noncovalently fluorescein-PEG functionalized C nanotubes at different pH values [(Nakayama-Ratchford et al. 2007), with permission]

nanoparticles functionalized with a hydrophobic zinc phthalocyanine (Camerin et al. 2016). Also, water-soluble gold nanoparticles functionalized with a thiolated PEG were synthesized (Penon et al. 2015) in addition to water-soluble, PEGylated multifunctional antibody–porphyrin gold nanoparticles for targeted photodynamic therapy (Penon et al. 2017). Furthermore, water-soluble gold nanoparticles entrapped within PEG-modified polyethyleneimine were prepared and have potential use for blood pool and tumor computed tomography (CT) imaging (Zhou et al. 2014). In all of these cases, PEGylation achieved an increase in the water solubility of the nanoparticles.

On the other hand, a study has described the development of a novel PEG-b-PAMA, poly(ethylene glycol)-block-poly((2-*N*,*N*-dimethylamino)ethyl methacrylate) water-soluble polymer that was used to modify the surface of gold nanoparticle rendering them completely dispersible under physiological conditions with high dispersion stability (Miyamoto et al. 2008).

Metal oxide nanoparticles, such as alumina, silica, iron oxide, gadolinium oxide, and cerium oxide, can function as controlled drug-delivery systems and magnetic contrast agents, in addition to therapeutic and diagnostic materials in cancer. The use of iron oxide nanoparticles, also known as magnetite nanoparticles such as $Fe_3O_4$ and substituted iron oxides ($MFe_2O_4$ where M = Co, Mn, Zn or Mg) for biomedical applications, has been studied extensively (Laurent et al. 2008). For example, iron oxide nanoparticles have demonstrated a potential for various nano-biomedical applications, including the hyperthermia treatment of tumor cells, magnetic drug targeting, and magnetic contrast age (Pankhurst et al. 2003; Lu et al. 2007; Salunkhe et al. 2014; Kallumadil et al. 2009; Banerjee and Chen 2007).

In several cases, magnetic nanoparticles were successfully PEGylated to enhance their aqueous solubility. Excellent stability and solubility were obtained in aqueous dispersions and physiological media (Umut 2013). In one of the reported cases, magnetic iron oxide nanoparticles were PEGylated, and their solubility was noticed to increase as a function of the molecular weight of PEG from 550 to 5000. Unfortunately, the increase in solubility was associated with an increase in particle size in addition to a loss of the magnetic properties of the nanoparticles. Another study showed that the solubilization kinetics of PEGylated supramagnetic iron oxide nanoparticles was dependent on the conformation of the used PEG, even though the extent of solubility was comparable. Dendritic PEGs resulted in a significantly higher rate of solubilization than linear PEGs. The former nanoparticles were suspended in water within seconds, whereas the latter nanoparticles required hours. Additionally, some of the nanoparticles functionalized with linear PEG exhibited substantially reduced aqueous solubility after freeze-drying. The different solubility behavior between linear and dendritically stabilized nanoparticles was attributed to the differences in the conformation of the dry shell of the nanoparticles. Upon drying, the linear polymer chains of neighboring particles are likely to interdigitate, resulting in smaller interparticle separations and consequently increasing van der Waals interactions, hence, the reduced dispersion rate. Contrastingly, the dendritic conformation forms a more compact shell (Gillich et al. 2013).

### 5.4.4 Water-Soluble PEGylated Dendrimers

Dendrimers are nanostructures composed of hyper-branched polymers of homogeneous composition with an overall compact globular structure formed by layers emanating from the center to the surface (Aranda-Lara et al. 2021).

Dendrimers have favorable applications in biomedical and pharmaceutical areas, such as bioimaging and tissue engineering, in addition to drug and gene delivery (D'Emanuele and Attwood 2005; Kimura et al. 2000). Nevertheless, several limitations, including hydrophobicity, restrict their use. In general, PEGylation of the terminal groups of dendrimers can overcome these limitations, including their solubility. It can modify the solubility profile of dendrimers, making them a more soluble system (Sideratou et al. 2001; Gajbhiye et al. 2007; Quintana et al. 2002; Lee et al. 2005; Luo et al. 2002). PEG chains that are coupled to dendrimers are proposed to form unimolecular micelles (Yang et al. 2004), thus aiding in the solubilization of the dendrimers.

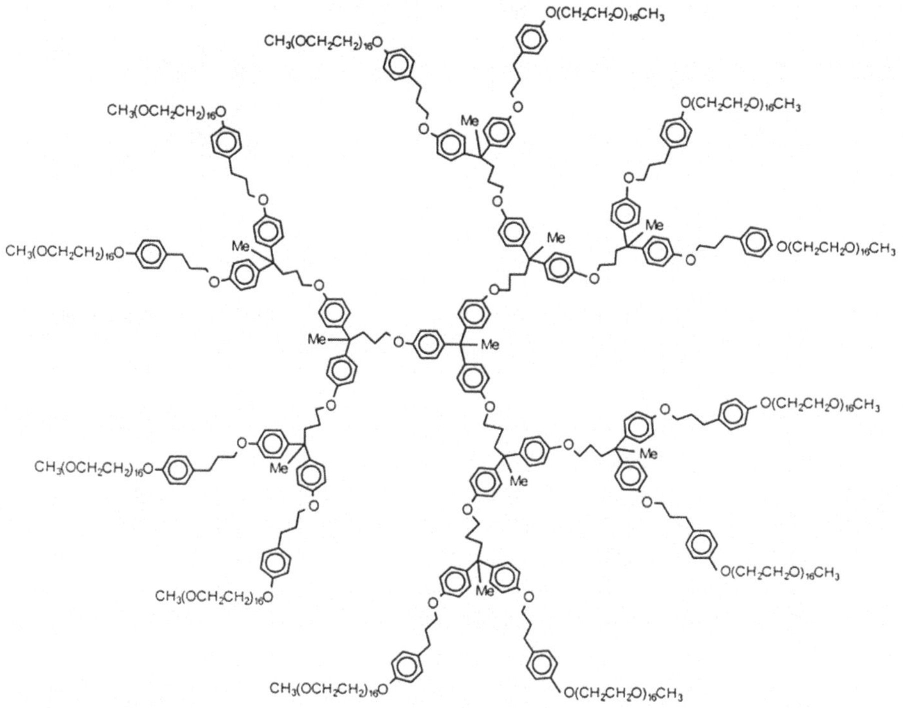


**Fig. 5.7** Water-soluble PEGylated 4,4-bis(4-hydroxy phenyl)pentanol dendrimer in the form of unimolecular micelle [(Liu et al. 2000), with permission]

Covalently PEG-functionalized Poly-amidoamine (PAMAM) dendrimers were found to possess desirable characteristics such as nonimmunogenicity, nonantigenicity, noncytotoxicity, biocompatibility, and high water solubility (Barraza et al. 2016; Jiang et al. 2010; Luong et al. 2016; Ryan et al. 2008; Thakur et al. 2015). Similarly, the hydrophobic dendritic hypercores made from 4,4-bis(4-hydroxy phenyl) pentanol were coupled with PEG as a hydrophilic shell, thus forming water-soluble dendritic unimolecular micelles (Liu et al. 2000) (Fig. 5.7). Moreover, diamino butane poly(propylene imine) dendrimers were PEGylated to enhance their hydrophilicity and modify their encapsulating properties and stabilizing ability (Sideratou et al. 2001).

### 5.4.5 Water-Soluble PEGylated Polymeric Micelles

Polymeric micelles are spherical nanoparticles that are thermodynamically formed from copolymers. The hydrophobic ends of the copolymers intertwine and form a nucleus, whereas the hydrophilic ends are exposed on the surface. Polymeric micelles have potential pharmaceutical and biomedical applications that are similar to those of dendrimers (Aranda-Lara et al. 2021).

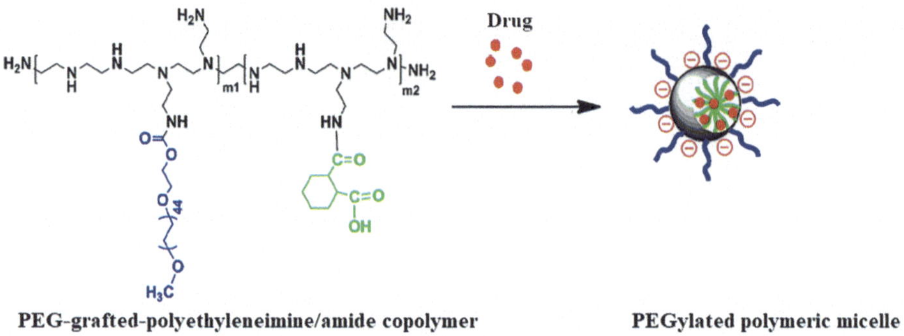


**Fig. 5.8** Formation of PEGylated polymeric micelles [(Liu et al. 2014), reproduced with permission]

Block copolymers containing PEG have been widely used and considered highly soluble in water. Coupling of PEG with polycationic polymers such as poly(L-lysine), polyspermine, and polyethylenimine has been done to improve their solubility, consequently forming PEG-coupled copolymers that are reported to be capable of forming self-assembling highly soluble complexes (Bhadra et al. 2002; Rackstraw et al. 2002) in the form of polymeric micelles. PEG-poly(lactic-*co*-glycolic acid) (PEG-PLGA) block copolymer forms an amphiphilic polymeric micelle in which PEG exists as a hydrophilic corona while PLGA serves as a hydrophobic core (Mozar and Chowdhury 2018). PEG and arginine-grafted reducible poly(disulfide amine) self-assembling micelles were also prepared (Nam et al. 2012). These polymeric micelles were suitable candidate nanocarriers for the delivery of the hydrophobic drug paclitaxel. PEGylated poly(β-benzyl L-aspartate) polymeric micelles were also prepared and showed a completely transparent aqueous solution (La et al. 1996).

In the other direction, a novel PEGylated polymer micelle formed from an amphiphilic block copolymer, PEG-b-poly(4-vinylbenzylphosphonate), spontaneously self-assemble in aqueous solutions into nanoparticles and was designed for bioimaging applications (Kamimura et al. 2011). Other novel PEGylated polymeric micelles (Fig. 5.8) composed of an amphiphilic copolymer and PEG-grafted-polyethyleneimine/amide were described for anticancer drug delivery (Liu et al. 2014).

## 5.5 Hydrated or Hydrophilic PEGylated Drug Nanocarriers

### 5.5.1 Hydrated PEGylated Lipid Nanocarriers

The degree of hydration on the membrane surface of liposomes plays a crucial role in their aggregation. The liposome shell membrane should be satisfactorily hydrated to reduce liposome aggregation and phagocytic cell uptake. The early attempts to increase liposomal membrane hydration were performed in the 1980s. This was accomplished by coating the membrane surface with hydrophilic polymers (Kraft et al. 2014). In 1995, the US Food and Drug Administration (FDA) approved the first liposome drug-delivery system for human use. It was marketed under the name of Doxil and constituted of PEGylated liposome-encapsulated doxorubicin (James 1995, p. 201).

Surface modification of lipid nanoparticles such as liposomes by PEGylation is performed in a similar manner to PEGylation of proteins (Beauchamp et al. 1984). PEGylated liposomes, also referred to as sterically stabilized or stealth liposomes, were first described by Allen and Chonn (1987). PEG is estimated to occupy an additional hydration thickness of 5 nm on the surface of the liposomes, which also depends on the length and density of the PEG polymer (Woodle et al. 1992) without significantly modifying the overall charge property of liposome membranes (Kraft et al. 2014).

### 5.5.2 Hydrophilic PEG-Coated Zein Nanocarriers

Zein is the main protein from corn and possesses a hydrophobic character. It is insoluble in water and classified as a GRAS substance (Irache and González-Navarro 2017; Penalva et al. 2015). Zein nanoparticles have the potential to be used for oral drug delivery. Nevertheless, the surface of zein nanoparticles is considered hydrophobic, thus denying them mucus-permeating properties. In an attempt to PEGylate the surface of these nanoparticles with PEG 35000, their surface hydrophobicity was significantly reduced, and the enhancement in hydrophilicity was associated with increased mobility in pig intestinal mucus. The hydrophobicity of the surface of the nanoparticles was inversely proportional to the PEG:zein ratio, as shown in Fig. 5.9. For example, the hydrophobicity of the nanoparticles prepared at a PEG:zein ratio of 5% and those prepared at + a ratio of 75% were 60% and 25% of that of the bare nanoparticles, respectively (Reboredo et al. 2021).

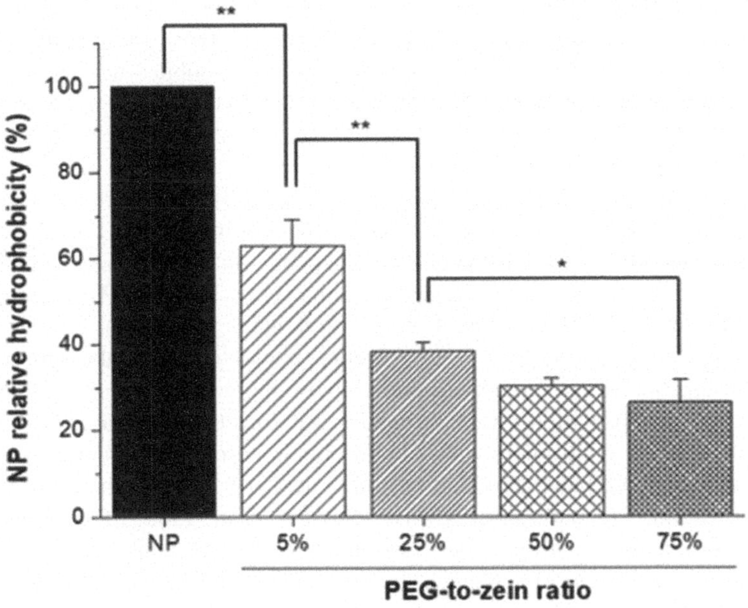


**Fig. 5.9** Surface hydrophobicity of PEGylated zein nanoparticles, normalized relative to that of the bare nanoparticles, as a function of the PEG:Zein ratio [(Reboredo et al. 2021), with permission]

## 5.6 Drug Solubilization Propensity of PEGylated Nanocarriers

As mentioned earlier, good water solubility is a prerequisite for the therapeutic utility of bioactive. Poor aqueous solubility limits the application or delivery of hydrophobic drugs as the release of the drug will be inadequate, consequently decreasing its bioavailability and therapeutic effect (Yiyun and Tongwen 2005; Shadrack et al. 2015; Prajapati et al. 2009; Sacchetti and Nejati 2012). It is estimated that 40% of the new chemical entities have low water solubility, which limits their clinical utility (Savjani et al. 2012; Sareen et al. 2012). There are various approaches to increase the solubility and bioavailability of drugs, such as comminution (particle size reduction), salt formation, complexation, co-crystallization, metastable forms, surfactants, and co-solvents. PEGylation of drugs and drug nanocarriers has emerged and is employed to enhance drug solubility (Fig. 5.3). Concerning this topic, most of the discussion hereby is dedicated to dendrimers as the literature mainly focuses on the PEGylation of this type of nanocarriers for modifying the solubility of hydrophobic drugs.

### 5.6.1 Survey of Nanocarriers Utilized in Hydrophobic Drug Solubility Enhancement

Complexation of drugs with dendrimers has been performed to enhance their solubility (Shadrack et al. 2018). In the late 1970s and early 1980s, the discovery of dendrimers (Vögtle 2003; Tomalia et al. 2012) provided the basis for utilizing PAMAM dendrimers in biomedical applications as solubility enhancers of poorly soluble actives (Shadrack et al. 2015; Prajapati et al. 2009). PEGylation of dendrimers can additionally increase the solubility of hydrophobic drugs carried on the dendrimers, among other improved properties (Gajbhiye et al. 2007; Luong et al. 2016; Yuan et al. 2010; Barraza et al. 2016; Jiang et al. 2010; Ryan et al. 2008; Thakur et al. 2015; Diaz et al. 2018).

Generally, dendrimer's ability to enhance drug solubility arises from their spherical shape, very low polydispersity, and accessible terminal groups (Gajbhiye et al. 2007). Their solubility enhancement effect depends on several factors, including their core (Hawker et al. 1993), branching units (Ooya et al. 2003), terminal end groups (Beezer et al. 2003), generation (Devarakonda et al. 2004), concentration, and pH (Asthana et al. 2005). At the molecular interaction level, the hydrophobic drug solubilization results from hydrophilic–hydrophobic interaction, ionic interaction, and drug encapsulation within the dendritic architecture (Sideratou et al. 2001). On the other hand, dendrimers with a hydrophobic core and hydrophilic terminal groups, such as PEGylated ones, demonstrate micellar behavior, which further aids in solubilization (Sideratou et al. 2001; Newkome et al. 1985; Stevelmans et al. 1996). Additionally, the drug can also be attached to PEG (Gajbhiye et al. 2007). PEGylation of dendrimers to enhance solubilization is considered superior to other surface modification polymers and approaches since it imparts a list of desirable properties (Fig. 5.3). Accordingly, the majority of the researchers utilized PEGylated dendrimers to achieve higher solubility associated with other enhanced properties such as biocompatibility (Liu et al. 1999; Nam et al. 2009; El-Sayed et al. 2002; Wolinsky and Grinstaff 2008; Wei et al. 2006) and controlled release ability (Tekade et al. 2009).

Many researchers have investigated the utility of PEGylated dendrimers in solubility enhancement and delivery of hydrophobic drugs. Water-soluble PEGylated 4,4-bis(4-hydroxy phenyl) pentanol dendrimers were found to solubilize drugs by forming dendritic unimolecular micelles. This effect, shown in Fig. 5.10, was demonstrated by pyrene solubilization in aqueous solution. The solubility of pyrene in water is very low ($8.0 \times 10^{-7}$ M), and a dramatic increase to $2.85 \times 10^{-4}$ M (356-fold) in an aqueous solution of $1.00 \times 10^{-4}$ M of generation 3 of the formed dendritic micelle was observed (Liu et al. 2000). Pyrene, as a probe, in addition to betamethasone valerate and betamethasone dipropionate, as active drug ingredients, was solubilized in PEGylated diamino butane-64-poly(propylene imine) dendrimers (Sideratou et al. 2001). The PEGylated dendrimers showed superior drug solubility to that of the bare dendrimers. Besides, PEGylated PAMAM dendrimers resulted in solubility enhancement of simvastatin of 33 folds in comparison to a 23 and 17.5 fold increase in solubility brought about by amine and hydroxyl functionalized dendrimers (Kulhari et al. 2011).

Table 5.2 shows the observed effects on the solubility of these drugs, along with other examples.

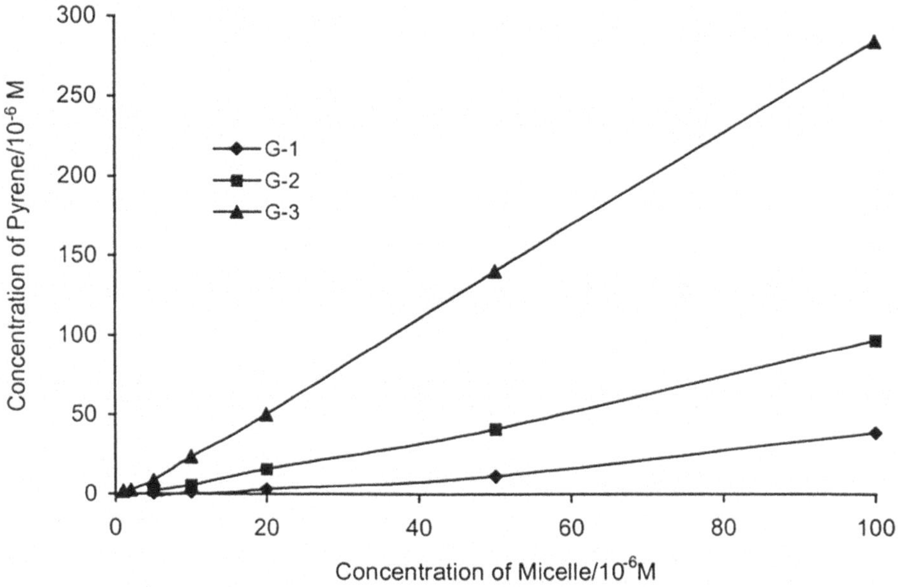


**Fig. 5.10** Pyrene solubilization by PEGylated 4,4-bis(4-hydroxy phenyl) pentanol dendrimer as a function of dendrimer generation and concentration [(Liu et al. 2000), with permission]

**Table 5.2** PEGylated dendrimer mediated solubility enhancement of drugs

| Dendrimer | Drug/probe | Solubility of drug/probe | | | References |
|---|---|---|---|---|---|
| | | Alone | In parent dendrimer | In PEGylated dendrimer | |
| 4,4-bis(4-hydroxy phenyl) pentanol dendrimers | Pyrene | $8.0 \times 10^{-7}$ M | – | $2.85 \times 10^{-4}$ M | Liu et al. (2000) |
| Poly(propylene imine) dendrimers | Imatinib | The fifth generation PEGylated dendrimer increased the aqueous solubility of the drug (no solubility data was mentioned) | | | Karthikeyan and Vijayarajkumar (2015) |
| Diaminobutane-64-poly(propylene imine) dendrimers | Pyrene | – | $2.15 \times 10^{-6}$ M | $5.40 \times 10^{-5}$ M | Sideratou et al. (2001) |
| | Betamethasone valerate | – | $2.95 \times 10^{-5}$ M | $3.85 \times 10^{-4}$ M | |
| | Betamethasone dipropionate | – | $1.84 \times 10^{-5}$ M | $2.56 \times 10^{-4}$ M | |
| PAMAM dendrimers | Pyrene | $6.44 \times 10^{-7}$ M | $6.8 \times 10^{-7}$ M | $1.52 \times 10^{-6}$ M | Yang et al. (2004) |
| | Silybin | 1280 μmol $L^{-1}$ | 3380 μmol $L^{-1}$ | 6300 μmol $L^{-1}$ | Diaz et al. (2018) |
| | Simvastatin | 33.4 μM/L | – | 1093.25 μM/L | Kulhari et al. (2011) |
| | Paclitaxel | 0.3 μg/mL | – | 3.2 mg/mL[a] | Khandare et al. (2006) |
| | Probucol | 5 ng $mL^{-1}$ | –[b] | ≈6.5 ng $mL^{-1}$ | Qi et al. (2015) |
| | 5-Fluorouracil | Enhanced solubilization was evident by 12 folds improved entrapment efficiency | | | Bhadra et al. (2003) |
| | Methotrexate Adriamycin | Improved encapsulation ability | | | Kojima et al. (2000) |
| Lysine dendrimers | Artemether | Improved entrapment efficiency | | | Bhadra et al. (2005) |

| | | | | | |
|---|---|---|---|---|---|
| Citric acid dendrimers | 5-Amino salicylic acid<br>Pyridine<br>Mefenamic acid<br>diclofenac | Improved entrapment efficiency | | | Namazi and Adeli (2005) |
| Polyglycerol dendrimers | Paclitaxel | 0.0003 mg/mL | – | 2.305 mg/mL | Ooya et al. (2003) |

[a] Solubility of paclitaxel–bis (PEG) conjugate was 2.5 mg/mL

[b] Solubility of probucol in $NH_2$–functionalized dendrimers was around 2.8 $ng{\cdot}mL^{-1}$

PEG 4000-modified liposomes were designed to increase the aqueous solubility of quercetin, which is an anticancer drug that also possesses antioxidant, antithrombotic, antihypertensive, angio-protective, and anti-inflammatory effects. Quercetin has a low water solubility of 1.53 ± 0.27 μg/mL, and the PEGylated liposomes increased this value by 2.2 folds (Demirbolat et al. 2022). In another attempt, PEGylated liposomes were prepared to enhance the solubility of paclitaxel, but the PEGylated formulation showed similar solubility to that of the parent liposomes (Yang et al. 2007b).

### 5.6.2 Proposed Mechanisms of Enhanced Drug Solubilization Mediated By PEGylation of Nanocarriers

The fundamental mechanisms of the solubility enhancement effect of PEG were mentioned earlier (refer to Sect. 5.1.3). Regarding enhanced solubility of drugs carried on nanoparticles, several mechanisms are recognized. The occurrence of these mechanisms depends on the designed PEGylated nanosystem.

Generally, in non-PEGylated dendrimers, the drug could be entrapped inside the core of the dendritic structure. Noncovalent bonding, such as hydrogen bonds, hydrophobic interactions, and electrostatic interactions, is involved in the entrapment process. Interaction of the drug with the periphery of the dendrimer could also occur via covalent bond formation (Choudhary et al. 2017; Menjoge et al. 2010; Milhem et al. 2000; Bhadra et al. 2005). The former effect may further induce the formation of unimolecular micelle or the so-called "dendritic box" (Menjoge et al. 2010). The overall result of these effects will be an enhancement of the solubility of the loaded drug.

It has been proposed that PEGylation of the dendrimers will add to the above-mentioned effects by providing a PEG coat in which the drug can also be solubilized (Sideratou et al. 2001; Yang et al. 2004) by interaction with the surface arms of PEG (Diaz et al. 2018; Kojima et al. 2000). This effect is possibly mediated by the enhanced complexation between the drug and the dendrimer by steric and electronic effects provided by the additional functional groups of PEG (Bhadra et al. 2003). On the other hand, PEG results in the formation of unimolecular micelles capable of increasing the solubilization of the hydrophobic drug (Yang et al. 2004; Gajbhiye et al. 2007; Bhadra et al. 2003; Liu et al. 2000). Illustrative examples of enhanced drug solubilization by dual effects of the formation of unimolecular micelles and enhanced complexation with the PEG coat in PEGylated dendrimers are shown in Figs. 5.11 and 5.12. The formation of multimolecular micelles from PEGylated dendrimers was also suggested to play a role in drug solubility enhancement (Wang et al. 2022) as shown in Fig. 5.13.

PEGylated polymeric micelles work similarly to PEGylated dendrimers. Solubilization by the formation of polymeric or copolymeric PEG micelles occurs when macromolecular PEG polymers or copolymers are added above their corresponding CAC, as they readily self-assemble to form micellar-like, nanosized, thermodynamically stable systems with core–shell structures (Hagan et al. 1995;

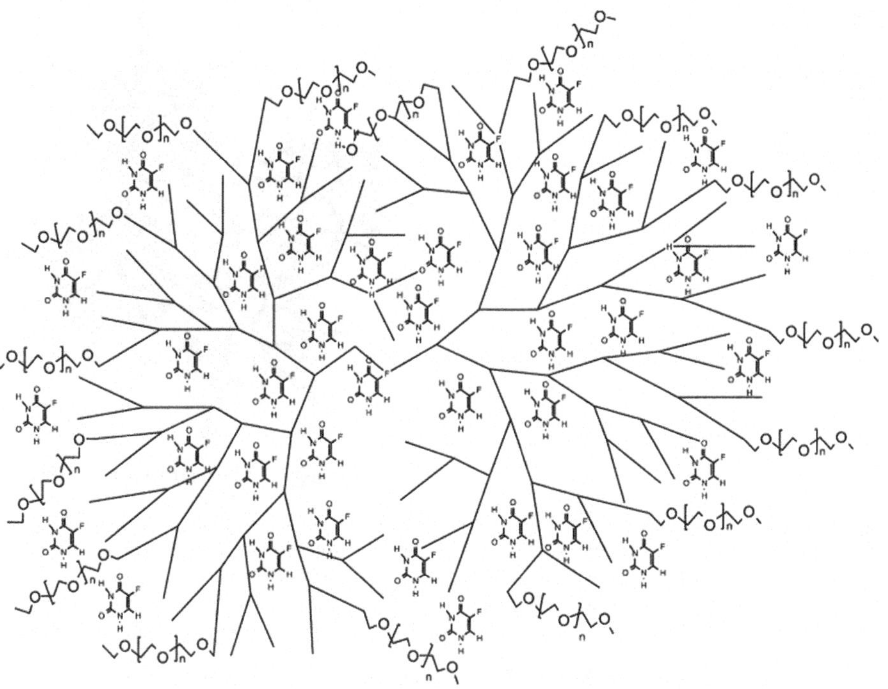

**Fig. 5.11** Solubilization of 5-fluorouracil in PEGelated dendrimers [(Bhadra et al. 2003), with permission]

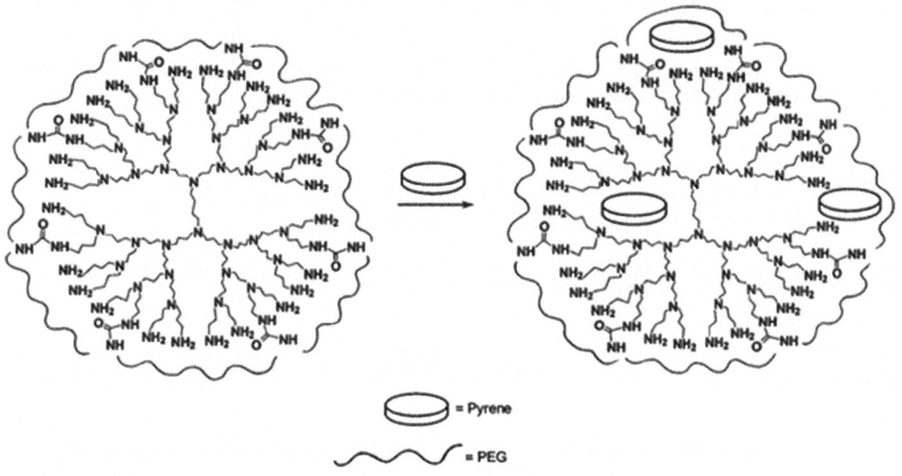


**Fig. 5.12** Solubilization of pyrene in PEGelated dendrimers [(Sideratou et al. 2001), with permission]

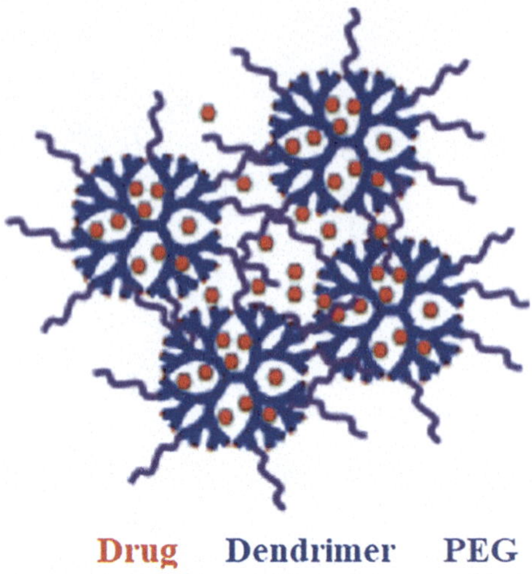


**Fig. 5.13** Drug solubilization in PEGelated dendrimers by the formation of multimolecular micelles [(Wang et al. 2022), reproduced with permission]

Kazunori et al. 1993; Otsuka et al. 2003). These micellar systems can increase the solubilization of hydrophobic drugs (Otsuka et al. 2003).

### 5.6.3 Added Advantages of Drug Solubility Enhancement by Nanocarriers PEGylation

An added advantage of dendrimer PEGylation in solubility enhancement is the improved drug loading. The encapsulation of the drug in the dendrimers will increase since the PEG chains will aid in loading additional drug molecules within PEG chains. For example, the enhanced encapsulation of 5-fluorouracil using PEGylated PAMAM dendrimers was reported (Bhadra et al. 2003).

### 5.6.4 Factors Affecting Drug Solubilization Propensity of PEGylated Dendritic Nanocarriers

#### 5.6.4.1 PEG Chain Length/Molecular Weight

In general, longer PEG chains, i.e., chains with higher molecular weight, are expected to exert a better solubility enhancement effect. Attachment of shorter PEG arms provides little improvement of drug solubility in comparison to the parent dendrimers.

In generation three PEGyated PAMAM dendrimers, the PEG arm length was found to significantly affect pyrene water solubility (Fig. 5.14). The short PEG arms (PEG 750) did not substantially change solubility compared to the parent dendrimers. It was hypothesized that these short chains produce a relatively open cavity due to weak interaction. Increasing the PEG arm length to 2000 Da led to a noticeable increase in pyrene solubility. This effect was attributed to forming a network of a thick

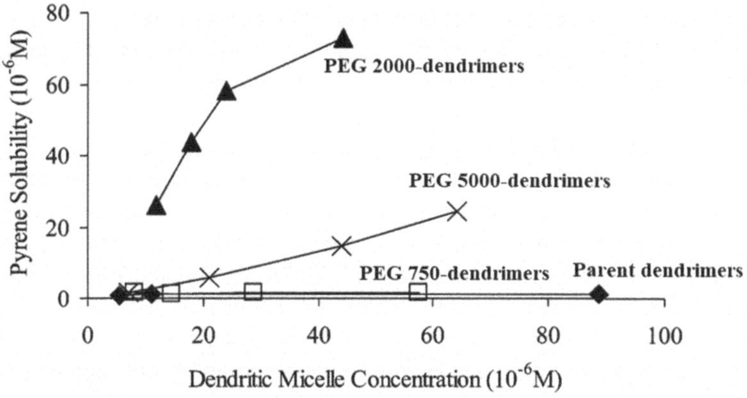


**Fig. 5.14** Effect of PEG chain length and dendrimer concentration on pyrene solubility [(Yang et al. 2004), reproduced with permission]

PEG layer at the surface of the dendrimer, within which pyrene can bind. These longer PEG chains gave each unimolecular micelle an isolated space to encapsulate the drug. However, PEG 5000 showed a lower capability for solubility enhancement compared to that of PEG 2000, and this could be a result of the higher tendency of the formed unimolecular micelles to aggregate, thus shielding some of the potential sites available for drug binding because of PEG chains entanglement (Yang et al. 2004).

Partially PEGylated generation four PAMAM dendrimers were used to enhance the water solubility of silybin. The longer PEG chain (PEG 2000) resulted in a higher increase in silybin solubility compared to the shorter chain (PEG 550) (Diaz et al. 2018). Similarly, PAMAM dendrimers PEGylated with PEG 2000 were found to encapsulate higher number of molecules of adriamycin and methotrexate compared to those PEGylated with PEG 550. This effect was justified by the inability of the shorter PEG chain to cover and stabilize the drug in addition to the ability of the longer chains to modify the conformation of the dendrimer chain, making the inner space of the dendrimer larger (Kojima et al. 2000).

#### 5.6.4.2 Extent/Density/Substitution Percentage of PEGylation

Increasing the PEGylation extent is supposed to provide better drug solubility enhancement. Nevertheless, the number of PEG chains attached to the surface of the dendrimer should be carefully controlled in order to preserve the flexibility of the nanosystem, which is an essential characteristic of using dendrimers in drug delivery. Moreover, crowded PEG arms may result in steric hindrance that reduces further coupling with other small molecules of interest to the dendrimer surface. To illustrate the effect of PEGylation extent, diamino butane-64-poly(propylene imine) dendrimers were PEGylated at two different levels. The densely PEGylated dendrimers showed a better ability to solubilize pyrene, betamethasone valerate, and betamethasone dipropionate (Sideratou et al. 2001). Paclitaxel solubility was also increased with increasing the generation of polyglycerol dendrimers, generations 3, 4, and 5 (Ooya et al. 2003).

#### 5.6.4.3 Generation and Concentration of the Dendrimer

Similar to non-PEGylated dendrimers, the ability of the PEGylated derivatives to enhance hydrophobic drug solubility increases with increasing dendrimer generation and concentration.

As shown in Fig. 5.10, PEGylated 4,4-bis(4-hydroxy phenyl) pentanol dendrimers of generation three were found superior to generation two in solubilizing pyrene. The latter was also better at solubilizing pyrene than generation one. Similarly, the effect of increasing the corresponding dendrimer concentration on the enhanced drug solubility is evident (Liu et al. 2000). PEGylated lysine dendrimers of generation 5 solubilized 0.411 ± 0.11 g of artemether/g of dendrimers compared to 0.35 ± 0.08 g/g of generation 4 dendrimers (Bhadra et al. 2005). The solubility of pyrene increases with increasing concentration of generation 3 PEGyated PAMAM dendrimers, as shown in Fig. 5.14 (Yang et al. 2004).

It is worth mentioning that even though higher generations of the dendrimers are associated with higher drug solubilization of higher generation as they have a higher tendency to entrap more hydrophobic drug molecules, PEGylation of the lower generation dendrimers may cause an enhanced drug solubility compared to the parent dendrimers. PEGylation will induce stretching out of the surface sites of the dendrimers as a result of the PEG arms. Consequently, the dendritic inner space will expand to the maximum in the aqueous phase providing a space for drug binding (Yang et al. 2004).

## 5.7 Conclusion and Future Perspective

The wide range of promising potential pharmaceutical and biomedical applications of PEGylated nanocarriers provides many advantageous criteria. The advantageous enhanced solubility of these systems and the drugs they carry necessitate further efforts to specifically approach solubility issues of different types of nanocarriers. Pharmaceutical solubilization of nanocarriers by PEGylation compromises a rich area of research. Within the same context, protein PEGylation as a means to solubilize enzymes and proteins for facilitated nanoparticle production has also emerged.

This chapter discussed the amount of knowledge available in the literature pertaining to different aspects of solubility and solubility-related issues of PEGylated nanocarriers. In most of the illustrated research, general approaches were employed with no focus on detailed aspects. Specific focus on the factors affecting the solubilization, in addition to the relation between the PEGylated nanocarriers' solubility and the other advantages offered by them, is worthy of investigation.

**Disclosures** There is no conflict of interest and disclosures associated with the manuscript.

## References

Abuchowski A, McCoy JR, Palczuk NC, van Es T, Davis FF (1977) Effect of covalent attachment of polyethylene glycol on immunogenicity and circulating life of bovine liver catalase. J Biol Chem 252:3582–3586

Albanese A, Tang PS, Chan WC (2012) The effect of nanoparticle size, shape, and surface chemistry on biological systems. Annu Rev Biomed Eng 14:1–16

Alivisatos P (2004) The use of nanocrystals in biological detection. Nat Biotechnol 22:47–52

Allen T, Chonn A (1987) Large unilamellar liposomes with low uptake into the reticuloendothelial system. FEBS Lett 223:42–46

Aranda-Lara L, García BEO, Isaac-Olivé K, Ferro-Flores G, Meléndez-Alafort L, Morales-Avila E (2021) Drug delivery systems-based dendrimers and polymer micelles for nuclear diagnosis and therapy. Macromol Biosci 21:2000362

Asthana A, Chauhan AS, Diwan PV, Jain NK (2005) Poly(amidoamine) (PAMAM) dendritic nanostructures for controlled sitespecific delivery of acidic anti-inflammatory active ingredient. AAPS PharmSciTech 6:E536–E542

Awasthi V, Garcia D, Klipper R, Goins BA, Phillips WT (2004) Neutral and anionic liposome-encapsulated hemoglobin: effect of postinserted poly(ethylene glycol)-distearoylphosphatidylethanolamine on distribution and circulation kinetics. J Pharmacol Exp Ther 309:241–248

Bailey FE (1990) Alkylene oxides and their polymers. CRC Press, Boca Raton

Balasubramanian K, Burghard M (2010) Carbon nanotubes: methods and protocols. Springer, Berlin

Banerjee SS, Chen D-H (2007) Magnetic nanoparticles grafted with cyclodextrin for hydrophobic drug delivery. Chem Mater 19:6345–6349

Barraza LF, Jiménez VA, Alderete JB (2016) Methotrexate complexation with native and PEGylated PAMAM-G4: effect of the PEGylation degree on the drug loading capacity and release kinetics. Macromol Chem Phys 217:605–613

Beauchamp C, Daddona PE, Menapace DP (1984) Properties of a novel PEG derivative of calf adenosine deaminase. In: Purine metabolism in man-IV: part A: clinical and therapeutic aspects; regulatory mechanisms. Springer, Berlin

Beezer A, King A, Martin I, Mitchel J, Twyman L, Wain C (2003) Dendrimers as potential drug carriers; encapsulation of acidic hydrophobes within water soluble PAMAM derivatives. Tetrahedron 59:3873–3880

Bhadra D, Bhadra S, Jain P, Jain N (2002) Pegnology: a review of PEG-ylated systems. Pharmazie 57:5–29

Bhadra D, Bhadra S, Jain S, Jain N (2003) A PEGylated dendritic nanoparticulate carrier of fluorouracil. Int J Pharm 257:111–124

Bhadra D, Bhadra S, Jain N (2005) PEGylated peptide-based dendritic nanoparticulate systems for delivery of artemether. J Drug Deliv Sci Technol 15:65–73

Bottini M, Rosato N, Bottini N (2011) PEG-modified carbon nanotubes in biomedicine: current status and challenges ahead. Biomacromolecules 12:3381–3393

Bruchez M Jr, Moronne M, Gin P, Weiss S, Alivisatos AP (1998) Semiconductor nanocrystals as fluorescent biological labels. Science 281:2013–2016

Camerin M, Moreno M, Marín MJ, Schofield CL, Chambrier I, Cook MJ, Coppellotti O, Jori G, Russell DA (2016) Delivery of a hydrophobic phthalocyanine photosensitizer using PEGylated gold nanoparticle conjugates for the in vivo photodynamic therapy of amelanotic melanoma. Photochem Photobiol Sci 15:618–625

Carrea G, Riva S (2000) Properties and synthetic applications of enzymes in organic solvents. Angew Chem Int Ed 39:2226–2254

Castellanos IJ, Al-Azzam W, Griebenow K (2005) Effect of the covalent modification with poly(ethylene glycol) on α-chymotrypsin stability upon encapsulation in poly(lactic-co-glycolic) microspheres. J Pharm Sci 94:327–340

Castillo B, Méndez J, Al-Azzam W, Barletta G, Griebenow K (2006) On the relationship between the activity and structure of PEG-α-chymotrypsin conjugates in organic solvents. Biotechnol Bioeng 94:565–574

Castillo B, Solá RJ, Ferrer A, Barletta G, Griebenow K (2008) Effect of PEG modification on subtilisin Carlsberg activity, enantioselectivity, and structural dynamics in 1,4-dioxane. Biotechnol Bioeng 99:9–17
Chen J, Spear SK, Huddleston JG, Rogers RD (2005) Polyethylene glycol and solutions of polyethylene glycol as green reaction media. Green Chem 7:64–82
Cheng Y, Samia AC, Meyers JD, Panagopoulos I, Fei B, Burda C (2008) Highly efficient drug delivery with gold nanoparticle vectors for in vivo photodynamic therapy of cancer. J Am Chem Soc 130:10643–10647
Cheng Y, Meyers JD, Agnes RS, Doane TL, Kenney ME, Broome AM, Burda C, Basilion JP (2011a) Addressing brain tumors with targeted gold nanoparticles: a new gold standard for hydrophobic drug delivery? Small 7:2301–2306
Cheng Y, Meyers JD, Broome A-M, Kenney ME, Basilion JP, Burda C (2011b) Deep penetration of a PDT drug into tumors by noncovalent drug-gold nanoparticle conjugates. J Am Chem Soc 133:2583–2591
Chiou WL, Riegelman S (1971) Pharmaceutical applications of solid dispersion systems. J Pharm Sci 60:1281–1302
Choi J-S, Jo B-W (2004) Enhanced paclitaxel bioavailability after oral administration of pegylated paclitaxel prodrug for oral delivery in rats. Int J Pharm 280:221–227
Choudhary S, Gupta L, Rani S, Dave K, Gupta U (2017) Impact of dendrimers on solubility of hydrophobic drug molecules. Front Pharmacol 8:261
Choudhury H, Pandey M, Chin PX, Phang YL, Cheah JY, Ooi SC, Mak K-K, Pichika MR, Kesharwani P, Hussain Z (2018) Transferrin receptors-targeting nanocarriers for efficient targeted delivery and transcytosis of drugs into the brain tumors: a review of recent advancements and emerging trends. Drug Deliv Transl Res 8:1545–1563
D'Emanuele A, Attwood D (2005) Dendrimer–drug interactions. Adv Drug Deliv Rev 57:2147–2162
D'Souza AA, Shegokar R (2016) Polyethylene glycol (PEG): a versatile polymer for pharmaceutical applications. Expert Opin Drug Deliv 13:1257–1275
Davis FF, van Es T, Palczuk NC (1979) Non-immunogenic polypeptides. Google Patents
Demirbolat GM, Erdoğan Ö, Coşkun GP, Çevik Ö (2022) PEG4000 modified liposomes enhance the solubility of quercetin and improve the liposome functionality: in vitro characterization and the cellular efficacy. Turk J Chem 46:1011–1023
Devarakonda B, Hill RA, De Villiers MM (2004) The effect of PAMAM dendrimer generation size and surface functional group on the aqueous solubility of nifedipine. Int J Pharm 284:133–140
Diaz C, Guzmán J, Jiménez VA, Alderete JB (2018) Partially PEGylated PAMAM dendrimers as solubility enhancers of silybin. Pharm Dev Technol 23:689–696
Elbert DL, Hubbell JA (1996) Surface treatments of polymers for biocompatibility. Annu Rev Mater Sci 26:365–394
Elhissi A, Ahmed W, Hassan IU, Dhanak V, D'Emanuele A (2012) Carbon nanotubes in cancer therapy and drug delivery. J Drug Deliv 2012:837327
El-Sayed M, Ginski M, Rhodes C, Ghandehari H (2002) Transepithelial transport of poly(amidoamine) dendrimers across Caco-2 cell monolayers. J Control Release 81:355–365
Erogbogbo F, Yong K-T, Roy I, Xu G, Prasad PN, Swihart MT (2008) Biocompatible luminescent silicon quantum dots for imaging of cancer cells. ACS Nano 2:873–878
Freichels H, Jérôme R, Jérôme C (2011) Sugar-labeled and PEGylated (bio) degradable polymers intended for targeted drug delivery systems. Carbohydr Polym 86:1093–1106
Fung W, Porter J, Bailon P (1997) Strategies for the preparation and characterization of polyethylene glycol (PEG) conjugated pharmaceutical proteins. Polym Prepr 38:565–566
Gajbhiye V, Vijayaraj Kumar P, Kumar Tekade R, Jain N (2007) Pharmaceutical and biomedical potential of PEGylated dendrimers. Curr Pharm Des 13:415–429
Gao X, Guo L, Li J, Thu HE, Hussain Z (2018) Nanomedicines guided nanoimaging probes and nanotherapeutics for early detection of lung cancer and abolishing pulmonary metastasis: critical appraisal of newer developments and challenges to clinical transition. J Control Release 292:29–57

Gillich T, Acikgöz C, Isa L, Schlüter AD, Spencer ND, Textor M (2013) PEG-stabilized core–shell nanoparticles: impact of linear versus dendritic polymer shell architecture on colloidal properties and the reversibility of temperature-induced aggregation. ACS Nano 7:316–329

González M, Vaillard SE, Grau RJ (2011) Pegylation: an overview and recent advances reported in the patent literature. Rec Patents Chem Eng 4:241–264

Gorain B, Choudhury H, Pandey M, Kesharwani P, Abeer MM, Tekade RK, Hussain Z (2018) Carbon nanotube scaffolds as emerging nanoplatform for myocardial tissue regeneration: a review of recent developments and therapeutic implications. Biomed Pharmacother 104:496–508

Greenwald RB, Pendri A, Bolikal D (1995) Highly water soluble taxol derivatives: 7-polyethylene glycol carbamates and carbonates. J Org Chem 60:331–336

Gullapalli RP, Mazzitelli CL (2015) Polyethylene glycols in oral and parenteral formulations—a critical review. Int J Pharm 496:219–239

Hadidi N, Kobarfard F, Nafissi-Varcheh N, Aboofazeli R (2011) Optimization of single-walled carbon nanotube solubility by noncovalent PEGylation using experimental design methods. Int J Nanomedicine 6:737–746

Hagan S, Davis S, Illum L, Davies M, Garnett M, Taylor D, Irving M, Tadros TF (1995) Estimation of the poly(ethylene glycol) chain length of L-polylactide-polyethylene glycol in aqueous dispersions using viscoelastic measurements. Langmuir 11:1482–1485

Hanif M, Ameer N, Ahmad Q-U-A, Aziz M, Mahmood K, Ramzan N, Abdur Rahman HM (2022) Improved solubility and corneal permeation of PEGylated curcumin complex used for the treatment of ophthalmic bacterial infections. PLoS One 17:e0258355

Harris JM (1992) Poly(ethylene glycol) chemistry: biotechnical and biomedical applications. Springer Science & Business Media, Berlin

Hawker CJ, Wooley KL, Fréchet JM (1993) Unimolecular micelles and globular amphiphiles: dendritic macromolecules as novel recyclable solubilization agents. J Chem Soc Perkin Trans 1:1287–1297

He Y, Kang ZH, Li QS, Tsang CHA, Fan CH, Lee ST (2009) Ultrastable, highly fluorescent, and water-dispersed silicon-based nanospheres as cellular probes. Angew Chem Int Ed 48:128–132

Hernaiz M, Sánchez-Montero J, Sinisterra J (1997) Influence of the nature of modifier in the enzymatic activity of chemical modified semipurified lipase from Candida rugosa. Biotechnol Bioeng 55:252–260

Ho RJY, Gibaldi M (2013) Biotechnology and biopharmaceuticals, 2nd edn. Wiley, New York

Hong C-Y, You Y-Z, Pan C-Y (2006) A new approach to functionalize multi-walled carbon nanotubes by the use of functional polymers. Polymer 47:4300–4309

Howard MD, Jay M, Dziubla TD, Lu X (2008) PEGylation of nanocarrier drug delivery systems: state of the art. J Biomed Nanotechnol 4:133–148

Hussain Z, Thu HE, Amjad MW, Hussain F, Ahmed TA, Khan S (2017a) Exploring recent developments to improve antioxidant, anti-inflammatory and antimicrobial efficacy of curcumin: a review of new trends and future perspectives. Mater Sci Eng C 77:1316–1326

Hussain Z, Thu HE, Ng S-F, Khan S, Katas H (2017b) Nanoencapsulation, an efficient and promising approach to maximize wound healing efficacy of curcumin: a review of new trends and state-of-the-art. Colloids Surf B: Biointerfaces 150:223–241

Hussain Z, Arooj M, Malik A, Hussain F, Safdar H, Khan S, Sohail M, Pandey M, Choudhury H, Ei Thu H (2018) Nanomedicines as emerging platform for simultaneous delivery of cancer therapeutics: new developments in overcoming drug resistance and optimizing anticancer efficacy. Artif Cells Nanomed Biotechnol 46:1015–1024

Hussain Z, Khan S, Imran M, Sohail M, Shah SWA, De Matas M (2019) PEGylation: a promising strategy to overcome challenges to cancer-targeted nanomedicines: a review of challenges to clinical transition and promising resolution. Drug Deliv Transl Res 9:721–734

Inada Y, Takahashi K, Yoshimoto T, Ajima A, Matsushima A, Saito Y (1986) Application of polyethylene glycol-modified enzymes in biotechnological processes: organic solvent-soluble enzymes. Trends Biotechnol 4:190–194

Irache JM, González-Navarro CJ (2017) Zein nanoparticles as vehicles for oral delivery purposes. Fut Med 12:1209
Israelachvili J (1997) The different faces of poly(ethylene glycol). Proc Natl Acad Sci 94:8378–8379
Jain DC, Mathur AK, Gupta MM, Singh AK, Verma RK, Gupta AP, Kumar S (1996) Isolation of high artemisinin-yielding clones of Artemisia annua. Phytochemistry 43:993–1001
James J (1995) DOXIL approved by FDA. AIDS Patient Care 9:306
Jiang Y-Y, Tang G-T, Zhang L-H, Kong S-Y, Zhu S-J, Pei Y-Y (2010) PEGylated PAMAM dendrimers as a potential drug delivery carrier: in vitro and in vivo comparative evaluation of covalently conjugated drug and noncovalent drug inclusion complex. J Drug Target 18:389–403
Jokerst JV, Lobovkina T, Zare RN, Gambhir SS (2011) Nanoparticle PEGylation for imaging and therapy. Nanomedicine 6:715–728
Joralemon MJ, McRae S, Emrick T (2010) PEGylated polymers for medicine: from conjugation to self-assembled systems. Chem Commun 46:1377–1393
Kallumadil M, Tada M, Nakagawa T, Abe M, Southern P, Pankhurst QA (2009) Corrigendum to "Suitability of commercial colloids for magnetic hyperthermia" [J. Magn. Magn. Mater. 321 (2009) 1509-1513]. J Magn Magn Mater 321:3650–3651
Kamimura M, Kanayama N, Tokuzen K, Soga K, Nagasaki Y (2011) Near-infrared (1550 nm) in vivo bioimaging based on rare-earth doped ceramic nanophosphors modified with PEG-b-poly(4-vinylbenzylphosphonate). Nanoscale 3:3705–3713
Karakoti AS, Das S, Thevuthasan S, Seal S (2011) PEGylated inorganic nanoparticles. Angew Chem Int Ed 50:1980–1994
Karra N, Benita S (2012) The ligand nanoparticle conjugation approach for targeted cancer therapy. Curr Drug Metab 13:22–41
Karthikeyan R, Vijayarajkumar P (2015) PEGylated nanoarchitechture mediated solubility enhancement of tyrosine-kinase inhibitor. Inventi Rapid: Novel Excipients 2:1–4
Kazunori K, Masayuki Y, Teruo O, Yasuhisa S (1993) Block copolymer micelles as vehicles for drug delivery. J Control Release 24:119–132
Khan S, Imran M, Butt TT, Shah SWA, Sohail M, Malik A, Das S, Thu HE, Adam A, Hussain Z (2018) Curcumin based nanomedicines as efficient nanoplatform for treatment of cancer: new developments in reversing cancer drug resistance, rapid internalization, and improved anticancer efficacy. Trends Food Sci Technol 80:8–22
Khandare JJ, Jayant S, Singh A, Chandna P, Wang Y, Vorsa N, Minko T (2006) Dendrimer versus linear conjugate: influence of polymeric architecture on the delivery and anticancer effect of paclitaxel. Bioconjug Chem 17:1464–1472
Kimura M, Kato M, Muto T, Hanabusa K, Shirai H (2000) Temperature-sensitive dendritic hosts: synthesis, characterization, and control of catalytic activity. Macromolecules 33:1117–1119
Knop K, Hoogenboom R, Fischer D, Schubert US (2010) Poly(ethylene glycol) in drug delivery: pros and cons as well as potential alternatives. Angew Chem Int Ed 49:6288–6308
Kojima C, Kono K, Maruyama K, Takagishi T (2000) Synthesis of polyamidoamine dendrimers having poly(ethylene glycol) grafts and their ability to encapsulate anticancer drugs. Bioconjug Chem 11:910–917
Kraft JC, Freeling JP, Wang Z, Ho RJ (2014) Emerging research and clinical development trends of liposome and lipid nanoparticle drug delivery systems. J Pharm Sci 103:29–52
Kulhari H, Pooja D, Prajapati S, Chauhan AS (2011) Performance evaluation of PAMAM dendrimer based simvastatin formulations. Int J Pharm 405:203–209
Kwon OH, Imanishi Y, Ito Y (1999) Catalytic activity and conformation of chemically modified subtilisin Carlsberg in organic media. Biotechnol Bioeng 66:265–270
La SB, Okano T, Kataoka K (1996) Preparation and characterization of the micelle-forming polymeric drug indomethacin-incorporated poly(ethylene oxide)–poly(β-benzyl L-aspartate) block copolymer micelles. J Pharm Sci 85:85–90
Lacerda L, Bianco A, Prato M, Kostarelos K (2006) Carbon nanotubes as nanomedicines: from toxicology to pharmacology. Adv Drug Deliv Rev 58:1460–1470

Laurent S, Forge D, Port M, Roch A, Robic C, Vander Elst L, Muller RN (2008) Magnetic iron oxide nanoparticles: synthesis, stabilization, vectorization, physicochemical characterizations, and biological applications. Chem Rev 108:2064–2110

Lee CC, Mackay JA, Fréchet JM, Szoka FC (2005) Designing dendrimers for biological applications. Nat Biotechnol 23:1517–1526

Li Z, Ruckenstein E (2004) Water-soluble poly(acrylic acid) grafted luminescent silicon nanoparticles and their use as fluorescent biological staining labels. Nano Lett 4:1463–1467

Liu M, Kono K, Fréchet JM (1999) Water-soluble dendrimer–poly(ethylene glycol) starlike conjugates as potential drug carriers. J Polym Sci A Polym Chem 37:3492–3503

Liu M, Kono K, Fréchet JM (2000) Water-soluble dendritic unimolecular micelles: their potential as drug delivery agents. J Control Release 65:121–131

Liu Z, Sun X, Nakayama-Ratchford N, Dai H (2007) Supramolecular chemistry on water-soluble carbon nanotubes for drug loading and delivery. ACS Nano 1:50–56

Liu A, Sun K, Yang J, Zhao D (2008) Toxicological effects of multi-wall carbon nanotubes in rats. J Nanopart Res 10:1303–1307

Liu GY, Li M, Zhu CS, Jin Q, Zhang ZC, Ji J (2014) Charge-conversional and pH-sensitive PEGylated polymeric micelles as efficient nanocarriers for drug delivery. Macromol Biosci 14:1280–1290

Lu AH, Salabas EEL, Schüth F (2007) Magnetic nanoparticles: synthesis, protection, functionalization, and application. Angew Chem Int Ed 46:1222–1244

Luo D, Haverstick K, Belcheva N, Han E, Saltzman WM (2002) Poly(ethylene glycol)-conjugated PAMAM dendrimer for biocompatible, high-efficiency DNA delivery. Macromolecules 35:3456–3462

Luong D, Kesharwani P, Deshmukh R, Amin MCIM, Gupta U, Greish K, Iyer AK (2016) PEGylated PAMAM dendrimers: enhancing efficacy and mitigating toxicity for effective anticancer drug and gene delivery. Acta Biomater 43:14–29

Madani SY, Naderi N, Dissanayake O, Tan A, Seifalian AM (2011) A new era of cancer treatment: carbon nanotubes as drug delivery tools. Int J Nanomedicine 6:2963–2979

Martinez-Oharriz M, Martin C, Goñi M, Rodriguez-Espinosa C, Tros-Ilarduya M, Zornoza A (1999) Influence of polyethylene glycol 4000 on the polymorphic forms of diflunisal. Eur J Pharm Sci 8:127–132

Md S, Kuldeep Singh JKAP, Waqas M, Pandey M, Choudhury H, Habib H, Hussain F, Hussain Z (2019) Nanoencapsulation of betamethasone valerate using high pressure homogenization–solvent evaporation technique: optimization of formulation and process parameters for efficient dermal targeting. Drug Dev Ind Pharm 45:323–332

Menjoge AR, Kannan RM, Tomalia DA (2010) Dendrimer-based drug and imaging conjugates: design considerations for nanomedical applications. Drug Discov Today 15:171–185

Meyers JD, Cheng Y, Broome AM, Agnes RS, Schluchter MD, Margevicius S, Wang X, Kenney ME, Burda C, Basilion JP (2015) Peptide-targeted gold nanoparticles for photodynamic therapy of brain cancer. Part Part Syst Charact 32:448–457

Michalet X, Pinaud FF, Bentolila LA, Tsay JM, Doose S, Li JJ, Sundaresan G, Wu A, Gambhir S, Weiss S (2005) Quantum dots for live cells, in vivo imaging, and diagnostics. Science 307:538–544

Milhem O, Myles C, McKeown N, Attwood D, D'Emanuele A (2000) Polyamidoamine starburst® dendrimers as solubility enhancers. Int J Pharm 197:239–241

Mirkin CA, Taton TA (2000) Semiconductors meet biology. Nature 405:626–627

Miyamoto D, Oishi M, Kojima K, Yoshimoto K, Nagasaki Y (2008) Completely dispersible PEGylated gold nanoparticles under physiological conditions: modification of gold nanoparticles with precisely controlled PEG-b-polyamine. Langmuir 24:5010–5017

Mohapatra A, Uthaman S, Park I-K (2019) Polyethylene glycol nanoparticles as promising tools for anticancer therapeutics. In: Polymeric nanoparticles as a promising tool for anti-cancer therapeutics. Elsevier, Amsterdam, pp 205–231

Mozar FS, Chowdhury EH (2018) Impact of PEGylated nanoparticles on tumor targeted drug delivery. Curr Pharm Des 24:3283–3296

Muralidharan P, Mallory E, Malapit M, Hayes D Jr, Mansour HM (2014) Inhalable PEGylated phospholipid nanocarriers and PEGylated therapeutics for respiratory delivery as aerosolized colloidal dispersions and dry powder inhalers. Pharmaceutics 6:333–353

Nakayama-Ratchford N, Bangsaruntip S, Sun X, Welsher K, Dai H (2007) Noncovalent functionalization of carbon nanotubes by fluorescein–polyethylene glycol: supramolecular conjugates with pH-dependent absorbance and fluorescence. J Am Chem Soc 129:2448–2449

Nam HY, Nam K, Hahn HJ, Kim BH, Lim HJ, Kim HJ, Choi JS, Park J-S (2009) Biodegradable PAMAM ester for enhanced transfection efficiency with low cytotoxicity. Biomaterials 30:665–673

Nam K, Nam HY, Kim P-H, Kim SW (2012) Paclitaxel-conjugated PEG and arginine-grafted bioreducible poly(disulfide amine) micelles for co-delivery of drug and gene. Biomaterials 33:8122–8130

Namazi H, Adeli M (2005) Dendrimers of citric acid and poly(ethylene glycol) as the new drug-delivery agents. Biomaterials 26:1175–1183

Nayak AK, Panigrahi PP (2012) Solubility enhancement of etoricoxib by cosolvency approach. ISRN Phys Chem 2012:1–5

Ndlovu ST, Ullah N, Khan S, Ramharack P, Soliman M, De Matas M, Shahid M, Sohail M, Imran M, Shah SWA (2019) Domperidone nanocrystals with boosted oral bioavailability: fabrication, evaluation and molecular insight into the polymer-domperidone nanocrystal interaction. Drug Deliv Transl Res 9:284–297

Newkome GR, Yao Z, Baker GR, Gupta VK (1985) Micelles. Part 1. Cascade molecules: a new approach to micelles. A [27]-arborol. J Org Chem 50:2003–2004

Nie H, Guo W, Yuan Y, Dou Z, Shi Z, Liu Z, Wang H, Liu Y (2010) PEGylation of double-walled carbon nanotubes for increasing their solubility in water. Nano Res 3:103–109

Ooya T, Lee J, Park K (2003) Effects of ethylene glycol-based graft, star-shaped, and dendritic polymers on solubilization and controlled release of paclitaxel. J Control Release 93:121–127

Otsuka H, Nagasaki Y, Kataoka K (2003) PEGylated nanoparticles for biological and pharmaceutical applications. Adv Drug Deliv Rev 55:403–419

Owens DE III, Peppas NA (2006) Opsonization, biodistribution, and pharmacokinetics of polymeric nanoparticles. Int J Pharm 307:93–102

Pankhurst QA, Connolly J, Jones SK, Dobson J (2003) Applications of magnetic nanoparticles in biomedicine. J Phys D Appl Phys 36:R167

Pasut G, Veronese FM (2012) State of the art in PEGylation: the great versatility achieved after forty years of research. J Control Release 161:461–472

Penalva R, Esparza I, Larraneta E, González-Navarro CJ, Gamazo C, Irache JM (2015) Zein-based nanoparticles improve the oral bioavailability of resveratrol and its anti-inflammatory effects in a mouse model of endotoxic shock. J Agric Food Chem 63:5603–5611

Penon O, Patiño T, Barrios L, Nogues C, Amabilino DB, Wurst K, Pérez-García L (2015) A new porphyrin for the preparation of functionalized water-soluble gold nanoparticles with low intrinsic toxicity. ChemistryOpen 4:127–136

Penon O, Marín MJ, Russell DA, Pérez-García L (2017) Water soluble, multifunctional antibody-porphyrin gold nanoparticles for targeted photodynamic therapy. J Colloid Interface Sci 496:100–110

Pinaud F, Michalet X, Bentolila LA, Tsay JM, Doose S, Li JJ, Iyer G, Weiss S (2006) Advances in fluorescence imaging with quantum dot bio-probes. Biomaterials 27:1679–1687

Prajapati RN, Tekade RK, Gupta U, Gajbhiye V, Jain NK (2009) Dendimer-mediated solubilization, formulation development and in vitro–in vivo assessment of piroxicam. Mol Pharm 6:940–950

Prato M, Kostarelos K, Bianco A (2008) Functionalized carbon nanotubes in drug design and discovery. Acc Chem Res 41:60–68

Qi R, Li Y-Z, Chen C, Cao Y-N, Yu M-M, Xu L, He B, Jie X, Shen W-W, Wang Y-N (2015) G5-PEG PAMAM dendrimer incorporating nanostructured lipid carriers enhance oral bioavailability and plasma lipid-lowering effect of probucol. J Control Release 210:160–168

Quintana A, Raczka E, Piehler L, Lee I, Myc A, Majoros I, Patri AK, Thomas T, Mulé J, Baker JR (2002) Design and function of a dendrimer-based therapeutic nanodevice targeted to tumor cells through the folate receptor. Pharm Res 19:1310–1316

Rackstraw BJ, Stolnik S, Davis SS, Bignotti F, Garnett MC (2002) Development of multicomponent DNA delivery systems based upon poly(amidoamine)–PEG co-polymers. Biochim Biophys Acta 1576:269–286

Radi L, Fach M, Montigny M, Berger-Nicoletti E, Tremel W, Wich PR (2016) Methods of protein surface PEGylation under structure preservation for the emulsion-based formation of stable nanoparticles. MedChemComm 7:1738–1744

Ravelli D, Montanaro S, Tomasi C, Galinetto P, Quartarone E, Merli D, Mustarelli P, Fagnoni M (2012) One-step decatungstate-photomediated PEGylation of single-walled carbon nanotubes. ChemPlusChem 77:210–216

Ravelli D, Merli D, Quartarone E, Profumo A, Mustarelli P, Fagnoni M (2013) PEGylated carbon nanotubes: preparation, properties and applications. RSC Adv 3:13569–13582

Reboredo C, González-Navarro C, Martínez-Oharriz C, Martínez-López A, Irache J (2021) Preparation and evaluation of PEG-coated zein nanoparticles for oral drug delivery purposes. Int J Pharm 597:120287

Roberts M, Bentley M, Harris J (2002) Chemistry for peptide and protein PEGylation. Adv Drug Deliv Rev 54:459–476

Rosso-Vasic M, Spruijt E, van Lagen B, De Cola L, Zuilhof H (2008) Alkyl-functionalized oxide-free silicon nanoparticles: synthesis and optical properties. Small 4:1835–1841

Ryan SM, Mantovani G, Wang X, Haddleton DM, Brayden DJ (2008) Advances in PEGylation of important biotech molecules: delivery aspects. Expert Opin Drug Deliv 5:371–383

Sacchetti M, Nejati E (2012) Prediction of drug solubility in lipid mixtures from the individual ingredients. AAPS PharmSciTech 13:1103–1109

Salunkhe AB, Khot VM, Pawar S (2014) Magnetic hyperthermia with magnetic nanoparticles: a status review. Curr Top Med Chem 14:572–594

Sanchez Armengol E, Unterweger A, Laffleur F (2022) PEGylated drug delivery systems in the pharmaceutical field: past, present and future perspective. Drug Dev Ind Pharm 48:129–139

Sareen S, Mathew G, Joseph L (2012) Improvement in solubility of poor water-soluble drugs by solid dispersion. Int J Pharm Invest 2:12

Savić R, Eisenberg A, Maysinger D (2006) Block copolymer micelles as delivery vehicles of hydrophobic drugs: micelle–cell interactions. J Drug Target 14:343–355

Savjani KT, Gajjar AK, Savjani JK (2012) Drug solubility: importance and enhancement techniques. ISRN Pharm 2012:195727

Shadrack DM, Mubofu EB, Nyandoro SS (2015) Synthesis of polyamidoamine dendrimer for encapsulating tetramethylscutellarein for potential bioactivity enhancement. Int J Mol Sci 16:26363–26377

Shadrack DM, Swai HS, Munissi JJ, Mubofu EB, Nyandoro SS (2018) Polyamidoamine dendrimers for enhanced solubility of small molecules and other desirable properties for site specific delivery: insights from experimental and computational studies. Molecules 23:1419

Sherman MR, Saifer MP, Williams LD, Hershfield MS, Kelly SJ (2004) Aggregate-free urate oxidase for preparation of non-immunogenic polymer conjugates

Shi J, Kantoff PW, Wooster R, Farokhzad OC (2017) Cancer nanomedicine: progress, challenges and opportunities. Nat Rev Cancer 17:20–37

Sideratou Z, Tsiourvas D, Paleos CM (2001) Solubilization and release properties of PEGylated diaminobutane poly(propylene imine) dendrimers. J Colloid Interface Sci 242:272–276

Stepankova V, Bidmanova S, Koudelakova T, Prokop Z, Chaloupkova R, Damborsky J (2013) Strategies for stabilization of enzymes in organic solvents. ACS Catal 3:2823–2836

Stevelmans S, van Hest J, Jansen J, van Boxtel D, de Brabander-van den Berg E, Meijer E (1996) Synthesis, characterization, and guest–host properties of inverted unimolecular dendritic micelles. J Am Chem Soc 118:7398–7399

Sudeep P, Page Z, Emrick T (2008) PEGylated silicon nanoparticles: synthesis and characterization. Chem Commun:6126–6127

Suk JS, Xu Q, Kim N, Hanes J, Ensign LM (2016) PEGylation as a strategy for improving nanoparticle-based drug and gene delivery. Adv Drug Deliv Rev 99:28–51

Sweet DM, Kolhatkar RB, Ray A, Swaan P, Ghandehari H (2009) Transepithelial transport of PEGylated anionic poly(amidoamine) dendrimers: implications for oral drug delivery. J Control Release 138:78–85

Tamura A, Oishi M, Nagasaki Y (2009) Enhanced cytoplasmic delivery of siRNA using a stabilized polyion complex based on PEGylated nanogels with a cross-linked polyamine structure. Biomacromolecules 10:1818–1827

Tekade RK, Kumar PV, Jain NK (2009) Dendrimers in oncology: an expanding horizon. Chem Rev 109:49–87

Thakur S, Kesharwani P, Tekade RK, Jain NK (2015) Impact of pegylation on biopharmaceutical properties of dendrimers. Polymer 59:67–92

Thomas A, Müller SS, Frey H (2014) Beyond poly(ethylene glycol): linear polyglycerol as a multifunctional polyether for biomedical and pharmaceutical applications. Biomacromolecules 15:1935–1954

Tilley RD, Warner JH, Yamamoto K, Matsui I, Fujimori H (2005) Micro-emulsion synthesis of monodisperse surface stabilized silicon nanocrystals. Chem Commun:1833–1835

Tomalia DA, Christensen JB, Boas U (2012) Dendrimers, dendrons, and dendritic polymers: discovery, applications, and the future. Cambridge University Press, Cambridge

Tran S, Degiovanni P-J, Piel B, Rai P (2017) Cancer nanomedicine: a review of recent success in drug delivery. Clin Transl Med 6:1–21

Umut E (2013) Surface modification of nanoparticles used in biomedical applications. Mod Surf Eng Treat 20:185–208

Veronese F, Largajolli R, Boccu E, Benassi C, Schiavon O (1985) Surface modification of proteins activation of monomethoxy-polyethylene glycols by phenylchloroformates and modification of ribonuclease and superoxide dismutase. Appl Biochem Biotechnol 11:141–152

Vllasaliu D, Fowler R, Stolnik S (2014) PEGylated nanomedicines: recent progress and remaining concerns. Expert Opin Drug Deliv 11:139–154

Vögtle F (2003) Dendrimers II: architecture, nanostructure and supramolecular chemistry. Springer, Berlin

Wang J, Li B, Qiu L, Qiao X, Yang H (2022) Dendrimer-based drug delivery systems: history, challenges, and latest developments. J Biol Eng 16:1–12

Wattendorf U, Merkle HP (2008) PEGylation as a tool for the biomedical engineering of surface modified microparticles. J Pharm Sci 97:4655–4669

Wei H, Zhang X-Z, Cheng H, Chen W-Q, Cheng S-X, Zhuo R-X (2006) Self-assembled thermo- and pH responsive micelles of poly(10-undecenoic acid-bN-isopropylacrylamide) for drug delivery. J Control Release 116:266–274

WILEY-VCH (2012) Polyethylene glycol [MAK value documentation, 1998]. The MAK-collection for occupational health and safety. Wiley-VCH, New York

Williams LD, Hershfield MS, Kelly SJ, Saifer MG, Sherman MR (2003) PEG-urate oxidase conjugates and use thereof. Google Patents

Wolinsky JB, Grinstaff MW (2008) Therapeutic and diagnostic applications of dendrimers for cancer treatment. Adv Drug Deliv Rev 60:1037–1055

Woodle M, Matthay K, Newman M, Hidayat J, Collins L, Redemann C, Martin F, Papahadjopoulos D (1992) Versatility in lipid compositions showing prolonged circulation with sterically stabilized liposomes. Biochim Biophys Acta 1105:193–200

Xia H-J, Zhang Z-H, Jin X, Hu Q, Chen X-Y, Jia X-B (2013) A novel drug–phospholipid complex enriched with micelles: preparation and evaluation in vitro and in vivo. Int J Nanomedicine 8:545–554

Xu Z, Li Y, Zhang B, Purkait T, Alb A, Mitchell BS, Grayson SM, Fink MJ (2015) Water-soluble PEGylated silicon nanoparticles and their assembly into swellable nanoparticle aggregates. J Nanopart Res 17:1–16

Yang H, Morris JJ, Lopina ST (2004) Polyethylene glycol–polyamidoamine dendritic micelle as solubility enhancer and the effect of the length of polyethylene glycol arms on the solubility of pyrene in water. J Colloid Interface Sci 273:148–154

Yang J, Lee CH, Ko HJ, Suh JS, Yoon HG, Lee K, Huh YM, Haam S (2007a) Multifunctional magneto-polymeric nanohybrids for targeted detection and synergistic therapeutic effects on breast cancer. Angew Chem Int Ed 46:8836–8839

Yang T, Cui F-D, Choi M-K, Cho J-W, Chung S-J, Shim C-K, Kim D-D (2007b) Enhanced solubility and stability of PEGylated liposomal paclitaxel: in vitro and in vivo evaluation. Int J Pharm 338:317–326

Yin QQ, Wu L, Gou ML, Qian ZY, Zhang WS, Liu J (2009) Long-lasting infiltration anaesthesia by lidocaine-loaded biodegradable nanoparticles in hydrogel in rats. Acta Anaesthesiol Scand 53:1207–1213

Yiyun C, Tongwen X (2005) Dendrimers as potential drug carriers. Part I. Solubilization of non-steroidal anti-inflammatory drugs in the presence of polyamidoamine dendrimers. Eur J Med Chem 40:1188–1192

Yuan Q, Yeudall WA, Yang H (2010) PEGylated polyamidoamine dendrimers with bis-aryl hydrazone linkages for enhanced gene delivery. Biomacromolecules 11:1940–1947

Zhao B, Hu H, Yu A, Perea D, Haddon RC (2005) Synthesis and characterization of water soluble single-walled carbon nanotube graft copolymers. J Am Chem Soc 127:8197–8203

Zhou B, Zheng L, Peng C, Li D, Li J, Wen S, Shen M, Zhang G, Shi X (2014) Synthesis and characterization of PEGylated polyethylenimine-entrapped gold nanoparticles for blood pool and tumor CT imaging. ACS Appl Mater Interfaces 6:17190–17199

# PEGylation as a Tool to Alter Immunological Properties of Nanocarriers

6

Akhilesh Tiwari, Sourajyoti Goswami, Megha Joshi, Sanyam Gandhi, Pranay Soni, Muktika Tekade, Shubham Ramdas Mule, and Rakesh Kumar Tekade

**Abstract**

Nanocarriers hold immense potential for targeted drug delivery, but their therapeutic efficacy is often hindered by immunological interactions. The covalent attachment of polyethylene glycol chains, known as PEGylation, has emerged as a promising approach to modifying the immunological properties of nanocarriers. PEGylation offers several advantages, including steric stabilization, reduced protein adsorption, and decreased recognition by the immune system, resulting in prolonged circulation time and improved biodistribution. By shielding nanocarriers from immune recognition, PEGylation can mitigate undesired immune

A. Tiwari (✉) · S. Goswami · P. Soni
Department of Pharmacy, Indira Gandhi National Tribal University, Amarkantak, Anuppur, Madhya Pradesh, India

M. Joshi
PHC, Amarkantak, Amarkantak, Madhya Pradesh, India

S. Gandhi
Takeda Pharmaceuticals, Boston, MA, USA

M. Tekade
School of Pharmacy, Devi Ahilya Vishwavidyalaya, Takshila Campus, Indore, Madhya Pradesh, India

School of Pharmaceutical Sciences, Lovely Professional University, Phagwara, Punjab, India

S. R. Mule · R. K. Tekade
National Institute of Pharmaceutical Education and Research (NIPER) Ahmedabad, An Institute of National Importance, Government of India, Department of Pharmaceuticals, Ministry of Chemicals and Fertilizers, Palaj, Opp. Air force station, Gandhinagar, Gujarat, India


R. K. Tekade, N. K. Jain (eds.), *PEGylated Nanocarriers in Medicine and Pharmacy*, https://doi.org/10.1007/978-981-97-7316-9_6

responses, such as rapid clearance by the mononuclear phagocyte system or induction of pro-inflammatory reactions. This chapter hereby highlights the nuances of the PEGylation technique while emphasizing the altered immunological properties of the PEGylated nanocarriers, which positively impact the various aspects of drug delivery.

**Keywords**

Nanocarriers · Polyethylene glycol (PEGylation) · Immune system · Drug delivery system

## 6.1 Introduction

Nanocarriers are microscopic structures that are designed to carry and deliver drugs, genes, and other therapeutic agents to specific target sites in the body. They comprise various materials, including proteins, lipids, and polymers, and are typically nanoparticles with diameters ranging from 10 nm to a few hundred nanometers (Rawat et al. 2006). These are essential tools in drug delivery due to their advantages of targeted drug delivery, improved bioavailability, and reduced toxicity. Nanoparticles such as liposomes, dendrimers, and polymeric micelles can be engineered to encapsulate drugs and deliver them directly to the site of interest, such as cancer cells or inflammatory tissues (Parhi et al. 2012). They can also protect the drugs from degradation and clearance by the immune system, prolonging their circulation time in the bloodstream (Lombardo et al. 2019). However, nanocarriers pose a challenge in terms of immunogenicity. The immune system is designed to recognize and eliminate foreign materials from the body, and nanoparticles can trigger an immune response that may result in the clearance of the carriers or cause adverse effects (Dobrovolskaia and McNeil 2007). The immune system can recognize the surface properties, size, and charge of the nanoparticles and mount an immune response against them, leading to inflammation, cytotoxicity, and allergic reactions (Qin et al. 2020).

Several strategies have been developed to overcome the immunogenicity challenge, such as surface modifications with stealth materials that can reduce interaction with the immune system (Fu et al. 2021). Researchers are also exploring the use of biocompatible and biodegradable materials that can be quickly metabolized or eliminated by the body, reducing the risk of immunogenicity. Overall, the immunogenicity of nanocarriers is a critical consideration when developing novel drug delivery systems, and researchers need to carefully balance the benefits and risks of nanocarriers in drug delivery (Cheng et al. 2021). Modifying nanocarriers can significantly improve drug delivery by addressing some of the challenges of traditional drug delivery methods.

Nanocarriers can be engineered to selectively target specific tissues or cells within the body (Cheng et al. 2021). This can increase the concentration of the drug at the desired site while reducing exposure to non-target tissues and thus minimizing side effects. Nanocarriers can be designed to release drugs in a controlled manner over a prolonged period, which can improve drug efficacy and reduce dosing frequency (Lee and Yeo 2015). Nanocarriers can be engineered to improve drug solubility, stability, and bioavailability. This can increase drug absorption, distribution, and elimination, leading to better therapeutic outcomes (Din et al. 2017). Nanocarriers can protect drugs from enzymatic degradation, clearance by the immune system, and premature elimination from the body, thus increasing the drug's half-life and bioavailability. Nanocarriers can be used to deliver multiple drugs or diagnostic agents simultaneously, allowing for synergistic effects and improved treatment outcomes.

## 6.2 Introduction to PEGylation

PEGylation is a technique that involves covalently attaching polyethylene glycol (PEG) to a nanocarrier surface (Howard et al. 2008). This modification has been extensively used to alter the immunological properties of nanocarriers, including nanoparticles, liposomes, and micelles. PEGylation as a tool to modify the immunological properties of nanocarriers has dramatically impacted the field of drug delivery and nanomedicine. It can potentially revolutionize the treatment of many diseases (Vllasaliu et al. 2014). Before PEGylation, the use of nanocarriers for drug delivery faced significant challenges, including rapid clearance from the bloodstream, poor targeting efficiency, and immune reactions (Mitchell et al. 2021). PEGylation has helped to overcome many of these challenges by altering the immunological properties of nanocarriers and improving their stability and biocompatibility.

The use of PEGylated nanocarriers has several advantages over traditional drug delivery methods. PEGylation can increase the circulation time of nanocarriers in the bloodstream, which can enhance their accumulation at the target site and improve their therapeutic efficacy. PEGylation can also reduce the clearance of nanocarriers by the immune system, enhancing their safety and reducing the risk of adverse reactions (Su and Kang 2020). In addition, PEGylation can improve the targeting efficiency of nanocarriers by reducing non-specific binding to tissues and cells. This can increase the specificity and efficacy of drug delivery and reduce the dose required for therapeutic effect.

### 6.2.1 Properties of PEG

PEG is a water-soluble, synthetic, and non-toxic polymer with repeating ethylene oxide units. It is a linear, flexible, and hydrophilic polymer that can be synthesized with different molecular weights (MW) and chain lengths. PEG is commonly used in various applications, including pharmaceuticals, biotechnology, and cosmetics (Su and Kang 2020). PEG is highly soluble in water and can form stable aqueous solutions with a wide range of concentrations. It is also miscible with many organic solvents, making it a versatile polymer for various applications. PEG has a low toxicity profile and is generally considered safe for human use, although some individuals may experience allergic reactions (Kolate et al. 2014).

The properties of PEG are highly dependent on its molecular weight and chain length; as the molecular weight of PEG increases, its viscosity, melting point, and boiling point also increase. Longer PEG chains tend to have greater hydrophilicity, solubility, and biocompatibility than shorter chains (Zheng et al. 2023). PEG has several unique properties, making it a valuable polymer for various applications. PEG is widely used in biomedical applications because of its excellent biocompatibility, low toxicity, and non-immunogenicity. Being highly hydrophilic, PEG serves as an ideal polymer for enhancing the solubility and bioavailability of poorly soluble drugs. PEG is a flexible polymer that can adopt a range of conformations, allowing it to adapt to different environments and surfaces. Outer PEG coating on the surface of nanocarriers renders them non-reactive toward many biomolecules. Hence, PEG is an ideal coating material to overcome the immune recognition of such nanocarriers.

### 6.2.2 Mechanisms of PEGylation

The mechanisms of PEGylation involve the creation of a covalent bond between the PEG molecule and the target molecule. This can be achieved through several methods, such as chemical crosslinking, enzymatic conjugation, and click chemistry techniques. Chemical crosslinking involves using a reactive group, such as a carboxyl or amine group, on the PEG molecule to form a covalent bond with a corresponding reactive group on the target molecule (Sinz 2006) whereas, enzymatic conjugation involves the use of an enzyme, such as transglutaminase or microbial transglutaminase, to catalyze the formation of a covalent bond between the PEG molecule and a specific amino acid residue on the target molecule (Fontana et al. 2008). Click chemistry, on the other hand, utilizes a small molecule linker that can react with both the PEG molecule and the target molecule to form a covalent bond between them (Moses and Moorhouse 2007).

Once the PEG molecule is covalently attached to the target molecule, it can modify the properties of the molecule in several ways. For example, the PEG molecule can sterically hinder interactions between the target molecule and other proteins or cells, which can increase the half-life of the molecule in the bloodstream. Additionally, the PEG molecule can increase the hydrophilicity of the target molecule, which can improve its solubility and reduce its aggregation (Harris et al. 2001). The mechanism of action of PEGylation in altering the immunological properties of nanocarriers is based on the "stealth effect" (Wen et al. 2023).

The PEG chains create a steric barrier around the nanocarriers, which prevents recognition and uptake by the reticuloendothelial system (RES)—a network of cells and tissues that play a role in immune defense and clearance of foreign particles from the blood. The RES includes macrophages, dendritic cells, and Kupffer cells, which can recognize and phagocytose foreign particles (Hadjesfandiari and Parambath 2018). This can improve the therapeutic efficacy and safety of drugs and nanocarriers for various biomedical applications (Caliceti and Veronese 2003). By preventing recognition by the immune system and the RES, PEGylation can increase the circulation time of the molecule in the blood, leading to improved pharmacokinetics and pharmacodynamics. The prolonged circulation time allows for enhanced drug delivery to the target site, increasing the drug's efficacy and reducing the required dose (Tian et al. 2022). Moreover, PEGylation can also reduce the immunogenicity of the molecule by masking its surface antigens, which can reduce the risk of immune reactions, such as antibody formation. This can improve the safety and tolerability of the drug or protein (Baker et al. 2010).

The mechanism of PEGylation can be broadly described in the following steps.

#### 6.2.2.1 Activation of PEG

PEG is usually activated by converting one end of the molecule into a reactive group, such as an amine or a thiol, which can then attach to the target molecule (Zalipsky 1995). This process is often called "PEGylation chemistry" and is usually accomplished using a coupling agent or a cross-linker. The choice of coupling agent or cross-linker will depend on the type of functional group used to activate the PEG and the chemistry of the target molecule (Gupta et al. 2019). Once the PEG has been activated, it can be conjugated to the target molecule through a covalent bond, resulting in a PEGylated product with improved properties (Saito et al. 2003).

#### 6.2.2.2 Conjugation of PEG to the Target Molecule

The activated PEG is then attached to the target molecule through a covalent bond, usually using a coupling agent or a cross-linker (Wildling et al. 2011). Once the PEG has been activated by converting one end of the molecule into a reactive group, it can be attached to the target molecule through a covalent bond (Ebner et al. 2007). This is typically achieved using a coupling agent or a cross-linker, which allows for forming a stable and durable bond between the PEG and the target molecule (Jia and Li 2015). The choice of coupling agent or cross-linker will depend on various factors, such as the chemistry of the target molecule, the type of reactive group on the PEG, and the desired length and configuration of the PEG chain (Ebner et al. 2007). The coupling reaction is typically carried out under carefully controlled conditions to ensure the formation of a high-quality PEGylated product with the desired properties.

### 6.2.3 Advantages and Disadvantages of PEGylation

PEGylation has several advantages and disadvantages, which should be considered when deciding whether to use this approach for a particular application. Therefore, The PEGylation strategy usually involves striking a balance between its advantages and disadvantages to suit the desired end outcome. The advantages and disadvantages of PEGylation are listed in Fig. 6.1.

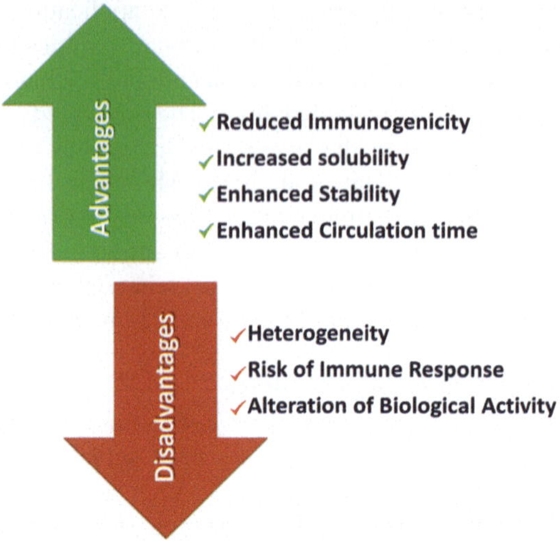


**Fig. 6.1** Advantages and disadvantages of PEGylation

#### 6.2.3.1 Advantages of PEGylation

PEGylation of nanocarriers offers a variety of advantages over their non-PEGylated counterparts. The most relevant benefits of PEGylation from the pharmaceutical perspective are mentioned below.

##### Increased Solubility

PEGylation can increase the solubility of a molecule in aqueous solutions, making it easier to administer or use in various applications. PEG chains are highly hydrophilic, exhibiting a strong affinity to water or aqueous solvents (Le Dévédec et al. 2013). PEGylation of poorly soluble compounds, including proteins and peptides, can enhance their water solubility and render them more convenient from a formulation perspective.

##### Improved Stability

PEGylation can increase the stability of a molecule, making it less prone to degradation or denaturation. The PEG chains around the PEGylated molecules form a steric barrier and prevent the molecules from aggregating (Suk et al. 2016). This steric barrier improves the stability of PEGylated moieties by preventing their aggregation. On the other hand, surrounding PEG chains reduce the exposure of PEGylated moieties to the surrounding metabolizing enzymes (Totten et al. 2019). This strategy is beneficial for drugs or proteins prone to degradation in biological fluids.

##### Reduced Immunogenicity

PEGylation can reduce the immunogenicity of a molecule, making it less likely to trigger an immune response when administered to humans. The coating of PEG around the molecules of interest helps in hiding their antigenic determinants, which ultimately reduces the immune recognition of such molecules, thereby significantly reducing their immunogenicity (Shi et al. 2022).

#### Enhanced Circulation Time

PEGylation can increase the overall circulation time of the proteins or drugs upon administration. PEGylation increases the overall molecular weight of the proteins or drugs, which reduces their renal clearance (Du et al. 2020). Reduced renal clearance coupled with declined immune recognition owing to PEGylation combined enhances the circulation time of the PEGylated molecules.

### 6.2.3.2 Disadvantages of PEGylation

Despite all the lucrative advantages that the technique of PEGylation offers, it still presents some disadvantages. Careful consideration and thorough examination of these disadvantages are recommended while adapting the strategic PEGylation approach for proteins, peptides, and drugs of interest. A few leading disadvantages of PEGylation are mentioned in following section.

#### Heterogeneity

PEGylation is a complex process necessitating the careful monitoring of various process parameters such as reaction temperature, reaction time, molecular weight of PEG, and pH. PEGylation reactions are time-consuming and involve forming heterogeneous final products over a monosubstituted product. This further makes the downstream processing cumbersome and increases the cost of production. Non-specific coupling reactions usually lead to the formation of heterogeneous products. However, this can be minimized by site-specific PEGylation of proteins and peptides, which specifically carries out PEGylation at the particular amino acid residue of the protein or peptide under consideration.

#### Risk of Immune Response

Although PEG is generally considered safe for human use, some individuals may develop an immune response to PEGylated molecules, leading to allergic reactions or other adverse effects. Multiple reports have suggested the formation of immediate and delayed anti-PEG antibodies in individuals after exposure to PEG. De Groot et al. have reported instances of anaphylactic shock with a marketed PEGylated contrast agent (De Groot et al. 2004). The cases of urticaria are also reported by Perez et al. using polysorbate 80, a PEG-containing polymer (Pérez-Pérez et al. 2011).

#### Alteration of Biological Activity

PEGylation can alter the biological activity of a molecule, either positively or negatively, depending on the specific application. The steric hindrance created by PEG may interfere with the interaction of the PEGylated molecules and their target, thereby reducing the desired pharmacological response. The drugs or proteins requiring cellular internalization to demonstrate their pharmacological activity may suffer significantly due to PEGylation. This effect can be prominently observed in the case of PEGylated enzymes, wherein the interaction between the therapeutic enzyme and the substrate is severely affected due to PEGylation.

Overall, PEGylation is a valuable tool for modifying the properties of molecules for various applications. However, the advantages and disadvantages of PEGylation should be carefully considered, along with other factors, such as the specific application, target molecule, and regulatory requirements.

## 6.3 Immunological Properties of Nanocarriers

Nanocarriers are engineered structures that are used to deliver drugs, genes, or other therapeutic agents to specific sites in the body. While nanocarriers have shown great promise as drug-delivery vehicles, they can also have immunological properties that must be carefully considered. To understand the immune response to the externally administered nanocarriers, it is essential first to understand the immune system. The immune system is a complex network of cells, tissues, and organs that protect the body against pathogens and foreign substances. It is responsible for identifying and eliminating harmful invaders while maintaining tolerance to self-antigens (Chaplin 2010). The immune system can be divided into two main parts—the innate immune system and the adaptive immune system. The innate immune system is the first line of defense against invading pathogens. It includes physical barriers, such as the skin and mucous membranes, as well as cellular components, such as phagocytic cells (e.g., macrophages and neutrophils) and natural killer (NK) cells. These cells can recognize and eliminate foreign invaders through non-specific mechanisms, such as phagocytosis, cytokine production, and complement activation (Anaya et al. 2013).

The adaptive immune system is a more specialized and targeted response that develops over time in response to specific pathogens. It includes B, T, and antigen-presenting cells (APCs). B cells produce antibodies that specifically recognize and neutralize foreign antigens, while T cells recognize and destroy infected or abnormal cells. APCs, such as dendritic cells and macrophages, present antigens to T cells to initiate an immune response (Den Haan et al. 2014). The humoral immunity and cell-mediated immunity are two branches of adaptive immunity. Humoral immunity involves the production of antibodies by B cells, which can neutralize pathogens in the bloodstream or other extracellular spaces.

In contrast, cell-mediated immunity consists of activating T cells, which can destroy infected or abnormal cells. The immune system can also remember previous exposure to pathogens, allowing for a faster and more efficient response upon subsequent encounters. This is the basis for vaccination, which involves the introduction of a non-pathogenic form of a pathogen to elicit an immune response and establish memory (Cooper and Eleftherianos 2017).

Depending on the therapeutic application, nanocarriers can play an essential role in modulating the immune response, either by enhancing or by suppressing immune responses. The immune response to nanocarriers can depend on several factors, including their size, shape, surface chemistry, and cargo.

### 6.3.1 Enhancing Immune Responses

Nanocarriers can be designed to enhance immune responses by delivering antigens, adjuvants, or immunomodulatory agents to antigen-presenting cells (APCs), such as dendritic cells and macrophages. This can activate T and B cells, generating an adaptive immune response against infectious diseases or cancer. For example, nanocarriers can be used as a platform for cancer vaccines to deliver tumor-associated antigens and adjuvants to APCs, which can enhance the immune response to cancer. Castro et al. developed chitosan/poly(γ-glutamic acid) nanoparticles (Ch/γ-PGA NPs) to modulate the inflammatory profile of macrophages and dendritic cells to impair their ability to promote cancer evasion (Castro et al. 2017). Ch/γ-PGA NPs demonstrated the development of immunostimulatory phenotype of dendritic cells, thereby enhancing the expression of co-stimulatory molecules such as CD86, CD40 and secretion of pro-inflammatory cytokines such as IL-6, IL-12p40, and TNF-α. Ch/γ-PGA NPs also promoted the macrophages to a pro-inflammatory profile, decreasing the expression of CD163. These developments observed using Ch/γ-PGA NPs inhibited the antigen-presenting cells' ability to invade colorectal cancer.

### 6.3.2 Suppressing Immune Responses

Nanocarriers can also be designed to suppress immune responses, such as treating autoimmune diseases or transplant rejection. This can be achieved by delivering immunosuppressive agents, such as siRNA or small molecule drugs, to immune cells, such as T cells or dendritic cells, which can reduce their activation and proliferation. Tang et al. developed PEG-b-poly(D,L-lactide-*co*-glycolide) (PEG-PLGA) nanoparticles and encapsulated cyclosporine A (CsA) into them (Tang et al. 2012). T-cell assay was performed to assess the suppression of T-cell proliferation and production of inflammatory cytokines. Both free CsA and CsA encapsulated PEG-PLGA NPs demonstrated dose-dependent production of inflammatory cytokines and suppression of T-cell proliferation. The release kinetics of CsA from CsA encapsulated PEG-PLGA NPs (55.6% release on day 1) suggested their ability to maintain the therapeutic concentration of CsA in vivo. This PEGylated NP-based delivery system for the delivery of immunosuppressive drugs presents the potential application of PEGylated techniques for the targeted delivery of immunosuppressive drugs.

### 6.3.3 Immune Evasion

Nanocarriers can also be designed to evade the immune system by using stealth coatings, such as PEG, or by mimicking the properties of natural cells, such as erythrocytes or platelets. This can increase the circulation time of nanocarriers and reduce their recognition and clearance by the immune system.

## 6.4 Immunological Barriers to Nanocarrier-Mediated Drug Delivery

PEGylation is the procedure of adhering PEG chains to a nanocarriers' surface, such as liposomes or nanoparticles. By inhibiting immune system detection and clearance, PEG chains can lengthen the duration that nanocarriers are in circulation in the bloodstream. Nanocarriers are foreign substances, and when they are introduced into the body, they can elicit an immune response. The immune system can recognize nanocarriers as foreign and try to eliminate them. This immune response can limit the effectiveness of nanocarriers as drug-delivery systems (Zolnik et al. 2010). By disguising the surface of the nanocarriers and preventing the immune system from recognizing them, adding PEG chains to nanocarriers can aid in lowering the immunological response. The "stealth effect" is the name given to this phenomenon. The stealth effect can prolong the circulatory circulation duration of nanocarriers, allowing for their accumulation in the target tissue or organ (Zalba et al. 2022). However, prolonged circulation of PEGylated nanocarriers can also lead to the formation of anti-PEG antibodies in some patients. These antibodies can recognize and eliminate PEGylated nanocarriers, limiting their effectiveness as drug delivery systems. The formation of anti-PEG antibodies may also lead to hypersensitivity reactions in some patients (Zhang et al. 2016a).

Nanocarrier-mediated drug delivery has shown great potential for the treatment of various diseases. However, there are several immunological barriers that can limit their efficacy and safety. The immune system can recognize nanocarriers as foreign invaders, which can trigger an immune response. This can lead to the activation of immune cells, such as macrophages and dendritic cells, and the production of inflammatory cytokines (Liu et al. 2017). When nanocarriers are recognized by the immune system, they can be coated with opsonins, such as antibodies and complement proteins. This opsonization can promote their uptake by phagocytic cells, such as macrophages, which can limit their circulation time and therapeutic efficacy (Gamucci et al. 2014).

Furthermore, nanocarriers can be cleared from the body by the reticuloendothelial system (RES), which includes the liver and spleen. The rate of clearance can depend on the size, shape, surface charge, and surface chemistry of the nanocarrier (Duan and Li 2013). Some nanocarriers can be immunogenic, meaning they can stimulate an immune response even without an antigen. This can be due to adjuvants, contaminants, or impurities in the nanocarrier (Zarreen Simnani et al. 2023).

Although the nanocarriers offer a variety of advantages for drug delivery purposes, their utility is often limited by the immunological responses that are elicited upon their administration. Some strategies, such as surface modification and PEGylation, can be undertaken to overcome the immunological barriers associated with nanocarriers.

### 6.4.1 Strategies to Overcome Immunological Barriers

There are several strategies that can be employed to overcome immunological barriers in nanocarrier-mediated drug delivery, including the use of biocompatible materials and surface modifications. Nanocarrier surface modifications possess a distinct advantage as they allow non-biocompatible materials to be delivered without eliciting an immune response. Surface modifications include using macromolecules, membranes from endogenous cells, receptors, and steric polymers. The aim of utilizing the surface modification strategies is to synthesize the stealth nanocarriers, which are undetectable to the immune system, allowing them to reach the target site and exert their desired therapeutic response. The commonly utilized surface modification strategies are summarized in Table 6.1.

#### 6.4.1.1 PEGylation

PEGylation, as described earlier, refers to the method of coating the surface of nanocarriers with PEG. Various proteins, such as complement proteins, plasma proteins, and antibodies, are known to aggregate on the surface of foreign bodies and lead to protein corona formation, which marks the initial step in the process of immune recognition. PEG has shown a potential to hinder the formation of protein corona on the surface of nanocarriers by forming a steric barrier (Schöttler et al. 2016). Additionally, PEG-coated nanocarriers can be designed to smartly shed the PEG coating in response to a stimulus upon reaching the target site. Qiao et al. successfully developed a nanosystem by combining matrix metalloprotease-2 sensitive peptides with HLAH, a pH-sensitive peptide, and PEG (Qiao et al. 2017). The PEG ensured the enhanced circulation time for the compound, whereas, upon reaching the acidic tumor microenvironment, the PEG coating was released, leading to the delivery of the cytotoxic compound to the target site.

Although PEGylation possesses various merits over non-PEGylated systems, it still presents certain limitations. PEG is susceptible to oxidative degradation, which

**Table 6.1** Strategies to overcome immunological barriers

| Surface modification strategy | Coating material | Mechanism of immune evasion | References |
|---|---|---|---|
| PEGylation | PEG | Form steric barrier and prevent protein corona formation | Schöttler et al. (2016) |
| Coating with cell membranes | Cell membranes derived from RBCs, leukocytes | Mimic circulatory cells and increases circulation time | Parodi et al. (2013) |
| Coating with carbohydrates | Heparin, hyaluronic acid, polysialic acid, glucosamine, and dextran | Mimic cell surface and prevent protein corona formation | Bellido et al. (2015) |
| Coating with proteins | Albumin, CD47 ligand, zwitterion of lysine, and glutamic acid | Form steric barrier and prevent protein corona formation | Nowinski et al. (2014) |

will lead to the formation of reactive oxygen species, and repeated administration of the PEGylated system might also lead to the formation of anti-PEG antibodies (Sung et al. 2010; Verhoef et al. 2014). However, compared to other polymers potentially used to meet the same purpose, PEG is preferred choice.

#### 6.4.1.2 Cell Membranes

Using cell membranes as a coating material for the nanocarriers is another area of research that primarily focuses on creating biomimetic systems to evade the immune response. The initial investigations in this area involved the development of nanocarriers coated with the membrane of red blood cells. This method exhibited promising applications as the nanocarriers prepared by membrane coating retained their original structure with enhanced circulation time and reduced clearance (Hu et al. 2011). The coating with cell membranes allows multiple administrations of the developed formulation without developing a significant immune response (Hu et al. 2011; Piao et al. 2014).

Apart from RBCs, the membranes of various leukocytes have been utilized to demonstrate immune evasion by the nanocarriers. Leukocyte-derived membranes showed immune evasion of the nanocarriers and provided unique functions to the nanocarriers, such as receptor-ligand affinity and diapedesis (Parodi et al. 2013). The nanocarriers coated with platelet membranes showed stealth properties and the ability to adhere to the damaged blood vessels (Hu et al. 2015). Using cell membranes to coat the nanocarriers' surface opened new avenues of opportunities in engineered biomimetic nanosystems. Future investigations in this domain are expected to develop synthetic biomimetic membranes with specific properties and targeting abilities.

#### 6.4.1.3 Carbohydrates

Macromolecules have been thoroughly investigated to develop coated nanocarriers that can remain significantly undetectable by the immune system. Heparin and hyaluronic acid are the natural glycosaminoglycan polysaccharides found on the surface of various cells. This imparts the ability to trick the immune system and prevents the development of an immunogenic response against the coated nanocarriers (Bellido et al. 2015; Peer et al. 2003; Toole 2004). Monosaccharides such as polysialic acid have demonstrated similar functions by shielding the nanocarriers and allowing targeted delivery of nanocarriers (Wilson et al. 2014; Fernandes and Gregoriadis 2001). A precursor to glycosaminoglycans, glucosamine, is another carbohydrate that reduces complement activation and thereby minimizes immunogenic response to the coated nanocarriers (Thasneem et al. 2013). Besides the above-mentioned polysaccharides and monosaccharides, dextran has also demonstrated the ability to evade complement activation (Thomas et al. 2011).

#### 6.4.1.4 Proteins

Like carbohydrates, the presence of various proteins, as an integral part of anatomical and physiological aspects of human beings, makes them a suitable choice for coating nanocarriers. Albumin, an abundantly present plasma protein, has been widely utilized as a coating material owing to its ability to shield the nanocarriers and form a steric barrier, which prevents the plasma proteins from forming a protein corona around the nanocarriers (Gulati et al. 2017, 2018). CD47 ligand is another commonly used coating material found on the surface of RBCs and is known to induce M1 phagocytic activity (Rodriguez et al. 2013). Zwitterions can also be utilized to coat the surface of nanocarriers effectively. Zwitterionic coating consisting of lysine and glutamic acid repeated chains has demonstrated significant promise as a coating material by preventing protein corona formation on the surface of nanocarriers (Nowinski et al. 2014).

## 6.5 Effects of PEGylation on In Vivo Behavior of Nanocarriers

PEGylation plays a crucial role in tailoring the in vivo behavior of nanocarriers, profoundly impacting their efficacy as drug delivery systems. PEGylation imparts stealth properties to the nanocarriers, thereby preventing their immune recognition and enabling them to remain in systemic circulation for longer. Various effects of PEGylation on the in vivo behavior of nanocarriers are discussed below.

### 6.5.1 Pharmacokinetics and Biodistribution of Nanocarriers

PEGylation can significantly affect the pharmacokinetics (PK) and biodistribution of nanocarriers. By attaching PEG chains to the surface of nanocarriers, the physicochemical properties of the nanocarriers can be altered, leading to changes in their PK and biodistribution. One of the main effects of PEGylation is the extension of the circulation time of nanocarriers in the bloodstream. PEG chains on the surface of nanocarriers can create a steric barrier that reduces their recognition and uptake by the reticuloendothelial system (RES), resulting in increased systemic circulation time. The longer circulation time can allow nanocarriers to accumulate in the target tissue or organ and improve their therapeutic efficacy (Suk et al. 2016).

Another effect of PEGylation is the reduction of non-specific interactions between nanocarriers and biological components, such as serum proteins, cell membranes, and extracellular matrix components. This can decrease the rate of clearance and uptake by the RES, which can lead to improved biodistribution and reduced accumulation in non-target organs (Donahue et al. 2019). The size of PEG chains and their density on the surface of nanocarriers can also impact PK and biodistribution. Higher PEG density on the surface of nanocarriers can lead to a more significant reduction of RES uptake, but it may also lead to faster kidney clearance. Conversely, lower PEG density may increase non-specific interactions, leading to rapid clearance by the RES (Perry et al. 2012).

### 6.5.2 Modulation of the Immune Response

PEGylation can also modulate the immune response by reducing the recognition of nanocarriers by the immune system and altering the interaction between nanocarriers and immune cells. As mentioned earlier, nanocarriers can elicit an immune response, which can reduce their efficacy as drug delivery systems. By adding PEG chains to the surface of nanocarriers, the immune response can be reduced due to the "stealth effect." The PEG chains can create a hydrophilic layer on the surface of nanocarriers, which can reduce their recognition and uptake by immune cells, such as macrophages and dendritic cells (Liu et al. 2017).

Moreover, PEGylation can also affect the cytokine profile of the immune response. Some studies have shown that PEGylation can lead to a shift toward an anti-inflammatory cytokine profile, such as interleukin 10 (IL-10) and transforming growth factor-beta (TGF-β), which can reduce inflammation and promote tissue repair. This shift in cytokine profile can be beneficial for treating inflammatory diseases, such as rheumatoid arthritis (Naing et al. 2016). However, prolonged circulation of PEGylated nanocarriers can also lead to the formation of anti-PEG antibodies, which can elicit an immune response and limit the effectiveness of PEGylated nanocarriers. In some cases, these anti-PEG antibodies can also lead to hypersensitivity reactions.

Therefore, the immunomodulatory effects of PEGylation need to be carefully considered in developing nanomedicines. The PEGylation strategy needs to be optimized to achieve the desired immune response while minimizing the risk of adverse effects, such as forming anti-PEG antibodies (Xia et al. 2021).

### 6.5.3 Immunogenicity of PEGylated Nanocarriers

Although PEGylation can reduce the immunogenicity of nanocarriers, it can also lead to the formation of anti-PEG antibodies, which can affect the safety and efficacy of PEGylated nanocarriers. The formation of anti-PEG antibodies is a concern because it can lead to accelerated clearance of PEGylated nanocarriers from the bloodstream and reduce their circulation time. This can decrease the therapeutic efficacy of PEGylated nanocarriers (Li et al. 2018). Several factors, including the size and shape of PEG chains, the density of PEG chains on the surface of nanocarriers, the route of administration, the frequency of administration, and the immune status of the patient, may influence the formation of anti-PEG antibodies. For example, smaller PEG chains may be more immunogenic than larger ones, and higher densities of PEG chains on the surface of nanocarriers may increase the likelihood of antibody formation (Kozma et al. 2020). Furthermore, it has been reported that some patients with pre-existing PEG-specific antibodies may be at an increased risk of developing hypersensitivity reactions to PEGylated nanocarriers. Hypersensitivity reactions can range from mild to severe, including anaphylaxis, and can limit the clinical application of PEGylated nanocarriers (Sellaturay et al. 2021).

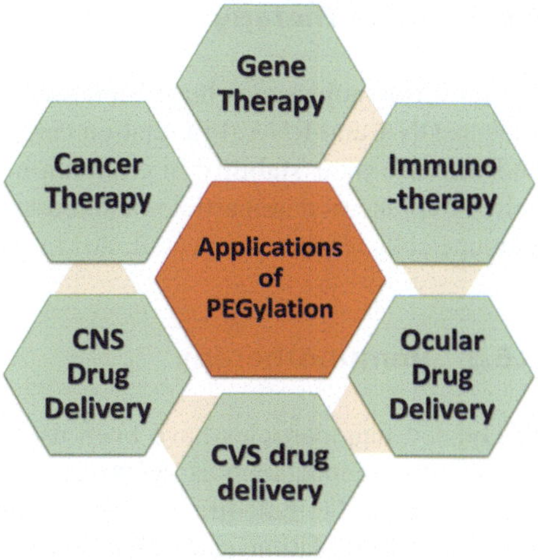


**Fig. 6.2** Applications of PEGylated nanocarriers

While PEGylation can reduce the immunogenicity of nanocarriers, it can also lead to the formation of anti-PEG antibodies, which can impact the safety and efficacy of PEGylated nanocarriers. Careful consideration of the immunogenicity of PEGylated nanocarriers and optimization of the PEGylation strategy is necessary to minimize the risk of adverse immune responses.

## 6.6 Applications of PEGylated Nanocarriers in Drug Delivery

Considering the advantages PEGylation offers, the PEGylated moieties are expected to possess a diverse spectrum of applications in drug delivery. PEGylated nanocarriers have been widely investigated as drug delivery systems for various therapeutic applications. The applications of PEGylated nanocarriers are elaborated in Fig. 6.2.

### 6.6.1 Cancer Therapy

PEGylated liposomes have been approved for treating various cancers, including ovarian and multiple myeloma. PEGylation can increase the circulation time of liposomes and enhance their accumulation in tumors through the enhanced permeability and retention (EPR) effect. Furthermore, PEGylation can reduce the toxicity and immunogenicity of liposomes (Fulton and Najahi-Missaoui 2023).

### 6.6.2 Gene Therapy

PEGylated nanoparticles have been used for gene therapy to deliver nucleic acids, such as DNA and RNA. PEGylation can protect the nucleic acids from degradation and increase their stability in circulation. Moreover, PEGylation can enhance the cellular uptake of nanoparticles and reduce their interaction with serum proteins and immune cells (Hatakeyama et al. 2011).

### 6.6.3 Immunotherapy

PEGylated nanoparticles have been used for immunotherapy to deliver immunomodulatory agents, such as cytokines and Toll-like receptor agonists. PEGylation can increase the half-life of nanoparticles and improve their bioavailability. Moreover, PEGylation can reduce the toxicity and immunogenicity of immunomodulatory agents (Debele et al. 2020).

### 6.6.4 Central Nervous System (CNS) Drug Delivery

PEGylated nanoparticles have been investigated for the delivery of drugs to the CNS, which is challenging due to the blood-brain barrier (BBB). PEGylation can enhance the penetration of nanoparticles through the BBB and improve the distribution of drugs in the brain. Furthermore, PEGylation can reduce the clearance of nanoparticles by the RES in the liver and spleen (Crawford et al. 2016).

### 6.6.5 Pulmonary Drug Delivery

PEGylated nanoparticles have been investigated for pulmonary drug delivery, a non-invasive route of administration for treating respiratory diseases. PEGylation can increase the retention time of nanoparticles in the lungs and enhance their penetration through the mucus layer. Moreover, PEGylation can reduce the clearance of nanoparticles by the immune system in the lungs (Almeida and Souto 2007).

### 6.6.6 Ocular Drug Delivery

PEGylated nanocarriers have shown potential in ocular drug delivery, allowing for sustained release of drugs to the eye. These nanocarriers can improve the bioavailability of drugs, prolong therapeutic effects, and enhance patient compliance in treating ocular diseases (Tsai et al. 2018).

### 6.6.7 Cardiovascular (CVS) Drug Delivery

PEGylated nanocarriers can be utilized to deliver drugs for treating cardiovascular diseases, such as atherosclerosis or hypertension. The nanocarriers can target specific sites in the vasculature, improving drug efficacy and reducing off-target effects (Mohamed et al. 2022). PEGylated nanocarriers have shown great potential as drug delivery systems for various therapeutic applications. The PEGylation strategy must be carefully optimized to achieve the desired pharmacokinetic and biodistribution properties while minimizing the risk of adverse reactions.

## 6.7 Challenges and Future Perspectives

PEGylation is a complex technique often involving multistep synthesis methods. Achieving high conjugation efficiency and uniform distribution of PEG chains on the nanocarrier surface is crucial for consistent performance. PEGylation involves the use of PEG chains of varying lengths. The choice of PEG chain length can influence the pharmacokinetics, biodistribution, and clearance of the nanocarriers. Achieving consistent and controlled PEGylation with precise chain lengths can be challenging, and variations in chain length can lead to unpredictable behavior and efficacy. These issues highlight the manufacturing challenges associated with PEGylated products. Hence, developing efficient PEGylation methods that ensure uniform coverage and minimal batch-to-batch variability is an ongoing challenge (Rabanel et al. 2014). The stability of the PEG coating is another concern related to the PEGylated molecules. PEG chains can undergo degradation or detachment from the nanocarriers' surface over time, potentially impacting the stability and performance of PEGylated nanocarriers. Strategies to improve the stability and prevent premature shedding of PEG chains are being explored (Sun et al. 2017).

PEGylation can limit the interaction of nanocarriers with target cells or tissues, as the PEG chains create a steric barrier. Overcoming this challenge involves designing strategies to enhance targeting and cellular penetration while maintaining the benefits of PEGylation. PEGylated moieties can further be conjugated with targeting ligands to improve the targeting efficiency of the designed nanocarriers. Few individuals may develop antibodies against PEG, leading to reduced efficacy and increased clearance of PEGylated nanocarriers. Developing strategies to overcome immune responses and minimize PEG-related immunogenicity is an active area of research (McSweeney et al. 2019). Despite the challenges associated with the PEGylation technique, it remains a most lucrative area of research endeavors related to drug delivery. Researchers are exploring innovative PEGylation methods, such as site-specific or controlled PEGylation, which allow for precise control over the location and density of PEG chains. These techniques aim to optimize the benefits of PEGylation while minimizing its potential drawbacks. Current developments in the formulation of PEGylated nanocarriers focus on designing nanocarriers with integrated functionalities beyond PEGylation. These may include incorporating targeting ligands, stimuli-responsive components, or imaging agents, thereby expanding the capabilities of nanocarriers for personalized and targeted therapy (Ekladious et al. 2019).

Researchers are investigating alternative polymers or coatings that can provide similar benefits to PEGylation while addressing some of its challenges. These include zwitterionic polymers, biomimetic coatings, or self-assembled monolayers, which offer improved stability, reduced immunogenicity, and enhanced targeting capabilities (Amoozgar and Yeo 2012). The future of nanocarrier-based drug delivery lies in combination therapies, where multiple therapeutic agents, such as drugs, siRNA, or immunotherapies, are co-delivered using PEGylated nanocarriers. This approach enables synergistic effects and improved treatment outcomes (Zhang et al. 2016b). Furthermore, with advancements in nanotechnology and PEGylation techniques, the future of nanocarriers lies in personalized medicine. Tailoring nanocarriers to specific patient profiles, disease characteristics, and therapeutic requirements holds excellent potential for improving treatment efficacy and minimizing side effects (Sakamoto et al. 2010).

While challenges exist, the field of PEGylation and nanocarrier-based drug delivery continues to evolve rapidly. With ongoing research, innovative techniques, and a better understanding of nanocarrier behavior, the future looks promising for overcoming current challenges and harnessing the full potential of PEGylated nanocarriers in clinical applications.

## 6.8 Conclusion

Nanocarriers are lucrative drug delivery vehicles, considering the unique advantages they offer. However, the immunogenicity concerns associated with nanocarriers constitute a significant obstacle in their applications. Various surface modification strategies have been explored to tackle the immune recognition of nanocarriers. One of the most prominent approaches is surface modification of nanocarriers with PEG, also known as PEGylation. The PEGylation has been regarded as a widely utilized strategy to improve the performance of nanocarriers in drug delivery. It offers benefits such as enhanced stability, prolonged circulation time, improved biocompatibility, and reduced immunogenicity.

Research in PEGylation continues to advance, exploring alternative coatings, site-specific PEGylation, controlled release systems, immunogenicity mitigation, combination therapies, advanced characterization techniques, and standardization. These future directions aim to overcome the limitations of PEGylation, expand its applications, optimize its performance, and ensure its safe and effective translation into clinical use. As research progresses, it is expected that PEGylation-based technologies will continue to play a significant role in drug delivery, biotechnology, and diagnostics. Through ongoing efforts and advancements, the field of PEGylation holds promise for addressing current challenges and further improving the effectiveness and safety of nanocarriers in various therapeutic applications.

## References

Almeida AJ, Souto E (2007) Solid lipid nanoparticles as a drug delivery system for peptides and proteins. Adv Drug Deliv Rev 59:478–490

Amoozgar Z, Yeo Y (2012) Recent advances in stealth coating of nanoparticle drug delivery systems. Wiley Interdiscip Rev Nanomed Nanobiotechnol 4:219–233

Anaya J-M, Shoenfeld Y, Rojas-Villarraga A, Levy R, Cervera R (2013) Autoimmunity: from bench to bedside. El Rosario University Press, Bogota

Baker MP, Reynolds HM, Lumicisi B, Bryson CJ (2010) Immunogenicity of protein therapeutics: the key causes, consequences and challenges. Self Nonself 1:314–322

Bellido E, Hidalgo T, Lozano MV, Guillevic M, Simón-Vázquez R, Santander-Ortega MJ, González-Fernández Á, Serre C, Alonso MJ, Horcajada P (2015) Heparin-engineered mesoporous iron metal-organic framework nanoparticles: toward stealth drug nanocarriers. Adv Healthc Mater 4:1246–1257

Caliceti P, Veronese FM (2003) Pharmacokinetic and biodistribution properties of poly(ethylene glycol)-protein conjugates. Adv Drug Deliv Rev 55:1261–1277

Castro F, Pinto ML, Silva AM, Pereira CL, Teixeira GQ, Gomez-Lazaro M, Santos SG, Barbosa MA, Gonçalves RM, Oliveira MJ (2017) Pro-inflammatory chitosan/poly(γ-glutamic acid) nanoparticles modulate human antigen-presenting cells phenotype and revert their pro-invasive capacity. Acta Biomater 63:96–109

Chaplin DD (2010) Overview of the immune response. J Allergy Clin Immunol 125:S3–S23

Cheng Z, Li M, Dey R, Chen Y (2021) Nanomaterials for cancer therapy: current progress and perspectives. J Hematol Oncol 14:85

Cooper D, Eleftherianos I (2017) Memory and specificity in the insect immune system: current perspectives and future challenges. Front Immunol 8:539

Crawford L, Rosch J, Putnam D (2016) Concepts, technologies, and practices for drug delivery past the blood–brain barrier to the central nervous system. J Control Release 240:251–266

De Groot MC, Van Zwieten-Boot BJ, Van Grootheest AC (2004) [Severe adverse reactions after the use of Sulphur hexafluoride (SonoVue) as an ultrasonographic contrast agent]. Ned Tijdschr Geneeskd 148: 1887–1888

Debele TA, Yeh CF, Su WP (2020) Cancer immunotherapy and application of nanoparticles in cancers immunotherapy as the delivery of immunotherapeutic agents and as the immunomodulators. Cancers (Basel) 12:3773

Den Haan JMM, Arens R, Van Zelm MC (2014) The activation of the adaptive immune system: cross-talk between antigen-presenting cells, T cells and B cells. Immunol Lett 162:103–112

Din FU, Aman W, Ullah I, Qureshi OS, Mustapha O, Shafique S, Zeb A (2017) Effective use of nanocarriers as drug delivery systems for the treatment of selected tumors. Int J Nanomedicine 12:7291–7309

Dobrovolskaia MA, McNeil SE (2007) Immunological properties of engineered nanomaterials. Nat Nanotechnol 2:469–478

Donahue ND, Acar H, Wilhelm S (2019) Concepts of nanoparticle cellular uptake, intracellular trafficking, and kinetics in nanomedicine. Adv Drug Deliv Rev 143:68–96

Du B, Jiang X, Huang Y, Li S, Lin JC, Yu M, Zheng J (2020) Tailoring kidney transport of organic dyes with low-molecular-weight PEGylation. Bioconjug Chem 31:241–247

Duan X, Li Y (2013) Physicochemical characteristics of nanoparticles affect circulation, biodistribution, cellular internalization, and trafficking. Small 9:1521–1532

Ebner A, Wildling L, Kamruzzahan ASM, Rankl C, Wruss J, Hahn CD, Hölzl M, Zhu R, Kienberger F, Blaas D, Hinterdorfer P, Gruber HJ (2007) A new, simple method for linking of antibodies to atomic force microscopy tips. Bioconjug Chem 18:1176–1184

Ekladious I, Colson YL, Grinstaff MW (2019) Polymer–drug conjugate therapeutics: advances, insights and prospects. Nat Rev Drug Discov 18:273–294

Fernandes AI, Gregoriadis G (2001) The effect of polysialylation on the immunogenicity and antigenicity of asparaginase: implication in its pharmacokinetics. Int J Pharm 217:215–224

Fontana A, Spolaore B, Mero A, Veronese FM (2008) Site-specific modification and PEGylation of pharmaceutical proteins mediated by transglutaminase. Adv Drug Deliv Rev 60:13–28

Fu D, Wang Z, Tu Y, Peng F (2021) Interactions between biomedical micro-/nano-motors and the immune molecules, immune cells, and the immune system: challenges and opportunities. Adv Healthc Mater 10:2001788

Fulton MD, Najahi-Missaoui W (2023) Liposomes in cancer therapy: how did we start and where are we now. Int J Mol Sci 24:6615

Gamucci O, Bertero A, Gagliardi M, Bardi G (2014) Biomedical nanoparticles: overview of their surface immune-compatibility. Coatings 4:139–159

Gulati NM, Pitek AS, Steinmetz NF, Stewart PL (2017) Cryo-electron tomography investigation of serum albumin-camouflaged tobacco mosaic virus nanoparticles. Nanoscale 9:3408–3415

Gulati NM, Pitek AS, Czapar AE, Stewart PL, Steinmetz NF (2018) The in vivo fates of plant viral nanoparticles camouflaged using self-proteins: overcoming immune recognition. J Mater Chem B 6:2204–2216

Gupta V, Bhavanasi S, Quadir M, Singh K, Ghosh G, Vasamreddy K, Ghosh A, Siahaan TJ, Banerjee S, Banerjee SK (2019) Protein PEGylation for cancer therapy: bench to bedside. J Cell Commun Signal 13:319–330

Hadjesfandiari N, Parambath A (2018) Stealth coatings for nanoparticles: polyethylene glycol alternatives. In: Engineering of biomaterials for drug delivery systems. Woodhead Publishing, Sawston

Harris JM, Martin NE, Modi M (2001) Pegylation: a novel process for modifying pharmacokinetics. Clin Pharmacokinet 40:539–551

Hatakeyama H, Akita H, Harashima H (2011) A multifunctional envelope type nano device (MEND) for gene delivery to tumours based on the EPR effect: a strategy for overcoming the PEG dilemma. Adv Drug Deliv Rev 63:152–160

Howard MD, Jay M, Dziubla TD, Lu X (2008) PEGylation of nanocarrier drug delivery systems: state of the art. J Biomed Nanotechnol 4:133–148

Hu C-MJ, Zhang L, Aryal S, Cheung C, Fang RH, Zhang L (2011) Erythrocyte membrane-camouflaged polymeric nanoparticles as a biomimetic delivery platform. Proc Natl Acad Sci U S A 108:10980–10985

Hu C-MJ, Fang RH, Wang K-C, Luk BT, Thamphiwatana S, Dehaini D, Nguyen P, Angsantikul P, Wen CH, Kroll AV, Carpenter C, Ramesh M, Qu V, Patel SH, Zhu J, Shi W, Hofman FM, Chen TC, Gao W, Zhang K, Chien S, Zhang L (2015) Nanoparticle biointerfacing by platelet membrane cloaking. Nature 526:118–121

Jia Y, Li J (2015) Molecular assembly of Schiff Base interactions: construction and application. Chem Rev 115:1597–1621

Kolate A, Baradia D, Patil S, Vhora I, Kore G, Misra A (2014) PEG—versatile conjugating ligand for drugs and drug delivery systems. J Control Release 192:67–81

Kozma GT, Shimizu T, Ishida T, Szebeni J (2020) Anti-PEG antibodies: properties, formation, testing and role in adverse immune reactions to PEGylated nano-biopharmaceuticals. Adv Drug Deliv Rev 154–155:163–175

Le Dévédec F, Strandman S, Hildgen P, Leclair G, Zhu XX (2013) PEGylated bile acids for use in drug delivery systems: enhanced solubility and bioavailability of itraconazole. Mol Pharm 10:3057–3066

Lee JH, Yeo Y (2015) Controlled drug release from pharmaceutical nanocarriers. Chem Eng Sci 125:75–84

Li B, Wang F, Gui L, He Q, Yao Y, Chen H (2018) The potential of biomimetic nanoparticles for tumor-targeted drug delivery. Nanomedicine (Lond) 13:2099–2118

Liu Y, Hardie J, Zhang X, Rotello VM (2017) Effects of engineered nanoparticles on the innate immune system. Semin Immunol 34:25–32

Lombardo D, Kiselev MA, Caccamo MT (2019) Smart nanoparticles for drug delivery application: development of versatile nanocarrier platforms in biotechnology and nanomedicine. J Nanomater 2019:3702518

McSweeney MD, Price LSL, Wessler T, Ciociola EC, Herity LB, Piscitelli JA, Dewalle AC, Harris TN, Chan AKP, Saw RS, Hu P, Jennette JC, Forest MG, Cao Y, Montgomery SA, Zamboni WC, Lai SK (2019) Overcoming anti-PEG antibody mediated accelerated blood clearance of PEGylated liposomes by pre-infusion with high molecular weight free PEG. J Control Release 311–312:138–146

Mitchell MJ, Billingsley MM, Haley RM, Wechsler ME, Peppas NA, Langer R (2021) Engineering precision nanoparticles for drug delivery. Nat Rev Drug Discov 20:101–124

Mohamed NA, Marei I, Crovella S, Abou-Saleh H (2022) Recent developments in nanomaterials-based drug delivery and upgrading treatment of cardiovascular diseases. Int J Mol Sci 23:1404

Moses JE, Moorhouse AD (2007) The growing applications of click chemistry. Chem Soc Rev 36:1249–1262

Naing A, Papadopoulos KP, Autio KA, Ott PA, Patel MR, Wong DJ, Falchook GS, Pant S, Whiteside M, Rasco DR, Mumm JB, Chan IH, Bendell JC, Bauer TM, Colen RR, Hong DS, Van Vlasselaer P, Tannir NM, Oft M, Infante JR (2016) Safety, antitumor activity, and immune activation of pegylated recombinant human INTERLEUKIN-10 (AM0010) in patients with advanced solid tumors. J Clin Oncol 34:3562–3569

Nowinski AK, White AD, Keefe AJ, Jiang S (2014) Biologically inspired stealth peptide-capped gold nanoparticles. Langmuir 30:1864–1870

Parhi P, Mohanty C, Sahoo SK (2012) Nanotechnology-based combinational drug delivery: an emerging approach for cancer therapy. Drug Discov Today 17:1044–1052

Parodi A, Quattrocchi N, Van De Ven AL, Chiappini C, Evangelopoulos M, Martinez JO, Brown BS, Khaled SZ, Yazdi IK, Enzo MV, Isenhart L, Ferrari M, Tasciotti E (2013) Synthetic nanoparticles functionalized with biomimetic leukocyte membranes possess cell-like functions. Nat Nanotechnol 8:61–68

Peer D, Florentin A, Margalit R (2003) Hyaluronan is a key component in cryoprotection and formulation of targeted unilamellar liposomes. Biochim Biophys Acta 1612:76–82

Pérez-Pérez L, García-Gavín J, Piñeiro B, Zulaica A (2011) Biologic-induced urticaria due to polysorbate 80: usefulness of prick test. Br J Dermatol 164:1119–1120

Perry JL, Reuter KG, Kai MP, Herlihy KP, Jones SW, Luft JC, Napier M, Bear JE, Desimone JM (2012) PEGylated PRINT nanoparticles: the impact of PEG density on protein binding, macrophage association, biodistribution, and pharmacokinetics. Nano Lett 12:5304–5310

Piao J-G, Wang L, Gao F, You Y-Z, Xiong Y, Yang L (2014) Erythrocyte membrane is an alternative coating to polyethylene glycol for prolonging the circulation lifetime of gold nanocages for photothermal therapy. ACS Nano 8:10414–10425

Qiao Z-Y, Zhao W-J, Gao Y-J, Cong Y, Zhao L, Hu Z, Wang H (2017) Reconfigurable peptide nanotherapeutics at tumor microenvironmental pH. ACS Appl Mater Interfaces 9:30426–30436

Qin M, Du G, Sun X (2020) Biomimetic cell-derived nanocarriers for modulating immune responses. Biomater Sci 8:530–543

Rabanel JM, Hildgen P, Banquy X (2014) Assessment of PEG on polymeric particles surface, a key step in drug carrier translation. J Control Release 185:71–87

Rawat M, Singh D, Saraf S, Saraf S (2006) Nanocarriers: promising vehicle for bioactive drugs. Biol Pharm Bull 29:1790–1798

Rodriguez PL, Harada T, Christian DA, Pantano DA, Tsai RK, Discher DE (2013) Minimal "self" peptides that inhibit phagocytic clearance and enhance delivery of nanoparticles. Science 339:971–975

Saito G, Swanson JA, Lee K-D (2003) Drug delivery strategy utilizing conjugation via reversible disulfide linkages: role and site of cellular reducing activities. Adv Drug Deliv Rev 55:199–215

Sakamoto JH, Van De Ven AL, Godin B, Blanco E, Serda RE, Grattoni A, Ziemys A, Bouamrani A, Hu T, Ranganathan SI, De Rosa E, Martinez JO, Smid CA, Buchanan RM, Lee S-Y, Srinivasan S, Landry M, Meyn A, Tasciotti E, Liu X, Decuzzi P, Ferrari M (2010) Enabling individualized therapy through nanotechnology. Pharmacol Res 62:57–89

Schöttler S, Becker G, Winzen S, Steinbach T, Mohr K, Landfester K, Mailänder V, Wurm FR (2016) Protein adsorption is required for stealth effect of poly(ethylene glycol)- and poly(phosphoester)-coated nanocarriers. Nat Nanotechnol 11:372–377

Sellaturay P, Nasser S, Ewan P (2021) Polyethylene glycol-induced systemic allergic reactions (Anaphylaxis). J Allergy Clin Immunol Pract 9:670–675
Shi D, Beasock D, Fessler A, Szebeni J, Ljubimova JY, Afonin KA, Dobrovolskaia MA (2022) To PEGylate or not to PEGylate: immunological properties of nanomedicine's most popular component, polyethylene glycol and its alternatives. Adv Drug Deliv Rev 180:114079
Sinz A (2006) Chemical crosslinking and mass spectrometry to map three-dimensional protein structures and protein-protein interactions. Mass Spectrom Rev 25:663–682
Su S, Kang PM (2020) Recent advances in nanocarrier-assisted therapeutics delivery systems. Pharmaceutics 12:837
Suk JS, Xu Q, Kim N, Hanes J, Ensign LM (2016) PEGylation as a strategy for improving nanoparticle-based drug and gene delivery. Adv Drug Deliv Rev 99:28–51
Sun Q, Zhou Z, Qiu N, Shen Y (2017) Rational design of cancer nanomedicine: nanoproperty integration and synchronization. Adv Mater 29
Sung H-J, Luk A, Murthy NS, Liu E, Jois M, Joy A, Bushman J, Moghe PV, Kohn J (2010) Poly(ethylene glycol) as a sensitive regulator of cell survival fate on polymeric biomaterials: the interplay of cell adhesion and pro-oxidant signaling mechanisms. Soft Matter 6:5196–5205
Tang L, Azzi J, Kwon M, Mounayar M, Tong R, Yin Q, Moore R, Skartsis N, Fan TM, Abdi R, Cheng J (2012) Immunosuppressive activity of size-controlled PEG-PLGA nanoparticles containing encapsulated cyclosporine A. J Transplant 2012:896141
Thasneem YM, Sajeesh S, Sharma CP (2013) Glucosylated polymeric nanoparticles: a sweetened approach against blood compatibility paradox. Colloids Surf B: Biointerfaces 108:337–344
Thomas SN, Van Der Vlies AJ, O'Neil CP, Reddy ST, Yu SS, Giorgio TD, Swartz MA, Hubbell JA (2011) Engineering complement activation on polypropylene sulfide vaccine nanoparticles. Biomaterials 32:2194–2203
Tian Y, Gao Z, Wang N, Hu M, Ju Y, Li Q, Caruso F, Hao J, Cui J (2022) Engineering poly(ethylene glycol) nanoparticles for accelerated blood clearance inhibition and targeted drug delivery. J Am Chem Soc 144:18419–18428
Toole BP (2004) Hyaluronan: from extracellular glue to pericellular cue. Nat Rev Cancer 4:528–539
Totten JD, Wongpinyochit T, Carrola J, Duarte IF, Seib FP (2019) PEGylation-dependent metabolic rewiring of macrophages with silk fibroin nanoparticles. ACS Appl Mater Interfaces 11:14515–14525
Tsai CH, Wang PY, Lin IC, Huang H, Liu GS, Tseng CL (2018) Ocular drug delivery: role of degradable polymeric nanocarriers for ophthalmic application. Int J Mol Sci 19:2830
Verhoef JJ, Carpenter JF, Anchordoquy TJ, Schellekens H (2014) Potential induction of anti-PEG antibodies and complement activation toward PEGylated therapeutics. Drug Discov Today 19:1945–1952
Vllasaliu D, Fowler R, Stolnik S (2014) PEGylated nanomedicines: recent progress and remaining concerns. Expert Opin Drug Deliv 11:139–154
Wen P, Ke W, Dirisala A, Toh K, Tanaka M, Li J (2023) Stealth and pseudo-stealth nanocarriers. Adv Drug Deliv Rev 198:114895
Wildling L, Unterauer B, Zhu R, Rupprecht A, Haselgrübler T, Rankl C, Ebner A, Vater D, Pollheimer P, Pohl EE, Hinterdorfer P, Gruber HJ (2011) Linking of sensor molecules with amino groups to amino-functionalized AFM tips. Bioconjug Chem 22:1239–1248
Wilson DR, Zhang N, Silvers AL, Forstner MB, Bader RA (2014) Synthesis and evaluation of cyclosporine A-loaded polysialic acid-polycaprolactone micelles for rheumatoid arthritis. Eur J Pharm Sci 51:146–156
Xia W, Tao Z, Zhu B, Zhang W, Liu C, Chen S, Song M (2021) Targeted delivery of drugs and genes using polymer nanocarriers for cancer therapy. Int J Mol Sci 22:2830
Zalba S, Ten Hagen TLM, Burgui C, Garrido MJ (2022) Stealth nanoparticles in oncology: facing the PEG dilemma. J Control Release 351:22–36
Zalipsky S (1995) Functionalized poly(ethylene glycols) for preparation of biologically relevant conjugates. Bioconjug Chem 6:150–165

Zarreen Simnani F, Singh D, Patel P, Choudhury A, Sinha A, Nandi A, Kumar Samal S, Verma SK, Kumar Panda P (2023) Nanocarrier vaccine therapeutics for global infectious and chronic diseases. Mater Today 66:371–408

Zhang P, Sun F, Liu S, Jiang S (2016a) Anti-PEG antibodies in the clinic: current issues and beyond PEGylation. J Control Release 244:184–193

Zhang RX, Wong HL, Xue HY, Eoh JY, Wu XY (2016b) Nanomedicine of synergistic drug combinations for cancer therapy—strategies and perspectives. J Control Release 240:489–503

Zheng X, Bao Y, Huang A, Yu L, Qin G (2023) Temperature dependence of thermophysical properties of polyethylene glycol in solid/liquid phase change region. J Chem Thermodyn 180:107022

Zolnik BS, González-Fernández AF, Sadrieh N, Dobrovolskaia MA (2010) Minireview: nanoparticles and the immune system. Endocrinology 151:458–465

# PEGylated Nanocarrier as a Promising Tool for Site-Specific Delivery of Therapeutics

7

Sweta Acharya, Niyati Lad, Aniket Navale, Simranjit Kaur, Aprameya Ganesh Prasad, and Rakesh Kumar Tekade

**Abstract**

Nanocarrier systems have gained popularity for drug delivery and diagnostic purposes, offering targeted and site-specific transport of highly efficacious medications. These systems connect biological and physical sciences, allowing for the treatment of various diseases and disorders. They can avoid severe side effects, improve target specificity, and reduce doses. Modifying carrier properties, such as targeted delivery, regulated distribution, and shielding from biological milieus, can further enhance their benefits. Several targeted nanocarrier systems such as nano-crystals, lipid nanoparticles (NPs), PEGylated polymeric nanocarriers, nanofibers, quantum dots, liposomes, dendrimers, micelles, protein-based nanoparticles, and metal-based nanoparticles are well explored for their benefits over conventional drug delivery systems. This chapter specifically highlights the different types of PEGylated nanocarriers and their uses in targeted drug delivery.

**Keywords**

PEGylated nanocarriers · Liposomes · Cancer · Brain disorders · Pulmonary disorders · Targeted delivery · EPR effect · RES system

S. Acharya · N. Lad · A. Navale · S. Kaur · R. K. Tekade (✉)
National Institute of Pharmaceutical Education and Research (NIPER) Ahmedabad, An Institute of National Importance, Government of India, Department of Pharmaceuticals, Ministry of Chemicals and Fertilizers, Palaj, Opp. Air force stations, Gandhinagar, Gujarat, India

A. G. Prasad
Department of Chemical and Biomolecular Engineering, Johns Hopkins University, Baltimore, MD, USA


R. K. Tekade, N. K. Jain (eds.), *PEGylated Nanocarriers in Medicine and Pharmacy*, https://doi.org/10.1007/978-981-97-7316-9_7

## Abbreviations

| | |
|---|---|
| Anti-hMAM | Anti-human mammaglobin |
| BBB | Blood brain barrier |
| Bcl-2 | B-cell lymphoma-2 |
| BPQDs | Black phosphorus quantum dots |
| BSA | Bovine serum albumin |
| CaP | Calcium phosphate |
| CAT | Catalase |
| ChNPs | Chitosan nanoparticles |
| CNTs | Carbon nanotubes |
| DDS | Drug delivery systems |
| DMARDs | Disease-modifying anti-rheumatic drugs |
| DMMA-P-DOX/LAP | 2,3-Dimethylmaleic-anhydride-poly(ethylene glycol)-$\varepsilon$-poly-L-lysine-doxorubicin/lapatinib polymeric |
| DOPE | 1,2-Dioleyl-*sn*-glycero-3-phosphoethanolamine |
| DOX | Doxorubicin |
| DPI | Dry powder inhaler |
| EAE | Encephalomyelitis |
| EGFP | Enhanced green fluorescent protein |
| EGFR | Epidermal growth factor receptor |
| EPR | Enhanced permeability and retention |
| ERs | Estrogen receptors |
| FA | Folic acid |
| FMN | Formononetin |
| GBM | Glioblastoma multiforme |
| GCs | Glucocorticoids |
| GEM | Gemcitabine |
| GFLG | Glycylphenylalanylleucylglycine |
| GO | Graphene oxide |
| GOD | Glucose oxidase |
| GSH | Glutathione |
| HA | Hyaluronic acid |
| Has | Human serum albumin |
| HEL | Hen egg-white lysozyme |
| HER2 | Human epidermal growth factor receptor 2 |
| HPMA | Hydroxy propyl methacrylamide-methacid |
| IA | Intra-articular |
| LRP | Low-density lipoprotein receptor |
| mAbs | Monoclonal antibodies |
| MCF-7 | Michigan cancer foundation-7 |
| Mcl-1 | Myeloid cell leukemia-1 |
| MMP | Mitochondrial membrane potential |
| MWNTs | Multi-walled carbon nanotubes |
| NCs | Nanocarriers |

| | |
|---|---|
| NGO | Nanosized graphene oxide |
| NIR | Near-infrared |
| NPs | Nanoparticles |
| NSAIDs | Nonsteroidal anti-inflammatory drugs |
| NSCLC | Non-small-cell lung carcinoma |
| OA | Osteoarthritis |
| OPSS-PEG-ProG | Orthopyridyldisulfide-polyethylene glycol-succinimidyl valerate |
| PAMAM | Polyamidoamine |
| PCL | Polycaprolactone |
| PDCs | Polymer-drug conjugates |
| PdNP | Palladium nanoparticle |
| PDT | Photodynamic therapy |
| PEG | Poly-ethylene glycol |
| PEGMnCaP | Pegylated $Mn^{2+}$-doped calcium phosphate |
| PEI | Polyethyleneimine |
| PGA | Polyglycolic acid |
| PIC | Polyion complex |
| PLA | Polylactic acid |
| PLGA | Poly(lactic-co-glycolic acid) |
| POEGMEMA | Poly(oligo(ethylene glycol) methyl ether methacrylate) |
| PPI | Polypropyleneimine |
| ProA | Protein-A |
| ProG | Protein-G |
| PTT | Photothermal therapy |
| PTX | Paclitaxel |
| QDs | Quantum dots |
| RES | Reticulo-endothelial system |
| RGD | Arginyl-glycyl-aspartic acid-peptide |
| RME | Receptor-mediated endocytosis |
| Sct | Salmon calcitonin |
| SiRNA | Small interfering RNA |
| SPIONs | Superparamagnetic iron oxide nanoparticles |
| SWNTs | Single-walled carbon nanotubes |
| Tf | Transferrin |
| VEGF | Vascular endothelial growth factor |

## 7.1 Introduction

The use of nanocarrier systems has been widely explored for the past few years for the delivery of drugs as well as for diagnostic purposes. Also, it is advancing from academic research to the field of commercial success and clinical use. The general availability of several nano-drug delivery systems serves as evidence of this. When it comes to treating chronic human diseases, nanotechnology has many advantages because of the targeted and site-specific transportation of highly efficient medications. These systems have been shown to connect the dots between biological and physical sciences (Patra et al. 2018). These systems are used for the delivery of an extensive array of compounds, such as chemotherapeutics, biological molecules, nucleic acids, enzymes, and peptides for the treatment of various diseases and disorders. With their application, severe side effects associated with anticancer drugs such as multiple organ failure, thrombocytopenia, and anemia can be avoided along with better target specificity and dose reduction. They are capable of passing across the smallest channels and biological barriers such as cell membranes, which are crucial for the transport of medications to intracellular areas of action, as well as epithelial and endothelial barriers and they are also capable of evading the phagocytes (Vllasaliu et al. 2014).

Because of their small size, usually in nanometers, they can remain in the bloodstream for a longer time. Additional significant benefits for drug delivery can be obtained by modifying the carrier's properties, including targeted delivery, regulated or stimuli-responsive distribution, and shielding of the drug from biological milieus. They have the capability to control the release of loaded drugs, due to their unique capability to biodegrade, heat sensitivity of structural elements, and pH, making them suitable for the administration of drugs or molecules (Farjadian et al. 2019). The following section expounds on their advantages over conventional drug systems.

## 7.2 Drawbacks of Conventional Drug Delivery Systems

One of the biggest challenges in the management and treatment of diseases is getting the effective drug to the desired spot. Major drawbacks of conventional drug delivery systems such as through the oral, buccal, sublingual, rectal, and subcutaneous routes are poor selectivity, insufficient bio-distribution, off-target effect, large molecular size, inability to cross the biological membranes, damage to healthy cells, and limited efficacy. Some physiological barriers of the body also hinder drug activity such as the blood-brain barrier by restricting the entry of the active component (Wilczewska et al. 2012). The majority of conventional systems have a large first explosion of drug release that occurs right after drug administration, and they also have a low tendency for drug solubility. Also, there is difficulty in eliminating the remnants of such systems, which can lead to the patient's body containing hazardous non-biodegradable substances. Targeted drug delivery systems can help to overcome the aforementioned limitations. The drug is carried to the site of action in controlled drug delivery systems, minimizing its impact on sensitive tissues and unfavorable side effects, especially in cancer treatment and brain-associated disorders. Additionally, they increase drug concentration in target tissues and shield them from quick oxidation or removal, thus reducing the dosage of the drug (Farjadian et al. 2019).

## 7.3 Nanocarrier-Based Targeted Drug Delivery

The fields of nanotechnology and nanomedicine have undergone a revolution as the number of nano-based medicinal compounds has increased dramatically since 1980. These innovative nano-based materials can either function as therapeutic agents or can be utilized to deliver active components to specific cells or tissues, with increased bioavailability as well as target specificity, ability to cross biological membranes, and prolong drug release. They can also escape endothelium at inflammatory regions, epithelium, malignancies, or micro-capillaries simply due to their small stature. Therapeutic compounds can be shielded from enzymatic breakdown by nucleases and proteases using an adaptation of this procedure. Numerous studies have shown that using nanoparticles (NPs) as a medication, delivery strategy has an array of benefits over conventional methods. The creation of nanoparticles using biodegradable materials enables prolonged drug release at the target site over days or even weeks (Singh and Lillard 2009).

These systems are used to enhance efficacy and reduce undesired side effects. The commonly marketed targeted nanocarrier system includes nanocrystals, lipid nanoparticles, PEGylated polymeric nanocarriers, nanofibers, quantum dots (QDs), liposomes, dendrimers, micelles, protein-based nanoparticles, and metal-based nanoparticles. The field of nanomedicine heavily relies on PEGylated nanocarriers (Vllasaliu et al. 2014). It has been widely known that PEGylation presents enormous potential for enhancing the efficacy of nanomedicines. This chapter will take a look at some of the PEGylated nanomedicines that are currently being investigated and highlight the benefits of PEGylation for drug delivery.

## 7.4 Methods of PEGylation

There are two approaches by which PEGylation can be achieved. Figure 7.1 is a schematic representation of the methods of PEGylation.

### 7.4.1 Covalent Approach

One of the methods for stabilizing proteins is PEG conjugation (PEGylation), which is commonly referred to as the covalent approach. Polyethylene glycol (PEG) is a hydrophilic, un-ionized, and safe polymer that offers proteins a steric barrier, improving the proteins' pharmacological properties. PEGylation also shields proteins in vitro from protease digestion and aggregation. However, covalent PEGylation necessitates a time-consuming and expensive chemical reaction to conjugate PEG to proteins (Kurinomaru and Shiraki 2015).

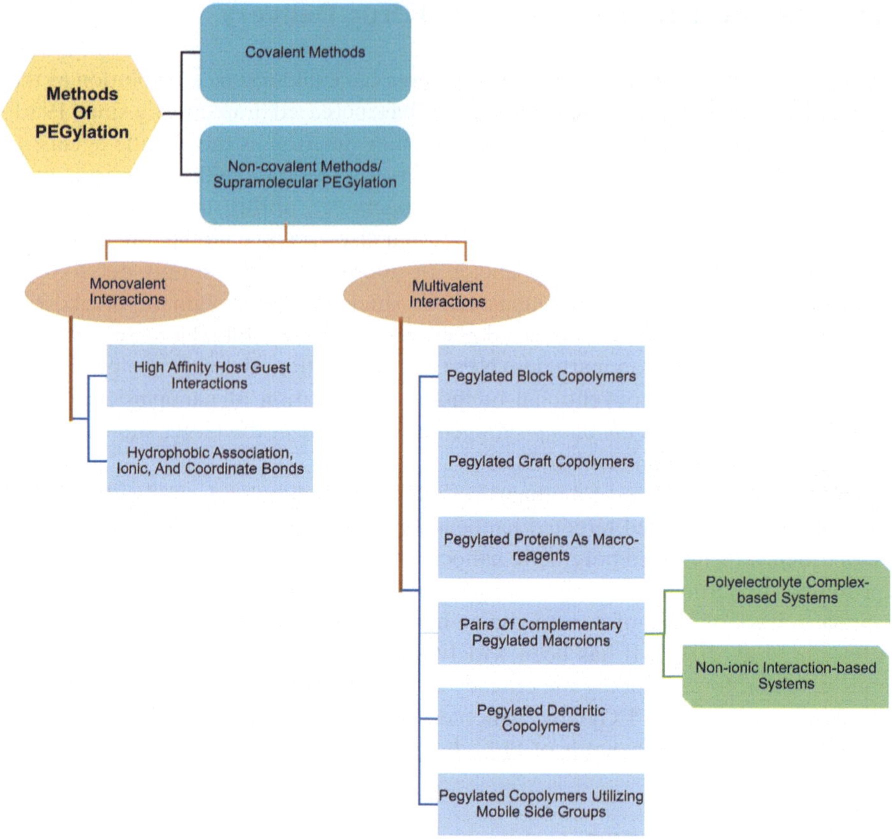


**Fig. 7.1** Various methods of PEGylation

### 7.4.2 Non-covalent Approach

A novel technique called non-covalent PEGylation avoids a chemical reaction between the PEG and the protein. It is based on the mechanisms of ionic interactions, protein polyelectrolyte complexes, hydrophobic interactions, or chelation. The main benefit of this method is that it prevents product loss that could result from additional purification procedures. However, a significant drawback of this strategy is that the protein is leached while being stored (Belén et al. 2019).

Protein stabilization using non-covalent PEGylation, also referred to as supramolecular PEGylation, has been proposed as an alternative technique. For non-covalent PEGylation, functional PEG derivatives that bind to proteins are frequently designed. Many PEG derivatives were conjugated with hydrophobic ligands, which prevented salmon calcitonin and lysozyme from aggregating. PEG derivatives have also been created that are conjugated with sugars, biotin, and nitrilotriacetic acid (Kurinomaru and Shiraki 2015; Andrianov 2023). Additionally, because of the

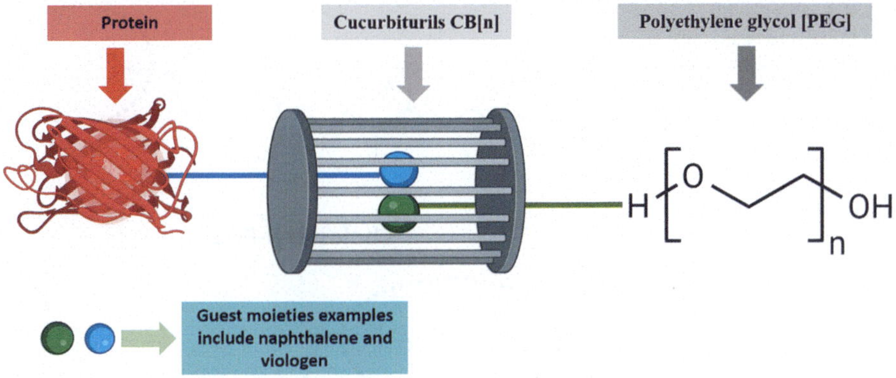


**Fig. 7.2** Representation of Cucurbiturils (CB[n]) mediated protein PEGylation utilizing guest moieties such as viologen and naphthalene-modified components

transitory nature of the PEG chain's bond with the protein, non-covalent PEGylation has been proposed to lower the danger of an immune response and facilitate straightforward approval by health authorities. The idea of supramolecular PEGylation goes beyond shielding proteins with ensuing drug release, much like releasable covalent PEGylation or other tunable drug release carriers. It stems from a defense mechanism, that involves dynamic protein modification mediated by different host-guest or polyelectrolyte exchange reactions. PEGylation that is non-covalent avoids site-specific attachment, enabling effective pharmacokinetic modulation using monovalent and multivalent strategies (Andrianov 2023).

#### 7.4.2.1 PEGylation Via Monovalent Interactions

The idea of functionalized PEG being attached to a protein surface at a single point is similar to how traditional PEGylation technology operates. As listed below, these are divided into different approaches.

##### High-Affinity Host-Guest Interactions

This method only works with proteins that have N-terminal aromatic residues, or the protein would need to be chemically altered. The protein surface must be treated with a small molecule recognition moiety in the latter scenario, but reactions involving end-functionalized PEG macromolecules will still be required. It is also common to refer to PEGylation involving host-guest interaction as "reversible" or "dynamic" PEGylation, implying the release of protein under specific circumstances in the absence of lysis of covalent bonds. The host-guest complexes that cause PEG-protein binding can be compared to *supramolecular "handcuffs"* that bind the components together as shown in Fig. 7.2.

##### Hydrophobic Association, Ionic, and Coordinate Bonds

Hen egg-white lysozyme (HEL) and salmon calcitonin (sCT) are two of several examples of proteins that have been stabilized against aggregation using

**Fig. 7.3** PEGs with functionalized structures that have a range of end groups that can interact with proteins non-covalently

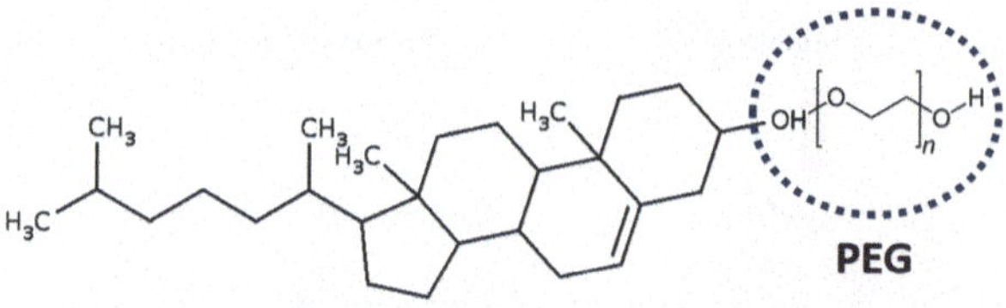


Cholesteryl-PEG

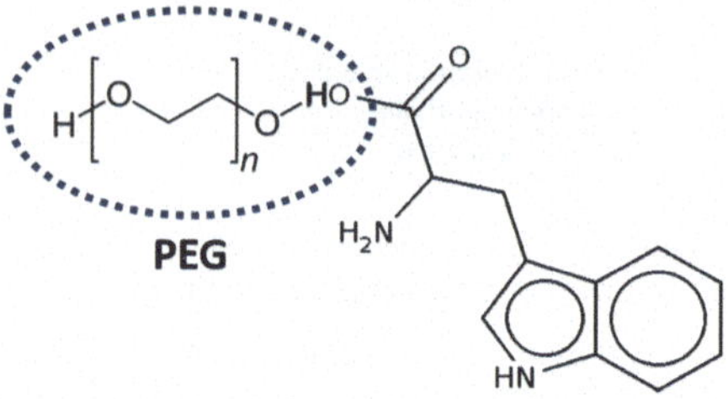


L-Tryptophan modified with PEG

mono-functionalized PEGs that contain benzyl-, phenyl butylamino-, cholesteryl-, dansyl-, and L-tryptophan. Some of the PEG functionalized derivatives are shown below in Fig. 7.3.

A unified site ionic bond or hydrophobic association drove the formation of complexes. The aggregation of sCT was lowered for up to 70 h by both dansyl and tryptophan derivatives. A benzyl-derivative of PEG caused some protein degradation, whereas cholesteryl functionalized PEG completely prevented HEL from aggregating. According to reports, PEGs that have been functionalized with phenylbutylamino, tryptophan, and dansyl extended the lag phase of protein aggregation and decreased its rate. Non-covalent PEGylation outperformed native protein by a factor of 3–4 as regards in vivo half-life, solubility, and bioavailability (Andrianov 2023).

#### 7.4.2.2 PEGylation Via Multivalent Interactions

Macroions are flexible and can locate oppositely charged surfaces on the protein—charge anisotropy model of interactions. This characteristic as well as entropy changes brought on by counterion release have both been shown to favor protein-polyelectrolyte interactions. Certain multivalent interactions utilize the “bind and slide” mechanism, where a counter-partner can switch between binding sites along a polymer chain. It is conceivable that the latter will advance the development of dynamic PEGylation, in which the protein’s active site is periodically made accessible to interact with its substrate molecule without chemical cleavage of the protein complex.

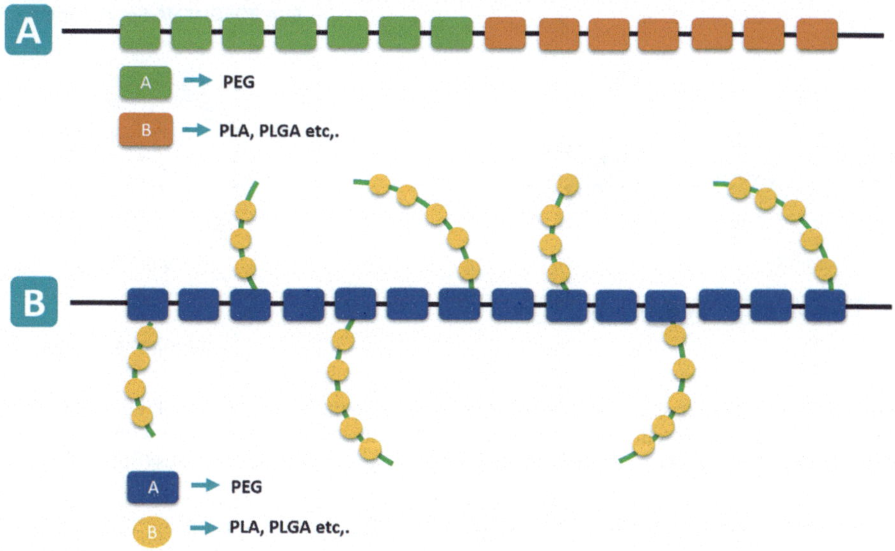


**Fig. 7.4** (**a**) PEGylated block copolymers. (**b**) PEGylated graft copolymers

### PEGylated Block Copolymers

Two or more distinct polymer chains joined by covalent bonds typically make up block copolymers as shown in Fig. 7.4 below. They can be effortlessly customized by mixing biodegradable polymers like polylactic acid (PLA), polyglycolic acid (PGA), poloxamers, polycaprolactone (PCL), poly (lactic-*co*-glycolic acid) (PLGA), and poly(ethylene glycol) (PEG). Block copolymers endowed with customized characteristics, chemical compositions, and molecular weights (MW) are now much easier to design, thanks to recent improvements in synthetic methods and thus been, widely used in the creation of nanomaterials to improve drug efficacy, decrease drug toxicity, and provide long-term therapeutic options. Other applications include precise programming of the drug release profile and surface modification with targeting ligands. Drug delivery is not the only application for block copolymers; they can also be used for tissue engineering, medical devices, and wound dressing (Agrahari and Agrahari 2018). The main distinction between a block copolymer and a graft copolymer is that the former has blocks of repeating units while the latter has branches of repeating units.

### PEGylated Graft Copolymers

Graft copolymers are made up of a main polymer chain called the backbone, to which one or more polymer sidechains are covalently joined, creating the branches as shown in Fig. 7.4. Typically, the branches' and backbone's chemical nature and composition are different (Sadeghi and Sayaf 2014; Madhu 2018). The complexation of two graft copolymers of Dextran Sulfate and PEGylated Pentosan Polysulfate with Keratinocyte Growth Factor-2 was used to investigate the role of

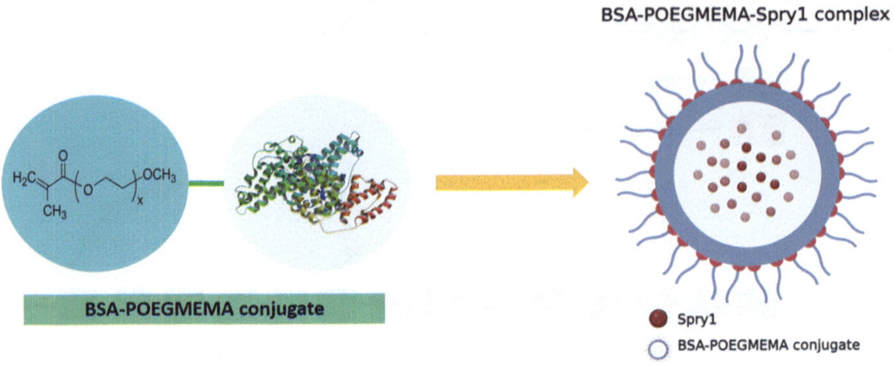


**Fig. 7.5** Schematic representation of the combination of the angiogenesis inhibitor Sprouty1 (Spry1) with poly(oligo (ethylene glycol) methyl ether methacrylate) modified with maleimide (MI-POEGMEMA) and bovine serum albumin (BSA) to create complex micelles (Spry1)

macromolecular architecture. An increase in the protein's melting temperature revealed that protein binding was decreased by both polymers' ability to stabilize the protein.

### PEGylated Proteins as Macro-reagents

PEGylated macro reagents are reagents that contain polyethylene glycol (PEG) chains attached to large molecules or structures, such as proteins, antibodies, liposomes, or nanoparticles. Protein-protein interactions as a basis for non-covalent PEGylation were also investigated. Covalent attachment of negatively charged bovine serum albumin (BSA) with poly(oligo(ethylene glycol) methyl ether methacrylate), POEGMEMA results in a "macro-reagent." This was tested for its ability to bind lysozyme and Sprouty 1 (Spry1), two positively charged proteins as shown in Fig. 7.5 below. The $IC_{50}$ value of Sprouty 1 was assessed against breast cancer cells and was found to be decreased as a result of this "protein-protein"-mediated modification. The complex's high anticancer effectiveness was also demonstrated by its ability to restrain the development of 3D MCF-7 multicellular tumor spheroids. While PEGylated protein mediator technology has some benefits like lowering the design complexity, it nevertheless requires the mediating protein to be covalently modified. The range of suitable interacting proteins is also restricted.

### Pairs of Complementary PEGylated Macroions

#### Polyelectrolyte Complex-Based Systems

Proteins can be coassembled with two PEGylated polyions that have opposing charges. This may have the benefit of increasing the environmental stability of polyelectrolyte complexes and multilayer coatings. The pathway typically starts with the complexation of the protein with the first polyion and is followed by the addition of an oppositely charged second polyelectrolyte. A pair of interacting non-ionic macromolecules may also be used to create a ternary polymer-protein-polymer

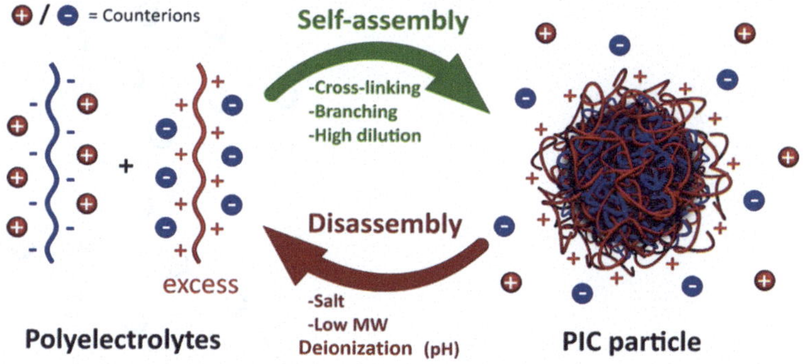


**Fig. 7.6** Counterion release after polyion self-assembly into PIC nanoparticles is depicted (charged spheres). In this instance, too much polycation in the mixture causes a neutral core to form, which is encircled by too much cationic material. (Adapted with permission from Insua et al. 2016 under the terms of the Creative Commons CC-BY license (https://s100.copyright.com/AppDispatchServlet?publisherName=ELS&contentID=S0014305716301860&orderBeanReset=true) © 2016 The Authors. Published by Elsevier Ltd)

assembly. In a solution, colloidal polyion complex (PIC) particles made of oppositely charged polyions can self-assemble as shown in Fig. 7.6 below.

#### Non-ionic Interaction-Based Systems

The co-assembly of the insulin hormone into a three-component complex during a dual-stage non-covalent process for PEGylation of insulin peptide was reported. First, insulin was complexed with zinc-chelated poly(aspartic acid-*co*-aspartglucosamine-*co*-aspartnitrilotriacetic acid) via the peptide's histidine groups. This was followed by the addition of a block copolymer of poly(ethylene glycol)-b-poly(aspartic acid-*co*-aspartamidophenylboronic acid) containing PEG. This produced PEG shell-stabilized submicron-sized (122–187 nm diameter) micelles (Andrianov 2023).

### PEGylated Dendritic Copolymers

The fundamental structure of telodendrimers is a diblock copolymer AB made up of a linear polymer (A) like PEG, a poly(ester), or a poly(amide); and a hyperbranched dendron like PAMAM-, polyester-, and polyamine (B). A schematic representation of telodendrimer is shown in Fig. 7.7. Most examples of telodendrimers that have been published are made of a hydrophilic linear polymer, like PEG, coupled with dendrons, like poly(lysine) or poly(ester), functionalized with hydrophobic moieties. Dendritic copolymers were utilized as yet another fascinating application of PEGylation that is completely dependent on non-covalent interactions. To create telodendrimers composed of a series of 11 copolymers, different combinations of charged functionalities were combined, including hydrophobic functionalities like heptadecanoic acid, cholesterol, and tocopherol, cationic functionalities like arginine and lysine, and anionic functionalities like oxalic acid.

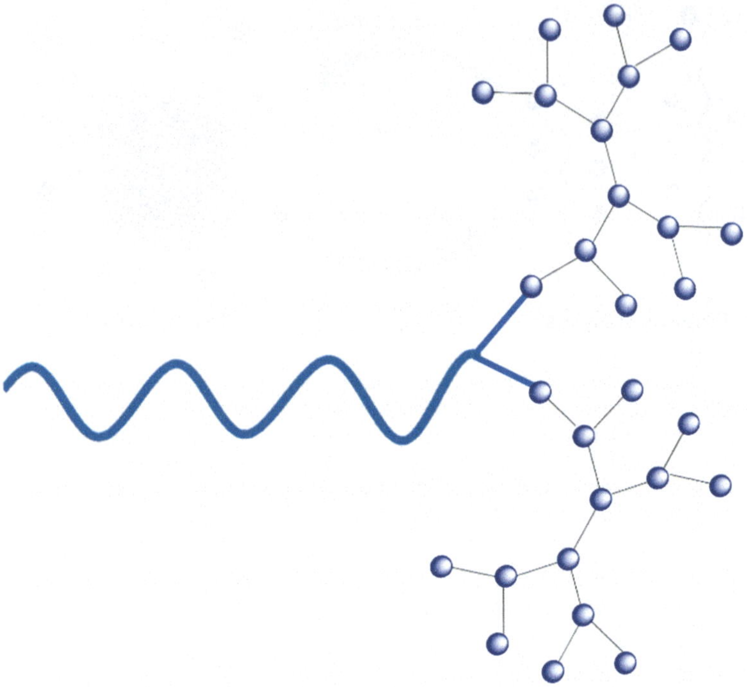

**Fig. 7.7** Typical representation of a telodendrimer

Utilizing isothermal titration calorimetry, electrophoresis, bio-layer interferometry, Forster resonance energy transfer spectroscopy, and other computational techniques, their interactions with proteins were investigated. The resultant nanoparticles loaded with protein were about 30 nm in size and effectively penetrated the cells. Furthermore, the enhanced permeability and retention (EPR) effect and enhanced peptide retention in an orthotopic brain tumor allowed the non-covalently bound nano-assemblies to effectively deliver proteins to xenografted tumors. Telodendrimers' dual multivalent ionic and hydrophobic functionalities have additive effects on protein binding affinity. Additionally, their hydrophobic groups stabilize protein binding in aqueous solutions by acting as annealing moieties (Andrianov 2023).

**PEGylated Copolymers Utilizing Mobile Side Groups**

Mobile side groups on PEGylated copolymers result in flexible and dynamic side chains on the polymer backbone. These side groups can interact with each other or with the environment, creating different shapes and structures of the copolymers. Polyrotaxane structural design, featuring cationic moieties and PEG grafts, significantly enhances polyelectrolyte-protein complexation (PEG-PRX). A diagrammatic representation of the PEG-PRX is shown in Fig. 7.8. PEG-PRX outperforms

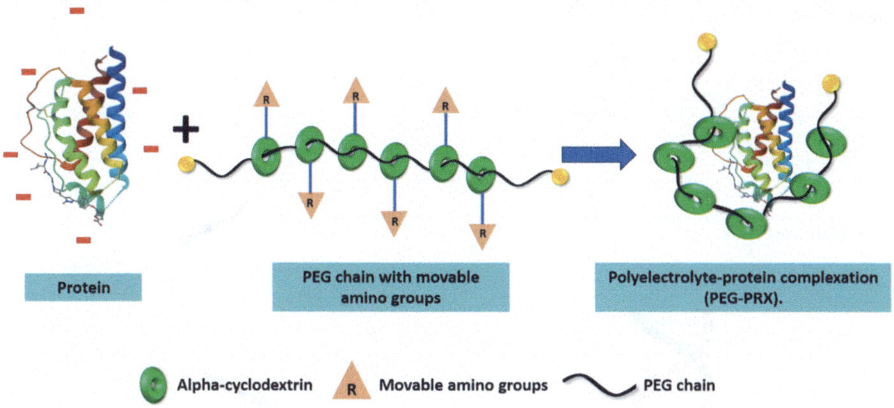


**Fig. 7.8** Copolymers of PEG with mobile side groups

PEG-DEX in protein uptake, despite non-covalent complexes with proteins like lysozyme, panitumumab, and insulin (Andrianov 2023).

## 7.5 Various Targeting Strategies

### 7.5.1 Active Targeting

EPR effect or passive targeting of the drugs alone cannot guarantee a successful cancer treatment since the drug may not reach the desired site in the intended or therapeutic concentrations, leading to therapeutic insufficiency of the drug. For an active targeting strategy, the cancer cells having many receptors that are overexpressed on their surface favor meeting enhanced demand of micronutrients and other factors required for the growth and survival of the cancer cells (Russell-Jones et al. 2004). For active targeting of the drugs, the backbones of the drug substances can be covered with a variety of ligands such as proteins, hormones, vitamins, and growth factors that are recognized by the cancer cells (Nateghian et al. 2016). For the cell-specific uptake, the surfaces of NPs are coated with specific ligands that can easily recognize and bind to the cell surfaces, as shown in Fig. 7.9 (Gajbhiye et al. 2020). The drug enters the tumor cell in one of two ways, depending on the linker molecule. The first one is receptor-mediated internalization of the entire prodrug by endocytosis and subsequent degradation by endosomal/lysosomal pathway or the second way is receptor-independent internalization of the drug into targeted cells after extracellular cleavage (Mishra et al. 2016). Table 7.1 below contains various active targeting strategies found in literature, along with the target receptors and intended use.

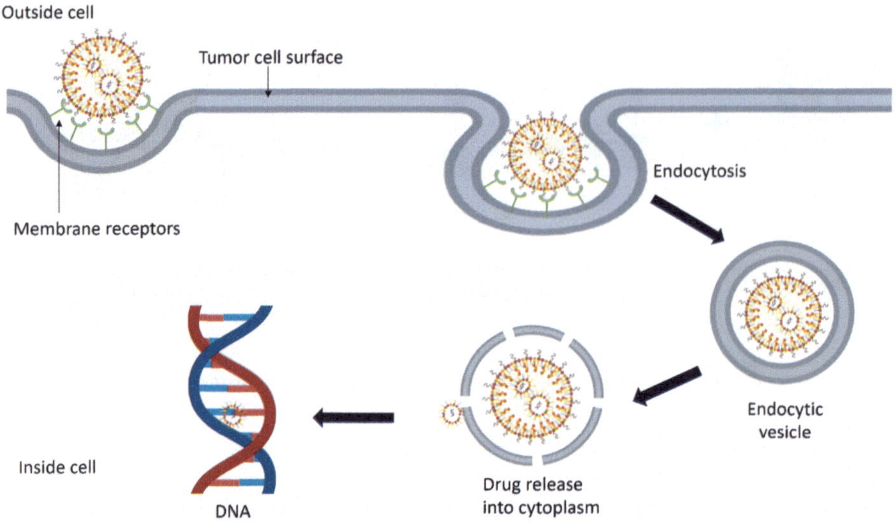


**Fig. 7.9** Active targeting of PEGylated nanocarriers

**Table 7.1** Various active targeting strategies

| Target receptor | Ligand | Spacer | Use | References |
|---|---|---|---|---|
| Several receptors | Antibodies and peptides | PLA-PEG | Targeting of thiol group containing drugs | Betancourt et al. (2009) |
| Microtubule receptors | Paclitaxel | PEG, PAMAM | Anti-cancer therapy | Rompicharla et al. (2019) |
| HER2 | Adenovirus | PEG | Metastatic cancer therapy | Kim et al. (2011) |
| Transferrin | Several anti-cancer drugs | PEG | Cancer therapy | Choi et al. (2010) |
| Folate receptors | Doxorubicin | PEG | Cancer therapy | McNeeley et al. (2007) |

## 7.5.2 Passive Targeting

Passive targeting in the case of cancer tissue is mainly due to the EPR effect (Fig. 7.10). The EPR effect consists of two mechanisms by which the tumor vasculature shows a leaky nature, viz.

1. Enhanced permeation due to the increased pore size on the tumor tissue surface.
2. Increased retention of the substances in the tumor region because of the compromised lymphatic drainage system.

The EPR effect, along with the characteristic cancer features such as increased acidity and unique tumor microenvironment can also serve as a guide to developing the tumor-targeted formulations. The EPR effect leads to the increased hydraulic conductivity of the tumor tissue, which enables macromolecules to freely cross

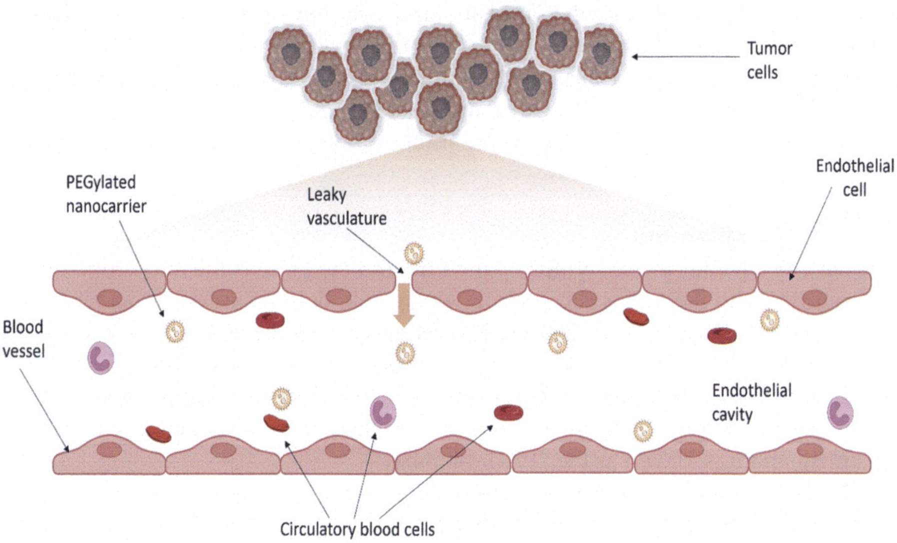


**Fig. 7.10** Passive targeting of PEGylated nanocarriers

through it. Normal tissues have lesser permeability, so the formulation may not accumulate in non-tumor regions (Mozar and Chowdhury 2018). PEGylation enhances the drugs' solubility, molecular mass, size, and serum stability (Mishra et al. 2016). Nanoparticles are more suitable for administration for tumor delivery because they can use the leaky vasculatures and compromised lymphatic drainage system of the solid tumors and hence enable them to passively accumulate into the tumor sites specifically. This also increases the retention time of the nanoparticles inside the tumor.

The PEG-modified nanoparticles possess advantages over uncoated nanoparticles such as:

1. Increased biocompatibility.
2. Diminished biological response.
3. Enhanced stability.
4. Delayed clearance by the MPS system (Wongpinyochit et al. 2015).

The size of the drug carrier has been controlled to achieve EPR-mediated delivery of the drugs, as the size of the molecules to permeabilize the vasculature varies from 200 to 800 nm. The EPR approach for targeting the drugs to tumor requires the macromolecular drugs and the drug delivery systems to possess the property of long circulation, i.e., stay in the body for longer periods without getting eliminated, so that the drug gets accumulated in the desired targeted organ. A simple approach to keep the drug at the target site is to mask it with the help of PEG moiety or with certain water-soluble polymers. Table 7.2 below includes some important applications of PEG moieties for passive targeting (Torchilin 2011).

The PEGylation approach for these purposes is successfully employed for liposomes, although it is thought to work with other drug carriers too. The important merit of prolonging the circulation of the drug and the carriers of the drug in the body is that it helps to maintain the desired therapeutic drug concentration at the target site. This approach is also useful for areas with limited blood supply and for ligand-mediated targeting of the drug molecules. It can also help to attain a sufficient amount of the drug in areas where it takes more time to achieve the targeted concentration (Kang et al. 2020).

#### 7.5.2.1 PEG Dilemma

The surface aqueous phase formed by the PEG moiety inhibits the interaction of the gene carrier with the tumor cell surface. As a result, cellular uptake is decreased to a large extent. Furthermore, PEGylation also improves the stability of nanoparticles, which results in poor endosomal escape because of the membrane fusion and the degradation of the cargoes in lysosomes or the digestive compartments. This issue sometimes arises with the use of the PEG moiety for gene delivery in cancer treatment and it is known as the "PEG dilemma."

The PEG dilemma can be successfully reduced by:

1. Using specific ligands, i.e., active targeting.
2. Cleavage of PEG from the carrier system.
3. Speeding up fusion or disruption of the membrane (Hatakeyama et al. 2011).

#### 7.5.2.2 Challenges in Drug Delivery by the EPR Effect

There are some challenges in the delivery of PEGylated nano-formulations through passive targeting mode (Nakamura et al. 2016), some of which are indicated in Fig. 7.11.

**Table 7.2** Key applications of PEG moiety for passive targeting

| Drug substance | Spacer moiety | Applications |
|---|---|---|
| Protein and peptide macromolecules | PEG | Anti-cancer properties by evading the MPS |
| L-asparaginase | PEG | Increase the half-life, for the treatment of lymphoma and leukemia |
| Interferon | PEG | Immuno-stimulator |
| Anti-cancer drugs | PEG, HPMA | Cancer therapy by using paclitaxel and doxorubicin |
| Doxorubicin | PEG | Liposomes to treat hepatocellular carcinoma and T-cell lymphoma |
| Paclitaxel | PEG | Increase in AUC and decrease in half-life |
| DNA-related products | PEG | Tumor-specific delivery of siRNA |

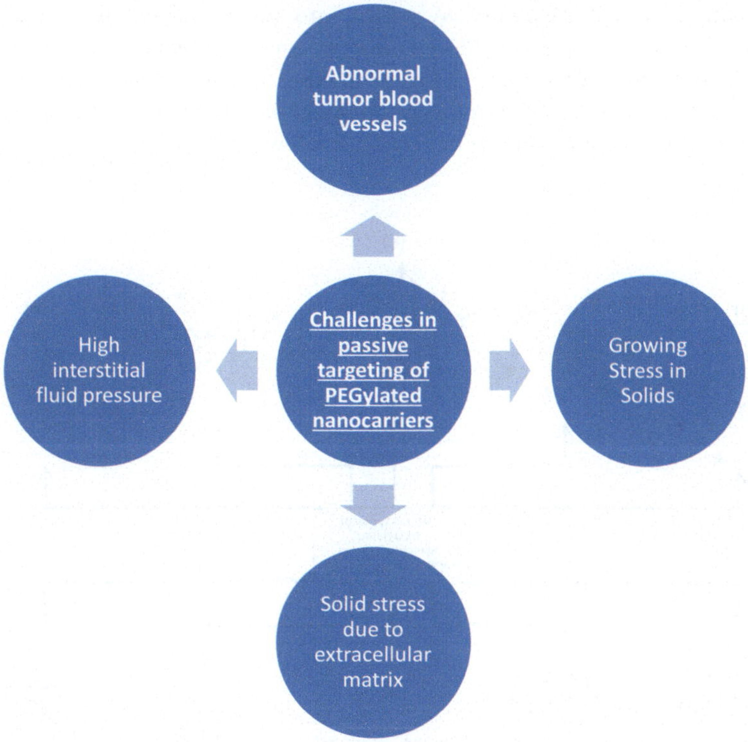


**Fig. 7.11** Challenges in passive targeting of PEGylated nanocarriers

## 7.6 PEGylated Nanocarriers and Their Types Used for Targeted Drug Delivery

The drugs that are administered systemically need to remain in circulation for a substantially longer period of time to accumulate in the targeted tissues at concentrations that are appropriate. But proteins and peptides are quickly broken down and removed from the blood; therefore, methods for extending circulation duration are required. Also, the reticuloendothelial system (RES) is a significant barrier for nanocarrier systems as it clears them out before reaching the site of action. This system quickly removes nanoparticles from circulation, especially hydrophobic particles, which causes their build-up in the liver, spleen, or bone marrow (Vllasaliu et al. 2014).

One such method involves coating the therapeutics' surface with an inert polymer that prevents interactions with the constituents of the bloodstream and confers it with stealth qualities. PEG is the most commonly utilized stealth polymer for drug delivery. PEG has been incorporated into drug delivery systems to improve the pharmacokinetic profile of both macromolecular therapies and particulate formulations. This increases the stability, and biocompatibility profile of the nanoparticle along with increased capability to infuse into the biological membranes and decreases the toxicity of drugs. PEGylation prevents aggregation, opsonization, as well as phagocytosis of nanoparticles by protecting the surface and extending circulation time (Suk et al. 2016). The diagrammatic representation of different types of PEGylated nanocarriers is shown in Fig. 7.12 below.

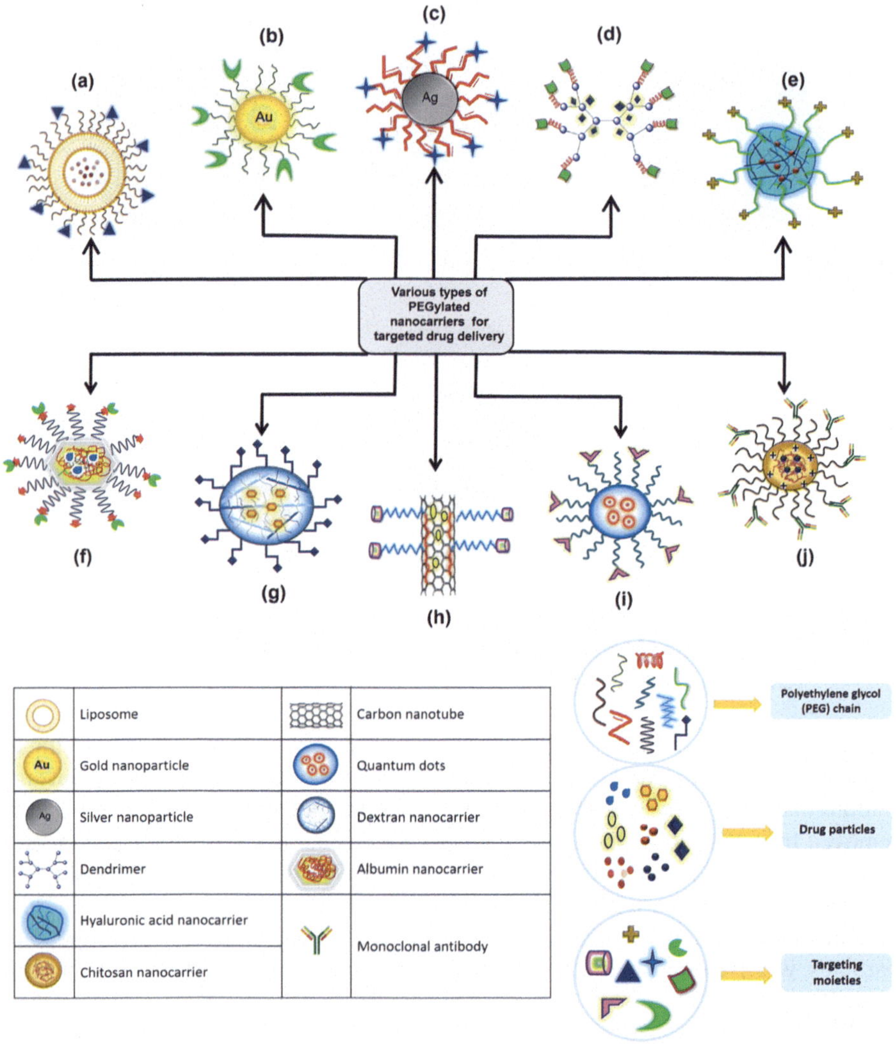


**Fig. 7.12** Various PEGylated nanocarriers designed to enhance drug delivery and targeting. The types shown include (**a**) PEGylated liposomes, (**b**) gold and (**c**) silver nanoparticles, (**d**) dendrimers, (**e**) hyaluronic acid, (**f**) chitosan, (**g**) dextran nanocarriers, (**h**) carbon nanotubes, (**i**) quantum dots, and (**j**) antibody-conjugated albuminnano carriers

### 7.6.1 PEGylated Liposomes for Targeted Drug Delivery

PEGylated liposomes as shown in Fig. 7.12a are recognized to be very promising nanocarriers for drug delivery applications, particularly for chemotherapeutics. PEGylated liposomes were used in an intriguing investigation to conceal and unmask a targeted ligand coupled to liposomes. It was possible to protect folate-modified liposomes against clearance from the RES, extend their circulation, and passively target tumors by using the cysteine-cleavable phospholipid-PEG5000. Experiments on tumor absorption in mice with artificially produced brain tumors (targeted liposomes containing DOX were administered, following a cysteine infusion) have verified the improved transport of folate-anchored targeted liposomes.

Kizelsztein et al. prepared PEGylated liposomes with entrapped antioxidant tempamine, which was tested for its ability to arrest the development of experimental autoimmune encephalomyelitis (EAE) in mice. According to the study, over 3% of the PEGylated liposome injection dose made it to the EAE mice's brains. The results obtained indicated that PEGylated liposomes reduced the severity and duration of EAE in this animal model (Kizelsztein et al. 2009).

Ghosh et al. in their study co-loaded vincristine in the clinically available PEGylated liposomal DOX and targeted against triple negative breast cancer and non-small cell lung cancer both in vitro and in vivo. It was concluded that the formulation significantly enhanced G2/M phase cell cycle arrest with consequent apoptosis and decreased cell viability in both tumor cell lines. In comparison to liposomal DOX, this carrier showed comparable acute toxicity, pharmacokinetic, and tissue distribution characteristics with a significantly higher rate of tumor regression. Thus, the treatment efficacy of doxorubicin and vincristine combined in clinically utilized PEGylated liposomal formulations against both cancers was dramatically enhanced (Ghosh et al. 2021).

### 7.6.2 PEGylated Gold Nanoparticle for Targeted Drug Delivery

Gold NPs (AuNPs) have unique optical, electronic, sensing, and biochemical properties and have been potentially applied for medical imaging, drug delivery, and tumor therapy in the early detection, diagnosis, and treatment (Kong et al. 2017). Smart drug delivery systems with gold nanoparticles are recent approaches for targeted therapy of life-threatening diseases such as cancer and cardiovascular diseases. Stimuli-responsive on-demand release of therapeutic agents at the diseased site can significantly limit serious adverse effects (Refaat et al. 2021). Due to their unique near-infrared (NIR) plasmon resonance, AuNPs are promising tools for early-stage cancer diagnosis and photothermal therapy. Their action is mediated by light-induced localized hyperthermia that in turn results in cell death and tumor inhibition (Sun et al. 2013). To achieve targeting for cancer diagnosis and therapy, it needs to be ensured that the activity of ligands such as antibodies (Abs) is stably retained (Day et al. 2010). For example, immunoglobulins (IgGs) have been immobilized on AuNPs at the Fc region mediated by a cofactor. This yields highly efficient and versatile immunoconjugated AuNPs. Ordered antibody immobilization can be achieved by IgG-binding proteins such as Protein-A (ProA) and Protein-G (ProG). Sun et al. (2013) successfully anchored in immobilizing IgGs on the AuNPs via PEGylation with orthopyridyldisulfide-polyethylene glycol-succinimidyl valerate, (OPSS-PEG-ProG), as shown in Fig. 7.12b. Efficient ablation upon laser irradiation was observed when the HER-2 antigen on SK-BR-3 breast cancer cells was targeted with IgGs. Based on ELISA, a 75% retention of anti-HER-2 binding activity of ProG was observed after PEGylation. The results indicated that localized hyperthermia of the AuNP/light interaction led to a loss in the cell membrane integrity when the nanoparticles were bound to the cell membrane.

### 7.6.3 PEGylated Silver Nanoparticles for Targeted Drug Delivery

Silver nanoparticles (AgNPs) have received particular attention due to their potential utilization in the treatment and diagnosis of cancer as well as in antibiotic resistance. However, some bacteria often show resistance towards AgNPs. To overcome this, PEGylated AgNPs have been investigated. Zhao et al. synthesized stable PEGylated AgNPs loaded with graphene oxide (GO) nanocomposite (GO-PEG-Ag) as shown in Fig. 7.12c, with strong antibacterial properties, high biocompatibility, and effectiveness over time. The addition of PEG to the nanocomposite significantly increased its stability in physiological fluids. Consequently, even after centrifugation, the GO-PEG-Ag may readily distribute in different media, which would significantly aid practical applications. The antibacterial test findings show that the GO-PEG-Ag demonstrated potent antibacterial activity towards bacteria (*E. coli* and *S. aureus*) carrying the MCR-1 (mobilized colistin resistance) gene encoding resistance to colistin, which is regarded as a last resort antibiotic against multidrug-resistant bacteria (Zhao et al. 2017a, b). In another study, Abdelfattah et al. constructed DOX-loaded PEGylated AgNPs and it was reported that the prepared nanoparticles showed sustained release with fewer side effects (Abdelfattah et al. 2022).

### 7.6.4 PEGylated Dendrimer for Targeted Drug Delivery

Dendrimers provide an ideal platform for the delivery of bioactive agents as they represent a well-defined, highly branched nanoscale architecture with defined molecular weight and availability in multiple generations determined by the number of branches built around the core with several modifiable surface groups. However, the real potential of dendrimers is limited by toxicity considerations. PEGylated dendrimer-mediated drug delivery overcomes the shortcoming of dendrimer reticuloendothelial system (RES) uptake, and drug leakage, and enhances the solubilization. Drugs with poor water solubility can be loaded in the internal pockets of dendrimers. On the other hand, its peripheral shell can be modified with a variety of surface functional groups. Several targeting ligands and moieties can be conjugated to these functional groups (Thakur et al. 2015).

Biotin (vitamin H) has shown great promise in cancer therapy. Biotin receptors are overexpressed by proliferating malignant cells to meet their biotin uptake. This affinity has been investigated for the development of several biotin-conjugated nanocarriers. This strategy results in enhanced drug uptake by multiple cancer cells. Rompicharla et al. (2019) developed PEGylated biotin-modified multifunctional Poly(amidoamine) (G4 PAMAM) dendrimers for effective delivery of Paclitaxel (PTX) (refer Fig. 7.12d), a chemotherapeutic drug, specifically to the cancer cells by following active-targeting approach. G4 PAMAM has been PEGylated and tagged with Biotin, an essential micronutrient for cellular functions. To avoid the toxicity of dendrimers due to their cationic nature and to prolong the systemic circulation of the conjugate, PEG has been attached. PTX was covalently linked to the surface of the G4 PAMAM dendrimer using a succinate linker. This study confirmed that biotin-tagged conjugate displayed superior penetration, and inhibition of the growth of cancer cells in comparison to the treatment with non-targeted conjugate and free drug.

Glioblastoma multiforme (GBM) is a serious form of brain cancer that is detrimental to the anatomy and physiology of the brain. GBM-bearing brain expresses an extensively large number of low-density lipoprotein receptors (LRP) on the luminal endothelial plasma membranes. Angiopep-2, an LRP ligand, has been reported to possess high perfusion capability and brain permeability in mice. Based on this theory, Parashar et al. (2018) designed PEGylated polypropyleneimine (PPI) dendrimers functionalized with Angiopep-2. PEG-2000 modification was intended to neutralize the positively charged dendrimer surface. Paclitaxel-loaded PEGylated dendrimers were then delivered to the brain glioma by receptor-mediated endocytosis (RME).

Further conjugation of these multifunctional nanocarriers with Angiopep-2 (ANG-PEG-PPP) resulted in the drug-loaded dendrimers demonstrating a superior anti-glioma effect when compared to the free drug. This was ascribed to enhanced drug delivery across the blood-brain barrier (BBB). Authors have developed PEGylated biotin-modified multifunctional Poly(amidoamine) (G4 PAMAM) dendrimers for effective delivery of paclitaxel (PTX) (Refer Fig. 7.12d), a chemotherapeutic drug, through active targeting.

Cancer cells overexpressing vitamin uptake receptors have been targeted by biotin-functionalized dendrimers. A succinate linker was used to covalently bind PTX to the surface of the G4 PAMAM dendrimer. The dendrimers were then PEGylated to extend systemic circulation as well as to minimize toxicity stemming from their cationic nature. Glioblastoma multiforme (GBM) is a serious form of brain cancer that is detrimental to the anatomy and physiology of the brain. GBM-bearing brain expresses an extensively large number of low-density lipoprotein receptors (LRP) on the luminal endothelial plasma membranes. Angiopep-2, an LRP ligand, has been reported to possess high perfusion capability and brain permeability in mice.

Based on this theory, Parashar et al. (2018) designed PEGylated polypropyleneimine (PPI) dendrimers functionalized with Angiopep-2. PEG-2000 modification was intended to neutralize the positively charged dendrimer surface. Paclitaxel-loaded PEGylated dendrimers were then delivered to the brain glioma by receptor-mediated endocytosis (RME). Further conjugation of these multifunctional nanocarriers with Angiopep-2 (ANG-PEG-PPP) resulted in the drug-loaded dendrimers demonstrating a superior anti-glioma effect when compared to the free drug. This was ascribed to enhanced drug delivery across the blood-brain barrier (BBB).

### 7.6.5 PEGylated Hyaluronic Acid for Targeted Drug Delivery

Hyaluronic acid (HA) is a naturally occurring polysaccharide present in the extracellular matrix alongside synovial fluids of the body. It is capable of binding to different cancer cells which overexpress CD44, a HA receptor. Because of this property, it has been used as a targeting moiety for cancer treatment. However, due to the cellular uptake by phagocytic cells of RES, it gets accumulated in the liver and the efficacy of HA-nanoparticles is reduced. To overcome this problem, Choi et al. synthesized PEGylated HA-nanoparticles labeled with cyanine 5.5 fluorescent dye. The results obtained indicated that the PEGylation of HA-NPs (Fig. 7.12e) decreased their cellular absorption in vitro, and more nanoparticles were absorbed by cancer cells that overexpressed the HA receptor CD44 compared to regular fibroblast cells. Following intravenous injections of Cy5.5-labeled PEGylated HA-nanoparticles into healthy mice, ex vivo findings of the organs utilizing optical imaging technology showed that PEGylation successfully decreased liver uptake of nanoparticles and extended their circulation time, and the brightest fluorescent signals were found in the tumor site suggesting their promising tumor targetability (Choi et al. 2011).

Zhang et al. developed a cleavable PEGylated hyaluronic acid nano-drug delivery system (HA-mPEG2k-DOX) that utilizes a tumor microenvironment pH-responsive imine bond for enhancing active tumor targeting, tumor cell uptake efficiency, and circulation duration of doxorubicin (DOX). The results obtained from this study indicated that HA-mPEG2k-DOX may self-assemble into stable nanoparticles. Additionally, the pH-responsive, cleavable PEG shell may be removed under weakly acidic conditions, thereby promoting the cellular uptake of HA-DOX nanoparticles in CD44-positive CT26 cells (Zhang et al. 2020).

### 7.6.6 PEGylated Albumin for Targeted Drug Delivery

Formulators have gained interest in human serum albumin (HSA)-based nanoparticles after the approval of paclitaxel-bound nanoparticle Abraxane® by USFDA, to treat patients with metastatic cancer of breast and non-small-cell lung carcinoma (NSCLC) (Zhao et al. 2015). Albumin, which carries drugs for binding and transport in the body, is suitable for use in these formulations due to its high capacity for loading nucleotides, biodegradability, non-toxicity, and consistent production procedure. Additionally, albumin is employed in gene therapy. Targeted drug delivery approaches to the tumor can enhance anticancer efficacy and improve the safety profile by sparing healthy tissues.

Recently, tumor-targeted monoclonal antibodies (mAbs) such as cetuximab, trastuzumab, and anti-CD3 mAb as well as other targeted ligands (including folate, RGD, and apolipoproteins), have been bound to the surfaces of albumin nanoparticles, further functionalizing them, enhancing drug uptake in tumor cells (Altintas et al. 2013). The epidermal growth factor receptor (EGFR) is highly overexpressed on the surface of tumor cells. Altintas et al. developed a glutaraldehyde crosslinked albumin core that has a surface modified with bifunctional PEG 3500 and the single variable domain of antibody-(Ega1) against the epidermal growth factor receptors. The multikinase inhibitors 17,864 have been loaded into the core of the nanoparticles and are retained into the particles via Lx-based platinum coordinal linkage-which couples the drug due to methionine residue of albumin and released in a reductive environment of the cytosol, where the 17,864 is released from the platinum linker. Seventeen thousand eight hundred sixty four is an analogue of sunitinib. Multikinase inhibitors inhibit the activity of several receptor tyrosine kinases by competitive inhibition with ATP, and are thus able to intervene in proliferation and survival of tumor cells and angiogenesis. Multikinase inhibitors inhibit the activity of several receptor tyrosine kinases by competitive inhibition with ATP, and thus are able to intervene in proliferation and survival of tumor cells and angiogenesis.

A study reveals that in comparison to PEGylated nanoparticles, nanoparticles modified by EGa1-PEG demonstrated a 40-fold higher affinity for EGFR-overexpressing cancer cells. By clathrin-mediated endocytosis, the nanoparticles loaded with 17,864-Lx were taken up by the tumor cells followed by lysosomal degradation. EGa1-directed nanoparticles, when administered intracellularly, successfully release the kinase inhibitor and reduce tumor cell growth; however, no anti-tumor effect was observed on 14C cells upon administration of the non-targeted formulation (Altintas et al. 2013)

The utilization of monoclonal antibodies (mAbs) and transferrin is widespread as targeting ligands for actively targeted nanomedicine formulations. In the family of epidermal growth factor receptors, the human epidermal growth factor receptor 2 (HER2) is overexpressed in the stomach, ovary, lung, and breast cancers (Tai et al. 2010). It comprises a tyrosine kinase domain inside the cell as well as an external ligand-binding domain. It might be argued that HER2 is an ideal marker for precise delivery to tumor cells because it is an accessible cell surface receptor that is overexpressed in both original tumors and metastatic locations. The anti-HER2 monoclonal antibody 1F2 recognizes the extracellular domain of HER2 exclusively. Kouchakzadeh et al. (2013) formulated a PEGylated human serum albumin (HSA) nanoparticle that is linked to a thiolated 1F2 mAb and allows for the incorporation of many drugs and nucleotides. For 1F2-modified nanoparticles, PEGylation can produce the maximum cellular absorption.

According to different studies, brain endothelial cells with expressed transferrin receptors mediate the endocytosis of transferrin by the cells. Most polar nucleosides, such as azidothymidine, an HIV-1 chemotherapeutic drug, do not easily cross the blood-brain barrier. Mishra et al. (2006), formulated transferrin-anchored PEGylated albumin nanoparticles with azidothymidine loading (Tf-PEG-Nps) (Fig. 7.12f). By employing PEG to modify the surface, surface-active RES interception and sequestration of nanoparticles might be prevented.

### 7.6.7 PEGylated Dextran for Targeted Drug Delivery

Dextran, a polymer composed of glucose residues primarily connected by 1,6-bonds, serves as a drug transporter. Nevertheless, there are specific challenges associated with dextran such as high molecular weight dextran that has been implicated in immunogenicity. The dextran molecule grafted with short PEG residues serves as a better approach than long-chain dextran as a carrier system as shown in Fig. 7.12g. After attaching several PEG residues, the dextran core may still include free reactive groups that are ideal for the attaching or single-point modification of a variety of medicines, such as proteins and peptides, to produce their long-circulating forms. Lukyanov et al. synthesized PEGylated dextran and from the biodistribution studies, it was reported that Dextran's liver absorption is reduced and its circulation half-life is markedly extended by PEG alteration. The amount of PEG residues added determines the extension of the circulation time. Further, it was concluded that the remaining reactive groups, such as amino groups, in the PEGylated dextrans, make them ideal long-circulating transporters for therapeutic as well as diagnostic substances (Lukyanov et al. 2004).

In another study, Naeye et al. synthesized PEGylated dextran nanogel for the delivery of siRNA. It was concluded from the study that PEGylated nanogels were readily absorbed by HuH-7 human hematoma cells and A431 human epithelial carcinoma cells. In addition, siRNA-loaded PEGylated nanogels significantly reduced enhanced green fluorescent protein (EGFP) levels in a human hepatoma cell line (HuH-7_EGFP) while being non-toxic to the cells (Naeye et al. 2010).

### 7.6.8 PEGylated Carbon Nanotube for Targeted Drug Delivery

Carbon nanotubes (CNTs) (Fig. 7.12h) have acquired a lot of attention in the biomedical field since they were discovered by Sumio Iijima in 1991 because of their peculiar framework and characteristics which include high aspect ratios, sizable surface areas, rich surface chemical functionalities, and size stability on the nanoscale (Son et al. 2016). A single graphene sheet can be utilized to create single-walled carbon nanotubes (SWNTs), while multiple graphene sheets can be utilized to create multi-walled carbon nanotubes (MWNTs) (Elhissi et al. 2012).

Additionally, studies have reported that non-spherical nanocarriers (such as CNTs) are preserved in the lymph nodes for a sustained duration as compared to the globular nanocarriers (e.g., liposome) (Bottini et al. 2011; Elhissi et al. 2012). The tiny nanoneedle mechanism allows for easy passage of functionalized CNTs (f-CNTs) across the plasma membrane with the help of an energy-dependent, endosome-mediated manner. This is because CNTs can be attached with large targeting moieties to offer sustained/controlled release behavior with cellular targeting.

It is a proven technique for treating cancer by stimulating apoptosis in cancer cells with the help of mitochondrial targeting. Many cancer cells develop anti-apoptotic properties by upregulating anti-apoptotic proteins, such as B-cell lymphoma-2 (Bcl-2), Bcl-XL, and myeloid cell leukemia-1 (Mcl-1), which prolongs their continuity. Bcl-2 proteins have been a potential therapeutic target for the development of methods to eradicate cancer cells. Due to their favorable pharmacokinetic, safety, and superior cellular internalization profiles, PEGylated Carbon Nanotubes (CNTs) have been used as a promising platform for delivering a variety of drugs, including small molecules and biomacromolecules. These anti-apoptotic proteins are susceptible to being inhibited by some small molecules like ABT737. Kim et al. (2017a, b) created a non-covalently bonded PEG-coated carbon nanotube-ABT737 nanodrug that enhances mitochondrial targeting and boosts the effectiveness of therapy against lung cancer. Apoptosis was brought on by the PEG-CNT-ABT737 nanodrug which was built up in the mitochondria of A549 non-small cell lung cancer cells. The proposed mechanism was abruption of mitochondrial membrane potential (MMP), lowering the expression of Bcl-2, and production of intracellular ROS generation resulting in enhanced efficacy against lung cancer.

In another study by Mehra and Jain (2015), it was reported that the nuclear hormone receptor superfamily's estrogen receptors (ERs) are overexpressed in malignant cells. After connecting specifically to each receptor by endocytosis or a tiny nanoneedle mechanism, MWCNTs tethering (Estrogen) likely distributes doxorubicin (DOX) more effectively into the malignant cells. Due to the upregulation of estrogen receptors (ERs) on human breast MCF-7 cells, estrone-anchored nanotube formulation was more readily absorbed than free DOX. Similarly, the elevated cancer-targeting propensity of the estrone-anchored MWCNTs formulations was also validated by the pharmacokinetic and boosted anti-tumor actions.

### 7.6.9 PEGylated Quantum Dots for Targeted Drug Delivery

The utility of quantum dots (QDs) (Fig. 7.12i) for biological applications is predicated on stably dispersing the particles in aqueous media. Their hydrophobicity and toxicity, however, limit their application in bioimaging (Xiao et al. 2014). To overcome this challenge, the surface of the QDs can be appropriately modified or they can be encapsulated in a suitable polymer. Hydrophobic drugs have been routinely delivered using PEG as it is not only highly soluble in organic solvents but is also biodegradable and biocompatible. Targeted drug delivery has been explored to treat inflammatory bowel disease. However, due to the lack of specific receptors, this strategy has resulted in suboptimal efficacy as well as intolerable toxicity. Epithelial cells and infiltrating immune cells in the colon have been reported to over-express CD98 upon inflammation.

Xiao et al. (2014), developed nanoparticles loaded with CD98 Fab′-conjugated quantum dots (Fab′-NPs). CD98 antibodies were digested and reduced to generate Fab′–SH. Next NPs were coated with a hetero-bifunctional cross-linker (MAL-PEG-NHS), yielding maleimide-functionalized PEG-NPs (MAL-PEGNPs). Finally, MAL-PEG-NPs were reacted with Fab′–SH to obtain the final NPs (Fab0-NPs). In vitro experiments revealed that the internalization of Fab′-NPs into cells occurred via CD98-mediated endocytosis. The study also demonstrated superior cellular uptake efficiency.

Black phosphorus quantum dots (BPQDs) have been recently discovered and they offer promising biomedical applications. Moreover, phosphorus, which constitutes roughly 1% of human body weight, is not only an essential macronutrient but is also known for its biocompatibility. A nanoplatform based on BPQDs has been designed for targeted tumor therapy. Wang et al. (2020), developed a novel FA (folic acid)-PEG@BPQD@DOX formulation for targeting the folate receptor (FR) in 293T mouse xenograft. Briefly, reactive oxygen and heat were produced in the BPQD upon laser irradiation, which further resulted in cell damage. Meanwhile, an increase in temperature led to drug release from the nanocomposite. Therefore, an excellent tumoricidal effect was demonstrated by this BP-based drug delivery system that combined chemotherapy, targeted synergistic photodynamic therapy (PDT), and photothermal therapy (PTT).

### 7.6.10 PEGylated Chitosan for Targeted Drug Delivery

Among all other polymeric nanoparticles, chitosan-based nanoparticles are the most effective, economical, and eco-friendly. Chitosan nanoparticles (ChNPs) (Fig. 7.12j) have attracted immense attention for biomedical applications due to their cationic characteristics, electrostatic interactions, biodegradability, and biocompatibility. A biological macromolecule called chitosan has a variety of bioremediation, anticancer, and drug carrier capabilities. Numerous studies have shown that ChNPs are a potential targeted medication delivery mechanism for the treatment of cancer. Many growth factors or growth factor receptors, including the vascular endothelial growth factor (VEGF) family, the epidermal growth factor (EGF) receptor family, and the transferrin receptor are overexpressed on the surface of tumor cells in cancer (Herdiana et al. 2021).

Chitosan nanoparticles (NPs) were modified at the surface with PEG to provide stealth character and transferrin (Tf) for the active targeting function by transferrin receptor-mediated endocytosis to promote drug delivery to cancer cells. It was discovered that cancer cells took up NP-PEG-Tf more readily than non-targeted cells.

Recent developments in monoclonal antibodies for targeted drug delivery represent a significant advancement in the biomedical profession, particularly in the treatment of breast cancer (Nag et al. 2016). To improve longer blood circulation times, higher solubility rates, better RES escape capabilities, and less antigenicity, doxorubicin-loaded chitosan nanoparticles were treated with polyethylene glycol. Additionally, labeling these nano-systems with breast cancer-specific monoclonal antibodies (mAbs) like anti-human mammaglobin (Anti-hMAM) and anti-human epidermal growth factor (Anti-HER2) has been a promising method to increase the specificity and sensitivity levels of the developed CSNPs system, potentially improving the survival rate and quality of life of breast cancer patients (Helmi et al. 2021).

## 7.7 Site Specific Delivery of Therapeutics

Site-specific drug delivery is the specialized form of drug delivery in which the drug is targeted to the desired target site in optimum therapeutic concentrations at a pre-defined rate. This avoids potential off-target adverse effects ranging from acute to chronic effects followed by coma and death. Site-specific delivery of the drug can be achieved through some of the following approaches:

1. Utilizing a carrier moiety in which the drug can be loaded, this can be done to safeguard the drug in other parts of the body.
2. Ligand-mediated targeting for release of the active drug only at the desired site.
3. The drug can be chemically or physically modified for targeted release, e.g., Prodrug approach (Hassanzadeh 2021).

The site-specific delivery of the drugs can be accomplished by using the PEGylation strategy for various disorders as follows.

### 7.7.1 Brain Disorders

In the fight against deadly brain diseases, targeted drug delivery via the blood-brain barrier (BBB) is a challenging task. The main obstacles include rapid blood flow, RES absorption, and nanocarrier excretion. PEGylation has proven to be beneficial for the targeted delivery of drugs to the brain. It is generally known that PEGylation can lengthen the period that nanocarriers remain in the bloodstream by preventing RES uptake, which is essential for enhancing the uptake of nanocarriers into the brain. For the targeted delivery of molecules that bind to the brain, PEGylation also serves as a linker. This carrier system is used to deliver proteins, peptides, and drugs of high molecular weight to the brain (Gajbhiye et al. 2020).

PEGylated dendrimers are also employed for targeted brain delivery. Santos and coauthors used various generations of cationic poly(amido amine) (PAMAM) linked with PEG for the treatment of stroke. BBB model's integrity was maintained in vitro, and rat primary astrocytes or bEnd.3 (bEnd.3 is a mouse brain cell line derived from BALB/c mice) cells were not negatively affected by PEGylated dendrimers, and no cytotoxicity was observed. After 24 h of injection, PEGylated dendrimers were found in the ischemic cortex neurons of animals suffering from focal brain ischemia, demonstrating their ability to serve as a delivery method during stroke (Santos et al. 2018).

Further Zhang et al. reported that DOX-loaded (Arginyl-glycyl-aspartic acid-peptide) RGD-PEG PAMAM dendrimer can target integrin receptors that are overexpressed in brain cancer (glioma). When tested on C6 glioma cells, it was found that the dendrimers were more effective than non-targeted PEGylated PAMAM. RGD-PEG-PAMAM accumulated in the tumor more significantly in mice with C6 brain tumors than in mice with unmodified dendrimers or free medicines. Also, the survival rate in mice treated with the modified formulation (RGD-PEG-PANAM) was significantly higher than the other groups (Zhang et al. 2011).

### 7.7.2 Pulmonary Disorders

Inhalation-based drug delivery systems using nanocarriers have become the subject of in-depth research. These drug delivery technologies offer strong, adaptable tools for specifically targeting and treating pulmonary disorders.

Kolte et al. prepared PEGylated nanoparticles of poly(lactic-*co*-glycolic acid) (PLGA) and cationic polymer polyethyleneimine (PEI) for delivering pDNA. PEGylation of nanoparticles resulted in decreased toxicity and improved cellular absorption and pDNA expression. It further enhanced the mucus barrier penetration of nanoparticles while preventing pulmonary macrophage absorption. In the next step, PEGylated composite NPs were lyophilized into a dry powder inhaler (DPI) and combined with lactose carrier particles. This increased the aerosolization characteristics and lung deposition with no alteration in the pDNA bioactivity. Thus it can be concluded from the study that local administration of pDNA to pulmonary tissue for efficient treatment of fatal lung illnesses may be made possible through this integrated method (Kolte et al. 2017).

Kim et al. studied the enhanced anti-cancer activity of PEGylated erlotinib for the treatment of non-small cell lung cancer. The formulation showed significant cytotoxicity in HCC-827 and NCI-H358 and also exhibited a better remission of tumor, indicating a strong therapeutic efficacy. In comparison to the control and free erlotinib groups, mice treated with the PEGylated nano formulation showed tumor reduction by five and twofolds, respectively. Based on these findings, it can be concluded that the prepared formulation offers a promising drug delivery strategy for the treatment of lung cancer (Kim et al. 2017a, b).

### 7.7.3 Cancer

The PEGylation approach can be employed in the formulations meant for tumor suppression. The extent of PEG coating required for tumor accumulation has been clearly stated by Mozar *and* Chowdhury. Low-density PEGylation creates a "mushroom state" (Fig. 7.13), that is susceptible to increased opsonization (engulfment of any foreign material, xenobiotics, or biomolecules by opsonin proteins) and cellular absorption and can eventually demonstrate higher tumor accumulation and enhanced intra-tumor drug effectiveness. However, a denser PEG results in less opsonization

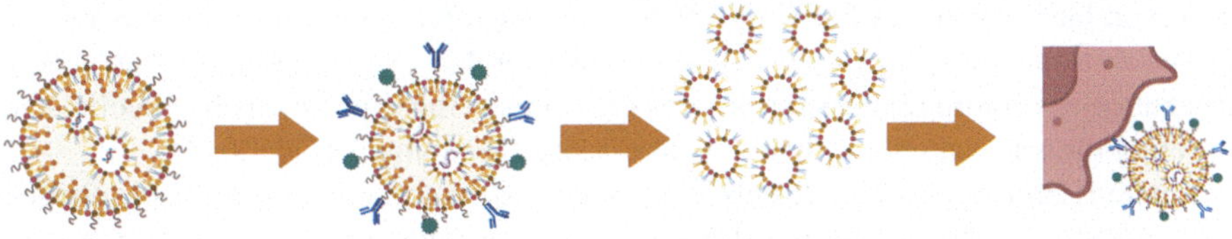


**Fig. 7.13** Formation of low-density "mushroom state" by PEG moiety

and cellular uptake ultimately resulting in less intratumor accumulation (Mozar and Chowdhury 2018).

In one study, Guo et al. reported the effectiveness of 2,3-dimethylmaleic-anhydride-poly(ethylene glycol)-$\varepsilon$-poly-L-lysine-doxorubicin/lapatinib polymeric (DMMA-P-DOX/LAP) nanoformulation against breast cancer. After being released from DMMA-P-DOX/LAP, the DOX and LAP significantly reduced the tumor in the MCF-7 (breast cancer cell line) model, and the penetration of active drugs inside the tumor tissues was effectively enhanced (Guo et al. 2020). Xin et al. synthesized paclitaxel (PTX) loaded methoxy poly(ethylene glycol)-poly(ε-caprolactone) (MPEG-PTMC) nanoparticles and investigated for their potential anticancer properties against glioblastoma multiform. The concentration of PEGylated nanoparticles was significantly increased in the brain tissues after 12 h of administration. The survival rate in the animals treated with MPEG-PTMC was higher than in the other groups (Xin et al. 2010).

Lee et al. studied the effect of albumin-based PEGylated nanoparticles loaded with PTX (HSA-PEG/PTX) on breast cancer. The spherical HSA-PEG/PTX nanoparticles have shown effective cellular delivery and resulted in cytotoxicity in multiple breast cancer cells. The HSA-PEG-based nanoparticles have further demonstrated sustained systemic circulation over 96 h and improved intra-tumoral retention in a mice model of breast cancer, leading to a noteworthy anticancer effect along with an extended lifespan of the animals (Lee et al. 2018).

### 7.7.4 Inflammatory Disorders

Inflammation is the first response of the body's immune system, and excessive or persistent inflammation causes chronic inflammatory diseases such as rheumatoid arthritis and inflammatory bowel disease (Cao et al. 2015). Activated macrophages greatly influence in initiation, maintenance, and pathogenesis of this disease by producing several pro-inflammatory cytokines. Suppression of pro-inflammatory cytokines and mediators from activated macrophages has been investigated as a key strategy for the treatment of inflammatory diseases. There are many Disease-modifying anti-rheumatic drugs (DMARDS), which are being used for treatment.

However, lack of efficacy and drug-related adverse effects are important reasons for the discontinuation of treatment in patients with inflammatory diseases. The development of new biological therapies seeks to address these problems by specifically targeting the pathogenic mechanisms of disease (Mcdonnell et al. 2013).

Folate (FA) receptors are highly overexpressed on these activated macrophages but are limited in normal tissues. FA conjugate was considered a therapeutic target drug for the population of pathologic cells (Pirmardvand Chegini et al. 2018). Chandrasekar et al. (2007) synthesized folate targeted PEG conjugates of anionic poly(amidoamine) (PAMAM) generation 3.5 dendrimer (G3.5 PAMAM) loaded with indomethacin as a targeted drug delivery system to treat inflammation. The conjugates revealed that approximately 7, 11, and 20 folate-PEG moieties were attached to each molecule of G3.5 PAMAM dendrimer. Drug loading was significantly increased in the folate PEG conjugates when compared with the nonconjugated dendrimer. The folate-PEG conjugates had shown reduced uptake by RES organs and in particular, the exposure to stomach was nearly one-tenth of the nonconjugated dendrimer, indicating the limited gastric-related side effects.

### 7.7.5 Bone Disorders

Osteoarthritis (OA) is a chronic and irreversible degenerative disease characterized by synovial inflammation and cartilage destruction (Xiong et al. 2021). Although many clinical therapeutics like nonsteroidal anti-inflammatory drugs (NSAIDs), glucocorticoids (GCs), and other drug treatments are effective strategies for OA, there are still some shortcomings that need to be overcome such as frequent injection, gastrointestinal, and cardiovascular risks, and potential overdose. Formononetin (FMN) is a phytoestrogen purified from natural herbal plants (e.g., *Astragalus membranaceus*, *Trifolium pretense*). It was reported to have pharmacological effects including anti-inflammatory capacity by suppressing IL-6 and TNF-α in neuroinflammatory rats and effectively antagonized proteoglycan loss by decreasing the expression of Matrix metalloprotease (MMP)-3, MMP-13, and attenuating oxidative stress. However, the bioavailability of FMN is low because it has poor water solubility and can hardly penetrate through the dense matrix of cartilage. Besides, with no specific targeting, FMN may be rapidly cleared in the joint. Sometimes, the retention time after intra-articular (IA) injection is short and repeated articular injection is inevitable. Thus, it is imperative to increase the water solubility and cartilage-targeting effect of FMN to improve its pharmacological effects (Xiong et al. 2021).

Polymer-drug conjugates (PDCs) are uniform-sized nanoparticles formed by linking hydrophilic polymers with drugs. It is one of the effective strategies in drug synthesis to enhance drug solubility and efficacy. Polyethylene glycol (PEG) with negligible toxicity and immunogenicity has been widely applied in PDCs, and it has increased drug solubility and improved cell growth (Xiong et al. 2021).

For cartilage-targeting, peptide-mediated (e.g., chondrocyte-affinity peptide CAP, anti-inflammatory peptide KAFAK, and RGD-modified) delivery systems have been used for modification of free drug molecules. Xiong et al. (2021) synthesized a nanosized amphiphilic polymer-drug conjugate (PEG-CollBP-FMN, PCFMN) for OA therapy, which was prepared by PEGylation of FMN followed by coupling with cartilage-targeting peptide (CollBP) to increase the bioavailability of FMN. PCFMN showed higher drug solubilization than unmodified FMN after PEGylation, demonstrating the effectiveness of PEGylation. PEGylation can enhance the pharmacokinetic properties of drugs. With excellent hydrophilic ability, PEG chains grafted on nanoparticles generate a sufficiently thick hydrated cloud that strongly prevents NPs from aggregation. As evidenced by the cellular uptake analysis, the CollB-peptide conjunction endowed PCFMN with a high cell penetration efficiency. Thus, PCFMN, a cartilage-targeting, and PEGylated nanodrug, had a more remarkable effect on down-regulating MMP-13 preventing cartilage degradation, and improving solubility. This contributed to enhanced anti-arthritic effects both in vitro and in vivo indicating that a targeted nano-drug design is a promising therapeutic strategy for OA (Xiong et al. 2021).

### 7.7.6 Blood Disorders

Hemophilia A and B involve a decrease in coagulation factors such as factor VII which results in prolonged bleeding. The PEGylation approach helps to prolong the half-life of drug or drug-loaded carriers in the blood. The brand name drug "Esperoct" includes factor VIII in glyco-PEGylated form for the treatment of hemophilia A. This was found to increase the drug's half-life by 1.6 times (Zhou et al. 2015).

Apart from that, PEGylated nanocarriers find applications in leukemia. Here PLGA is modified with mPEG for modulating its hydrophilicity as well as to render the nanoparticles with stealth properties to evade the RES and prolong systemic circulation. These biodegradable and biocompatible nanoparticles have been extensively studied for applications in controlled release. Following intravenous administration, the residence time for PLGA-mPEG nanoparticles was reported to be significantly longer than that for PLGA nanoparticles which were rapidly cleared from circulation. This has been attributed to the steric barrier provided by PEG which minimizes the opsonization of the particles.

## 7.8 Stimuli-Sensitive Nanocarriers

To improve drug delivery specificity, efficacy, and biological activity, stimuli-sensitive nanoparticles have been logically conceived and designed based on tissue pathology, tumor microenvironment, and intracellular compartments. Overall, the stimuli that the nanocarriers can respond to are the following: internal stimuli such as redox potential, pH, hypoxia, $H_2O_2$, ATP, and specific enzymes; external stimuli can include thermal, magnetic and electric field, ultrasound, and light (Fig. 7.14). The stimuli-sensitive functions make it easier to release drugs on demand or under

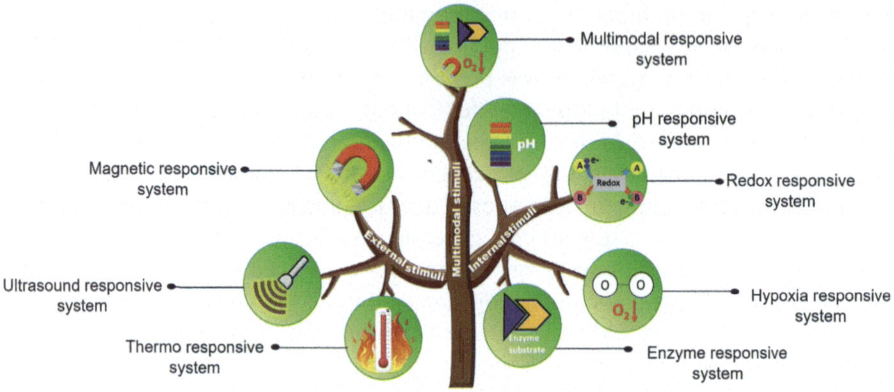


**Fig. 7.14** Various stimuli responsive strategies

control, promote site-specific accumulation, activate drugs or probes, expose ligands, activate nanoparticles, convert charges, signal in particular locations, detect unique pathological factors or molecules, and thus widely used as theragnostic agents. Furthermore, external forces, or stimuli, may also be able to alert nanocarriers to changes in their biological performance. For instance, an external magnetic field may enhance the tumor's magnetic nanocarrier accumulation to rise. Additionally, the stimuli might be used to activate specific prodrug-formulated nanocarriers biologically in sick areas or cells to provide precision therapy (Mi 2020).

## 7.8.1 External-Responsive Nanocarriers

### 7.8.1.1 Ultrasound-Responsive PEGylated Nanocarriers

Researchers have developed PLGA-PEG nanoparticles as suitable vehicles for the efficient delivery of miRNAs using ultrasound and microbubble approaches. This type of delivery system was specifically used for the delivery of plasmid vectors expressing a specific promotor called survivin on their surface, for the treatment of cancer. This novel approach overcomes the limitations associated with miRNA delivery into the body. Researchers have developed a formulation, which upon administration into the body, followed by the ultrasound treatment of particular frequency, releases miRNA for anti-cancer therapy (Devulapally et al. 2018). There are many challenges associated with systemic delivery of siRNA and the most troublesome challenge is the rapid degradation of naked siRNAs. siRNAs, when administered as complex formulation, suffer from aggregation issues thereby getting engulfed by the macrophages in the body.

To combat these issues, scientists have invented "SiPlex," a complex of siRNA and liposomes. The most effective means of delivery of SiPlex is by coupling it with PEG moiety and administering it as a microbubble formulation. Ultrasound radiation-triggered release of these microbubbles was significantly higher than the unaltered PEG-siPlexes. This SiPlex enters the cells on exposure to ultrasound and entry is more efficient than the formulations without PEG moiety. Also, the developed PEG-siPlexes showed greater efficacy in gene silencing, as compared to unmodified drugs. One more advantage of this system is that it only releases the siRNA in the presence of ultrasound, so the time and space for the release of siRNA in the body can be precisely controlled with this approach (Vandenbroucke et al. 2008).

#### 7.8.1.2 Thermal-Responsive PEGylated Nanocarriers

Researchers have explored the PEGylation approach for inducing photothermal damage to tumor cells when exposed to light, known as photothermal therapy. Various nanomaterials including nanostructures, carbon nanotubes, and nanographene have been investigated for the delivery of anti-cancer agents by using the photothermal approach. The PEG-modified nanocarriers possess very strong optical absorption properties when they are subjected to NIR radiation of frequency in the NIR window. Specifically, an NIR laser of 808 nm and a power density of 1 $W/cm^2$ was applied and it resulted in a rapid increase in the temperature in a concentration-dependent manner. This makes for an efficient stimuli-responsive approach to the delivery of anti-cancer medications (Farani et al. 2020).

Doxorubicin suffers from a change in the crystallization state at various temperatures. Doxil®, a stealth liposomal formulation of doxorubicin, was studied for this thermal effect. Various crystalline states of the drug can be easily confirmed by differential scanning calorimetry (DSC). To study the effect of temperature on Doxorubicin, scientists have studied the Doxil® formulation in comparison with unloaded liposomes. Techniques like microcalorimetry provide information about the changes in thermal behavior of the drug under different heating programmed in an orderly fashion.

For this study, PEGylated Doxorubicin liposomes were prepared and upon application of heat for a prolonged period of time (2 h at 80 °C), irreversible thermal behavior of the drug was analyzed. This indicates the leakage of the drug from the formulation on prolonged heat treatment (Perinelli et al. 2017). Researchers have developed PEGylated NGO (Nanosized Graphene Oxide), by a one-step green reduction reaction. This PEGylated NGO is useful in photothermally controllable drug delivery systems. This formulation, upon laser irradiation with an 808 nm laser, shows a 14-fold increase in drug release. This PEGylated NGO was found to show decreased cell viability, loss of mitochondrial membrane potential, and suppression of tumor cell growth. Its effect on the induction of apoptosis has been studied in various animal tumor models. PEGylation of NGOs has also increased their stability on prolonged storage (Chen et al. 2014).

#### 7.8.1.3 Magnetic Responsive PEGylated Nanocarriers

Scientists have developed a Doxorubicin formulation for anti-cancer applications by formulating it with PEG-coated superparamagnetic iron oxide nanoparticles (SPIONs) for improved drug delivery to cancer cells. SPIONs are magnetic-active materials, popularly used in many magnetic-responsive drug delivery systems as they get activated upon the application of a magnetic field. In this study, nanoparticles were loaded with the drug, using a pre-formed DOX-$Fe^{2+}$ complex, reversible at lower tumorous pH. Doxorubicin-loaded nanoparticles were found to have a hydrodynamic size around 60 nm and a zeta potential of zero at the physiological pH. These parameters favored increased stability as well as decreased elimination. At physiological pH, when compared with the plain drug administration, the liposomes show 60% drug release. The in vitro cytotoxicity of the formulation on MCF-7 breast cancer cells was equivalent to that of free DOX solution. DOX was reversibly bound to the SPION surface and PEGylation influenced the release and stability of the liposomal formulation (Gautier et al. 2012).

PEGylated liposomes were also studied by the Vannier group, wherein, they developed citrated SPIONS (Superparamagnetic iron oxide nanoparticles), i.e., C-SPIONS and the side effects were studied after PEG coating. The cardiotoxicity and hepatotoxicity were observed when the liposomes were coated with PEG moiety. The negative charges of the citrate group prevent liposomal aggregation and enhance the stability of the formulation. The stealth nature imparted by the PEG moiety shows improved release characteristics under the application of a magnetic field as well as improved stability properties. PEGylated C-SPIONS were also found to show enhanced doxorubicin efficacy and delivery.

Dual pH/magnetic responsive PEGylated $Fe_3O_4$ DOX nanoparticles have been developed to overcome limitations of the drug such as non-selective distribution and side effects. Citrate-coated nanoparticles were prepared and then functionalized with the PEG moiety to obtain PEGylated nanoparticles. The nanoparticles were roughly spherical in shape and showed strong magnetism as compared to the neat DOX nanoparticles. They also showed high drug loading capacity, pH, and magnetic field-responsive drug release (Ji et al. 2018).

## 7.8.2 Internal-Responsive Nanocarriers

### 7.8.2.1 pH-Responsive Systems

Solid tumors have a unique microenvironment that is characterized by several factors such as elevated temperatures, elevated expression of certain enzymes, a redox potential biased toward reduction, and acidic pH, usually around 6.5. The low extracellular pH of solid tumors is due to their preference for anaerobic respiration. To deliver drugs to tumors, pH-responsive NPs have been extensively researched. Changes in pH trigger drug release in these NPs. This can be achieved by the incorporation of protonatable groups or the formation of acid-labile bonds. pH-responsive NPs have been designed using protonatable/ionizable groups. Acidic pH results in charge reversal or protonation of the incorporated functional groups which disturbs the nanocarrier's hydrophilic-hydrophobic equilibrium. This imbalance disassembles the NP structure subsequently releasing the encapsulated cargo. Some of the commonly used ionizable groups include amino, sulfonate, imidazolyl, and carboxyl groups. Drug is released from these NPs through mechanisms that depend on

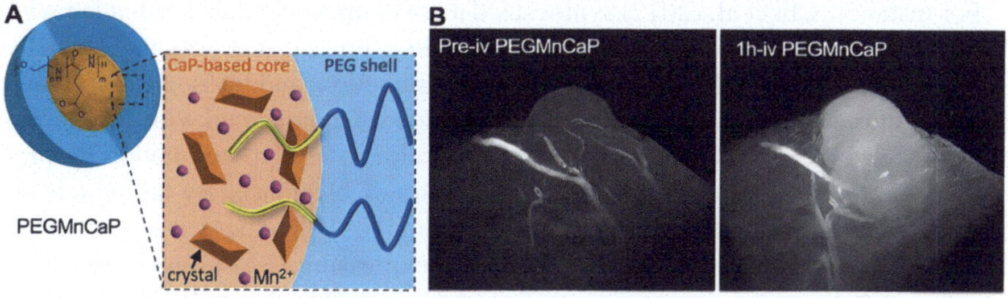


**Fig. 7.15** The pH-responsive PEGMnCaP nanocarriers with contrast amplification ability have been developed for MR imaging of tumor malignancy. (**a**) The composition and characterization of $Mn^{2+}$-doped PEGMnCaP. (**b**) PEGMnCaP specifically enhanced the contrast in C26 tumors for three-dimensional (3D) MR imaging. (Adapted with permission from Mi 2020 under CC BY 4.0 DEED Attribution 4.0 International (https://creativecommons.org/licenses/by/4.0/))

the acid dissociation constant ($pK_a$) of the introduced functional group such as aggregation, precipitation, or dissociation (Alsawaftah et al. 2022).

In order to effectively deliver anticancer drugs, one study proposes the formulation of pH-responsive PEGylated palladium nanoparticles (PdNP) as a drug nanocarrier system. To synthesize the nanocarrier, doxorubicin (DOX) had to be conjugated via hydrazone interaction to the surface of PEGylated PdNPs. It was shown that HeLa cancer cells took up the NPs preferentially. Studies on in vitro apoptosis have demonstrated that PEGylated PdNPs loaded with DOX induce cell death. In addition to their low toxicity, the unloaded PEGylated PdNPs were biocompatible. NPs loaded with DOX showed a strong tumoricidal effect in HeLa tumor xenograft models (Shanthi et al. 2015). pH-responsive PEGylated nanocarriers have also been utilized in MR imaging. Mi et al. have developed $Mn^{2+}$-doped, polymer hybrid calcium phosphate (CaP) nanocarriers (PEGMnCaP). This platform finds application in MR-based tumor imaging as it is endowed with intratumoral pH-triggered contrast amplification. High relaxivities can be obtained upon binding of $Mn^{2+}$ with surrounding proteins This enables specific and sensitive amplification of contrast in tumors resulting in accurate 2D and 3D MR imaging (Fig. 7.15) (Mi 2020).

#### 7.8.2.2 Redox-Responsive Systems

Redox-responsive delivery systems with disulfide bonds have been well studied by many researchers. Disulfide bonds can be easily broken down by reducing glutathione into sulfhydryl groups, which causes the degradation of carriers and facilitates the release of cargo. The disulfide bond can be cleaved by the tumoral intracellular glutathione (GSH, 2–10 mM), a strong biological reducing agent while remaining stable in a predominantly oxidizing extracellular space, where GSH concentration is much lower, 2–20 μM (Chi et al. 2017). Since cancer cells have four times the amount of GSH compared to normal cells, GSH is utilized to create redox-responsive nano preparations.

For instance, Chi et al. (2017) synthesized a novel liposome that was loaded with doxorubicin (DOX) and showed cytoplasmic drug release that was triggered by GSH due to disulfide linker reduction, PEG de-protection, and liposome membrane destabilization. Chol-SSmPEG/HA is a sheddable PEG conjugated with cholesterol through a disulfide (eSSe) bond. PEGylation prolongs in vivo circulation by preventing the reticuloendothelial system's (RES) quick clearance. By taking advantage of ligand-receptor interaction, hyaluronic acid (HA) has emerged as a desirable ligand for intracellular delivery to tumors overexpressing CD44, and it has been demonstrated to improve anticancer efficacy. Once inside the cells, PEG detachment would cause DOX to be released in response to intracellular GSH.

With many studies on the redox-responsiveness of disulfide bonds going on, diselenide bonds are attracting much attention as well. Diselenide bonds have similar reduction sensitivity and redox-responsive ability as that of disulfide bonds. As the Se–Se bond and C–Se bond have lower bond energy than the of S–S bonds, a more sensitive redox-responsive delivery system can be designed with the use of diselenide bonds in tumor therapy (Guo et al. 2018). Diselenide-rich amphiphilic diblock copolymers were fabricated to encapsulate two antitumor drugs with synergistic therapeutic effects, Camptothecin (CPT) and Doxorubicin (DOX), then visible light-induced dynamic exchange of diselenide bonds in the hydrophobic core promoted the formation of core crosslinking micelles, CPT/DOX-CCM. Upon intravenous administration, hand-in-hand release of CPT and DOX was accelerated significantly in tumor's redox microenvironments.

#### 7.8.2.3 Enzyme-Responsive Systems

Enzyme-responsive drug delivery systems (DDS) are designed to respond to the biocatalytic activity of enzymes, resulting in macroscopic changes in their physicochemical characteristics. The development of these responsive elements for DDS largely depends on the regulation and/or dysregulation of enzymes within the intracellular environment and their involvement in various biological and metabolic processes (Raza et al. 2019).

Certain enzymes that are overexpressed are associated with the pathophysiology of many diseases. Enzyme-cleavable peptides can be used to create enzyme-driven nanocapsule deshielding, allowing the drug to be released selectively into cancerous tissues or tumor cells. Although most nanocapsules cannot be used for intracellular drug release, enzyme-responsive delivery systems have been designed specifically for cancer treatment. Such targeted delivery approaches can offer superior tumor inhibition and minimize toxicity in normal tissues. As enzymes are secreted at a specific time and location, their structural characteristics provide excellent substrate specificity (Zhang et al. 2017). Enzyme-triggered nanocarriers (NCs) can be created by modifying their surface, which react catalytically with enzymes overexpressed in the extracellular environment of cancer cells (Kaushik et al. 2022).

Among the most useful enzymes for developing innovative drug delivery systems (DDS) are proteases, which are commonly elevated in infectious disorders like cancer. Glycylphenylalanylleucylglycine (GFLG) is an oligopeptide that has been incorporated in multiple drug delivery systems as an enzyme-sensitive spacer. It is

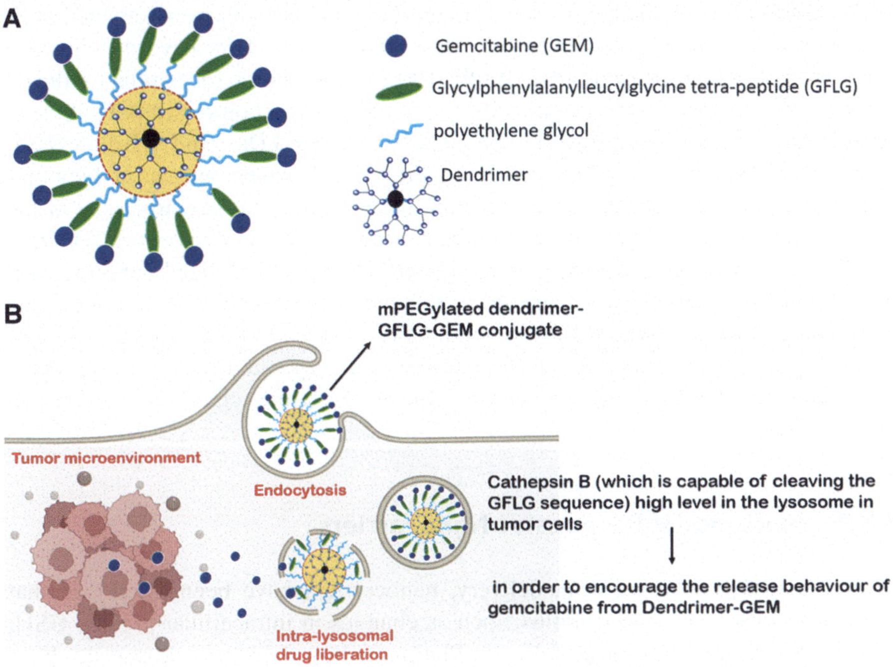


**Fig. 7.16** (**a**) Schematic representation of the preparation of lysine peptide dendrimer-GFLG-gemcitabine conjugate (dendrimer-GEM) and (**b**) tumor uptake of the conjugated nanoparticles is mediated by the EPR effect and enzyme-responsive drug release

preferentially cleaved by the lysosomal cysteine protease cathepsin B, which is highly expressed in most tumor cells, including breast cancer cells.

In response to this, Zhang et al. (2017), developed an efficient nanoscale chemotherapeutic prodrug, a PEGylated lysine peptide dendrimer-gemcitabine conjugate (Dendrimer-GEM) based nanoparticle that is responsive to the enzyme cathepsin B. The dendrimer and gemcitabine (GEM) are conjugated through a highly effective click reaction using GFLG tetrapeptide. The resulting nanoparticles release the drug significantly faster in the tumor microenvironment in response to higher cathepsin B concentration (Fig. 7.16).

#### 7.8.2.4 Hypoxia-Responsive Systems

Solid tumors often suffer from hypoxia, a condition characterized by low oxygen pressure due to irregular blood flow and high interstitial fluid pressures. Hypoxic regions in tumors can contribute to disease progression, resistance to chemotherapy and radiotherapy, and relapse. Unfortunately, drug carriers often cannot penetrate these hypoxic niches, making it challenging to treat tumors in areas such as the breast, pancreas, cervix, rectum, head, and neck. Among other causes, resistance to therapy is often induced by poor drug delivery to the hypoxic regions.

To address this challenge, hypoxia imaging agents and hypoxia-activated anti-cancer prodrugs have been developed. In a study, researchers have developed a nanocarrier PEG-azobenzene-PEI-DOPE (PAPD) that allows for hypoxia-induced siRNA uptake as well as gene silencing. This nanocarrier comprises polyethylene glycol 2000, azobenzene, polyethyleneimine (PEI) (1.8 kDa), and 1,2-dioleyl-*sn*-glycero-3-phosphoethanolamine (DOPE) units. Azobenzene acts as a hypoxia-responsive linker and undergoes a reduction in the tumor microenvironment of the hypoxic niches. Researchers utilized azobenzene as a hypoxia-responsive bioreductive linker for hypoxia-targeted delivery of siRNA from PEGylated nano preparations upon PEG removal/cleavage. siRNA delivery was evaluated by linking the azobenzene unit to PEG2000 at one end. The other end was conjugated to PEI (1.8 kDa)-DOPE to obtain PAPD. PEG2000 was incorporated due to its hydrophilicity and to stabilize the nanocarrier during systemic circulation (Kulkarni et al. 2018).

### 7.8.3 Multimodal Responsive Nanocarriers

In multimodal responsive drug delivery, nanocarriers have been developed that respond to more than one stimulus, such as changes in intracellular pH and GSH, which help improve the drug delivery. Platinum drug delivery nanocarriers have been developed to react to intracellular GSH and low pH. While GSH induces separation, controlled drug release is triggered by the low pH. Several responsive nanoparticles have shown great potential in achieving prolonged systemic circulation, enhanced tumor uptake and penetration, cellular internalization as well as endosomal escape. For instance, nanocarriers can disassemble at the acidic pH in the tumor microenvironment. GSH then activates the platinum prodrugs, promoting penetration to treat pancreatic tumors that are known to have low permeability (Mi 2020).

Scientists have successfully developed glucose-sensitive vesicles with immobilized glucose oxidase (GOD) using pH-sensitive polymers. The GOD catalyzes glucose to gluconic acid, which then alters the structure of the vesicles and releases their contents. Researchers have extensively studied GOD-immobilized polymer vesicles, particularly Gu and colleagues. They employed pH-sensitive diblock polymers, including PEG and ketal-modified polyserine (PEG-poly(Ser-Ketal)), to encapsulate recombinant human insulin, GOD, and catalase (CAT) to prepare glucose-sensitive vesicles (Fig. 7.17).

Ketal, an acid-labile group, undergoes acidic hydrolysis when gluconic acid is produced from glucose oxidization, leading to the formation of PEG-polyserine. PEG-poly serine shows high aqueous solubility and it causes the dissociation of vesicles which subsequently release insulin. To prolong the release of insulin following subcutaneous administration, vesicles were combined with a thermoresponsive and biodegradable polymer called Pluronic-127. This mixture was reported to rapidly form a stable hydrogel with evenly dispersed nanovesicles. The

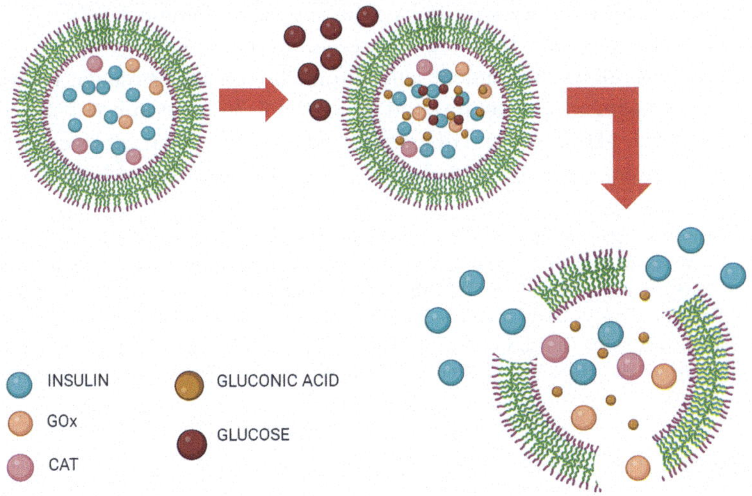


**Fig. 7.17** Chemical structure of pH-sensitive copolymer PEG-poly(Ser-Ketal) and schematic representation of enzyme-based glucose-responsive nanovesicle

nanovesicles were found to be highly compatible with living tissue and effectively regulated blood glucose levels in the normal range for up to 5 days (Zhao et al. 2017a, b).

## 7.9 Conclusion

Nanocarriers are often PEGylated to achieve specialized drug delivery mechanisms with improved target specificity, targeted delivery, regulated drug distribution, and enhanced drug efficacy in vivo. PEG has been incorporated into drug delivery systems to improve the pharmacokinetic profile of both macromolecular therapies and particulate formulations. This increases the stability, and biocompatibility profile of the nanoparticle along with increased capability to infuse into the biological membranes and decreases the toxicity of drugs. Out of the two main PEGylation approaches, non-covalent modification is generally preferred because of the additional benefits of the purification and prevention of drug loss. The PEGylated nanocarriers hence formed could be targeted to the desired body site by active (with the help of specialized receptors) and passive (e.g., EPR effect in cancer) targeting mechanisms. Several PEGylated targeted nanocarrier systems such as polymeric nanocarriers, nanofibers, quantum dots, liposomes, dendrimers, micelles, protein-based nanoparticles, and metal-based nanoparticles are well explored for their benefits over conventional drug delivery systems. These systems are employed mainly for the treatment of pulmonary, brain, bone disorders, and gastrointestinal disorders, and are widely explored in cancer.

To improve drug delivery specificity, efficacy, and biological activity, stimuli-sensitive nanoparticles have been logically conceived and designed based on tissue pathology, tumor microenvironment, and intracellular compartments. Overall, the stimuli that the nanocarriers can respond to are the following: internal stimuli such as redox potential, pH, hypoxia, $H_2O_2$, ATP, and specific enzymes; external stimuli can include thermal, magnetic, and electric fields, ultrasound, and light.

**Acknowledgments** The author, R.K.T, acknowledges the Department of Pharmaceuticals, Ministry of Chemicals and Fertilizers, India, for supporting the drug discovery and formulation research at NIPER Ahmedabad. R.K.T also acknowledges the Department of Science and Technology, Government of India, for a Core Research Grant funding (File No. CRG/2021/005402) and also acknowledges the Indian Council of Medical Research (ICMR), New Delhi, for the grant File Id: 2021-14161 and grant File Id: IIRP-2023-4849/F1 for supporting research in RKT lab.

## References

Abdelfattah A, Aboutaleb AE, Abdel-Aal ABM, Abdellatif AAH, Tawfeek HM, Abdel-Rahman SI (2022) Design and optimization of PEGylated silver nanoparticles for efficient delivery of doxorubicin to cancer cells. J Drug Deliv Sci Technol 71:103347

Agrahari V, Agrahari V (2018) Advances and applications of block-copolymer-based nanoformulations. Drug Discov Today 23:1139–1151

Alsawaftah NM, Awad NS, Pitt WG, Husseini GA (2022) pH-responsive nanocarriers in cancer therapy. Polymers (Basel) 14:936. https://doi.org/10.3390/polym14050936

Altintas I, Heukers R, van der Meel R, Lacombe M, Amidi M, van Bergen En Henegouwen PMP, Hennink WE, Schiffelers RM, Kok RJ (2013) Nanobody-albumin nanoparticles (NANAPs) for the delivery of a multikinase inhibitor 17864 to EGFR overexpressing tumor cells. J Control Release 165(2):110–118

Andrianov AK (2023) Noncovalent PEGylation of protein and peptide therapeutics. Wiley Interdiscip Rev Nanomed Nanobiotechnol 15:e1897

Belén LH, Rangel-Yagui CDO, Beltrán Lissabet JF, Effer B, Lee-Estevez M, Pessoa A, Castillo RL, Farías JG (2019) From synthesis to characterization of site-selective PEGylated proteins. Front Pharmacol 10:1450

Betancourt T, Byrne JD, Sunaryo N, Crowder SW, Kadapakkam M, Patel S et al (2009) PEGylation strategies for active targeting of PLA/PLGA nanoparticles. J Biomed Mater Res Part A 91(1):263–276

Bottini M, Rosato N, Bottini N (2011) PEG-modified carbon nanotubes in biomedicine: current status and challenges ahead. Biomacromolecules 12:3381. https://doi.org/10.1021/bm201020h

Cao J, Naeem M, Noh J-K, Lee EH, Yoo J-W (2015) Dexamethasone phosphate-loaded folate-conjugated polymeric nanoparticles for selective delivery to activated macrophages and suppression of inflammatory responses. Macromol Res 23:485–492

Chandrasekar D, Sistla R, Ahmad FJ, Khar RK, Diwan PV (2007) Folate-coupled poly(ethyleneglycol) conjugates of anionic poly(amidoamine) dendrimer for inflammatory tissue-specific drug delivery. J Biomed Mater Res A 82A:92–103

Chen J, Liu H, Zhao C, Qin G, Xi G, Li T, Wang X, Chen T (2014) One-step reduction and PEGylation of graphene oxide for photothermally controlled drug delivery. Biomaterials 35:4986–4995

Chi Y, Yin X, Sun K, Feng S, Liu J, Chen D, Guo C, Wu Z (2017) Redox-sensitive and hyaluronic acid functionalized liposomes for cytoplasmic drug delivery to osteosarcoma in animal models. J Control Release 261:113–125. https://doi.org/10.1016/j.jconrel.2017.06.027

Choi CHJ, Alabi CA, Webster P, Davis ME (2010) Mechanism of active targeting in solid tumors with transferrin-containing gold nanoparticles. Proc Natl Acad Sci 107(3):1235–1240

Choi KY, Min KH, Yoon HY, Kim K, Park JH, Kwon IC, Choi K, Jeong SY (2011) PEGylation of hyaluronic acid nanoparticles improves tumour targetability in vivo. Biomaterials 32:1880–1889

Day ES, Bickford LR, Slater JH, Riggall NS, Drezek RA, West JL (2010) Antibody-conjugated gold-gold sulfide nanoparticles as multifunctional agents for imaging and therapy of breast cancer. Int J Nanomedicine 5:445–454

Devulapally R, Lee T, Barghava-Shah A, Sekar TV, Foygel K, Bachawal SV, Willmann JK, Paulmurugan R (2018) Ultrasound-guided delivery of thymidine kinase–nitroreductase dual therapeutic genes by PEGylated-PLGA/PEI nanoparticles for enhanced triple negative breast cancer therapy. Nanomedicine 13:1051–1066

Elhissi AMA, Ahmed W, Hassan IU, Dhanak VR, D'Emanuele A (2012) Carbon nanotubes in cancer therapy and drug delivery. J Drug Deliv 2012:1–10. https://doi.org/10.1155/2012/837327

Farani MR, Khadiv-Parsi P, Riazi GH, Ardestani MS, Rad HS (2020) PEGylation of graphene/iron oxide nanocomposite: assessment of release of doxorubicin, magnetically targeted drug delivery and photothermal therapy. Appl Nanosci 10:1205–1217

Farjadian F, Ghasemi A, Gohari O, Roointan A, Karimi M, Hamblin MR (2019) Nanopharmaceuticals and nanomedicines currently on the market: challenges and opportunities. Nanomedicine (Lond) 14:93–126

Gajbhiye K, Pawar A, Mahadik K, Gajbhiye V (2020) PEGylated nanocarriers: a promising tool for targeted delivery to the brain. Colloids Surf B: Biointerfaces 187:110770

Gautier J, Munnier E, Paillard A, Hervé K, Douziech-Eyrolles L, Soucé M, Dubois P, Chourpa I (2012) A pharmaceutical study of doxorubicin-loaded PEGylated nanoparticles for magnetic drug targeting. Int J Pharm 423:16–25

Ghosh S, Lalani R, Maiti K, Banerjee S, Bhatt H, Bobde YS, Patel V, Biswas S, Bhowmick S, Misra A (2021) Synergistic co-loading of vincristine improved the chemotherapeutic potential of pegylated liposomal doxorubicin against triple-negative breast cancer and non-small cell lung cancer. Nanomedicine 31:102320

Guo X, Cheng Y, Zhao X, Luo Y, Chen J, Yuan W-E (2018) Advances in redox-responsive drug delivery systems of tumor microenvironment. J Nanobiotechnol 16:74

Guo Z, Sui J, Ma M, Hu J, Sun Y, Yang L, Fan Y, Zhang X (2020) pH-responsive charge switchable PEGylated ε-poly-l-lysine polymeric nanoparticles-assisted combination therapy for improving breast cancer treatment. J Control Release 326:350–364

Hassanzadeh P (2021) The biomedical significance of multifunctional nanobiomaterials: the key components for site-specific delivery of therapeutics. Life Sci 277:119400

Hatakeyama H, Akita H, Harashima H (2011) A multifunctional envelope type nano device (MEND) for gene delivery to tumours based on the EPR effect: a strategy for overcoming the PEG dilemma. Adv Drug Deliv Rev 63:152–160

Helmi O, Elshishiny F, Mamdouh W (2021) Targeted doxorubicin delivery and release within breast cancer environment using PEGylated chitosan nanoparticles labeled with monoclonal antibodies. Int J Biol Macromol 184:325–338. https://doi.org/10.1016/j.ijbiomac.2021.06.014

Herdiana Y, Wathoni N, Shamsuddin S, Joni IM, Muchtaridi M (2021) Chitosan-based nanoparticles of targeted drug delivery system in breast cancer treatment. Polymers (Basel) 13:1717. https://doi.org/10.3390/polym13111717

Insua I, Wilkinson A, Fernandez-Trillo F (2016) Polyion complex (PIC) particles: preparation and biomedical applications. Eur Polym J 81:198–215

Ji F, Zhang K, Li J, Gu Y, Zhao J, Zhang J (2018) A dual pH/magnetic responsive nanocarrier based on PEGylated Fe3O4 nanoparticles for doxorubicin delivery. J Nanosci Nanotechnol 18:4464–4470

Kang H, Rho S, Stiles WR, Hu S, Baek Y, Hwang DW, Kashiwagi S, Kim MS, Choi HS (2020) Size-dependent EPR effect of polymeric nanoparticles on tumour targeting. Adv Healthc Mater 9:1901223

Kaushik N, Borkar SB, Nandanwar SK, Panda PK, Choi EH, Kaushik NK (2022) Nanocarrier cancer therapeutics with functional stimuli-responsive mechanisms. J Nanobiotechnol 20:152. https://doi.org/10.1186/s12951-022-01364-2

Kim PH, Sohn JH, Choi JW, Jung Y, Kim SW, Haam S, Yun CO (2011) Active targeting and safety profile of PEG-modified adenovirus conjugated with herceptin. Biomaterials 32(9):2314–2326

Kim SW, Kyung Lee Y, Yeon Lee J, Hee Hong J, Khang D (2017a) PEGylated anticancer-carbon nanotubes complex targeting mitochondria of lung cancer cells. Nanotechnology 28:465102. https://doi.org/10.1088/1361-6528/aa8c31

Kim J, Ramasamy T, Choi JY, Kim ST, Youn YS, Choi H-G, Yong CS, Kim JO (2017b) PEGylated polypeptide lipid nanocapsules to enhance the anticancer efficacy of erlotinib in non-small cell lung cancer. Colloids Surf B: Biointerfaces 150:393–401

Kizelsztein P, Ovadia H, Garbuzenko O, Sigal A, Barenholz YJJON (2009) Pegylated nanoliposomes remote-loaded with the antioxidant tempamine ameliorate experimental autoimmune encephalomyelitis. J Neuroimmunol 213:20–25

Kolte A, Patil S, Lesimple P, Hanrahan JW, Misra A (2017) PEGylated composite nanoparticles of PLGA and polyethylenimine for safe and efficient delivery of pDNA to lungs. Int J Pharm 524:382–396

Kong FY, Zhang JW, Li RF, Wang ZX, Wang WJ, Wang W (2017) Unique roles of gold nanoparticles in drug delivery, targeting and imaging applications. Molecules 22:1445. https://doi.org/10.3390/molecules22091445

Kouchakzadeh H, Shojaosadati SA, Tahmasebi F, Shokri F (2013) Optimization of an anti-HER2 monoclonal antibody targeted delivery system using PEGylated human serum albumin nanoparticles. Int J Pharm 447:62–69. https://doi.org/10.1016/j.ijpharm.2013.02.043

Kulkarni P, Haldar MK, Karandish F, Confeld M, Hossain R, Borowicz P, Gange K, Xia L, Sarkar K, Mallik S (2018) Tissue-penetrating, hypoxia-responsive echogenic polymersomes for drug delivery to solid tumors. Chem Eur J 24:12490–12494. https://doi.org/10.1002/chem.201802229

Kurinomaru T, Shiraki K (2015) Noncovalent PEGylation of l-asparaginase using PEGylated polyelectrolyte. J Pharm Sci 104:587–592

Lee JE, Kim MG, Jang YL, Lee MS, Kim NW, Yin Y, Lee JH, Lim SY, Park JW, Kim J, Lee DS, Kim SH, Jeong JH (2018) Self-assembled PEGylated albumin nanoparticles (SPAN) as a platform for cancer chemotherapy and imaging. Drug Deliv 25:1570–1578

Lukyanov AN, Sawant RM, Hartner WC, Torchilin VP (2004) PEGylated dextran as a long-circulating pharmaceutical carrier. J Biomater Sci Polym Ed 15:621–630

Madhu (2018) Difference between block and graft copolymer. https://www.differencebetween.com/difference-between-block-and-graft-copolymer/#:~:text=The%20key%20difference%20between%20block,other%20via%20covalent%20chemical%20bonds. Accessed 24 2023

Mcdonnell T, Ioannou Y, Rahman A (2013) PEGylated drugs in rheumatology—why develop them and do they work? Rheumatology 53:391–396

McNeeley KM, Annapragada A, Bellamkonda RV (2007) Decreased circulation time offsets increased efficacy of PEGylated nanocarriers targeting folate receptors of glioma. Nanotechnology 18(38):385101

Mehra NK, Jain NK (2015) One platform comparison of estrone and folic acid anchored surface engineered MWCNTs for doxorubicin delivery. Mol Pharm 12:630–643. https://doi.org/10.1021/mp500720a

Mi P (2020) Stimuli-responsive nanocarriers for drug delivery, tumor imaging, therapy and theranostics. Theranostics 10:4557. https://doi.org/10.7150/thno.38069

Mishra V, Mahor S, Rawat A, Gupta PN, Dubey P, Khatri K, Vyas SP (2006) Targeted brain delivery of AZT via transferrin anchored pegylated albumin nanoparticles. J Drug Target 14:45–53. https://doi.org/10.1080/10611860600612953

Mishra P, Nayak B, Dey R (2016) PEGylation in anti-cancer therapy: an overview. Asian J Pharm Sci 11:337–348

Mozar FS, Chowdhury EH (2018) Impact of PEGylated nanoparticles on tumour targeted drug delivery. Curr Pharm Des 24:3283–3296

Naeye B, Raemdonck K, Remaut K, Sproat B, Demeester J, De Smedt SC (2010) PEGylation of biodegradable dextran nanogels for siRNA delivery. Eur J Pharm Sci 40:342–351

Nag M, Gajbhiye V, Kesharwani P, Jain NK (2016) Transferrin functionalized chitosan-PEG nanoparticles for targeted delivery of paclitaxel to cancer cells. Colloids Surf B Biointerfaces 148:363–370. https://doi.org/10.1016/j.colsurfb.2016.08.059

Nakamura Y, Mochida A, Choyke PL, Kobayashi H (2016) Nano drug delivery: is the enhanced permeability and retention effect sufficient for curing cancer? Bioconjug Chem 27:2225–2238

Nateghian N, Goodarzi N, Amini M, Atyabi F, Khorramizadeh MR, Dinarvand R (2016) Biotin/folate-decorated human serum albumin nanoparticles of docetaxel: comparison of chemically conjugated nanostructures and physically loaded nanoparticles for targeting of breast cancer. Chem Biol Drug Des 87:69–82

Parashar AK, Gupta AK, Jain NK (2018) Synthesis and characterization of Agiopep-2 anchored PEGylated poly propyleneimine dendrimers for targeted drug delivery to glioblastoma multiforme. J Drug Deliv Ther 8:74–79

Patra JK, Das G, Fraceto LF, Campos EVR, Rodriguez-Torres MDP, Acosta-Torres LS, Diaz-Torres LA, Grillo R, Swamy MK, Sharma S, Habtemariam S, Shin H-S (2018) Nano-based drug delivery systems: recent developments and prospects. J Nanobiotechnol 16:71

Perinelli DR, Cespi M, Bonacucina G, Rendina F, Palmieri GF (2017) Heating treatments affect the thermal behaviour of doxorubicin loaded in PEGylated liposomes. Int J Pharm 534:81–88

Pirmardvand Chegini S, Varshosaz J, Taymouri S (2018) Recent approaches for targeted drug delivery in rheumatoid arthritis diagnosis and treatment. Artif Cells Nanomed Biotechnol 46:502–514

Raza A, Rasheed T, Nabeel F, Hayat U, Bilal M, Iqbal HMN (2019) Endogenous and exogenous stimuli-responsive drug delivery systems for programmed site-specific release. Molecules 24:1117. https://doi.org/10.3390/molecules24061117

Refaat A, del Rosal B, Palasubramaniam J, Pietersz G, Wang X, Moulton SE, Peter K (2021) Near-infrared light-responsive liposomes for protein delivery: towards bleeding-free photothermally-assisted thrombolysis. J Control Release 337:212–223. https://doi.org/10.1016/j.jconrel.2021.07.024

Rompicharla SVK, Kumari P, Bhatt H, Ghosh B, Biswas S (2019) Biotin functionalized PEGylated poly(amidoamine) dendrimer conjugate for active targeting of paclitaxel in cancer. Int J Pharm 557:329–341. https://doi.org/10.1016/j.ijpharm.2018.12.069

Russell-Jones G, McTavish K, McEwan J, Rice J, Nowotnik D (2004) Vitamin-mediated targeting as a potential mechanism to increase drug uptake by tumours. J Inorg Biochem 98:1625–1633

Sadeghi GMM, Sayaf M (2014) Nanostructure formation in block copolymers. In: Nanostructured polymer blends. Elsevier, Amsterdam

Santos SD, Xavier M, Leite DM, Moreira DA, Custódio B, Torrado M, Castro R, Leiro V, Rodrigues J, Tomás H, Pêgo AP (2018) PAMAM dendrimers: blood-brain barrier transport and neuronal uptake after focal brain ischemia. J Control Release 291:65–79

Shanthi K, Vimala K, Gopi D, Kannan S (2015) Fabrication of a pH responsive DOX conjugated PEGylated palladium nanoparticle mediated drug delivery system: an in vitro and in vivo evaluation. RSC Adv 5:44998–45014. https://doi.org/10.1039/c5ra05803a

Singh R, Lillard JW (2009) Nanoparticle-based targeted drug delivery. Exp Mol Pathol 86:215–223

Son KH, Hong JH, Lee JW (2016) Carbon nanotubes as cancer therapeutic carriers and mediators. Int J Nanomedicine 11:5163. https://doi.org/10.2147/IJN.S112660

Suk JS, Xu Q, Kim N, Hanes J, Ensign LM (2016) PEGylation as a strategy for improving nanoparticle-based drug and gene delivery. Adv Drug Deliv Rev 99:28–51

Sun X, Zhang G, Keynton RS, O'Toole MG, Patel D, Gobin AM (2013) Enhanced drug delivery via hyperthermal membrane disruption using targeted gold nanoparticles with PEGylated protein-G as a cofactor. Nanomedicine 9:1214–1222. https://doi.org/10.1016/j.nano.2013.04.002

Tai W, Mahato R, Cheng K (2010) The role of HER2 in cancer therapy and targeted drug delivery. J Control Release 146:264. https://doi.org/10.1016/j.jconrel.2010.04.009

Thakur S, Kesharwani P, Tekade RK, Jain NK (2015) Impact of PEGylation on biopharmaceutical properties of dendrimers. Polymer (Guildf) 59:67–92. https://doi.org/10.1016/J.POLYMER.2014.12.051

Torchilin V (2011) Tumour delivery of macromolecular drugs based on the EPR effect. Adv Drug Deliv Rev 63:131–135

Vandenbroucke RE, Lentacker I, Demeester J, De Smedt SC, Sanders NN (2008) Ultrasound assisted siRNA delivery using PEG-siPlex loaded microbubbles. J Control Release 126:265–273

Vllasaliu D, Fowler R, Stolnik SJEOODD (2014) PEGylated nanomedicines: recent progress and remaining concerns. Expert Opin Drug Deliv 11:139–154

Wang J, Liang D, Qu Z, Kislyakov IM, Kiselev VM, Liu J (2020) PEGylated-folic acid-modified black phosphorus quantum dots as near-infrared agents for dual-modality imaging-guided selective cancer cell destruction. Nanophotonics 9:2425–2435. https://doi.org/10.1515/nanoph-2019-0506

Wilczewska AZ, Niemirowicz K, Markiewicz KH, Car HJPR (2012) Nanoparticles as drug delivery systems. Pharmacol Rep 64:1020–1037

Wongpinyochit T, Uhlmann P, Urquhart AJ, Seib FP (2015) PEGylated silk nanoparticles for anticancer drug delivery. Biomacromolecules 16:3712–3722

Xiao B, Yang Y, Viennois E, Zhang Y, Ayyadurai S, Baker MT, Laroui H, Merlin D (2014) Glycoprotein CD98 as a receptor for colitis-targeted delivery of nanoparticles. J Mater Chem B 2:1499–1508. https://doi.org/10.1039/c3tb21564d

Xin H, Chen L, Gu J, Ren X, Wei Z, Luo J, Chen Y, Jiang X, Sha X, Fang X (2010) Enhanced anti-glioblastoma efficacy by PTX-loaded PEGylated poly(ε-caprolactone) nanoparticles: in vitro and in vivo evaluation. Int J Pharm 402:238–247

Xiong W, Lan Q, Liang X, Zhao J, Huang H, Zhan Y, Qin Z, Jiang X, Zheng L (2021) Cartilage-targeting poly(ethylene glycol) (PEG)-formononetin (FMN) nanodrug for the treatment of osteoarthritis. J Nanobiotechnol 19:197

Zhang L, Zhu S, Qian L, Pei Y, Qiu Y, Jiang Y (2011) RGD-modified PEG–PAMAM–DOX conjugates: in vitro and in vivo studies for glioma. Eur J Pharm Biopharm 79:232–240

Zhang C, Pan D, Li J, Hu J, Bains A, Guys N, Zhu H, Li X, Luo K, Gong Q, Gu Z (2017) Enzyme-responsive peptide dendrimer-gemcitabine conjugate as a controlled-release drug delivery vehicle with enhanced antitumor efficacy. Acta Biomater 55:153–162. https://doi.org/10.1016/j.actbio.2017.02.047

Zhang X, Zhao M, Cao N, Qin W, Zhao M, Wu J, Lin DJBS (2020) Construction of a tumour microenvironment pH-responsive cleavable PEGylated hyaluronic acid nano-drug delivery system for colorectal cancer treatment. Biomater Sci 8:1885–1896

Zhao M, Lei C, Yang Y, Bu X, Ma H, Gong H, Liu J, Fang X, Hu Z, Fang Q (2015) Abraxane, the nanoparticle formulation of paclitaxel can induce drug resistance by up-regulation of P-gp. PLoS One 10:e0131429. https://doi.org/10.1371/journal.pone.0131429

Zhao L, Wang L, Zhang Y, Xiao S, Bi F, Zhao J, Gai G, Ding J (2017a) Glucose oxidase-based glucose-sensitive drug delivery for diabetes treatment. Polymers (Basel) 9:255. https://doi.org/10.3390/polym9070255

Zhao R, Lv M, Li Y, Sun M, Kong W, Wang L, Song S, Fan C, Jia L, Qiu S, Sun Y, Song H, Hao R (2017b) Stable nanocomposite based on PEGylated and silver nanoparticles loaded graphene oxide for long-term antibacterial activity. ACS Appl Mater Interfaces 9:15328–15341

Zhou T, Zhao X, Liu L, Liu P (2015) Preparation of biodegradable PEGylated pH/reduction dual-stimuli responsive nanohydrogels for controlled release of an anti-cancer drug. Nanoscale 7:12051–12060

# PEGylated Nanocarriers for Gene Therapy

# 8

Lopamudra Mishra, Lakshmi Kumari, Yash Sharma, Kanak Chahar, Satyam Khare, Preeti Patel, Dilpreet Singh, and Balak Das Kurmi

**Abstract**

The developed disease and mutations in treatment need emerging strategies to tackle physiological and biopharmaceutical challenges. Many drugs/therapies are available for symptomatic relief; however, they are sometimes only partially effective and have a number of dangerous adverse effects. New therapeutic approaches must therefore be developed to overcome complications and target permanent cures for specific diseases. Permanent and targeted clinical care has been significantly impacted by the production of nanotechnology-based products. The use of nanotechnology has aided in the creation of extremely disease-selective gene therapy systems, which will be able to deliver the gene in the body to correct the abnormality and recover the genetic function. Carriers like liposomes, dendrimers, and polymer nanomaterials have recently gained attention as possible drug delivery systems for various complex physiological diseases, decreasing toxicity and increasing biocompatibility. PEGylated nanocarriers are

L. Mishra · L. Kumari · S. Khare
Department of Pharmaceutics, ISF College of Pharmacy, Moga, Punjab, India

Y. Sharma · K. Chahar
Department of Pharmaceutical Quality Assurance, ISF College of Pharmacy, Moga, Punjab, India

P. Patel
Department of Pharmaceutical Chemistry, ISF College of Pharmacy, Moga, Punjab, India

D. Singh
University Institute of Pharma Sciences, Chandigarh University, Gharuan, Mohali, India

B. D. Kurmi (✉)
Department of Pharmaceutical Quality Assurance, ISF College of Pharmacy, Moga, Punjab, India

Department of Pharmaceutics, ISF College of Pharmacy, Moga, Punjab, India


R. K. Tekade, N. K. Jain (eds.), *PEGylated Nanocarriers in Medicine and Pharmacy*, https://doi.org/10.1007/978-981-97-7316-9_8

significantly the most cutting-edge techniques for non-invasive gene delivery technologies. When compared to non-PEGylated complexes, PEGylated nanocarriers showed superior biodistribution, enhanced safety profiles, and effective gene transfer in vector-based nanoformulations. Additionally, compared to polymeric systems PEGylation improves the gene expression and stability in a system that has been progressively studied in pre-clinical settings. In the present chapter, we compiled the utilization of PEGylated nanocarriers along with different formulation approaches for gene therapy in an extensive set of disease conditions and barriers.

**Keywords**

Gene therapy · CNS · CVS · Vectors · PEGylation · Liposomes

## 8.1 Introduction

### 8.1.1 Background on Gene Therapy

Gene therapies entail the effective intracellular delivery of certain genomic materials (transgene) inside specific cells to produce a pharmacological effect by replacing or repairing an existing abnormality or giving the cells a new genetic function (Nayerossadat et al. 2012; Stone 2010). To turn off a particular gene or restore a specific gene function, various gene delivery-based systems may be used in gene therapy. Gene therapy's ultimate objective is to replace the malfunctioning or missing gene with a single dose of the appropriate base material (Katare and Aeri 2010).

In 1989, tumor-infiltrating lymphocytes underwent the first human gene transfer, and the first human gene therapy was performed on the adenosine deaminase (ADA) gene in 1990 to treat SCID (severe combined immunodeficiency defect) patients (Aiuti et al. 2017). Due to its immense therapeutic potential to cure a variety of genetic illnesses by introducing novel genes (DNA and RNA) into target cells together with transgene expression, gene therapy has expanded quickly (Cao et al. 2019). Moreover, the direct in vivo delivery of certain naked therapeutic genes is unfeasible and rife with difficulties, such as the susceptibility to prone gene degradation with nucleases in plasma having non-specificity towards the targeted cells, and the inability of entering negatively charged genes into negatively charged cellular membranes (Razi Soofiyani et al. 2013). In fact, just six gene therapies have been approved in the west despite around 2600 gene therapeutics were finished or been tested in clinical studies (Ginn et al. 2018). The European Medicines Agency (EMA) have granted conditional marketing authorization to Glybera®, an AAV-based treatment for familial lipoprotein lipase impairment, in 2012.

The US Food and Drug Administration (FDA) licensed IMLYGIC®, a herpes simplex virus type 1 that have been genetically altered, in 2015 for the local treatment of melanoma patients' unresectable lesions. The European Medicines Agency (EMA) approved Strimvelis® in 2016 as a treatment for severe combined immunodeficiency brought on by adenosine deaminase deficiency (ADA-SCID). The FDA authorized three gene therapies in 2017. Both KYMRIAH® and YESCARTA® are genetically altered autologous CAR T cell immunotherapies that are directed against CD19; they are both approved for the treatment of non-Hodgkin lymphoma, and KYMRIAH® is also approved by regulatory authorities to treat acute lymphoblastic leukemia. AAV-based gene therapy called LUXTURNA® is intended to treat biallelic RPE65 mutation-related retinal degeneration (Anguela and High 2019). There are mainly two basic subcategories of gene therapy: somatic and germline. Germline gene therapy has a lot of potential; however, it cannot currently be used because of ethical constraints. Somatic cells are the only cells in humans that have been altered by gene therapy thus far (Goncalves and Paiva 2017). Genetic material can generally be transferred by a vector that can be defined as the vehicle which is helpful to deliver particularly the desired gene. The ideal vector could provide an accurate or precise dosage of genetic material to every target cell, enabling the production of the gene product without hazardous side effects (Wang et al. 2019).

An ideal vector must be non-immunogenic, safe, and able to transport a gene to a certain cell type. It should also be able to accommodate foreign genes of a suitable size and achieve the amount and duration of trans-genic expression necessary to rectify the deficiency (Davis and Cooper 2007). Transduction is usually the transfer of a gene using a viral vector, whereas transfection describes the transfer of a gene using a non-viral vector. Although, there are several viral as well as nonviral vectors for gene delivery as depicted in Fig. 8.1, the prime utilization in a therapy depend on genetic material being transported on the cell membrane and ultimately inside the cell nucleus (Ramamoorth and Narvekar 2015).

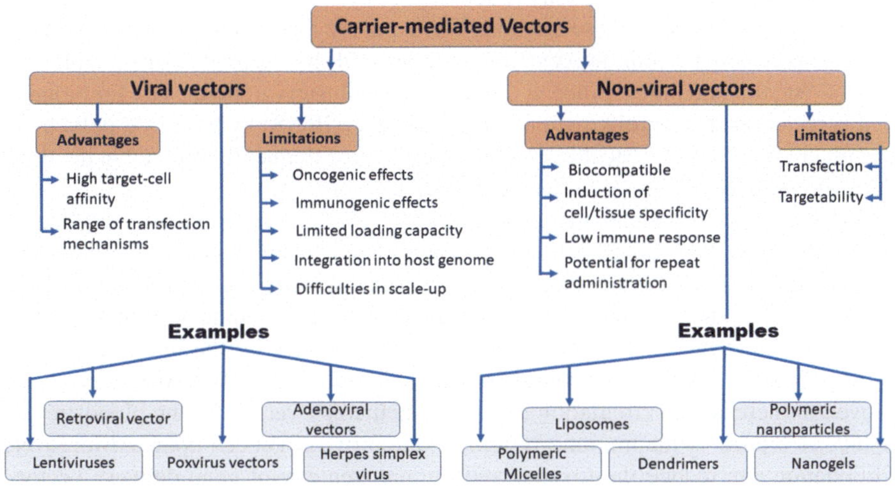


**Fig. 8.1** Various types of vectors used for gene therapy along with its advantages and limitations

## 8.2 Challenges for Gene Delivery and PEGylation as a Solution

Efficient as well as safe delivery of genes into the target cells remains a significant challenge. There are several challenges associated with gene delivery, including poor stability, low transfection efficiency, rapid clearance by the immune system, and off-target effects. To overcome these challenges, various gene delivery vectors developed rationally, which includes viral and non-viral vectors (Wang et al. 2013). Retrovirus (RV), adenovirus (AD), adeno-associated virus (AAV), and lentivirus (LV) are examples of viral vectors that often demonstrate a wide variety of species and tissue tropism and require for straightforward preparation and administration techniques. They also exhibit persistent gene expression and excellent transduction efficiency (Boulaiz et al. 2005). However, they possess a number of disadvantages includes high cost, limited target cell selectivity, high toxicity, immunogenicity, and mutagenicity, as well as the inability to transfer large-sized genes.

Circumventing the immune response to a particular vector is wide known principle challenge which induces an immune response, especially, adenovirus and AAV, particularly express the immunogenic epitopes inside the organism (Razi Soofiyani et al. 2013). Despite appearing to be less effective than viral approaches, nonviral delivery mechanisms like nanocarriers have inherent benefits including flexibility and safety. For the creation of nanocarriers along with minimal toxicity, high cell selectivity, and high stability, nanoparticle complexes modified by functional molecules such as peptides, proteins, integrins, lectins, and antibodies has been developed (Wang et al. 2022). There are several kinds of nanoparticles (NPs) that can be used as nanocarriers, including micelles, dendrimers, inorganic or metal NPs, quantum dots (QDs), liposomes, and polymeric NPs. These nanocarriers must circulate inside the blood stream for as long as feasible in order to reach target tissues at appropriate concentrations (Edis et al. 2021).

An example of one of these methods is coating the therapeutic's surface with the help of inert polymer that prevents interactions with bloodstream constituents and gives it "stealth" qualities. Since, it often has a long history of being safe for humans and is regarded as generally recommended as safe (GRAS) by the FDA, polyethylene glycol (PEG) is the "stealth" polymer that is most frequently utilized in the drug delivery industry (Padin-Gonzalez et al. 2022).

PEGylation, the attachment of polyethylene glycol (PEG) onto the surface of gene delivery vectors, have been shown to upregulate the efficiency as well as safety of gene delivery in order to improve its pharmacokinetic and pharmacodynamic properties (Lu and Zhang 2018). PEGylation can increase the stability of gene delivery vectors inside the bloodstream, reduce clearance by the immune system, and enhance their ability to penetrate target cells (Suk et al. 2016). Also, it has been proven to increase the circulation time of gene therapy vectors in the bloodstream, reduce clearance by the immune system, and improve target cell uptake. Moreover, PEGylation can reduce the toxicity and immunogenicity of gene delivery vectors (Suk et al. 2016; Chen et al. 2016).

In the 1970s, Davis and colleagues developed the term "PEGylation," which they defined as the attachment of polyethylene glycol (PEG) with drug molecules (Hamidi et al. 2006). Preparing conjugates of protein/peptide drugs with PEGs with varying degrees of polymerization was the major goal of this procedure. Later, the goal is to change the pharmacokinetic profile as well in the in vivo course of the bound drug in the body of host following systemic administration (Veronese and Pasut 2005; Roberts et al. 2020).Talking about chemistry of PEGs, PEGs are amphiphilic polymers made of ethylene oxide repeating units that are comparatively chemically inert. PEG molecules come in a wide range of molecular weights and are accessible commercially. PEGs are named depending on the amount of ethylene oxide units present in the polymer chain. PEGs may be arranged in a variety of ways and are primarily divided into two classes: linear polymers and branched polymers. Additionally, PEGs can be categorized into two groups: PEGs with free hydroxyl (–OH) groups at both ends and PEGs with one or two methoxylated end group(s) (i.e., –OH replaced by –OCH3) (Hamidi et al. 2006).

Nanocarriers are often utilized to carry genes and medicinal molecules, with or without coupling of ligands. Because of their unique targeting mechanism and regulated drug release profile, these carriers are highly selectable and efficient (Edis et al. 2021). The reticuloendothelial system (RES) removes nanocarriers from the circulation, which is their main disadvantage. RES absorption, stability, immunogenicity, drug leakage, hydrophobicity and hemolytic toxicity, are the key reasons why nanocarriers' employment is constrained in the biomedical area, despite the fact that they have demonstrated a wide variety of uses (Su and Kang 2020). PEGylation of nanocarriers is the only method available to bypass the restrictions. PEGylation is the process of coating or conjugating polyethylene glycol (PEG) with nanocarrier systems.

To increase biocompatibility and get around the opsonins of nanocarriers, PEGylation have been an absolute boon in the nanomedicine field (Verhoef and Anchordoquy 2013). PEGylation, however, may protect the targeting ligand and hinder the targeting of a particular cell, but un-PEGylated nanocarriers show fast clearance from the blood circulation. Because of this, there are less opportunities for targeted nanocarriers to engage with or be recognized by the appropriate receptor site, which will ultimately result in a less successful therapeutic outcome (Li and Huang 2010). Additionally, PEGylation causes nanocarriers to expand in size and molecular weight, which might change their biological capabilities. The charge, size, biocompatibility, contact with macrophages, and circulation duration of nanocarriers can all be affected by the molecular weight as well as density of PEG chains. When compared to PEG or targeting ligand conjugation alone, PEGylated targeted nanocarriers (PEG + targeting ligand) have demonstrated improved targeting outcomes at clinical level (van Vlerken et al. 2007). PEGylation is applicable to reduce the immunogenicity and prolong the circulation of various viral gene vectors such as adenovirus and associated virus (Wonganan and Croyle 2010)..

The development of an ideal PEGylated nanocarrier for systemic usefulness in the early 1980s and 1990s was prompted by the reproducibility of protein PEGylation which was used for generating efficient circulating, and subsequently more effective intravenous therapies for drug and gene delivery (Klibanov et al. 1990). The mononuclear phagocyte system (MPS) cells quickly remove NPs from systemic circulation after they are recognized as foreign objects, preventing accumulation in target cells and tissues (Ahmed et al. 2005). However, PEG coatings on NPs protect the surface from aggregation, phagocytosis, and opsonization, prolonging circulation time, much like what was seen with PEGylated proteins.

Doxil®, the first PEGylated nanoparticle (NP) product to receive FDA approval, was introduced in 1995. With a drug half-life of around 72 h along with circulation half-life of 36 h, Doxil "Stealth®" liposomes improves doxorubicin bioavailability about 90-fold in 1 week after injection compared to free drug (Gabizon et al. 2003; Laginha et al. 2005). Since then, PEGylation has established itself as a staple in NP formulation. PEG chains presented on NPs produce a hydrated cloud with a sizable excluded volume which can sterically prevent NPs from interacting with nearby NPs or blood components because of their hydrophilic nature. Additionally, the flexibility of PEG's conformational range makes it thermodynamically unfavorable for outside substances to penetrate the PEG corona (Lu and Zhang 2018). Ultimately, gene delivery is still a major problem in the field of gene therapy. PEGylation, on the other hand, has become a potential method for enhancing the effectiveness and security of gene delivery vectors. The area of gene therapy has the potential to undergo a revolution with the introduction of PEGylated gene delivery vectors, opening the door for the creation of safer and more effective gene treatments.

## 8.3 Types of PEGylated Nanocarriers

Number of nanocarriers, including liposomes, dendrimers, polymeric nanoparticles, polymeric micelles, metal-organic frameworks have achieved clinical authorization for the efficient delivery of a wide range of therapies, making them promising instruments for the targeted drug delivery for the treatment of cancer and many other disorders. They have the benefits of being biocompatible, comparatively harmless, and naturally able to shield the encapsulated payload from enzymatic degradation. Other unfavorable circumstances and advantages and disadvantages are given in Table 8.1 (Taher et al. 2023). Additionally, by changing the pharmacokinetic profile of the medications, they found to improve the therapeutic efficacy of the encapsulated pharmaceuticals and reduce their toxicity. Due to their combined benefits, nanoscale drug carriers have been approved for use in several clinical settings with a range of small molecule therapies, most recently siRNA. However, worries about potential carrier toxicity occasionally prevent the broad use of nanocarrier-based treatments in medicine (Gajbhiye et al. 2020).

In rare instances in animal trials, it has been observed that nanocarriers with specific physical and chemical properties stimulate host-immune system, leading to several side effects and/or a lack of beneficial effects of the medicine that is encapsulated. For instance, the size of the generated nanocarriers' particles or surface electrostatic charges are crucial in defining such immune responses. Immune cells in these trials effectively identify the nanoparticles as alien substances, leading to

**Table 8.1** Advantages and disadvantages of PEGylated nanocarriers

| Nanocarriers | Advantages | Disadvantages | References |
|---|---|---|---|
| Liposomes | Liposomes offered numerous advantages in delivering genes into the cells<br>Liposomes can be complexed either with negatively or positively charged molecules<br>Liposome offers a huge degree of protection to DNA from degradative processes<br>Liposomes are enabled to carry large pieces of DNA, as big as a chromosome<br>Liposomes must be targeted into specific cells or tissues | Production cost of liposomes is high<br>Leakage and fusion of encapsulated complexed drug may be possible<br>Sometimes phospholipids present on liposomes undergoes oxidation and hydrolysis-like reactions<br>Shorter half-life | Suk et al. (2016) |
| Micelle | Polymeric micelles possess several unique features that favor their applicability for drug delivery in cancer. The main advantage of polymeric micelles is their capability to solubilize several solubility compromised drugs within the core, thus improving their bioavailability | Polymeric micelles mainly show two weaknesses: one of which is low payload of drugs and the other is less stability in aqueous medium. Hence, this part pledged with drug loading capacity of polymeric micelles | Shiraishi and Yokoyama (2019) |
| Nanogels | Nanogels have shown excellent biocompatibility with a high loading capacity for different guest molecules as they possess hydrophilic properties and due to their unique physical properties, nanogels showed distinct advantages over other types of nanomaterials for various biomedical utilities | One of the major challenges during the formation of nanogels with the help of such polymers is to limit the particle size, that requires fine-tuning of the polymer concentrations or various environmental parameters includes temperature, pH, and ionic strength | Vijayan et al. (2017) |
| Inorganic nanoparticles | Inorganic nanoparticles possess several properties such as non-toxic, hydrophilic, biocompatible, and shows high stability when compared to organic materials. These drug delivery systems are designed for improving drug efficacy and reduced down the adverse effects have been evolved by the development of various novel inorganic nanoparticles | There are several issues shown by inorganic nanoparticles including long-term stability, low drug loading efficiency, and also brings batch-to-batch variations. Moreover, non-biodegradability as well as long-term toxicity are the limitations of inorganic nanocarriers | Karakoti et al. (2011) |

(continued)

**Table 8.1** (continued)

| Nanocarriers | Advantages | Disadvantages | References |
|---|---|---|---|
| Polymeric nanoparticles | Polymeric nanoparticles offered huge advantages over conventional systems which may include biocompatibility, biodegradability, easy to fabricate, non-immunogenic, non-toxic, and site-specific targeting to organs or tissues | The limitations of these polymeric nanoparticles include toxic degradation, toxic monomers aggregation, residual material associated with them, as well as toxic degradation process | Suk et al. (2016) |

the multilayered immune responses (Geiger et al. 2018). Additionally, it has been noted that the nanoparticles interact with circulating protein levels, including complement proteins, which is the prime part of humoral immune response, causing the mononuclear phagocyte system (MPS) to quickly eliminate them from body. The innate immune system's mononuclear phagocyte system (MPS), which is made up of dendritic cells (DCs), monocytes, and macrophages, is essential for the phagocytosis of pathogens during all stages of the immune response. When considered collectively, these immune system interactions having the potential to affect the in vivo fate of administered nanocarriers and may reduce their therapeutic efficacy.

### 8.3.1 PEGylated Liposome

The most effective nanocarriers that have achieved commercial success are liposomes. The Greek terms "lipos" and "soma," which translate to "fat" and "body," majorly, are the source of the word "liposome." A liposome is a globular lipid bilayer that ranges in size from 50 to 1000 nm. Cholesterol and phospholipids are used to make it. After that, polyethylene glycol (PEG)-based liposomes were created to improve their clinical acceptability (Mohamed et al. 2019).

PEGylation can be done in two different ways: by combining PEG-lipids with liposomal dispersion before liposome formation (post-insertion approach) or by adding PEG-lipids to the composition of lipids prior to liposome formation. The effectiveness of liposomal PEGylation is influenced by both the dimension and saturation density of the PEG. Although very long PEG chains result in a dramatic drop-in transfection activity, very small PEG molecules cannot stop protein absorption and speed up blood circulation. Typically, medium-length PEG molecules are utilized to modify liposomes. The covering density increases with the molecular PEG-lipid/lipid composition ratio (Jain and Nahar 2010). Figure 8.2 illustrates the de-PEGylation system, where PEG is getting cleaved from the liposomes after the liposomes reached the target site in order to improve endosomal escape of carrier.

Liposome PEGylation prolongs circulation duration by limiting liposome-macrophage interaction and MPS uptake. Additionally, PEG-liposomes are known to assemble in tumor foci. The steric barrier among PEG-liposomes and macrophages and attachment to blood proteins determine how PEG-liposomes interact

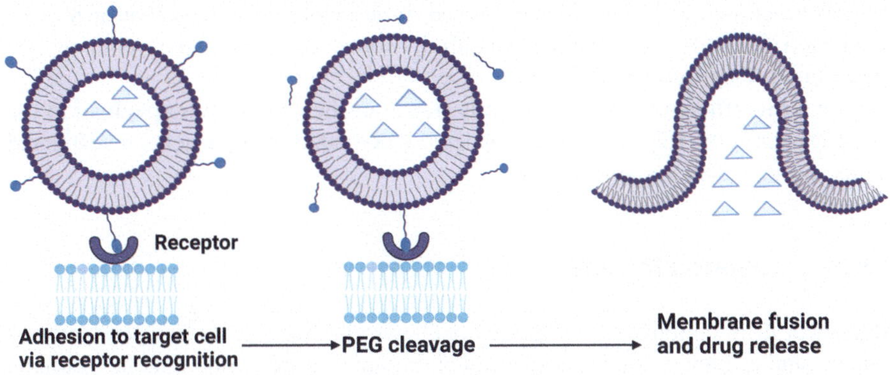


**Fig. 8.2** De-PEGylation technology in liposomes. PEG molecules on the liposome's surface are cleaved once the liposome binds to the suitable target cell through specific identification of the receptor by the ligand. Gene delivery is made more effective by the release of PEG, which speeds up membrane fusion of the liposome and liposome breakdown

with them. In general, steric barrier stability and circulation time are higher in liposomes that have a brush-like covering. PEGylated liposomes, on the other hand, exhibit noticeably decreased cellular uptake or endosomal escape, which lessens the overall effect of gene silencing (de Souza Guimaraes et al. 2022).

Recurrent PEG-liposome injection is thought to cause fast blood clearance, which dramatically reduces the drug's therapeutic impact. Activated B-cells generate IgM antibodies following the initial injection. The subsequent delivery results in drug-IgM binding, complement system activation, and macrophage liposome capture (Koudelka et al. 2016). This phenomenon is an incredibly quick immunological response that happens 10 min after the initial intravenous administration. Additionally, it appears that spleen is the primary generator of anti-PEG IgM. It should be noted that, the rate of immune response activation is higher when nucleic acids are used to stimulate the immune response and that the release of IgM antibodies is dosage dependent. However, because of MPS cell saturation, the third and subsequent injections of PEG-liposomes do not have such a dramatic effect (Wang et al. 2023).

Shroff et al. created PEGylated liposomes with the fibronectin-mimetic peptide-amphiphile PR_b integrated into their bilayer, which may target a variety of cancer cells which can overexpress, including the MDA-MB-231 breast cancer cells utilized in this investigation. To maximize the therapeutic potential of doxorubicin through PEGylation and active targeting to cancer cells, we have encapsulated it inside liposomes (Shroff and Kokkoli 2012). The findings demonstrate that PR_b--functionalized stealth liposomes could bind to MDA-MB-231 cells with high specificity, and the binding would be modulated by peptide concentration. When the intracellular trafficking of the doxorubicin liposomes was investigated, it was shown that most of them were likely in the early endosomes minutes after administration, while after a longer length of time, they have been collected in the late endosomes and lysosomes. Particularly at higher doxorubicin concentrations, it was discovered that the functionalized liposomes delivered more cytotoxicity than that of nontargeted and GRGDSP-functionalized stealth liposomes and were just as lethal as free doxorubicin (Demirgoz et al. 2008).

In order to efficiently and specifically transport their therapeutic payload to the breast cancer cells, the PR_b-functionalized PEGylated nanoparticles investigated in this study present a viable approach.

### 8.3.2 PEGylated Micelles

Anticancer drug-containing polymeric micelles were first created independently by Kataoka and Kabanov. Anti-cancer medications are integrated into micelles with the help of chemical conjugation or physical entrapment. Clinical experiments are being conducted on several micelles, and the polymer-micelle system is developing. Micelles are also utilized in the system for delivering genes. A PIC can self-assemble from a cationic polymer and DNA. In order to establish a complex with DNA, a block or graft copolymer made of hydrophilic-cationic polymers, including PEG-poly(L-lysine) and PEG-poly[*N*-(2-aminoethyl)-2-aminoethyl] aspartamide (PEG-PAsp(DET)), is typically utilized. Numerous in vitro and in vivo tests have been conducted to show the impact of PEG shielding. PEG shielding, for instance, considerably reduces the accessibility and utilization of biomacromolecules, such as serum proteins, and attenuates the charged state of the polyplex, which inhibits the aggregation.

The Kataoka group has conducted substantial research on the creation of micelles utilized for gene delivery systems and has created a number of block copolymers, including PEG-PLL and PEG-PAsp(DET) (Torchilin 2007). Because its cationic component has been discovered to decay more quickly than PEG-PLL under various physiological settings, PEG-PAsp(DET) is the one that is largely studied as a potential carrier of genes. Due to the increase in cytotoxicity of the cationic polymeric with molecular weight, biodegradability of the cationic polymer is crucial in minimizing toxicity. By reducing electrostatic contact, the breakdown of cationic polymers enables the release of plasmid DNA. PEG-PAsp(DET)'s effectiveness as a gene carrier has been assessed in vivo (Cagel et al. 2017).

A variety of polyion complex (PIC) micelles were created by Gao et al., and PEGylated in a range of ratios (PEG2k and PEG550). To clarify the effect of PEGylation patterns on the biodistribution of micelles, comprehensive research was conducted on the murine macrophage absorption, plasma protein adsorption, and in vivo biodistribution using iodine-125 as the tracer. They showed that the reticuloendothelial system (RES) cleared PEGylated micelles with shorter hydrophilic PEG chains mixed quickly on the surface, and that single PEG2k PEGylated micelles would effectively extend blood circulation time and increase their deposition over tumor sites (Shiraishi and Yokoyama 2019; Gao et al. 2013). The goal of the current work is to increase the comprehension of the PEGylation technique and create the best nanocarriers for medication administration and imaging applications.

Camptothecin (CPT), a weakly soluble anticancer medication and by Mu et al., prepared its micelles made from the blend of poly (ethylene glycol)-phosphatidyl ethanolamine conjugate (PEG-PE) and D-tocopherol poly ethylene glycol 1000 succinate (TPGS) (Mu et al. 2005). When compared to previously described PEG-PE-only micelles, the solubilization of CPT by the mixture of micelles ware found more effective. The developed formulation i.e. CPT-loaded mixed micelles, remain stable throughout storage as well as on dilution also, the cytotoxicity of the mixed micelles containing CPT was noticeably greater when compared to the free drug against the number of cancer cells. For poorly soluble medicines and imaging applications, PEG-PE/TPGS mixed micelles may be used as therapeutic nanocarriers with increased solubilization capability (Mu et al. 2005; Gao et al. 2008).

### 8.3.3 PEGylated Nanogels

Nanoparticles made of a polymers network and cross-linked chemically or physically are referred to as nanogels. In PEGylated nanogel particles with a polyamine core that has undergone chemical cross-linking and a PEG surface 94. The PEGylated nanogels exhibit greater durability against very diluted and high salt environments than those of self-assembled nanocarriers like liposomes and micelles due to their covalently cross-linked polyamine gel core (Tamura et al. 2011). *In vitro*, studies suggested that it was possible to stabilize nucleic acids and effectively deliver them inside the cells with success; but when nanogel was injected intravenously inside the bloodstream, it tended to mix with anionic serum proteins in order to produce substantial aggregates. Although, the non-fouling characteristics of the nanogel surface needs to be enhanced in order to use nanogels as a gene delivery mechanism for systemic applicability (Chen et al. 2018).

Nanogels may to some extent expose amino groups, that connect to serum proteins and cells that are outside the particle due to their loosely cross-linked gel structure. Moreover, the PEGylated nanogels that had a cross-linked density of 5% demonstrated the lowest toxicity (LD50 >200 mg/kg), that is quite sufficient for in vivo utilities. In reality, the nanogel caused severe hemolysis when the cross-linking density is of 1%. It was necessary to further raise the PEG corona density in order to further enhance the bioinert nature of nanogels. This stable nanoparticle is useful for researching how PEGylation alters the physicochemical parameters of the nanoparticle surface and affects in vivo pharmacokinetics characteristics (Gupta et al. 2015). High PEG-density nanogels can be created using the new post-PEGylation procedure, that refers to a quaternized reaction involving the amine in the nanogel core as well as the bromobenzyl-terminated PEG, in order to create long-circulation nanogels (Fig. 8.2). In comparison to nanogels without post-PEGylation, those that had it considerably increased the blood circulation time. This work unequivocally shows how PEGylation of nanoparticles affects their biodistribution (Vijayan et al. 2017).

Gold nanoparticles (GNPs) prepared by Nakamura et al. in poly-[2-(*N*,*N*-diethylamino) ethyl methacrylate] (PEAMA) gel core, the PEGylated nanogel including GNPs synthesized at pH 6 and 60 °C temperature, (PEGylated GNG) was shown the maximum GNP-loading capacity (Nakamura et al. 2010). PEGylated GNG showed an outstanding photothermal efficacy ($\Delta T = 7.7$ °C) when exposed to an Ar ion ($Ar^+$) laser at a fluence of 39 W $cm^{-2}$ for 6 min (14 kJ $cm^{-2}$). It should be noted that PEGylated GNG demonstrated non-cytotoxicity within the total absence of irradiation with $Ar^+$ laser (480 mg $mL^{-1}$: >90% cell viability), while pronounced cytotoxicity ($IC_{50}$ = 110 μg $mL^{-1}$) were noticed for PEGylated GNG according to irradiation supplied with $Ar^+$ laser at a fluence of 26 W $cm^{-2}$ for at least 5 min (7.8 kJ $cm^{-2}$), considering heat-generation from GNPs in the cells, and these resulted in selective and non-invasive cancer Photodynamic therapy (PTT). Consequently, a potential nanomedicine for cancer PTT would be PEGylated GNG, that has a high GNP-loading capacity (Miyamoto et al. 2008).

### 8.3.4 PEGylated Inorganic Nanoparticles

The topic of medicinal delivery has also sparked interest in inorganic nanoparticles. These inorganic nanoparticles, including silicon oxide, calcium phosphate, iron oxide, and gold are simple to manufacture with adjustable sizes and are simple to functionalize. Because of their low toxicity, gold nanoparticles have been extensively exploited in gene delivery systems; nonetheless, inorganic nanoparticles are often unstable and may prove hazardous in biological systems. Therefore, it is anticipated that surface modification will increase biological stability and compatibility (Karakoti et al. 2011). Inorganic nanoparticles can be made more stable and toxic-free by using PEG to modify their surfaces. Thiol groups, for instance, make good anchors for gold nanoparticles.

For the purpose of gene delivery in vivo, Kawano et al. coupled electroporation with PEG-modified cationic gold nanoparticles. By reducing chloroauric acid ($HAuCl_4$) with sodium borohydride ($NaBH_4$) in the presence of 2-aminoethanol and PEG-SH, they created cationic gold nanoparticles. Plasmid DNA can be attached to resulting PEG-modified cationic gold nanoparticle. Mice were intravenously injected with DNA complexes including PEG-modified nanoparticles, and electroporation was used to confirm gene expression (Chopra 2004). One issue is that under physiological settings, de-PEGylation of inorganic nanoparticles caused by mono-end-functionalized PEG frequently leads to particle aggregation. This issue might be solved by alternative methods of stabilizing gold nanoparticles, such as numerous anchoring among gold nanoparticle as well as functionalized PEG like penta-ethylene hexamine-ended PEG (N6-PEG). Other than gold nanoparticles, multiple anchoring technique can also be used to stabilize the inorganic nanoparticles.

For mesoporous silicon (PSi), a particular dual PEGylation (DPEG) technique was created by Nissinen et al., and put through in vitro as well as in vivo testing (Nissinen et al. 2016). The half-life of the nanoparticles was increased from 1 to 241 min by the DPEG coating, which drastically altered their in vivo behavior. Additionally, uncoated nanoparticles would quickly deposit in the liver, whereas the spleen was where the coated particles accumulated. The particle's protein coronas varied greatly from one another. While the coated nanoparticles had proteins that could inhibit cellular uptake, the uncoated particles have significantly more amount of proteins adsorbed, particularly liver and immune active proteins (Nakki et al. 2015). These factors, coupled with the agglomeration seen in blood circulation, were determined to be the root causes of the variations in behavior in vivo. By inserting superparamagnetic oxide of iron nanocrystals inside the pores of the particles, which made dynamic imaging possible, the bio-fate of the particles was monitored by magnetic resonance imaging. The current study's findings open the door to further developmental strategies of the porous inorganic system for delivery in the sense of active targeting because the carriers are easily modifiable chemically, enabling magnetically targeted distribution and diagnostics (Sarparanta et al. 2012).

### 8.3.5 PEGylated Polymeric Nanoparticles

Polymer nanoparticles (PNPs) are tiny polymeric nanospheres (NS) and nanocapsules. Matrix particles, or those whose entire mass is solid, include nanospheres. The surface of the sphere or the particle matrix may have molecules adsorbed on it. Polymer nanoparticles have increasingly been expanding and they play important roles in a broad range of areas, such as photonics, electronics, conducting materials, sensors, healthcare, biotechnology, environmental control, and environmental technology. Biodegradable nanoparticles are extensively employed in biotechnology and medicine to enhance the therapeutic efficacy of various medications. The solubility, bioavailability, retention duration, effectiveness, specificity, tolerability, and drug therapeutic index values of medicines that have been nano encapsulated in PNPs are all increased. PNPs can be functionalized to accomplish the so-called "intelligent targeting," i.e., targeted distribution to certain cells, tissues, or organs (Lukasiewicz et al. 2021).

Several biodegradable polymers including polylactic acid (PLA), polycaprolactone (PCL), polyglycolic acid (PGA), and polylactide-*co*-glycoside (PLGA) are being studied for efficient use in drug delivery systems. Numerous medications have been effectively encapsulated in PNPs to increase their bioactivity, bioavailability, and regulate delivery. Diseases include cancer, AIDS, insulin resistance, malaria, prion infections, and tuberculosis are the principal areas of application. When creating new drug delivery systems, PNP characteristics including toxicity, biocompatibility, biodistribution, and immunogenicity are critical. It is widely known that the criteria that affect these properties are the size, particles charges, and surface changes. The effects of possible nanocarriers that on cells of the immune system, which act as the body's first line of defense against dangers from the outside, are particularly crucial to study (Xin et al. 2011).

It is essential to develop carriers that, to phagocytic cells like macrophages, are undetectable ("the stealth property"). Immune system cells called macrophages play a role in inflammation. When activated, macrophages phagocyte harmful substances like scavengers. When macrophages are exposed to pathogenic particulates, which may also be nanoparticles, this process of activation takes place. Because nanoparticles are perceived as foreign and adequately ingested and digested by phagocytic cells, macrophages represent the first obstacle in the route of pharmaceutical nanocarriers to their destinations. Therefore, one of the primary objectives in developing novel methods for drug delivery is to reduce the absorption of PNPs by macrophages (Gajbhiye et al. 2020).

Paclitaxel (PTX)-loaded PEG-poly (trim ethylene carbonate) (MPEG-PTMC) NPs were investigated for their potential anticancer properties against glioblastoma multiforme by Xin et al. in 2010. Twelve hours after the intravenous injection, there was a considerably greater concentration of PEGylated NPs in the tissues of brain tumors. Additionally, PTX-loaded PEGylated NPs showed much greater in vivo anticancer effects than saline or Taxol, which resulted in a longer than the mean duration of survival for mice treated with them (27 days) (Xin et al. 2010). Angiopep-anchored PEG-PCL NPs (ANG-PEG-PCL NPs) were created by a similar team of researchers to take advantage of the overexpressed low-density lipoprotein receptor-related protein 1 (LRP1) seen over the BBB as well as glioma cells. On U87 MG glioma cells, PTX-loaded ANG-PEG-PCL NPs showed better inhibitory effects. The research also showed that the intracranial U87 MG glioma tumor-bearing in vivo model accumulated more targeted NPs (Xin et al. 2011).

## 8.4 Applications of PEGylated Nanocarriers for Gene Delivery

### 8.4.1 Cancer

Cancer is one of the mysterious and terrifying disease which is the uncontrolled proliferation. Multicellular organisms have been shown to be affected by cancer for more than 200 million years, and the cancer evidence in the progenitors of contemporary humans dates back around million years. However, in contrast to various infectious diseases, parasitic diseases, as well as many environmental diseases, this disease is not likely brought on by an outside factor. In cancer, human cells being having a sense of their uncontrolled growth, and which further changes into pathogenic organisms or the nuclei of tumors, often serves as its various agents of destruction (Yin et al. 2021; Dart 2022).

Cancers are legitimately known to be "genetic diseases" because genetic mutations play a significant part in the disease. Mutations are mainly responsible for causing and elevating the outcomes of the cancer, which can be considered as a major factor for induction of cancer. This is because external agents carry the huge potential to dismiss cellular functions and produce mutations, they also play greater role in the genesis of cancer (Ma et al. 2022; Chakravarty and Solit 2021).

Cancer gene therapy is a novel area of research promised with several potential treatments. The phrase "gene therapy" is known as a broad spectrum of medical procedures that utilizes genetic material to alter cells (either in vitro or in vivo) to treat a disease. Several preclinical and in vitro animal models applicable to test a wide variety of gene therapy drugs have been demonstrated for astounding success.

Gene therapy is utilized for development of cancer vaccines, induction of target viruses to cancer cells for lysis and death, reducing the blood supply towards the tumor, and induce genes into the cancer cells which further causes death or restoration of normal cellular phenotype, and showed to improve survival in various models of lung cancer (Giamas 2020; Giamas and Gagliano 2022).

Pegylated nanocarriers have demonstrated encouraging outcomes in the treatment of cancer. Figure 8.3 represents multifunctional pegylated nanocarriers for cancer therapy. These nanocarriers can encapsulate medications and deliver them precisely to cancer cells with the least chances of harm to healthy cells. These nanocarriers have been "pegylated," or have had polyethylene glycol (PEG) chains added to their surface. Because of the complexity of cancer biology and off-site toxicity with multidrug resistance, the conventional treatment regimen for use in cancer having a single chemotherapeutic agent falls far short of clinical expectations and is also linked to low Quality of Life (Gajbhiye et al. 2020; Jain and Nahar 2010).

The PEGylation of nanocarriers would shield both compounds from the harsh gastro-intestinal tract (GIT) GIT environment and subsequently encourage intestinal absorption through the lymphatic route into the systemic circulation This alteration can assist prevent their clearance and promote their accumulation in tumors by increasing their stability, lengthening the time they spend in the bloodstream, and decreasing their detection by the immune system. Many anticancer medications, including paclitaxel, doxorubicin, and cisplatin, have been delivered via pegylated nanocarriers. For added cancer cell selectivity, they might be functionalized with targeted ligands like antibodies or peptides. Pegylated nanocarriers

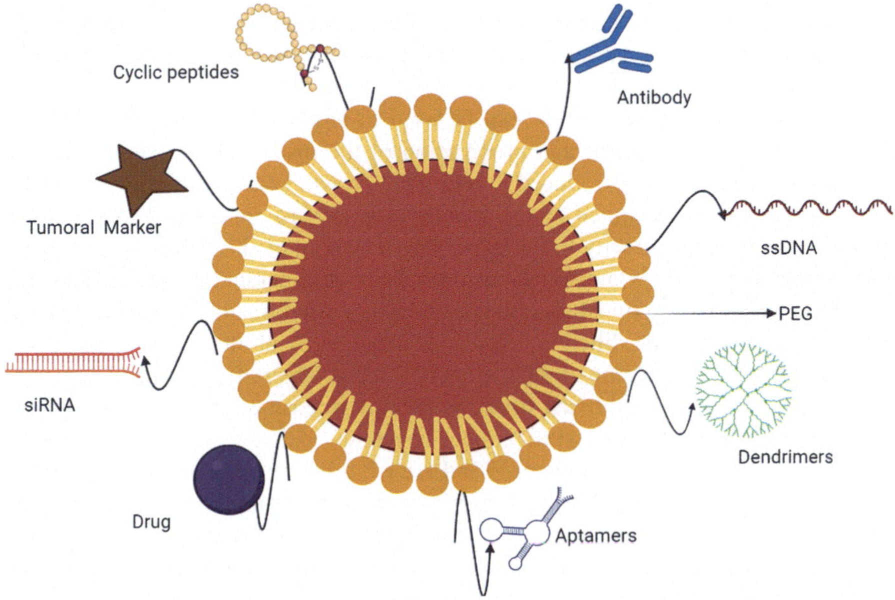


**Fig. 8.3** Multifunctional PEGylated nanocarriers for gene delivery in cancer cell

have been shown in studies to improve the therapeutic index of anticancer medicines by lowering their toxicity and boosting their efficacy (Koide et al. 2009; Wang et al. 2015a). They can also get around several drawbacks of conventional chemotherapy, which includes low solubility, non-specific distribution, and drug resistance. Pegylated nanocarriers have a lot of potential for the development of safer and more effective cancer treatments.

Further study is required, though, to improve their layout, assess their long-term security, and deal with potential difficulties including immunogenicity and drug resistance. Due to their capacity to safeguard and distribute therapeutic nucleic acids to cancer cells, PEGylated nanocarriers are an effective tool for cancer gene therapy. Mostly made of biodegradable polymers, these nanocarriers have a stable nanoparticle structure. The nanocarrier in cancer gene therapy can transfer different nucleic acids, like as plasmid DNA, small interfering RNA (siRNA), or microRNA (miRNA), which can control the expression of genes in cancer cells. For instance, miRNA can be used to restore the expression of tumor suppressor genes after miRNA has been used to quiet the expression of oncogenes (Muralidharan et al. 2014; Xue et al. 2015). In Table 8.1, different kinds of PEGylated nanocarriers and their clinical status is described.

#### 8.4.1.1 Breast Cancer

PEGylated nanocarriers are a promising approach for gene therapy in breast cancer. Gene therapy involves the introduction of genes into cells to treat or prevent disease. However, the induction of genes to specific cells inside the body can be challenging. Nanocarriers are small particles which can encapsulate and induce therapeutic genes to targeted cells in the body. PEGylation, which is the process of attaching polyethylene glycol (PEG) onto the surface of the nanocarrier, increases their circulation time in the body and improves their stability (Braden et al. 2014; Liu et al. 2022).

In breast cancer, gene therapy can be used to target specific genes which are involved in the development and progression of the disease. For example, tumor suppressor genes can be introduced to retard the growth and spread of cancer cells. Nanocarriers can be designed to specifically target breast cancer cells and deliver the therapeutic genes directly to the tumor site (Dastjerd et al. 2022).

Several types of PEGylated nanocarriers have been generated for gene therapy in breast cancer, including liposomes, polymeric nanoparticles, and dendrimers. These nanocarriers could be modified to enhance their targeting ability, enhance their stability, and control the release of the therapeutic genes. Overall, PEGylated nanocarriers offer a promising approach for gene therapy in case of breast cancer. While more research is needed to optimize their design and delivery, they have the huge potential to improve the effectiveness and reduce down the side effects of current breast cancer treatments. Several studies have demonstrated the potential of PEGylated nanocarriers for gene therapy in breast cancer. For example, one study used PEGylated liposomes to deliver a gene that inhibits tumor growth to breast cancer cells in vitro, resulting in a significant reduction in tumor cell viability. Another study used PEGylated nanoparticles to deliver a gene that enhances the immune response to breast cancer cells in a mouse model, resulting in a significant reduction in tumor growth (Dastjerd et al. 2022; Bottai et al. 2017; Sahu and Pattanayak 2020).

Overall, the utilization of pegylated nanocarriers for gene therapy in breast cancer holds great promise as a targeted and effective treatment strategy. However, future research is needed to optimize the design as well as delivery of these nanocarriers, and significantly evaluate their safety and efficacy in clinical trials.

#### 8.4.1.2 Lung Cancer

Lung cancer is still the emerging cause of cancer death all over the world. More people will die from lung cancer than from breast, prostate, and colon cancers put together. Before the fifth decade of life, lung cancer is quite uncommon; subsequently, the risk rises with age. Women are less impacted than men. Remarkably, only 15% of smokers acquire lung cancer despite smoking being the exposure most closely linked to the disease (it is thought to be the cause of 80–90% of cases), pointing to a hereditary vulnerability (Bade and Dela Cruz 2020; Nasim et al. 2019).

PEGylated nanocarriers have also been investigated as an efficient delivery system for gene therapy in lung cancer. Lung cancer is one of the leading causes of cancer-related deaths around worldwide, and traditional treatment options including chemotherapy and radiation therapy often have limited efficacy and can cause significant side effects. Gene therapy using PEGylated nanocarriers offers a promising strategy for targeted and effective management of lung cancer (Shahbazi et al. 2023).

One of the emerging challenges in the treatment of lung cancer is the ability of cancer cells to develop resistance to chemotherapy and other treatments. Gene therapy using nanocarriers can be used to deliver genes that sensitize cancer cells to chemotherapy or radiation therapy, making these treatments more effective. PEGylation of the nanocarriers can improve their stability and circulation time, allowing for targeted delivery of therapeutic genes into the cancer cells while minimizing toxicity to healthy tissues (Lara-Guerra and Roth 2016).

Several studies have shown the potential of PEGylated nanocarriers for gene therapy in lung cancer. For example, one study used PEGylated nanoparticles to deliver a gene that sensitizes cancer cells to radiation therapy in a mouse model of lung cancer, resulting in an efficient reduction in tumor growth. Another study used PEGylated liposomes to deliver a gene that inhibits tumor growth in lung cancer cells in vitro. In addition to sensitizing cancer cells to traditional treatments, gene therapy using nanocarriers can also be used to deliver genes that improves the immune response against cancer cells. For example, a study using PEGylated nanoparticles to deliver a gene that enhances the immune response in a mouse model of lung cancer which resulted in a significant reduction in tumor growth (Nair et al. 2020).

Overall, the use of pegylated nanocarriers for gene therapy in lung cancer holds great promise as a targeted and effective treatment approach. However, more research is required to optimize the design and delivery of these nanocarriers, and to evaluate their safety and efficacy in clinical trials.

#### 8.4.1.3 Colon Cancer

Together with breast, prostate, and lung cancers, colon cancer is also one of the most common tumors worldwide and is regarded as a major cause of death. Being the second most leading cause of cancer-related death, colon cancer is a major global health concern. There are drawbacks and certain side effects which are associated with conventional cancer treatments such as surgery, radiation, and chemotherapy. As a result, existing cancer treatment methods must be updated, and new methods must take their place. PEGylated nanocarriers for gene therapy present a potential approach for the focused and efficient treatment of colon cancer. The stability as well as circulation time of the nanocarriers can be increased through PEGylation, enabling the targeted delivery of therapeutic genes to cancer cells with the least amount of harm to normal tissues (Pawlik 2018; Otani et al. 2019; Zwacka and Dunlop 1998).

The capability of cancer cells to acquire resistance to chemotherapy and other treatments is one of the difficulties in treating colon cancer. Nanocarrier-based gene therapy can be used to transfer genes that make cancer cells more susceptible to chemotherapy or radiation therapy, increasing the efficacy of these therapies. For instance, in a study utilizing PEGylated nanoparticles to carry a gene that makes colon cancer cells more susceptible to chemotherapy, the tumor growth in a mouse model of colon cancer was significantly reduced.

Pishavar et al. (2020), in their study, they have studied about 5% and 3% of PAMAM primary amine and performed its substitution for transmitting of the plasmid encoding IL-12 gene. In colon cancer cells, the characteristics of modified PAMAMs incorporating size, surface charge density, cytotoxicity, and transfection efficiency were examined. According to an in vitro investigation, this modified carrier was able to double the amount of IL-12 production when compared to the unmodified PAMAM. An effective and secure non-viral IL-12 gene for colon cancer immunogens therapy may be produced by enhancing the polymer hydrophobic balance as well as by modulating the surface positive charge (Pishavar et al. 2020).

#### 8.4.1.4 Brain Cancer

The most prevalent form of brain cancer, glioblastoma, is one of the most fatal and severe neoplasms and is known to be very invasive. The majority of brain cancer patients still have a poor prognosis, and the median survival time rarely surpasses 16 months despite breakthroughs in surgery and medicine. The blood-brain barrier (BBB), which protected the central nervous system by acting as a semi-permeable membrane, greatly hinders drug transport to the brain (Tang et al. 2019).

PEGylated nanocarriers have been studied as a potential drug delivery system for gene therapy in brain cancer. Brain cancer, including gliomas and other types of brain tumors, are particularly challenging to treat due to the blood-brain barrier, which can prevent therapeutic agents from reaching the brain (Mehrabian et al. 2022).

PEGylation of nanocarriers can help to overcome this barrier by improving the stability and circulation time of the nanocarriers, allowing them to pass the blood-brain barrier and deliver therapeutic genes directly to the brain tumor cells. Additionally, PEGylation can reduce the clearance of the nanocarriers by the immune system, allowing for more effective and sustained delivery of therapeutic genes.

Several studies have demonstrated the emerging potential of PEGylated nanocarriers for gene therapy in brain cancer. For example, one study used PEGylated liposomes to deliver a gene that sensitizes brain tumor cells to chemotherapy, resulting in significant reduction in tumor growth in a mouse model of brain cancer. Another study used PEGylated nanoparticles to deliver a gene that inhibits tumor growth and increases the sensitivity of brain tumor cells to radiation therapy. In addition to sensitizing tumor cells to traditional treatments, gene therapy using nanocarriers can also be used to deliver genes that enhance the immune response against brain tumor cells. For example, a study using PEGylated nanoparticles to deliver a gene that enhances the immune response in a mouse model of brain cancer which results in a significant reduction in tumor growth (Bruun et al. 2015).

Overall, the use of PEGylated nanocarriers for gene therapy in brain cancer shows great promise as a targeted and potential treatment option. However, more research is needed to optimize the design and delivery of these nanocarriers and evaluate their safety and efficacy in clinical trials. Additionally, further studies are needed to determine the most effective therapeutic genes to deliver via PEGylated nanocarriers for the treatment of brain cancer (Wang et al. 2015b) (Table 8.2).

### 8.4.2 Autoimmune Diseases

Autoimmune diseases refer to a group of disorders in which the immune system mistakenly attacks the body's own healthy tissues and organs, leading to inflammation and damage. These diseases can affect various parts of the body, including

**Table 8.2** Clinical status of PEGylated nanocarriers for cancer therapy

| Types of nanocarriers | Drug loaded | Indication | Status | References |
|---|---|---|---|---|
| PEGylated liposomes | DOX | Alleviated systemic toxicity with increase in plasma circulation half-life and tumor accumulation | Approved | Hofheinz et al. (2005) |
| PEGylated polymeric micelles | Dox | Accumulation of drug in solid tumor | Phase 1 clinical trails | Matsumura et al. (2004) |
| PEGylated polymeric nanoparticles | Cisplatin | Reduction in tumor growth | Preclinical trail | Mattheolabakis et al. (2009) |
| PEGylated polymeric micelles | PTX | Alleviated systemic toxicity, prolonged plasma circulation half-life, increased tumor accumulation and improving in anticancer efficacy | Approved in South Korea | Werner et al. (2013) |
| PEGylated lipid nanoparticles | Docetaxel | Prolonged plasma half-life of drugs along with substantial accumulation at the tumor site | Preclinical trails | Khalid et al. (2006) |

joints, skin, muscles, blood vessels, and internal organs. Some examples of autoimmune diseases such as rheumatoid arthritis, multiple sclerosis, lupus, type 1 diabetes, inflammatory bowel disease, psoriasis, and autoimmune thyroid disease. These conditions may have a significant impact on a person's quality of life and may require ongoing medical care and management. The exact causes of autoimmune diseases are not fully understood, but it is assumed that a combination of genetic, environmental, and lifestyle factors may play a role. Treatment options may include medications to overcome inflammation and suppress the immune system, as well as lifestyle changes such as stress reduction and dietary modifications. It is necessary for individuals with autoimmune diseases to work closely with their healthcare providers to develop an appropriate management plan (Wang et al. 2015c; Wójcik et al. 2021).

The treatment of autoimmune diseases probably involves medications to reduce inflammation and to suppress the immune system. The type of medication used will depend on the specific autoimmune disease being treated, as well as the severity and location of the symptoms. Some common medications used to treat autoimmune diseases include: Nonsteroidal anti-inflammatory drugs (NSAIDs), corticosteroids, Disease-modifying antirheumatic drugs (DMARDs) (Beheshti et al. 2022; Afzal et al. 2017).

While these medications can be effective in managing symptoms and slowing disease progression, they may also have significant limitations and potential side effects. For example, long-term consumption of corticosteroids can increase the risk of infections, osteoporosis, and other complications. Biologic drugs can be expensive and may increase the risk of infections and other side effects. Additionally, while medications can help manage symptoms, they may not cure the underlying autoimmune disease. Lifestyle modifications such as stress reduction, exercise, and a healthy diet may also play a role in managing autoimmune diseases and improving overall health. It is beneficial for individuals with autoimmune diseases to work closely with their healthcare providers to develop an appropriate management plan that considers the benefits and limitations of various treatment options (Lim et al. 2019; An et al. 2020).

Gene therapy is a promising area of research for the management of autoimmune diseases. The goal of gene therapy is to alter the genetic material of a patient's cells to correct the underlying cause of a disease. In the case of autoimmune diseases, gene therapy could be used to modify the patient's immune cells to reduce the overactive immune response that causes the disease. One approach involves using gene editing approaches such as CRISPR to edit the genes responsible for the immune response. Another approach is to modify the patient's T cells or other immune cells to target and destroy the cells that are attacking the body's own tissues (Leung et al. 2010; Shu et al. 2015).

While the efficiency of gene therapy for treating autoimmune diseases is exciting, there are still significant challenges and limitations that must be overcome before it can become a widely available treatment option. Some of these challenges include safety concerns, delivery of therapeutics, and specificity. Despite these challenges, gene therapy holds significant potential as a treatment strategy for autoimmune diseases. Continued research and development in this field could lead to new and innovative therapies that offer more effective and targeted treatment options for individuals living with autoimmune diseases. PEGylated nanocarriers are a promising approach for delivering gene therapy to specific immune cells in autoimmune diseases. These nanocarriers are made up of biocompatible materials, such as lipids or polymers, and could be modified with PEG to increase their circulation time and reduce clearance by the immune system (Xue et al. 2015; Delogu et al. 2009).

In the context of autoimmune diseases, PEGylated nanocarriers could be applicable to deliver therapeutic genes to modify the patient's immune cells and reduce the overactive immune response that causes the disease. The PEG coating on the nanocarrier could help protect the therapeutic genes from degradation and immune clearance, while also increasing their specificity for certain immune cells. Pegylated nanocarriers for gene therapy has been studied in animal models of autoimmune diseases and showed promising results in reducing disease severity and improving outcomes. For example, in a mouse model of rheumatoid arthritis, PEGylated nanocarriers were shown to deliver therapeutic genes specifically to the inflamed joint tissues, leading to a reduction in inflammation and joint damage. One advantage of using PEGylated nanocarriers for gene therapy in autoimmune diseases is their ability to target specific immune cells, which can help reduce the risk of systemic immunosuppression and other side effects associated with non-targeted therapies. Additionally, PEGylated nanocarriers may be able to overcome some of the delivery challenges associated with traditional gene therapy approaches (Ren et al. 2019).

### 8.4.3 Inflammatory Disorders

Chronic, persistent inflammation plays a wide role in the clinical development of advanced atherosclerotic lesions. After being administered to mice, Col IV-targeted PLGA nanoparticles encasing Ac2-26 (capable of simulating the pro-resolving effects of annexin A1) were investigated for therapeutic effectiveness in chronic and advanced atherosclerosis. Col IV-Ac2-26 PLGA NPs significantly reduced plaque necrosis suppressed oxidative stress, and advanced plaque properties, indicating that targeted delivery of a resolution-mediating peptide activated receptors over myeloid cells to stabilize advanced atherosclerotic lesions (Fredman et al. 2015). Super paramagnetic iron oxide nanoparticles included in PLGA nanoparticles can also be employed for the treatment of joint inflammation, suggesting a possible way to prevent the generation of inflammatory responses in joint illnesses (Gu et al. 2013). Reduced interchange with the mononuclear phagocyte system has been seen in PEGylated PLGA nanoparticles, which results in reduced development of immunological responses (Ikoba et al. 2015).

Crohn's disease and ulcerative colitis are two examples of the chronic relapsing GI disorders that fall under the umbrella label of "inflammatory bowel disease" (IBD). Mucosal inflammatory cycles that come and go are a defining aspect of Ulcerative colitis (UC) and crohn's disease (CD). Both disorders frequently impact the whole colon, necessitating colon-specific medication administration to treat inflammatory bowel disease (IBD) (Hua et al. 2015). To test its application in an experimental model of ulcerative colitis, chitosan (CS)-modified PLGA nanospheres (NS) with a nuclear factor kappa B (NF-kB) decoy oligonucleotide (ODN) oral delivery system were created. Dextran sulphate sodium-induced diarrhea was significantly improved by decoy ODN-loaded CS-PLGA NS, and bloody feces and myeloperoxidase activity were also significantly increased. These results suggest that CS-PLGA NS might be a useful method for colon-specific oral decoy ODN delivery in UC (Tahara et al. 2011).

Targeted drug delivery devices of nano range with potential to particularly aggregate in inflamed mucosal tissues might lessen adverse medication responses in association with oral administration of anti-inflammatory as well as immunosuppressive pharmaceuticals for the treatment of IBD. Different PLGA micro- and nanoparticles with PEG-surface functionalization demonstrated enhanced translocation and deposition in inflamed mucosa and recognized as an avant-garde method of treating IBD (Lautenschläger et al. 2013).

### 8.4.4 Cardiovascular Diseases

Despite notable clinical advancements, cardiovascular diseases (CVDs), which include a variety of heart and blood vessel illnesses as well as stroke, continue to be the emerging cause of death in the United States. As per estimates from the World Health Organization (WHO), about 25 million deaths worldwide will be attributable to CVDs by 2030 (Lloyd-Jones et al. 2010; Tucka et al. 2012). For use in CVDs, numerous nanostructures-based drug delivery systems, particularly those made of biodegradable PLGA, are being developed. These systems feature a variety of sizes, shapes, as well as surface functionalization, as well as a huge range of electrostatic charges and bio-molecular conjugations (Cristallini et al. 2016; Pascual-Gil et al. 2017).

Controlled imaging and medication delivery in order to treat atherosclerosis, myocardial infarction, restenosis, and other cardiovascular disorders have been one of the key applications of nanotechnology in the cardiovascular investigation. An unfavorable consequence of endovascular procedures is restenosis, which is the narrowing of a blood vessel and results in constrained blood flow. Numerous medications, such as cytotoxins, inhibitors of the growth of smooth muscle cells (such as paclitaxel, etoposides, cytarabine, and doxorubicin), immunomodulators (such as steroids and bisphosphonates), inhibitors of the receptor for platelet-derived growth factors (PDGF), such as tyrphostin, antibiotics (such as fumagillin), and gene therapy, are used to prevent this restenosis. These biomolecules and genetic components are enclosed in polymeric nanoparticles to create regulated release patterns while protecting them from unintended enzymatic breakdown. By creating PLGA-based nanoparticles that are coated with AGL 2043 and AG1295, two specific PDGF receptor protein tyrosine kinase blockers, restenosis is prevented in rats with balloon-injured carotid arteries (Godin et al. 2010).

Life-threatening disorders include peripheral artery thrombosis, ischemic stroke, deep vein thrombosis, ischemic stroke, myocardial infarction, and pulmonary embolism resulting from the development of a thrombus in the blood circulatory system. For the therapy of these illnesses, numerous plasminogen activators have been developed. Plasminogen activator-loaded PLGA and PEG copolymer microspheres showed that they can successfully retain medication and release it at the target spot in concentrations of more than 4 mg/mL. The enhanced thrombolysis is demonstrated by pressure-driven permeation into the clot's interior. However, fibrin clot pores exhibit resistance to the admission of carriers 1 m or larger into the clot interior when there is no hydrodynamic pressure. Therefore, chitosan (CS) and CS-GRGD coated, PLGA NPs loaded with PA were created, and their capacity for thrombolysis was examined in a blood clot-occluded model. The shortest time range for clot lysis was shown by PLGA/CS NPs, while the highest percentage of digested clots was shown by PLGA/CS-GRGD NPs. By demonstrating persistent adhesion and aggregation of the clot font and interior, these nanoparticles displayed enhanced penetration (Vyas and Vaidya 2009).

Depletion of myocardiocytes and fibrosis result in myocardial ischemia (MI), which causes cardiac dysfunction. Delivery of recombinant protein or DNA forms of proangiogenic cytokines, such as growth factors includes vascular endothelial, hepatocyte, or fibroblast growth factor 1 and 2 (VEGF), (HGF), and (FGF-1 and -2), increases neovascularization of ischemic heart tissue and reduces dysfunction. Moreover, issues regarding growth factor delivery have resulted in the administration of synthetic compounds that can increase the release of endogenous proangiogenic cytokines. By polymerizing ONO-1301 with PLGA, a slow-release version of the prostacyclin agonist ONO-1301 (SR-ONO) were created. The results of the current investigation demonstrated that ONO1301 greatly accelerated with recovery of perfusion in the rat's hind leg and boosted endogenous HGF production. However, epicardial injection of SR-ONO in case of swine chronic ischemia enhanced local wall motion of ventricles, encouraged collateral development, and decreased left ventricle hypertrophy. The tissue can progressively absorb PLGA without suffering any harm. The oral administration of SR-ONO has not been related to any side effects such as diarrhoea, hypotension, or tachycardia (Iwata et al. 2009).

To combat hypoxia and its aftereffects, angiogenic therapy with vascular endothelial growth factor (VEGF) is an effective approach. It has been demonstrated that PLGA particles which are loaded with VEGF are an efficient method for delivering cytokines to a rat myocardial ischemia model. This strategy might be investigated further for various clinical investigations (Formiga et al. 2010). The compelling evidence supporting the contribution of oxidative stress to MI supports the utilization of antioxidants. Because of its function in the mitochondrial electron transport chain, Coenzyme Q10 (CoQ10) seems like a good choice to treat MI; however, because of its poor biopharmaceutical qualities, there is a need to find effective delivery methods. CoQ10 was given to rats with MI after being encapsulated in PLGA-based nanoparticles, which improved the draggability of CoQ10. Ejection fraction measurements were made before and after 3 months of treatment to examine cardiac function. The ejection fraction significantly improved after 3 months, according to the results (Simón-Yarza et al. 2013).

Early reperfusion has been demonstrated to be a conventional method for reduction in myocardial infarction size in MI patients to achieve improved clinical outcomes. Myocardial ischemia-reperfusion (IR) injury is described as the paradoxical death of cardiomyocytes caused by reperfusion of coronary arteries. Inflammatory mediators, in particular monocytes, are significant pathogenesis-related factors, making them prospective therapy targets for IR injury. Irbesartan, which is an angiotensin receptor blocker (ARB), was incorporated into PLGA nanoparticles and examined in a mouse model having IR injury. A single IV dosage of PLGA nanoparticles demonstrated improved distribution inside the mouse heart's myocardium and monocytes. Irbesartan nanoparticles have demonstrated efficacy in inhibiting monocyte recruitment in the IR heart and greatly reduced down the size of infarct. These results support their use as a therapeutic treatment for myocardial IR injury (Nakano et al. 2016).

Current developments in interventional cardiology using PCI and various revascularization techniques have improved the symptoms of CVDs. However, the atherosclerotic disease continues to be the emerging cause of death in the world. To enhance patient prognosis, new treatments must be developed that can interfere with underlying biological pathways responsible for disease pathogenesis. The development of several genetic models, such as that of atherosclerotic heart disease, has given rise to the crucial concept of using gene therapy to create novel treatments for CVDs. A hallmark of atherosclerosis development is inflammation of the arterial wall involving monocytes. As a result, monocyte chemoattractant protein-1 (MCP-1) and its receptor chemokine receptor 2 (CCR2) are promising target sites for gene therapy in the prevention of monocyte-regulated inflammation in atherosclerosis. 7ND, a mutant MCP-1 lacking the N-terminal 7 amino acids, is capable of binding with CCR2 and inhibiting MCP-1-mediated monocyte chemotaxis. A suitable gene delivery system has been created using PLGA. In mice with hypercholesterolemia, gene treatment with the 7ND plasmid encased in PLGA nanoparticles inhibited atherosclerosis. Finally, gene therapy uses a cutting-edge NP-regulated gene delivery system that targets MCP-1/CCR2 signals which is an efficient therapeutic method for the treatment of cardiovascular diseases (Matoba and Egashira 2011).

### 8.4.5 Ocular Diseases

Despite the fact that the eyes are generally accessible, multiple hurdles in effective drug delivery prevent potential treatment of the various blindness disorders which affect the eyes. The common frequent method of eye administration is topical drops, but due to quick clearance as well as poor absorption, less than 5% of the administered dose reached in intraocular tissues (Järvinen et al. 1995). NPs have been investigated as a method of extending the residence time and penetration of medications delivered to the ocular surface. Giannavola and colleagues investigated the use of PEG which works as a mucoadhesive to increase contacts between acyclovir-loaded

PLA NPs and the eye's surface (Giannavola et al. 2003). Despite being mostly described as a mucoinert surface coating thus far, PEG has a wide history of application in promoting mucoadhesion by interpenetration and/or hydrogen bonding with the mucus (Wang et al. 2008).

Furthermore, one more recent research by Schopf and colleagues which suggested to applied topically to the surface of the rabbit eye, loteprednol etabonate (LE) delivery in to the cornea as well as retina was increased by nanoparticles coated with a dense, mucoinert PEG surface coating when compared to nanoparticles without having a mucoinert PEG surface coating. The fact that nanoparticles (240 nm) and not microparticles (>1 m) benefited from the mucoinert surface coating suggests that the nanoparticles (LE-MPP) were able to enter the membrane-bound eye mucus layer, resulting in a prolonged residence time and efficient drug delivery. Then, in a pigmented rabbit model of increased retinal vascular permeability, they showed that topical LE-MPP dramatically reduced vascular leakage (Schopf et al. 2015).

Mun and colleagues sought to better understand the cornea's ability to block fluorescently labelled nanoparticles (NPs). They delivered thiol-functionalized silica nanoparticles (thiol NPs) and silica particles incorporated with 750 or 5000 Da PEG to bovine eyes that were purchased from a butcher 1 day before use. The different NP solutions were exposed to the central cornea of the eyes while they were housed in a beaker with a Franz cell donor chamber on top. They discovered that the thiolated silica NPs (21–45 nm) stuck onto the cornea's surface even when the epithelium had been removed before exposure. Similar findings were obtained for the 750 Da PEG-functionalized silica particles (27–54 nm). When the epithelium was removed before the procedure, the silica NPs functionalized with 5000 Da PEG (43–69 nm) penetrated the stroma but not the intact cornea. The stroma is more of an aqueous layer with small resistance to transcorneal permeation, whereas the corneal epithelium acts as the primary barrier restricting drug absorption in the eye. This is why they hypothesized that the difference in penetration between the corneal epithelium and the stroma was caused by the different barrier characteristics of the tissue layers (Mun et al. 2014).

Injections into an eye are frequently utilized to get over the several obstacles in order to delivery in front of the eye, especially for medication administration to the rear of the eye. Even though the vitreous gel still presents a considerable obstacle to efficient retinal administration, intravitreal injections are frequently employed to deliver inside the retina. Sanders and colleagues investigated how PEGylation affected DNA NPs for retinal gene delivery. They reasoned that PEG coatings can boost the mobility and vitreous penetration, would likely bring them closer to the target cells for the transfection, but they would not like to compromise the DNA NPs' efficacy of transfection. By adding DNA to cationic liposomes made of DOTAP and DOPE lipids, they created DNA NPs. The DSPE-PEG lipids in the "pre-PEGylated" lipoplexes were afterwards coated with DNA.

In order to create the "post-PEGylated" lipoplexes, non-PEGylated lipoplexes were first created and then treated with ceramide-C8-PEG lipids. Within 30 min of the animal's death, they retrieved the vitreous from its eyes and added the different lipoplexes. When the non-PEGylated lipoplexes were injected into the vitreous, they discovered that they greatly aggregated. While the 17 mol% PEG pre-PEGylated lipoplexes looked stable and evenly dispersed following injection inside the isolated vitreous, the addition of 4 mol% PEG in pre-PEGylated lipoplexes did give stability. Non-PEGylated and 4 mol% pre-PEGylated lipoplexes were found to be immobilized in the vitreous whereas the 17 mol% pre-PEGylated lipoplexes were shown to be mobile by fluorescence recovery after photobleaching (FRAP). The pre-PEGylation (using DSPE-PEG) significantly reduced the in vitro transfection of retinal pigment epithelium (RPE) cells, which were later discovered. The researchers went on to show that although the post-PEGylated lipoplexes retained stability in the vitreous, they were internalized by RPE cells similar to that of non-PEGylated lipoplexes. Thus, they showed that it was able to create DNA lipoplexes wrapped in PEG to increase vitreous stability while maintaining in vitro RPE cell transfection (Sanders et al. 2007).

### 8.4.6 Drug Delivery to Central Nervous System (CNS)

The frequency of neurodegenerative illnesses has grown due to population ageing and their widespread distribution. Alzheimer's disease (AD), which affects more than 35 million people globally, and by 2050, that number is projected to be double (Gregori et al. 2015). Because the blood-brain barrier (BBB) majorly prevents drugs from reaching the central nervous systems, present therapies are ineffective and new drug delivery mechanisms must be developed. Particularly PLGA-based polymeric nanoparticles have proven useful for this purpose (Fazil et al. 2012; Vilella et al. 2015).

In order to demonstrate that loperamide-loaded PLGA nanoparticles may effectively cross the BBB, their surfaces were functionalized with the help of monoclonal antibody for active targeting towards the transferrin receptor (Fornaguera et al. 2015a). A further neurodegenerative condition is Parkinson's disease. To obtain a neuroprotective effect in parkinsonism brought on by reactive oxygen species (ROS), poly (lactic-*co*-glycolic acid) (PLGA) nanoparticles were produced (Jo et al. 2015). Lorazepam-loaded PLGA nanoparticles were successfully created utilizing the nanoprecipitation process. Drug release characteristics in vitro were discovered to be almost identical to sheep nasal mucosa ex vivo penetration studies. Through an in vitro cell viability assay, the safety of NPs was assessed using the Vero cell line. Lzp-PLGA-NPs were radiolabeled with Technitium-99 m in order to create a nose-to-brain bio-distribution route in rats models using scintigraphy imaging (Sharma et al. 2014). Despite intensive investigation into cerebral vasospasm therapy, the fate of different patients with subarachnoid hemorrhage (SAH) remained disastrous. An important underlying mechanism for the development and maintenance of early brain damage (EBI) is inflammation after SAH. Curcumin is enclosed in PLGA for increasing its solubility-related concentration and provide a potent anti-neuroinflammatory reaction in the affected CNS. In a mouse model of SAH, delivery of nanocurcumin decreased leukocyte chemotaxis and subsequent inflammation, indicating that the herb's nanoparticles would be clinically effective to reduce the apoptosis brought on by SAH (Chang et al. 2015).

Resveratrol is a polyphenolic molecule that is abundant in grapes, blueberries, mulberries, cranberries, and peanut sprouts. It has anti-aging, cardioprotective, anticancer, and protective benefits against PD features. Its encapsulation in PLGA nanoparticles prevents it from being converted from its *trans* form into its inactive *cis* form, increasing its bioavailability and prolonging the anti-Parkinson effects shown in a rat model for up to 4 days (Ganesan et al. 2015). One of the most major neurodegenerative conditions that causes dementia is Alzheimer's disease (AD). The main characteristic of this illness, which causes the production of extracellular aggregates, is the deposition of amyloid (A). Curcumin exhibits anti-amyloidogenic activity in order to inhibit the formation of fresh A aggregates and dispersing the ones that are already present. Because curcumin cannot effectively penetrate the BBB, the brain absorbs less of it. Curcumin was therefore enclosed in PLGA NPs to increase its BBB-crossing capacity. Hippocampal cell culture in vitro investigations have not revealed any harm. Further evidence that curcumin loaded PLGA NPs are viable carriers for effective AD treatment comes from the considerable decrease in aggregation caused by these NPs (Barbara et al. 2017).

In various animal models of Alzheimer's disease, siRNAs and miRNAs are frequently delivered via PLGA nanoparticles. In order to enhance the characteristics of nanoparticles, various chemicals, proteins, and nucleic acids can be introduced into the PLGA matrix. Emerging nanotechnology-based delivery technologies may be employed for miRNA treatment in illnesses of the central nervous system (Shi 2015). One other study has shown that PLGA nanoparticles may be used to create the most effective and secure Alzheimer's disease vaccine. In present work, PLGA nanoparticles were created to encapsulate an A-1-15 amino acid-containing peptide for the subcut or intranasal immunization of Balb/c mice. Elicited Abs titers against complete A have been seen in mice (Puras et al. 2011).

Demyelination progresses over time in the central nervous system disorder known as multiple sclerosis (MS). Insulation, metabolic support for axons, and assistance with electrical signal transmission are all functions of the myelin sheath. Demyelination causes permanent neurodegeneration and loss of electrical conduction, which calls for an efficient therapeutic strategy. Leukemia inhibitory factor (LIF), which is a cytokine that promotes self-immunological tolerance, is a promyelination factor. LIF is delivered using PLGA nanoparticles that have been surface functionalized with antibodies against NG2 expressed on oligodendrocyte precursor cells (OPCs). This promotes the differentiation of OPC precursor cells inside the mature oligodendrocytes that can repair myelin. Both degrees of myelin repair were seen, and the increased number of myelinated axons and enhanced myelin thickness per axon imply that this method may be used to treat multiple sclerosis (MS) (Rittchen et al. 2015).

PLGA has been used to create GAL-loaded nanoparticles with improved encapsulation effectiveness, a regulated drug release profile, a non-toxic effect, and the preservation of GAL's pharmacological activity, lengthening the duration of its therapeutic benefits. These improved GAL-loaded nanoparticles are a potential medication delivery method for neurological disorders (Fornaguera et al. 2015b). Additionally, nanopatterned scaffolds made of PLGA has been created to increase the therapeutic potential of stem cells in treating illnesses such as brain damage and neurodegenerative disorders. Neuronal stem cells (NSCs) are being stimulated to extend their neurites and are being controlled in their development using biophysical signals provided by nanotopographical characteristics. Biodegradable PLGA polymeric substrates with nanoscale topographic features were created with DOPA (dihydroxyphenylalanine) coating for guided neurite outgrowth and improved differentiation of human neural stem cells (hNSC). With the addition of nerve growth factor, hNSC differentiation was significantly enhanced, demonstrating the joint impact of biochemical as well as physical cues on stem cell differentiation (Yang et al. 2015).

## 8.5 Conclusion

Due to their capacity to shield genetic material from deterioration and enhance its reception into cells, PEGylated nanocarriers are a potential technique for gene transfer. Future developments in this area can be anticipated as researchers attempt to improve the functionality and efficiency of PEGylated nanocarriers for gene therapy applications. How to combat the immunological response that can be brought on by PEGylated nanoparticles is one of the major issues that researchers will need to address. Finding ways to lessen this response will be a key area of attention as it may restrict their efficacy and lead to negative patient reactions. The use of PEGylated nanocarriers for targeted gene therapy, in which particular genes are delivered to particular cells or tissues in the body, will also be investigated in future studies. This might be accomplished by altering the nanocarrier's surface with targeting molecules that recognize and bind to particular cell types. As a tool for gene therapy, PEGylated nanocarriers show great potential, and as scientists seek to solve the remaining issues and refine the technology for clinical application, we may anticipate seeing further advancements in this area. Before PEGylated nanocarriers can be extensively used as a gene therapy tool, extensive safety and efficacy studies must be carried out because, like any new technology, there are still numerous unresolved issues and possible dangers.

## References

Afzal W et al (2017) Autoimmune neutropenia updates: etiology, pathology, and treatment. South Med J 110(4):300–307. https://doi.org/10.14423/smj.0000000000000637

Ahmed M et al (2005) Combined radiofrequency ablation and adjuvant liposomal chemotherapy: effect of chemotherapeutic agent, nanoparticle size, and circulation time. J Vasc Interv Radiol 16(10):1365–1371. https://doi.org/10.1097/01.RVI.0000175324.63304.25

Aiuti A, Roncarolo MG, Naldini L (2017) Gene therapy for ADA-SCID, the first marketing approval of an ex vivo gene therapy in Europe: paving the road for the next genera-

tion of advanced therapy medicinal products. EMBO Mol Med 9(6):737–740. https://doi.org/10.15252/emmm.201707573
An HJ et al (2020) Sarcopenia in autoimmune and rheumatic diseases: a comprehensive review. Int J Mol Sci 21(16):5678. https://doi.org/10.3390/ijms21165678
Anguela XM, High KA (2019) Entering the modern era of gene therapy. Annu Rev Med 70:273–288. https://doi.org/10.1146/annurev-med-012017-043332
Bade BC, Dela Cruz CS (2020) Lung Cancer 2020: epidemiology, etiology, and prevention. Clin Chest Med 41(1):1–24. https://doi.org/10.1016/j.ccm.2019.10.001
Barbara R et al (2017) Novel Curcumin loaded nanoparticles engineered for blood-brain barrier crossing and able to disrupt Abeta aggregates. Int J Pharm 526(1–2):413–424. https://doi.org/10.1016/j.ijpharm.2017.05.015
Beheshti SA et al (2022) CAR Treg: a new approach in the treatment of autoimmune diseases. Int Immunopharmacol 102:108409. https://doi.org/10.1016/j.intimp.2021.108409
Bottai G et al (2017) Progress in nonviral gene therapy for breast cancer and what comes next? Expert Opin Biol Ther 17(5):595–611. https://doi.org/10.1080/14712598.2017.1305351
Boulaiz H et al (2005) Non-viral and viral vectors for gene therapy. Cell Mol Biol (Noisy-le-grand) 51(1):3–22
Braden AM, Stankowski RV, Engel JM, Onitilo AA et al (2014) Breast cancer biomarkers: risk assessment, diagnosis, prognosis, prediction of treatment efficacy and toxicity, and recurrence. Curr Pharm Des 20(30):4879–4898. https://doi.org/10.2174/1381612819666131125145517
Bruun J et al (2015) Investigation of enzyme-sensitive lipid nanoparticles for delivery of siRNA to blood-brain barrier and glioma cells. Int J Nanomedicine 10:5995–6008. https://doi.org/10.2147/ijn.S87334
Cagel M et al (2017) Polymeric mixed micelles as nanomedicines: achievements and perspectives. Eur J Pharm Biopharm 113:211–228. https://doi.org/10.1016/j.ejpb.2016.12.019
Cao Y et al (2019) Recent advances in chitosan-based carriers for gene delivery. Mar Drugs 17(6):381. https://doi.org/10.3390/md17060381
Chakravarty D, Solit DB (2021) Clinical cancer genomic profiling. Nat Rev Genet 22(8):483–501. https://doi.org/10.1038/s41576-021-00338-8
Chang CZ et al (2015) Curcumin, encapsulated in nano-sized PLGA, down-regulates nuclear factor κB (p65) and subarachnoid hemorrhage induced early brain injury in a rat model. Brain Res 1608:215–224. https://doi.org/10.1016/j.brainres.2015.02.039
Chen J et al (2016) Production and clinical development of nanoparticles for gene delivery. Mol Ther Methods Clin Dev 3:16023. https://doi.org/10.1038/mtm.2016.23
Chen Y et al (2018) In vitro evaluation of anti-aggregation and degradation behavior of PEGylated polymeric nanogels under in vivo like conditions. Macromol Biosci 18(1). https://doi.org/10.1002/mabi.201700127
Chopra A (2004) Pegylated bis(4-(N-(2-naphthyl)phenylamino)phenyl)-fumaronitrile (NPAPF) nanoparticles. In: Molecular imaging and contrast agent database (MICAD). National Center for Biotechnology Information (US), Bethesda
Cristallini C et al (2016) Micro- and macrostructured PLGA/gelatin scaffolds promote early cardiogenic commitment of human mesenchymal stem cells in vitro. Stem Cells Int 2016:7176154. https://doi.org/10.1155/2016/7176154
Dart A (2022) Cancer's sensory experience. Nat Rev Cancer 22(8):433. https://doi.org/10.1038/s41568-022-00492-z
Dastjerd NT et al (2022) Gene therapy: a promising approach for breast cancer treatment. Cell Biochem Funct 40(1):28–48. https://doi.org/10.1002/cbf.3676
Davis PB, Cooper MJ (2007) Vectors for airway gene delivery. AAPS J 9(1):E11–E17. https://doi.org/10.1208/aapsj0901002
de Souza Guimaraes M et al (2022) PEG-grafted liposomes for L-asparaginase encapsulation. Pharmaceutics 14(9):1819. https://doi.org/10.3390/pharmaceutics14091819
Delogu LG et al (2009) Conjugation of antisense oligonucleotides to PEGylated carbon nanotubes enables efficient knockdown of PTPN22 in T lymphocytes. Bioconjug Chem 20(3):427–431. https://doi.org/10.1021/bc800540j

Demirgoz D, Garg A, Kokkoli E (2008) PR_b-targeted PEGylated liposomes for prostate cancer therapy. Langmuir 24(23):13518–13524. https://doi.org/10.1021/la801961r
Edis Z et al (2021) Nanocarriers-mediated drug delivery systems for anticancer agents: an overview and perspectives. Int J Nanomedicine 16:1313–1330. https://doi.org/10.2147/IJN.S289443
Fazil M et al (2012) Nanotherapeutics for Alzheimer's disease (AD): past, present and future. J Drug Target 20(2):97–113. https://doi.org/10.3109/1061186x.2011.607499
Formiga FR et al (2010) Sustained release of VEGF through PLGA microparticles improves vasculogenesis and tissue remodeling in an acute myocardial ischemia-reperfusion model. J Control Release 147(1):30–37. https://doi.org/10.1016/j.jconrel.2010.07.097
Fornaguera C et al (2015a) PLGA nanoparticles prepared by nano-emulsion templating using low-energy methods as efficient nanocarriers for drug delivery across the blood-brain barrier. J Control Release 211:134–143. https://doi.org/10.1016/j.jconrel.2015.06.002
Fornaguera C et al (2015b) Galantamine-loaded PLGA nanoparticles, from nano-emulsion templating, as novel advanced drug delivery systems to treat neurodegenerative diseases. Nanoscale 7(28):12076–12084. https://doi.org/10.1039/c5nr03474d
Fredman G et al (2015) Targeted nanoparticles containing the proresolving peptide Ac2-26 protect against advanced atherosclerosis in hypercholesterolemic mice. Sci Transl Med 7(275):275ra20. https://doi.org/10.1126/scitranslmed.aaa1065
Gabizon A, Shmeeda H, Barenholz Y (2003) Pharmacokinetics of pegylated liposomal doxorubicin: review of animal and human studies. Clin Pharmacokinet 42(5):419–436. https://doi.org/10.2165/00003088-200342050-00002
Gajbhiye KR et al (2020) PEGylated nanocarriers: a promising tool for targeted delivery to the brain. Colloids Surf B Biointerfaces 187:110770. https://doi.org/10.1016/j.colsurfb.2019.110770
Ganesan P et al (2015) Recent trends in the development of nanophytobioactive compounds and delivery systems for their possible role in reducing oxidative stress in Parkinson's disease models. Int J Nanomedicine 10:6757–6772. https://doi.org/10.2147/ijn.S93918
Gao Y, Li LB, Zhai G (2008) Preparation and characterization of Pluronic/TPGS mixed micelles for solubilization of camptothecin. Colloids Surf B Biointerfaces 64(2):194–199. https://doi.org/10.1016/j.colsurfb.2008.01.021
Gao H et al (2013) The impact of PEGylation patterns on the in vivo biodistribution of mixed shell micelles. Int J Nanomedicine 8:4229–4246. https://doi.org/10.2147/IJN.S51566
Geiger BC et al (2018) Cartilage-penetrating nanocarriers improve delivery and efficacy of growth factor treatment of osteoarthritis. Sci Transl Med 10(469):eaat8800. https://doi.org/10.1126/scitranslmed.aat8800
Giamas G (2020) Cancer gene therapy: vision and strategy for the new decade. Cancer Gene Ther 27(3–4):115. https://doi.org/10.1038/s41417-020-0169-8
Giamas G, Gagliano T (2022) Cancer gene therapy 2020: highlights from a challenging year. Cancer Gene Ther 29(1):1–3. https://doi.org/10.1038/s41417-021-00340-6
Giannavola C et al (2003) Influence of preparation conditions on acyclovir-loaded poly-d,l-lactic acid nanospheres and effect of PEG coating on ocular drug bioavailability. Pharm Res 20(4):584–590. https://doi.org/10.1023/a:1023290514575
Ginn SL et al (2018) Gene therapy clinical trials worldwide to 2017: an update. J Gene Med 20(5):e3015. https://doi.org/10.1002/jgm.3015
Godin B et al (2010) Emerging applications of nanomedicine for the diagnosis and treatment of cardiovascular diseases. Trends Pharmacol Sci 31(5):199–205. https://doi.org/10.1016/j.tips.2010.01.003
Goncalves GAR, Paiva RMA (2017) Gene therapy: advances, challenges and perspectives. Einstein (Sao Paulo) 15(3):369–375. https://doi.org/10.1590/S1679-45082017RB4024
Gregori M, Masserini M, Mancini S (2015) Nanomedicine for the treatment of Alzheimer's disease. Nanomedicine (Lond) 10(7):1203–1218. https://doi.org/10.2217/nnm.14.206
Gu W et al (2013) Nanotechnology in the targeted drug delivery for bone diseases and bone regeneration. Int J Nanomedicine 8:2305–2317. https://doi.org/10.2147/ijn.S44393
Gupta P et al (2015) Quercetin conjugated poly(beta-amino esters) nanogels for the treatment of cellular oxidative stress. Acta Biomater 27:194–204. https://doi.org/10.1016/j.actbio.2015.08.039

Hamidi M, Azadi A, Rafiei P (2006) Pharmacokinetic consequences of pegylation. Drug Deliv 13(6):399–409. https://doi.org/10.1080/10717540600814402

Hofheinz RD et al (2005) Liposomal encapsulated anti-cancer drugs. Anti-Cancer Drugs 16(7):691–707. https://doi.org/10.1097/01.cad.0000167902.53039.5a

Hua S et al (2015) Advances in oral nano-delivery systems for colon targeted drug delivery in inflammatory bowel disease: selective targeting to diseased versus healthy tissue. Nanomedicine 11(5):1117–1132. https://doi.org/10.1016/j.nano.2015.02.018

Ikoba U et al (2015) Nanocarriers in therapy of infectious and inflammatory diseases. Nanoscale 7(10):4291–4305. https://doi.org/10.1039/c4nr07682f

Iwata H et al (2009) Local delivery of synthetic prostacycline agonist augments collateral growth and improves cardiac function in a swine chronic cardiac ischemia model. Life Sci 85(5–6):255–261. https://doi.org/10.1016/j.lfs.2009.06.002

Jain NK, Nahar M (2010) PEGylated nanocarriers for systemic delivery. Methods Mol Biol 624:221–234. https://doi.org/10.1007/978-1-60761-609-2_15

Järvinen K, Järvinen T, Urtti A (1995) Ocular absorption following topical delivery. Adv Drug Deliv Rev 16(1):3–19

Jo DH et al (2015) Size, surface charge, and shape determine therapeutic effects of nanoparticles on brain and retinal diseases. Nanomedicine 11(7):1603–1611. https://doi.org/10.1016/j.nano.2015.04.015

Karakoti AS et al (2011) PEGylated inorganic nanoparticles. Angew Chem Int Ed Engl 50(9):1980–1994. https://doi.org/10.1002/anie.201002969

Katare DP, Aeri V (2010) Progress in gene therapy: a review. IJTPR 1(33):e41

Khalid MN et al (2006) Long circulating poly(ethylene glycol)-decorated lipid nanocapsules deliver docetaxel to solid tumors. Pharm Res 23(4):752–758. https://doi.org/10.1007/s11095-006-9662-5

Klibanov AL et al (1990) Amphipathic polyethyleneglycols effectively prolong the circulation time of liposomes. FEBS Lett 268(1):235–237. https://doi.org/10.1016/0014-5793(90)81016-h

Koide H et al (2009) Elucidation of accelerated blood clearance phenomenon caused by repeat injection of PEGylated nanocarriers. Yakugaku Zasshi 129(12):1445–1451. https://doi.org/10.1248/yakushi.129.1445

Koudelka S et al (2016) Liposomal nanocarriers for plasminogen activators. J Control Release 227:45–57. https://doi.org/10.1016/j.jconrel.2016.02.019

Laginha KM et al (2005) Determination of doxorubicin levels in whole tumor and tumor nuclei in murine breast cancer tumors. Clin Cancer Res 11(19 Pt 1):6944–6949. https://doi.org/10.1158/1078-0432.CCR-05-0343

Lara-Guerra H, Roth JA (2016) Gene therapy for lung cancer. Crit Rev Oncog 21(1–2):115–124. https://doi.org/10.1615/CritRevOncog.2016016084

Lautenschläger C et al (2013) PEG-functionalized microparticles selectively target inflamed mucosa in inflammatory bowel disease. Eur J Pharm Biopharm 85(3 Pt A):578–586. https://doi.org/10.1016/j.ejpb.2013.09.016

Leung PS et al (2010) Gene therapy in autoimmune diseases: challenges and opportunities. Autoimmun Rev 9(3):170–174. https://doi.org/10.1016/j.autrev.2009.10.004

Li SD, Huang L (2010) Stealth nanoparticles: high density but sheddable PEG is a key for tumor targeting. J Control Release 145(3):178–181. https://doi.org/10.1016/j.jconrel.2010.03.016

Lim JA et al (2019) Development of the clinical assessment scale in autoimmune encephalitis. Ann Neurol 85(3):352–358. https://doi.org/10.1002/ana.25421

Liu L et al (2022) PEGylation effectively improves anti-breast cancer efficiency of heat shock protein gp96 inhibitory polypeptide. Sheng Wu Gong Cheng Xue Bao 38(9):3363–3378. https://doi.org/10.13345/j.cjb.220067

Lloyd-Jones D et al (2010) Heart disease and stroke statistics—2010 update: a report from the American Heart Association. Circulation 121(7):e46–e215. https://doi.org/10.1161/circulationaha.109.192667

Lu X, Zhang K (2018) PEGylation of therapeutic oligonucletides: from linear to highly branched PEG architectures. Nano Res 11(10):5519–5534. https://doi.org/10.1007/s12274-018-2131-8

Lukasiewicz S et al (2021) Polycaprolactone nanoparticles as promising candidates for nanocarriers in novel nanomedicines. Pharmaceutics 13(2):191. https://doi.org/10.3390/pharmaceutics13020191

Ma F, Laster K, Dong Z (2022) The comparison of cancer gene mutation frequencies in Chinese and U.S. patient populations. Nat Commun 13(1):5651. https://doi.org/10.1038/s41467-022-33351-4

Matoba T, Egashira K (2011) Anti-inflammatory gene therapy for cardiovascular disease. Curr Gene Ther 11(6):442–446. https://doi.org/10.2174/156652311798192888

Matsumura Y et al (2004) Phase I clinical trial and pharmacokinetic evaluation of NK911, a micelle-encapsulated doxorubicin. Br J Cancer 91(10):1775–1781. https://doi.org/10.1038/sj.bjc.6602204

Mattheolabakis G et al (2009) In vivo investigation of tolerance and antitumor activity of cisplatin-loaded PLGA-mPEG nanoparticles. Eur J Pharm Biopharm 71(2):190–195. https://doi.org/10.1016/j.ejpb.2008.09.011

Mehrabian A et al (2022) Preparation, characterization, and biodistribution of glutathione PEGylated nanoliposomal doxorubicin for brain drug delivery with a post-insertion approach. Iran J Basic Med Sci 25(3):302–312. https://doi.org/10.22038/ijbms.2022.60306.13369

Miyamoto D et al (2008) Completely dispersible PEGylated gold nanoparticles under physiological conditions: modification of gold nanoparticles with precisely controlled PEG-b-polyamine. Langmuir 24(9):5010–5017. https://doi.org/10.1021/la703813f

Mohamed M et al (2019) PEGylated liposomes: immunological responses. Sci Technol Adv Mater 20(1):710–724. https://doi.org/10.1080/14686996.2019.1627174

Mu L, Elbayoumi TA, Torchilin VP (2005) Mixed micelles made of poly(ethylene glycol)-phosphatidylethanolamine conjugate and d-alpha-tocopheryl polyethylene glycol 1000 succinate as pharmaceutical nanocarriers for camptothecin. Int J Pharm 306(1–2):142–149. https://doi.org/10.1016/j.ijpharm.2005.08.026

Mun EA et al (2014) On the barrier properties of the cornea: a microscopy study of the penetration of fluorescently labeled nanoparticles, polymers, and sodium fluorescein. Mol Pharm 11(10):3556–3564. https://doi.org/10.1021/mp500332m

Muralidharan P et al (2014) Inhalable PEGylated phospholipid nanocarriers and PEGylated therapeutics for respiratory delivery as aerosolized colloidal dispersions and dry powder inhalers. Pharmaceutics 6(2):333–353. https://doi.org/10.3390/pharmaceutics6020333

Nair J et al (2020) Translatable gene therapy for lung cancer using Crispr CAS9—an exploratory review. Cancer Gene Ther 27(3–4):116–124. https://doi.org/10.1038/s41417-019-0116-8

Nakamura T et al (2010) Large payloads of gold nanoparticles into the polyamine network core of stimuli-responsive PEGylated nanogels for selective and noninvasive cancer photothermal therapy. Nanoscale 2(5):739–746. https://doi.org/10.1039/b9nr00329k

Nakano Y et al (2016) Nanoparticle-mediated delivery of Irbesartan induces cardioprotection from myocardial ischemia-reperfusion injury by antagonizing monocyte-mediated inflammation. Sci Rep 6:29601. https://doi.org/10.1038/srep29601

Nakki S et al (2015) Improved stability and biocompatibility of nanostructured silicon drug carrier for intravenous administration. Acta Biomater 13:207–215. https://doi.org/10.1016/j.actbio.2014.11.019

Nasim F, Sabath BF, Eapen GA (2019) Lung cancer. Med Clin North Am 103(3):463–473. https://doi.org/10.1016/j.mcna.2018.12.006

Nayerossadat N, Maedeh T, Ali PA (2012) Viral and nonviral delivery systems for gene delivery. Adv Biomed Res 1:27. https://doi.org/10.4103/2277-9175.98152

Nissinen T et al (2016) Tailored dual PEGylation of inorganic porous nanocarriers for extremely long blood circulation in vivo. ACS Appl Mater Interfaces 8(48):32723–32731. https://doi.org/10.1021/acsami.6b12481

Otani K et al (2019) Colon cancer with perforation. Surg Today 49(1):15–20. https://doi.org/10.1007/s00595-018-1661-8

Padin-Gonzalez E et al (2022) Understanding the role and impact of poly (ethylene glycol) (PEG) on nanoparticle formulation: implications for COVID-19 vaccines. Front Bioeng Biotechnol 10:882363. https://doi.org/10.3389/fbioe.2022.882363

Pascual-Gil S et al (2017) Cytokine-loaded PLGA and PEG-PLGA microparticles showed similar heart regeneration in a rat myocardial infarction model. Int J Pharm 523(2):531–533. https://doi.org/10.1016/j.ijpharm.2016.11.022

Pawlik TM (2018) Colon cancer. Surg Oncol Clin N Am 27(2):xiii–xiv. https://doi.org/10.1016/j.soc.2017.11.013

Pishavar E et al (2020) Cholesterol-conjugated PEGylated PAMAM as an efficient nanocarrier for plasmid encoding interleukin-12 immunogene delivery toward colon cancer cells. Biotechnol Prog 36(3):e2952. https://doi.org/10.1002/btpr.2952

Puras G et al (2011) Encapsulation of Aβ(1-15) in PLGA microparticles enhances serum antibody response in mice immunized by subcutaneous and intranasal routes. Eur J Pharm Sci 44(3):200–206. https://doi.org/10.1016/j.ejps.2011.07.010

Ramamoorth M, Narvekar A (2015) Non viral vectors in gene therapy—an overview. J Clin Diagn Res 9(1):GE01-6. https://doi.org/10.7860/JCDR/2015/10443.5394

Razi Soofiyani S et al (2013) Gene therapy, early promises, subsequent problems, and recent breakthroughs. Adv Pharm Bull 3(2):249–255. https://doi.org/10.5681/apb.2013.041

Ren H et al (2019) Role of liposome size, surface charge, and PEGylation on rheumatoid arthritis targeting therapy. ACS Appl Mater Interfaces 11(22):20304–20315. https://doi.org/10.1021/acsami.8b22693

Rittchen S et al (2015) Myelin repair in vivo is increased by targeting oligodendrocyte precursor cells with nanoparticles encapsulating leukaemia inhibitory factor (LIF). Biomaterials 56:78–85. https://doi.org/10.1016/j.biomaterials.2015.03.044

Roberts TC, Langer R, Wood MJA (2020) Advances in oligonucleotide drug delivery. Nat Rev Drug Discov 19(10):673–694. https://doi.org/10.1038/s41573-020-0075-7

Sahu R, Pattanayak SP (2020) Strategic developments & future perspective on gene therapy for breast cancer: role of mTOR and Brk/ PTK6 as molecular targets. Curr Gene Ther 20(4):237–258. https://doi.org/10.2174/1566523220999200731002408

Sanders NN et al (2007) Wanted and unwanted properties of surface PEGylated nucleic acid nanoparticles in ocular gene transfer. J Control Release 122(3):226–235. https://doi.org/10.1016/j.jconrel.2007.05.004

Sarparanta M et al (2012) Intravenous delivery of hydrophobin-functionalized porous silicon nanoparticles: stability, plasma protein adsorption and biodistribution. Mol Pharm 9(3):654–663. https://doi.org/10.1021/mp200611d

Schopf LR et al (2015) Topical ocular drug delivery to the back of the eye by mucus-penetrating particles. Transl Vis Sci Technol 4(3):11. https://doi.org/10.1167/tvst.4.3.11

Shahbazi R et al (2023) Design and optimization various formulations of PEGylated niosomal nanoparticles loaded with phytochemical agents: potential anti-cancer effects against human lung cancer cells. Pharmacol Rep 75(2):442–455. https://doi.org/10.1007/s43440-023-00462-8

Sharma D et al (2014) Formulation and optimization of polymeric nanoparticles for intranasal delivery of lorazepam using Box-Behnken design: in vitro and in vivo evaluation. Biomed Res Int 2014:156010. https://doi.org/10.1155/2014/156010

Shi J (2015) Regulatory networks between neurotrophins and miRNAs in brain diseases and cancers. Acta Pharmacol Sin 36(2):149–157. https://doi.org/10.1038/aps.2014.135

Shiraishi K, Yokoyama M (2019) Toxicity and immunogenicity concerns related to PEGylated-micelle carrier systems: a review. Sci Technol Adv Mater 20(1):324–336. https://doi.org/10.1080/14686996.2019.1590126

Shroff K, Kokkoli E (2012) PEGylated liposomal doxorubicin targeted to alpha5beta1-expressing MDA-MB-231 breast cancer cells. Langmuir 28(10):4729–4736. https://doi.org/10.1021/la204466g

Shu SA et al (2015) Gene therapy for autoimmune disease. Clin Rev Allergy Immunol 49(2):163–176. https://doi.org/10.1007/s12016-014-8451-x

Simón-Yarza T et al (2013) Functional benefits of PLGA particulates carrying VEGF and CoQ10 in an animal of myocardial ischemia. Int J Pharm 454(2):784–790. https://doi.org/10.1016/j.ijpharm.2013.04.015
Stone D (2010) Novel viral vector systems for gene therapy. Viruses 2(4):1002–1007. https://doi.org/10.3390/v2041002
Su S, Kang PM (2020) Recent advances in nanocarrier-assisted therapeutics delivery systems. Pharmaceutics 12(9):837. https://doi.org/10.3390/pharmaceutics12090837
Suk JS et al (2016) PEGylation as a strategy for improving nanoparticle-based drug and gene delivery. Adv Drug Deliv Rev 99(Pt A):28–51. https://doi.org/10.1016/j.addr.2015.09.012
Tahara K et al (2011) Oral nuclear factor-κB decoy oligonucleotides delivery system with chitosan modified poly(D,L-lactide-co-glycolide) nanospheres for inflammatory bowel disease. Biomaterials 32(3):870–878. https://doi.org/10.1016/j.biomaterials.2010.09.034
Taher M et al (2023) PEGylated liposomes enhance the effect of cytotoxic drug: a review. Heliyon 9(3):e13823. https://doi.org/10.1016/j.heliyon.2023.e13823
Tamura M et al (2011) In vitro and in vivo characteristics of core-shell type nanogel particles: optimization of core cross-linking density and surface poly(ethylene glycol) density in PEGylated nanogels. Acta Biomater 7(9):3354–3361. https://doi.org/10.1016/j.actbio.2011.05.027
Tang W et al (2019) Emerging blood-brain-barrier-crossing nanotechnology for brain cancer theranostics. Chem Soc Rev 48(11):2967–3014. https://doi.org/10.1039/c8cs00805a
Torchilin VP (2007) Micellar nanocarriers: pharmaceutical perspectives. Pharm Res 24(1):1–16. https://doi.org/10.1007/s11095-006-9132-0
Tucka J, Bennett M, Littlewood T (2012) Cell death and survival signalling in the cardiovascular system. Front Biosci (Landmark Ed) 17(1):248–261. https://doi.org/10.2741/3925
van Vlerken LE, Vyas TK, Amiji MM (2007) Poly(ethylene glycol)-modified nanocarriers for tumor-targeted and intracellular delivery. Pharm Res 24(8):1405–1414. https://doi.org/10.1007/s11095-007-9284-6
Verhoef JJ, Anchordoquy TJ (2013) Questioning the use of PEGylation for drug delivery. Drug Deliv Transl Res 3(6):499–503. https://doi.org/10.1007/s13346-013-0176-5
Veronese FM, Pasut G (2005) PEGylation, successful approach to drug delivery. Drug Discov Today 10(21):1451–1458. https://doi.org/10.1016/S1359-6446(05)03575-0
Vijayan VM, Shenoy SJ, Muthu J (2017) Octreotide-conjugated fluorescent PEGylated polymeric nanogel for theranostic applications. Mater Sci Eng C Mater Biol Appl 76:490–500. https://doi.org/10.1016/j.msec.2017.03.125
Vilella A et al (2015) Endocytosis of nanomedicines: the case of glycopeptide engineered PLGA nanoparticles. Pharmaceutics 7(2):74–89. https://doi.org/10.3390/pharmaceutics7020074
Vyas SP, Vaidya B (2009) Targeted delivery of thrombolytic agents: role of integrin receptors. Expert Opin Drug Deliv 6(5):499–508. https://doi.org/10.1517/17425240902878002
Wang YY et al (2008) Addressing the PEG mucoadhesivity paradox to engineer nanoparticles that "slip" through the human mucus barrier. Angew Chem Int Ed Engl 47(50):9726–9729. https://doi.org/10.1002/anie.200803526
Wang W et al (2013) Non-viral gene delivery methods. Curr Pharm Biotechnol 14(1):46–60
Wang C et al (2015a) Accelerated blood clearance phenomenon upon cross-administration of PEGylated nanocarriers in beagle dogs. Int J Nanomedicine 10:3533–3545. https://doi.org/10.2147/ijn.S82481
Wang F et al (2015b) Efficient RNA delivery by integrin-targeted glutathione responsive polyethyleneimine capped gold nanorods. Acta Biomater 23:136–146. https://doi.org/10.1016/j.actbio.2015.05.028
Wang L, Wang FS, Gershwin ME (2015c) Human autoimmune diseases: a comprehensive update. J Intern Med 278(4):369–395. https://doi.org/10.1111/joim.12395
Wang D, Tai PWL, Gao G (2019) Adeno-associated virus vector as a platform for gene therapy delivery. Nat Rev Drug Discov 18(5):358–378. https://doi.org/10.1038/s41573-019-0012-9
Wang H et al (2022) Mechanisms and challenges of nanocarriers as non-viral vectors of therapeutic genes for enhanced pulmonary delivery. J Control Release 352:970–993. https://doi.org/10.1016/j.jconrel.2022.10.061

Wang S et al (2023) Liposomes for tumor targeted therapy: a review. Int J Mol Sci 24(3):2643. https://doi.org/10.3390/ijms24032643

Werner ME et al (2013) Preclinical evaluation of Genexol-PM, a nanoparticle formulation of paclitaxel, as a novel radiosensitizer for the treatment of non-small cell lung cancer. Int J Radiat Oncol Biol Phys 86(3):463–468. https://doi.org/10.1016/j.ijrobp.2013.02.009

Wójcik P et al (2021) Oxidative stress and lipid mediators modulate immune cell functions in autoimmune diseases. Int J Mol Sci 22(2):723. https://doi.org/10.3390/ijms22020723

Wonganan P, Croyle MA (2010) PEGylated adenoviruses: from mice to monkeys. Viruses 2(2):468–502. https://doi.org/10.3390/v2020468

Xin H et al (2010) Enhanced anti-glioblastoma efficacy by PTX-loaded PEGylated poly(varepsilon-caprolactone) nanoparticles: in vitro and in vivo evaluation. Int J Pharm 402(1–2):238–247. https://doi.org/10.1016/j.ijpharm.2010.10.005

Xin H et al (2011) Angiopep-conjugated poly(ethylene glycol)-co-poly(epsilon-caprolactone) nanoparticles as dual-targeting drug delivery system for brain glioma. Biomaterials 32(18):4293–4305. https://doi.org/10.1016/j.biomaterials.2011.02.044

Xue HY et al (2015) Lipid-based nanocarriers for RNA delivery. Curr Pharm Des 21(22):3140–3147. https://doi.org/10.2174/1381612821666150531164540

Yang K et al (2015) Biodegradable nanotopography combined with neurotrophic signals enhances contact guidance and neuronal differentiation of human neural stem cells. Macromol Biosci 15(10):1348–1356. https://doi.org/10.1002/mabi.201500080

Yin W et al (2021) Cancer and stem cells. Exp Biol Med (Maywood) 246(16):1791–1801. https://doi.org/10.1177/15353702211005390

Zwacka RM, Dunlop MG (1998) Gene therapy for colon cancer. Hematol Oncol Clin North Am 12(3):595–615. https://doi.org/10.1016/s0889-8588(05)70010-1

# PEGylated Nanocarrier System for Nucleic Acid Delivery

# 9

Akhilesh Tiwari, Sreeharsha Nagaraja, Rakesh Sahu, and Muktika Tekade

**Abstract**

The transport of nucleic acids to host sites must be successful in creating gene therapy techniques. Due to their capacity to shield nucleic acids from deterioration and improve their cellular absorption and bioavailability, PEGylated nanocarrier systems have demonstrated significant potential as delivery technologies. This manuscript outlines creating and evaluating a PEGylated nanocarrier system for transporting nucleic acids. The nanocarrier comprised a lipid-based core for encasing nucleic acids and a polyethylene glycol (PEG) coating to increase stability and decrease immunogenicity. The nanocarrier may effectively encapsulate and shield the nucleic acids from nuclease deterioration. Moreover, it displayed in vitro minimal cytotoxicity and good cellular uptake and transfection

A. Tiwari (✉)
Department of Pharmacy, Indira Gandhi National Tribal University, Amarkantak, Anuppur, Madhya Pradesh, India

S. Nagaraja
Department of Pharmaceutics, Vidya Siri College of Pharmacy, Bengaluru, Karnataka, India

Department of Pharmaceutical Sciences, College of Clinical Pharmacy, King Faisal University, Al-Ahsa, Saudi Arabia

R. Sahu
National Institute of Pharmaceutical Education and Research (NIPER) Ahmedabad, An Institute of National Importance, Government of India, Department of Pharmaceuticals, Ministry of Chemicals and Fertilizers, Palaj, Opp. Air force station, Gandhinagar, Gujarat, India

M. Tekade
School of Pharmacy, Devi Ahilya Vishwavidyalaya, Takshila Campus, Indore, Madhya Pradesh, India

School of Pharmaceutical Sciences, Lovely Professional University, Phagwara, Punjab, India


R. K. Tekade, N. K. Jain (eds.), *PEGylated Nanocarriers in Medicine and Pharmacy*, https://doi.org/10.1007/978-981-97-7316-9_9

efficiency. In vivo tests revealed that considerable gene activation in the liver was caused by the nanocarrier's ability to effectively carry nucleic acids to target cells in a mouse model.

**Keywords**

PEGylated · Nanocarrier · Nucleic acid delivery

## 9.1 Introduction

Therapies based on nucleic acids have shown promise in treating various illnesses, such as cancer, viral infections, and genetic abnormalities. Unfortunately, nucleic acid treatments have difficulty translating into the clinic due to their poor pharmacokinetics, bioavailability, and lack of adequate and secure delivery mechanisms. Due to their capacity to shield nucleic acids from deterioration, improve cellular absorption, and enable intracellular transport, nanocarrier systems, such as liposomes, polymeric nanoparticles, and dendrimers, have been intensively explored as possible delivery vehicles for nucleic acids (Sridharan and Gogtay 2016; Shtykalova et al. 2023).

The quick removal of nucleic acids from the circulation by the reticuloendothelial system (RES), which restricts their accumulation at the target location, is one of the main obstacles to developing efficient nanocarrier systems for nucleic acid delivery. Increased circulation duration has been achieved via polyethylene glycol (PEG)ylation, which involves the covalent attachment of PEG chains to the surface of nanocarriers by reducing their recognition by RES cells. PEGylation can also enhance the stability and biocompatibility of nanocarriers and reduce their toxicity and immunogenicity.

Attaching PEG to a molecule is known as PEGylation, sometimes known as polyethylene glycosylation. The pharmaceutical industry frequently employs this technique to enhance the pharmacokinetic characteristics of medications, including extending their duration in circulation and lowering their immunogenicity. Many medications, including interferons, enzymes, and antibodies, have been proven to be safer and more effective after being PEGylated. The PEG molecule is biocompatible, non-toxic, and non-immunogenic, making it an ideal candidate for drug modification. One example of a PEGylated drug is Pegasys chronic hepatitis B, and C are managed with peginterferon alfa-2a. The addition of PEG to interferon alfa-2a improves its pharmacokinetic properties and allows for less frequent dosing compared to non-PEGylated interferon alfa-2a. Another example is Adagen (pegademase bovine), a PEGylated enzyme used to treat adenosine deaminase deficiency. The PEGylation of the enzyme increases its half-life and reduces its immunogenicity, allowing for a longer-lasting and more effective treatment (Veronese and Pasut 2005).

PEGylation is a process that joins the polymer PEG strands to molecules, most often peptides, proteins, and antibody fragments. It can increase the efficacy and safety of numerous medicines. The physicochemical characteristics, such as conformational changes, electrostatic binding, hydrophobicity, are changed. These physiological and physical modifications let the therapeutic ingredient stay in the body for longer. Also, it can alter the distribution and absorption patterns and impact how well the medicinal component binds to cell receptors. Making a molecule heavier through PEGylation can confer numerous significant pharmacological advantages over the unaltered version. These advantages include improved drug solubility, decreased dosage frequency with potential for reduced toxicity without sacrificing efficacy, extended circulation life, and higher drug absorption. PEGylated versions might also qualify for patent protection due to their enhanced stability and defense against proteolytic degradation (Guo et al. 2022).

This chapter will explain PEGylated nanocarrier methods for delivering nucleic acids, including their design, synthesis, and applications. The advantages and limitations of PEGylation and the factors affecting PEGylated nanocarrier systems' efficiency and safety will be discussed. Finally, the current and prospects for nucleic acid therapeutics of PEGylated nanocarrier systems will be outlined.

## 9.2 Advantage of PEGylation

The developments in nanotechnology over the past several decades could significantly affect the field of drug delivery and diagnostic imaging. Although they have also been developed as controlled release systems for water-soluble drugs, colloidal drug delivery systems (CDDS) are most frequently used to deliver highly lipophilic drugs and drugs that are unstable in biological environments (for example, proteins, nucleic acids, and peptides). Polymeric nanoparticles, liposomes (self-assembled lipid bilayers), micelles (self-assembled amphiphilic molecules), and dendrimers are examples of solid lipid nanoparticles (SLNs, dispersions of solid lipids) (repeatedly branched spherical polymers) are examples of systems that are typically included in CDDS (Fig. 9.1). These nanostructures provide several benefits over traditional delivery methods for many medications (Howard et al. 2008a, b).

First, their small size and targeting skills enable them to deliver substances to specific locations, increasing local concentrations, and lowering systemic toxicity. The treatment of neurological illnesses, which is hampered by the difficulty of many medications to enter the BBB (BLOOD BRAIN BARRIER), and cancer therapy, in which toxic side effects are frequently manifested, are two significant therapeutic areas where this may have an impact. Moreover, these nanostructures might shield medicines from enzymatic deterioration.

Lastly, it has been demonstrated that nanocarriers generally have low toxicity levels, high stability levels, and increased medication solubility. They could also be functionalized to create release systems that are time- or space-controlled. The same is true for delivering radionuclidic and contrast-enhancing drugs using nano carrivers to improve diagnostic imaging.

Nanocarriers are removed from the bloodstream with the opsonization of the particles and subsequent absorption of the RES by cells. Nevertheless, interactions with proteins adsorbed on the surface of the nanoparticles appear necessary for

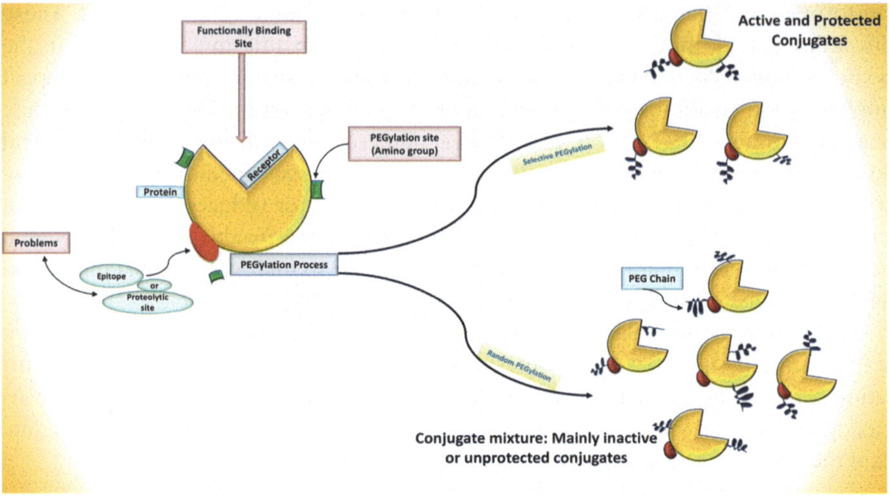


**Fig. 9.1** PEGylation process

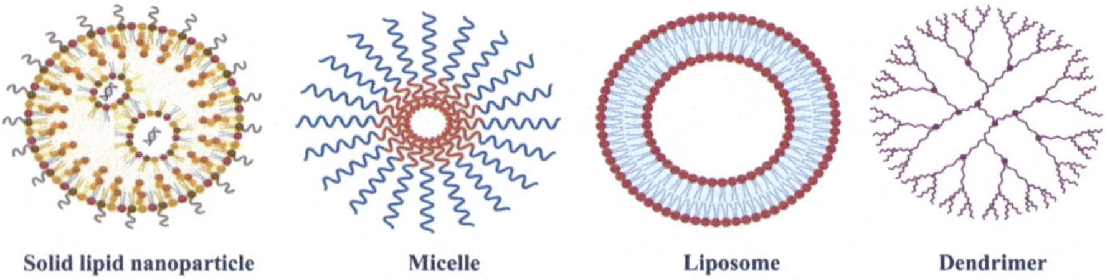


**Fig. 9.2** Colloidal delivery systems

macrophage absorption of nanocarriers in most species proteins called opsonin (Fig. 9.2).

Colloidal drug delivery systems are nanoscale carriers that improve drug solubility, stability, and bioavailability. These systems include liposomes, micelles, nanoparticles, and nanoemulsions, which encapsulate or bind drugs for targeted delivery. By protecting drugs from degradation, enabling controlled release, and enhancing site-specific accumulation, colloidal systems reduce side effects and increase therapeutic efficacy. They are widely used in cancer therapy, gene delivery, and the treatment of infectious and inflammatory diseases. The small solid spheres represent the hydrophilic head groups of amphiphilic molecules, and the long chains are lipophilic tails depending on their size, charge, stiffness, or hydrophobicity, including complement protein C3b, immunoglobulins G and M, fibronectin, C-reactive protein, −2-glycoprotein, and apolipoproteins. They need to be made less likely to be recognized by these opsonin tropical substitutes for conventional distribution methods. The most popular technique for reducing protein adsorption and introducing "stealthiness" is PEGylation. A number of explanations have been

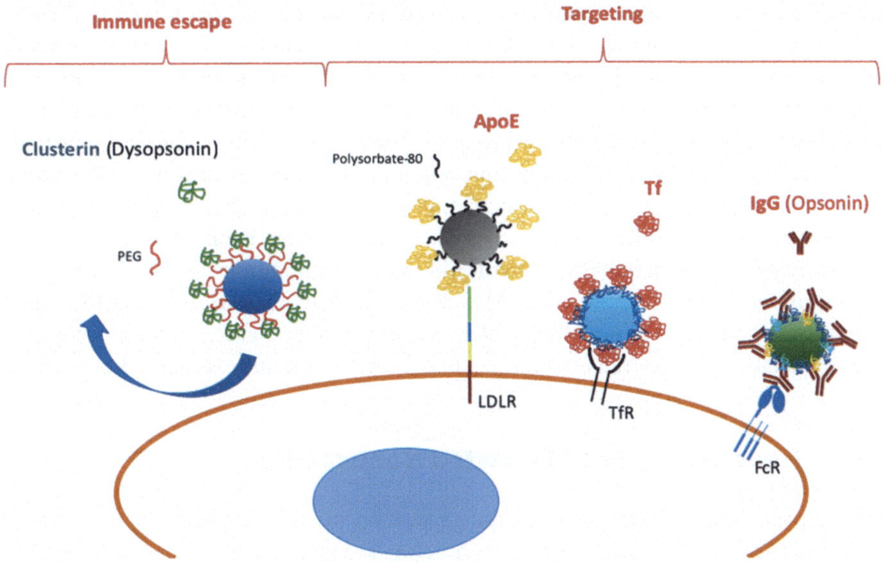


**Fig. 9.3** Possible mechanisms governing PEG's ability to stop protein adsorption. Adapted from Creative Commons Attribution (CC BY) © 2019 license (http://creativecommons.org/licenses/by/4.0/). (Howard et al. 2008a, b)

put out to explain why PEG makes nanocarriers more stealthy, including reduced surface charge, hydrophilicity, molecular flexibility, and non-immunogenicity (Howard et al. 2008a, b).

PEG has been proposed as a way to reduce the interfacial energy needed for protein binding, albeit has not been thoroughly investigated. According to Jeon et al., PEG chains have an extended shape when they are free in solution due to their hydrophilicity and flexibility. A repulsive force is produced by this transition to a higher energy shape, which can inhibit protein interaction. Several researchers say exhibiting long-circulating properties involves drawing in the suitable proteins rather than preventing protein adsorption. Several investigations by Moghimi and Patel provide credence to the presence of opsonin which is particular to the liver and spleen and may result in a changed distribution profile. Two serum components were shown to be permanently linked to poloxamine 908 (four PEG polymers coupled to four polys (propylene gly-phthalcol) in early experiments (PPG); the polyester all joined by an ethylene diamine moiety (Fig. 9.3).

## 9.3 PEGylated Nanocarriers as Therapeutic Delivery System

PEGylation, which involves applying poly (ethylene glycol) (PEG) to a surface through adsorption, grafting, or entrapment techniques, has emerged as the preferred technique for enhancing nanocarriers' ability to be biocompatible with increased stealth. The features that PEG-coated surfaces convey, such as their resistance to the reticuloendothelial system (RES) uptake, have not been replicated by other substances (Veronese and Mero 2008a, b). Nonetheless, there is currently little general agreement on several related issues. Scientists interested in using this technology frequently find themselves with more questions than answers because of the vast differences in systems, components, procedures, and accompanying findings. Secondly, PEG seems to have contradictory characteristics, acting, depending on the circumstance, as both a mucoadhesive polymer and a barrier to protein adsorption. Second, while numerous research studies have been conducted to clarify how PEG endows nanocarriers with long-circulating properties, further work is still required to guide its application. Most PEGylation research of several PEGylated systems was compared to non-PEGylated systems. Because each situation is unique, it can be challenging to transform the data is transformed into valuable knowledge.

## 9.4 Development of PEGylated Nanocarrier

In the 1970s, they were swiftly followed by several investigations and advancements made by other teams, as listed in Table 9.1. The US FDA has cleared PEG, a non-toxic, non-immunogenic polymer polyethylene glycol, for internal usage. The researchers took advantage of PEG's unique features. The benefit of PEG is that it has a wide range of molecular weight species with low polydispersity, exceptional flexibility, excellent solubility in both aqueous and organic environments, high hydration that increases its hydrodynamic volume, and favorable biological properties. 431 PEG-covalently bound compounds obtain each of these characteristics.. Since a water cloud surrounds the polymer, proteins conjugated with PEG become more soluble and resistant due to their increased size (Fig. 9.1), and antibodies, proteolytic enzymes, and cells are ultra-filtered less slowly by kidneys (Veronese and Mero 2008a, b). Proteins are delicate molecules that are easily denatured and inactivated. Finding a binding chemistry gentle enough not to harm the protein was the first challenge to attain success with PEGylation, which involves joining PEG to proteins. There is a chemical process for gently converting the terminal hydroxyl group of the PEG into one that combines with the leftovers of protein amino acids. The amino groups in proteins have been primarily exploited for conjugation because they are always present, commonly found near the protein surface, and are accessible to the solvent. Cysteine thiol residues are also very beneficial when present because they permit site-specific conjugation and are infrequently seen in proteins; moreover, it is possible that these alterations were proven to be essential for maintaining the protein's biological function (Veronese and Mero 2008a, b).

Developing a less polydisperse product species with a specific structure—the branching form being the most efficient—was another important aspect of enhancing PEG properties. It is evident in this regard that the evolution of PEGylation chemistry corresponds to the use and efficiency of biological treatments (Table 9.1). PEGylation has contributed to the success of various therapeutic proteins and oligonucleotides, including some items that have become blockbusters (Table 9.1). Moreover, because conjugated proteins dissolve in some organic solvents, they can

**Table 9.1** Background TO PEGylations

| Decade | PEGs | General observation | Application | Reference |
|---|---|---|---|---|
| In 2000 era | Used in enzymatic coupling reactions, Coupling for disulfides, Continuous PEGs, PEGs star | More strict regulatory requirements, detailed chemical and biological characterization of conjugates, and combination of generic engineering and PEGylation in the design and discovery of new drugs. | PEGylation of non-protein drugs, oligonucleotide PEGylation 3, and cell PEGylation. | Veronese and Mero (2008a, b) |
| From 1990 to 2000 era | PEGs as branched, NHS-PEGs, OPSS-PEGs | PEGlyted drug selectivity and marketing improved. | Hormones, anticancer drug targeting, cytokines | |
| In 1980–1990 era | Aldehyde-PEGs, carbonyl imidazole PEGs, Succinimidyl carbonate- PEGs etc. | Absence of diols results in reduced polydispersed site-specific conjugations. | Substitution treatment of enzymes. | |
| In 1970–80 | PEG-succinimidyl succinate PEG-chloro triazine | Lack of selectivity in the immunogenic autocratic starting material, extremely polydisperse peg | Study investigations, biocatalyst enzymatic alterations. | |

AA amino acids, *NHS* N-hydroxy succinimide, *OPSS* ortho-pyridyl disulfide, *PEG* polyethylene glycol

be employed as new biocatalysts because PEG is an amphiphilic polymer (both hydrophilic and hydrophobic). What makes this scenario intriguing are the unexpected catalytic roles of lipase or chymotrypsin in synthetic processes. A PEG-protein conjugate must consider several unique variables. In the following sections, several aspects are discussed with typical illustrations of polymer-modified proteins. To obtain more detailed documentation, the reader is suggested to study several recent evaluations (Suk et al. 2016).

### 9.4.1 Chemistry of PEGylation

A substance, usually a medication or a protein, is regulated by covalently adding PEG. Extending the molecule's circulation time in the body enhances its stability, solubility, and therapeutic efficacy. The process of PEGylation contains the reaction of a reactive group on a PEG molecule on the target molecule, generally an amine, thiol, or carboxylic acid. The PEG molecule is typically activated with a functional group like succinimidyl ester or maleimide. PEG-maleimide conjugation to a cysteine residue on a protein illustrates a PEGylation reaction. The maleimide functional

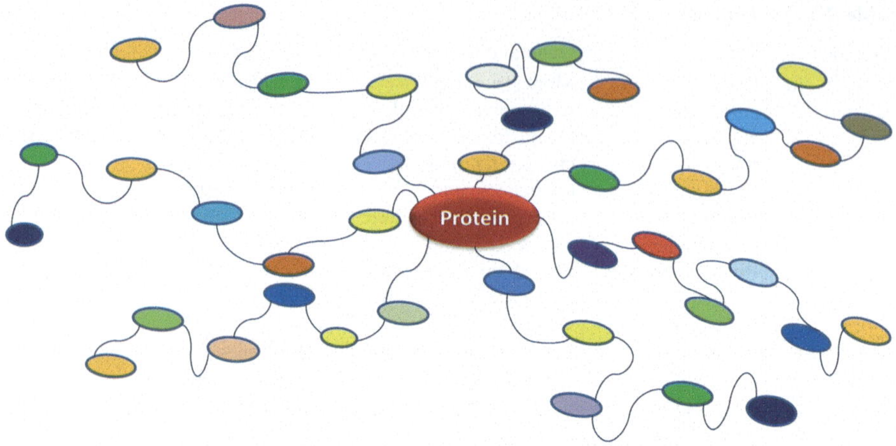


**Fig. 9.4** Schematic representation of protein with bound polyethylene glycol

group on PEG and the thiol group on cysteine interact in this reaction to generate a thioether bond, which results in a strong covalent connection between the protein and PEG. The conjugation of PEG-succinimidyl ester to an amine group on a medicinal molecule is another illustration of PEGylation chemistry. The primary amine group on the drug molecule and the succinimidyl ester functional group on PEG combine in this reaction to produce an amide bond (Veronese and Mero 2008a, b) (Fig. 9.4). A protein with bound polyethylene glycol is shown schematically. The ether-oxygen groups of the polymer work along with the surrounding circles to simulate a cloud of water. This schematic depicts a protein conjugated with polyethylene glycol (PEG), where PEG chains are attached to the protein surface. The ether-oxygen groups of the PEG are represented, along with surrounding circles, simulating a hydrated "cloud" of water molecules. This hydration layer stabilizes the protein, reduces immunogenicity, and improves solubility, making PEGylation a valuable technique in drug development and therapeutic protein delivery.

#### 9.4.1.1 Amino Acid Modifications

PEGylation is a common method for enhancing the pharmacokinetics and biodistribution of nanocarriers such as liposomes, nanoparticles, and micelles. In PEGylation, which entails covalently gluing PEG chains to the surface of the nanocarrier, PEG (polyethylene glycol) is a biocompatible and hydrophilic polymer that is frequently utilized (Veronese and Mero 2008a, b).

Amino groups are another functional group that can be used for modifying nanocarriers. Amino groups can be introduced onto the surface of nanocarriers by using amine-containing molecules, such as polyamines or amino acids, during the preparation of the nanocarrier. Once the amino groups are present on the upper surface of the nanocarrier, they can be used for the covalent conjugation of PEG chains or other molecules.

Collecting amino groups to the overhead of nanocarriers can provide several advantages over traditional PEGylation. For example, amino groups can be used to attach targeting ligands, such as peptides or antibodies, to the surface of the nanocarrier. This allows targeting the delivery of the nanocarrier to particular cells or tissues, improving the therapeutic efficacy and reducing off-target effects.

In addition, amino groups can be used for pH-sensitive drug release from the nanocarrier. By attaching pH-sensitive moieties, the medicine can be released in reaction to pH variations thanks to the amino groups on the surface of the nanocarrier, such as in the tumor's acidic environment (Veronese and Mero 2008a, b).

Overall, adding amino groups to nanocarriers creates a flexible framework for creating pH-sensitive and tailored drug delivery systems (Suk et al. 2016).

#### 9.4.1.2 Conjugation of Histidine to the Imidazole Group

The pKa value of histidine's secondary amine group is lower than that of its primary amine. A properly activated PEG can be conjugated at this level with a pH close to neutral. Nevertheless, because the acetate-histidine link is unstable, the conjugate functions like a prodrug and releases the free protein over time, as with Peginterferon-2b (Veronese and Mero 2008a, b).

#### 9.4.1.3 Cysteine Thiol Residue Conjugation

PEG-cysteine reagents can create stable thiol-ethers or disulfides, such as maleimide and vinyl sulfone (e.g., PEG-pyridyl disulfide), to conjugate proteins. The latter results in the production of a protein-PEG conjugate that may be a bifunctional thiol-reactive PEG that reacts with a substance produced under reducing circumstances and might also be present to build a new body. Free cysteine residues stable bridges with a polymer attached are highly uncommon because of this amino acid conjugation to the carboxylic group. When present, arginine often participates in catalysis or the formation of disulfide bridges. Proteins employed for conjugation contain it. Granulocyte colony-stimulating factor (also known as sugars CSF or G-hydroxyl residues) is one instance where the conjugal and a free cysteine are partly accessible. Direct coupling of PEG-$NH_2$ to these would cause the amines' natural protein to respond to activated protein carboxylic ion that is only conceivable after partially reversible denaturation of the groups is not feasible. Alternatively, several examples exist of the protein itself or another nearby protein molecule having cysteine inserted by genetic engineering at a valuable spot of the protein sequence connections within or between molecules.

The research of Zalipsky and Meno-Rudolph led to an innovative approach that involved using PEG hydrazide, an amino donor that is reactive at low pH. Just recently, a cunning plan was developed to take advantage, although no practical use has been discovered to date. The cysteines are harmed when disulfide bridges are present. There are not many instances of arginine-reactive PEGs; these are mainly based on PEG with a bifunctional thiol-reactive PEG (reported in proprietary literature only) reacts with a substance produced under reducing circumstances and might also be present to produce a new body. Free cysteine residues stable bridges with a polymer attached are highly uncommon because of this amino acid (Veronese and Mero 2008a, b) (Fig. 9.5).

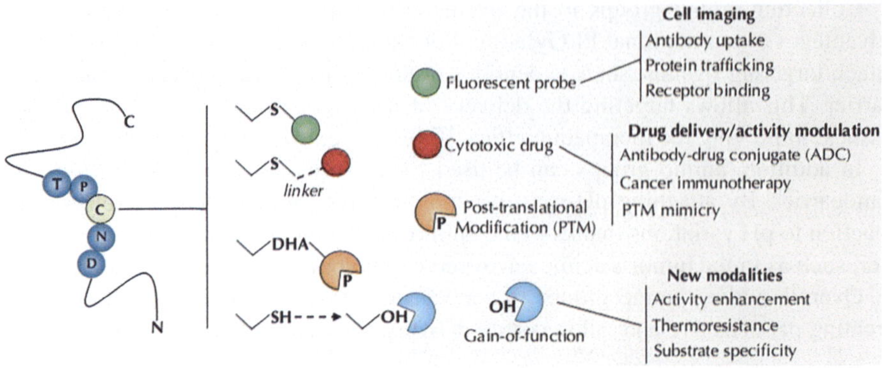


**Fig. 9.5** Schematic illustrates protein modification strategies targeting cysteine residues, including disulfide bond formation, thiol-ene and thiol-yne reactions, maleimide conjugation, and oxidative modifications. These approaches allow site-specific protein labeling, stabilization, and conjugation with drugs or probes. Applications in biomedicine and biotechnology include targeted drug delivery, protein engineering, biosensor development, and therapeutic antibody modification, enhancing protein stability, functionality, and therapeutic efficacy.  (Pessatti et al. 2021).

#### 9.4.1.4 Conjugation to Arginine, Carboxylic Groups, and Sugar or Hydroxyl Residues

As they would react with the amine's natural protein, as in the case of CSF, where a free cysteine is partially exposed, direct binding of PEG-NH2 to activated protein carboxylation takes place only after partially reversible denaturation of the groups. On the other hand, the protein itself or a nearby protein molecule can produce multiple examples. Genetic engineering is used to insert cysteine into a useful location in the protein sequence connections within or between molecules. The research of Zalipsky and Meno-Rudolph led to an innovative approach that involved using PEG hydrazide, an amino donor that is reactive at low pH. How a creative technique was just recently developed to take advantage, although there has yet to be any practical use. The cysteines are harmed by disulfide bridges when they are present. There are not many instances of arginine-reactive PEGs (reported in private literature only, where free cysteine is first produced via a reduction process); they are often based on PEG with vicinal dicarboxylic groups, albeit the approach has not been successful due to a lack of selectivity.

#### 9.4.1.5 Conjugation of Glutamine Residues

While the high glutamine amide and a large number of diol impurities in the samples cannot be precisely reacted chemically, conjugation may be carried out enzymatically by polydispersity, the presence in the polymer of distinct species of transglutaminases, either specific or non-specific. Heterogeneity in the finished product was caused by these two elements: a component found in all living things and obtained from microbial sources. In the first instance, covalent protein collects and combines with the transamidase process between the protein's PEG-NH2, which acts as the nucleophile donor and glutamine. The following mass is diverse. It took not only a specification, very high mass polymer (30–40 kDa). Shearwater sequence around the glutamine resides but also the sale of a pure, low polydisperse, highly selective PEGylation, as we just discovered—polymers, Inc.—to fix this issue later, at the end of the 1990s a connection between the site of a certain conjugation and the flexibility of the protein chain. APEG conjugation of glutamines to the lysine group can also be catalyzed by the same enzyme. Pharmaceutical grade PEG should have a polydispersity of between 1.01 and 1.1 for products with a 5 kDa to 50 kDa molecular weight (Veronese and Mero 2008a, b). Keep in mind that the value is 1.00 for a monodisperse product.

### 9.4.2 Releasable PEGs

Releasable PEG nanocarriers are a nanoscale drug delivery system that utilizes polyethylene glycol (PEG) chains to increase the circulation time of the drug in the body and protect it from immune system recognition. These nanocarriers are designed to release the drug payload at a specific target site, such as a tumor, in a controlled and sustained manner.

One approach to achieving this controlled release is to use stimuli-responsive PEGs that change their properties, such as solubility or molecular weight, in reaction to certain triggers, such as pH changes, temperature changes, or the presence of enzymes. The drug payload may be released from the nanocarrier due to this alteration in characteristics.

For example, tumors have a lower pH than healthy tissues, and their acidic environment can cause PEGs to release the therapeutic payload. Similarly, heat sensitivity is created to deliver the pharmacological payload in response to increased temperature found in inflamed or cancerous tissues (Veronese and Mero 2008a, b).

Releasable PEGs nanocarriers have improved drug solubility and stability, longer circulation time, less toxicity, and focused drug administration, which are only a few benefits over conventional drug delivery methods. These systems could increase the effectiveness and safety of numerous drugs, including chemotherapeutic agents, hormones, and gene therapies.

## 9.5 Nanocarrier Mechanism of Delivery to Nucleic Acid

Nanocarriers are delivery vehicles that can deliver various types of cargo, including nucleic acids, to particular cells or tissues inside the body. The mechanism of delivery of nanocarriers to nucleic acids involves several steps:

Nanocarrier uptake: The target cells must take up the nanocarriers to deliver their cargo. This can occur through a variety of mechanisms, including endocytosis, phagocytosis, or receptor-mediated uptake.

***Endosomal escape:*** Once the nanocarriers are inside the cell, they are typically sequestered within endosomes. To reach the nucleus, the nanocarriers must go away

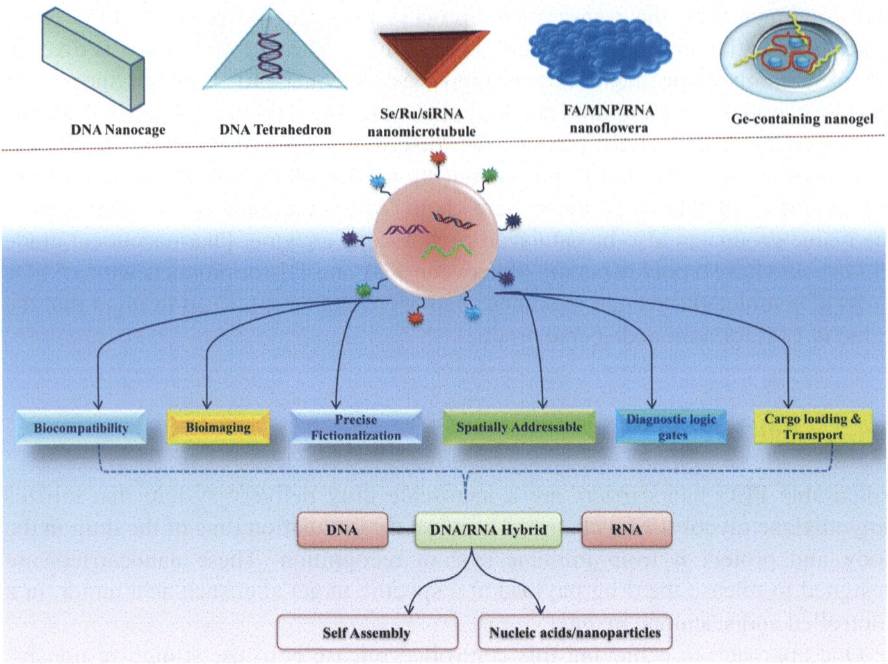


**Fig. 9.6** Nanocarrier mechanism of delivery to nucleic acid: Nanocarriers deliver nucleic acids by encapsulating or binding them in protective structures, safeguarding them from degradation. They enter target cells via endocytosis or membrane fusion. Once inside, nanocarriers release the nucleic acids, which travel to the nucleus or cytoplasm. Common nanocarriers include lipid nanoparticles, polymers, and dendrimers, designed to improve stability, targeting specificity, and transfection efficiency for therapeutic applications such as gene therapy or RNA interference

from the endosome and enter the cytoplasm. This can be achieved through various mechanisms, such as the proton sponge effect, photochemical internalization, or other pH-sensitive strategies.

***Nuclear localization:*** Finally, the nanocarriers must be able to reach the nucleus and deliver their cargo to the nucleic acids within. This can be achieved through nuclear localization signals or other targeting strategies.

Once the nanocarriers have delivered their cargo to the nucleic acids, they can be used for various purposes, including gene therapy (RNA interference). The choice of nanocarrier and delivery mechanism will depend on the specific application and the properties of the target cells and nucleic acids (Fig. 9.6).

## 9.6 Release of Drug from PEGylated Nanocarriers

The most popular technique for giving drug nanocarriers stealth characteristics is now “PEGylation.” The neutrality, hydrophilicity, flexibility, and ability to hydrate the PEG moiety in nanoparticles generate a steric barrier that prevents opsonin proteins from adhering. Surfaces of PEGylated particles may be accomplished by straightforward or adsorption by covalently attaching PEG to functional groups that have been activated on the particle's exterior. Moreover, PEG molecules have been modified to accomplish the controlled release of medicinal chemicals trapped in them and enhance their absorption by certain targets (such as cancers). Various characterization approaches have been developed due to the increase in PEGylation, and mathematical modeling is now being used more frequently to direct formulation development. This study provides an overview of PEGylation theories, PEGylation techniques, PEGylation particle characterization, PEGylation-related mathematical modeling, and how PEGylation can improve nanocarrier drug delivery systems to give an overview of the direction that nanocarrier PEGylation and nanomedicine will take the existing accomplishments and shortcomings of PEGylation are assessed (Sur et al. 2019a, b).

Targeted and customized drug delivery systems were developed to address the drawbacks of traditional dosage forms. A revolutionary drug delivery technology was in great demand at the time. One such method of targeted drug delivery is nanoparticles. Colloidal nanoparticles that deliver medications site specifically. Nanoparticles come in many forms, but polymeric nanoparticles are unique. One of the subjects being investigated the most at the moment is polymeric nanoparticles. Their main benefits include better therapeutic efficacy, longer clearance times, less toxicity, and greater control over size. Dendrimers, polymeric micelles, ligand-based nanoparticles, PEGylated nanoparticles, and other types of polymer-based nanoparticles are only a few examples(Sur et al. 2019a, b).

## 9.7 Challenges with Nanocarriers for RNA Delivery

An essential biological macromolecule in nature, RNA conveys genetic information and directs the creation of proteins. It is a single-strand transcript created using one DNA strand as a template and can be used to construct and modify the general structure of RNA nanocarriers. The double helix structure's restrictions can be overcome by RNA nanotechnology, which can create a wide range of structures and many kinds of circular-structured modules. Many researchers have become interested in using RNA interference processes brought on by small-molecule RNA in the therapy of tumors in recent years. Some enzymes and antisense siRNA generate a complex that increases the quantity of double-stranded RNA the enzyme produces until the mRNA is relocated while simultaneously degrading similar single-stranded RNA. For tumor therapy, specific dsRNA may be delivered into cells using RNA-based vectors, which are more effective, more exact, and quicker than conventional approaches to target genes selectively. They may be adapted to different requirements. Gene silencing, drug delivery, and biological imaging rely on RNA self-assembly and RNA decoration on nanoparticles (NPs). To overcome the drawbacks of chemotherapy, researchers have developed techniques for making nanocarriers with high selectivity and low cytotoxicity. To address this problem, scientists have combined RNA with polyethylene glycol, inorganic NPs, glue bundles, carbon nanospheres, peptides, proteins, and other compounds to make RNA composite nanomaterials. High selectivity, stability, and resistance to deterioration are all features of these nanomaterials. The addition of RNA overcomes the limitations of conventional NPs (Xue et al. 2015; Veronese and Pasut 2005).

### 9.7.1 Nanocarriers for RNA Delivery Based on Lipid and the Difficulties

Considering that lipid-based nanocarriers are favorable for the transport of RNA, several issues need to be fixed before this class of delivery devices performs optimally in vivo or the clinical situation. Several issues with the lipid nanocarriers, which are ostensibly safe, result in subpar toxicity profiles of drugs (Xue et al. 2015).

#### 9.7.1.1 Cationic Lipid Toxicology

To increase RNA encapsulation and stability, cationic lipids are employed in RNAi therapies; however, they also have serious toxicity problems. A lipoplex made of cationic lipid molecules can irritate an exposed cell at low concentrations, leading to cytoplasm vacuolization, decreased mitoses, and cell shrinkage. Cell lysis and necrosis may start when the lipoplex level is high enough. The interaction of the cationic groups with biological enzymes like protein kinase C may potentially result in cell damage. Several genes involved in cell apoptosis were discovered to be affected by oligofectamine, which raised the probability of early apoptotic cell death (Xue et al. 2015; Veronese and Pasut 2005).

#### 9.7.1.2 RNA Nanocarriers Based on Lipids

The production of tumor necrosis factor, interferon 6, interleukin 6, and 12, leukopenia and thrombocytopenia, and mouse mortality have all been linked to lipoplexes as systemic toxins. More labile, biodegradable linkers may be utilized to lessen toxicity; however, doing so may affect stability and transfection effectiveness (Xue et al. 2015; Veronese and Pasut 2005).

#### 9.7.1.3 Concerns with PEGylation

Especially in those lipid carriers of RNA, lipid materials naturally interact efficiently with cell surfaces in a non-specific manner. While creating lipid nanocarriers for nucleic acids, PEGylation (surface coating with polyethylene glycol groups) is a crucial common procedure, although it has downsides. The RNA-loaded nanocarriers must efficiently enter the target cells by endocytosis for RNAi-mediated gene silencing to be effective, and the RNA payload must leave the endosomal compartment and enter the cytoplasm. The "accelerated blood clearance (ABC) phenomenon" refers to the ability of empty PEGylated liposomes given orally to speed up the clearance of subsequent injections of liposomes (Xue et al. 2013). The generation of anti-PEG IgM, following activation of the complement system, and rapid capture were responsible for this event made by Kupffer cells from liposomes. When stabilized plasmid lipid particles were administered in many doses, it was similarly discovered that the targeted tumor had less expression, and the liver had more expression in the later doses (Xue et al. 2015; Veronese and Pasut 2005).

### 9.7.2 Strategies to Overcome the Challenges

Several strategies have been explored to maintain the benefits of lipid-based RNA nanocarriers while addressing their drawbacks (Xue et al. 2015, Veronese and Pasut 2005).

#### 9.7.2.1 Modification of Lipids

Cationic lipids mainly cause the toxicity of lipid-based RNA nanocarriers, and effective engineering and alteration of the lipid molecule structure can enhance transfection efficiency. A well-known example is DOTAP, which has a hydrophobic hydrocarbon backbone, a linker region, and a cationic head. It has been investigated how to increase transfection effectiveness without increasing toxicity by mixing multiple alkyl chains into the same lipid, employing an asymmetrical backbone, maximizing alkyl chain length, and leveraging steroid hydrophobic domains. One should use caution when selecting the "least hazardous" cationic lipids for in vivo or clinical purposes because most of the reported lipid toxicity knowledge is based on in vitro experiments (Xue et al. 2015; Veronese and Pasut 2005).

#### 9.7.2.2 Utilization of Lipids Not Catatonic

Non-cationic lipids are often used in RNA nanocarriers, although they may decrease the RNA's stability, encapsulation, and transfection efficacy. To boost many properties, such as endosomal escape and physical stability, and decrease toxicity, neutral or anionic lipids can be added. Yet, these lipids might not always lessen toxicity. For instance, DOTAP coupled with cholesterol in a nanosystem killed less non-specific cells than DOTAP alone. Lipid mixtures must be carefully considered if developing a less dangerous nanocarrier is the primary objective (Xue et al. 2015; Veronese and Pasut 2005).

#### 9.7.2.3 Nano Formulations of RNA with High Potency

Researchers developed nano formulations that could suppress gene expression at extremely low siRNA dosage levels by combinatorial production and screening of lipid-like materials. In other research, replacing DOTAP with the newly created cationic lipid DLinMC3DMA in the lipid nanoparticles significantly increased the potency of siRNA. From 10 mg siRNA/kg body weight to 0.005 mg, the IC50 was decreased. The authors hypothesized that the improvement is caused by employing ionizable cationic lipids with enhanced endosomal bilayer destabilizing properties and improved pKa values (Xue et al. 2015; Veronese and Pasut 2005).

#### 9.7.2.4 Hybrid Lipid and Polymeric Materials Nanocarrier Integration

While polymeric materials often have strong affinities for giant molecules like RNA, lipid-based nanoparticles are skilled at managing drug release rates and have good biocompatibility and immunogenicity. Hybrid nanoparticles combining lipids and polymers have been produced to enhance RNA encapsulation and delivery, lessen cellular toxicity, and achieve high cell transfection efficiencies. Recently, Ewe and Aigne (2014) created lip polyplexes for RNA transport. Inhalation delivery of lip polyplexes containing siRNA and DNA is possible (Xue et al. 2015, Veronese and Pasut 2005).

## 9.8 RNA Lipid Nanoparticle

Lipid nanoparticles have been shown to be the most effective RNA nanocarriers so far. Several approaches have been included to solve various problems with RNA delivery, such as restricted in vivo circulation, ineffective transfection, and intolerable toxicity. It is made up of cationic lipids, PEG-conjugated lipids, and neutral lipids such as cholesterol and 1,2-distearoylsn-glycero-3-phosphocholine. It is an electron-dense complex structure with a solid core that differs from liposomes in terms of its physicochemical characteristics. A thorough evaluation of LNP performance in vivo and clinical settings revealed no appreciable complement activation, release of pro-inflammatory cytokines, delayed coagulation, or other alterations in hematological parameters. No evidence of hepatotoxicity was seen in the phase I clinical study employing LNP delivering siRNA against apolipoprotein B, although flu-like symptoms are probably related to the immunostimulant. One of the participants given the highest treatment dosage level showed signs of a high siRNA payload. Thus, the manufacturer (Tekmira) chose to end the experiment early. LNPs have great promise for delivering RNA. More toxicology data, particularly those pertaining to immunogenicity, are required to support the therapeutic utility of these medicines (Xue et al. 2015, Veronese and Pasut 2005).

## 9.9 Clinical Trials for Nucleic Acid Delivery of Nanoparticles

Many clinical trials and experiments have been conducted to assess the security and effectiveness of nanoparticle-based nucleic acid delivery. Examples of clinical trials with citations are as follows:

1. Phase I/II trial of RNAi treatment targeting VEGF and KSP in patients with advanced solid tumors: In this study, the effectiveness and safety of an RNAi-based therapy administered via nanoparticles in patients with advanced solid tumors were assessed. The research showed that the treatment was both safe and well-tolerated, and there was proof of therapeutic efficacy.
2. Age-related macular degeneration: Phase I trial of siRNA targeting the RTP801 gene: This study assessed the effectiveness and safety of a siRNA-based treatment administered via nanoparticles in individuals with age-related macular degeneration. The study showed that the treatment was secure and well-tolerated and that biological activity was present.

3. Human anti-TNF mRNA given through polymer in patients with rheumatoid disease: In this study, individuals with rheumatoid arthritis received an mRNA-based treatment that was administered via nanoparticles. The therapy's safety and effectiveness were assessed. The research showed that the treatment was both safe and well-tolerated, and there was proof of therapeutic efficacy.
4. Phase I investigation evaluating the safety and effectiveness of an RNAi-based treatment administered utilizing nanoparticles to patients with hematological malignancies. This trial focused on the M2 isoform of pyruvate kinase. The research showed that the treatment was both safe and well-tolerated, and there was proof of therapeutic efficacy.
5. Phase I research of an mRNA vaccine for the Zika virus based on lipid nanoparticles: In this study, healthy people were given the vaccine to assess its immunogenicity and safety. The study proved that the vaccination was secure and well-tolerated and that immune responses were present.

## 9.10 Merits and Prospects of a Nanocarrier System

These clinical trials describe the efficiency of nucleic acid delivery using nanoparticles for various therapeutic applications.

1. Nanocarrier systems, also known as nanoparticles, are microscopic particles that can be designed to transport drugs, genes, or other therapeutic agents to specific targets in the body. These systems have the potential to promote drug effectiveness and lessen side effects, leading to better patient outcomes. Here are some of the merits and prospects of nanocarrier systems, supported by relevant citations.
2. Improved drug delivery: Nanocarriers can encapsulate drugs and transport them to specific tissues or cells, improving drug efficacy and reducing toxicity. For example, liposomes have been employed to deliver doxorubicin to cancer cells, resulting in better therapeutic outcomes and fewer side effects (Wang et al. 2019).
3. Targeted therapy: Target cell receptors can be bound explicitly to using nanocarriers functionalized with targeting ligands, such as antibodies or peptides. This might increase the precision and potency of medicine delivery. For instance, siRNA was delivered to tumor cells via a targeted nanocarrier technology, which enhanced survival in animal models and inhibited tumor development (Shi et al. 2017).
4. Increased bioavailability: Drugs' bioavailability can be increased by using nanocarriers to shield them from the body's natural processes of breakdown and clearance. For instance, polymeric nanoparticles have been utilized to improve the oral bioavailability of medications with low solubility, like curcumin (Li et al. 2019).
5. Regulated release: Drugs can be released from nanocarrier systems in a controlled way, for as in reaction to a particular trigger (for instance, temperature, pH, or enzymes). The pharmacokinetics may be enhanced, and toxicity may be decreased. A pH-sensitive nanoparticle system, for example, was created to

deliver anticancer medications specifically to tumor cells, improving treatment efficacy and minimizing adverse effects (Jhaveri and Torchilin 2014).
6. Multipurpose platforms: Nanocarrier systems may be designed to perform various tasks, including imaging and treatment. Iron oxide nanoparticles, for instance, have been employed as both MRI contrast agents and medication delivery systems (Sun et al. 2018).

In conclusion, nanocarrier systems have the potential to boost bioavailability, permit targeted and controlled drug release, and improve drug delivery. They may also be designed to perform various tasks, giving them flexible platforms for imaging and treatment.

## 9.11 The Creation of Nucleic Acid-Based Nanocarriers and Their Use in Nanobiology Delivery Systems

In nanobiology, nucleic acid-based nanocarriers like DNA and RNA have been thoroughly investigated for their potential as medication delivery methods. These nanocarriers are capable of cellular absorption, targeted distribution to certain cells or tissues, and protection from degradation for therapeutic compounds. Below are some examples of current studies on the creation and use of nucleic acid-based nanocarriers.

1. DNA origami nanocarriers: DNA is used as a structural material in the DNA origami method, which produces very accurate nanoscale forms. These structures can be loaded with medicinal agents and functionalized with targeted ligand molecules for targeted drug delivery. For example, Li et al. (2018a, b) constructed DNA origami nanocarriers that were functionalized with folate for targeted delivery to cancer cells and loaded with doxorubicin for chemotherapy.
2. RNA nanocarriers: RNA molecules can also be used as nanocarriers and have the advantage of being easier to synthesize and modify than DNA. For example, Guo et al. constructed RNA nanocarriers that were functionalized with targeting ligands and loaded with siRNA for gene silencing. Another study by Zhou et al. (2018) used RNA nanocarriers to deliver CRISPR-Cas9 gene editing machinery to target cells.
3. Hybrid DNA/RNA nanocarriers: Hybrid nanocarriers that combine the advantages of DNA and RNA have also been developed. For example, Liu et al. constructed hybrid DNA/RNA nanocarriers that were functionalized with targeting ligands and loaded with siRNA for gene silencing.
4. Aptamer-conjugated nanocarriers: Aptamers are short nucleic acid sequences with a high affinity and selectivity for binding to particular targets. To allow targeted distribution to certain cells or tissues, they can be coupled to nanocarriers. For instance, Liu et al. developed aptamer-conjugated DNA nanocarriers to deliver doxorubicin to prostate cancer cells precisely.

Overall, nucleic acid-based nanocarriers have great potential for drug delivery in nanobiology, and continued research in this area is expected to lead to new and improved therapies for a range of diseases.

## 9.12 PEGylated Nanocarrier System for Nucleic Acid Delivery for Anti-PEG IgM Production

Plasmid DNA (pDNA) and small interfering RNA (siRNA), for example, must be delivered systemically, and this needs carriers that are safe, efficient, and can overcome the pharmacokinetic restrictions of nucleic acids. The potential for producing nucleic acid-containing lipoplexes using an efficient cationic liposome is enormous (Abu Lila and Ishida 2019a, b).

In the delivery technique of gene therapy, most lipoplexes are PEG-treated for in vivo stability and extended circulation (PEGylation). Nevertheless, when given to the same animal again at predefined intervals, PEGylated liposomes lost their long-circulating properties. This unexpected and undesirable occurrence is known as the accelerated blood clearance (ABC) phenomenon. PEGylated liposomes' initial dose resulted in anti-PEG IgM, which has been discovered as a significant cause of ABC (Abu Lila and Ishida 2019a, b).

## 9.13 Conclusion

PEGylated nanocarrier systems are, in summary, promising techniques for delivering nucleic acids. Enhancing the stability, biocompatibility, and pharmacokinetics of nanoparticles with PEG can result in more effective drug administration with lower toxicity. PEGylated nanocarriers can also overcome additional biological barriers that prevent the transfer of nucleic acids, such as endosomal escape and cellular absorption. Despite these benefits, more investigation is required to improve the design and formulation of PEGylated nanocarriers for the distribution of nucleic acids and to comprehend their interactions with biological systems. In general, PEGylated nanocarrier systems show promise as a method for creating therapeutic nucleic acids that are both secure and efficient.

## References

Abu Lila AS, Ishida T (2019a) Anti-PEG IgM production via a PEGylated nanocarrier system for nucleic acid Delivery. In: Ogris M, Sami H (eds) Nanotechnology for nucleic acid delivery, vol 1943. Springer, New York, pp 333–346. https://doi.org/10.1007/978-1-4939-9092-4_22

Abu Lila AS, Ishida T (2019b) Anti-PEG IgM production via a PEGylated nanocarrier system for nucleic acid delivery. Methods Mol Biol 1943:333–346

Ewe A, Aigne A (2014) Nebulization of liposome–polyethyleneimine complexes (lipo polyplexes) for DNA or siRNA delivery: physicochemical properties and biological activity. Eur J Lipid Sci Technol 116:1195–1204

Guo Y, Cao X, Zheng X, Abbas SJ, Li J, Tan W (2022) Construction of nanocarriers based on nucleic acids and their applications in nanobiology delivery systems. Natl Sci Rev 9(5):nwac006. https://doi.org/10.1093/nsr/nwac006

Howard MD, Jay M, Dziubla TD, Lu X (2008a) PEGylation of Nanocarrier drug Delivery systems: state of the art. J Biomed Nanotechnol 4(2):133–148. https://doi.org/10.1166/jbn.2008.021

Howard MD, Jay M, Dziubla TD, Xiuling L (2008b) PEGylation of nanocarrier drug delivery systems: state of the art. J Biomed Nanotechnol 4(2):133–148

Jhaveri AM, Torchilin VP (2014) Multifunctional polymeric nanoparticles for delivery of drugs and siRNA. Front Pharmacol 5:77

Li X, Sun AN, Liu YJ et al (2018a) Amphiphilic dendrimer-engineered nanocarrier systems for co-delivery of siRNA and paclitaxel to matrix metalloproteinase-rich tumors for synergistic effect. NPG Asia Mater 10:238

Li J et al (2018b) DNA origami as a carrier for circumvention of drug resistance. J Am Chem Soc 140(46):15486–15489

Li X, Xu X, Zhang X, Li S, Wang Y, Chen Y, Zhang Q (2019) Nanocarrier-mediated co-delivery of resveratrol and curcumin for synergistic enhancement of anti-inflammatory and antioxidant effects. J Nanobiotechnol 17(1):2151

Pessatti TB, Terenzi H, Bertoldo JB (2021) Protein modifications: from chemoselective probes to novel biocatalysts. Catalysts 11:1466. https://doi.org/10.3390/catal11121466

Shi J, Kantoff PW, Wooster R, Farokhzad OC (2017) Cancer nanomedicine: progress, challenges, and opportunities. Nat Rev Cancer 17(1):20–37

Shtykalova S, Deviatkin D, Freund S, Egorova A, Kiselev A (2023) Non-viral carriers for nucleic acids Delivery: fundamentals and current applications. Life (Basel) 13(4):903. https://doi.org/10.3390/life13040903. PMID: 37109432; PMCID: PMC10142071

Sridharan K, Gogtay NJ (2016) Therapeutic nucleic acids: current clinical status. Br J Clin Pharmacol 82(3):659–672. https://doi.org/10.1111/bcp.12987

Suk JS, Xu Q, Kim N, Hanes J, Ensign LM (2016) PEGylation as a strategy for improving nanoparticle-based drug and gene delivery. Adv Drug Deliv Rev 99:28–51. https://doi.org/10.1016/j.addr.2015.09.012

Sun C, Lee JS, Zhang M (2018) & magnetic nanoparticles in MR imaging and drug delivery. Adv Drug Deliv Rev 136:64–73

Sur S, Rathore A, Dave V, Reddy KR, Chouhan RS, Sadhu V (2019a) Recent developments in functionalized polymer nanoparticles for efficient drug delivery system. Nano-Structures Nano-Objects 20:100397. https://doi.org/10.1016/j.nanoso.2019.100397

Sur S, Rathore A, Dave V, Reddy KR, Chouhan RS, Sadhu V (2019b) Recent developments in functionalized polymer nanoparticles for efficient drug delivery system. Nano Structures Nano Objects 20:100397

Tenchov R, Bird R, Curtze AE, ACS QZ (2021) Lipid nanoparticles from liposomes to mRNA vaccine Delivery, a landscape of research diversity and advancement. Nano 15(11):16982–17015. https://doi.org/10.1021/acsnano.1c04996

Veronese FM, Mero A (2008a) The impact of PEGylation on biological therapies. BioDrugs 22(5):315–329. https://doi.org/10.2165/00063030-200822050-00004

Veronese FM, Mero A (2008b) The impact of PEGylation on biological therapies. BioDrugs 22:315–329

Veronese FM, Pasut G (2005) PEGylation, successful approach to drug delivery. Drug Discov Today 10(21):1451–1458

Wang M, Zhao X, Zhu D, Liu T, Liang Y, Zhang J, Xu S (2019) Improving the therapeutic effect of doxorubicin

Xue HY, Narvikar M, Zhao JB et al (2013) Lipid encapsulation of cationic polymers in hybrid nanocarriers reduces their non-specific toxicity to breast epithelial cells. Pharm Res 30(2):572

Xue HY, Guo P, Wen W-C, Wong HL (2015) Lipid-based nanocarriers for RNA delivery. Curr Pharm Des 21:3140

Zhou J et al (2018) A Cas9-based toolkit to program gene expression in vitro and in vivo. Cell 169(7):1429–1438

# PEGylated Nanocarriers for Protein and Peptide Delivery

# 10

Tejas Girish Agnihotri, Vasu Peddinti, Shyam Sudhakar Gomte, Biswajit Rout, and Aakanchha Jain

**Abstract**

PEGylation is the biochemical process of modifying bioactive molecules with polyethylene glycol (PEG), which endows proteins/peptides, antibodies, and vesicles with several desirable properties that are used for the therapy. Protein and Peptide drug delivery systems are considered novel drug delivery systems and have immense therapeutic potential. As a result, using proteins and peptides as therapeutic agents is thought to be a promising strategy for treating a variety of diseases. However, the main limitation with protein and peptide drugs is poor permeation across intestinal mucous barriers and they tend to get rapidly degraded and cleared from the bloodstream. The PEGylation process increases the efficiency of therapeutic proteins by protecting them from proteolytic enzymes. The other advantages include site-specific drug delivery, increased circulation time in the blood, decreased immunogenicity, and prevention or decreased uptake by the reticuloendothelial system. PEGylation of such drugs results in improved physicochemical properties such as an increase in hydrophilicity, size, and molecular weight, changes in conformation, and steric hindrance of intermolecular interactions. This book chapter discusses about protein and peptide drug delivery and their importance in drug delivery, along with PEGylation and its importance, types of PEGylated nanocarriers, the importance of PEGylation in protein and peptide drug delivery, and characterization of PEGylated nanocarriers. The limitations and strategies to overcome the limitations of PEGylated nanocarriers have also been discussed at length.

T. G. Agnihotri · V. Peddinti · S. S. Gomte · B. Rout · A. Jain (✉)
Department of Pharmaceutics, National Institute of Pharmaceutical Education and Research (NIPER)-Ahmedabad, Gandhinagar, Gujarat, India
e-mail: aakanchha.jain@niperahm.res.in



R. K. Tekade, N. K. Jain (eds.), *PEGylated Nanocarriers in Medicine and Pharmacy*, https://doi.org/10.1007/978-981-97-7316-9_10

**Keywords**
Proteins · Peptides · PEGylation · Nanocarriers · Physicochemical properties

## 10.1 Introduction: Proteins and Peptides

Proteins and peptides, the building blocks of life have been implicated in various cellular functions such as gene expression, regulation, enzymatic activity, signal transduction, cell apoptosis, immunity, etc. They have also been known to take part in a myriad of diseases like cancer, hypertension, metabolic disorders, and neurological disorders to name a few. As a result, they form an indispensable part of the therapeutic regimen to counteract those disorders. Owing to their specific physiological properties, proteins and peptides held a special place in the biopharmaceutical market rendering them one of the most effective drug deliveries (Jain et al. 2013). Therapeutic peptides usually consist of amino acids of molecular weight ranging from 500–5000 Da, which have been arranged in a specific order (Henninot et al. 2018; Wang et al. 2022). The synthesis of insulin in 1921 paved the way for the further development of peptides, which culminated in the approval of more than 80 therapeutic peptides globally. With the advancement of protein purification and synthesis, structural elucidation, genetic sequencing, and biotechnological tools, peptide development has taken a giant stride from the 1950s to the 1990s. During this period, apart from natural peptides, synthetic vasopressin, oxytocin, and estrogen have emerged. With the arrival of the twenty-first century, peptide drug development ushered into a new phase with rapid progress in structural biology, recombinant biologics, and new synthetic and analytic tools considerably speeding up the process. Since 2000, 33 non-insulin peptide medicines have been authorized globally (Wang et al. 2022).

Proteins and peptides are not generally recommended for oral administration for the fear of instability in the gastrointestinal tract (GIT). The size and hydrophilicity of peptides and proteins further complicate the process of absorption leading to low bioavailability (Deb et al. 2019; Goldberg and Gomez-orellana 2003). Other routes of administration also are not devoid of any side effects and accompanied by allergic reactions, scarring, pain, and infection for intravenous delivery. The transdermal delivery also poses absorption challenges owing to the presence of a skin barrier and impeding the entry of especially hydrophilic drugs having a molecular weight greater than 500 Da (Oberli et al. 2016).

Proteins and peptides are also prone to metabolic reactivity and usually have low bioavailability. Further, peptide oral bioavailability is restricted by a breakdown in the GI tract as well as their inability to pass the epithelial barrier. These medications often have high molecular weight peptides, poor lipophilicity, and charged functional groups, which make them difficult to absorb. Other factors including greater metabolism, protease degradation, opsonization, conformational changes, non-covalent complexation with plasma proteins, etc. could amount to low bioavailability of therapeutic peptides and proteins. Structural modification of peptides could offer a plausible solution to solve the aforementioned challenges, PEGylation, being one of them (Bruno et al. 2014). This chapter will emphasize the role of PEGylation in different nanocarriers along with a brief overview of methods of PEGylation. The limitations of PEGylated nanocarriers and the subsequent challenges to overcome those limitations have also been illustrated in this chapter.

## 10.2 Role of PEGylation in Nanocarriers

Drug delivery using nanocarriers, with or without ligand coating or coupling, is very common. Because of their unique targeting mechanism and controlled drug release profile, these carriers have gained immense popularity. However, the main disadvantage of nanocarriers is that the reticuloendothelial system (RES) removes them from the bloodstream. RES uptake, immune response, stability, drug loss, and hydrophobicity are the main reasons why nanocarriers use is restricted in the biomedical field, despite the fact that they have demonstrated a wide range of applications. PEGylation of nanocarriers can be used to get around the aforementioned restrictions. PEGylation is the process of coating or conjugating polyethylene glycol (PEG) with nanocarrier systems (Gajbhiye et al. 2020) (Fig. 10.1). PEG coatings on nanoparticles (NPs) protect the surface from aggregation, opsonization, and phagocytosis, thereby increasing the circulation time. For instance, Doxil®, the first PEGylated liposomal product approved by the FDA, was introduced in 1995. With a drug half-life of 72 h and a circulation half-life of 36 h, Doxil "Stealth®" liposomes enhanced doxorubicin bioavailability about 90 times at 1 week after injection compared to the free drug (Ahmed et al. 2005; Gabizon et al. 2003; Laginha et al. 2005). Due to the increased permeability and retention (EPR) effect, long-circulating PEGylated NPs provide an excellent opportunity to target tumors following intravenous administration (Agnihotri et al. 2023; Matsumura and Maeda 1986). The chance of interactions between the ligands presented on the NP surface and the corresponding receptors on cell surfaces rises with enhanced circulation time. However, the inclusion of ligands increases formulation complexity and may compromise the

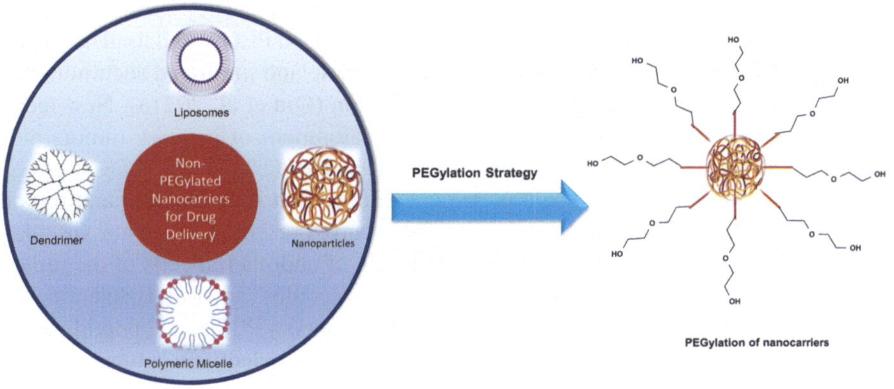


**Fig. 10.1** Schematic representation of PEGylation of various nanocarriers

"stealth" characteristics offered by surface PEG coatings (Moghimi et al. 2001). PEGylation has been employed to decrease the immune response and increase the circulation time of synthetic NPs as well as viral gene vectors like adenovirus and adeno-associated viruses (Kim et al. 2012; Lee et al. 2005). Table 10.1 represents the overview of few research findings regarding PEGylated nanocarriers-based protein/peptide delivery.

### 10.2.1 PEGylated Nanocarriers

PEGylated nanocarriers have gained widespread acceptance due to their distinct targeting mechanism as well as their controlled drug release profile. PEGylated NPs, liposomes, dendrimers, and polymeric micelles have been discussed in detail in the following section.

#### 10.2.1.1 Polymeric NPs

NPs are solid particles or particulate dispersions with sizes between 10–1000 nm. These NPs offer a number of benefits, including biocompatibility and the ability to control the rate and time of degradation with drug release. However, the use of bare polymeric NPs is limited by RES uptake, drug loss, hemolysis, immune response, and cytotoxicity. These restrictions are typically overcome by the PEGylation of NPs (Janrao et al. 2022; Nag et al. 2016). Rheumatoid arthritis (RA) is an autoimmune disorder that causes joint inflammation as well as immune system impairment. Inflammation-fighting responses from the patient would be triggered by pro-resolving mediators that target the inflammation-resolving pathway. Ac2–26, a 25-amino acid peptide obtained from Annexin A (a pro-resolving mediator), has demonstrated its effectiveness in treating inflammatory disorders. The low bioavailability of Ac2–26 peptides, on the other hand, limits their efficacy in vivo. To address this limitation, Qin et al. formulated PEGylated lipid nanoparticles (LDNPs) by co-assembling L-ascorbyl palmitate (L-AP) and N-(carbonyl methoxypolyethylene glycol-2000)-1,2-distearoyl-snglycero-3 phosphoethanolamine (DSPE-PEG2k) to encapsulate and deliver Ac2–26 peptides to arthritis rats. They exhibited excellent stability and biocompatibility. Ac2–26 peptide-loaded PEGylated lipid nanoparticles (ADNPs) showed longer in vivo circulation time and improved accumulation in inflamed regions after intravenous administration (Qin et al. 2021a). New techniques or technologies that may prevent the development of primary tumors and metastatic spread are particularly important since metastasis is the main cause of mortality in cancer patients. There is strong evidence that inflammation and tumor growth are closely related. Inflammation is frequently present in the tumor vasculature, which may enhance angiogenesis. The lining of endothelial cells of the tumor vasculature has an overexpression of adhesion molecules, such as E-selectin. To target E-selectin, Hao et al. chose a peptide and designed a PEGylated peptide-drug conjugate (PEGylated PDC). This prodrug consists of E-selectin as a tumor vascular endothelial target, E-selectin binding peptide as a targeting ligand, and the drug SN38 was combined with a spacer that could break and release SN38 in the tumor

**Table 10.1** Overview of research findings concerning PEGylated peptide/protein-based nanocarriers

| Sr. No. | Type of PEGylated nanocarriers | Enclosed protein/ peptide | Functions of protein/peptide | Benefits of PEGylation | Ref. |
|---|---|---|---|---|---|
| 1 | PEGylated dendrimers of polyglutamic acid | Nattokinase | Thrombolytic activity | PEGylation protects the enzyme activity of nattokinase and increases its systemic circulation time. | Zhang et al. (2017) |
| 2 | Deoxycholic acid conjugated PEGylated polyhydroxy butyrate nanoparticles | Insulin | Antihyperglycemic activity | Improved hydrophilic characteristic of polyhydroxy butyrate and cellular uptake of nanocarriers. | Chaturvedi et al. (2015) |
| 3 | PEGylated albumin-based poly-ionic micelles | Sprouty-1 protein | Control cancer development, proliferation, and metastasis | Improved anticancer efficacy and cytotoxicity of Sprouty-1 protein. | Jiang et al. (2016) |
| 4 | PEGylated Nano-niosomes | CDC-20 siRNA | Inhibit the overexpression of CDC-20 protein and thereby Inhibit the proliferation of cancer cells | Increased siRNA protection against nuclease enzyme | Hemati et al. (2019) |
| 5 | PEGylated liposomes | Insulin | Hypoglycemic activity | Increased insulin stability in GIT with prolonged hypoglycemic effect. | Iwanaga et al. (1997) |
| 6 | Peptide-conjugated PEGylated liposomes | Parathyroid hormone (PTH) analogous (Teriparatide) | Used for the treatment of osteoporosis | Increased blood circulation time and stability of the formulation | Salave et al. (2023) |
| 7 | Octreotide-conjugated PEGylated liposomes | Lipocalin-2 siRNA | Lessen the overexpression of lipocalin-2 protein in breast cancer | Improved circulation time, biocompatibility, and liposomal stability. | Gote and Pal (2021) |
| 8 | PEGylated Ferrisilicate nanoparticles | Insulin | Hypoglycemic effect | Enhanced insulin release from nanoparticles. | Jermy et al. (2023) |

(continued)

**Table 10.1** (continued)

| Sr. No. | Type of PEGylated nanocarriers | Enclosed protein/ peptide | Functions of protein/peptide | Benefits of PEGylation | Ref. |
|---|---|---|---|---|---|
| 9 | PEGylated silicon nanoparticles | Insulin | Hypoglycemic effect | Increased GIT stability and mucoadhesive property of silicon nanoparticles | Andreani et al. (2014) |
| 10 | PEGylated lipid nanoparticles | KRAS targeting siRNA | Knocking down proto-oncogene (KRAS) expression | Increased circulation time of lipid nanoparticles | Sasayama et al. (2019) |
| 11 | PEGylated lipid nanoparticles | Ac2–26 (Annexin A protein equivalent) | Anti-inflammatory effect | Extended *in Vivo* circulation time and enhanced accumulation of nanocarrier at inflamed site | Qin et al. (2021b)) |
| 12 | Acid-sensitive PEGylated solid lipid nanoparticles | Tumor necrosis factor-α (TNF-α) siRNA | Decreased the formation of TNF-α-associated inflammatory cytokines | Increased delivery of siRNA at the targeted site | Aldayel et al. (2018) |

microenvironment's high glutathione (GSH) concentration. Due to the amphiphilic properties of this prodrug, known as PEG-Pep-SN38, NPs could form in aqueous conditions. These self-assembled NPs enabled the simultaneous death of activated endothelium cells and the surrounding tumor cells by enhancing drug concentration and retention at the tumor site. It was found that the formation of PEG-Pep-SN38 significantly improved the solubility in water and in vivo toxicity of SN38. PEGylated PDCs have several advantages over antibody-drug conjugates (ADCs), (1) high structural diversity because peptides may be chemically altered more often than antibodies; (2) low cost because PEGylated PDCs can be produced on a large scale using chemical methods; and (3) low immune response (Ahmad et al. 2021).

Targeted administration of an anti-inflammatory drug in osteoarthritis (OA) treatment has the ability to significantly reduce unwanted systemic adverse reactions and lower the required therapeutic dosage. A targeted, non-invasive drug delivery system was developed by McMasters et al. to lessen the inflammation in an ex vivo osteoarthritis model. For the making of the osteoarthritis model, cartilage plugs from 3-month-old bovine knee joints that were procured from a slaughterhouse within 24 h after the animal's death were collected. They were then washed three times in a medium which is serum-free and equilibrated in 5% FBS-supplemented media for 3 days. The removal of native aggrecan simulated OA conditions. To remove aggrecan, plugs were treated at 37 °C for about 3 h with 0.5% (w/v) trypsin in HBSS. To inactivate residual trypsin activity, plugs were rinsed three times in HBSS and incubated for 20 min in 20% FBS after trypsin treatment. In the plugs, inflammation was induced by administering 20 ng/mL IL-1β. Hollow thermoresponsive poly (N-isopropyl acrylamide) (pNIPAM) NPs were produced by degrading a cross-linked N,N′-bis(acryloyl)cystamine (BAC) core out of a non-degradable pNIPAM shell. In the shell, sulfated 2-acrylamido-2-methyl-1--propanesulfonic acid (AMPSA) was copolymerized to improve the passive loading of an anti-inflammatory cell-penetrating peptide (KAFAK) that inhibits mitogen-activated protein kinase (MK2). The hollow PEGylated NPs exhibited a decrease in hydrodynamic radius, increased hydrophilicity, increased uptake, and a higher loading of KAFAK compared to their solid NPs. The hollow NPs loaded more KAFAK and released a larger proportion of their cargo for an extended period of time than the solid NPs. The drug-loaded hollow PEGylated NPs were successful in decreasing pro-inflammatory interleukin-6 (IL-6) expression after interleukin-1 beta (IL-1 β) activation in bovine cartilage explants by releasing a therapeutically effective dose of KAFAK. These thermosensitive hollow PEGylated NPs offer an excellent platform for the delivery of peptide drugs into highly proteolytic settings, such as osteoarthritis (Mcmasters et al. 2017). Movileanu et al. developed magnetite nanoparticles ($Fe_3O_4$) stabilized with PEG and the surface was modified with folic acid (FA) to enable targeted internalization in cells expressing the folic acid receptors (FR). These NPs, which are functionalized with FA, may be used to deliver various therapeutics such as biologically active proteins, peptides, vaccines, etc. (Movileanu et al. 2022).

#### 10.2.1.2 Liposomes

Liposomes resemble cell membranes in that they have an aqueous core surrounded by a lipid bilayer. This distinction facilitates the fusion of liposomes with cell membranes and subsequent cellular uptake. Because liposomes are amphipathic, they can be used to encapsulate drugs that are both hydrophilic and hydrophobic. RES uptake results in the rapid clearance of liposomes from the circulation, which then builds up in the liver and spleen. Biocompatible PEG polymers have been used to sterically stabilize conventional liposomes. Conventional liposomes are PEGylated to significantly extend their half-life in circulation (Marzban et al. 2015; Tang et al. 2018). Orally given insulin must overcome the harsh gastrointestinal tract environment, cross the enteric epithelial barrier, and avoid the first pass effect before reaching the systemic circulation. To address this issue, Sarhadi et al. developed PEGylated liposomal insulin and modified it with B12 to increase the stability and absorption of insulin in the gastrointestinal environment. As compared to non-targeted liposomes, B12-targeted PEGylated liposomes significantly increased cellular uptake in Caco-2 cells, according to in vitro results (Sarhadi et al. 2022). Yamazoe et al. investigated the viability of using liposomes modified with dense PEG2000 as mucus-penetrating particles (MPPs) for oral administration of systemically absorbed peptides. The densely PEGylated liposomes showed mucus permeability in an in vitro artificial mucus model. Fluorescein isothiocyanate dextran (FD) was used as a model peptide drug to assess the in vivo oral absorption of liposomes in rats. PEGylated liposomes had a greater oral absorption than unmodified liposomes (Yamazoe et al. 2020). By taking advantage of PEGylation, Yazdi et al. prepared liposomal preparation by using phospholipids, which are having a high transition temperature (Tm) combined with folic acid. The prepared liposomal formulation showed improved stability and high penetrating ability. The cell uptake results showed that FA conjugation increased insulin uptake. The results of biodistribution also revealed that FA-targeted PEGylated liposomes had greater concentrations in the blood and liver and longer residence times in the stomach and intestine. The PEGylated liposomes conjugated with FA had effective benefits in increasing insulin levels and decreasing blood glucose, as shown by the anti-diabetic effects of the formulation in vivo. The results of this study suggested that employing phospholipids with high Tm, PEGylation, and targeting ligands could increase the effectiveness of liposome oral delivery (Yazdi et al. 2020). An effective PEGylated liposomal delivery system having two ligands with target selectivity and characteristics that would improve cellular uptake was developed by Golam Kibria et al. The lipid film hydration process was used to produce PEGylated liposomes (PEG-LP). On the PEG-LP, the cyclic RGD (Arg-Gly-Asp) peptide, a specific ligand with an affinity for integrin αvβ3, was attached. Octa arginine that has been stearylated (STR-R8) was added, which acts as a second ligand (R8/RGD-PEG-LP) that serves as a Cell Penetrating Peptide (CPP) to the surface of the RGD-PEG-LP. In cells expressing Integrin αvβ3, R8/RGD-PEG-LP indicated improved cellular uptake in addition to greater transfection efficiency. From this, they concluded that PEG-LP attached with dual ligands has a high capacity for internalizing PEG-LP effectively, and as a result, it would be a useful tool for the targeted delivery of peptides and anti-cancer drugs (Kibria et al. 2011). On a similar note, Kato et al. created a new adapter-lipid compound that permits several click-reaction-based cyclic peptide modifications. The cyclic RGDfK (cRGD) peptide was used as the cyclic peptide ligand, which targeted integrin αvβ3.They developed a new alkyne tagged lipid having distinct peptide spacer and it was combined with cRGD peptide with the help of click reaction, which resulted in cRGD complexed lipid with excellent water dispersibility for formulating cRGD-modified PEGylated liposomes utilizing the post-insertion method. The prepared PEGylated liposomes showed excellent cytotoxicity when they were loaded with anticancer drug doxorubicin (Kato et al. 2022) (Fig. 10.2).

#### 10.2.1.3 Dendrimers

Dendrimers are well-defined, highly branching, nanoscale macromolecules with numerous active groups that may bind to other functional moieties for active targeting. Despite many advantages dendrimers have for drug delivery, their toxic properties due to amine groups make their clinical translation difficult. PEG conjugation has been employed in order to decrease the toxicity of amine groups on the surface of dendrimers due to their nonimmunogenic characteristics and hydrophilic nature (Alibolandi et al. 2021). PEGylation and acetylation of polyamidoamine (PAMAM) dendrimers can improve biocompatibility, reduce cytotoxicity, and delay their clearance from the bloodstream. Molecular dynamics simulations have been carried out by Khamseh et al. to study the interaction of eptifibatide (EPT) with native acetyl-terminated G4-PAMAM dendrimers and its partially (25, 50, and 75%) PEGylated analogs. Chains with molecular weights of 500 and 1000 Da were used to examine

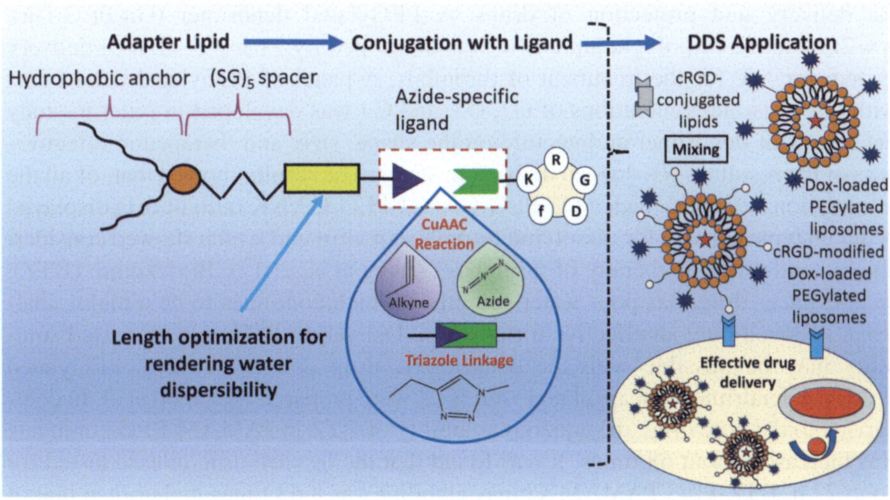


**Fig. 10.2** Schematic illustration of cRGD-modified PEGylated liposomes utilizing click reaction approach and doxorubicin as a model drug

the effect of PEG length. The G4-16PEG1000 system (1:7) was able to achieve its maximum loading capacity largely owing to the conjugation of 3 EPT molecules on the surface and the encapsulation of 4 EPT molecules in the internal cavities. G4-16PEG1000 is the most potential carrier for EPT in terms of loading capacity, preferred drug molecule binding sites, and release patterns (Badalkhani-Khamseh et al. 2023). Due to its less solubility in pharmaceutical solvents, 7-Ethyl-10-hydroxy-camptothecin (SN38) belongs to the camptothecin family with the highest biological activity, and cannot be used widely. PAMAM dendrimers can be used as an effective drug delivery system for molecules with low aqueous solubility. Mahmoudi et al. designed two CPPs (BR2 and CyLoP1) conjugated formulations of PEGylated PAMAM dendrimers having SN38 as a drug. Studies on cellular uptake proved that CPP-conjugated dendrimers showed higher values in a time-dependent manner. The majority of CPP-conjugated formulations significantly outperformed irinotecan in inhibiting tumor growth, according to in vivo studies, and all the formulations which were prepared showed increased survival rates. The majority of formulations had higher tumor accumulation than the commercial irinotecan formulation (positive control), according to bio-distribution studies by collecting whole tumors, various organs like kidneys, spleen, heart, and a part of the liver and lung. In summary, prepared dendrimeric formulations of SN38 that are combined with CPP showed effective tumor-inhibitory properties (Mahmoudi et al. 2019). Nattokinase (NK) has been used as a new-generation thrombolytic drug because of its high degree of safety, low cost, and few adverse effects. It is, however, easily affected by external environmental changes and may lose its enzymatic activity. The characteristics of peptide dendrimers, such as their surface functional groups, biodegradability, minimal toxicity, and biocompatibility make them an ideal vehicle for the delivery and protection of drugs. A PEGylated dendrimer (Gn-PEG-Gn), $n = 2,3,4$ made of polyglutamic acid was developed by Zhang et al. as a delivery system for NK for the treatment of thrombus. A panel of PEGylated dendrimers with three distinct generations of G2, G3, and G4 was developed in order to study the effects of dendrimer architecture on the shape, size, and therapeutic effectiveness of the resultant NK-loaded delivery systems. The results showed that, of all the formulations, the NK-loaded G3-PEG-G3 (G3-PEG-G3/NK ratio of 6:1) displayed better enzyme activity for dissolving thrombus in vitro and which showed considerable potential for the therapy of thrombus (Zhang et al. 2017). Bortezomib (BTZ), an anticancer drug, has poor water solubility, which continues to be a major challenge in developing an effective formulation. Delivering BTZ with increased solubility may be possible with the dendrimeric drug delivery system. PEGylated PAMAM dendrimers encapsulated with BTZ were prepared by Sahoo et al. In comparison to the pure drug, the aqueous solubility of BTZ in PAMAM-PEG conjugate was increased about 68 times. It was found that the in vitro drug release lasted for up to 72 h. BTZ-PEG-PAMAM's bioavailability was 8.63 times higher than that of the pure drug in an in vivo pharmacokinetic study on Sprague Dawley rats. BTZ-PEG-PAMAM's pharmacokinetic characteristics were superior to those of BTZ and other prepared formulations. In conclusion, the developed BTZ-PEG-PAMAM formulation produced meaningful results, and this approach may be explored further for improved delivery of BTZ (Sahoo et al. 2020).

#### 10.2.1.4 Polymeric Micelles

Polymeric micelles are composed of amphiphilic block or graft copolymers with hydrophilic and hydrophobic segments. These polymeric micelles have a hydrophobic inner core and a hydrophilic outer shell. Although there are many different types of amphiphilic copolymers, block copolymers create stable polymeric micelles with a characteristic core-shell structure. PEGylated polymeric micelles increase the aqueous solubility, stability, and circulation time of the drugs (Kamimura and Nagasaki 2014). A new ligand for colorectal neoplasia, LTTHYKL peptide, was added to PEGylated octadecyl lithocholate micelles by Khondee et al. to develop a targeted nanomedicine. Free rapamycin, rapamycin not labeled with micelles, and rapamycin micelles labeled with peptide were given intraperitoneally for 35 days to Apc mice that developed colonic adenomas. Adenoma regression in vivo was assessed using endoscopy. When compared to normal rapamycin micelles, the mean regression rate was considerably greater for peptide-labeled rapamycin micelles with $P < 0.01$. Results showed no other toxicities and peptide-labeled rapamycin micelles had less renal toxicity than the free drug. Together, these distinct targeted micelles offer a potential treatment for colorectal neoplasia that has systemic toxicity that is significantly lower and therapeutic efficacy that is comparable to the rapamycin-free drug (Khondee et al. 2015). Siyu Guan et al. developed two strategies for developing a multi-targeted delivery system by utilizing tumor characteristics. The first was the production of pH-dependent micelles, which enabled increased drug release by taking advantage of the tumor's acidic environment. The second

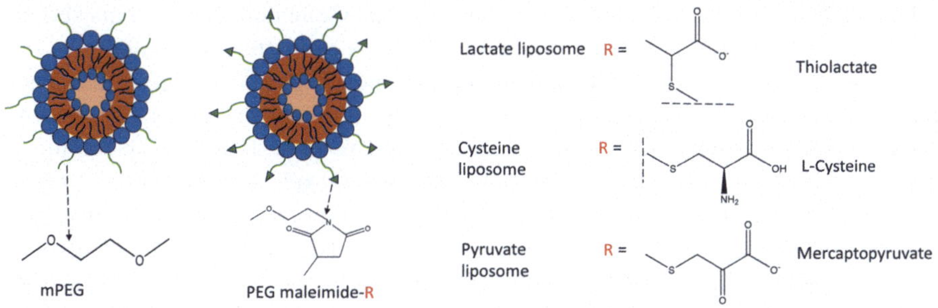


**Fig. 10.3** text

method involved using the tumors' surface-exposed phosphatidylserine (PS), which is usually found in the inner leaflet of healthy cells. Using PS as the target site, PS binding peptide (PSBP-6) was combined to pH-dependent mixed micelles consisting of poly (ethylene glycol)-b-poly (L-histidine) (PEG-PHIS) and poly (ethylene glycol)-b-poly (D, L-lactide) (PEG-PDLLA). After the micelles were successfully prepared, the anti-cancer drug paclitaxel was used to assess for drug loading capacity and encapsulation effectiveness. The results showed 7.9% and 83.5%, drug loading capacity and encapsulation efficiency, respectively. At pH 5.0, 6.5, and 7.4, the in vitro release of PTX from mixed micelles was 78.1, 56.8, and 51.4%, respectively, showing acid-triggered drug release. In comparison to unmodified mixed micelles, the micelles modified with PSBP-6 significantly increased in vitro cytotoxicity and showed increased cellular uptake in the HeLa cell lines. Additionally, studies on the pharmacokinetics, in vivo biodistribution, and fluorescence imaging of PSBP-6-PEG-PDLLA/PEG-PHIS mixed micelles demonstrated that they increased blood circulation time and promote tumor accumulation (Guan et al. 2020). Khan et al. described the recent accomplishments in the design and formulation of nanoscale pH-responsive micelles conjugated with CPPs to provide controlled drug release in the tumor cells. They have also explained about the drawbacks and future prospects of stimuli-sensitive drug delivery systems complexed with CPPs (Fig. 10.3). The summary of their study is the use of pH-sensitive polymeric micelles-based DDSs combined with peptides for controlled administration might be thought as a viable strategy to increase the therapeutic index of anticancer medications (Khan et al. 2021).

## 10.3 PEGylated Nanocarriers for Miscellaneous Applications

Under the realm of gene therapy, major advancements are expected in the use of siRNA as a therapeutic agent in the treatment of chronic illnesses and genetic abnormalities. However, delivering it, particularly via systemic ways, remains a difficulty. Since cells do not easily absorb siRNA, practical applications of siRNA rely heavily on the development of delivery mechanisms capable of delivering intact siRNA into the cytoplasm of the target cells. Gene delivery systems should be developed to preserve genetic materials from a premature breakdown in the systemic circulation while still transferring therapeutic genes to target cells effectively. While the PEGylation of these delivery methods preserves the siRNA, additional issues may arise, such as a reduction in gene transfection owing to decreased cellular absorption or restricted endosomal release (Ryan et al. 2008). To overcome this, pH-cleavable PEGylation of monomolecular siRNA complexes was proposed by DeRouchey and his team who found that PEG coating of siRNA-PEG complexes elicited biodegradability with hydrazone moiety or pH-sensitive disulfide linker between $K_{14}$ (14-lysine) and PEG (DeRouchey et al. 2008). As naked siRNA suffers from rapid degradation and siRNA complex tends to accumulate within blood or get captured by macrophages, PEGylation of siRNA is often exercised. However, PEGylation of siRNA does not correlate well with gene silencing efficiency, thus creating a great need for effectively delivering siRNA. Ultrasound coupled with microbubbles has been a popular choice, in the recent past for delivering siRNA. However, gene silencing efficiency is often low requiring a greater load of siRNA to be delivered when it is associated with ultrasound-microbubbles technique. Vandenbroucke et al. combined these two approaches of PEGylated siPlex (siRNA-liposome complex)-loaded microbubbles, which can be induced on exposure to ultrasound which exhibited greater gene silencing expressions both in terms of space and time-controlled settings (Vandenbroucke et al. 2008).

PEGylation also has a special place in pulmonary delivery. For instance, Qiu et al. PEGylated synthetic cationic peptide KL4 to deliver mRNA through the pulmonary route, which eventually formed a complex with mRNA at 10:1 ratio (w/w). The PEGylation of the peptide was done to serve two purposes. One is being able to overcome poor solubility issues owing to the presence of leucine residues present in peptide structure and the other is to impart steric hindrance, which further facilitates the stability against degrading enzymes. This complex was then formulated into dry powder formulation and evaluated on human lung epithelial cells and in animals. The results were promising enough to establish PEGylated peptide as an effective non-viral vector for the delivery of mRNA via a pulmonary route of administration (Qiu et al. 2019).

The therapeutic potential of PEGylated salmon calcitonin derivatives (PEG-sCT) was investigated by Youn et al. On administration of PEG-sCT intratracheally, it was subjected to a bioactivity study, proteolytic resistance determination, and pulmonary pharmacokinetic study. The results showed improved in vivo efficacy and prolonged half-life owing to their proteolytic stability and also showed an enhanced area under the curve when sCT was attached to 5 kDa molecular weight PEG as compared to the 1 and 2 kDa variants (Lee et al. 2009; Qin et al. 2022; Youn et al. 2008). However, the molecular weight of PEG (20 and 40 kDa) had a negligible effect on the therapeutic retention of anti-interleukin-17A and anti-interleukin-13 fragments of antibody in the lungs when administered through the intratracheal route. Instead, the location of delivery had a major role to play in the retention of antibody fragments in the lungs of mice (Patil et al. 2018).

Drug delivery to the central nervous system (CNS) has become a prime concern en route to the treatment of neurodegenerative disorders. In this context, a gold nano prism is an attractive option because of the special characteristics exhibited like no requirement of organic solvents, simple synthetic procedure, and biocompatibility. It can also exhibit a photothermal effect and show localized surface plasmon resonance in the near-infrared region, thus enjoying therapeutic applications in the cancer field, too. In the work done by Arellano et al., gold nano prisms were functionalized by PEG and angiopep-2 peptide, which facilitated BBB crossing in zebrafish larvae model. Angiopep-2 is known to bind low-density lipoprotein receptor-related protein-1 (LRP1), which further facilitated the entry of drugs across BBB through transcytosis. The study on SH-SY5Y neuroblastoma cell lines and in zebrafish larvae demonstrated no toxicity profile both in vitro as well as in vivo. Overall, this study holds great potential in treating CNS-related disorders (Tapia-arellano et al. 2021).

## 10.4 Limitations of PEGylated Nanocarriers

PEGylation, which has been extensively used to improve the physicochemical characteristics of nanocarriers, showed unfavorable patterns of a sudden release of the loaded drug before or right after administration, even before NPs adequately accumulate in target tissues. Additionally, when comparing PEGylated and non-PEGylated nanocarriers, it was found that the particle size enlargement by the PEGylation process elicited an opposite effect on the desired RES evasion and accumulation in tumor tissue by the EPR effect (Sebak 2018). For example, doxorubicin-loaded PEGylated liposomes released the drug more rapidly than their non-PEGylated liposomes, with 90% of the drug being released 3 hours after intravenous administration, whereas the liposomes without the drug were still circulating in the blood before reaching the target region. Therefore, it was wrong to presume that the loaded drug was released upon absorption by the target cells (Press 2017). Additionally, the PEGylation process and purification are challenging procedures that may result in the polymers' inherent polydispersity, which eventually leads to batch-to-batch variability (Rattan et al. 2017).

The hydrophilic polymer provides steric hindrance and conceals surface charge, as was previously described under the advantages of PEGylation, to inhibit the adsorption of tagging proteins (opsonization), which is necessary for absorption by RES macrophages. PEGylation has been proven to have a limited ability to prevent opsonization, and reports of eventual opsonization and RES clearance have been made time and time again. For instance, it was found in one study that PEGylation of thiolated gelatin reduced the mean residence time (MRT) and half-life of NPs in blood. In other words, circulating phagocytes and tissue macrophages cleared the bulk of NPs from the blood, and the NPs subsequently found in the liver and spleen (Xu et al. 2013). Another notable drawback of PEGylation is the lower intracellular uptake of PEGylated NPs compared to non-PEGylated ones. This occurs as a result of the hydrophilicity of the surfaces

**Table 10.2** Characterization techniques regarding PEGylated nanocarriers for protein and peptide delivery

| Sr. No. | Characterization technique | Peptide/protein | Remarks | Ref |
|---|---|---|---|---|
| 1 | HPLC | tTF-NGR | HPLC-based gel filtration separated N-terminal PEGylated protein rendering pure elution of only mono-PEGylated protein | Brand et al. (2015) |
| 2 | LC-MS | Oxytocin | Conjugation of N-amine PEG to formulation was confirmed by MALDI-TOF. The differentiation between positional isomers was determined confirmed by MALDI-TOF | Collins et al. (2016) |
| 3 | LC-MS | Filgrastim | With LC-MS technique, identification and detection of positional isomers were done for PEGylation. | Shekhawat et al. (2019) |
| 4 | DLS | Human growth hormone | Size and zeta potential of PEGylated hGH were determined by DLS technique and found that size was increased and zeta potential was decreased. | Khameneh et al. (2016) |
| 5 | DLS | L-asparaginase | DLS revealed increase in size (hydrodynamic diameter) of protein on PEGylation | Meneguetti et al. (2019) |
| 6 | NMR | Cetuximab | N-glycan analysis and structural elucidation of antibody by LC/MS/MS and NMR was determined. | Wiegandt and Meyer (2014) |
| 7 | ELISA | Pegasys | Pharmacokinetics of Pegasys and PEG-intron with IFN antibodies were evaluated by ELISA. | Bruno et al. (2007) |
| 8 | Computational modeling | Erythropoietin | Parameters of PEGylation were optimized for erythropoietin analogs by in silico approach. | Mirzaei et al. (2016) |
| 9 | Bioinformatics method | Interferon | Changes at molecular level induced by PEGylation of interferons were detected by molecular simulations | Xu et al. (2018) |

of the NPs and the steric hindrance produced by the projecting chains of PEG polymers (Mao et al. 2013). Another major concern for PEGylated nanocarriers is inadequacy of characterization techniques to completely understand the elucidation and fate of formulated system. Although a rapid progress has been made towards characterization techniques supported by research findings (Table 10.2), (Fig. 10.4) quantity of PEG on and in the core of NPs cannot be quantified precisely. At the same time, ligand modification would often render the conformational analysis of PEGylated nanocarriers complicated (Shi et al. 2021).

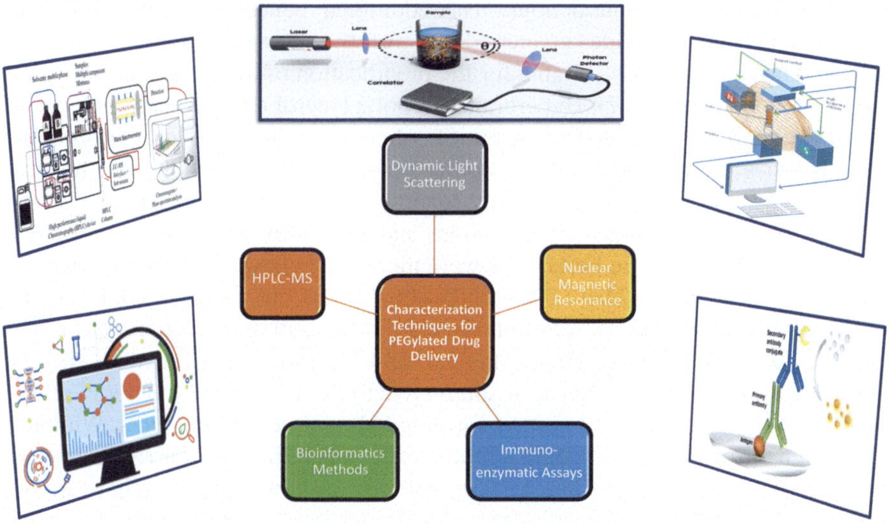


**Fig. 10.4** Overview of different characterization techniques for evaluating PEGylated nanocarriers

## 10.5 Strategies to Overcome the Limitations of PEGylated Nanocarriers

There are numerous methods available to get around the drawbacks of PEGylation, for example, controlling the various factors that can influence the outcomes of PEGylation, such as molecular weight of PEG, its density, and coating stability, and selecting appropriate polymers having low or no antigenicity (Wan et al. 2016). In contrast to methoxy-PEG (mPEG) and other unbranched PEG forms, and branched PEG, one study found that branched PEG was the least immunogenic and antigenic (Zhang et al. 2014). The issue of PEG's non-biodegradability might also be approached and solved, for example by combining PEG with segments of PLA or poly (glycolic acid) (PGA), to produce equivalent biodegradable copolymers (Saez-martinez et al. 2014; Sebak 2018; Ulbricht et al. 2014). Alternative nanocarrier coatings have been researched as a different strategy to solve the PEGylation dilemma. For instance, it has been found that NPs with zwitterionic coatings can avoid brain immune components and reduce the rate of non-specific adhesion for proteins and lipids without steric hindrance (Jenkins et al. 2016).

A different study used acetylation of dendrimers instead of PEGylation of dendrimers. This method provided the benefit of RES evasion without introducing steric hindrance, which might decrease the interaction between FA on the acetylated dendrimers and the folate receptor on the surface of the target cells, which could decrease the absorption of the NPs as in the case of FA-conjugated PEGylated dendrimers. Aside from decreasing the electrostatic interactions with serum proteins required for RES macrophage absorption, acetylation of the dendrimers' cationic surface renders them neutral. It is believed that acetylation reduces RES clearance via these electrostatic interactions. The additional benefit of acetylation over PEGylation is that it provides greater control over the size and polydispersity of the NPs, making it potentially suitable for the modification of different types of nanocarriers (Rattan et al. 2017). Hyperbranched polyglycerol (HPG) has shown promise in the delivery of neurotherapeutics when used in place of conventional PEG. Hyperbranched polyglycerol displayed more advantageous physicochemical characteristics, such as increased drug loading, increased NPs surface coverage, increased efficacy for ligand conjugation, and increased circulation time (Zhou et al. 2017). The magnetic $Fe_3O_4$ NPs were more stable in the cell culture medium due to the HPG coating. These coated NPs showed less macrophage uptake and no toxicity to mammalian cells. Additionally, HPG offers a platform for preparing NPs surfaces for active targeting (Hadjesfandiari and Parambath 2018). Lipids may be employed as a coating for NPs as an alternative to PEG to extend the duration that they remain in the circulatory system, forming hybrid lipid-polymer NPs. With outstanding in vitro and in vivo stability, biological compatibility, and biodegradability, hybrid NPs are a potential drug delivery platform. Additionally, they have a high drug loading, surface functionality, controlled particle size, and stimuli-responsive drug release characteristics (Bose et al. 2017; Date et al. 2017; Dong et al. 2018; Garg et al. 2018). To accomplish active targeting markers of self or target-specific ligands, such as FA or transferrin, may be introduced to the surface of the NPs (Kettiger et al. 2013). The arginine-glycine-aspartic acid (RGD) tripeptide has been found to enhance the selectivity, binding, and absorption of NPs by targeted cells by binding to integrin receptors that are expressed more on tumor cells. This method can specifically improve the steric hindrance problem caused by PEG polymers' limited uptake drawback (Hatakeyama et al. 2013). Compared to non-targeted ones, targeted PEGylated liposomes that target α5β1 in metastatic colon cancer have shown greater transfection efficiency. These liposomes formed by conjugation to PRb, a peptide that mimics fibronectin ("42. levine et al. 2017 Dual-ligand α5β1 and α6β4 integrin targeting enhances gene delivery and selectivity to cancer cells.pdf," n.d.). The similar targeting strategy might be used to enhance PEGylated nanocarrier-based drug delivery to the brain. For example, brain tumors exhibited greater accumulation of PEGylated gold NPs functionalized with transferrin-specific peptides (Sun et al. 2017). Paclitaxel was coupled to cyclo-[Arg-Gly-Asp-D-Phe-Lys] peptide in dual targeting, which enhances brain drug delivery by precisely interacting with integrin receptors overexpressed on glioma cells. The lipid-polymer hybrid nanoparticles were then loaded with this conjugate (Agrawal et al. 2014). In summary, the PEGylation technique has shown tremendous promise for improving the characteristics of drug delivery systems, leading to its use in a variety of pharmaceuticals. However, poor PEGylation process management has produced a number of drawbacks. Therefore, extreme care should be taken while using the PEGylation method.

## 10.6 Conclusion

The covalent bonding of peptides and proteins to PEG remains a popular method of choice for altering the pharmacokinetic and immunological properties of therapeutic molecules, as evidenced not only by the introduction of PEGylated drugs to the market, but also by the growing number of clinical studies that are currently in progress. The chemical versatility of polyethylene glycol derivatives allows the synthesis of a wide range of PEGylated protein structures, with a preference for target-specific amino acid residues located at the terminal ends (N or C-terminus) of the peptides or proteins of interest, which contributes to the formation of homogeneous and well-defined conjugates. The biological activity of the PEGylated molecule must be preserved by these site-specific changes. The novel methodologies based on novel in the form of enzyme ligation and bio-orthogonal chemistry-based techniques are being applied as part of the evolution of the science of PEGylation. PEG coatings will undoubtedly remain a cornerstone in the design of NPs for proteins, and peptides applications, allowing for further research into how the characteristics of PEG coatings affect NP biodistribution and removal from the body. Further research into the immunogenic qualities of PEG coatings as a function of molecular weight, functional group, surface density, NPs properties, dosage frequency, etc., will almost likely result in more effective peptide and protein products. Summarily, more research into understanding the impact of properties on the systemic administration of peptide/protein-based nanocarriers is warranted to supplement the existing landscape of PEGylated nanocarriers.

## References

Agnihotri TG, Alexander A, Agrawal M, Dubey SK, Jain A (2023) In vitro-in vivo correlation in nanocarriers: from protein corona to therapeutic implications. J Control Release 354:794–809. https://doi.org/10.1016/j.jconrel.2023.01.063

Agrawal U, Chashoo G, Sharma PR, Kumar A, Saxena A, Vyasa SP (2014) Accepted crt. Colloids Surfaces B Biointerfaces. https://doi.org/10.1016/j.colsurfb.2014.12.045

Ahmad K, Teng Y, Liu Z, Li J, Guo N, Yu P (2021) European journal of medicinal chemistry tumor vasculature-targeting PEGylated peptide-drug conjugate prodrug nanoparticles improve chemotherapy and prevent tumor. Eur J Med Chem 219:113430. https://doi.org/10.1016/j.ejmech.2021.113430

Ahmed M, Lukyanov AN, Torchilin V, Tournier H, Schneider AN, Goldberg SN (2005) Combined radiofrequency ablation and adjuvant liposomal chemotherapy: effect of chemotherapeutic agent, nanoparticle size, and circulation time. J Vasc Interv Radiol 16:1365–1371. https://doi.org/10.1097/01.RVI.0000175324.63304.25

Aldayel AM, O'Mary HL, Valdes SA, Li X, Thakkar SG, Mustafa BE, Cui Z (2018) Lipid nanoparticles with minimum burst release of TNF-α siRNA show strong activity against rheumatoid arthritis unresponsive to methotrexate. J Control Release 283:280–289. https://doi.org/10.1016/j.jconrel.2018.05.035

Alibolandi M, Shahriari M, Ramezani M (2021) Principal concept in PEGylated dendrimer-based cancer therapeutics. In: Dendrimer-based nanotherapeutics. Elsevier, Amsterdam, pp 183–202

Andreani T, Kiill CP, Luiza A, De Souza R, Fangueiro JF, Fernandes L, Doktorovová S, Santos DL, Garcia ML, Palmira M, Gremião D, Souto EB, Silva AM (2014) Surface engineering of silica nanoparticles for oral insulin delivery: characterization and cell toxicity studies. Colloids Surfaces B Biointerfaces 123:916. https://doi.org/10.1016/j.colsurfb.2014.10.047

Badalkhani-Khamseh F, Ebrahim-Habibi A, Hadipour NL, Behmanesh M (2023) PEGylated PAMAM dendrimers as eptifibatide nanocarriers: an atomistic view from molecular dynamics simulations. Chem Eng Sci 267:118283

Bose RJC, Ravikumar R, Karuppagounder V, Bennet D, Rangasamy S, Thandavarayan RA (2017) Lipid—polymer hybrid nanoparticle—mediated therapeutics delivery : advances and challenges. Drug Discov Today 00:1258. https://doi.org/10.1016/j.drudis.2017.05.015

Brand C, Fröhlich M, Ring J, Schliemann C, Kessler T, Mantke V, König S, Lücke M, Mesters RM, Berdel WE, Schwöppe C (2015) Tumor growth inhibition via occlusion of tumor vasculature induced by N-terminally PEGylated retargeted tissue factor tTF-NGR. Mol Pharm 12:3749–3758. https://doi.org/10.1021/acs.molpharmaceut.5b00508

Bruno R, Sacchi P, Scagnolari C, Torriani F, Maiocchi L, Patruno S, Bellomi F, Filice G, Antonelli G (2007) Pharmacodynamics of peginterferon alfa-2a and peginterferon alfa-2b in interferon-naïve patients with chronic hepatitis C: a randomized, controlled study. Aliment Pharmacol Ther 26:369–376. https://doi.org/10.1111/j.1365-2036.2007.03392.x

Bruno BJ, Miller GD, Lim CS (2014) NIH Public Access 4:1443–1467. https://doi.org/10.4155/tde.13.104.Basics

Chaturvedi K, Ganguly K, Kulkarni AR, Rudzinski WE, Krauss L, Nadagouda MN, Aminabhavi TM (2015) Oral insulin delivery using deoxycholic acid conjugated PEGylated polyhydroxybutyrate co-polymeric nanoparticles. Nanomedicine 10:1569–1583. https://doi.org/10.2217/nnm.15.36

Collins J, Kempe K, Wilson P, Blindauer CA, McIntosh MP, Davis TP, Whittaker MR, Haddleton DM (2016) Stability enhancing N-terminal PEGylation of oxytocin exploiting different polymer architectures and conjugation approaches. Biomacromolecules 17:2755–2766. https://doi.org/10.1021/acs.biomac.6b00919

Date T, Nimbalkar V, Kamat J, Mittal A, Mahato RI, Chitkara D (2017) NU SC. J Control Release. https://doi.org/10.1016/j.jconrel.2017.12.016

Deb PK, Al-attraqchi O, Chandrasekaran B, Paradkar A, Tekade RK (2019) Protein/peptide drug delivery systems : practical considerations in pharmaceutical product development, basic fundamentals of drug delivery. Elsevier, Amsterdam. https://doi.org/10.1016/B978-0-12-817909-3.00016-9

DeRouchey J, Schmidt C, Walker GF, Koch C, Plank C, Wagner E, Rädler JO (2008) Monomolecular assembly of siRNA and poly(ethylene glycol)-peptide copolymers. Biomacromolecules 9:724–732. https://doi.org/10.1021/bm7011482

Dong W, Wang X, Liu C, Zhang X, Zhang X, Chen X, Kou Y, Mao S (2018) Chitosan based polymer-lipid hybrid nanoparticles for oral delivery of enoxaparin. Int J Pharm 547:499. https://doi.org/10.1016/j.ijpharm.2018.05.076

Gabizon A, Shmeeda H, Barenholz Y (2003) Pharmacokinetics of pegylated liposomal doxorubicin: review of animal and human studies. Clin Pharmacokinet 42:419–436. https://doi.org/10.2165/00003088-200342050-00002

Gajbhiye KR, Pawar A, Mahadik KR, Gajbhiye V (2020) PEGylated nanocarriers: a promising tool for targeted delivery to the brain. Colloids Surfaces B Biointerfaces 187:110770. https://doi.org/10.1016/j.colsurfb.2019.110770

Garg NK, Tandel N, Jadon RS, Tyagi RK, Katare OP (2018) Lipid-polymer hybrid nanocarrier-mediated cancer therapeutics: current status and future directions. Drug Discov Today 23:1610. https://doi.org/10.1016/j.drudis.2018.05.033

Goldberg M, Gomez-orellana I (2003) Challenges for the oral delivery of macromolecules. Nat Rev Drug Discov 2:289. https://doi.org/10.1038/nrd1067

Gote V, Pal D (2021) Octreotide-targeted lcn2 sirna pegylated liposomes as a treatment for metastatic breast cancer. Bioengineering 8:44. https://doi.org/10.3390/bioengineering8040044

Guan S, Zhang Q, Bao J, Duan T, Hu R, Czech T (2020) European journal of pharmaceutics and biopharmaceutics phosphatidylserine targeting peptide-functionalized pH sensitive mixed micelles for enhanced anti-tumor drug delivery. Eur J Pharm Biopharm 147:87–101. https://doi.org/10.1016/j.ejpb.2019.12.012

Hadjesfandiari N, Parambath A (2018) 13. Stealth coatings for nanoparticles: polyethylene glycol alternatives, engineering of biomaterials for drug delivery systems. Elsevier, Amsterdam. https://doi.org/10.1016/B978-0-08-101750-0.00013-1

Hatakeyama H, Akita H, Harashima H (2013) Polyethyleneglycol: a classical but innovative material the polyethyleneglycol dilemma: advantage and disadvantage of PEGylation of liposomes for systemic genes and nucleic acids delivery to tumors. Biol Pharm Bull 36:892–899

Hemati M, Haghiralsadat F, Jafary F, Moosavizadeh S, Moradi A (2019) Targeting cell cycle protein in gastric cancer with CDC20siRNA and anticancer drugs (doxorubicin and quercetin) co-loaded cationic PEGylated nanoniosomes. Int J Nanomedicine 14:6575–6585. https://doi.org/10.2147/IJN.S211844

Henninot A, Collins JC, Nuss JM (2018) The current state of peptide drug discovery: back to the future? J Med Chem 61:1382. https://doi.org/10.1021/acs.jmedchem.7b00318

Iwanaga K, Ono S, Narioka K, Morimoto K, Kakemi M, Yamashita S, Nango M, Oku N (1997) Oral delivery of insulin by using surface coating liposomes. Improvement of stability of insulin in GI tract. Int J Pharm 157:73–80. https://doi.org/10.1016/S0378-5173(97)00237-8

Jain A, Jain A, Gulbake A, Shilpi S, Hurkat P (2013) Peptide and protein delivery using new drug delivery systems. Crit Rev Ther Drug Carrier Syst 30:293–329

Janrao C, Khopade S, Bavaskar A, Gomte SS, Agnihotri TG, Jain A (2022) Recent advances of polymer based nanosystems in cancer management. J Biomater Sci Polym Ed 0:1–62. https://doi.org/10.1080/09205063.2022.2161780

Jenkins SI, Weinberg D, Arwa F, Fer- AR, Yiu HHP, Telling ND, Roach P, Chari DM (2016) ‘Stealth’ nanoparticles evade neural immune cells but also evade major brain cell populations: implications for PEG-based neurotherapeutics. J Control Release 224:136. https://doi.org/10.1016/j.jconrel.2016.01.013

Jermy BR, Salahuddin M, Tanimu G, Dafalla H, Almofty S, Ravinayagam V (2023) Design and evaluation of Pegylated large 3D pore ferrisilicate as a potential insulin protein therapy to treat diabetic mellitus. Pharmaceutics 15:593. https://doi.org/10.3390/pharmaceutics15020593

Jiang Y, Lu H, Chen F, Callari M, Pourgholami M, Morris DL, Stenzel MH (2016) PEGylated albumin-based Polyion complex micelles for protein delivery. Biomacromolecules 17:808–817. https://doi.org/10.1021/acs.biomac.5b01537

Kamimura M, Nagasaki Y (2014) PEGylated polymer micelles for anticancer drug delivery carrier, colloid and interface science in pharmaceutical research and development. Elsevier, Amsterdam. https://doi.org/10.1016/B978-0-444-62614-1.00014-4

Kato N, Sato T, Fuchigami Y, Suga T, Geng L, Tsurumaru M, Hagimori M, Mukai H, Kawakami S (2022) Synthesis and evaluation of a novel adapter lipid derivative for preparation of cyclic peptide-modified PEGylated liposomes: application of cyclic RGD peptide. Eur J Pharm Sci 176:106239. https://doi.org/10.1016/j.ejps.2022.106239

Kettiger H, Schipanski A, Wick P, Huwyler J (2013) Engineered nanomaterial uptake and tissue distribution: from cell to organism. Int J Nanomedicine:3255–3269

Khameneh B, Saberi MR, Hassanzadeh-Khayyat M, Mohammadpanah H, Ghandadi M, Iranshahi M, Baratian A, Jaafari MR (2016) Evaluation of physicochemical and stability properties of human growth hormone upon enzymatic PEGylation. J Appl Biomed 14:257–264. https://doi.org/10.1016/j.jab.2016.06.002

Khan S, Vahdani Y, Hussain A, Haghighat S (2021) Polymeric micelles functionalized with cell penetrating peptides as potential pH-sensitive platforms in drug delivery for cancer therapy: a review. Arab J Chem 14:103264. https://doi.org/10.1016/j.arabjc.2021.103264

Khondee S, Rabinsky EF, Owens SR, Joshi BP, Qiu Z, Duan X, Zhao L, Wang TD (2015) Targeted therapy of colorectal neoplasia with rapamycin in peptide-labeled pegylated octadecyl lithocholate micelles. J Control Release 199:114–121. https://doi.org/10.1016/j.jconrel.2014.11.034

Kibria G, Hatakeyama H, Ohga N, Hida K, Harashima H (2011) Dual-ligand modification of PEGylated liposomes shows better cell selectivity and efficient gene delivery. J Control Release 153:141–148. https://doi.org/10.1016/j.jconrel.2011.03.012

Kim J, Kim P-H, Kim SW, Yun C-O (2012) Enhancing the therapeutic efficacy of adenovirus in combination with biomaterials. Biomaterials 33:1838–1850

Laginha KM, Verwoert S, Charrois GJR, Allen TM (2005) Determination of doxorubicin levels in whole tumor and tumor nuclei in murine breast cancer tumors. Clin Cancer Res 11:6944–6949. https://doi.org/10.1158/1078-0432.CCR-05-0343

Lee GK, Maheshri N, Kaspar B, Schaffer DV (2005) PEG conjugation moderately protects adeno-associated viral vectors against antibody neutralization. Biotechnol Bioeng 92:24–34

Lee KC, Chae SY, Kim TH, Lee S, Lee ES, Youn YS (2009) Intrapulmonary potential of polyethylene glycol-modified glucagon-like peptide-1s as a type 2 anti-diabetic agent. Regul Pept 152:101–107. https://doi.org/10.1016/j.regpep.2008.09.008

Levine et al (2017) Dual-ligand α5β1 and α6β4 integrin targeting enhances gene delivery and selectivity to cancer cells. J Control Release 251:24

Mahmoudi A, Reza M, Ramezanian N, Gholami L (2019) BR2 and CyLoP1 enhance in-vivo SN38 delivery using pegylated PAMAM dendrimers. Int J Pharm 564:77–89. https://doi.org/10.1016/j.ijpharm.2019.04.037

Mao Z, Zhou X, Gao C (2013) Influence of structure and properties of colloidal biomaterials on cellular uptake and cell functions. Biomater Sci 1:896–911. https://doi.org/10.1039/c3bm00137g

Marzban E, Hoda S, Ghiadi M, Khoshangosht M (2015) Colloids and surfaces B : biointerfaces optimizing the therapeutic efficacy of cisplatin PEGylated liposomes via incorporation of different DPPG ratios : in vitro and in vivo studies. Colloids Surfaces B Biointerfaces 136:885–891. https://doi.org/10.1016/j.colsurfb.2015.10.046

Matsumura Y, Maeda H (1986) A new concept for macromolecular therapeutics in cancer chemotherapy: mechanism of tumoritropic accumulation of proteins and the antitumor agent smancs. Cancer Res 46:6387–6392

Mcmasters J, Poh S, Lin JB, Panitch A (2017) Delivery of anti-inflammatory peptides from hollow PEGylated poly(NIPAM) nanoparticles reduces inflammation in an ex vivo osteoarthritis model. J Control Release 258:161. https://doi.org/10.1016/j.jconrel.2017.05.008

Meneguetti GP, Madalena P, Mariana K, Obreque T, Marcello C, Barbosa V, Monteiro G, Helena S, Farsky P, De Oliveira AM, Angeli CB, Ventura M, Pessoa-Junior A, Id CDOR, Santos JHPM, Obreque KMT, Barbosa CMV, Monteiro G, Farsky SHP, Marim de Oliveira A, Angeli CB, Palmisano G, Ventura SPM, Pessoa-Junior A, de Oliveira Rangel-Yagui C, Madalena P, Mariana K, Obreque T, Marcello C, Barbosa V, Monteiro G, Helena S, Farsky P, De Oliveira AM, Angeli CB, Ventura M, Pessoa-Junior A, Id CDOR (2019) Novel site-specific PEGylated L-asparaginase. PLoS One 14:1–19

Mirzaei H, Kazemi B, Bandehpour M, Shoari A, Asgary V, Ardestani MS, Madadkar-Sobhani A, Cohan RA (2016) Computational and nonglycosylated systems: a simpler approach for development of nanosized PEGylated proteins. Drug Des Devel Ther 10:1193–1200. https://doi.org/10.2147/DDDT.S98323

Moghimi SM, Hunter AC, Murray JC (2001) Long-circulating and target-specific nanoparticles: theory to practice. Pharmacol Rev 53:283–318

Movileanu C, Anghelache M, Turtoi M, Voicu G, Neacsu IA, Ficai D, Trusca R, Oprea O, Ficai A, Andronescu E (2022) Folic acid-decorated PEGylated magnetite nanoparticles as efficient drug carriers to tumor cells overexpressing folic acid receptor. Int J Pharm 625:122064

Nag M, Gajbhiye V, Kesharwani P, Jain NK (2016) Transferrin functionalized chitosan-PEG nanoparticles for targeted delivery of paclitaxel to cancer cells. Colloids Surfaces B Biointerfaces 148:363–370

Oberli MA, Schoellhammer CM, Langer R, Blankschtein D (2016) Future Challenges 5:843–857. https://doi.org/10.4155/tde.14.32.Ultrasound-enhanced

Patil HP, Freches D, Karmani L, Duncan GA, Ucakar B, Suk JS, Hanes J, Gallez B, Vanbever R (2018) Fate of PEGylated antibody fragments following delivery to the lungs: influence of delivery site, PEG size and lung inflammation. J Control Release 272:62–71. https://doi.org/10.1016/j.jconrel.2017.12.009

Press D (2017) Docetaxel-loaded PLGA and PLGA-PEG nanoparticles for intravenous application: pharmacokinetics and biodistribution profile. Int J Nanomedicine 12:935

Qin X, He L, Fan D, Liang W, Wang Q, Fang J (2021a) Title page key Laboratory of Advanced Technologies of materials, Ministry of Education. Asian J Pharm Sci 3:1. https://doi.org/10.1016/j.ajps.2021.03.001

Qin X, He L, Fan D, Liang W, Wang Q, Fang J (2021b) Targeting the resolution pathway of inflammation using Ac2–26 peptide-loaded PEGylated lipid nanoparticles for the remission of rheumatoid arthritis. Asian J Pharm Sci 16:483–493. https://doi.org/10.1016/j.ajps.2021.03.001

Qin L, Cui Z, Wu Y, Wang H, Zhang X, Guan J, Mao S (2022) Challenges and strategies to enhance the systemic absorption of inhaled peptides and proteins. Pharm Res 40:1037. https://doi.org/10.1007/s11095-022-03435-3

Qiu Y, Man RCH, Liao Q, Kung KLK, Chow MYT, Lam JKW (2019) E ff ective mRNA pulmonary delivery by dry powder formulation of PEGylated synthetic KL4 peptide. J Control Release 314:102–115. https://doi.org/10.1016/j.jconrel.2019.10.026

Rattan R, Bhattacharjee S, Zong H, Swain C, Siddiqui MA, Visovatti SH, Kanthi Y, Desai S, Pinsky DJ, Goonewardena SN (2017) Nanoparticle-macrophage interactions: a balance between clearance and cell-specific targeting. Bioorg Med Chem 25:4487–4496

Ryan SM, Mantovani G, Wang X, Haddleton DM, Brayden DJ (2008) Advances in PEGylation of important biotech molecules: delivery aspects. Expert Opin Drug Deliv 5:371–383. https://doi.org/10.1517/17425247.5.4.371

Saez-martinez V, Olalde B, Martinez-redondo D, Braceras I, Morin F (2014) J Bioactive Compatible Polym Biomed Appl 29:270. https://doi.org/10.1177/0883911514528597

Sahoo RK, Gothwal A, Rani S, Nakhate KT, Ajazuddin Gupta U (2020) PEGylated dendrimer mediated delivery of Bortezomib: drug conjugation versus encapsulation. Int J Pharm 584:119389. https://doi.org/10.1016/j.ijpharm.2020.119389

Salave S, Shinde SD, Rana D, Sahu B, Kumar H, Patel R, Benival D, Kommineni N (2023) Peptide engraftment on PEGylated Nanoliposomes for bone specific delivery of PTH (1-34) in osteoporosis. Pharmaceutics 15:608. https://doi.org/10.3390/pharmaceutics15020608

Sarhadi S, Moosavian SA, Mashreghi M, Rahiman N, Golmohamadzadeh S, Tafaghodi M, Sadri K, Chamani J, Jaafari MR (2022) B12-functionalized PEGylated liposomes for the oral delivery of insulin: in vitro and in vivo studies. J Drug Deliv Sci Technol 69:103141

Sasayama Y, Hasegawa M, Taguchi E, Kubota K, Kuboyama T, Naoi T, Yabuuchi H, Shimai N, Asano M, Tokunaga A, Ishii T, Enokizono J (2019) In vivo activation of PEGylated long circulating lipid nanoparticle to achieve efficient siRNA delivery and target gene knock down in solid tumors. J Control Release 311–312:245–256. https://doi.org/10.1016/j.jconrel.2019.09.004

Sebak AYAA (2018) Limitations of Pegylated nanocarriers: unfavourable physicochemical properties. Biodistribution Patterns Cell Subcell Fates 10:6–12

Shekhawat R, Shah CK, Patel A, Srinivasan S, Kapoor P, Patel S, Kumar S, Sonar S, More N, Joshi M, Patel J, Vachhani M, Kodaganti BP, Choavatiya U, Pushpaja A, Argade S, Nuwal N, Kumar M, Khambhampaty S (2019) Structural similarity, characterization of poly ethylene glycol linkage and identification of product related variants in biosimilar pegfilgrastim. PLoS One 14:e0212622. https://doi.org/10.1371/journal.pone.0212622

Shi L, Zhang J, Zhao M, Tang S, Cheng X, Zhang W, Li W, Liu X, Peng H, Wang Q (2021) Effects of polyethylene glycol on the surface of nanoparticles for targeted drug delivery. Nanoscale 13:10748–10764. https://doi.org/10.1039/d1nr02065j

Sun C, Ding Y, Zhou L, Shi D, Sun L, Webster TJ (2017) Noninvasive nanoparticle strategies for brain tumor targeting. Nanomedicine nanotechnology. Biol Med 13:2605–2621. https://doi.org/10.1016/j.nano.2017.07.009

Tang AX, Sun J, Ge T, Zhang K (2018) PT SC. Colloids Surfaces B Biointerfaces. 172:26. https://doi.org/10.1016/j.colsurfb.2018.08.022

Tapia-arellano A, Gallardo-toledo E, Ortiz C, Araya E, Sierpe R, Henríquez J, Feij CG, Kogan MJ (2021) Materials Science & Engineering C Functionalization with PEG / Angiopep-2 peptide to improve the delivery of gold nanoprisms to central nervous system: in vitro and in vivo studies. Mater Sci Eng 121:111785. https://doi.org/10.1016/j.msec.2020.111785

Ulbricht J, Jordan R, Luxenhofer R (2014) Biomaterials On the biodegradability of polyethylene glycol , polypeptoids and poly ( 2-oxazoline ) s. Biomaterials 35:4848–4861. https://doi.org/10.1016/j.biomaterials.2014.02.029

Vandenbroucke RE, Lentacker I, Demeester J, De Smedt SC, Sanders NN (2008) Ultrasound assisted siRNA delivery using PEG-siPlex loaded microbubbles. J Control Release 126:265–273. https://doi.org/10.1016/j.jconrel.2007.12.001

Wan X, Zhang J, Yu W, Shen L, Ji S, Hu T (2016) Effect of protein immunogenicity and PEG size and branching on the anti-PEG immune response to PEGylated proteins. Process Biochem 52:183. https://doi.org/10.1016/j.procbio.2016.09.029

Wang L, Wang N, Zhang W, Cheng X, Yan Z, Shao G, Wang X, Wang R, Fu C (2022) Therapeutic peptides: current applications and future directions. Signal Transduct Targeted Ther 7:48. https://doi.org/10.1038/s41392-022-00904-4

Wiegandt A, Meyer B (2014) Unambiguous characterization of N-glycans of monoclonal antibody cetuximab by integration of LC-MS/MS and 1H NMR spectroscopy. Ana Chem 86(10):4807–4814

Xu J, Gattacceca F, Amiji M, De Pharmacie F, Montpellier U, Flahault C (2013) Biodistribution and pharmacokinetics of EGFR-targeted Thiolated gelatin nanoparticles following systemic Administration in Pancreatic Tumor-Bearing Mice, vol 10, p 2031

Xu D, Smolin N, Shaw RK, Battey SR, Tao A, Huang Y, Rahman SE, Caylor ML (2018) Molecular insights into the improved clinical performance of PEGylated interferon therapeutics: a molecular dynamics perspective. RSC Adv 8:2315–2322. https://doi.org/10.1039/c7ra12480e

Yamazoe E, Fang J, Tahara K (2020) Oral mucus-penetrating PEGylated liposomes to improve drug absorption: differences in the interaction mechanisms of a mucoadhesive liposome. Int J Pharm 120148:120148. https://doi.org/10.1016/j.ijpharm.2020.120148

Yazdi JR, Tafaghodi M, Sadri K, Mashreghi M, Nikpoor AR, Nikoofal-Sahlabadi S, Chamani J, Vakili R, Moosavian SA, Jaafari MR (2020) Folate targeted PEGylated liposomes for the oral delivery of insulin: in vitro and in vivo studies. Colloids Surfaces B Biointerfaces 194:111203. https://doi.org/10.1016/j.colsurfb.2020.111203

Youn YS, Kwon MJ, Na DH, Chae SY, Lee S, Lee KC (2008) Improved intrapulmonary delivery of site-specific PEGylated salmon calcitonin: optimization by PEG size selection. J Control Release 125:68–75. https://doi.org/10.1016/j.jconrel.2007.10.008

Zhang F, Liu M, Wan H, Anti-peg THE, Peg HTO (2014) Discussion about several potential drawbacks of PEGylated therapeutic. Proteins 37:335–339

Zhang S, Gao C, Lü S, He J, Liu M, Wu C (2017) Colloids and surfaces B : biointerfaces synthesis of PEGylated polyglutamic acid peptide dendrimer and its application in dissolving thrombus. Colloids Surfaces B Biointerfaces 159:284–292. https://doi.org/10.1016/j.colsurfb.2017.08.009

Zhou Y, Peng Z, Seven ES, Leblanc RM (2017) Crossing the blood-brain barrier with nanoparticles. J Control Release 270:290. https://doi.org/10.1016/j.jconrel.2017.12.015

# PEGylation of Therapeutic Proteins and Peptides

# 11

Natasha Akojwar, Ankit Mishra, Pranali Mishra, Muktika Tekade, Shubham Ramdas Mule, and Rakesh Kumar Tekade

**Abstract**

PEGylation refers to the technique of modifying protein and peptide drugs with PEG, which offers numerous advantages over their unmodified counterparts in terms of physicochemical properties and therapeutic efficacy. This book chapter

N. Akojwar
Department of Pharmaceutical Sciences, Rashtrasant Tukdoji Maharaj Nagpur University, Nagpur, MS, India

A. Mishra (✉)
Department of Pharmaceutics, VNS Group of Institutions, Faculty of Pharmacy, Bhopal, MP, India

P. Mishra
Department of Pharmaceutics, VNS Group of Institutions, Faculty of Pharmacy, Bhopal, MP, India

Swami Vivekanand College of Pharmacy, Bhopal, MP, India

M. Tekade
School of Pharmacy, Devi Ahilya Vishwavidyalaya, Takshila Campus, Indore, Madhya Pradesh, India

School of Pharmaceutical Sciences, Lovely Professional University, Phagwara, Punjab, India

S. R. Mule
National Institute of Pharmaceutical Education and Research (NIPER), Ahmedabad, An Institute of National Importance, Government of India, Department of Pharmaceuticals, Ministry of Chemicals and Fertilizers, Palaj, Opp. Air Force Station, Gandhinagar, Gujarat, India

R. K. Tekade
National Institute of Pharmaceutical Education and Research (NIPER) Ahmedabad, An Institute of National Importance, Government of India, Department of Pharmaceuticals, Ministry of Chemicals and Fertilizers, Palaj, Opp. Air force station, Gandhinagar, Gujarat, India


R. K. Tekade, N. K. Jain (eds.), *PEGylated Nanocarriers in Medicine and Pharmacy*, https://doi.org/10.1007/978-981-97-7316-9_11

aims to equip the readers with detailed insights into objectives, methods, and end applications of PEGylation. The chemistry-related aspects of PEGylation, with special emphasis on the site of PEGylation, is the central theme of this piece of literature. Various critical factors affecting the process of PEGylation and its outcomes, such as the structure of PEG and size of PEG, have been emphasized in this manuscript. The applications of PEGylation in active and passive targeting and various FDA-approved PEGylated products are thoroughly discussed.

**Keywords**

PEG · PEGylation · Proteins · Peptides · PK profiling

## 11.1 Introduction

The advancements in the modern technologies of production, modification, and analytical advancements have helped in the development of protein and peptide drugs (PPDs) significantly during the past 10 years (Henninot et al. 2018). Countless therapeutic applications for natural and synthetic peptides have been discovered and explored. The application of PPDs as therapeutics has been viewed as an appealing strategy to tackle numerous diseases since the isolation of insulin in 1922 (Bliss 1982). Recent advancements in biotechnology and pharmaceutical sciences, especially recombinant technology, have made it feasible to develop PPDs commercially as promising therapeutics (Gupta et al. 2016; Seif et al. 2017). Protein and peptide medications are intriguing biological molecules in drug research because they offer great efficacy, minimal toxicity, and strong specificity compared to conventional chemically manufactured pharmaceuticals (Dimitrov 2012).

By far, the FDA has approved more than 240 PPDs. Other potential therapeutic PPD candidates undergo clinical trials (Chen et al. 2022). PPDs have been synthesized and modified utilizing both biological and chemical techniques. Although protein and peptide medications offer unrivaled benefits in treating disease conditions at the cellular and extracellular levels, they are not the ideal treatment choice owing to their molecular makeup. These macro-molecules are difficult to administer because of their large size, complex structure, hydrophilicity, immunogenicity, and susceptibility to enzymatic cleavage, leading to shorter half-life necessitating frequent administration (Joseph et al. 2017).

Researchers attempt various approaches to reconfigure these PPDs to address the above issues, and many of these techniques can drastically enhance their stability and safety and improve their pharmacokinetic profile while lowering the dose and frequency of administration. The novel design and delivery technologies assisted in overcoming the inherent limitations of PPDs, permitting the continuous expansion of this industry. The following section will review PEGylation as one of the strategies researchers employ to combat the shortcomings of protein and peptide drugs.

### 11.1.1 Clinical and Research Status of PPDs

From the 1980s to the 1990s, peptide products with 2–10 amino acids comprise a large portion of the amino acid composition of polypeptide drugs. However, as peptide synthesis and purification technology have advanced in the twenty-first century, polypeptide products with more than ten amino acids have gradually emerged, and the myriad of polypeptide drugs has expanded (Lau and Dunn 2018). According to global industry research on peptide treatments, peptide medicine sales surpassed 70 billion USD in 2019; this research forecasted a compound annual growth rate (CAGR) reaching 9.1% from 2016 to 2024. Over 100 peptide therapeutics have entered clinical trials, over 400 peptide-based drugs are in the pre-clinical research phase, and more than 70 peptide drug types have been certified globally for commercialization (Lee et al. 2019). On studying the intended uses of these licensed peptide medications, it appears that the most frequently targeted disease groups are cancer and metabolic disorders. It can be said that the robust expansion of this industry is probably due to the anticipated rise in the prevalence of malignancies and metabolic diseases (Li et al. 2021; Cheetham et al. 2016). Liraglutide (Victoza) and glucagon-like peptide 1 (GLP-1) are top-selling peptide medications for metabolic disorders, total sales of at least two billion USD annually. Over four billion USD in sales were also attributed to well-known peptide medications like leuprolide (Lupron), gosarelin (Zoladex), and somatostatin analogs like octreotide and lanreotide (Ibeanu et al. 2020).

Correspondingly, there is an increase in the variety of protein and peptide therapeutic molecular targets. G-protein-coupled receptors (GPCRs) are the most prominent extracellular targets, preceded by the natriuretic peptide receptor and cytokine receptor family. These three receptor families are the primary targets of 70–90% of the peptide medicines tested in clinical studies (Sriram and Insel 2018; Davenport et al. 2020). Thus, drug research currently focuses on identifying the role of proteins and peptides in the etiology of a disease; these proteins include enzymes, viral proteins, receptors, and channel molecules. Therefore, conducting an in-depth analysis of protein and peptide therapeutic targets is crucial for developing these novel medicines (Lagassé et al. 2017).

### 11.1.2 Factors Influencing PPDs' Short-Term Efficiency

Major drawbacks of these peptide and protein therapies include limited bioavailability and higher metabolic sensitivity. The GI tract's digestion of peptides and their failure to cross the epithelium limits their oral bioavailability (Alqahtani et al. 2019). These PPDs frequently have high molecular weights (MWs), minimal lipophilicity, and charged moiety within their structure, all hindering absorption (Han et al. 2019). These properties cause most orally given peptides to have low bioavailability (2%) and brief half-lives (30 min). The problem of absorption is resolved by intravenous (IV) or subcutaneous (SC) delivery of these therapeutic agents, but other factors, such as systemic proteolytic enzymes, rapid clearance, opsonization, structural rearrangements, separation of subunit protein molecules, complex formation with blood components, and destabilization of susceptible side-groups, limit the bioavailability of these therapeutics (Craik et al. 2013). Another challenge is that the PPDs are difficult to manufacture and store because of poor stability (Bruno et al. 2013).

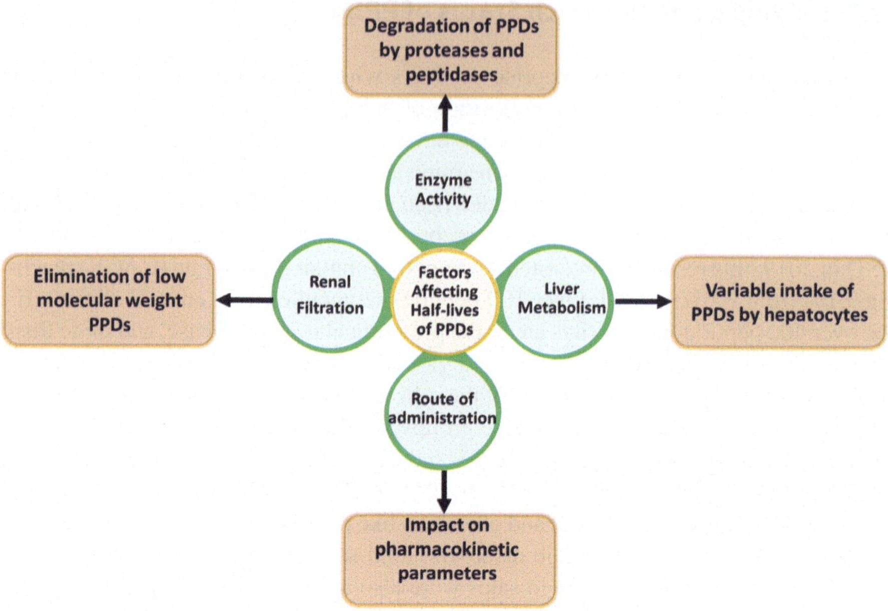


**Fig. 11.1** Various factors affecting the half-lives of protein and peptide drugs

Furthermore, every protein therapy has the potential to cause immunogenicity complications in patients, and antibody formation is one such concern that poses serious safety issues and lowers the efficacy of PPDs (Szlachcic et al. 2011; Fu et al. 2020). Above all, the short half-lives of the PPDs limit their potential application. The most common causes of the short half-life of PPDs are depicted in Fig. 11.1. The hepatic and renal tissues and bloodstream all contain significant numbers of peptidases and proteases, which cause the peptide or protein to be swiftly broken down by the enzyme's function. Renal filtration, on the other hand, leads to quick elimination of PPDs with low molecular weights. Liver metabolism is another factor influencing the half-life of PPDs. The particular sequence of amino acids of a polypeptide chain also significantly impacts their hepatocyte intake and the rate at which the liver metabolizes them; this uptake is a crucial step as the liver largely metabolizes them. Various drug delivery systems and administration routes may impact the drugs' pharmacokinetics (ADME) and metabolic processes, ultimately affecting the half-life of PPDs (Cao et al. 2020).

## 11.2 What Is PEGylation?

PEGylation is the non-covalent or covalent modification of other molecules utilizing polyethylene-glycol (PEG). This PEGylation produces interesting outcomes in PPDs. It has produced a revolutionary product line that has already hit the market, and others will be accessible soon. PEGylation enables the therapeutic use of PPDs with poor pharmacokinetics to be clinically applicable (Pasut and Veronese 2006).

Frank Davis's laboratories conducted early research on PEGylated proteins and enzymes in the late 1970s. Researchers today established the groundwork for protein PEGylation as a targeted delivery mechanism. In a series of experiments by Davis et al., cyanuric chloride was used to couple a methoxy-PEG to the amino acids in proteins. Their research showed that PEGylated proteins exhibited lower immunogenicity and longer circulatory half-lives. PEGylation has been a widely used technique in targeted drug delivery systems due to the recently discovered use of proteins in therapeutics (Dozier and Distefano 2015).

In PEGylation, proteins, peptides, antibodies, and vesicles are biochemically modified using PEG to give them numerous desired qualities that are useful for cell genetic modification or therapy. Nevertheless, PEGylation of proteins is a complicated procedure that can be accomplished using a variety of approaches (enzymatic, entrapping, physical, and chemical methods), depending on the type of the protein and the intended application. Inert PEG strings are covalently or non-covalently complexed with proteins of interest. Several PEGylated medications are used for conditions such as anemia, kidney dysfunction, multiple sclerosis, hemophilia, and malignancies (Awwad et al. 2018).

The increased hydrodynamic volume of the PEG–protein conjugate, which is a result of the potential of PEG to co-ordinate water molecules as well as from the flexibility of the PEG chain, can be used to explain why the physicochemical properties of a protein have significantly improved after coupling of PEG to a protein. As a result, the PEG protein conjugate has an apparent molecular weight that is roughly five to ten times more than the protein's nominal molecular weight (Gupta et al. 2019). Besides, PEG chains can wrap around a protein to cloak and protect it from exposure to the environment. Still, they can also affect how the protein interacts with other proteins, which affects how well the protein performs its biological function. Typically, in biological systems, these adverse impacts are counterbalanced by important positive consequences they offer, viz., enhanced blood circulation, increased drug bioavailability by evading the digestive system, reduced toxicity, decreased immunogenicity, and enhanced passive drug delivery (Jevsevar et al. 2010). Figure 11.2 gives detailed insights into the process of PEGylation.

### 11.2.1 Ideal Properties of PEG

PEG are inert, long-chain amphiphilic molecules synthesized by connecting ethylene oxide repeating units. Many possible PEG molecules can be made in various arrangements, such as linear or branched architectures, and with different molecular weights. The immunologic, pharmacokinetic, and pharmacodynamic characteristics of a target protein are altered by PEGylation, which can expand the protein's possible applications (Harris et al. 2001). Although PEG is typically considered a non-biodegradable polymer, certain studies have shown that it can be oxidatively broken down by various enzymes, including alcohol and aldehyde dehydrogenases and CYP450-dependent oxidases (Beranová et al. 1990). Reports presented that Alcohol dehydrogenases convert PEG chains under 400 Da in vivo into harmful metabolites. The elimination of longer PEG chains, utilized for PEGylate proteins, depends on their molecular mass

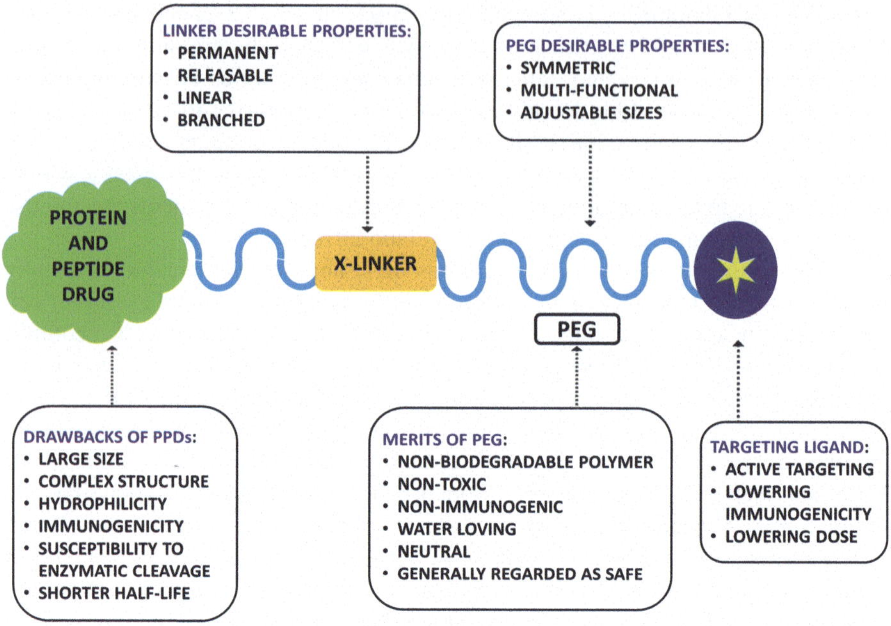


**Fig. 11.2** Overview of the process of PEGylation

and is not susceptible to metabolism. Renal filtration eliminates PEG and PEG–protein complexes with PEGs smaller than 20 kDa. In contrast, protein complexes with bigger PEG molecules are eliminated from the body through different pathways, such as hepatic uptake, immune system interaction, and proteolytic metabolism of the protein component of the conjugate (Caliceti and Veronese 2003).

PEG has a longstanding experience of being a non-toxic, non-immunogenic, water-loving, neutral, and non-degradable polymer and has been given the "generally regarded as safe" certification by the US FDA (Alconcel et al. 2011). PEGs and PEG reagents with wide polydispersity have been utilized; today, the acceptable standard for PEG reagents up to 30 kDa is polydispersity indices of about 1.05 (Wang et al. 2020a). A polydispersity around 1.1 might be appropriate for higher molecular weight forms, but the current trend is to use PEGs with narrower distributions. High doses of PEG as an intravenous therapeutic agent have been demonstrated to cause dose-dependent and molecular mass-dependent systemic immune toxic reactions. PEGylated therapies utilize 10,000 to 1000-fold lower amounts of PEG. Hence, therapeutic doses of PEGylated medications are generally significantly lower than those needed to cause PEG toxicity (Webster et al. 2009).

In toxicity studies, it has been shown that extremely high dosages of PEG–protein conjugates can cause nephron vacuolization, which is not accompanied by functional problems and goes away after the therapy (Rudmann et al. 2013). As a result, PEG–protein conjugates continue to be considered non-toxic, safe, and immunologically harmless. It has also been shown that the attachment of longer, branching PEGs is preferable to shorter, linear PEGs for reducing possible protein immunogenicity. PEG's minimal immunogenicity and the comparatively low doses of PEG-conjugates greatly limit the chance of an immunogenic reaction (Qi and Chilkoti 2015).

### 11.2.2 Advantages of PEGylation

The PPD's physicochemical characteristics are changed due to PEGylation, including its conformational, steric impediment, electrostatic interaction capabilities, and lipophilicity. The improvement in the therapeutic efficiency that PEGylation offers because of the several advantages over non-PEGylated products is summed up in this section. Figure 11.3 enlists the benefits of PEGylation. PEGylation expands the size and molecular mass of PPDs, boosts their water solubility, and improves their pharmacokinetics and pharmacodynamics. PEG polymer interacts with two or more water molecules for every ethylene glycol subunit when coupled to a protein (Delgado et al. 1992). This property produces a bulky hydrophilic shielding that effectively hides the conjugated protein from immune detection and enzymatic breakdown.

Additionally, the conjugated protein is five to ten times bigger and more soluble than an equivalent soluble protein of the same mass. The shield produced by the PEG moiety and linked water molecules significantly reduces renal clearance (Milla et al. 2012). This last characteristic seems to be especially beneficial for smaller proteins and peptides, which the kidneys filter quickly. To lengthen the body's residence period, PEGylation with a PEG mass of 40–50 kDa can be accomplished by conjugating either a single large PEG molecule or numerous smaller PEG molecules (Vargason et al. 2021).

PEGylation increases the plasma half-life of proteins and prevents phagocytic system clearance by producing a steric obstacle to opsonization. Proteases are hindered by PEG, which increases the proteins' resistance to proteolytic breakdown. PEGylation can maintain the protein structure and function this way (Veronese 2001). Additionally, PEGylation can mask antigenic regions of foreign proteins, preventing the development of antibodies directed against the same, thereby lowering their immunogenicity (Knop et al. 2010). Adenosine deaminase became the first PEGylated PPD commercialized and licensed by the FDA to manage immunodeficiency illnesses in 1990. Ever since, numerous conjugates have received approval

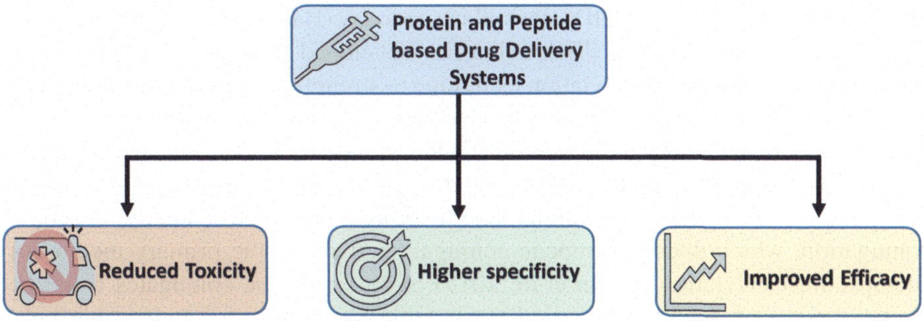


**Fig. 11.3** Advantages of PEGylation of proteins and peptides

and gone on the market, including epotein protein for treating renal failure, PEGylated interferon for treating hepatitis C, and granulated stimulating factor for treating neutropenia (Villegas et al. 2018).

Shorter half-lives in vivo, which lead to much lower potency as demonstrated in vitro, limit the biological agent's incredible potential. These frequently needed short-acting therapeutic drugs, specifically for chronic illnesses, can limit their application in the clinic. To enable better biological treatments, half-life augmentation approaches are now available in clinical settings (Swierczewska et al. 2015). The most well-established half-life extension technique today is PEGylation, which has been safe in humans for over two decades. It also helps reduce the dose and frequency of application, enhancing patient compliance. Nevertheless, it will be among the most cutting-edge systems used during the next 20 years.

### 11.2.3 Generations of PEGylation Process

The key trait of low molecular weight linear PEGs defines the first-generation PEG-conjugated therapeutic proteins. The activation of PEG is required before it is conjugated to a protein. This activation is done utilizing a functional group located on one of its terminal sides appropriate for interaction with an available active site on the protein (typically the amino groups). This activation could be achieved by reacting the PEG with a proper initiator or termination reagent (Swierczewska et al. 2015). Because of the availability to create reactive PEGs that stop the synthesis of cross-linked polypeptides, monomethoxy PEG (mPEG) is more appropriate for conjugating proteins (Monfardini et al. 1995). It is hypothesized that the PEG's amphiphilic property, ability to bind to water molecules, and flexibility of the backbone chain are related to its capacity to induce precipitation of proteins, reduce immunogenicity and antigenicity, keep proteins and cells away from surfaces, and prevent degradation by enzyme activity or mammalian cells.

Furthermore, irreversible conjugation is necessary to achieve the primary goal of first-generation PEGylation. Due to the removal of the PEG coating, the method's widespread use is restricted. Regulators approved several first-generation pegylated medications despite these restrictions. Pegaspargase (Oncaspar), a pegylated form of the enzyme asparaginase employed for leukemia, and pegademase (Adagen), a pegylated derivative of the enzyme adenosine deaminase for the treatment of severe combined immunodeficiency disease (SCID), are still in use today (Harris and Chess 2003).

The second generation of PEGylation has developed from the first generation, and it uses a site-specific PEGylation technique to strengthen the specificity between PEG with different protein moieties. Chemically stable bonds are indeed site-specifically created during covalent PEGylation, which is a most desirable and potentially economical method. Some techniques used to carry out site-specific PEGylation include thiol, N-terminal, and histidine markers. Constantly reversible conjugation, which does not impede conjugate activity, is the primary method of site-specific PEGylation (Belén et al. 2019). As a result, the conjugates could be liberated per a predetermined period attributable to the use of the cleavable linkages, which enable temporary anchoring of PEG molecules. However, a covalent approach has significant drawbacks; it may not always be practical and could take the longest to develop.

Like random PEGylation, multiple target-specific sites in the molecule may also form PEGylated species with different degrees of modification and isomerism positions (Zuma et al. 2022). Creating larger PEG polymers is a general objective of second-generation PEGylation techniques to enhance the pharmacokinetic and pharmacodynamic consequences observed following relatively low molecular mass PEGs. Some of the changes are significant, such as the PEGylation of interleukin-6 (IL-6), which caused a 100-fold increase in half-life and a 500-fold increase in thrombopoietin activity (Harris et al. 2001).

Using branching structures instead of the only linear structures present in first-generation PEGs is another advancement in second-generation PEG polymers. There have been developed branched PEGs with significantly higher molecular weights (up to 60 kDa or above) than the 12 kDa or fewer observed in first-generation PEGs. A branching PEG is much larger than a linear PEG with the same molecular mass (Fee 2007).

Additionally, branched PEGs are more effective at shielding the connected polypeptide drug again from immune cells and digestive enzymes, lowering its antigenicity and reducing the probability that it will be destroyed. To attain better potency and a circulation half-life focused on fast-acting, site-specific, and lower doses, third-generation PEGylation is being developed (Kursa et al. 2003). With an aim to establish longer-acting therapeutics, third-generation PEGylation seeks to reduce the struggle to find a balance between efficiency and circulatory half-life. Releasable PEG conjugates, a prodrug strategy invented by Enzon Pharmaceuticals, are one method. After being administered in vivo, the medication can be converted to its active form owing to specially designed linkers between PPD active molecules and PEG. Another strategy to reduce the API's steric barrier is constructing modifiable PEGylation regions on proteins using technologies like Ambrx's ReCODE and EuCODE to support therapeutic candidates with problematic PK characteristics (Manandhar et al. 2021).

## 11.3 Chemistry of PEGylation

Proteins and appropriately activated PEGylation reagents often interact chemically to PEGylate proteins. There are many chemical groups on the side chains of amino acids that might feasibly be used for the interaction with PEG, including disulfide (-S-S-) bonds, -NH2, -NH-, -COOH, -OH, and -SH groups. The location of attachment upon the protein, the nature of bonding, the size and form of the linker, and the PEG reagent are all factors that must be considered when discussing PEGylation, particularly when considering their modified properties (Zalipsky 1995).

### 11.3.1 Random PEGylation

In the 1970s, the historical background of "random" PEGylation reagents commenced with PEG-chlorotriazine. It proceeded with PEG-benzotriazole carbonate,

PEG-tosylate, PEG-nitrophenyl carbonate, PEG-carbonyl imidazole, succinimidyl succinate (SS-PEG), as well as succinimidyl carbonate PEG reagents (SC-PEGs). Although SS-PEG reagents form strong amide bonds with protein, the subsequent PEG-protein adducts are vulnerable to hydrolysis because of an additional ester linkage in the polymer's backbone. In addition to reacting with lysine residues, SC-PEG reagents also react with histidine residues, forming urethane linkages, which lead to hydrolytically weak bonds (Zalipsky et al. 1992). The weak connection could be advantageous when designing control release or prodrug compositions, but if conjugation instability is not desired, it could be a severe drawback (Warren et al. 2012).

Looking back at the development of PEGylation, it is apparent that, until recently, the preponderance of PEGylation reagents primarily targeted the amino groups on side chains of lysine residues in proteins, which is considered random PEGylation. Lysines are relatively numerous, polar amino acid residues typically found on the surface of proteins, making them susceptible to chemical interactions with PEG reagents. As a result, these reactions proceed rapidly and produce complex mixes of conjugates that vary in the quantity and location of the attached PEG chains (Cheng et al. 2019). These reactions are not confined to the amino groups of lysine residues; they also somewhat react with certain other protein nucleophiles, such as the side chains of serine, threonine, and tyrosine cysteine residues, the N-terminal amino groups and the imidazolyl nitrogens of histidine residues. Although the pH of the medium can influence the reaction to some degree, these conjugation processes typically result in complex PEGylation mixes (Maiser et al. 2014). Also, first-generation high molecular weight, pure monofunctional PEG reagents are challenging to construct and ineffective for protein conjugation because they have a 15% maximum diol concentration (Zhang et al. 2007).

### 11.3.2 Site-Specific PEGylation

The issues of diol contaminants, confinement to low-molecular-weight mPEG, instability of links, side-reactions, and lack of substitution selectivity have been avoided by site-specific PEGylation chemistry (Xu et al. 2021). N-terminal, serine, threonine, tryptophan, and cysteine PEGylations are examples of traditional, well-known methods for site-specific PEGylation reactions discussed in the following section.

#### 11.3.2.1 Amine Conjugation

mPEG-propionaldehyde is one of the earliest instances of second-generation chemistry. Given the PEG acetaldehyde's susceptibility to dimerization, mPEG-propionaldehyde is simpler to manufacture and employ. Because the N-terminal alpha-amine has a lower pKa than other nucleophiles, mPEG-propionaldehyde is mainly selected for it under acidic circumstances (about pH 5). The nucleophilicity of every amino acid residue plays a critical role in attaching electrophilic PEG to amino acid residue on protein. Only after the pH of the solution containing protein is close to or higher than the residue's pKa will a nucleophilic attack occur. As a result, each residue's reactivity is also influenced by its nearby amino acid residues. The heterogeneity often observed with lysine chemistry is considerably reduced, even though total selectivity is not seen. Because amino groups in proteins have a comparable propensity to PEG amines and may thus form cross-linked aggregates, reductive alkylation involving PEG-amine is challenging. In these circumstances, PEG-hydrazides are more advantageous. The amino groups in proteins are mostly protonated in an acidic environment (approximate pH 5). Still, because PEG-hydrazide is a weaker base than primary amines (pKa, about 3), the reaction is selective for the synthesis of PEG-hydrazone (Wang et al. 2018). Using an N-terminal serine or threonine, which may be oxidized by periodate into a glyoxylic derivative, is another strategy for site-specific conjugation (Gaertner and Offord 1996).

#### 11.3.2.2 Cysteine Conjugation

Because agents that preferentially react with cysteines have been created and because there are far fewer free cysteine residues on the surface of proteins than there are lysine residues, PEGylation targeting free cysteine in proteins is the primary method for site-specific modification (Gaertner and Offord 1996). Genetic engineering can insert one or more free cysteines into native proteins that do not contain any. This method can enable site-specific PEGylation in protein regions, minimizing biological activity loss but reducing immunogenicity (Byrne et al. 2021). This tactic has several drawbacks, however. For instance, the possibility of improper disulfide generation and protein dimerization increases when free cysteines are inserted by genetic engineering (Gunnoo and Madder 2016).

Various PEG derivatives, each with unique benefits and drawbacks, have been produced using cysteine residues. Starting with PEG-vinyl sulfone (PEG-VS) at mildly basic circumstances (pH 7–8), PEG-VS slowly reacts with thiols to create a robust thioether bond with the protein. The reaction will occur more rapidly if the pH is raised. While stable in aqueous systems, PEG-VS may interact with lysine residues at high pH levels (Morpurgo et al. 1996).

Unlike PEG-VS, the PEG-maleimide (PEG-MAL) is unstable in water and can initiate ring-opening reactions or the addition of water molecules to the double bond. The thioether bond between the PEG-MAL and proteins is stable, although hydrolysis can slowly cleave one of the amide links. However, it is more active in thiols even in slightly acidic pH 6–7 (Fontaine et al. 2015). Through nucleophilic substitution, PEG-iodoacetamide (PEG-IA) progressively interacts with free thiols to form a persistent thioether bond. To prevent the production of free iodine, which might react with other amino acids, the reaction should be carried out in a dark container with a small molar excess of PEG-IA [74]. Lastly, to create a disulfide bond with the protein, ortho-pyridyl disulfide-PEG (PEG-OPSS) can be used to interact with sulfhydryl groups in both acidic and basic environments (pH 3–10). These disulfide links are stable, except when transformed to thiols in a reducing environment (Kunstelj et al. 2013).

#### 11.3.2.3 Serine, Threonine, and Tryptophan Conjugation

A periodate oxidation reaction can produce glyoxylic groups by targeting the N-terminal positions of serines and threonines. The susceptibility of 1, 2- amino alcohols towards periodate oxidation affects this process (Kolate et al. 2014). The N-terminal serine residue was previously subjected to a site-specific PEGylation procedure, which involved oxime ligation of aminooxy and hydrazide PEG derivatives and sodium peroxide for oxidation. Interleukin-8 (IL-8), granulocyte colony-stimulating factor (G-CSF), and IL-1r preserved their biological activity after PEGylation.

Strong reaction conditions, which frequently cause protein denaturation in conjugation protocols, or decreased reaction yields are the main problems. Much focus has recently been placed on the bioorthogonal alteration of the side chains of aromatic amino acids. Tyrosine, in particular, is a desirable target for protein modification due to its relative rarity (as opposed to certain other sensitive endogenous amino acids like lysine) and capacity for alteration without changing charge. The alkylation reaction of tyrosine residues catalyzed by palladium is one technique for modifying proteins. This process modifies PPDs in an aqueous medium at room temperature using electrophilic intermediates such as π-allyl produced via allylic acetate and carbamate predecessors (Tilley and Francis 2006).

Other strategies for tyrosine-targeted posttranslational modifications using aniline derivatives were also developed, which allows the tyrosine's phenol group to react as a nucleophile. The diazonium coupling procedure has been discovered, in which the phenol of tyrosine residues in proteins is coupled to a diazonium salt made from aniline derivatives. Tyrosine residues have been targeted by diazonium in the polymer coupling of medically necessary proteins and peptides (Jones et al. 2012). Another method is the Mannich-type reaction, which allows for highly selective tyrosine residue alteration at the ortho position on the phenol group. At a pH range of 5.5–6.5, Mannich-type reactions occur under gentle circumstances with millimolar reagent concentration. Both techniques use aniline derivatives to target tyrosine in milder, biocompatible, and metal-free environments (Szijj et al. 2020).

Peptides with N-terminal tryptophan residues can be changed using the Pictet–Spengler procedure with an aldehyde in a solution of glacial acetic acid. The creation of a stable C-C bond occurs in just one step as a result of this reaction, which involves the oxidation of the N-terminal amino group to imine and the cyclic condensation of an aldehyde and the α-amine as well as the indole functionality of a tryptophan residue (Turecek et al. 2016). For example, researchers created folic acid (FA) and monoclonal antibody (mAb) conjugates utilizing a tryptophan (Trp)-selective process, which produced relatively consistent results when compared to other approaches. The mAb-FA conjugates significantly killed cancer cells expressing the folate receptor, proving that the conjugates still carried out the Fc region's primary purpose (Tagawa et al. 2020).

#### 11.3.2.4 Enzymatic Tools for Selective PEGylation

Enzymes have special selective, specific, and catalytic properties. They have not been used to their full potential in chemistry up to this point, but enormous strides could be made with this method soon (Sato 2002). Regarding the application of enzymes for selective PEGylation, one method, namely transglutaminase, has produced significant outcomes and sparked several PEGylation investigations. Recently,

PEG chains were covalently linked at the glutamine (Gln) protein residue's carboxamide group using the transglutaminase (TGase) enzyme (Mero et al. 2009). TGases are an enzyme type that can catalyze an acyl transfer between two proteins using the glutamyl group of glutamine as the acyl donor and a primary amine as the acyl acceptor, typically the ε-amino of lysine. It is important to emphasize that glutamine cannot be altered chemically without tagging other residues or impairing the structure of the protein. Because TGase has stringent constraints for the amino acid sequence and substrate flexibility, this enzymatic conjugation is selective (Fontana et al. 2008). Thus, just one or two of the many glutamines in a protein typically meet the requirements for TGase catalysis.

A new method known as glycopegylation for site-directed PEGylation that attaches PEG to O-glycans using glycosyltransferases is proposed. Proteins produced in Escherichia coli without glycosylation undergo enzymatic GalNAc glycosylation at specified serine and threonine residues. Then, sialic acid conjugated with PEG is transferred enzymatically to the GalNAc residues. The method was used with three therapeutic polypeptides that are currently used in clinical settings: granulocyte colony-stimulating factor (G-CSF), interferon-alpha2b (IFN-alpha2b), and granulocyte/macrophage colony-stimulating factor (GM-CSF) (Defrees et al. 2006). Because between 80–90% of eukaryotic PPDs are acetylated at the N-termini, the specific modification of the α-amine employing subtiligase is a powerful method in proteomics to enrich novel N-termini originating from proteolytic recognition and fragmentation. This benefit might be utilized for the novel application of selectively attaching PEG-modified peptides and proteins to increase conjugation effectiveness (Weeks and Wells 2020). Butelase 1, a productive asparagine and aspartate-specific cysteine-ligase, was discovered in the medicinal plant *Clitoria ternatea*. Butelase 1 was only recently identified, yet it has already been used for various things, including protein engineering and modification (Nguyen et al. 2015). Other enzymes, such as tubulin tyrosine ligase, N-myristoyl transferase, biotin ligase, and lipoid acid ligase, have also been used for protein bioconjugation.

## 11.4 Critical Parameters to Consider for Protein PEGylation

### 11.4.1 Binding Affinity

PEGylation may have a range of effects on a protein's binding affinity, from an increase to a decrease to having no effect at all. For instance, PEGylated Trypsin and Amino Oxidase exhibit higher activity when used with low-molecular-weight substrates. These findings were attributed to either a positive change in the three-dimensional structure or an improved affinity for the substrates due to the altered microenvironment the amphiphilic PEG produced (Monfardini et al. 1995). Pegaptanib is an aptamer that targets vascular endothelial growth factor (VEGF). This protein causes pathological neovascularization of the eye and increased vascular permeability, contributing to a particular type of macular degeneration. The first anti-VEGF aptamer used to bind the protein displayed the typical aptamer drawback of a limited half-life caused by nuclease degradation and a fast kidney clearance rate. Conjugation to a 40 kDa branching PEG increased the plasma half-life and lengthened tissue retention while reducing binding affinity by a factor of four. PEGylation allows delayed vitreous humor diffusion, increasing effectiveness and lowering systemic exposure (Ng et al. 2006). A humanized antitumor necrosis factor (TNF) monoclonal antibody marketed as certolizumab pegol is a PEGylated Fab fragment that binds to and neutralizes membrane-bound or soluble membrane TNFa. In this instance, the conjugation with a branching PEG allows for the retention of recognition ability, a long blood circulation period, and perhaps even a decrease in immunogenicity.

Furthermore, because the Fab segment is targeted for site-specific PEGylation at a location distant from the antigen-binding site, the compound retains biologic activity while exhibiting the same affinity for TNF as the original antibody. PEG2 maleimide with a molecular weight of 40 kDa serves as the reagent for the conjugation, which takes place at a single, free thiol cysteine residue. In April 2008, the FDA authorized certolizumab pegol to treat people with moderate to severe Crohn's disease (Schreiber 2011). G120K, in which a lysine was put in place of glycine, was the first growth hormone mutant to be investigated as an antagonist. G120K, like the growth hormone itself, had a relatively short half-life but binds the growth hormone receptor with great affinity while blocking signal transmission. Its plasma circulation time was increased by performing a randomized conjugation with PEG 5 kDa. This conjugation resulted in a protein with a low affinity for the receptor and a half-life of roughly 100 hours. Further mutations were added to lessen the number of possible PEG binding sites to address this (Fishburn 2008).

### 11.4.2 Altered Biological Activity

PEGylation can occasionally change the biological features of the protein; for example, PEGylated cholesterol oxidase and esterase swap specificity from total cholesterol to HDL cholesterol (Sugiuchi et al. 1995). This concept has been exploited in developing a commercial kit for HDL cholesterol assay. After being PEGylated, IL-15 competes with other ligands for receptor binding (Pettit et al. 1997). Pegfilgrastim's self-regulating pharmacokinetics illustrates how PEGylation changes biological characteristics (Finck et al. 2020). Similarly, the PEGylation of growth hormone altered the activity from agonist to antagonist (Yowell and Blackwell 2002).

### 11.4.3 Physicochemical Modifications

PEGylation causes numerous physicochemical changes in biomolecules. Since PEG comprises ethylene oxide subunits that absorb three water molecules, the hydrodynamic volume has increased (Bailon and Won 2009). Furthermore, aggregation may result from charge alteration (such as acylation), which is more apparent when assessed after incubating at higher temperatures. Less aggregation occurs when the charge on the alpha-amino group is preserved after PEGylation. PEGylation may also conceal charges and glycosylation capabilities, reducing the ability of RES and hepatocytes to phagocytose PEG-induced epitope shielding on proteins may lessen immunogenicity and proteolytic breakdown (Baumann et al. 2014).

### 11.4.4 PEG Size

The only polymer used for several years was 5 kDa in size and was also end-capped by the methoxyl group and terminated by the hydroxyl group. Despite it becoming evident that the size of the polymer has a significant impact on assessing the biological behavior of PEG conjugates, this has not been the case for many years. This decision was made since the polymer is simple to activate, and conjugation may be carried out not simply by professional polymer scientists but also in biological laboratories (Plesner et al. 2011).

Also, it was understood that the mass of attached PEG was crucial in determining how long it remained in the blood. One large PEG chain, or several tiny PEG chains, can be added to the protein to achieve the necessary mass. Due to the strong polydispersity and considerable amount of diol contaminants in the samples, conjugation was troublesome for a very long time creating heterogeneity in the finished product (Wang et al. 2020a). Commercializing a purified, low polydisperse, activated, extremely high-weight polymer (30–40 kDa) that was far more ideal for medical applications helped solve this issue later in the 1990s. Pharmaceutical PEG should range in polydispersity from 1.01 for a 5 kDa product to 1.1 for a 50 kDa product, considering that the value is 1.00 for a monodisperse product.

### 11.4.5 PEG Structure

PEGs can be created in linear, branching, Y-shaped geometries, or multi-arm. It was discovered that the shape of the connected protein is just as crucial to its biological function as its weight or the binding chemistry (Veronese et al. 1997). The "Y"-shaped branched PEG substantially enhances the PEG shielding function on a protein surface, making it more successful at defending the conjugated protein against proteolytic enzymes and antibodies. The increased steric hindrance hinders the PEG from reaching the enzyme's active site or other less accessible locations vital to biological activity. Enzyme active sites are frequently hidden in clefts inaccessible to bulky molecules. By utilizing hindered branched PEGs, this property can be used to prevent the active site from being modified during the PEGylation procedure. Compared to linear PEGs, branching PEGs have limited access to buried sites due to their structure, which helps preserve the enzymatic function. Also, the "umbrella-like" design of branching PEGs shields the proteins from deterioration to a larger extent and provides a higher level of surface protection per point of attachment (Veronese et al. 1997).

## 11.5 Purification of PEGylated Proteins

PEGylated proteins must be purified to extract the end product from complex mixtures. It is necessary to separate the specific PEG–protein conjugate from unreacted proteins, proteins that have been over-PEGylated, unreacted PEG reagent, and additional reagents that will eventually be incorporated into the PEGylation mixture. Typically, one seeks to take advantage of variances in physicochemical features while constructing a bioseparation method. Variability in charge, hydrodynamic radii, hydrophobicity, and, in some circumstances, affinity is used for the separation of the target PEG-protein. The complexity of the PEGylation mixture often determines how effectively the purification procedure produces the necessary homogeneity of the product (Veronese et al. 1997).

Column chromatography, particularly size exclusion chromatography (SEC) and ion-exchange chromatography (IEC), is used to purify most PEGylated proteins. The small amount of sample that can be loaded and the low mobile-phase flow velocity are constraints on SEC, a technique focused on the solute difference in size. As a result, it is typically utilized for analytical separations. IEC is centered around charge differences, with uncharged PEG being able to protect the protein's intrinsic charge to a greater extent as PEGylation levels rise (Yoshimoto and Yamamoto 2012). Unfortunately, the separation speed is slowed down by the diffusion limitation of column-based IEC based on porous particle chromatography. PEGylated proteins typically have a binding capability on IEC media that is weaker than native proteins. Both charge shielding and an increase in hydrodynamic radius, which reduces diffusivity, have been used to explain this phenomenon. While this makes it simple to separate native proteins and PEGylated proteins practically, it is challenging to fractionate multiple PEGylated forms.

A purified protein sample that has been PEGylated poses two different categories of purifying difficulties. In the first, PEG-proteins are separated from other reaction products, such as but not exclusively unreacted PEGs and proteins. The second involves sub-fractionating PEG-proteins according to positional isomerism and the amount of PEGylation in both. Making such separations may seem simple, but several PEG polymer-related issues might make it more difficult (Fee and Van Alstine 2006).

## 11.6 PK Profiling

The two most crucial elements that affect the concentration of PEG-modified therapeutics in blood circulation are molecular weight and injection site. The elimination rates from the injection point are intraperitoneal (IP) > subcutaneously (SC) > intramuscularly (IM) [109]. The drug delivered as IM injection forms a reservoir at the injection site and progressively diffuses into the bloodstream; to a smaller degree, the same is true for the drug administered SC. As PEG's molecular mass increases, the elimination rate for SC and IM reduces. The pharmacokinetic characteristic of the PEG adducts is significantly influenced by the mode of delivery (Veronese and Mero 2008).

A two-compartment model of biodistribution suggests that the relatively small PEG translocates readily between circulation to extravascular cells and back again through diffusion. In contrast, large PEG translocates much more poorly (Hamidi et al. 2006). On the other hand, relatively high-weight PEG circulates in the blood for a prolonged period while urine clearance is lowered. Regardless of molecular weight, PEG accumulates in tissues and organs (such as muscle, skin, bone, and, to a greater extent, the liver). PEG uptake increases by the Küpffer cells when the weight reaches 50 kDa. Small and linear PEG are distributed all over the body with a high distribution volume, whereas branched PEG is distributed to a lesser extent, mainly in the liver and spleen (Witt et al. 2001).

The primary motivation for protein modification, especially PEGylation, is regulating protein PK properties, focusing on the elimination and half-life extension. Rodents, usually rats that are big enough to provide the time course sampling necessary for PK profile determination, are typically used for the initial screening of PK properties. A universal analytical approach for identifying the conjugate in complicated biological materials is required to accurately and satisfactorily quantify the concentration of the PEG protein conjugate over time in blood sera. Thus, the preferred technique continues to be the protein-specific ELISA, which is typically sold commercially and has antibodies targeted to the conjugated protein (Kozma et al. 2020). Since these ELISAs can only detect the protein portion, they are less sensitive to PEGylated proteins than their counterparts that are not PEGylated. This reduced sensitivity is one of their fundamental limitations.

Reduced sensitivity results from PEGs' tendency to conceal crucial amino acid positions for receptor binding, which reduces the strength of the interaction between the receptor and the target antibody. The affinity for the antibodies is typically reduced by large PEGs and multi-PEGs connected to protein, which causes a less steep dose–response curve. It is crucial to utilize the same pure PEG-conjugate for the standard plot to prevent errors in the concentration of PEG–protein conjugates that are determined. Thus, the standard curve obtained from purified PEG-conjugate should be used to compute the final quantities of PEG-conjugates in blood serum. Instead, the un-PEGylated protein should not be utilized as a comparison .

## 11.7 Passive and Active Targeting

The targeting mechanisms of polymeric therapeutic PPDs might be passive or active. This property is mainly exploited in the treatment of cancer. Passive targeting is demonstrated by the EPR-induced accumulation of PEG-modified drugs inside permeable tumors. EPR's passive targeting is enhanced by active targeting, which also increases selectivity. By receptor-mediated endocytosis, targeting moieties attached to the polymer backbone can further distinguish between healthy and malignant cells.

The effectiveness of passive targeted drug delivery techniques mostly depends on the concentration gradient between intracellular and extracellular areas. With leaky vasculature and inadequate lymphatic drainage, PEG conjugates take advantage of the increased permeation and retention (EPR) phenomenon demonstrated by tumors and accumulate in the pathophysiological microenvironment of tumor arteries (Greish 2007). This size-dependent effect cannot be explored with low-molecular-weight proteins that rapidly extravasate and cause systemic toxicity. PEGylation enhances solubility, molecular mass, size, and serum stability. PEGylation is regarded as one of the most remarkable techniques for the passive targeted delivery of anticancer treatments for all these reasons.

The idea behind active drug targeting is to couple drug molecules to targeting agents (antibodies, ligands, etc.) to specifically engage with the structures on the cell surface and deliver an anticancer agent where needed. The targeting and linker molecules on the prodrug govern how it will behave within the body. Also, based on the linker molecules, the drug can enter the tumor cell in one of two ways: either receptor-mediated endocytosis, which involves internalizing the entire prodrug for later destruction by the endosomal/lysosomal route, or receptor-independent internalization, which occurs following extracellular dissociation of the prodrug (Filpula and Zhao 2008). Figure 11.4 represents the mechanism of internalization of the PEGylated PPDs.

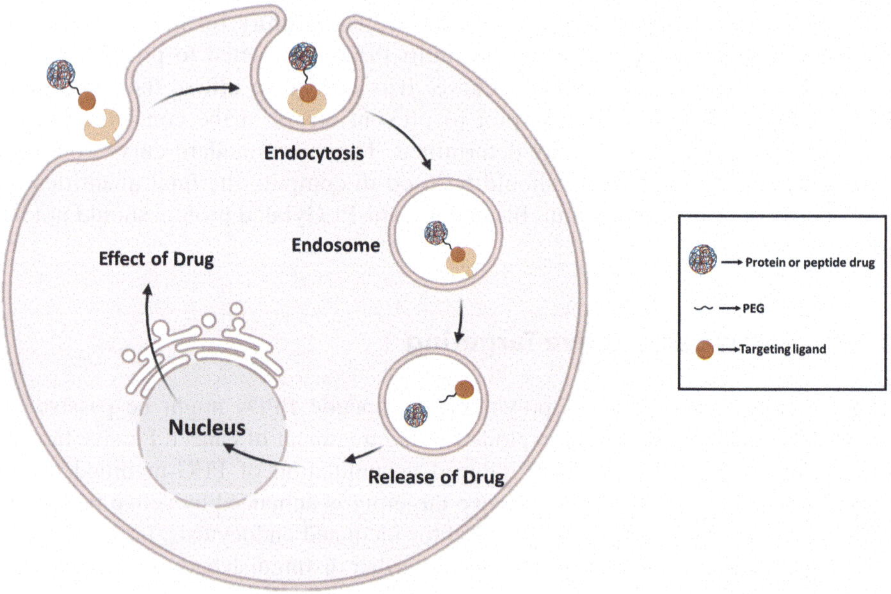


**Fig. 11.4** The mechanism of internalization of PEGylated PPDs

## 11.8 Analytical Methods for Characterization

Analysis techniques for PEG protein adducts and PEG reagents are needed at different phases of the development process to create a safe and effective PEGylated therapy. To successfully develop PEGylated therapies, properly characterizing PEG reagents, which serve as a critical raw material, is imperative. PEG reagent quality can vary significantly depending on molecular weights, polydispersity, impurity presence, and activation degree. NMR is often employed to determine the functional groups both qualitatively and quantitatively, and it is the release method chosen by all manufacturers to determine the terminal activity of activated PEGs. A derivatization process is required to achieve UV absorbance because most PEG reagents are UV transparent and non-fluorescent. Reversed-phase (RP)-HPLC, for instance, can be used to test methoxy-PEG aldehyde after it has been derivatized with 4-aminobenzoic acid (Zhang et al. 2014b). The variation in molecular mass of PEG chains can be determined using reverse phase chromatography and Size Exclusion Chromatography in conjunction with corona detection mode, enabling impurity detection in the final PEG reagent. The molecular weight of PEG, which affects the protein's final half-life and directly impacts the bioavailability of PEGylated medicines, is another aspect of PEG that needs to be carefully managed (Wang et al. 2020b). The same technique is employed to identify polydispersity and the major peak fraction in PEG.

The complete characterization of the PEG-protein conjugates during the PEGylated therapeutics production process is a very challenging job. Beginning with the evaluation of PEGylated reaction mixtures, it continues with the analysis of each fraction throughout purification and ends with thorough final product characterization. The characterization of PEGylated PPDs is affected by the reality that the associated PEG significantly modifies the protein's properties. As was already mentioned, the most noticeable effects of PEGylation are a higher hydrodynamic volume and larger molecular size. Many techniques, including SEC, electrophoretic techniques, light scattering, and mass spectrometry, can be used to measure the molecular weight of proteins and PEG-protein conjugates (Caserman et al. 2009). Peptide mapping and MS are utilized to identify and quantify PEGylation sites via analyzing PEGylated and unPEGylated counterparts as well as to characterize impurities that are not always resolved and detected using simpler techniques (Kemptner et al. 2010).

## 11.9 FDA-Approved PEGylated Products

PEGylated medications treat conditions like cancer, chronic renal disease, hepatitis, hemophilia, and gastrointestinal issues. FDA approved the sale of ADAGEN, a PEGylated protein Enzon Pharmaceuticals produced in March 1990. Several PEGylated PPDs have followed in ADAGEN's footsteps since its release, and many more are currently through clinical trials or other phases of research. Neulasta and Cimzia, two of the FDA-approved medications, reached \$3.221 billion and \$1.953 billion in sales in 2019. It is important to note that all medications approved by the FDA involve methoxypolyethylene glycol. Table 11.1 represents the marketed PEGylated PPDs.

**Table 11.1** Marketed PEGylated products

| Trade Name | Company | PEGylated entity | Indications | Average MW of PEGs | Year of approval | Reference |
|---|---|---|---|---|---|---|
| Rolvedon | Spectrum Pharmaceuticals | G-CSF | Febrile neutropenia | 3.4 kDa | 2022 | Schwartzberg et al. (2020) |
| Stimufend | Fresenius Kabi | G-CSF | Neutropenia | 20 kDa | 2022 | Panda et al. (2023) |
| Fylnetra | Amneal pharmaceuticals LLC | G-CSF | Neutropenia | 20 kDa | 2022 | Humphreys et al. (2022) |
| BESREMi | PharmaEssentia Corp | Interferon | Polycythemia vera | 40 kDa | 2021 | Okikiolu et al. (2023) |
| Skytrofa | Ascendis | Human growth hormone | Growth hormone deficiency | 4 × 10 kDa | 2021 | Maniatis et al. (2022) |
| Empaveli | Apellis | Pentadecapeptide | Paroxysmal nocturnal hemoglobinuria (PNH) | 40 kDa | 2021 | Shah et al. (2022) |
| Nyvepria | Pfizer Inc. | G-CSF | Neutropenia associated with chemotherapy | 20 kDa | 2020 | Yang et al. (2021) |
| Esperoct | Novo Nordisk | Recombinant antihemophilic factor | Hemophilia A | 40 kDa | 2019 | Ezban et al. (2020) |
| Ziextenzo | Sandoz | G-CSF | Infection during chemotherapy | 20 kDa | 2019 | (Cornes et al. 2020) |
| Udenyca | Coherus biosciences | G-CSF | Infection during chemotherapy | 20 kDa | 2018 | Webster et al. (2020) |
| Palynziq | BioMarin Pharmaceutical | Recombinant phenylalanine ammonia lyase | Phenylketonuria | ~ 9 X 20 kDa | 2018 | Patrawala et al. (2020) |
| Revcovi | Leadiant bioscience | Recombinant adenosine deaminase | ADA-SCID | 80 kDa | 2018 | Carbonaro-Sarracino et al. (2021) |
| Fulphila | Mylan GmbH | G-CSF | Infection during chemotherapy | 20 kDa | 2018 | Selby et al. (2021) |
| Asparlas | Servier pharma | L-asparaginase | Leukemia | 31–39 × 5 kDa | 2018 | Bender et al. (2021) |

## 11.10 Obstacles and Pitfalls of PEGylation

PEGylation has recently gained popularity as a post-production modification technique for enhancing protein therapeutics' biological effectiveness and physicochemical characteristics. Given the usefulness and safety of this approach, which several commercial medications have already demonstrated for more than 10 years, PEGylation is anticipated to be used to modify more potential therapeutic proteins. However, given the trend towards using branch and high molecular weight PEGs, the non-biodegradability of PEG could pose a significant constraint for the subsequent generation of protein therapeutics, which are used at high concentrations and over an extended period.

Protein PEGylation has progressed from the first to the second generation, and current attempts to use the third generation are being made to boost efficacy. PEG polymeric size and orientation concerning protein conjugates can significantly change the characteristics of the resulting product. Polydispersity value, site-specific PEGylation, and degree of PEGylation are further disadvantages. Similarly, the method primarily blocking proteolytic enzymes from destroying PEGylated protein can prevent a substrate from accessing the protein's active site. Hence, active-site protective agents are utilized to avoid such troubles and other issues. However, PEGylation can still occur close to the protected region.

PEGylation therapy has adversely affected patients by entering the vasculature, resulting in dermatitis, mucositis, and hands and foot syndrome (HFS) (Najem et al. 2014). Additional disadvantages have been reported in biotechnology and nanomedicine systems, wherein receptor interaction is reduced because of the steric hindrance provided by the disordered PEG chain. In in vivo trials, enzymes like alcohol dehydrogenase and cytochrome P450 can gradually shorten the chain length. The 40 kDa branched PEG version is currently the greatest PEG molecular weight used for protein conjugation.

Due to the PEGylation PPDs improved bioavailability, thermal and physical stability, decreased immunogenicity, increased half-lives, and optimized pharmacokinetic and pharmacodynamic characteristics, these compounds have shown promise in the field of biomedicine (Schellekens et al. 2013). However, the extensive usage of PEGylation is somewhat constrained due to anti-PEG, vacuoles associated with PEGs, and other drawbacks. On the other hand, this approach should be applied more frequently and appropriately. To address these issues, we must first investigate the various modified process parameters for multiple PEGs to determine the ideal level of modification and the final product's molecular weight (Zhang et al. 2014a).

## 11.11 Conclusion

Many facets of PEGylation technology have been discussed in this review. They include the creation of PEG reagents, PEGylation processes, purifications, and studying PEG–protein conjugates. PEGylation is an established and tried technology that has already produced various FDA-approved medicines, attesting to its viability and safety. Since its inception, PEGylation has mostly been utilized to control the life cycles of already-existing therapeutic proteins. Since every protein is unique, each PEG moiety must be tailored individually to the desired therapeutic molecule. The impact on pharmacokinetic and pharmacodynamic qualities depends heavily on the length and form of each PEG moiety. In addition, homogeneity and pyrogenicity requirements for pegylated drugs must be met, and activation and binding methods must be repeatable. Notwithstanding these difficulties, methods for PEGylating proteins and peptides have greatly improved recently. In the near future, we may anticipate and sincerely hope that the numerous studies conducted by academic and commercial researchers will successfully address the issues with PEG-small drug conjugates, allowing the introduction of these novel products to the market as well.

**Acknowledgments** The author, RKT, acknowledges the Department of Pharmaceuticals, Ministry of Chemicals and Fertilizers, India, for supporting the drug discovery and formulation research at NIPER Ahmedabad. RKT also acknowledges the Department of Science and Technology, Government of India, for a Core Research Grant funding (File No. CRG/2021/005402) and also acknowledges the Indian Council of Medical Research (ICMR), New Delhi, for the grant File Id: 2021-14161 and grant File Id: IIRP-2023-4849/F1 for supporting research in RKT lab.

## References

Alconcel SNS, Baas AS, Maynard HD (2011) Fda-approved poly(ethylene glycol)–protein conjugate drugs. Polym Chem 2:1442–1448

Alqahtani AD, O'connor D, Domling A, Goda SK (2019) Strategies for the production of long-acting therapeutics and efficient drug delivery for cancer treatment. Biomed Pharmacother 113:108750

Awwad S, Ginn C, Brocchini S (2018) The case for protein Pegylation. Woodhead Publishing, London

Bailon P, Won CY (2009) Peg-modified biopharmaceuticals. Expert Opin Drug Deliv 6:1–16

Baumann A, Tuerck D, Prabhu S, Dickmann L, Sims J (2014) Pharmacokinetics, metabolism and distribution of Pegs and Pegylated proteins: quo Vadis? Drug Discov Today 19:1623–1631

Belén LH, Rangel-Yagui CDO, Beltrán Lissabet JF, Effer B, Lee-Estevez M, Pessoa A, Castillo RL, Farías JG (2019) From synthesis to characterization of site-selective pegylated proteins. Front Pharmacol 10:1

Bender C, Maese L, Carter-Febres M, Verma A (2021) Clinical utility of Pegaspargase in children, adolescents and young adult patients with acute lymphoblastic leukemia: a review. Blood Lymphat Cancer 11:25–40

Beranová M, Wasserbauer R, Vancurová D, Stifter M, Ocenásková J, Mára M (1990) Effect of cytochrome P-450 inhibition and stimulation on intensity of polyethylene degradation in microsomal fraction of mouse and rat livers. Biomaterials 11:521–524

Bliss M (1982) Banting's, Best's, and Collip's accounts of the discovery of insulin. Bull Hist Med 56:554–568

Bruno BJ, Miller GD, Lim CS (2013) Basics and recent advances in peptide and protein drug delivery. Ther Deliv 4:1443–1467

Byrne SA, Bedding MJ, Corcilius L, Ford DJ, Zhong Y, Franck C, Larance M, Mackay JP, Payne RJ (2021) Late-stage modification of peptides and proteins at cysteine with Diaryliodonium salts. Chem Sci 12:14159–14166

Caliceti P, Veronese FM (2003) Pharmacokinetic and biodistribution properties of poly(ethylene glycol)-protein conjugates. Adv Drug Deliv Rev 55:1261–1277

Cao S-J, Lv Z-Q, Guo S, Jiang G-P, Liu H-L (2020) An update—prolonging the action of protein and peptide drugs. J Drug Deliv Sci Technol 61:102124

Carbonaro-Sarracino DA, Chun K, Clark DN, Kaufman ML, Jin X, Wang X, Kohn DB (2021) Gene delivery using Aav8 in vivo for disease stabilization in a bimodal gene therapy approach for the treatment of Ada-deficient Scid. Mol Ther Methods Clin Dev 20:765–778

Caserman S, Kusterle M, Kunstelj M, Milunović T, Schiefermeier M, Jevsevar S, Porekar VG (2009) Correlations between in vitro potency of polyethylene glycol-protein conjugates and their chromatographic behavior. Anal Biochem 389:27–31

Cheetham AG, Keith D, Zhang P, Lin R, Su H, Cui H (2016) Targeting tumors with small molecule peptides. Curr Cancer Drug Targets 16:489–508

Chen G, Kang W, Li W, Chen S, Gao Y (2022) Oral delivery of protein and peptide drugs: from non-specific formulation approaches to intestinal cell targeting strategies. Theranostics 12:1419–1439

Cheng F, Yang J, Schwaneberg U, Zhu L (2019) Rational surface engineering of an arginine deiminase (an antitumor enzyme) for increased Pegylation efficiency. Biotechnol Bioeng 116:2156–2166

Cornes P, Gascon P, Vulto AG, Aapro M (2020) Biosimilar Pegfilgrastim: improving access and Optimising practice to supportive care that enables cure. BioDrugs 34:255–263

Craik DJ, Fairlie DP, Liras S, Price D (2013) The future of peptide-based drugs. Chem Biol Drug Des 81:136–147

Davenport AP, Scully CCG, De Graaf C, Brown AJH, Maguire JJ (2020) Advances in therapeutic peptides targeting G protein-coupled receptors. Nat Rev Drug Discov 19:389–413

Defrees S, Wang ZG, Xing R, Scott AE, Wang J, Zopf D, Gouty DL, Sjoberg ER, Panneerselvam K, Brinkman-Van Der Linden EC, Bayer RJ, Tarp MA, Clausen H (2006) Glycopegylation of recombinant therapeutic proteins produced in Escherichia Coli. Glycobiology 16:833–843

Delgado C, Francis GE, Fisher D (1992) The uses and properties of peg-linked proteins. Crit Rev Ther Drug Carrier Syst 9:249–304

Dimitrov DS (2012) Therapeutic proteins. Methods Mol Biol 899:1–26

Dozier JK, Distefano MD (2015) Site-specific Pegylation of therapeutic proteins. Int J Mol Sci 16:25831–25864

Ezban M, Hansen M, Kjalke M (2020) An overview of Turoctocog alfa Pegol (N8-Gp; Esperoct(®) ) assay performance: implications for Postadministration monitoring. Haemophilia 26:156–163

Fee CJ (2007) Size comparison between proteins Pegylated with branched and linear poly(ethylene glycol) molecules. Biotechnol Bioeng 98:725–731

Fee CJ, Van Alstine JM (2006) Peg-proteins: reaction engineering and separation issues. Chem Eng Sci 61:924–939

Filpula D, Zhao H (2008) Releasable Pegylation of proteins with customized linkers. Adv Drug Deliv Rev 60:29–49

Finck B, Tang H, Civoli F, Hodge J, O'kelly H, Vexler V (2020) Pharmacokinetic and Pharmacodynamic equivalence of Pegfilgrastim-Cbqv and Pegfilgrastim in healthy subjects. Adv Ther 37:4291–4307

Fishburn CS (2008) The pharmacology of Pegylation: balancing Pd with Pk to generate novel therapeutics. J Pharm Sci 97:4167–4183

Fontaine SD, Reid R, Robinson L, Ashley GW, Santi DV (2015) Long-term stabilization of Maleimide-thiol conjugates. Bioconjug Chem 26:145–152

Fontana A, Spolaore B, Mero A, Veronese FM (2008) Site-specific modification and Pegylation of pharmaceutical proteins mediated by transglutaminase. Adv Drug Deliv Rev 60:13–28

Fu K, March K, Alexaki A, Fabozzi G, Moysi E, Petrovas C (2020) Immunogenicity of protein therapeutics: a lymph node perspective. Front Immunol 11:11

Gaertner HF, Offord RE (1996) Site-specific attachment of functionalized poly(ethylene glycol) to the amino terminus of proteins. Bioconjug Chem 7:38–44

Greish K (2007) Enhanced permeability and retention of macromolecular drugs in solid tumors: a Royal Gate for Targeted Anticancer Nanomedicines. J Drug Target 15:457–464

Gunnoo SB, Madder A (2016) Chemical protein modification through cysteine. Chembiochem 17:529–553

Gupta V, Sengupta M, Prakash J, Tripathy BC (2016) Production of recombinant pharmaceutical proteins. In: Basic and applied aspects of biotechnology. Springer, Singapore, pp 77–101. https://doi.org/10.1007/978-981-10-0875-7_4

Gupta V, Bhavanasi S, Quadir M, Singh K, Ghosh G, Vasamreddy K, Ghosh A, Siahaan TJ, Banerjee S, Banerjee SK (2019) Protein Pegylation for cancer therapy: bench to bedside. J Cell Commun Signal 13:319–330

Hamidi M, Azadi A, Rafiei P (2006) Pharmacokinetic consequences of Pegylation. Drug Deliv 13:399–409

Han Y, Gao Z, Chen L, Kang L, Huang W, Jin M, Wang Q, Bae YH (2019) Multifunctional Oral delivery systems for enhanced bioavailability of therapeutic peptides/proteins. Acta Pharm Sin B 9:902–922

Harris JM, Chess RB (2003) Effect of Pegylation on pharmaceuticals. Nat Rev Drug Discov 2:214–221

Harris JM, Martin NE, Modi M (2001) Pegylation: a novel process for modifying pharmacokinetics. Clin Pharmacokinet 40:539–551

Henninot A, Collins JC, Nuss JM (2018) The current state of peptide drug discovery: Back to the future? J Med Chem 61:1382–1414

Humphreys SZ, Geller RB, Walden P (2022) Pegfilgrastim Biosimilars in us supportive oncology: a narrative review of administration options and economic considerations to maximize patient benefit. Oncol Ther 10:351–361

Ibeanu N, Egbu R, Onyekuru L, Javaheri H, Khaw PT, Williams GR, Brocchini S, Awwad S (2020) Injectables and depots to prolong drug action of proteins and peptides. Pharmaceutics 12:999

Jevsevar S, Kunstelj M, Porekar VG (2010) Pegylation of therapeutic proteins. Biotechnol J 5:113–128

Jones MW, Mantovani G, Blindauer CA, Ryan SM, Wang X, Brayden DJ, Haddleton DM (2012) Direct peptide bioconjugation/Pegylation at tyrosine with linear and branched polymeric Diazonium salts. J Am Chem Soc 134:7406–7413

Joseph M, Trinh H, Mitra A (2017) Peptide and protein-based therapeutic agents. Elsevier, Amsterdam

Kemptner J, Marchetti-Deschmann M, Siekmann J, Turecek PL, Schwarz HP, Allmaier G (2010) Gemma and Maldi-Tof Ms of reactive Pegs for pharmaceutical applications. J Pharm Biomed Anal 52:432–437

Knop K, Hoogenboom R, Fischer D, Schubert US (2010) Poly(ethylene glycol) in drug delivery: pros and cons as well as potential alternatives. Angew Chemie Int Ed 49:6288–6308

Kolate A, Baradia D, Patil S, Vhora I, Kore G, Misra A (2014) Peg—a versatile conjugating ligand for drugs and drug delivery systems. J Control Release 192:67–81

Kozma GT, Shimizu T, Ishida T, Szebeni J (2020) Anti-peg antibodies: properties, formation, testing and role in adverse immune reactions to Pegylated Nano-biopharmaceuticals. Adv Drug Deliv Rev 154-155:163–175

Kunstelj M, Fidler K, Skrajnar S, Kenig M, Smilović V, Kusterle M, Caserman S, Zore I, Porekar VG, Jevševar S (2013) Cysteine-specific Pegylation of Rhg-Csf via Selenylsulfide bond. Bioconjug Chem 24:889–896

Kursa M, Walker GF, Roessler V, Ogris M, Roedl W, Kircheis R, Wagner E (2003) Novel shielded transferrin-polyethylene glycol-Polyethylenimine/Dna complexes for systemic tumor-targeted gene transfer. Bioconjug Chem 14:222–231

Lagassé HA, Alexaki A, Simhadri VL, Katagiri NH, Jankowski W, Sauna ZE, Kimchi-Sarfaty C (2017) Recent advances in (therapeutic protein) drug development. *F1000res* 6:113

Lau JL, Dunn MK (2018) Therapeutic peptides: historical perspectives, current development trends, and future directions. Bioorg Med Chem 26:2700–2707

Lee AC-L, Harris JL, Khanna KK, Hong J-H (2019) A comprehensive review on current advances in peptide drug development and design. Int J Mol Sci 20:2383

Li CM, Haratipour P, Lingeman RG, Perry JJP, Gu L, Hickey RJ, Malkas LH (2021) Novel peptide therapeutic approaches for cancer treatment. Cells 10:2908

Maiser B, Dismer F, Hubbuch J (2014) Optimization of random pegylation reactions by means of high throughput screening. Biotechnol Bioeng 111:104–114

Manandhar M, Chun E, Romesberg FE (2021) Genetic code expansion: inception, development, commercialization. J Am Chem Soc 143:4859–4878

Maniatis AK, Casella SJ, Nadgir UM, Hofman PL, Saenger P, Chertock ED, Aghajanova EM, Korpal-Szczyrska M, Vlachopapadopoulou E, Malievskiy O, Chaychenko T, Cappa M, Song W, Mao M, Mygind PH, Smith AR, Chessler SD, Komirenko AS, Beckert M, Shu AD, Thornton PS (2022) Safety and efficacy of Lonapegsomatropin in children with growth hormone deficiency: enlighten trial 2-year results. J Clin Endocrinol Metab 107:E2680–E2689

Mero A, Spolaore B, Veronese FM, Fontana A (2009) Transglutaminase-mediated Pegylation of proteins: direct identification of the sites of protein modification by mass spectrometry using a novel monodisperse peg. Bioconjug Chem 20:384–389

Milla P, Dosio F, Cattel L (2012) Pegylation of proteins and liposomes: a powerful and flexible strategy to improve the drug delivery. Curr Drug Metab 13:105–119

Monfardini C, Schiavon O, Caliceti P, Morpurgo M, Harris JM, Veronese FM (1995) A branched Monomethoxypoly(ethylene glycol) for protein modification. Bioconjug Chem 6:62–69

Morpurgo M, Veronese FM, Kachensky D, Harris JM (1996) Preparation of characterization of poly(ethylene glycol) vinyl sulfone. Bioconjug Chem 7:363–368

Najem A, Deregnaucourt D, Ramdane S, Dridba M, Djouba F, Vercambre-Darras S (2014) Intertrigo-like dermatitis with Pegylated liposomal doxorubicin: diagnosis and management. J Clin Oncol 32:E104–E106

Ng EW, Shima DT, Calias P, Cunningham ET Jr, Guyer DR, Adamis AP (2006) Pegaptanib, a targeted anti-Vegf aptamer for ocular vascular disease. Nat Rev Drug Discov 5:123–132

Nguyen GK, Kam A, Loo S, Jansson AE, Pan LX, Tam JP (2015) Butelase 1: a versatile ligase for peptide and protein macrocyclization. J Am Chem Soc 137:15398–15401

Okikiolu J, Woodley C, Cadman-Davies L, O'sullivan J, Radia D, Garcia NC, Harrington P, Kordasti S, Asirvatham S, Sriskandarajah P, Saunders J, Saha C, Sanchez I, Delavallade H, Mclornan DP, Harrison CN (2023) Real world experience with Ropeginterferon alpha-2b (Besremi) in essential Thrombocythaemia and Polycythaemia Vera following exposure to Pegylated interferon alfa-2a (Pegasys). Leuk Res Rep 19:100360

Panda S, Singh PK, Mishra S, Mitra S, Pattnaik P, Adhikary SD, Mohapatra RK (2023) Indian Biosimilars and vaccines at crossroads-replicating the success of Pharmagenerics. Vaccines (Basel) 11:110

Pasut G, Veronese FM (2006) Pegylation of proteins as tailored chemistry for optimized bioconjugates. In: Satchi-Fainaro R, Duncan R (eds) Polymer therapeutics I. Springer, Berlin

Patrawala M, Kuruvilla M, Li H (2020) Successful desensitization of Pegvaliase (Palynziq®) in a patient with phenylketonuria. Mol Genet Metab Rep 23:100575

Pettit DK, Bonnert TP, Eisenman J, Srinivasan S, Paxton R, Beers C, Lynch D, Miller B, Yost J, Grabstein KH, Gombotz WR (1997) Structure-function studies of interleukin 15 using site-specific mutagenesis, polyethylene glycol conjugation, and homology modeling. J Biol Chem 272:2312–2318

Plesner B, Fee CJ, Westh P, Nielsen AD (2011) Effects of peg size on structure, function and stability of Pegylated Bsa. Eur J Pharm Biopharm 79:399–405

Qi Y, Chilkoti A (2015) Protein-polymer conjugation-moving beyond Pegylation. Curr Opin Chem Biol 28:181–193

Rudmann DG, Alston JT, Hanson JC, Heidel S (2013) High molecular weight polyethylene glycol cellular distribution and peg-associated cytoplasmic Vacuolation is molecular weight dependent and does not require conjugation to proteins. Toxicol Pathol 41:970–983

Sato H (2002) Enzymatic procedure for site-specific Pegylation of proteins. Adv Drug Deliv Rev 54:487–504

Schellekens H, Hennink WE, Brinks V (2013) The immunogenicity of polyethylene glycol: facts and fiction. Pharm Res 30:1729–1734
Schreiber S (2011) Certolizumab Pegol for the treatment of Crohn's disease. Ther Adv Gastroenterol 4:375–389
Schwartzberg LS, Bhat G, Peguero J, Agajanian R, Bharadwaj JS, Restrepo A, Hlalah O, Mehmi I, Chawla S, Hasal SJ, Yang Z, Cobb PW (2020) Eflapegrastim, a long-acting granulocyte-Colony stimulating factor for the management of chemotherapy-induced neutropenia: results of a phase iii trial. Oncologist 25:E1233–E1241
Seif S, Planz V, Windbergs M (2017) Delivery of therapeutic proteins using electrospun fibers-recent developments and current challenges. Arch Pharm (Weinheim) 350:1700077
Selby C, Peyton-Thomas B, Eslami P (2021) Pegfilgrastim Biosimilars: where are we now? J Adv Pract Oncol 12:541–547
Shah S, Chamlagain R, Musalman ZH, Raj Adhikari Y, Chhetri S, Paudel S, Gundabolu K, Dhakal P (2022) Pegcetacoplan in paroxysmal nocturnal hemoglobinuria: a systematic review on efficacy and safety. Res Pract Thromb Haemost 6:E12781
Sriram K, Insel PA (2018) G protein-coupled receptors as targets for approved drugs: how many targets and how many drugs? Mol Pharmacol 93:251–258
Sugiuchi H, Uji Y, Okabe H, Irie T, Uekama K, Kayahara N, Miyauchi K (1995) Direct measurement of high-density lipoprotein cholesterol in serum with polyethylene glycol-modified enzymes and sulfated alpha-Cyclodextrin. Clin Chem 41:717–723
Swierczewska M, Lee KC, Lee S (2015) What is the future of Pegylated therapies? Expert Opin Emerg Drugs 20:531–536
Szijj PA, Kostadinova KA, Spears RJ, Chudasama V (2020) Tyrosine bioconjugation—an emergent alternative. Org Biomol Chem 18:9018–9028
Szlachcic A, Zakrzewska M, Otlewski J (2011) Longer action means better drug: tuning up protein therapeutics. Biotechnol Adv 29:436–441
Tagawa H, Maruyama K, Sasaki K, Konoue N, Kishimura A, Kanai M, Mori T, Oisaki K, Katayama Y (2020) Induction of Adcc By A folic acid–mab conjugate prepared by tryptophan-selective reaction toward folate-receptor-positive cancer cells. RSC Adv 10:16727–16731
Tilley SD, Francis MB (2006) Tyrosine-selective protein alkylation using pi-Allylpalladium complexes. J Am Chem Soc 128:1080–1081
Turecek PL, Bossard MJ, Schoetens F, Ivens IA (2016) Pegylation of biopharmaceuticals: a review of chemistry and nonclinical safety information of approved drugs. J Pharm Sci 105:460–475
Vargason AM, Anselmo AC, Mitragotri S (2021) The evolution of commercial drug delivery technologies. Nat Biomed Eng 5:951–967
Veronese FM (2001) Peptide and protein Pegylation: a review of problems and solutions. Biomaterials 22:405–417
Veronese FM, Mero A (2008) The impact of Pegylation on biological therapies. BioDrugs 22:315–329
Veronese FM, Caliceti P, Schiavon O (1997) Branched and linear poly(ethylene glycol): influence of the polymer structure on enzymological , pharmacokinetic, and immunological properties of protein conjugates. J Bioact Compat Polym 12:196–207
Villegas MR, Baeza A, Vallet-Regí M (2018) Nanotechnological strategies for protein delivery. Molecules 23:1008
Wang L, Zhou W, Wang Q, Xu C, Tang Q, Yang H (2018) An injectable, dual responsive, and self-healing hydrogel based on oxidized sodium alginate and hydrazide-modified poly(Ethyleneglycol). Molecules 23:546
Wang J, Deng T, Liu Y, Chen K, Yang Z, Jiang ZX (2020a) Monodisperse and Polydisperse Pegylation of peptides and proteins: a comparative study. Biomacromolecules 21:3134–3139
Wang T, Guo Y, He Y, Ren T, Yin L, Fawcett JP, Gu J, Sun H (2020b) Impact of molecular weight on the mechanism of cellular uptake of polyethylene glycols (Pegs) with particular reference to P-glycoprotein. Acta Pharm Sin B 10:2002–2009

Warren K, Bent R, Wolters PL, Prager A, Hanson R, Packer R, Shih J, Camphausen K (2012) A phase 2 study of Pegylated interferon A-2b (peg-intron(®)) in children with diffuse intrinsic pontine glioma. Cancer 118:3607–3613

Webster R, Elliott V, Park BK, Walker D, Hankin M, Taupin P (2009) Peg and peg conjugates toxicity: towards an understanding of the toxicity of peg and its relevance to Pegylated biologicals. In: Veronese FM (ed) Pegylated protein drugs: basic science and clinical applications. Birkhäuser, Basel

Webster J, Scott J, Smith H, Wieland D, Donaldson J, Catasus C, Smith R (2020) Udenyca has equivalent efficacy to the Pegfilgrastim originator in breast cancer patients receiving highly Myelosuppressive chemotherapy. J Clin Oncol 38:E19273–E19273

Weeks AM, Wells JA (2020) Subtiligase-catalyzed peptide ligation. Chem Rev 120:3127–3160

Witt KA, Huber JD, Egleton RD, Roberts MJ, Bentley MD, Guo L, Wei H, Yamamura HI, Davis TP (2001) Pharmacodynamic and pharmacokinetic characterization of poly(ethylene glycol) conjugation to met-Enkephalin analog [D-Pen2, D-Pen5]-Enkephalin (Dpdpe). J Pharmacol Exp Ther 298:848–856

Xu L, Kuan SL, Weil T (2021) Contemporary approaches for site-selective dual functionalization of proteins. Angew Chem Int Ed Engl 60:13757–13777

Yang J, Liu R, Granghaud A, Zaidi O, Stephens J (2021) Biosimilar Pegfilgrastim may offer affordable treatment options for patients in France: a budget impact analysis on the basis of clinical trial and real-world data. J Med Econ 24:665–674

Yoshimoto N, Yamamoto S (2012) Pegylated protein separations: challenges and opportunities. Biotechnol J 7:592–593

Yowell SL, Blackwell S (2002) Novel effects with polyethylene glycol modified pharmaceuticals. Cancer Treat Rev 28 Suppl A:3–6

Zalipsky S (1995) Functionalized poly(ethylene glycols) for preparation of biologically relevant conjugates. Bioconjug Chem 6:150–165

Zalipsky S, Seltzer R, Menon-Rudolph S (1992) Evaluation of a new reagent for covalent attachment of polyethylene glycol to proteins. Biotechnol Appl Biochem 15:100–114

Zhang J, Zhao Y, Su Z, Ma G-H (2007) Synthesis of Monomethoxy poly(ethylene glycol) without diol poly(ethylene glycol). J Appl Polym Sci 105:3782–3786

Zhang F, Liu MR, Wan HT (2014a) Discussion about several potential drawbacks of Pegylated therapeutic proteins. Biol Pharm Bull 37:335–339

Zhang H, Wilson J, Zhang J, Luo Y (2014b) Characterization of potential degradation products in a Pegylating reagent 20 Kda Monomethoxy polyethylene glycol Propionaldehyde by Rp-Hplc, Apci-Ms and Nmr. J Pharm Biomed Anal 89:221–226

Zuma LK, Gasa NL, Makhoba XH, Pooe OJ (2022) Protein Pegylation: navigating recombinant protein stability, aggregation, and bioactivity. Biomed Res Int 2022:8929715

# PEGylated Nanocarriers for Diagnostic Applications

# 12

Naveen Gupta, Ankit Mishra, Pran Kishore Deb, and Muktika Tekade

**Abstract**

Polyethylene glycol, due to its "stealth" characteristics and biocompatibility, is frequently used in the administration of drugs and nanomaterials. PEGylation enables biomaterials and particle delivery systems to circumvent the immune system and extend circulation lifetimes. To diagnose various diseases, substantial attempts are being made to create alternative imaging techniques that can improve the signal or produce high positive contrast for effective molecular imaging. The development of PEGylated nanoparticles as contrast agents in imaging technology has made it possible to gain precise cellular and molecular imaging, detect drug delivery, particularly to tumoral areas, and provide information for adequate surgical excision of solid tumors. PEGylated nanocarriers have been identified as potential candidates for the targeted treatment and imaging of malignant tumors. Peptides or antibodies can be conjugated to the surface of PEGylated nanocarriers to directly target tumor cells and potentially impair their

N. Gupta (✉)
Patel Institute of Pharmacy, Madhyanchal Professional University, Bhopal, Madhya Pradesh, India

A. Mishra
Department of Pharmaceutics, VNS Group of Institutions, Faculty of Pharmacy, Bhopal, Madhya Pradesh, India

P. K. Deb
Department of Pharmaceutical Sciences and Technology, Birla Institute of Technology (BIT), Ranchi, Jharkhand, India

M. Tekade
School of Pharmacy, Devi Ahilya Vishwavidyalaya, Takshila Campus, Indore, Madhya Pradesh, India

School of Pharmaceutical Sciences, Lovely Professional University, Phagwara, Punjab, India


R. K. Tekade, N. K. Jain (eds.), *PEGylated Nanocarriers in Medicine and Pharmacy*, https://doi.org/10.1007/978-981-97-7316-9_12

active signaling pathways. They improve magnetic resonance imaging contrast, helping physicians to track anatomical, physiological, and molecular changes in a disease as treatment progresses. PEGylated nanocarriers are effective imaging tools for tumor diagnosis, follow-up, and disease monitoring, which help in the overall clinical management of tumors.

**Keywords**

PEGylation · PEGylated · Nano carriers · Diagnostic · Imaging

## 12.1 Introduction

Polyethylene glycol (PEG) is an aqueous, safe, and biocompatible polymer approved by the Food and Drug Administration (FDA). PEG generally finds applications in intravenous, oral, and dermal delivery routes for human use. PEG with high molecular weight are viscous and colorless liquids, and larger molecular weight PEGs are waxy liquids. They are miscible with glycols, soluble in water, alcohol, acetone, and chloroform, and insoluble in ether. Due to its purported "stealth" qualities and biocompatibility, PEG is frequently used in pharmaceutical delivery systems. PEG has demonstrated promise in delaying renal clearance and extending circulation lives. The prevailing consensus is that PEGylation enables biomaterials and particulate delivery systems to avoid the immune system by extending circulation lifetimes. The process by which PEGylation prolongs circulation durations is widely thought to include a considerable reduction in opsonization, which accounts for its "stealth" behavior.

Studies have examined reduced protein binding to nanoparticles containing PEGylated components and associated the amount of protein adsorption (as a measure of opsonization) with half-life in the distribution, both of which agree with this theory. In fact, PEG is believed to conceal surface charge on nanoparticles (as can be observed by a near-neutral zeta potential) and form a hydrophilic barrier that sterically inhibits protein adsorption. Numerous in vitro studies have clearly shown that PEGylation can reduce macrophage absorption, and it is hypothesized that this effect is the one that allows PEG to lengthen circulation durations in vivo (Verhoef and Anchordoquy 2013).

A drug delivery system must be in the bloodstream for sufficient time to reach the target region. Through the reticuloendothelial system (RES), plasma proteins known as opsonins can bind circulating drug delivery devices, namely nanocarriers, and eliminate them from circulation in seconds to minutes. Giving these drug delivery systems a stealth shielding on their surface stops opsonins from identifying these particles, restricting phagocytosis by the RES cells and extending the time it takes for blood to circulate throughout the body from minutes to hours or days, as represented in Fig. 12.1. To achieve such stealth-shielding and long-circulation of pharmaceuticals or delivery devices, polyethylene glycol modification has arisen as a frequent tactic. PEG is also called polyethylene oxide (PEO) when the molecular weight is larger than 20 kDa. PEGylation term is used to describe the covalent

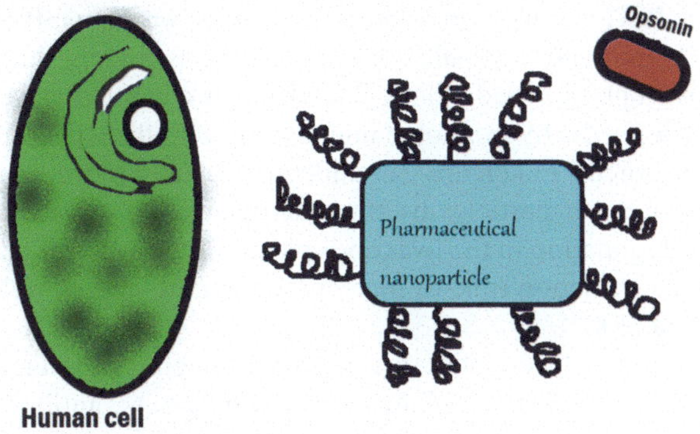


**Fig. 12.1** Prevention of opsonization by surface modification of nanocarriers with PEG

binding, non-covalent entrapment, or adsorption of PEG onto a product and is a common term for PEG modification (Van Vlerken et al. 2007).

A highly effective strategy for enhancing therapeutic drug systemic distribution is the PEGylation of nanoparticles (NPs). PEGylation of NPs has also been employed as a method to cross-extracellular barriers related to alternative administration methods, such as distribution to the mucosa or delivery to the brain. PEG coatings will undoubtedly remain an integral component in the design of NPs for drug and gene delivery applications, enabling more research to understand how the characteristics of PEG coating affect NPs biodistribution and removal from the body.

As more NP-based products are being used in the clinical setting, it is unclear what role the emerging PEG alternatives will play. However, further research into the potential immunogenic properties of PEG coatings as a function of molecular weight, functional group, surface density, NPs core properties, dosing frequency, etc., will undoubtedly result in more effective products (Suk et al. 2016).

### 12.1.1 Modification of the PEG Surface and Long-Circulation Properties

The typical structure of PEG is HO-$(CH_2CH_2O)$n-$CH_2CH_2$-OH, which includes a chemically inert polyether backbone and terminal hydroxyl groups that can be triggered for conjugation to various polymers and medicinal products. Poloxamers and poloxamines, which are amphiphilic block co-polymers made of blocks of hydrophilic PEG and hydrophobic polypropylene oxide (PPO), are additionally PEG derivatives frequently used for surface adsorption or entrapment modification. Covalent methods, such as grafting PEG chains onto the surface of the polymeric NPs, or co-polymers, in which PEG is covalently bonded to another type of polymer, can modify the surface of the NPs. PEG has the benefits of being non-toxic and non-immunogenic, which led to the FDA of the United States approving it for use in individuals internally, a listing of inactive substances for parenteral and oral use, as well. Considering PEG is not biodegradable, it is available in a wide range of molecular weights up to several million daltons, which affects the rate at which the body can eliminate it.

PEG chains with a larger molecular weight shift from urine to fecal excretion. The protective (stealth) action of PEG is mainly due to the formation of a dense, hydrophilic cloud of long flexible chains on the surface of the colloidal particles that reduces the hydrophobic interactions with the RES. The tethered and/or chemically anchored PEG chains can undergo spatial conformations, thus preventing the opsonization of particles by the macrophages of the RES, which leads to preferential accumulation in the liver and spleen. PEG surface modification, therefore, enhances the circulation time of molecules and colloidal particles in the blood (Van Vlerken et al. 2007).

Today, PEGylation is crucial for imaging and diagnosing different body tissues. Macromolecular chelators such as polymers, antibodies, and recognition proteins are becoming more popular among chelators. PEGylation extends the period that paramagnetic chelates remain in the body. These chelates will then be eliminated from the body at slower rates through the kidney or liver than unmodified molecules, providing magnetic resonance to produce accurate images. PEG also serves as a linker between the targeting and diagnostic moieties, which significantly impacts the biodistribution pattern of radiodiagnostics. PEG might assist in collecting more precise pictures in the case of protein-targeted diagnostics by reducing background noise brought on by nonspecific protein–protein interaction (Veronese and Pasut 2005).

To increase the effectiveness of drug and gene delivery to target cells and tissues, PEG is frequently used to coat the surface of NPs, a process known as "PEGylation." The length of the systemic circulation is increased by PEG coatings on NPs, which protect the surface from opsonization, aggregation, and phagocytosis.

Systemically delivered medicines must circulate in the bloodstream for as long as possible to reach target tissues at sufficient concentrations. However, proteins and peptides rapidly break down and are removed from the blood; therefore, methods for prolonging circulation duration are required. One such way is to coat the therapeutic's surface with an inert polymer that prevents interactions with bloodstream constituents and gives it "stealth" qualities. Due to its long history of safety in human usage and FDA classification as Generally Regarded as Safe (GRAS), PEG is the most commonly utilized "stealth" polymer in the drug delivery industry (Li and Huang 2010).

Branched PEGs may provide steric repulsion due to the attachment of neighboring PEG molecules, resulting in empty spaces on the surface, just enough to attract the nonspecific binding of opsonin proteins. These molecular weights of PEG have been reported to be between 1500 and 5000. To get the highest stealth qualities, a high density of short-chain PEG molecules is required, yet chains that are too short may be too rigid to give the necessary flexibility (Karakoti et al. 2011).

### 12.1.2 Current Market Scenario

Many PEG-based medications have been developed, and a number of these have gained market acceptance. Due to vast clinical expertise, it has become possible to synthesize PEG prodrug conjugates with increased therapeutic efficacy while minimizing systemic toxicity. PEGylated products, such as PEGvisomant and certolizumab PEGol, demonstrated that PEGylated forms can be commercialized independently of the past commercialization of their non-PEGylated counterparts. In contrast, PEGylated drugs, such as PEGinterferon and PEGfilgrastim, have proven their cost-effectiveness in the market. The market for therapeutic PEGylated proteins appears to have strong long-term potential. Because of its enormous clinical advantages, PEGylation is fundamental in delivering and targeting drugs and other bioactives. PEGylation is essential for achieving cellular targetability (Banerjee et al. 2012).

Few well-known PEGylated products in the market include PEG-interferon α-2a, PEG-Intron (PEG-interferon alpha-2b) for hepatitis (Herndon et al. 2012), Somavert (PEG-human growth hormone receptor antagonist) for acromegaly (Alconcel et al. 2011), Cimzia (certolizumabPEGol, PEGylated antihuman TNF-alpha Fab) for rheumatoid arthritis (Lang 2008), Neulasta (PEGfilgrastim, PEG-recombinant human granulocyte colony stimulating factor analogue) for neutropenia associated with cancer chemotherapy (Bence et al. 2002), PEG-erythropoietin (EPO) for anemia (Rubio 2015), Adagen (PEG-adenosine deaminase) for immunodeficiency (Booth and Gaspar 2009), PEG-Hirudin for thrombosis, and Oncaspar (PEG asparaginase) for cancer treatment (Dinndorf et al. 2007).

More than 20 PEGylated nanocarriers have received clinical approval from the US FDA to date, including Doxil and Macugen, which are essential in managing cancer and neovascular age-related macular degeneration (Shi et al. 2021).

### 12.1.3 Ideal Properties of PEGylated Nanocarriers for Diagnostic Applications

PEG is an aqueous, safe, non-toxic, non-immunogenic, and non-antigenic polymer that is FDA-approved for human use. They are miscible with glycols, soluble in water, alcohol, acetone, and chloroform, and insoluble in ether. Due to its purported "stealth" qualities and biocompatibility, polyethylene glycol is frequently used in pharmaceutical delivery. The PEG-drug conjugates have several benefits, including an extended residence in the body, delaying renal clearance, a reduced breakdown rate by metabolic enzymes, and a decreased level of protein immunogenicity. The viscosity of PEG solution with lower molecular weight is low and increases with the increase of molecular weight. PEG's distinctive stability against oxidation, reduction, and decomposition by acids, bases, relatively high temperatures, hydrogen peroxide, and sodium borohydride distinguishes it from other materials. Because PEG contains both hydrophilic and hydrophobic groups, it is a preferred solvent for green synthesis. Because of its capacity to exhibit phase separation under regulated conditions. The stealth features offered by PEG coatings have enhanced the enormous potential of nanoparticles in therapy, diagnostic imaging, treatment, and disease prevention (Karakoti et al. 2011).

Nanoparticles with PEG coatings prevent surface aggregation, opsonization, and phagocytosis, extending circulation duration. Doxil, the first PEGylated NPs product to receive FDA approval, was introduced in 1995. With a drug half-life of 72 h and a circulation half-life of 36 h, Doxil "Stealth" liposomes enhanced doxorubicin bioavailability about 90-fold at 1 week after injection compared to free drug (Working et al. 1994).

The therapeutic effectiveness of systemic chemotherapy is insufficient, and the systemic toxicity restricts the dosages of injectable drugs, thus reducing their therapeutic efficacy. Creating a revolutionary system that permits early diagnosis and tailored therapy to address these shortcomings in present diagnostic and therapeutic techniques is essential.

## 12.2 PEGylated Nanocarriers in Diagnostic Applications

Recently, nanotechnology has drawn attention for its novel cancer diagnostic and therapeutic tool. These alternatives have been used to overcome the challenges that traditional diagnostic and therapeutic approaches offer. These difficulties include toxicity, rapid drug release, and non-specificity. According to research done over the years, nanocarriers improve the bioavailability, specificity, and accumulation of medications at the target location. Their easily adjustable physical and chemical features can be attributed to these advancements. They are frequently altered in size and surface texture to improve their accumulation at the target areas and overall targeting capabilities. Nanocarriers are good candidates for fluorescence-guided surgery and imaging methods due to their impressive fluorescent characteristics. Biomimetic NPs, which effectively interact with complicated biological systems and imitate the functions of biological components, are another potential class of NPs (Zhang et al. 2020).

Researchers are examining nanotechnology-based tools and techniques as viable potential for the early diagnosis and treatment of cancer. Numerous studies have been done on enhancing targeted drug delivery and imaging while preserving efficacy. PEGylated nanocarriers have been utilized in recent times for the diagnosis and imaging of many diseases, including tumors. PEGylated magnetic NPs demonstrating magnetic characteristics are frequently used in magnetic targeting, magnetic heating, and MRI contrast enhancement. Long-circulating NPs required for MRI and other theranostic applications are well-coated with PEGs (Illés et al. 2018).

A diagnostic tool that can characterize EPR is vital from a diagnostic standpoint. PEGylated NPs may be used to monitor drug release and long-term pharmacological efficacy, visualize and quantify biodistribution, and target site accumulation of NPs. Nanoparticles help characterize tumor angiogenesis, and very tiny NPs (less than 5 nm) can be utilized for molecular diagnostics for extravascular targets (Baetke et al. 2015).

The physicochemical characteristics of NPs make them suitable delivery systems for delivering drugs to brain tumors. The core or the surface of NPs can be coated with molecules, such as contrast agents or medications. As MRI contrast agents, numerous NPs have been developed with or without an organic material shell-like polyethylene glycol. NPs could be designed for imaging specific cell subpopulations, like endothelial cells or stem cells. NPs may advance the diagnosis, surgical treatment, and adjuvant therapy of brain tumors. NPs-based MRI contrast agents have the potential to make tumor areas visible that would not have been seen with conventional MRI, particularly at the tumor–brain interface (Cho et al. 2010).

Systemic chemotherapy exhibits an unfavorable biodistribution of anticancer medicines after systemic delivery, resulting in severe side effects, and has limited therapeutic efficacy. Additionally, systemic toxicity restricts the dosages of injectable drugs, thereby lowering their therapeutic efficacy. Creating a revolutionary system that permits early diagnosis and tailored therapy to address these shortcomings in present diagnostic and therapeutic approaches is vital. The development of a method that allows early identification and focused treatment to enhance current procedures that have a high failure rate of colon cancer leading to metastases is crucial for improved management of colon cancer.

Colon tumors were successfully examined utilizing an near-infrared fluorescence (NIRF) imaging technique using a theranostic carrier system based on PEGylated hyaluronic acid NPs bearing NIRF imaging dye. Because of their remarkable tumor-targeting abilities, they demonstrated significant inhibition of tumor growth with less systemic toxicity (Choi et al. 2012).

### 12.2.1 PEGylated Iron Oxide Nanoparticles

Polyethylene glycol has been extensively used on various nanoparticle systems to increase surface hydrophilicity and improve circulation half-life by decreasing interactions with blood proteins and mononuclear phagocyte system (MPS) cells. The use of PEGylated iron oxide NP ferrofluids in in vivo biomedical applications such as magnetic imaging resonance (MRI) contrast enhancement and cancer therapy by hyperthermia and/or targeted drug delivery is currently the subject of research. The magnetic characteristics of NPs could be utilized for drug delivery to improve the antitumor efficacy and lessen the systemic side effects of anticancer medications. Loaded PEGylated iron oxide NPs may be held in a cancer site and afterward released the drug there if an external localized magnetic field is applied. The surface of iron oxide NPs has been transformed using PEG, which has been proven to increase biocompatibility and blood circulation rates. PEG surface coating is utilized to prevent plasma adsorption and escape from the MPS (Shan et al. 2009).

Gadolinium, superparamagnetic iron oxides, ultrasmall (5–10 nm) superparamagnetic iron oxides, gadolinium-doped carbon nanotubes, quantum dots embedded paramagnetic micelles, and soft NPs are currently excellent magnetic contrast agents in the research field. With their magnetic properties and contrast, PEGylated iron oxide NPs have already succeeded in biomedical areas such as diagnostics as a probe (MRI scanning) for discovering diseases or issues within the brain, cardiovascular system, liver, blood arteries, and other essential organs. PEGylated iron oxide NPs have also been designed with advancements for carrying out multiple operations in single-stage imaging. In recent times, PEGylated iron oxide NP-based nanohybrids were modified for partial thromboplastin time (PTT) and hyperthermia, where even small concentrations can effectively increase heat generation at the tumor site and be employed for cellular treatments (Vallabani and Singh 2018).

PEGylated iron oxide nanoparticles with specific surface chemical characteristics have been widely exploited for several biological applications, including cell separation, tissue healing, targeted drug delivery, hyperthermia, and MRI contrast enhancement. Due to their low toxicity, size-dependent superparamagnetism, and biocompatibility with cells and tissue, superparamagnetic nature, PEGylated iron oxide NPs may, therefore, be able to increase contrast in MRI images more effectively than typical paramagnetic Gadolinium (Gd)-based contrast agents. Additionally, leaky vasculatures near tumors cause PEGylated iron oxide NPs with an appropriate particle size to accumulate in tumor sites through the enhanced permeability and retention (EPR) effect. Developing PEGylated iron oxide NPs with controlled size and shape-restricted size distribution is preferable to gain advantage from their high-quality and regular imaging property in MRI and precise tumor targeting (Yue-Jian et al. 2010).

PEGylated iron oxide NPs can replace the currently utilized gadolinium-based contrast agents in MRI, which have demonstrated poor blood lifetime and low proton relaxation efficiency. These limitations have led to greater administration doses, increasing human and environmental toxicity. Results indicate that these particles may be used for therapeutic applications and longitudinal imaging studies due to their high specific absorption rate (SAR) values. These biocompatible NPs are a potent tool for biomedical applications, primarily because of the current interest in non-invasive diagnostic assays based on real-time and long-term tracking and monitoring of tagged cells (i.e., tumor-related macrophages) (Lazaro-Carrillo et al. 2020). There are many different biomedical and bioengineering applications for PEGylated iron oxide NPs, which belong to the family of magnetic materials known as ferrimagnetic materials. There are different types of iron oxide-based NPs like maghemite (y-$Fe_2O_3$), magnetite ($Fe_3O_4$), and mixed ferrites ($MFe_2O_4$), where M = Co, Mn, Ni, or Zn). PEGylated iron oxide NPs are produced after surface modification. They can be used for magnetic resonance imaging, magnetic particle imaging (MPI), targeted delivery of drugs, proteins, antibodies, and nucleic acids, separation of biomolecules, hyperthermia, biosensing, and tissue repair. These current exhaustive applications are brought about by magnetic characteristics and the fact that they may be synthesized in various forms and sizes. When exposed to an external magnetic field, superparamagnetic iron oxide nanoparticles have significant magnetic moments, and even when the magnetic field is removed, there is no longer any magnetic moment. In preclinical and clinical trials, many iron oxide NPs have been examined, and a number of them have been commercialized.

### 12.2.2 PEGylated Magnetic Nano Assemblies in Diagnosis

Magnetic nanocrystals made in an organic phase can be employed as contrast agents because of their well-defined crystal structures and excellent magnetic sensitivity. To increase their colloidal stability in the aqueous phase, additional surface modification with a hydrophilic layer, for example, PEG, is needed because they are disseminated in the organic phase. PEG has a high hydrophilicity, a low cytotoxicity, and a high cell permeability. Specifically, it helps reduce nonspecific interactions with serum proteins by creating hydrogen bonds with water molecules, which extends the duration that the substance is in circulation in the blood. For PEG-based materials, there are two primary surface modification techniques. The first one is the exchange technique, in which the hydrophobic ligand on the magnetic nanocrystal (MNC) surface is replaced with PEG at elevated temperatures, and the second technique is the addition technique, in which a hydrophobic MNC's surface is covered with the emulsion technique using a PEG-based amphiphilic polymer as a surfactant (Fig. 12.2).

Using this addition technique, it is possible to load MNCs, fluorescent materials, and drugs together to create a multifunctional nanocomposite. It is challenging to obtain constantly sized NPs, particularly nanoclusters. Because it also triggers the transition between the super magnetic–ferrimagnetic transitions, growing the size of MNCs to increase the saturation magnetization has limitations. Instead, it has been suggested that creating magnetic nanoclusters successfully maintains superparamagnetic behavior with high magnetization.

Due to their high magnetic susceptibility, low coercive force, and high magnetic characteristics, magnetic NPs made up of several single MNCs are very appealing. Due to the PEG molecules on the particle surface (PEGylation effect), the PEGylated MNs displayed better durability in an aqueous phase for a longer period of time than those of a commercial MRI contrast agent, as well as an appropriate MRI contrast

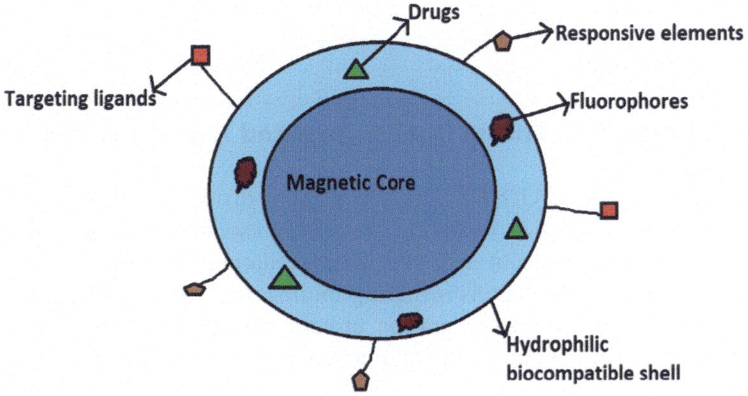


**Fig. 12.2** An illustration of a magnetic nanoparticle structure for targeting ligands, fluorophores, and responsive elements

effect due to the magnetic clustering effect. Additionally, in vivo tests confirm that PEGylated MNCs have a potential value as MRI agents for cancer diagnosis (Kang et al. 2019). An external magnetic field can influence magnetic NPs, indicating their ability to be site-specific. MNPs are an excellent choice for the simultaneous delivery of diagnostic and treatment, commonly known as theranostics; due to their imaging capabilities and magnetic characteristics, MNPs have a higher theranostic potential than liposomes or other polymer-based nanoparticles. Owing to their capacity to be simultaneously directed, visualized, and heated by external magnetic fields, MNPs are useful in theranostics (Alromi et al. 2021). Magnetic nanoparticles are being developed and employed in magnetic resonance imaging as contrast-enhancing agents for diagnosing a wide range of ailments, including cancer, cardiovascular disease, and neurological disorders.

Magnetic NPs that have been radioactively labeled have been shown to be highly effective at diagnosing cancer. Compared to conventional imaging methods, it has few key advantages. Due to the EPR effect, magnetic NPs labelled with radioactive tracers have precision targeting; they also have a high surface-to-volume ratio, enabling the use of high-density radioactive labels; and they are capable of complementary multimodal imaging, such as integrated imaging using various radionuclide nanoparticles.

Magnetic fluorescent nanocomposites that are superparamagnetic and fluorescent have been developed by integrating magnetic NPs with fluorescent elements. The efficiency of cell imaging is significantly increased when the magnetic fluorescent nanocomposites are introduced to living cells because the fluorescence can be monitored and programmed to be enriched at the desired area when an external magnetic field is applied.

In addition to having the general features of NPs, magnetic NPs also possess magnetic capabilities, which have recently attracted the attention of researchers in the field of nanomedicine. Due to their special characteristics, magnetics are extensively employed in biomedicine, drug-gene delivery, magnetic resonance imaging, molecular probes, tumor detection, tumor therapy, and other domains. Magnetic NPs will be utilized more frequently in tumor diagnosis and treatment as nanotechnology advances (Bian et al. 2021).

### 12.2.3 PEGylated Quantum Dots in Targeted Tissue Imaging

PEGylated quantum dots (QDs) are fluorescent NPs, distinguished by their extraordinary optical characteristics, including strong fluorescence emission, photostability, a compact emission spectrum, and a wide excitation wavelength. They are an intriguing option for applications involving bioimaging due to such features. The PEG coating helps them escape plasma protein adsorption and RES clearance, which prolongs the blood circulation of quantum dots in the body and increases the possibility that they will reach to their target area successfully. PEGylated QDs is currently a dynamic delivery system for diagnostic applications (Fig. 12.3).

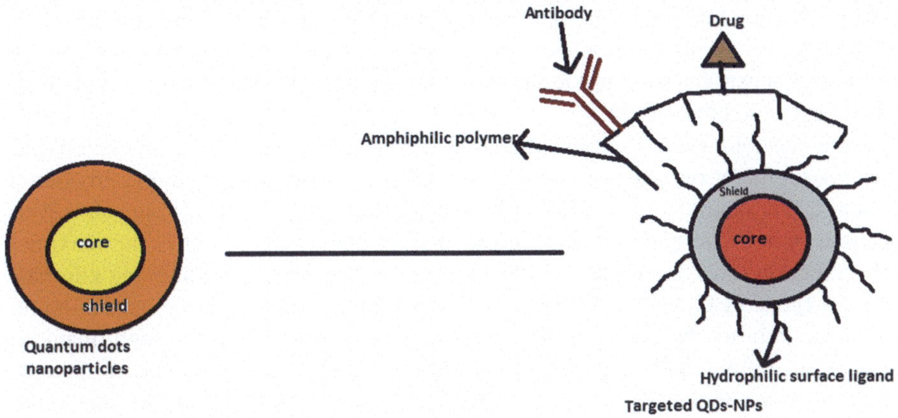


**Fig. 12.3** Targeting of quantum dots

Long-term dynamic imaging is a perfect application for PEGylated QDs because of their optical characteristics, specifically their capacity to emit fluorescence in the near-infrared (NIR) range and their resistance to photobleaching. As a result, they could be utilized in fluorescence-guided surgery (FGS) to speed up tumor removal and produce negative surgical margins and therefore enhancing the prognosis of a patient. In surgical oncology, FGS is a real-time intraoperative procedure, which is designed to highlight tumors during surgical resection to speed up tumor excision and provide negative tumor margins. The prognosis of a patient must be improved by complete tumor excision with negative margins. Additionally, PEGylated QDs have the ability to generate fluorescence in the near-infrared spectrum, which is of significant relevance to fluorescence-guided surgery (Dirheimer et al. 2022).

They provide enhanced fluorescence intensity, superior photobleaching resistance, size-tunable light emission, and the capacity to generate different fluorescent colors from a single excitation source. The optimum surface for coupling with a range of targets, such as antibodies, peptides, and several other small molecules, is offered by QDs. They might thus present effective alternatives to the more sensitive and focused cancer molecular targeting and bioimaging detection methods currently in use. Cancer imaging has evolved greatly since Gao intravenously administered PEG-coated QDs functionalized with antibodies to prostate-specific membrane antigen. Furthermore, the ability to resist bleaching for extended periods enables the collection of sharp, well-contrasted images that are particularly helpful for 3D optical sectioning of tumors and the environment around them because bleaching of fluorophores hinders the accurate reconstruction of 3D structures.

Additionally, their long-term stability, excellent brightness, wider and continuous excitation spectrum, and deep penetration make them the perfect choice for in vivo cancer diagnostics and bio-imaging. Some of the goals of creating new and improved cancer diagnostics and imaging probes have already been realized because of the rapid advancement of QDs technology. The active targeting of tumors is now achievable due to their characteristics and effective conjugation with biomolecules.

Despite the potential and large application of QDs in cancer imaging and detection to date, there are still obstacles to be cleared in terms of improving specificity, enhancing sensitivity, and reducing QDs toxicity before clinical applications may advance (Liang et al. 2021).

Aptamer-conjugated PEGylated QDs is recent advancement for targeting epidermal growth factor receptor variant III for fluorescence imaging of glioma. In glioma patients, the degree of resection plays a substantial predictive role. However, because tumors are naturally infiltrative, it is challenging to identify the maximal safe resection level. Recently, FGS, a novel technique for safely removing gliomas, has emerged. PEGylated QDs have a number of exceptional benefits, including chemical stability, broad absorption with narrow emission spectra, high quantum fluorescence yield, and water solubility. QDs have significantly increased biocompatibility when their shell is made of ZnS and coated with PEG. Additionally, the PEG coating helps them to evade plasma protein adsorption and RES clearance, which increases the likelihood that they will reach their intended area.

One of the most well-known near-infrared imaging probes was made using PEGylated Silver Sulphide QDs with special optical properties. They showed much-improved blood flow and tumor accumulation in vivo when compared to the original ones, making them great candidates for use as tumor imaging probes.

High-resolution whole-body blood vessel imaging of living mice was accomplished, and near-infrared imaging was also used to examine the biodistribution of QDs. The wide excitation range, resistance to photobleaching, and lack of toxic heavy metal components that PEGylated Silver Sulphide QDs display are all crucial for bioimaging. In addition to these fundamental qualities, surface features were essential for in vivo applications since they may directly affect the blood flow and biodistribution of the NPs. PEG grafting is one of the most popular methods for extending nanoparticle circulation and enhancing tumor accumulation by lowering nonspecific binding to proteins and cells. As a result, PEGylated Ag2S QDs offer a uniform structure for in vivo imaging using quantum dots (Lu et al. 2020).

### 12.2.4 PEGylated Gold NPs for Imaging Applications

PEGylated Gold nanoparticles (PEG-AuNPs) are highly biocompatible and have a variety of potential biological applications. PEG-AuNPs, which have had their surface modified with PEG, can be used to conjugate them with a variety of biological molecules. Unique features of gold nanoparticles include their small size, excellent biocompatibility, low toxicity, basic surface chemistry, and ease of surface modification. PEG-AuNPs are up-and-coming prospects for biomedical use with a wide range of biological applications, including as biosensors and drug delivery vectors for chemotherapy and radiation therapy for cancer.

However, due to absorption and rapid clearance by the RES, specifically macrophages, administration of nanoparticles to target tumor locations is typically constrained. To prevent the absorption of opsonin proteins, PEG-coated NPs (PEGylation) can provide a hydrophilic protective layer consequently reducing macrophage recognition and clearance and extending their duration in circulation. Regardless of the method of delivery, in vivo investigations have shown that PEGylated NPs have minimal adverse impact on various cell lines. PEGylation has been recognized as a particularly efficient and biocompatible way for decreasing renal clearance (Huang et al. 2014) (Fig. 12.4).

PEG-AuNPs have a negative charge, various biomolecules, including pharmaceuticals, genes, and targeting ligands, that can easily functionalize them. Surface plasmon resonance (SPR) bands are present in PEG-AuNPs, which also have an ultra-small size, a macroscopic quantum tunnelling effect, and a distinct surface effect. PEG-AuNPs confirm the most promising material for a variety of biomedical applications, such as biosensing, molecular imaging, drug carriers, and so on (Kong et al. 2017).

Synthesized PEG-AuNPs have highly unique plasmonic features that enable the detection of various compounds using SER spectroscopy, which can be done on solid spots or in liquid droplets. The SER spectra obtained using this new class of nanoparticles on various molecules of interest (methylene blue, rhodamine 6G, doxorubicin, and 5-fluorouracil) are highly reproducible due to their distinctive plasmonic properties, making them excellent candidates for further use as SERS substrates (Niţică et al. 2018).

Due to the unique properties of AuNPs, they have long been thought of as a possible tool for cancer diagnostics and delivery of drug applications. Among these characteristics are a high surface area-to-volume ratio, stable nature, surface plasmon resonance, surface chemistry, and ease of synthesis. Additionally, the non-toxic and non-immunogenic properties of AuNPs as well as their EPR effect offer additional benefits by making it simple for drugs to penetrate and accumulate at tumor sites. AuNPs are used mainly in ultrasensitive detection and imaging-based therapeutic approaches necessary for treating fatal diseases like cancer because of their optical qualities. Numerous varieties of AuNPs, including gold nanorods, nanocages, nanostars, nanocubes, and nanospheres, have been developed for human cancer and cell biology.

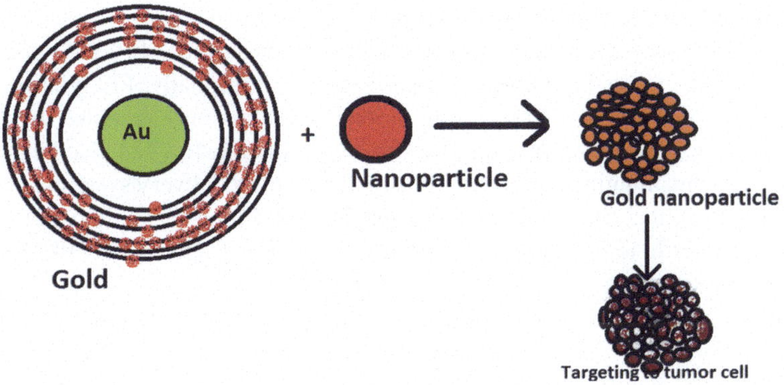


**Fig. 12.4** Gold nanoparticles targeting tumor cells

AuNPs' optical characteristics depend on SPR. The SPR is a mechanism that involves the resonance of gold electrons in response to an incoming radiation, which causes them to simultaneously absorb and scatter light. Photoimaging can help identify tumors in their early stages and direct them towards precise surgical intervention. Having a clear understanding of where the tumor ends and the healthy tissue begins is one of the main issues facing surgeons today. Throughout a surgical procedure, the surgeons must decide how much of the tumor must be removed; if they are too conservative, they risk leaving some tumor cells behind, and if they are too liberal, they risk removing healthy tissues that may be crucial. The majority of tumors come back over time because most treatments are highly conservative.

The tumors are only recognized when they reach a particular threshold since MRI and Computed Tomography (CT) scans have limitations and can only detect tumors above a size of several millimeters or roughly ten million cells. Gold nanoparticles are targetedly injected into the tumor, where they precisely bond to the cancer cells and scatter (shine), making it simpler for the surgeons to distinguish between the tumor and healthy cells. This revolutionary method of cancer treatment is called photoimaging. Due to their bioinertness and capacity to boost spatial and temporal resolution for imaging, gold nanoparticles (nanorods, nanocages, and nanoshells) are regarded as the best photo-imaging nanoparticles for cancer therapies currently on the market (Singh et al. 2018).

### 12.2.5 PEGylated PLGA Nanocarriers in Diagnostic Applications

Poly(D, L-lactic-co-glycolic acid) (PLGA) has been extensively used among the range of organic or inorganic nanomaterials available for the development of drug delivery systems for cancer therapy and detection because of its biocompatibility and biodegradability.

Since lactic acid and glycolic acid, which are organic, non-toxic substances that can eventually be broken down into water and carbon dioxide, may be produced when PLGA and PLA are broken down, they have been utilized more frequently than other materials in the production of nanocarriers. PLGA additionally gained FDA approval for usage in medical applications.

The different targeting moieties can be added to the surface of PLGA-based drug delivery systems to enable them to target tumors. Additionally, PEGylation of the outer surface of the PLGA-based drug delivery systems may increase their blood circulation time. The imaging properties for tumor diagnostics can also be acquired by integrating the imaging agents into PLGA-based drug delivery systems. Potential candidates for cancer therapies and diagnostics, called "cancer theranostics" can be developed by co-loading therapeutic drugs and imaging agents into PLGA-based drug delivery systems.

By applying external stimuli such as photo, ultrasound, magnetic field, and radiofrequency to the targeted areas, remotely triggered cancer therapy can enable the selective and accurate removal of tumors and controlled release of chemotherapeutics. Additionally, remotely triggered cancer therapy can control when and how long the treatment lasts, allowing precise treatment and reducing systemic toxicity. In addition to hydrophilic therapeutic drugs and imaging agents, therapeutic hydrophobic medicines and imaging agents can also be encapsulated into PLGA NPs by using the water-in-oil-in-water (W/O/W) double emulsion method. This may enhance the targeting and bioavailability of hydrophobic drugs and imaging agents and ultimately lead to the accomplishment of cancer theranostics. A promising anticancer strategy that can selectively and accurately eliminate solid tumors and remotely control the drug release is using outside factors to initiate cancer therapy (Shen et al. 2020).

PLGA NPs delivered intravenously are rapidly eliminated by the RES and are easily opsonized. A new class of amphiphilic block co-polymer nanoplatform, PEG-PLGA NPs, has been developed through conjugating PEG with PLGA to enhance the characteristics of PLGA nanoparticles and achieve long-term therapeutic benefits. PLGA-based contrast agent composite nanomaterials' biocompatibility is substantially improved by PEGylation. Through the EPR effect, PEG-PLGA NPs modified with ligands can target specific receptors on the tumor surface, facilitating tumor targeting. However, more study is needed to determine the biocompatibility, toxicity, and safety of these nanocarriers to ensure their clinical applicability. Studies must continue searching for ways to customize NPs to recognize tumors as new cancer-specific target molecules are constantly being found. This strategy could potentially open new avenues for effective cancer treatment and diagnostic imaging (Zhang et al. 2022).

## 12.3 Role of PEGylated Nanocarriers in Diagnosis of Diseases

Coating the surface of NPs with PEG, or "PEGylation," is a commonly used approach for improving the diagnostic and imaging application employed for various critical diseases.

### 12.3.1 PEGylated Nanocarriers in Cancer Tissue Diagnosis

In the United States, cancer is the second greatest cause of morbidity and mortality, and it is predicted that rates will rise for at least few years to come. Successfully delivering innovative therapeutic medicines to the target site while preventing adverse side effects from systemic treatment is a significant challenge. The EPR effect has become widely recognized due to the particular physiology of tumors, which includes fenestrated vasculature and inadequate lymphatic drainage. By this process, the tumor–neovasculature's wide gaps between adjacent endothelial cells enable passive targeting of the tumor site. At the same time, inadequate lymphatic outflow increases the retention of macromolecular therapies within the tumor mass.

The ability of the therapies to reach their designated cellular and intracellular target locations while minimizing accumulation and action at nonspecific areas is crucial for the success of anticancer therapies. PEG surface modification of nanoparticulate carriers has emerged as a strategy to improve hydrophobic drug solubility, extend circulation time, reduce nonspecific uptake, and enable specific tumor-targeting through the EPR effect.

Additionally, PEG modification has become a platform for incorporating active targeting ligands, offering the drug and gene carriers unique tumor-targeting abilities through a flexible tether. Longer circulation improves the possibility that the nanocarriers will reach the tumor interstitium, where it may be possible to accumulate a more considerable portion of the given dose than the tumor would experience when medications are administered parenterally without a carrier (Van Vlerken et al. 2007). Due to PEG's ability to allow for prolonged bloodstream circulation of the NPs followed by passive accumulation in the tumor tissue, it has been widely researched for the production of tumor-targetable NPs The PEG surface, in particular, makes it possible for NPs to bypass the RES and minimizing their clearance from the liver site. Future brain tumor treatment is likely to be significantly impacted by nanotechnology. Especially, NPs could revolutionize brain research, tumor imaging, surgery, and adjuvant therapies. MRI is one of the most advanced nanotechnology applications for diagnosing brain tumors.

As MRI contrast agents, several NPs have been developed. Iron oxide (IO) crystals have produced NP-based contrast agents with an organic shell-like PEG. The primary advantage of NPs-based materials is that they can give more accurate information regarding the size of a tumor. Both gadolinium-based contrast agents and NPs-based contrast agents increase tumors by traveling through regions with damaged blood-brain barrier (BBB), where they change the magnitude of the MRI signal. IO-based NPs can be used to find improved MRI. Additionally, IO-based NPs have a propensity to persist longer inside the tumor and more precisely delineate tumor margins than freely diffusing gadolinium chelates. NPs also have the ability to target molecular tumors very specifically. Developing NPs with specific cellular imaging capabilities, such as stem cells or endothelial cells, is feasible. As tumor-specific contrast agents, several IO NPs now under research appear promising. NPs-based MRI contrast agents have the potential to make tumor areas visible that would not have been seen with conventional MRI, particularly at the tumor–brain interface. Furthermore, NPs may eventually enhance the accuracy of brain tumor resection (Orringer et al. 2009).

### 12.3.2 PEGylated Nanocarriers in Diagnosis of Glioma–Brain Tumor

The most prevalent and deadly malignant brain tumor is glioma. Due to a lack of effective treatment, glioblastoma (GBM) patients often only survive for 12–15 months after diagnosis. The most successful treatment for glioma patients is surgical resection, and maximizing safe resection is essential in improving the prognosis of patients with low or high-grade gliomas. However, the infiltrative nature of the tumors makes it challenging to strike a balance between maximum cytoreduction and preservation of normal brain tissue. It is believed that small glioma cells that infiltrate the region around the primary tumor are the cause of over 90% of recurrent tumors, which manifest within a margin of 2–3 cm from the primary site. It has been established that one of the most significant predictors of overall survival and progression-free survival and neurological prognosis is the degree of resection.

Thus, maximum safe resection becomes tremendously essential. Therefore, a technique that can delineate the boundary of tumors to realize maximum safe resection and protect the normal brain tissues is urgently needed. It has been proved that intraoperative fluorescein is used to guide the surgical resection and can improve the overall survival and progression-free survival of patients. Although research suggests that fluorescence, such as sodium fluorescein, 5-aminolevulinate acid, and others, can be used to guide the removal of gliomas, their clinical utility is constrained by problems such as low quantum fluorescence yield and poor photostability.

To deliver chemotherapy medications to the brain, various NP compositions have been studied. Most of these formulations made use of polymers that complied with the strict standards necessary to be approved for use in biological applications. Nanoparticle technology has improved with the optimization of drug loading, encapsulation effectiveness, and release profile during the past few years. Brain cancer imaging has been made easier by using NPs with ligands attached to their surfaces. To shield NPs from RES and blood protein interactions, PEGylation of NPs has been frequently used in medication delivery.

PEGylated QDs are fluorescent nanocarriers with many notable benefits in the field of bio-imaging for therapeutic and diagnostic purposes. Compared to the standard fluorescent materials, maximum safe resection becomes extremely important since it is essential for improving the prognosis of patients with low- or high-grade gliomas by surgical resection. However, both grade gliomas tend to penetrate deeply into the parenchyma around them, so resecting them is challenging. Therefore, it is crucial to develop a method that makes it possible to see the tumor's border to carry out the safest possible resection and safeguard the healthy brain tissues. First, we have to search for a biomarker that is only expressed in tumors and not in healthy organs.

In contrast to normal tissues, epidermal growth factor receptor (EGFRvIII) had only been found in gliomas, lung cancer, breast cancer, squamous cell carcinoma of the head and neck, and colorectal cancer. EGFRvIII is a tumor-specific antigen or one that is exclusively expressed in tumors and not in normal tissues. EGFRvIII is, therefore, the perfect target for the treatment of gliomas. Second, the antibody EGFRvIII, like many drugs, is too large to pass across the BBB. Thus, a small chemical that can cross the BBB and specifically bind to EGFRvIII is required. PEGylated QDs can be used to mark the EGFRvIII (Tang et al. 2017).

Another study suggests that the oncogene BMI-1, a central protein in the polycomb group, could be an entirely novel therapeutic target for GBM. PTC209, a BMI-1 inhibitor, has been incorporated into a PLGA-PEG NP coupled with CD133 antibody (Nano-PTC209) to increase effectiveness and decrease toxicity. NPs significantly increased apoptosis and significantly reduced cell viability in a dose-dependent manner. Additionally, they markedly reduced cells' capacity for migration and increased the production of reactive oxygen species (Poonaki et al. 2021).

### 12.3.3 PEGylated Nanocarriers in Diagnosis of Gastrointestinal Tract Imaging

Molecular imaging techniques have a profound impact on the diagnosis and prognosis of disorders, including those affecting the gastrointestinal tract. Colorectal cancer is the second biggest leading cause of death, closely followed by stomach cancer for mortality. The keys to effective treatments for gastrointestinal cancer are early detection and fast intervention. Gastrointestinal endoscopy is an invasive procedure; gas-barium double-contrast imaging is resistant to gastrointestinal cancer in its early stages, which increases the likelihood of mistaken diagnoses. The non-invasive methods used to diagnose a variety of disorders with the advancement of medical examination technology include MRI, CT, single-photon emission computed tomography (SPECT), ultrasonography (US), positron emission tomography (PET), as well as optical imaging techniques.

Designing and synthesizing contrast agents with strong imaging properties that are appropriate for gastrointestinal imaging is therefore crucial. Additionally, it is more realistic to diagnose and predict GI disease by doping other materials to do various imaging modes. PEGylated NPs performed better on CT scans and absorbed more X-ray than iodinated contrast material at the same concentrations. The NPs might also be applied to MRI in the presence of Gd. PEGylated nanomaterials had shown excellent biocompatibility and low systemic toxicity in both in vitro and in vivo toxicity tests. PEGylated NPs are excellent techniques as CT/MRI dual-modal contrast agents for in vivo imaging of the GI tract. These NPs showed improvement in CT imaging compared to the commercially available iodine contrast agents (Cui et al. 2020).

The two varieties of MRI contrast agents now in use are $T_1$ and $T_2$ contrast agents. Gadolinium and Mn are primarily used in $T_1$ contrast agents. $Fe_3O_4$ NPs with a superparamagnetic field are $T_2$ agents. Targetable contrast agents are a crucial component of molecular imaging, which could significantly improve the precision and accuracy of monitoring. Various $Mn_3O_4$-based contrast agents have been reported for tumor imaging, including MRI with a mix of fluorescence. A Mn-based contrast agent surface coated with PEG exhibits good results for MRI of stomach cancer monitoring. The MRI effects were instantly markedly amplified by these PEGylated NPs, which could effectively concentrate in stomach cancer tissues and cells. Additionally they have the potential to serve as a novel MRI contrast agent for the postoperative monitoring of gastric cancer (Li et al. 2020).

An efficient oral MRI contrast agent for gastrointestinal tract imaging is NPs of gadolinium-incorporated Prussian blue with PEG coating. It exhibits high $r_1$ relaxivity (i.e., high sensitivity) and excellent temporal stability, with contrast enhancement characteristics remaining nearly unchanged when passing throughout the GI tract. Other commonly employed oral contrast agents lack the ability to pass through cell membranes to serve as cellular MRI probes, have modest relaxivity values, and exhibit poor temporal stability when administered to the GI tract. Based on its applicability as MRI contrast agent, it is tempting to hypothesize that gadolinium-incorporated Prussian blue presents the potential for developing a sensitive cellular MRI probe for early cancer detection in the GI tract. These PEGylated NPs have very high $r_1$ relaxivity and can rapidly penetrate the cell (Perera et al. 2016).

## 12.4 Conclusion and Future Resolutions

PEGylated components are typically used to prevent opsonization and protein binding in serum for therapeutic proteins and NPs to show extended circulation periods. It has been shown to be a practical scaffold for incorporating active targeting ligands and influencing intracellular target localization. PEG has significantly developed drug delivery systems, particularly for detecting, diagnosing, and treating tumors. PEG's adaptability will make it possible to investigate new applications and make ongoing advancements to anticancer therapies and diagnosis. Preclinical and clinical trials for more PEGylated proteins and short peptides are anticipated to begin soon. There are also numerous PEGylated nanoparticle medicinal and diagnostic compounds in active development. Accelerating the clinical translation of innovative PEGylated agents, including carbon nanotubes, nanodiamonds, microbubbles, dendrimers, and other nanoparticles, will require using existing technology and creating new techniques in the diagnostic application.

A number of key requirements, such as non-toxicity, non-immunogenicity, target selectivity, and biodegradability, need to be fulfilled by nanomedicines in order for them to be commercially successful for the diagnostic application. The production of multifunctional PEG molecules that can be grafted with anti-biofouling compounds, fluorescent agents, and other functional polymers is critical in developing PEGylated NPs. PEGylated nanocarriers have a promising future in biomedical research and diagnostics, particularly in the fields of disease diagnosis, early detection, cellular and deep tissue imaging, drug/gene delivery, and multifunctional therapies. These efforts would essentially simplify, accelerate, and minimize the invasiveness of the diagnosing processes. PEGylated nanocarriers would be an appealing material for diagnostic and imaging applications in the following years and could alter the normal business practices of pharmaceutical companies..

## References

Alconcel SN, Baas AS, Maynard HDJPC (2011) Fda-approved poly (ethylene glycol)–protein conjugate. Drugs 2:1442–1448

Alromi DA, Madani SY, Seifalian A (2021) Emerging application of magnetic nanoparticles for diagnosis and treatment of cancer. Polymers 13:4146

Baetke SC, Lammers T, Kiessling F (2015) Applications of nanoparticles for diagnosis and therapy of cancer. Br J Radiol 88:20150207

Banerjee SS, Aher N, Patil R, Khandare J (2012) Poly(ethylene glycol)-prodrug conjugates: concept, design, and applications. J Drug Deliv 2012:103973

Bence AK, Adams VR, Piascik PJJOTAPA (2002) Pegfilgrastim: A new therapy to prevent neutropenic fever. J Am Pharm Assoc (Wash) 42:806–808

Bian X, Guo T, Zhang J, Xia J, Feng X, Wang F, Lin M, Tian W (2021) The magnetic nanomaterial biofunctions in cancer diagnosis and therapy. J Nanomater 2021:9968166

Booth C, Gaspar HB (2009) Pegademase bovine (peg-Ada) for the treatment of infants and children with severe combined immunodeficiency (Scid). Biologics 3:349–358

Cho EC, Glaus C, Chen J, Welch MJ, Xia Y (2010) Inorganic nanoparticle-based contrast agents for molecular imaging. Trends Mol Med 16:561–573

Choi KY, Jeon EJ, Yoon HY, Lee BS, Na JH, Min KH, Kim SY, Myung SJ, Lee S, Chen X, Kwon IC, Choi K, Jeong SY, Kim K, Park JH (2012) Theranostic nanoparticles based on Pegylated hyaluronic acid for the diagnosis, therapy and monitoring of colon cancer. Biomaterials 33:6186–6193

Cui F-Z, Liu J-H, Liu Y, Yuan B-Y, Gong X, Yuan Q-H, Gong T-T, Wang L (2020) Synthesis of pegylated Bagdf5 nanoparticles as efficient Ct/Mri dual-modal contrast agents for gastrointestinal tract imaging. Chin J Anal Chem 48:1004–1011

Dinndorf PA, Gootenberg J, Cohen MH, Keegan P, Pazdur R (2007) Fda drug approval summary: Pegaspargase (Oncaspar) for the first-line treatment of children with acute lymphoblastic leukemia (all). Oncologist 12:991–998

Dirheimer L, Pons T, Marchal F, Bezdetnaya L (2022) Quantum dots mediated imaging and phototherapy in cancer spheroid models: State of the art and perspectives. Pharmaceutics 14:2136

Herndon TM, Demko SG, Jiang X, He K, Gootenberg JE, Cohen MH, Keegan P, Pazdur R (2012) U.S. food and drug administration approval: Peginterferon-alfa-2b for the adjuvant treatment of patients with melanoma. Oncologist 17:1323–1328

Huang YC, Yang YC, Yang KC, Shieh HR, Wang TY, Hwu Y, Chen YJ (2014) Pegylated gold nanoparticles induce apoptosis in human chronic myeloid leukemia cells. Biomed Res Int 2014:182353

Illés E, Szekeres M, Tóth IY, Farkas K, Földesi I, Szabó Á, Iván B, Tombácz E (2018) Pegylation of superparamagnetic iron oxide nanoparticles with self-organizing Polyacrylate-peg brushes for contrast enhancement in Mri diagnosis. Nanomaterials (Basel) 8:776

Kang B, Lim J, Son H-Y, Choi Y, Kang T, Jung J, Huh Y-M, Haam S, Lim E-K (2019) Pegylated magnetic Nano-assemblies as contrast agents for effective T2-weighted MR imaging. Nanomaterials (Basel) 9:410

Karakoti AS, Das S, Thevuthasan S, Seal S (2011) Pegylated Inorganic nanoparticles. Angew Chem Int Ed Engl 50:1980–1994

Kong FY, Zhang JW, Li RF, Wang ZX, Wang WJ, Wang W (2017) Unique roles of gold nanoparticles in drug delivery, targeting and imaging applications. Molecules 22:1445

Lang L (2008) Fda approves Cimzia to treat Crohn's disease. Gastroenterology 134:1819

Lazaro-Carrillo A, Filice M, Guillén MJ, Amaro R, Viñambres M, Tabero A, Paredes KO, Villanueva A, Calvo P, Del Puerto Morales M, Marciello M (2020) Tailor-made peg coated iron oxide nanoparticles as contrast agents for long lasting magnetic resonance molecular imaging of solid cancers. Mater Sci Eng C Mater Biol Appl 107:110262

Li SD, Huang L (2010) Stealth nanoparticles: high density but Sheddable peg is a key for tumor targeting. J Control Release 145:178–181

Li K, Li P, Wang Y, Han S (2020) Manganese-based targeted nanoparticles for postoperative gastric cancer monitoring via magnetic resonance imaging. Front Oncol 10:601538

Liang Z, Khawar MB, Liang J, Sun H (2021) Bio-conjugated quantum dots for cancer research: detection and imaging. Front Oncol 11:749970

Lu F, Ju W, Zhao N, Zhao T, Zhan C, Wang Q, Fan Q, Huang W (2020) Aqueous synthesis of Pegylated ag(2)S quantum dots and their in vivo tumor targeting behavior. Biochem Biophys Res Commun 529:930–935

Niţică Ş, Moldovan AI, Toma V, Moldovan CS, Berindan-Neagoe I, Ştiufiuc G, Lucaciu CM, Ştiufiuc R (2018) Pegylated gold nanoparticles with interesting Plasmonic properties synthesized using an original, rapid, and easy-to-implement procedure. J Nanomater 2018:5954028

Orringer DA, Koo YE, Chen T, Kopelman R, Sagher O, Philbert MA (2009) Small solutions for big problems: the application of nanoparticles to brain tumor diagnosis and therapy. Clin Pharmacol Ther 85:531–534

Perera VS, Chen G, Cai Q, Huang SD (2016) Nanoparticles of gadolinium-incorporated Prussian blue with peg coating as an effective Oral Mri contrast agent for gastrointestinal tract imaging. Analyst 141:2016–2022

Poonaki E, Ariakia F, Jalili-Nik M, Shafiee Ardestani M, Tondro G, Samini F, Ghasemi S, Sahab-Negah S, Gorji A (2021) Targeting Bmi-1 With Plga–Peg nanoparticle-Containing Ptc209 Modulates the behavior of human glioblastoma stem cells and cancer Cells. Cancer Nanotechnol 12:5

Rubio T (2015) Summaries of safety labeling changes approved by Fda—boxed warnings highlights October–December 2014, vol 72. Oxford University Press, Oxford, p 343

Shan X, Yuan Y, Liu C, Tao X, Sheng Y, Xu F (2009) Influence of peg chain on the complement activation suppression and longevity in vivo prolongation of the Pcl biomedical nanoparticles. Biomed Microdevices 11:1187–1194

Shen X, Li T, Xie X, Feng Y, Chen Z, Yang H, Wu C, Deng S, Liu Y (2020) PLGA-based drug delivery systems for remotely triggered cancer therapeutic and diagnostic applications. Front Bioeng Biotechnol 8:381

Shi L, Zhang J, Zhao M, Tang S, Cheng X, Zhang W, Li W, Liu X, Peng H-S, Wang Q (2021) Effects of polyethylene glycol on the surface of nanoparticles for targeted drug delivery. Nanoscale 13:10748

Singh P, Pandit S, Mokkapati V, Garg A, Ravikumar V, Mijakovic I (2018) Gold nanoparticles in diagnostics and therapeutics for human cancer. Int J Mol Sci 19:1979

Suk JS, Xu Q, Kim N, Hanes J, Ensign LM (2016) Pegylation as a strategy for improving nanoparticle-based drug and gene delivery. Adv Drug Deliv Rev 99:28–51

Tang J, Huang N, Zhang X, Zhou T, Tan Y, Pi J, Pi L, Cheng S, Zheng H, Cheng Y (2017) Aptamer-conjugated Pegylated quantum dots targeting epidermal growth factor receptor variant iii for fluorescence imaging of glioma. Int J Nanomedicine 12:3899–3911

Vallabani NVS, Singh S (2018) Recent advances and future prospects of iron oxide nanoparticles in biomedicine and diagnostics. 3 Biotech 8:279

Van Vlerken LE, Vyas TK, Amiji MM (2007) Poly(ethylene glycol)-modified Nanocarriers for tumor-targeted and intracellular delivery. Pharm Res 24:1405–1414

Verhoef JJ, Anchordoquy TJ (2013) Questioning the use of Pegylation for drug delivery. Drug Deliv Transl Res 3:499–503

Veronese F, Pasut G (2005) Veronese, F.M. & Pasut, G. Pegylation, successful approach to drug delivery. Drug Discov Today 10:1451–1458

Working PK, Newman MS, Huang SK, Mayhew E, Vaage J, Lasic DD (1994) Pharmacokinetics, biodistribution and therapeutic efficacy of doxorubicin encapsulated in stealth® liposomes (Doxil®). J Liposome Res 4:667–687

Yue-Jian C, Juan T, Fei X, Jia-Bi Z, Ning G, Yi-Hua Z, Ye D, Liang G (2010) Synthesis, self-assembly, and characterization of peg-coated iron oxide nanoparticles as potential Mri contrast agent. Drug Dev Ind Pharm 36:1235–1244

Zhang M, Du Y, Wang S, Chen B (2020) A review of biomimetic nanoparticle drug delivery systems based on cell membranes. Drug Des Devel Ther 14:5495–5503

Zhang D, Liu L, Wang J, Zhang H, Zhang Z, Xing G, Wang X, Liu M (2022) Drug-Loaded Peg-Plga nanoparticles for cancer treatment. Front Pharmacol 13:990505

# Multifunctional PEGylated Nanoparticles in Theragnosis

# 13

Akhilesh Tiwari, Muktika Tekade, Shubham Ramdas Mule, Girish Meravanige Basavarajappa, and Rakesh Kumar Tekade

**Abstract**

Target-specific therapeutic modalities solve shortcomings associated with traditional medication administration and therapy. The success of liposomes, polymeric nanoparticles (NPs), dendrimers, micelles, protein-drug conjugates, and other nanoplatforms in improving the pharmacokinetics and biodistribution of the therapeutic after systemic administration has led to promising results in clinical settings and market traction. NPs have been the gold standard for medication administration for millennia; however, this conventional method still faces problems with drug solubility, sustained payload release, and getting the necessary treatment to the targeted sickness site. NPs-mediated drug administration and targeting are further improved by encasing nanoformulations in a stealth coating made of hydrophilic polymers authorized by the FDA, such as PEG, chitosan, polyacrylamides, etc. The NPs are protected by this surface layer against opso-

A. Tiwari (✉)
Department of Pharmacy, Indira Gandhi National Tribal University, Amarkantak, Anuppur, Madhya Pradesh, India

M. Tekade
School of Pharmacy, Devi Ahilya Vishwavidyalaya, Takshila Campus, Indore, Madhya Pradesh, India

School of Pharmaceutical Sciences, Lovely Professional University, Phagwara, Punjab, India

S. R. Mule · R. K. Tekade
National Institute of Pharmaceutical Education and Research (NIPER) Ahmedabad, An Institute of National Importance, Government of India, Department of Pharmaceuticals, Ministry of Chemicals and Fertilizers, Palaj, Opposite Air Force Station, Gandhinagar, Gujarat, India

G. M. Basavarajappa
Department of Biomedical Sciences, College of Medicine, King Faisal University, Al-Ahsa, Saudi Arabia



R. K. Tekade, N. K. Jain (eds.), *PEGylated Nanocarriers in Medicine and Pharmacy*, https://doi.org/10.1007/978-981-97-7316-9_13

nization, immune system recognition, and aggregation. The benefits and requirements for the PEGylation of NPs are discussed in this chapter. Here, the focus is on NPs as a means of drug administration and how the PEGylation of NPs enhances their therapeutic utility. Several PEGylated NPs that have demonstrated promising outcomes for both systemic and nonsystemic drug delivery have also been highlighted.

**Keywords**

PEG · PEGylated · Theragnostic · Nanoparticles · Pharmacokinetic · Biodistribution

## 13.1 Introduction

The term "theragnosis" in medicine refers to the simultaneous diagnosis and treatment of disorders. Combining therapeutic and diagnostic procedures makes this possible. It combines the fields of therapeutics and diagnostics to improve patient care and outcomes. This technique has drawn a lot of interest in the field of customized medicine, where therapies may be modified based on specific patient features. The term "theragnosis" refers to various methods and tools that make it possible to find specific biomarkers associated with particular disease conditions, provide customized medications, and track the efficacy of treatments in real time. Several theragnostic elements are being developed in recent years.

Genetic testing, also referred to as molecular diagnostic technologies, is crucial to the diagnosis since it can identify distinctive biomarkers connected to diseases. Genetic or epigenetic components may be included in these biomarkers. Several molecular diagnostic methods are used to find and assess these biomarkers, including polymerase chain reaction (PCR), next-generation sequencing (NGS), and microarray technologies (Arora et al. 2023). Breast cancer can be diagnosed with molecular diagnostic tools such as PCR, NGS, and gene profiling, which help identify altered biomarker expression levels and genetic mutations (Litton et al. 2019). Imaging technologies used for the noninvasive visualization and characterization of illnesses and advanced imaging methods are crucial components of theragnostic endeavors. Theragnosis frequently employs methods including magnetic resonance imaging (MRI), positron emission tomography (PET), computed tomography (CT), and ultrasound. These imaging techniques can reveal essential details of the location, severity, and treatment response of the disease (Weissleder and Pittet 2008).

In order to administer therapeutic medications to injured tissues with the fewest side effects on healthy tissues, theragnosis employs tailored drug delivery devices. The nanoparticles (NPs), liposomes, and other drug carriers can be functionalized with unique ligands that target disease-specific receptors or biomarkers. This targeted approach reduces systemic toxicity and enhances treatment efficacy. Theragnosis enables real-time treatment response monitoring, enabling any required adjustments to be made to therapy. Biomarkers, imaging techniques, and wearable technologies may all be used to assess the progression of a disease, the effectiveness of treatment interventions, and to help make well-informed decisions. The

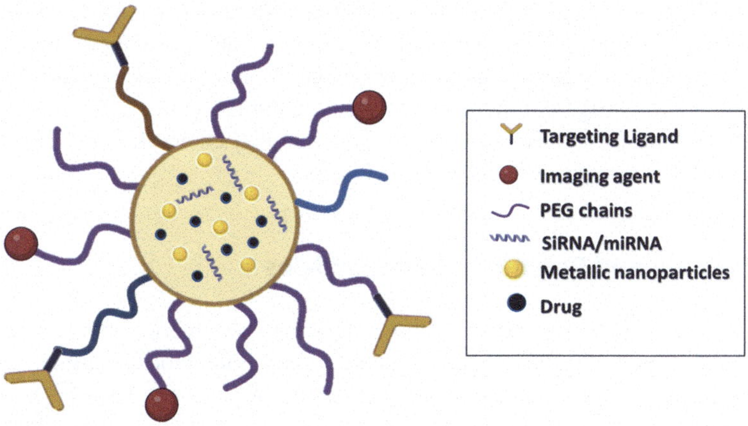


**Fig. 13.1** Multifunctional PEGylated NPs

identification of compounds having multiple uses that have both medicinal and diagnostic capabilities is referred to as theragnostic. For instance, medicines and imaging agents for disease visualization for specialized medical treatment can both be transported by diagnostic NPs. These agents assist individualized therapy by delivering medication and real-time disease information at the same time. The theragnosis concept can potentially revolutionize the healthcare sector by improving disease detection, treatment efficacy, and patient outcomes. By enabling doctors to base decisions on the specifics of each patient, it allows them to perform medicine more accurately and personally. The schematic representation of multifunctional PEGylated NPs is given in Fig. 13.1. In this chapter, the details of PEGylated NPs in the theragnostic domain are discussed thoroughly, emphasizing their therapeutic applications.

## 13.2 Development of Different PEGylated Nanoparticles for Theragnosis

PEGylated NPs have shown promise as theragnostic platforms since they combine therapy and diagnostics into one device. These NPs boost medication transport, increase therapeutic efficacy, and enable noninvasive imaging for diagnostics by fusing the benefits of nanotechnology with those of polyethylene glycol (PEG). Herein, a succinct summary of how various PEGylated NPs for theragnosis has been well elaborated.

### 13.2.1 PEGylated Lipid-Based NPs

Using lipid-based NPs like liposomes and solid lipid NPs (SLNs), theragnosis has been extensively studied. These NPs are more stable, have longer circulation times, and are less absorbed by the reticuloendothelial system (RES), thanks to the addition of PEG. PEGylated liposomes have been investigated for the combined delivery of therapeutic medicines and imaging agents. For instance, doxorubicin, a chemotherapeutic drug, and indocyanine green, a fluorescent dye, have been delivered together via PEGylated liposomes for combined chemotherapy and fluorescence imaging (Grabarnick Portnoy et al. 2021).

### 13.2.2 PEGylated Polymeric Nanoparticles

When it comes to design and functionalization, polymeric NPs are flexible. Polymeric NPs that have been PEGylated are more stable, biocompatible, and have longer circulation times. For theragnosis, a variety of polymers have been utilized, including poly (lactic-co-glycolic acid) (PLGA), polyethyleneimine (PEI), and polymeric micelles. It has been looked at using PEGylated PLGA NPs to simultaneously deliver medicinal and imaging agents. For instance, PEGylated PLGA NPs loaded with the fluorescent probes quantum dots and the chemotherapeutic drug paclitaxel have been created for combination therapy and fluorescence imaging (Li et al. 2001).

### 13.2.3 PEGylated Metal-Based Nanoparticles

Gold nanoparticles (AuNPs) and iron oxide nanoparticles (IONPs) are two examples of metal-based NPs with distinctive optical and magnetic properties appropriate for therapy and imaging. Metal-based NPs can be surface functionalized, which increases their stability and biocompatibility. For combination of photothermal therapy and photoacoustic imaging, PEGylated AuNPs have been used. For instance, PEGylated AuNPs have been used to administer paclitaxel and photothermal agents such as indocyanine green simultaneously for the purposes of imaging and synergistic therapy (Kim et al. 2006).

### 13.2.4 Multifunctional PEGylated Nanocarriers

Multifunctional PEGylated nanocarriers used in combination therapy have drawn much interest from the drug delivery community. These nanocarriers have several benefits, including increased drug solubility, longer circulation times, targeted administration, and the capacity to incorporate several therapeutic agents. Several research activities highlighting the utility of multifunctional PEGylated nanocarriers in managing various diseases have been undertaken in the past few years.

Interestingly, multifunctional PEGylated NPs that can simultaneously transport diagnostic modalities such as MRI contrast agents and anticancer medications have been well explored. The successful encapsulation of chemotherapeutic drugs and NPs within the nanocarriers with targeting capabilities such as receptor-mediated targeting or magnetic targeting has been demonstrated by many researchers independently. The multifunctional nanocarriers have demonstrated improved treatment efficacy and the potential for imaging-based real-time drug delivery monitoring. Yim et al. formulated pH-responsive PEGylated gold NPs, which were further conjugated with an anticancer drug, doxorubicin via citraconic anhydride linkage. The NPs were additionally labeled with radioisotopes $^{123}I$ or $^{124}I$ for detection using PET. The citraconic anhydride linkage degraded in response to the acidic conditions at cancerous cells, releasing the doxorubicin intracellularly. Clear PET images of the cellular accumulation of gold NPs were obtained along with the PET images of the clearance pattern of the gold NPs, and significant cell mortality was achieved due to the presence of doxorubicin (Yim et al. 2021).

Hence, it can be concluded that the development of PEGylated NPs for the theragnostic applications opens a new avenue of opportunities for managing various diseases. Persistent research in this domain is required to develop clinically applicable theragnostic modalities.

### 13.2.5 Targeted PEGylated Nanocarriers

A fundamental difficulty in the clinical treatment of cancer is the effective and site-specific administration of therapeutic medicines. Researchers have studied a range of liposomes, micelles, and polymeric NPs as nanoscale carriers to improve the bioavailability and pharmacokinetic properties of medications. One such process is the increased permeability and retention effect (EPR effect) (Wu 2021). It might be possible to improve treatment by attaining targeted delivery to the cancer cells or the tumor vasculature by conjugating targeting ligands onto nanocarriers. It has been demonstrated that the administration of receptor-targeted nanocarriers enhances therapeutic benefits both in vitro and in vivo. There have been studies on several different ligands, such as folate, transferrin, antibodies, peptides, and aptamers.

NPs can be made to perform a wide range of tasks, including intracellular medication release and imaging. It is crucial to remember that the choice of ligands for PEGylated nanocarriers' surface functionalization depends on the precise target tissues or cells and the intended therapeutic outcomes. Jin et al. developed PLGA-PEG NPs with further surface modification with the GE11 peptide, an EGFR-targeting peptide. The NPs so prepared were utilized to deliver curcumin to the breast cancer cells. Upon in vitro and in vivo examination of the NPs, it was reported that the phosphoinositide 3-kinase signaling was significantly reduced along with reduced viability of cancer cells and suppressed tumor burden compared to free curcumin and nontargeted NPs (Jin et al. 2017).

## 13.3 Chemistry of PEGylated NPs

The field of nanomedicine frequently employs PEGylated NPs, also referred to as PEG NPs, as a form of drug delivery platform. PEGylation is the technique of creating a PEG corona around NPs by covalently connecting PEG chains to the surface of the NPs (Li et al. 2014). The PEGylated NPs have many beneficial qualities, including increased stability, longer circulation times, less immunogenicity, and more effective drug delivery.

It is crucial to understand the chemistry behind developing and evaluating PEGylated NPs to formulate the PEGylated NPs with desired physicochemical properties and therapeutic outcomes. In this section, the chemistry of PEGylated NPs has been discussed thoroughly.

### 13.3.1 Surface Modification Chemistry

PEG chains are covalently attached to the NPs' surface through various chemical processes during the surface modification of NPs with PEG. Thiol-ene, amine-carboxylic acid, and maleimide-thiol reactions are typical techniques used for surface modification with PEG (Kharkar et al. 2016). In the amine-carboxylic reaction, the carboxylic group on the surface of NPs and the amine group in the structure of PEG react to form stable amide bonds (Rahme and Dagher 2019). The amine-carboxylic reactions are carried out using carbodiimide chemistry such as N-hydroxysuccinimide (NHS) or 1-ethyl-3-(3-dimethylaminopropyl) carbodiimide (EDC) (Ahmad et al. 2015).

In the thiol-maleimide reaction, maleimide-modified NPs are conjugated with thiol-functionalized PEG to form the stable thioether bond (Martínez-Jothar et al. 2018). These processes enable the attachment of PEG molecules with various functional groups and lengths to the surface of NPs. The method of PEGylation is primarily governed by reaction parameters such as temperature, pH, and reaction time (Suk et al. 2016). Serious consideration of all these factors must be given to obtain the final product with the desired attributes.

### 13.3.2 Polymer Chemistry

The creation, evaluation, and subsequent conjugation of PEG polymers to the surface of NPs all depend heavily on polymer chemistry. PEG polymers with various MWs and structural types are used in the PEGylation of NPs. The preferred characteristics of the NPs, such as stability, drug-loading capacity, and release kinetics, will determine which PEG polymer is used (Suk et al. 2016; Zielińska et al. 2020). The higher molecular weight PEGs offer better steric hindrance and enhance stealth properties (Salmaso and Caliceti 2013). This ultimately increases the circulation time of NPs in the bloodstream. However, the higher MWs of PEG affect the tissue penetration of the PEG-modified NPs, which necessitates the thorough determination of polymer selection criteria. The functional groups on the surface of PEG, such as thiol, amine, maleimide, and carboxyl, govern the choice of chemistry for further modification of NPs with PEG. For instance, using thiol maleimide chemistry, thiol-terminated PEG can be conjugated with maleimide-modified NPs.

### 13.3.3 Synthesis of NPs

PEGylated NPs generation depends heavily on the chemistry involved in the synthesis of NPs. Various techniques are used to create NPs with diverse sizes, shapes, and surface qualities, including nanoprecipitation, emulsion/solvent evaporation, and self-assembly. The effectiveness and stability of the PEGylated NPs are influenced by the chosen method of NP synthesis. Hence, the selection of the process of synthesis of PEGylated NPs is a crucial aspect to consider to generate NPs with desired pharmacokinetic attributes (Goldberg et al. 2007).

### 13.3.4 Characterization Techniques

PEGylated NPs are characterized by utilizing a variety of analytical techniques. These include transmission electron microscopy (TEM), which gives insights into the morphological characteristics of the NPs, and dynamic light scattering (DLS) is used to measure the particle size and zeta potential of the NPs. Fourier-transform infrared spectroscopy (FTIR), a well-known technique for functional group detection, can be utilized to determine the surface modification of the NPs and the conjugation of PEG with the desired entity of interest (Younis et al. 2021). Nuclear magnetic resonance (NMR) is a preferred technique for detecting the conjugation of the PEG with the molecules of interest. The NPs can be further characterized by X-ray diffraction analysis, which determines the crystallinity of the prepared formulation (Mourdikoudis et al. 2018). In the case of metallic PEGylated NPs, the identification and determination of the metallic content of the NPs is necessary. This can be achieved using energy-dispersive X-ray (EDX) analysis and inductively coupled plasma-mass spectrometry (ICP-MS).

## 13.4 Mechanism of Enhanced Drug Delivery Through PEGylated NPs

PEGylation is the term used to describe the process of attaching PEG chains to NPs' surfaces. Utilizing PEGylated NPs to enhance drug pharmacokinetics and therapeutic efficacy is known to enhance the therapeutic aspects of drug delivery by various mechanisms. Some crucial outcomes associated with the PEGylation of NPs that significantly impact the in vivo behavior of the prepared PEGylated NPs are shown. The mechanisms of enhanced drug delivery through PEGylated NPs are shown in Fig. 13.2.

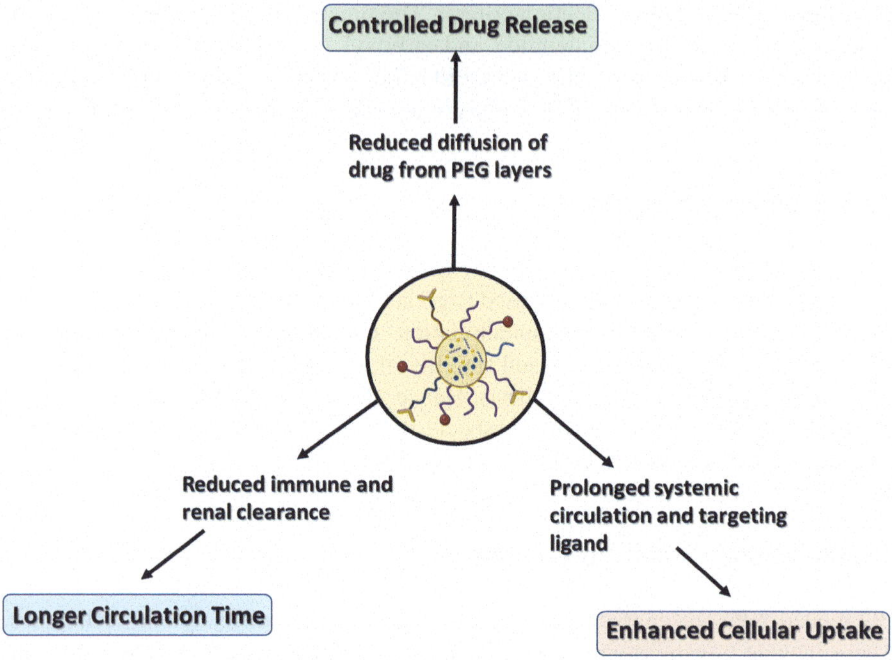


**Fig. 13.2** Mechanism of enhanced drug delivery through PEGylated NPs

### 13.4.1 Longer Circulation Time

PEGylation reduces the RES's ability to recognize and opsonize NPs, giving them a stealth effect. The hydrophilic PEG chains form a steric barrier that delays opsonization and subsequent clearance by the RES (Gref et al. 1994). This prolongs the circulation period of NPs in the bloodstream. The effectiveness of medication delivery is improved by the accumulation of NPs at the target site due to the lengthened circulation duration.

The PEG coating on the surface of NPs reduces nonspecific interactions with plasma proteins and cell surfaces, limiting absorption by macrophages and other immune cells, resulting in decreased clearance and enhanced biodistribution (Alexis et al. 2008). This decrease in nonspecific interactions promotes the availability of NPs at the target site and slows down fast clearance. The EPR effect, which passively permits PEGylated NPs to collect in tumor tissues with leaky vasculature, further improves drug delivery to solid tumors (Maeda et al. 2009).

### 13.4.2 Enhanced Cellular Uptake

By removing numerous obstacles, PEGylation can enhance the cellular uptake of NPs. PEGylation is a proven technique to lessen the serum proteins' ability to adhere to the surface of NPs, which can impede cellular uptake (Veronese and Harris 2002). Depending on the ligands connected to the PEG chains, the presence of PEG on the surface of NPs can also promote endocytosis by engaging with particular cell surface receptors, such as the asialoglycoprotein receptor or folate receptor (Torchilin 2007). This active targeting strategy enhances the specificity and effectiveness of drug delivery to specific cell types. PEGylation combined with further surface modification with the targeting ligand will enhance the cellular uptake multifold owing to the improved circulation time due to PEGylation and receptor-mediated cellular uptake due to the targeting ligand.

### 13.4.3 Controlled Drug Release

PEGylation can alter how quickly medications are released from NPs. Because PEG chains are hydrophilic, they prevent medication from diffusing out of the NPs matrix, producing a prolonged release profile. PEG layers on the surface of NPs prevent the rapid diffusion of the drug, thereby prolonging the release rate. To fine-tune the release rate and obtain the desired therapeutic results, the size and structure of the PEG chains can be changed to achieve the desired drug release pattern.

## 13.5 Application of Theragnosis in Different Therapies

A new area of medicine known as theragnostic combines therapy and diagnostics, intending to create individualized treatments based on the unique features of each patient's ailment. It entails using diagnostic tools to pinpoint specific molecular targets or biomarkers, followed by administering therapeutic therapies explicitly made for those targets. Some of the theragnostic use in various therapy are accompanied by pertinent sources.

### 13.5.1 Cancer Theragnostic

The goal of cancer prognosis is the creation of patient-specific tailored medicines. Specific tumor biomarkers can be seen and measured utilizing molecular imaging techniques like PET or single-photon emission computed tomography (SPECT). This makes it possible to choose the best treatments, including radiotherapy or tailored drug administration. Choi et al. developed the PEG-hyaluronic acid (HA) NPs to target CD44 receptors overexpressing cancer cells. The NPs were further labeled with cyanine 5.5 for the fluorescence-based detection of the localization of the NPs (Choi et al. 2010). It was reported that a significant number of the administered NPs were localized at the tumor site, which suggests the high tumor targetability of the prepared NPs. These results suggest the possible application of the PEGylated NPs for the theragnosis of cancer.

### 13.5.2 Cardiovascular Theragnostic

Cardiovascular theragnostics aim to create individualized treatments for heart failure and atherosclerosis, two cardiovascular illnesses. Determining the extent of disease progression and prospective therapy targets is possible using noninvasive imaging methods like CT or MRI. Gene therapy and targeted drug delivery are examples of therapeutic approaches. Ren et al. formulated the PEG-PLGA NPs loaded with miR-30b-5p for targeted therapy of cardiac failure. Echocardiography, HE/Masson staining, and immunofluorescence were utilized to assess the effects of the therapy on the cardiac structure and function. Western blotting and reverse transcription-polymerase chain reaction (RT-PCR) were done to evaluate the expression of markers associated with inflammation and cardiac hypertrophy. The miR-30b-5p-loaded PEG-PLGA NPs improved cardiac function, myocardial injury, and regulated markers related to inflammation and cardiac hypertrophy (Ren et al. 2021).

### 13.5.3 Theranostics for Neurological Conditions

This field focuses on the detection and treatment of neurological conditions like brain tumors, Parkinson's disease, and Alzheimer's disease. Functional MRI (fMRI) and PET are the two examples of molecular imaging techniques that can be used to identify specific molecular targets or gauge the effectiveness of a treatment. Therapeutic approaches include gene therapy, deep brain stimulation, and targeted drug delivery. Liu et al. developed PEG-poly(lactic) acid NPs and imparted surface modification using B6 peptides to target clathrin-mediated and lipid-raft-mediated endocytosis. The prepared NPs were loaded with the neuroprotective peptide NAPVSIPQ (NAP) for the management of the Alzheimer's disease. B6-decorated NPs demonstrated higher accumulation in brain capillary endothelial cells. The administration of the NPs in mouse models of AD showed improvement in learning functions and amelioration of hippocampal neuronal loss and cholinergic disruption. This suggests the utility of PEGylated NPs in managing neurological conditions (Liu et al. 2013).

## 13.6 Diagnosis and Therapy Through PEGylated NPs

PEGylated NP-based diagnosis and treatment have drawn much interest from the medical community. Considering PEG's advantages, such as longer circulation time and targeting ability with surface modification with targeting ligands, the PEGylated NPs offer significant promise as a theragnostic platform for various diseases. Hereby, the uses and advantages of PEGylated NPs in treatment and diagnostics have been thoroughly discussed in this section.

### 13.6.1 Cancer Diagnosis and Therapy

PEGylated NPs have been thoroughly investigated for cancer treatment and diagnostics. They have many benefits, including increased tumor accumulation, extended circulation duration, and regulated medication release. Several investigations in this regard are reported with promising results. Choi et al. developed PEG-conjugated hyaluronic acid NPs to deliver anti-cancer drugs doxorubicin and camptothecin. The developed NPs were evaluated for their efficacy against the cancer cells SCC7 and MDA-MB-231 and the animal tumor model. In tumor-bearing mice, the PEGylated NPs demonstrated selective uptake into the tumor site after systemic administration. This can be attributed to PEG-mediated prolonged circulation time and hyaluronic acid-mediated CD44 receptor targeting. No significant increase in tumor size was observed for up to 35 days, suggesting anti-cancer activity (Choi et al. 2011).

### 13.6.2 Imaging and Contrast Agents

PEGylated NPs can be customized with imaging agents to serve as contrast agents in a variety of imaging modalities, including CT, MRI, and fluorescence imaging (Mulder et al. 2009). Zhou et al. developed PEGylated AuNPs for the imaging of tumor sites. The developed nanosystems explicitly accumulated in the tumor site owing to the EPR effect and demonstrated better X-ray attenuation properties than the marketed small molecular iodinated contrast agent and prolonged half-decay time. The CT imaging has been successfully done using the above-developed AuNPs, which hints towards the utility of PEGylated systems as contrast agents for imaging applications (Zhou et al. 2014).

### 13.6.3 Gene Therapy and RNA Delivery

PEGylated NPs have also demonstrated potential in gene therapy and RNA delivery applications. They can effectively transfer genetic material and RNA-based therapies to target cells, enabling gene editing and regulation. Santos et al. have developed liposomes of 1,2-distearoyl-sn-glycero-3-phosphoethanolamine-poly(ethylene glycol) (DSPE-PEG) with further surface modification with hexapeptide for the delivery of SiRNA. The developed liposomes demonstrated increased uptake by SCLC cells and HMEC-1 microvascular endothelial cells. PEG, in this case, can ensure enhanced circulation time in vivo, whereas hexapeptide has been used for targeting purposes (Santos et al. 2010).

## 13.7 Multifunctional PEGylated Nanoparticles in Theragnosis

In theragnosis, which is the fusion of therapeutic and diagnostic activities within a single platform, PEGylated NPs are frequently employed. These NPs have more excellent stability, longer circulation times, and better drug delivery, among other benefits. A concise overview of multifunctional PEGylated NPs in theragnosis, along with relevant examples, is given as follows.

### 13.7.1 PEGylation for Enhanced Stability

By minimizing nonspecific interactions with biological components, PEGylation improves stability by adding PEG chains to NPs. This alteration prevents opsonization, which can cause the RES to clear substances quickly (Owens 3rd and Peppas 2006). PEGylation has been routinely used to increase the stability and circulation time of theragnostic NPs. The PEG chains around the PEGylated NPs create a steric barrier that prevents the aggregation of the NPs thereby imparting higher stability to them. Adjusting the zeta potential of the NPs to meet the stability criteria can also be employed along with steric stabilization via PEG to attain additional stability.

### 13.7.2 Prolonged Circulation Time

Longer circulation times are also a result of PEGylation, which is referred to as the "stealth effect." Because the hydrophilic PEG chains provide a steric barrier that shields against immune system identification and uptake, NPs can circulate for a longer time (Blanco et al. 2015). Conversely, PEGylation increases the overall molecular weight of the PEGylated moieties, thereby reducing renal clearance (Gupta et al. 2019). This characteristic is particularly essential for theragnostic NPs since it enhances the possibility of accumulating at the target region. Hence, increased molecular weight as well as reduced immune clearance due to PEG modification are jointly responsible for the prolonged circulation time.

### 13.7.3 Improved Drug Delivery

Multifunctional PEGylated NPs can effectively transport therapeutic materials, such as medicines or genes, to the appropriate location by encapsulating them. By enhancing stability during circulation and avoiding early drug release and degradation, the PEG coating enhances the overall pharmacokinetic behavior of the PEGylated NPs. Furthermore, targeting ligands can be functionalized onto the surface of PEGylated NPs to enable selective identification and binding to the intended cells or tissues (Peer et al. 2007). This active targeting increases the effectiveness of medicine delivery. Nanocarriers that target specific receptors can be used to treat diseases that show overexpression of certain receptors. For instance, folate receptors and CD44 receptors have been widely exploited for the targeted therapy of various cancers (Kesharwani et al. 2021).

### 13.7.4 Diagnostic Functionality

Theragnostic NPs combine drug delivery with diagnostic capabilities, enabling simultaneous imaging and therapy. These NPs can transport imaging agents like fluorescent dyes or contrast agents in various imaging modalities such as MRI, CT, or optical imaging (Han et al. 2019). The diagnostic element aids in defining the target site, tracking therapy effectiveness, and directing therapeutic choices. Multifunctional PEGylated NPs have considerable potential for diagnosis because they combine better stability, prolonged circulation time, improved drug delivery, and diagnostic capabilities. However, theragnosis has many limitations that need to be addressed, even if it offers a lot of potential for individualized therapy and better patient outcomes.

## 13.8 Technical Challenges

Theragnosis entails creating cutting-edge imaging and sensing technologies that can precisely identify and track the course of disease. However, these methods have technological constraints, including sensitivity, resolution, and specificity, which can impact the precision and dependability of diagnostic data. To address these technical problems, more research and development is required. The following section enlists the various technical challenges associated with the clinical applications of the PEGylated NPs in theragnosis.

### 13.8.1 Limited Availability of Targeted Therapies

By locating certain molecular targets linked to diseases, theragnosis aims to provide targeted therapeutics. Targeted treatments are, however, currently scarce for many illnesses. More tailored treatment drugs that diagnostic data can lead must be developed for theragnosis to realize its full potential. Developing theragnostic modalities for all diseases is not practically possible. This puts reasonable restrictions on the broad-scale application of the theragnostic NPs.

### 13.8.2 Cost and Accessibility

The use of pricy diagnostic imaging techniques and specialized therapeutic drugs during the implementation of the theragnosis could considerably raise the expense of treatment (Mangalath et al. 2014). This might make theragnosis less available to patients, especially in places with limited access to healthcare resources. To guarantee wider access, efforts must be made to lower the price and increase the availability of theragnostic technology.

### 13.8.3 Ethical and Regulatory Considerations

The use of personal health information, patient privacy, and informed permission are all crucial ethical and legal issues that are raised by the theragnosis (Coletta et al. 2020). The integration of diagnostic and therapeutic activities also questions the proper use and interpretation of diagnostic data for treatment decisions. The widespread implementation of theragnostic methods depends on creating precise rules and laws to address these moral and legal concerns.

### 13.8.4 Limited Clinical Validation

Although encouraging preclinical and early clinical research has shown theragnosis' potential, additional clinical validation is required to prove its efficacy and utility in treating various diseases and patient demographics. Extensive clinical studies and meticulous evaluation are needed to assess the clinical advantages, financial viability, and long-term effects of theragnostic techniques (Fogel 2018). It is crucial to remember that theragnosis is a rapidly developing discipline, and future technological improvements and continuing research are expected to resolve some of these limits. Theragnosis has the potential to transform customized medicine and enhance patient care with ongoing innovation and cooperation..

## 13.9 Regulatory Requirements of Theragnosis

The term "theragnosis" refers to the fusion of treatment and diagnosis, in which therapeutic actions are individualized and guided by diagnostic findings. It entails the discovery of certain biological targets or biomarkers that assist in choosing the best course of action for every patient. To the best of our knowledge, theragnosis is not the subject of any specific regulatory requirements. Regulatory standards for diagnostics and personalized treatment may differ by nation or location. Through the Food and Drug Administration (FDA) in the United States, specifically the Center for Devices and Radiological Health (CDRH) and the Center for Biologics Evaluation and Research (CBER), controls diagnostic procedures and equipment. The FDA oversees diagnostic procedures, including those used in diagnosis, to verify their efficacy and safety (Genzen 2019).

The FDA classifies typically diagnostic tests as either in vitro diagnostic (IVD) tools or laboratory-developed tests (LDTs) (Sarata and Johnson 2014). LDTs are tests created and carried out within a single laboratory, whereas IVD devices are commercially distributed tests. For each of these categories, the FDA has a different regulatory process, such as premarket approval (PMA), 510(k) clearance, and the laboratory-developed test (LDT) policy (Sarata and Johnson 2014). Customized medication and diagnostics are subject to regulatory oversight in the European Union (EU). IVDs are governed by the in vitro Diagnostic Regulation (IVDR) 2017/746, which became effective in May 2022 (Sarata and Johnson 2014). The IVDR establishes standards for the reliability, efficiency, and clinical validity of IVDs.

It is vital to remember that regulatory requirements are subject to periodic changes. For the most up-to-date and correct information on the regulatory requirements for theragnosis in particular, it is recommended to keep a tab of all the regulatory changes.

## 13.10 Conclusion

The quest for the development of enhanced therapeutic and diagnostic techniques fuels the research endeavors in the realm of drug delivery. One of the most important outcomes of these efforts is the development of PEGylated NPs in theragnosis, a domain dealing with both therapeutics and diagnostics. PEGylated NPs demonstrate remarkable properties such as reduced immunogenicity, increased stability, and increased circulation time, enabling them to deliver drugs precisely, minimize the off-target effects, and maximize therapeutic efficacy.

Targeted PEGylated NPs thereby reduce the need for high drug doses, ultimately reducing the systemic toxicity of drugs. The unique surface properties of PEGylated NPs allow the attachment of various ligands, such as targeting moieties and imaging agents, allowing them to attain therapeutic as well as diagnostic functions simultaneously. This helps in the early detection of cancers, wherein an early intervention can bring significant patient outcomes in terms of survival rates. Despite the advantages that PEGylated NPs offer, they present themselves with many technical challenges, such as achieving precise control over the size of NPs and their surface properties, which significantly affect the pharmacokinetic behavior of NPs. Achieving multifunctionality while tackling the challenge of increased complexity is another prime challenge associated with developing PEGylated NPs. Furthermore, there is a need to have clear regulatory guidelines and standardized protocols for the development of theragnostic modalities such as multifunctional PEGylated NPs. Overall, the PEGylated NPs offer various advantages in the domain of theragnostic intervention of various diseases, which hints towards their potential clinical utility.

## References

Ahmad Z, Shepherd JH, Shepherd DV, Ghose S, Kew SJ, Cameron RE, Best SM, Brooks RA, Wardale J, Rushton N (2015) Effect of 1-Ethyl-3-(3-Dimethylaminopropyl) Carbodiimide and N-Hydroxysuccinimide concentrations on the mechanical and biological characteristics of cross-linked collagen Fibres for tendon repair. Regen Biomater 2:77–85

Alexis F, Pridgen E, Molnar LK, Farokhzad OC (2008) Factors affecting the clearance and biodistribution of polymeric nanoparticles. Mol Pharm 5:505–515
Arora N, Chaudhary A, Prasad A (2023) Editorial: methods and applications in molecular diagnostics. Front Mol Biosci 10:1239005
Blanco E, Shen H, Ferrari M (2015) Principles of nanoparticle design for overcoming biological barriers to drug delivery. Nat Biotechnol 33:941–951
Choi KY, Chung H, Min KH, Yoon HY, Kim K, Park JH, Kwon IC, Jeong SY (2010) Self-assembled hyaluronic acid nanoparticles for active tumor targeting. Biomaterials 31:106–114
Choi KY, Yoon HY, Kim J-H, Bae SM, Park R-W, Kang YM, Kim I-S, Kwon IC, Choi K, Jeong SY, Kim K, Park JH (2011) Smart Nanocarrier based on Pegylated hyaluronic acid for cancer therapy. ACS Nano 5:8591–8599
Coletta RD, Yeudall WA, Salo T (2020) Grand challenges in oral cancers. Front Oral Health 1:3
Fogel DB (2018) Factors associated with clinical trials that fail and opportunities for improving the likelihood of success: a review. Contemp Clin Trials Commun 11:156–164
Genzen JR (2019) Regulation of laboratory-developed tests: a clinical laboratory perspective. Am J Clin Pathol 152:122–131
Goldberg M, Langer R, Jia X (2007) Nanostructured materials for applications in drug delivery and tissue engineering. J Biomater Sci Polym Ed 18:241–268
Grabarnick Portnoy E, Andriyanov AV, Han H, Eyal S, Barenholz Y (2021) Pegylated liposomes remotely loaded with the combination of doxorubicin, quinine, and indocyanine green enable successful treatment of multidrug-resistant tumors. Pharmaceutics 13:13
Gref R, Minamitake Y, Peracchia MT, Trubetskoy V, Torchilin V, Langer R (1994) Biodegradable long-circulating polymeric Nanospheres. Science 263:1600–1603
Gupta V, Bhavanasi S, Quadir M, Singh K, Ghosh G, Vasamreddy K, Ghosh A, Siahaan TJ, Banerjee S, Banerjee SK (2019) Protein Pegylation for cancer therapy: bench to bedside. J Cell Commun Signal 13:319–330
Han X, Xu K, Taratula O, Farsad K (2019) Applications of nanoparticles in biomedical imaging. Nanoscale 11:799–819
Jin H, Pi J, Zhao Y, Jiang J, Li T, Zeng X, Yang P, Evans CE, Cai J (2017) Egfr-targeting Plga-peg nanoparticles as a curcumin delivery system for breast cancer therapy. Nanoscale 9:16365–16374
Kesharwani P, Chadar R, Sheikh A, Rizg WY, Safhi AY (2021) Cd44-targeted Nanocarrier for cancer therapy. Front Pharmacol 12:800481
Kharkar PM, Rehmann MS, Skeens KM, Maverakis E, Kloxin AM (2016) Thiol-Ene click hydrogels for therapeutic delivery. ACS Biomater Sci Eng 2:165–179
Kim J, Park S, Lee JE, Jin SM, Lee JH, Lee IS, Yang I, Kim J-S, Kim SK, Cho M-H, Hyeon T (2006) Designed fabrication of multifunctional magnetic gold nanoshells and their application to magnetic resonance imaging and photothermal therapy. Angew Chem Int Ed Engl 45:7754–7758
Li Y, Pei Y, Zhang X, Gu Z, Zhou Z, Yuan W, Zhou J, Zhu J, Gao X (2001) Pegylated Plga nanoparticles as protein carriers: synthesis, preparation and biodistribution in rats. J Control Release 71:203–211
Li Y, Kröger M, Liu WK (2014) Endocytosis of Pegylated nanoparticles accompanied by structural and free energy changes of the grafted polyethylene glycol. Biomaterials 35:8467–8478
Litton JK, Burstein HJ, Turner NC (2019) Molecular testing in breast cancer. Am Soc Clin Oncol Educ Book 39:E1–E7
Liu Z, Gao X, Kang T, Jiang M, Miao D, Gu G, Hu Q, Song Q, Yao L, Tu Y, Chen H, Jiang X, Chen J (2013) B6 peptide-modified peg-Pla nanoparticles for enhanced brain delivery of neuroprotective peptide. Bioconjug Chem 24:997–1007
Maeda H, Bharate GY, Daruwalla J (2009) Polymeric drugs for efficient tumor-targeted drug delivery based on Epr-effect. Eur J Pharm Biopharm 71:409–419
Mangalath U, Aslam SA, Abdul Khadar AH, Francis PG, Mikacha MS, Kalathingal JH (2014) Recent trends in prevention of Oral cancer. J Int Soc Prev Community Dent 4:S131–S138

Martínez-Jothar L, Doulkeridou S, Schiffelers RM, Sastre Torano J, Oliveira S, Van Nostrum CF, Hennink WE (2018) Insights into Maleimide-thiol conjugation chemistry: conditions for efficient surface functionalization of nanoparticles for receptor targeting. J Control Release 282:101–109

Mourdikoudis S, Pallares RM, Thanh NTK (2018) Characterization techniques for nanoparticles: comparison and complementarity upon studying nanoparticle properties. Nanoscale 10:12871–12934

Mulder WJ, Strijkers GJ, van Tilborg GA, Cormode DP, Fayad ZA, Nicolay K (2009) Nanoparticulate assemblies of Amphiphiles and diagnostically active materials for multimodality imaging. Acc Chem Res 42(7):904–914. https://doi.org/10.1021/ar800223c

Owens DE 3rd, Peppas NA (2006) Opsonization, biodistribution, and pharmacokinetics of polymeric nanoparticles. Int J Pharm 307:93–102

Peer D, Karp JM, Hong S, Farokhzad OC, Margalit R, Langer R (2007) Nanocarriers as an emerging platform for cancer therapy. Nat Nanotechnol 2:751–760

Rahme K, Dagher N (2019) Chemistry routes for copolymer synthesis containing peg for targeting, imaging, and drug delivery purposes. Pharmaceutics 11:11

Ren Y, Wang X, Liang H, He W, Zhao X (2021) Mechanism of Mir-30b-5p-loaded Peg-Plga nanoparticles for targeted treatment of heart failure. Front Pharmacol 12:745429

Salmaso S, Caliceti P (2013) Stealth properties to improve therapeutic efficacy of drug nanocarriers. J Drug Deliv 2013:374252

Santos AO, Da Silva LCG, Bimbo LM, De Lima MCP, Simões S, Moreira JN (2010) Design of peptide-targeted liposomes containing nucleic acids. Biochim Biophys Acta Biomemb 1798:433–441

Sarata AK and Johnson JA (2014) Regulation of clinical tests: In Vitro Diagnostic (Ivd) Devices, Laboratory Developed Tests (Ldts), and genetic tests

Suk JS, Xu Q, Kim N, Hanes J, Ensign LM (2016) Pegylation as a strategy for improving nanoparticle-based drug and gene delivery. Adv Drug Deliv Rev 99:28–51

Torchilin VP (2007) Targeted pharmaceutical Nanocarriers for cancer therapy and imaging. AAPS J 9:E128–E147

Veronese FM, Harris JM (2002) Introduction and overview of peptide and protein Pegylation. Adv Drug Deliv Rev 54:453–456

Weissleder R, Pittet MJ (2008) Imaging in the era of molecular oncology. Nature 452(7187):580–589. https://doi.org/10.1038/nature06917

Wu J (2021) The enhanced permeability and retention (Epr) effect: the significance of the concept and methods to enhance its application. J Pers Med 11:771

Yim MS, Hwang YS, Bang JK, Jung D-W, Kim JM, Yi G-R, Lee G, Ryu EK (2021) Morphologically homogeneous, Ph-responsive gold nanoparticles for non-invasive imaging of Hela cancer. Nanomedicine 34:102394

Younis U, Rahi AA, Danish S, Ali MA, Ahmed N, Datta R, Fahad S, Holatko J, Hammerschmiedt T, Brtnicky M, Zarei T, Baazeem A, Sabagh AE, Glick BR (2021) Fourier transform infrared spectroscopy vibrational bands study of Spinacia Oleracea and Trigonella Corniculata under biochar amendment in naturally contaminated soil. PLoS One 16:E0253390

Zhou B, Zheng L, Peng C, Li D, Li J, Wen S, Shen M, Zhang G, Shi X (2014) Synthesis and characterization of Pegylated Polyethylenimine-entrapped gold nanoparticles for blood Pool and tumor Ct imaging. ACS Appl Mater Interfaces 6:17190–17199

Zielińska A, Carreiró F, Oliveira AM, Neves A, Pires B, Venkatesh DN, Durazzo A, Lucarini M, Eder P, Silva AM, Santini A, Souto EB (2020) Polymeric nanoparticles: production, characterization, toxicology and ecotoxicology. Molecules 25:3731

# Reversible PEGylation of Nanocarriers 14

Mahima Mishra, Sweety Shah, Gagandeep Kaur,
Aniket Navale, Heet Jani, Vaishnavi Chinkure,
and Rakesh Kumar Tekade

**Abstract**

Nanocarriers have transformed drug delivery by providing targeted and regulated distribution of therapeutic agents, as a result increasing efficacy and decreasing side effects. Among the various approaches used to maximize the performance of nanocarriers, polyethylene glycol (PEG)ylation has attracted a lot of interest due to its capacity to improve biocompatibility and extend the duration of circulation in vivo. Conventional PEGylation, often result in unsatisfactory therapeutic payload delivery and less cellular absorption. A potential solution to these issues is reversible PEGylation, which preserves effective drug release with cellular incorporation while providing the advantages of extended circulation. This review examines the principles and uses of reversible PEGylation in drug delivery and nanocarrier design. At one point, the review also examines the effects of reversible PEGylation concerning the pharmacokinetics, biodistribution and colloidal stability of nanocarriers. A discussion on the translational opportunities and difficulties related to reversible PEGylation, including stability, scalability, and regulatory issues is also incorporated.

M. Mishra · S. Shah · G. Kaur · A. Navale · H. Jani · V. Chinkure
National Institute of Pharmaceutical Education and Research (NIPER) Ahmedabad, An Institute of National Importance, Government of India, Department of Pharmaceuticals, Ministry of Chemicals and Fertilizers, Palaj, Opp. Air force station, Gandhinagar, Gujarat, India

R. K. Tekade (✉)
National Institute of Pharmaceutical Education and Research (NIPER) Ahmedabad, An Institute of National Importance, Government of India, Department of Pharmaceuticals, Ministry of Chemicals and Fertilizers, Palaj, Opp. Air force station, Gandhinagar, Gujarat, India
e-mail: rakeshtekade@niperahm.res.in



R. K. Tekade, N. K. Jain (eds.), *PEGylated Nanocarriers in Medicine and Pharmacy*, https://doi.org/10.1007/978-981-97-7316-9_14

**Keywords**

Reversible PEGylation · Nanocarriers · Polyethylene glycol (PEG) · Stimuli-responsive polymers · Cleavable linkers · pH-sensitive

## 14.1 Introduction

### 14.1.1 Nanocarrier in Drug Delivery

Pharmaceutical nanocarriers are extremely versatile, nanosized drug delivery system comprising liposomes, nanotubes, nanocomplexes, niosomes, polymeric, lipidic, and inorganic nanoparticles, among many more. Drug delivery systems using nanocarriers offer a cutting-edge approach for enhancing the effectiveness and accuracy of medication delivery (Alshawwa et al. 2022). These nanoscale carriers are intended to carry and distribute pharmaceutical substances, such as medications or genetic material, to particular target locations within the body. The principal objective is to optimize the therapeutic outcome while mitigating adverse reactions and lowering the treatment's overall toxicity. While nanocarrier-based drug delivery systems seem promising, there are still issues to be resolved, such as concerns regarding their long-term safety, their ability to be scalable for mass production, their possible toxicity, and their uptake by the reticuloendothelial system (RES) (Sun et al. 2015a).

### 14.1.2 PEGylation of Nanocarriers and Its Role

Nanostructures like solid lipid nanoparticles, polymeric nanoparticles, liposomes, micelles, and dendrimers offer several benefits, including the ability to deliver drugs to particular sites, defense against enzymatic degradation, low toxicity, stability, enhanced drug solubility, and the possibility of controlled release (Veronese and Pasut 2005). When creating sophisticated drug delivery systems, one typical tactic is to coat the surface of nanocarriers with substances that improve therapeutic delivery behavior. Although the synthesis of nanocarriers has advanced, little is understood about how they behave in intricate biological systems (Shete et al. 2022). The creation of a protein corona, primarily made up of plasma proteins like albumin, causes alterations in the surfaces of nanocarriers upon exposure to biological fluids.

The properties of nanomedicines, including as size, surface charge, aggregation resistance, and hydrodynamic size, can all be strongly impacted by this corona. Thus, these modifications could affect biodistribution, targeting ability, and biocompatibility—underscoring the need to take the protein corona into account when developing "smart" nanomedicines (Vllasaliu et al. 2014). Opsonization and absorption by the RES frequently result in unaltered nanostructures having short plasma half-lives (Thakur et al. 2015).

One of the recommended techniques for improving nanocarrier biocompatibility is PEGylation, which is accomplished by adsorbing, grafting, or entrapping poly (ethylene glycol) (PEG) onto surfaces improves stealth qualities are conferred by this technique, which surpasses other chemicals' ability to inhibit RES absorption. "Stealth" property mentioned in the context of nanocarrier's surfaces, diminishes their recognition or uptake by the cellular immune system, leading to longer circulation time, reduced dosage and frequency, and superior site-selective delivery of drugs (Howard et al. 2008).

### 14.1.3 Reversible PEGylation and Its Significance

The selection of a PEGylation strategy depends on various factors, such as the target parameter, entity function, and structure (drug molecule or colloidal carrier). Permanent modification via stable covalent binding of PEG strands to the agent or particle surface and releasable PEGylation via the use of specially designed linkers responsive to hydrolysis, enzymes, or certain stimuli are the two most investigated approaches. But the primary barrier to PEGylation is the inactivation that frequently occurs when a PEG chain covalently attaches to a protein, particularly peptide medications (Shechter et al. 2005).

A useful delivery platform is provided by reversible PEG modification, which permits control over pharmacological action and active agent regeneration. The conjugate created with this method becomes a "prodrug," slowly and spontaneously hydrolyzing under physiological conditions to release the native active peptide or protein drug from the inert conjugation over an extended period of time (Shechter et al. 2005). In a recent study, researchers used polyamidoamine dendrimers that have been loaded with doxorubicin and PEGylated with both acid-insensitive succinic linkage and acid-sensitive cis-aconityl linkage (Zhu et al. 2010).

Another research group employed this phenomenon to enhance the circulatory lifespan of peptide and protein drugs. A novel PEG-IFNR2 conjugate, PEG40-FMS-IFNR2, was developed. This conjugate, formed through a slowly hydrolyzable bond, allows for the regeneration of native interferon R2 (IFNR2) under physiological conditions (Fig. 14.1). The in vitro regeneration rate indicated a half-life of 65 hours. The pharmacokinetic profile, simulated numerically, demonstrated prolonged in vivo maintenance and active-native IFNR2 regeneration, offering an advantage over current formulations in terms of accessibility to peripheral tissues. Utilizing this reversible PEGylation event has various benefits, some of which were already covered (Peleg-Shulman et al. 2004; Rieger et al. 2009).

Drug delivery techniques such as reversible PEGylation are essential because they enable the temporary attachment of PEG chains to peptides and proteins, controlled release of the active agent, maintenance of the therapeutic activity of the drug while prolonging its half-life in circulation, and improvement of the flexibility and adaptability of nanomedicines (Shechter et al. 2005).

**Fig. 14.1** Diagrammatic representation showing PEGylation and reversible PEGylation

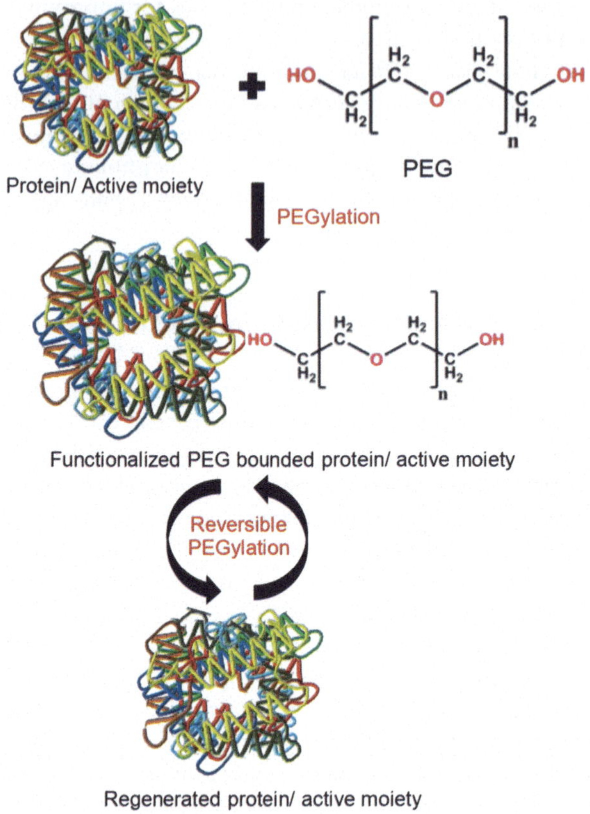


## 14.2 Reversible PEGylation Strategies

Reversible PEGylation allows for precise control over the PEGylation state, offering advantages in drug delivery, protein engineering, and bioconjugation. Reversible PEGylation is a novel technology that allows the release of native proteins, such as interferon-alpha 2 (IFN-alpha 2), over a prolonged period in vivo (Peleg-Shulman et al. 2004). PEGylation, or the covalent attachment of polyethylene glycol (PEG) chains, is a technique used to prolong the action of therapeutic proteins in the body. However, PEGylated peptides and proteins are often inactive due to their short-circulatory half-life (Peleg-Shulman et al. 2004). Reversible PEGylation aims to overcome the limitations of traditional PEGylation by using a combination of PEG and other components, such as 2-sulfo-9-fluorenylmethoxycarbonyl (FMS), to create a slowly hydrolyzable bond between the PEG and the protein (Peleg-Shulman et al. 2004). This allows the protein to retain its activity and be released gradually in the body, providing a sustained therapeutic effect.

For example, a study developed a PEG-IFNalpha2 conjugate, PEG (40)-FMS-IFNalpha2, which was capable of regenerating native interferon alpha 2 at a slow rate under physiological conditions. The in vitro rate of regeneration of native interferon was estimated to have a half-life of 65 hours, and the active IFNalpha2 levels peaked at 50 h, with substantial levels still being detected 200 hours after administration (Peleg-Shulman et al. 2004).

Another example is the use of a supramolecular polymer-based transformable material, PEG-NH2-PRX, which is a mixing-type PEGylation material that provides amino groups to interact with protein drugs on demand (Utatsu et al. 2021). This material has been shown to efficiently form complexes with proteins and improve their stability compared to other PEGylation materials (Utatsu et al. 2021). In summary, reversible PEGylation is a promising approach to improve the therapeutic efficacy of protein drugs by releasing native proteins in a controlled manner, overcoming the limitations of traditional PEGylation (Utatsu et al. 2021).

### 14.2.1 Reversible PEGylation Chemistry

The most basic linker design for releasable PEGylation could have an exposed ester bond between the drug and the polymer. However, simplicity in design does not deliver the needed customized release kinetics required for drug release control and personalized pharmacokinetics (Filpula and Zhao 2008). Furthermore, molecules frequently lack a functional hydroxyl or carboxyl group for ester formation. Many biologically active compounds have additional functional groups, such as amines and thiols, which necessitate more complex linker architectures. The present stage of proficiency in linker design for releasable PEGylation is a result of a recent focused study on prodrug chemistry (Filpula and Zhao 2008).

### 14.2.2 Linkers for Reversible PEGylation

PEG chains can be temporarily attached to biomolecules like proteins or peptides with the help of **reversible PEGylation linkers**. These linkers then permit the PEG chains to be cleaved or detached under specific circumstances. These linkers' chemistry can vary, but they frequently contain functional groups or cleavable bonds that react to variations in pH, temperature, or other external conditions. The chemistry of a few popular varieties of reversible PEGylation linkers is shown in Fig. 14.2.

Aromatic and aliphatic linkers are two broad categories of chemical linkers or spacers used in various applications, including molecular biology, drug design, and materials science.

#### 14.2.2.1 Aromatic Linkers

Contains one or more aromatic rings in their structure which are characterized by a closed loop of alternating single and double bonds, such as benzene rings. Aromatic linkers often provide rigidity to the molecule due to the conjugated pi-electron system in the aromatic ring. They are generally planar and can participate in pi-pi stacking interactions with other aromatic molecules. They can have distinct

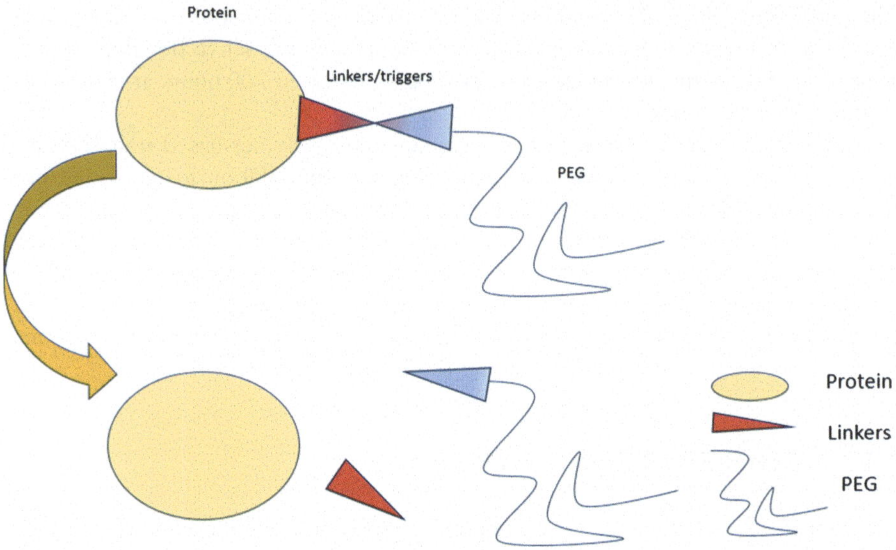


**Fig. 14.2** Attributes of releasable PEGylation. Diagram of key elements in rPEGylation that include a trigger segment that initiates the linker degradation pathway and the release of the original protein as verified by multiple analytical methods

electronic properties and may be involved in electronic or photophysical processes (Fig. 14.3).

#### 14.2.2.2 Aliphatic Linkers

They are characterized by open-chain structures, lacking the closed-loop of alternating single and double bonds found in aromatic rings. Aliphatic linkers often contain saturated carbon-carbon bonds, such as in alkyl chains. Aliphatic linkers are typically flexible and less rigid compared to aromatic linkers. They do not have the same pi-pi stacking interactions that aromatic rings can form and are often used to increase the flexibility and mobility of attached functional groups (Fig. 14.3).

### 14.2.3 Cleavage Linkers

Cleavable linkers in reversible PEGylation are used to attach PEG to a therapeutic molecule, such as a protein or peptide, and enable its release under specific conditions (Gupta et al. 2019). These linkers allow for the temporary attachment of PEG, which can be cleaved under certain biological conditions, leading to the release of the active molecule. Examples of cleavable linkers include enzyme-cleavable ester or amide bonds, disulfide bonds, and other specific chemical moieties that can be cleaved under physiological conditions. Reversible PEGylation using cleavable linkers has been studied for its potential in drug delivery and cancer therapy (Gupta et al. 2019).

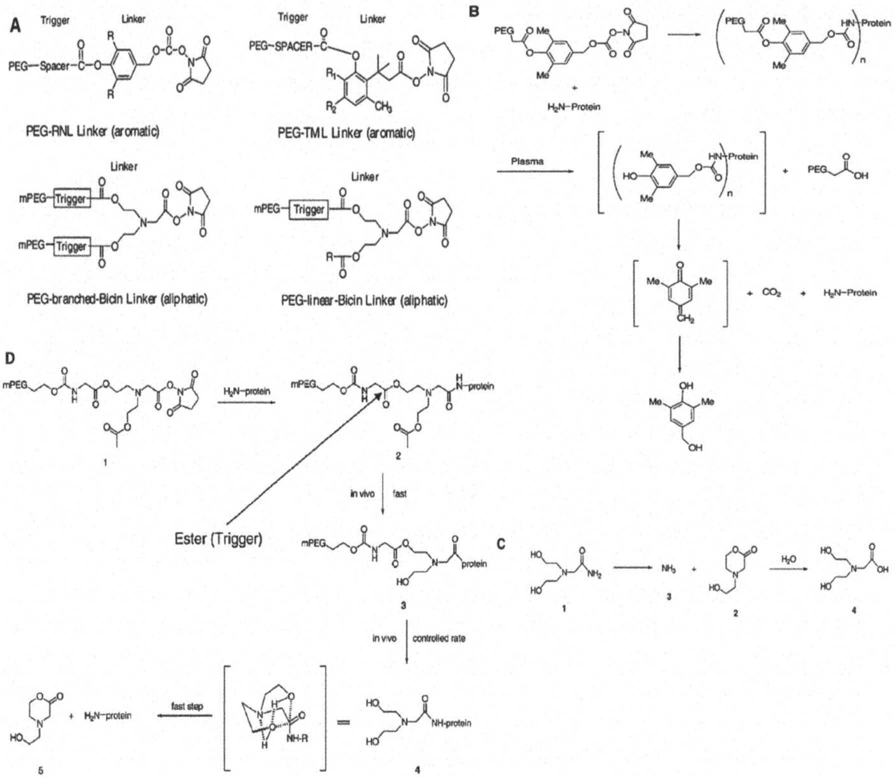


**Fig. 14.3** Chemistry of releasable linkers for PEGylation. Adapted with permission from (Filpula and Zhao 2008), Copyright 2007. (**a**) Prodrug designs of aromatic (BE: RNL or TML) and aliphatic (bicin) linkers include NHS leaving group, a trigger group for initiation of elimination reactions, and a linker segment attached to a spacer or directly to PEG. The linker segment may be rationally or empirically designed to provide steric hindrance or additional nucleophilic sites such that the release kinetics of the linker are controllable. (**b**) Reaction of an activated PEG of the BE linker type with a protein amine to form the bioconjugate. In plasma, the initial ester cleavage in the trigger element first discharges the PEG and complete release of the linker element from protein occurs rapidly. (**c**) Chemical basis of the bicin series of releasable linkers. Bis-N-2-hydroxyethylglycinamide (1) is subject to hydrolytic assistance from both hydroxyethyl side chains and cyclizes to a morpholinolactone (2), which is rapidly hydrolyzed to the acid, bicin (4). (**d**) Reaction of an activated PEG of the bicin linker type (1) with a protein amine to form the bioconjugate (2). In vivo, the acetyl group is rapidly lost, while the elimination reaction to release the PEG segment occurs at a moderate and controllable rate with the final release of the original protein predicted to occur via an intramolecular cyclization event

Cleavable linkers in PEGylation offer a versatile approach to controlling drug release kinetics, allowing for targeted and controlled release of the therapeutic molecule under specific biological conditions.

#### 14.2.3.1 Hydrolyzable Linkers

These linkers are designed to be cleaved by hydrolysis under specific conditions, such as changes in pH or enzymatic activity. They enable the release of the PEG from the therapeutic molecule under controlled biological environments.

Examples of hydrolyzable linkers in PEGylation include:

1. **Hydrolytically labile ester linkers**: These linkers are designed to be cleaved by hydrolysis under specific conditions, such as changes in pH or enzymatic activity. They enable the release of the PEG from the therapeutic molecule under controlled biological environments.
2. **Carbonate linkers**: A hydrolyzable carbonate linker can connect the drug payload to the PEG. This connection breaks under acidic or basic conditions, allowing for controlled drug release.
3. **Succinimide-based linkers**: These linkers contain a succinimide ring that can undergo a hydrolysis reaction when the linker-payload combination reacts with cysteine residues on other proteins. This reaction can open the succinimide ring, allowing the linker to undergo an elimination pathway and become completely stable. Seattle Genetics' new generation of linkers incorporate a primary amine adjacent to the maleimide, which promotes a much faster hydrolysis reaction.
4. **Glucuronic acid-based linkers**: Some hydrolyzable linkers replace the maleimide group, where the linker is cleaved when it enters the cancer cell, with a hydrophilic glucuronic acid moiety. Lysosomal compartments of cells have β-glucuronidase enzymes that can remove the glucuronic acid, which can improve the ADC's pharmacokinetic properties.

These hydrolyzable linkers in PEGylation offer a versatile approach to controlling drug release kinetics, allowing for targeted and controlled release of the therapeutic molecule under specific biological conditions.

#### 14.2.3.2 Enzymatically Cleavable Linkers

These linkers are designed to be cleaved by specific enzymes, allowing for targeted drug release at particular sites in the body. For example, protease-sensitive linkers can be used to achieve controlled release of the therapeutic molecule in response to enzymatic activity (Ghosal et al. 2021).

Examples of enzymatically cleavable linkers in PEGylation include:

1. **Dipeptide-based linkers**: Val-Cit and Phe-Lys are examples of dipeptide-based linkers that can be used in enzymatically cleavable PEGylation. These linkers can be cleaved by specific enzymes, allowing for controlled drug release (Mccombs and Owen 2015).
2. **PAB (p-amino benzyl alcohol)**: The dipeptide valine-citrulline combined with PAB is a popular enzymatic cleavage sequence. PAB is a self-immolative linker that can release the free drug when it is cleaved by enzymes (Mccombs and Owen 2015).

3. **Peptide linkers**: Creative Biolabs offer a class of enzymatically cleavable peptide linkers for conjugating antibodies and drugs. These linkers can be designed to be cleaved by specific enzymes, allowing for controlled drug release (Mccombs and Owen 2015).
4. **β-glucuronide linkers**: These linkers incorporate a hydrophilic sugar group that can be cleaved by β-glucuronidase. Once the sugar is cleaved from the phenolic backbone, self-immolation of the PAB group releases the free drug.

These enzymatically cleavable linkers in PEGylation offer a versatile approach to controlling drug release kinetics, allowing for targeted and controlled release of the therapeutic molecule under specific biological conditions (Mccombs and Owen 2015).

### 14.2.4 pH-Responsive PEGylation

It is a technique used in drug delivery systems to target specific cells or tissues. It involves the use of PEG chains that are attached to nanoparticles and can be designed to be pH sensitive (Rustad et al. 2022). PEGylation, which reverses after a therapeutic agent reaches the target cell, is an attractive feature for drug, protein, or nucleic acid delivery. PEG has been widely used to enhance the circulation time of nanocarriers in the bloodstream, improving drug delivery efficiency (Sun et al. 2015a). However, its nonspecific binding can hinder effective drug release at the target site. It also appears that the covalent binding of PEG to proteins results in steric interference, which shields conjugates from proteolysis and lowers the pace at which they are removed from the bloodstream by renal filtration and intracellular absorption (Tsubery et al. 2004). The pH-responsive reversible PEGylation is a groundbreaking advancement in drug delivery systems, allowing nanocarriers to be coated with PEG chains that selectively cleave in response to changes in pH levels. This allows for precise drug release at the desired location, reducing the risk of premature release.

The pH-responsiveness of these systems arises from the protonation of specific groups, leading to the destabilization of the nanoparticles and the release of the encapsulated drug (Fig. 14.4). Research has shown that pH-responsive PEGylation can affect release kinetics and cellular uptake in cancer cells, particularly in glioblastoma cells. pH-sensitive liposomes with cleavable PEGylation have been designed to target tumor cells, exhibiting a significantly charge shift and highly efficient phagocytosis by tumor cells while retaining long blood circulation time. Therefore, pH-responsive PEGylation represents a promising strategy for targeted drug delivery, especially in the context of cancer therapy (Ghosh and Dey 2020).

This method has applications in cancer therapy and inflammatory diseases, enhancing treatment efficacy while minimizing side effects (Nie et al. 2011). This innovative approach represents a significant step towards personalized medicine, paving the way for more effective and patient-friendly treatments. These merits are in agreement with the reports of Sun et al. They synthesized acid-sensitive PEGylated

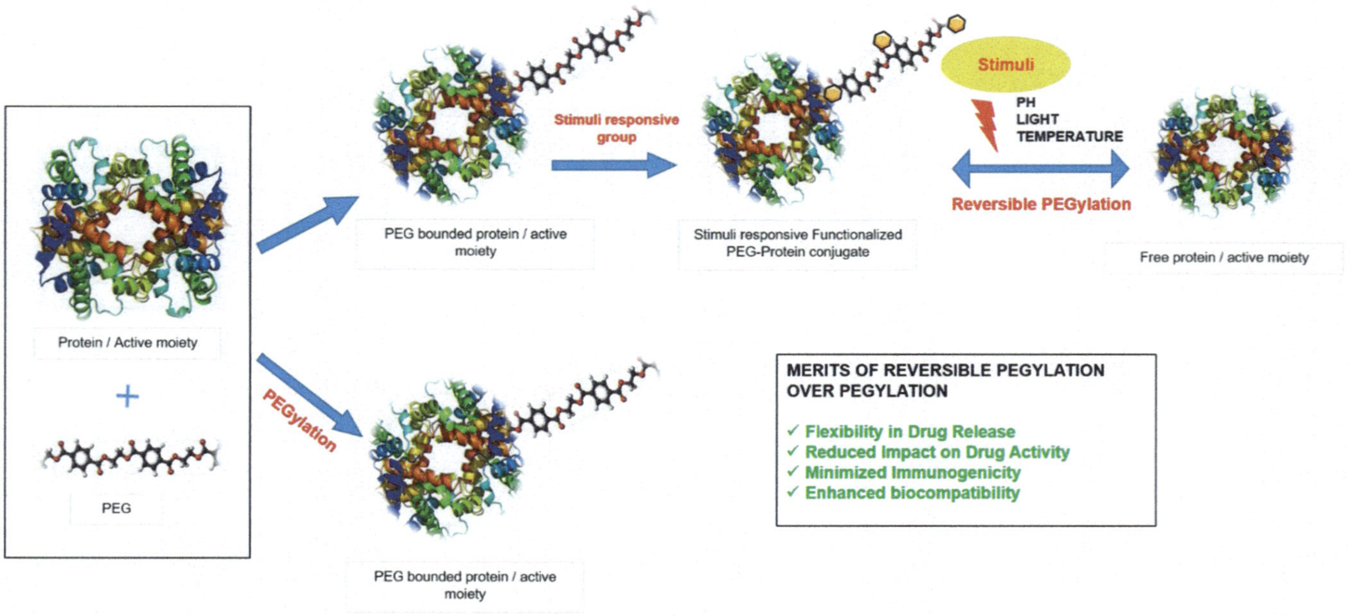


**Fig. 14.4** Diagrammatic representation showing reversible PEGylation strategies and its merit

DOX, which enhanced tumor accumulation and accelerated intracellular drug release in acidic conditions. This pH-responsive reversible DOX demonstrated prolonged circulation time in blood, enhanced tumor tissue accumulation, and accelerated intracellular drug release, suggesting potential for clinical chemotherapy of various malignancies. The study also showed no damage to the body (Sun et al. 2015b).

Another group of researchers use this phenomenon, they use PEG aldehyde-carboxypyridylhydrazone, N-hydroxysuccinimide esters (PEG-HZN-NHS) and synthesized for bioreversible surface shielding of DNA polyplexes. Both monofunctional and bifunctional mPEG-HZN-NHS were used, with mPEG-HZN-NHS shielded particles remaining shielded at pH 7.4. Luciferase gene transfections with epidermal growth factor (EGF)-containing polyplexes showed up to 16-fold enhancement in gene expression with reversibly shielded polyplexes compared to stably shielded polyplexes. Reversibly shielded polyplexes also mediated an enhanced tumor-specific in vivo transgene expression (Fella et al. 2008).

pH-responsive PEGylated nanocarriers offer several advantages in drug delivery, including:

1. **Targeted drug delivery**: pH-responsive nanocarriers can be designed to release drugs specifically in the acidic environment of tumors, leading to targeted drug delivery and reduced side effects (Alsawaftah et al. 2022).
2. **Improved drug stability**: PEGylation can improve the stability of drugs, protecting them from degradation and premature release (Li et al. 2022).
3. **Increased circulation time**: PEGylation can increase the circulation time of nanocarriers in the bloodstream, allowing for prolonged drug exposure and improved therapeutic efficacy (Karimi et al. 2016; Alsawaftah et al. 2022).
4. **Controlled drug release**: pH-responsive nanocarriers can be designed to release drugs in a controlled manner, allowing for sustained drug release and improved therapeutic outcomes (Alsawaftah et al. 2022).

However, there are also limitations to consider, such as reduced intracellular uptake of PEGylated nanoparticles and immunogenicity of PEG. Despite these limitations, pH-responsive PEGylated nanocarriers remain a promising strategy for targeted drug delivery, especially in the context of cancer therapy.

There are several drugs that have been successfully delivered using pH-responsive PEGylated nanocarriers. Some examples include:

(a) **Paclitaxel**: Acetal-linked PEGylated paclitaxel prodrugs have been developed to form free-paclitaxel-loaded pH-responsive micelles with high drug loading capacity and improved drug delivery (Li et al. 2022).

(b) **Doxorubicin**: pH-sensitive liposomes with a PEGylated pH-responsive poly (amino acid) (PAA) corona have been developed to release encapsulated doxorubicin in a pH-responsive manner, showing potential for tumor-specific drug delivery (Karimi et al. 2016).

#### 14.2.4.1 Proteasome Inhibitor MG132

Micelles composed of poly (ethylene glycol)-poly-aspartate copolymer with 45 nm particle size have been developed to deliver proteasome inhibitor MG132, showing low in vivo toxicity of micelles than free MG132 and prolonged circulation of drug-loaded micelles in the bloodstream (Karimi et al. 2016).

These examples demonstrate the potential of pH-responsive pegylated nanocarriers in delivering drugs for cancer therapy and other applications. pH-responsive PEGylation has been explored in various drug delivery systems, such as liposomes and poly (lactic-co-glycolic acid) (PLGA) nanoparticles, for cancer therapy. Here are some examples:

1. **pH-responsive liposomes with cleavable PEGylation**: These liposomes exhibit a significant charge shift and highly efficient phagocytosis by tumor cells while retaining long blood circulation time.
2. **pH-sensitive liposomes with a PEGylated pH-responsive poly (amino acid) (PAA) corona**: These liposomes display a high colloidal stability in blood and a pH-responsive release of encapsulated doxorubicin both in vitro and in vivo. The in vitro doxorubicin release showed a half-life of 3.4 h at pH 7.4, which increased to 21.6 h at pH 5.0. The in vivo antitumor efficacy was evaluated in mice bearing 4 T1 luciferase-labeled breast cancer, showing the potential of these pH-responsive liposomes for tumor-specific drug delivery.
3. **pH-responsive PEGylation of PLGA nanoparticles**: The effect of pH on the stability of PLGA nanoparticles with respect to the pH range of 7.4 (pH of cytoplasm) to 6.5 (pH of lysosome) was investigated. The study found that the pH of the medium does not affect the stability of the formulation, suggesting that the pH of the medium does not influence the degradation of the polymer (Hu et al. 2020).

These examples demonstrate the potential of pH-responsive PEGylation in cancer drug delivery, offering improved targeted delivery and reduced side effects.

### 14.2.5 Temperature-Responsive PEGylation

A clever tactic in drug delivery systems is temperature-responsive reversible PEGylation, in which medications or carriers are linked to PEG chains, which are then attached and detached in response to variations in temperature (Fig. 14.4). This technique enables PEGylation to be modulated reversibly, enabling a dynamic response to temperature changes. PEG chains cover the drug or carrier at lower temperatures and separate at higher temperatures, which may improve drug release and interactions with biological systems shown in Fig. 14.5. This reversible PEGylation that is temperature-responsive enhances the flexibility and effectiveness of drug delivery systems by providing a customized response to the physiological environment for better therapeutic results.

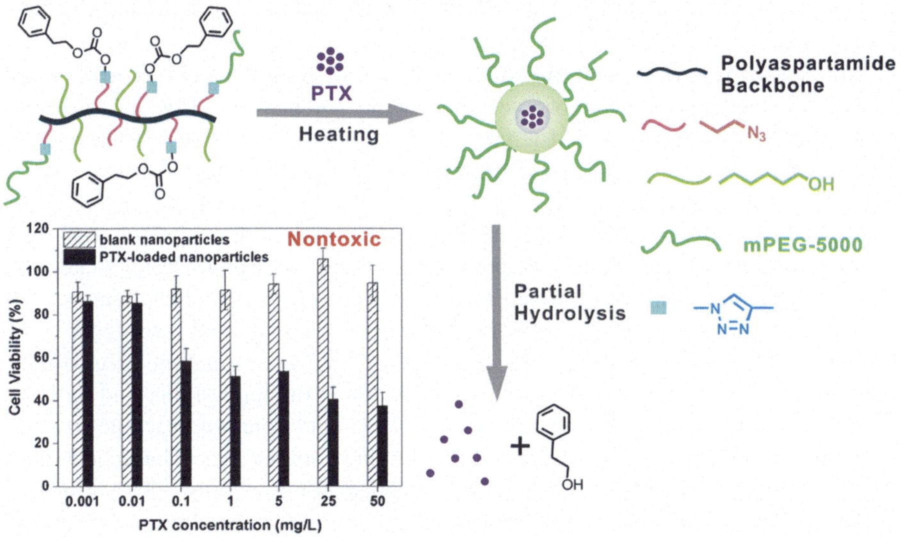


**Fig. 14.5** Temperature responsive nanoparticles based on PEGylated polyaspartamide derivative for drug delivery. Adopted from (Zhang and Jiang 2019) under Creative Commons Attribution (CC BY) license (http://creativecommons.org/licenses/by/4.0/) Copyright 2019

Rieger Jutta et al. reported this approach, and by using this, simple reversible addition-fragmentation chain-transfer polymerization (RAFT)-mediated aqueous dispersion polymerization process has successfully prepared PEGylated thermoresponsive micelles and coreshell nanogels such as double hydrophilic (poly (ethylene oxide)-b-poly (N, N-dimethyl acrylamide)). Excellent control over the aqueous dispersion polymerization of N, N-diethyl acrylamide (DEAAm) was achieved with the development of PEO-b-PDMAAm reversible chain transfer agents, which have a tunable length of the poly (N, N dimethyl acrylamide) (PDMAAm) segment. These agents were used in the radical polymerization of DEAAm in the presence of a crosslinker, resulting in thermoresponsive gel particles in one step.

The researchers found that a critical minimum length of the PDMAAm segment was necessary to obtain nanometric particles, which decreased with increasing length. This approach is a straightforward method for preparing PEO-coated nanogels with a limited number of reactants and high solids contents (Rieger et al. 2009). Polymers like poly (N-isopropyl acrylamide) (PNIPAM) is a temperature-sensitive polymer that can undergo a phase transition around body temperature. This property can be utilized to create PEGylated drug carriers where PEGylation is reversible upon changes in temperature.

One example of a temperature-responsive PEGylated polyaspartamide derivative is mPEG-PAAHP, which was synthesized by the click reaction. This derivative was characterized and confirmed using FTIR and 1H NMR spectroscopy (Zhang and Jiang 2019). In drug delivery, mPEG-PAAHP has shown potential as a promising system due to its apparent biocompatibility and temperature-responsive properties (Zhang and Jiang 2019).

Another study investigated the effect of PEGylation on the stability of thermo-responsive nanogels. It was observed that in the absence of a PEG coating, the nanogels aggregated at elevated temperatures (Motlaq et al. 2018). However, the addition of a PEG coating improved the heat responsiveness of the polymers, requiring a lower thermal dose to trigger drug release or gelation (Amin et al. 2022). Overall, temperature-responsive PEGylation is a promising approach that can enhance the performance of drug delivery systems and tissue engineering materials. Developing temperature-responsive PEGylated drugs faces several challenges.

One of the main challenges is the potential loss of activity and heterogeneity of PEGylated therapeutic proteins, which can affect their conformation, electrostatic binding, and hydrophobicity, leading to a reduction in binding affinity and biological activity. Additionally, the existence of anti-PEG antibodies and vacuoles related to PEGs can limit the extensive use of PEGylation in some cases (Zhang and Jiang 2019). Another challenge is the difficulty in achieving optimal drug release kinetics, as the temperature-responsive properties of the materials can be affected by various factors, such as the molecular weight and concentration of the PEGylated polymers (Amin et al. 2022). Furthermore, the stability of thermo-responsive nanogels can be affected by the absence of PEG coating, leading to aggregation at elevated temperatures (Motlaq et al. 2018). These factors need to be carefully considered when developing temperature-responsive pegylated drugs to ensure their efficacy and safety.

In summary, temperature-responsive PEGylation is a promising approach in drug delivery and tissue engineering, as it allows for the development of materials that exhibit controlled drug release or gelation in response to temperature changes. This strategy can improve the biocompatibility and stability of thermo-responsive materials, making them more effective for various applications.

### 14.2.6 Light-Responsive PEGylation

PEG chains are reversibly linked to and disengaged from medications or carriers in light-responsive reversible PEGylation, an inventive approach to drug delivery systems. This tactic makes use of photosensitive compounds, which are frequently added to PEG chains and which alter structurally when exposed to particular light wavelengths. This change initiates a reversible transformation of the PEGylation status, influencing the carrier's or drug's interaction with the environment. Light-responsive reversible PEGylation offers a viable path towards the creation of dynamic and flexible drug delivery systems with improved therapeutic precision by enabling precise and spatiotemporal control over drug release and bioavailability.

This strategy is in line with the increasing interest in creating intelligent and adaptable medication delivery systems to enhance patient outcomes. Artificial membranes that react to several stimuli at once can imitate biological channels and pumps. Customized conical nanopores with combined light- and pH-responsive features by grafting PEG-spiropyrans. According to studies on ionic transport, a nanopore opens in reaction to UV light and shuts in response to visible light. The gating property of the nanopore is facilitated by PEG-spiropyran self-assembly, which is also affected by the solvent. Water inhibits the photoswitchable molecule in its zwitterionic state, but conformational switching is encouraged by ethanol. The pH can regulate the selectivity of nanopores under UV radiation; an acidic pH is selective for anion transport, while a neutral pH is favorable for cation transport.

Using this method, adaptive systems that resemble biological pumps and channels can be designed (Ma et al. 2018). PEGylation increases the hydrophilicity of nanoparticles, reduces immune clearance, and prolongs circulation time (Son et al. 2019). Light-responsive properties can be achieved by incorporating photosensitive moieties into the polymeric structures of hydrogels (Xing et al. 2022).

Light-responsive hydrogels have been developed for various applications, such as:

1. **Controlled drug delivery**: Light-responsive hydrogels can be used to release drugs in a controlled manner, allowing for more precise drug delivery and reduced side effects (Xing et al. 2022).
2. **Biophotonic imaging**: Light-responsive nanoparticles can be used for biophotonic imaging, which enables real-time monitoring of drug release and distribution within the body (Son et al. 2019).
3. **Targeted therapy**: Light-responsive nanoparticles can be coated with PEG to enhance their hydrophilicity and reduce immune clearance, allowing for more selective targeting and improved drug delivery (Zhu et al. 2019).
4. **Selective targeting and reduced side effects**: The combination of PEGylation and light-responsiveness enables more selective targeting and reduced side effects, as it allows for precise drug delivery and release (Xing et al. 2022).
5. **Improved biocompatibility and reduced immune clearance**: Surface PEGylation of nanoparticles enhances their hydrophilicity, reduces immune clearance, and prolongs circulation time, leading to improved biocompatibility (Xing et al. 2022).

These advantages make light responsive PEGylation a promising strategy for controlled drug delivery, targeted therapy, and improved biomedical applications. Light responsive drug delivery system built through covalent incorporation of specific light-sensitive chemical groups with the aim to locally release cargo from the delivery system (Lehner and Hunziker 2012).

Some examples of light-responsive nanoparticles used in drug delivery include:

(a) **Spiropyran-based nanoparticles**: A nanoparticulate drug delivery system comprising spiropyran and PEGylated lipid has been developed (Fig. 14.6), where the light-sensitive switch enables particles to fluoresce and release drugs inside cells when exposed to light (Xing et al. 2022).

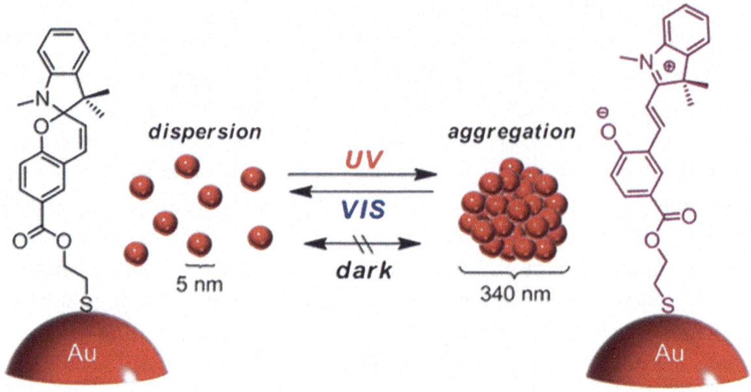


**Fig. 14.6** Spiropyran modified gold nanoparticles. Adapted with permission from Shiraishi et al. (2014), Copyright 2014, Elsevier

(b) **Photochemical reaction-based drug release**: Light-responsive nanoparticles have been used for photochemical reaction-based drug release, where the release of drugs is triggered by light exposure.
(c) **Gold nanoparticles**: Gold nanoparticles have been explored for ocular drug delivery through photothermal reactions. These nanoparticles can rapidly absorb energy from various light sources and release energy as heat, which can be used to trigger drug release (Abdelmohsen et al. 2023).
(d) **Light-responsive hydrogels**: Light-responsive hydrogels have been developed for controlled drug delivery, using light as a powerful trigger for controlled drug delivery systems. These systems can provide excellent manipulation of therapeutic agents in chemical and biological stimuli (Xing et al. 2022).

These light-responsive nanoparticles have been utilized in various drug delivery applications, including chemotherapy, immunotherapy, photodynamic therapy, gene therapy, wound healing, and ocular drug delivery. The use of light as a trigger for drug release offers precise control over the delivery of therapeutic agents, improving patient compliance and convenience.

Light-responsive PEGylation has some potential disadvantages, including:

(i) **Immunogenicity**: PEGylation can lead to hypersensitivity indirectly by side products formed during synthesis, and the existence of anti-PEG can cause vacuoles related to PEGs (Zhang et al. 2014).
(ii) **Disadvantages on liposomes**: PEGylation can impose certain disadvantages on liposomes, especially for the delivery of genes and nucleic acids in anticancer therapy (Mishra et al. 2016).

However, these disadvantages are not specific to light-responsive PEGylation and can be addressed by optimizing the modification degree and molecular weight of PEG, as well as designing new PEG reagents for site-directed PEGylation modification (Zhang et al. 2014).

## 14.3 Characterization of Reversible PEGylation

### 14.3.1 Analytical Techniques

Various analytical techniques are used to characterize reversible PEGylation in drug delivery systems. These techniques help researchers understand the nature of PEGylation, monitor changes, and optimize the design of drug carriers. Common techniques include Dynamic Light Scattering (DLS), Nuclear Magnetic Resonance (NMR), High-Performance Liquid Chromatography (HPLC), Fourier Transform Infrared Spectroscopy (FTIR), UV-Visible Spectroscopy, Mass Spectrometry (MS), Fluorescence Spectroscopy, and Zeta Potential Measurement. DLS measures particle size distribution, providing information about changes in hydrodynamic diameter of PEGylated nanoparticles. NMR spectroscopy is useful for studying the structure and composition of PEGylated compounds, identifying PEG chains, and determining their attachment sites.

HPLC separates and quantifies components in mixtures, analyzing PEGylated drugs and determining the degree of modification. FTIR helps identify chemical bonds and functional groups, confirming PEG on drug carriers and understanding the nature of bonds formed during PEGylation (Zalipsky et al. 2007). MS determines the molecular weight of PEGylated compounds, confirms PEG chain attachment, and assesses reversibility. Fluorescence spectroscopy studies fluorescence properties of labeled compounds, providing information about changes in the microenvironment of fluorophores. Zeta potential measurement evaluates the surface charge of nanoparticles, providing insights into the stability and surface characteristics of PEGylated carriers (Shechter et al. 2005). NMR is a fingerprint technique for the characterization of these systems. Also (Rieger et al. 2009) mentioned in their report that chemical structures of SAD-PEG-SAD and CAD-PEG-CAD which are reversible PEGylated complexes confirmed by proton nuclear magnetic resonance.

#### 14.3.1.1 Qualitative Characterization of Reversible PEGylation

**Zeta Potential**

Zeta potential can be determined based on the DLS principle. DLS measures the molecular radii of the samples and hence can be used to estimate the molecular weight of PEGylated proteins. It can also differentiate between linear and branched PEGs. The DLS method is popularly used to study the interactions in mono-PEGylated proteins.

The researchers have formulated the pH-reversible lipoplexes, and upon checked for the zeta potential, they found that the surface charge changes significantly when the amount of PEG incorporated into the liposomes is increased. The zeta potential without PEG coating was +27 mV; after incorporation of only 5% PEG, the zeta potential was changed to +11 mV. High surface charge on the liposomes might help in interacting with the receptors, but can decrease the stability due to aggregation (Nie et al. 2011). Fella et al. have made amine-reactive polyplexes. The polyplexes had high positive zeta potential (+25 mV). When the polyplexes were partly shielded with PEG moiety, zeta potential changed to +15 mV. But when the polyplexes are shielded with 30-fold higher concentration of PEG, the zeta potential was found to be below +5 mV (Fella et al. 2008).

**Hydrophobic Interaction Chromatography (HIC)**
HIC comparatively measures the hydrophilicity and lipophilicity of the particle. In this technique, two washes are provided, in which the first wash selectively washes the hydrophilic constituents away, whereas the second wash is responsible for elimination of comparatively more hydrophobic particles.

The "HIC Ratio" is calculated as follows:

$$HIC\ Ratio = Elution\ volume\ of\ sample\ /\ Elution\ volume\ of\ reference$$

The lower HIC ratio indicates the presence of hydrophobic particles and higher value indicates hydrophilic particles.

Researchers have used the hydrophobic interaction chromatography in the purification of mono-PEGylated antiepidermal growth factor receptors (EGFR). For the selective capture, a weak cation exchanger was used. Upon increasing the size of PEG from 5 to 30 kDa, the retention on HIC was increased. HIC resulted in the suppressed aggregation of the protein and allowed good resolution with high purity and product recovery. The purity was found to be increased to 98% (Moosmann et al. 2012).

**Near Infrared (NIR) Spectroscopy**
The NIR spectroscopic technique coupled with Principal Component Analysis (PCA) can also give the information about the hydrophilic/lipophilic balance of the compound. PCA is the statistical technique which measures the variance of the given set of data (Bista and Bruch 2008). NIR spectroscopy shows the upconversion spectra when the PEGylated nanoparticles were subjected to NIR laser with 980 nm diode excitation. This upconversion of the nanoparticles is because of the use of PEG moiety has been employed in the biolabeling in the biological window. Zhang et al. have prepared the pH-reversible PEG conjugate and the methods to characterize it, the absorbance was measured by UV-VIS-NIR spectrometer to determine the activity of PEGylated protein. The NIR data were used for the relative activity of the proteins added in various formulations (Zhang et al. 2020).

**Fourier Transform-Infrared Spectroscopy (FT-IR)**
In FTIR spectra, the presence of the PEGylation can be seen by the presence of CH2 and C-O-C peaks. This method is effectively used for various kinds of metallic as well as polymeric nanoparticles. The PEGylation of the nanoparticles can be

confirmed by the FTIR analysis. The characteristic peak for PEG moiety were observed at wavelength of 2884 $cm^{-1}$ (stretching vibration) and 1468 $cm^{-1}$ (bending vibration). Massoumi et al. have developed the pH-reversible polymeric nanocapsules and the FTIR spectra of neat PEG showed the stretching vibration of C-O at 1108 $cm^{-1}$, bending vibration of -CH2- groups at 1467 $cm^{-1}$ and aliphatic C-H shows the stretching vibration at 2950–2800 $cm^{-1}$. The hydroxyl or adsorbed water shows the vibration band at 3490 $cm^{-1}$ (Massoumi et al. 2020).

#### $^{13}C$-NMR

In NMR, the decrease in the CH and CH2 peaks indicates the PEGylation of the nanoparticles. NMR spectroscopy is mostly used for the PEGylated proteins, in which the PEGylation leads to alteration in the intensity of peaks. Herald's group performed the solid state $^{13}C$-NMR, and the spectrum clearly showed the resonance of methyl group at 17 ppm and OCH2 groups shown 70.3 ppm intensity, which clearly indicates that the groups were not resolved clearly. In the characterization of the diblock copolymer of chitosan-PEG, when the $^{13}C$-NMR spectra of the proposed diblock polymer was compared with the spectra of chitosan alone, one broad peak (68.6 ppm) and four narrow peaks (59.0, 63.4, 64.1, and 173 ppm) were found to be the characteristic peaks due to the presence of PEG micromere (Ganji and Abdekhodaie 2008).

#### Mass Spectrometry

The specialized instrumentation employing Liquid Metal Ion Gun (LMIG) and single-stage reflectron analyzer is used for the qualitative determination of the PEGylation in the given sample (Stigsnaes et al. 2007). Furthermore, TOF- SIMS data are obtained and interpreted with software. The fragment ions C2H2O2 and C2H5O+ indicate the presence of PEG moiety. MALDI-TOF mass spectrometry was used by scientists to identify the sites of PEGylation of eluted fraction. They have PEGylated various proteins. In the results, five distinct peaks were obtained at interval of 44 Da. The graph also gave the unreacted free biotin-PEG molecules. From the molecular weights of the fragments, it can be inferred that which site is utilized for the PEGylation among the possible seven amino groups. Higashi et al. have PEGylated bromelain, and the MALDI TOF spectrum was obtained showing specific peaks at 23,500 m/z values. In case of modified bromelain for reversible PEGylation approach, the peak was obtained at an m/z value of 24,400, suggesting the effective PEGylation of the parent moiety has been occurred (Higashi et al. 2020).

### 14.3.1.2 Quantitative Characterization of Reversible PEGylation

#### High-Performance Liquid Chromatography (HPLC)

HPLC can be coupled with techniques such as refractive index detector, evaporative light scattering detector, or mass spectrometry for the effective estimation of the degree of PEGylation. Simple HPLC methodology employing UV detector often fails due to the absence of chromophore in the PEG moiety. To obtain enhanced hypotensive effect, et al. have conjugated atrial natriuretic peptide (ANP) with PEG moiety. Furthermore, the HPLC analysis was performed using reverse phase-18 column and analyzed at the wavelength of 220 nm. The HPLC analysis was found useful to eliminate the free atrial natriuretic peptide from the prepared ANP-PEG conjugates and serves as the better means of purification of the PEG conjugates (Nesher et al. 2008).

#### Calorimetry

For the calorimetric quantification of the PEGylation, first the alkaline hydrolysis of nanoparticles is done and the concentration and localization of PEG chains is determined. For perfectly spherical nanoparticles, the surface density of PEG and the distance between the PEG chains can also be determined effectively. The calorimetric analysis of PEG conjugates in mainly employed in the case of the macromolecules. Researchers have carried out the calorimetric study of the conjugates of the PEG with myoglobin moiety and upon application of the thermodynamic models, the thermal denaturation of the conjugates is examined. If the denaturation has occurred, a single step state is observed. It is proposed from the calorimetric data that the exothermic processes occurring do not contribute largely to the starting of the thermodynamic curve (Pelosi et al. 2019).

#### X-Ray Photoelectron Spectroscopy (XPS)

This technique basically provides the elemental composition of the surface of nanoparticle. The extent of PEGylation caused can be assessed by comparing the spectrum obtained before and after PEGylation. This method is most useful in case of the metallic nanoparticles, but possess some limitations in case of polymeric nanoparticles. When various substrates were PEGylated with catechol-grafted PEG, XPS analysis of the PEGylated substrates shown the differences in the atomic composition of the surface. Quantitative analysis by using XPS revealed the immobilization of the PEG molecules. After PEGylation, the signals for Si in silicon have been decreased in intensity. For qualitative examination, it was evident that the signals for the substrate were decreased significantly after modification with PEG-catechol; however, the signals arising due to catechol and PEG were increased. The relationship between the XPS spectra values and the amount of PEG added can be established for the accurate determination of the PEGylation. Zhang et al. have examined different nanodiamond samples to identify the elements found in the nanodiamonds. XPS scan was performed in the range of 0–1100 eV. XPS studies revealed the presence of C, N, and O atoms (Zhang et al. 2012).

#### Nuclear Magnetic Resonance (NMR)

Recently, the NMR analysis is also found to be an effective methodology for estimating the PEGylation. In this, the integrals of 1H NMR PEG peak are compared with the peak of some standard compound to get an idea about the quantity of PEG attached to the surface of nanoparticles. The scientists have used the NMR spectroscopy to determine the degree of PEGylation of the complex macromolecules by using the NMR spectroscopy. The proton NMR spectroscopy of proteins was carried out to determine the number of methoxy PEG chains grafted to a protein, also known as the degree of PEGylation.

The proton NMR spectroscopy overcomes the biases caused to the chromatographic methods used in the determination of the degree of PEGylation. Spectroscopy was performed at 300 MHz for protons. For this, the biconjugate solution was transferred to the NMR tube, and also, DMSO was added as an internal standard, upon carrying out the NMR spectroscopy, it was found that the singlet at 3.69 ppm was observed for ethylene groups of PEG and a singlet 2.71 ppm peak was observed for methyl groups of DMSO (Utatsu et al. 2021).

#### TGA-DSC

The combination of Thermogravimetric analysis (TGA) and Differential scanning calorimetry (DSC) can be used to determine the amount of PEG grafted on the nanocarrier system and can be interpreted as comparison of the amount of PEG grafted on the surface to the total weight of the nanocarrier system. The samples are vacuum dried prior to analysis. Zhang et al. have prepared the nanodiamonds, and the TGA analysis have shown that the percent reduction in the weight of OH and COOH was 0.4% and 3.1%, respectively, which suggests that the functional groups are further attached to the surface of nanodiamond after oxidation. The TGA data also revealed that the thermal decomposition temperature of PEG is about 400 °C and after 500 °C, no PEG was remained on the surface (Zhang et al. 2012).

The scientists have PEGylated BSA protein and examined the thermal events by using DSC. DSC gives the maximum heat capacity of the sample. The key finding of the experiment that the apparent folding temperature, i.e., $T_{max}$ was affected by PEGylation, but not affected by the molecular weight of PEG. The apparent $T_{max}$ was lower for PEGylated BSA than for intact BSA, The apparent $T_{max}$ was found to be 82 °C for intact BSA protein and was 2 °C lower for the conjugate of PEG with BSA (Paolino et al. 2017).

## 14.3.2 In Vitro Characterization of Reversible PEGylation

In vitro methods are generally used for the exact determination of complement consumption, protein adsorption, macrophage uptake, and also for the release rate of the drug substance and its stability. Table 14.1 contains some of the important in vitro techniques along with their applications.

### 14.3.2.1 Protein Adsorption

For estimation of adsorption of proteins on the nanoparticle surface, the nanoparticles are diluted and vortexed. The size of nanoparticles is obtained in different sera by nanoparticle tracking analysis (NTA). For determination of collagen-bound nanoparticles, the confocal laser scanning images can be acquired after the polymerization of collagen, and the diffusion coefficient can then be determined by using Raster image correlation spectroscopy (RICS) (Filpula and Zhao 2008).

**Table 14.1** In vitro characterization of reversible PEGylation

| In vitro technique | Application |
|---|---|
| 1. Electrophoretic methods, Electron Microscopy, Gel Permeation Chromatography (GPC) | Study of PEGylated proteins and structure of proteins |
| 2. Surface PEO density | Determination of complement consumption |
| 3. Cell uptake studies | Uptake of PEGylated nano-carriers by RES |
| 4. Release studies | Extent of PEGylation required |

#### 14.3.2.2 Cellular Association

The cellular association can be determined by using flow cytometry. The particular kind of cells are first cultured in the suitable culture media. The cells are then seeded on 12-well plates, 100,000 cells per well of the plate. Once the log phase of the cellular growth is attained, the cells are incubated with PEGylated nanocarrier system and kept for 3 h. Then, the cells were detached and analyzed with flow cytometry. The cellular uptake is determined by the fluorescent intensity measurement of the nanoparticles relative to the fluorescent intensity obtained in the untreated population of cells (Filpula and Zhao 2008).

#### 14.3.2.3 Cellular Effects and Target Affinity

Upon the exposure to the PEGylated and non-PEGylated proteins, the short-term and long-term cellular events are monitored by microelectronic cell sensor array (MCSA). These measurements give increase in the mass over the time and allow easy calculation of doubling time of the given population. Specifically, the results are checked for the presence of any kind of acute, chronic, or cell proliferation effects (Hamidi et al. 2006).

#### 14.3.2.4 Retention of Biological Effect

The biological effect of PEGylation should be maintained for sufficient period of time for the successful development of suitable drug candidate. As PEGylation maintains the drug concentrations for comparatively longer times in the body, sometimes, a little decrease in the biological activities does not appear to be troublesome. Using surface plasmon resonance (SPR), the binding affinities of proteins and the PEG chains can be easily determined. Generally, PEGylation leads to little or no loss of affinity for the biological site (Tao et al. 2009).

#### 14.3.2.5 Bioactivity Assay

These assays can be used for the quantitative determination of the PEGylated molecules in serum samples. PEGylated enzymes are more popularly tested by this approach as many detectable substrates are available for the enzymes. Pharmacokinetic properties of PEGylated nanocarriers can be determined by the cytopathic effect assay (CPE) (Tao et al. 2009).

#### 14.3.2.6 Enzyme-Linked Immunosorbent Assay (ELISA)

ELISA can be effectively used for the measurement of the concentration of the PEGylated proteins in serum samples. This technique provides an idea about the effect of PEGylation on the immunogenicity of the proteins. In this, a specific antigen is coated with microliter plate wells to capture PEGylated antibodies in the sample, and then, the amount of antibody bound to the antigen is determined by adding a detection antibody. A direct ELISA methodology using the antigen-coated microliter plates can also be used for the pharmacokinetics determination of the PEGylated nanocarriers. By using the competition assay methodology, direct ELISA can be used to measure the concentration of PEGylated compounds. In this case, the PEGylated nanocarrier is coated in the wells of a microtiter plate and the serum concentration of the PEGylated nanocarrier can be estimated from the degree of binding of the specific antibody to the plate coated with the compound (Filpula and Zhao 2008).

#### 14.3.2.7 Sandwich ELISA

As the simple ELISA methodology requires only one specific antibody for the detection of PEGylated compound, it cannot usually differentiate between the PEGylate and non-PEGylated proteins or nanocarriers. So, the sophisticated methodologies are currently used, for example, the indirect or Sandwich ELISA which makes the use of two antibodies, one to capture the analyte on the solid surface and the other to effectively determine the concentration of the detected analyte. In order to get detected, the analyte must possess at least two antigen-binding sites, that can be considered as the distinct binding epitopes present on the same analyte. For multimeric analytes, the same epitope can be used as the detection antibodies that bind to the same molecule. Sandwich ELISA is used to measure the concentration of the PEGylated nanocarriers in the complex biological samples. This is a more sensitive technique than direct and competitive ELISA and often considered with due priority when the analyte is present in low concentration (Liu et al. 2020).

#### 14.3.2.8 Anti-PEG ELISA

ELISA can be made more accurate and sensitive when the antibodies that specifically bind to PEG are used. For the generation of PEG-specific antibodies, a laboratory animal is infected with PEG linked proteins. The polyclonal antibodies thus obtained can be effectively employed for the specific binding of PEGylated nanocarriers. As the PEG possess very low immunogenicity, the animal is exposed to the PEG attached protein multiple times to attain the required extent of immunogenicity (Ozer et al. 2022).

### 14.3.3 In Vivo Characterization of Reversible PEGylation

#### 14.3.3.1 In Vivo Blood Circulation Half-Life

The pharmacokinetic and tissue distribution studies are the popular kinds of in vivo assessment for the PEGylated nanocarriers. If the PEGylation is achieved to the sufficient extent, it should increase the circulation time of the formulation in blood. It should also cause the reduction in the uptake by the liver as compared to the non-PEGylated particles (Howard et al. 2008). The main purpose of the PEGylation of the nanocarriers is to increase their residence time inside the body. The circulation half-life of the PEGylated nanoparticles can be measured using mice as an animal model.

In the estimation, it is evident that the nanoparticles having lower PEG surface coverage will have lowest circulation time and the blood circulation time will then increase as the concentration of PEG on surface is increased. The circulation half-life was determined in female BALB/c nude mice. Mice aged 6 weeks are housed into four groups in different cages, while maintaining the temperature and relative humidity. The animals were then anesthetized by fentanyl and midazolam subcutaneous injection. Blood samples are collected from the saphenous vein, and the blood collection was carried out before injection and after 10 min, 30 min, and 1, 2, 4, 6, and 24 h after the injection of the formulation. Collected samples were then diluted by heparin and then vortexed and then centrifuged. The fluorescence of the supernatant was measured by excitation at 535 nm and emission at 620 nm wavelength using a spectrophotometer (Böttger et al. 2016).

#### 14.3.3.2 Radiolabeling

This is a very sensitive method for the determination of the pharmacokinetics of the PEG moiety and the PEGylated compounds. In this method, a radioisotope is chemically included in the PEG or PEGylated molecules and then administered in the small amounts. The radionucleotide atom keeps emitting the radiation which can be then detected by gamma counter to monitor the emissions of the gamma rays and the X-rays. The radioactive emissions can also be estimated by using the scintillation counter, which give an idea about α and β particle emissions. For this purpose, many radioisotopes have been used by the researchers. In the case of PEGylated liposomes and micelles, the radiolabeling is often useful for the pharmacokinetic and biodistribution studies. The radioisotopes are generally attached to the surface of a liposome or can be trapped inside the liposome. The main approaches for labeling of the liposomal surface:

(a) Anchoring of the metallic radioisotopes on the surface of the liposomes.
(b) Covalent linkage of the radiolabeled chelating agents to the surface of the liposomes.

The pharmacokinetic estimation of the PEGylated peptides can be successfully carried out by the incorporation of radioiodine, if the polypeptide backbone contains tyrosine residues (Cheng et al. 2012; Gaberc-Porekar et al. 2008).

## 14.4 Application of Reversible PEGylation

The PEGylation approach has been extensively employed for reducing the immunogenic properties, increasing enzymatic activities and most importantly, to prolong clearance time (Kodera et al. 1998). The various applications of the reversible PEGylation can be classified broadly as therapeutic and pharmaceutical applications.

(a) Therapeutic applications: Reversible PEGylation is being successfully employed to modify the pharmacokinetic properties of immunological preparations, anticancer agents, and antibiotics. Recently, it was also found to be useful in enzyme replacement therapy, RBC substitution, and some oxygen toxicity diseases.
(b) Pharmaceutical applications: PEGylation aids the drug delivery, and the liposomes are frequently PEGylated to impart the stealth properties. Proteins are PEGylated for increasing their stability. The PEGylation approach also finds its applications in targeted delivery of therapeutic agents.

### 14.4.1 Therapeutic Applications

Although the scientists have made the proteins having high compatibility with the host immune system, these proteins often produce the negative immunogenic properties. So, PEGylated proteins can serve as the better means which do not affect the immunity of the person. PEG stimulates the production of the antibodies, and it itself does not induce the immunogenicity. PEG conjugation does not only cause of the suppression of the hyperactive immune system, but also switches the immunogenic types of the proteins into the tolerogenic proteins, and hence, they are considered suitable for immunogenic protein administration. Severe attempts have been made to screen such immunogenic proteins and have checked for their binding with PEG moiety. The strategy for immunogenic protein PEGylation mainly involves:

(a) Identification of the conjugation site.
(b) Masking of the immunogenic sites of protein by PEG conjugates.
(c) Removal of the conjugate by immune system.

Then, conjugation sites are often indicated by the presence of the lysine and serine, which, in due process of PEGylation, get substituted with cysteines. To decrease the immunogenicity of the protein, the antigenic epitope of the protein is masked with the hydrated PEG cloud. The improved immunological nature of this protein is also depicted by the enhanced pharmacokinetic properties of the protein. The half-life of the coated protein will also be prolonged because of the decrease clearance of the proteins by immune system and also by glomerular filtration. After its action, the conjugates and free PEG present in the circulation is removed by the IgG and IgM antibodies (Shi et al. 2022).

#### 14.4.1.1 Anticancer Activity

Many enzymes possess the anticancer activity. Enzymes mainly cause depletion of the nutrients required by the cancer cell growth, so the starvation of cancer cells is caused without affecting the normal cells. Enzymes also face the similar issue like other proteins including the short half-life inside the body and immunogenic nature. An enzyme asparaginase is popularly used for the treatment of leukemia, but possess serious issue of causing immunogenic reactions which lead to the allergies and anaphylactic shock. To combat the issues with asparaginase, PEGylated asparaginase was made by Enzon Inc. These PEGylated enzymes not only show decreased immunogenicity, but also protect the enzymes from the degrading enzymes and also prolong their circulation half-life (Higashi et al. 2020).

#### 14.4.1.2 Antibiotic Administration

The antibiotics show limitations in use as anticancer drugs due to their low solubilities and side effects. The conjugates of antibiotics with PEG moiety can effectively alter the solubility, permeability, and the distribution characteristics of the antibiotics in order to achieve the selective drug delivery with decreased side effects and improved targeting at the desired sites of action. Scientists have successfully formulated the polymeric micelles of the antibiotics employing PEG moiety to overcome the issues associated and achieve more patient-friendly therapy (Alavi et al. 2022).

#### 14.4.1.3 Enzyme-Replacement Therapy

Sometimes, a disease state is associated with the alterations in the normal physiological levels of certain enzymes in the body. The enzyme activity maybe decreased or completely lost, indicating a disease state. Enzon Inc. made mPEG-adenosine deaminase, i.e., ADAGEN for the treatment of severe combined immunodeficiency disease (SCID). SCID causes partial or total dysfunction of immune system and after receiving the formulation, patients show improved immune responses (Yadav and Dewangan 2021).

#### 14.4.1.4 Red Blood Cell Substitution

There are many diseases which are caused due to blood transfusions. To overcome this issue, researchers have developed an alternative to natural blood for blood transfusions. Artificial blood can supply the increasing demand of blood and also has increased half-life than the natural blood. The artificial blood is generally comprised of the perfluorocarbons, also known as "white bloods," which resemble the natural blood and are chemically related to the synthetic oils.

Red bloods are made by modification of the naked hemoglobin from RBCs, so as to carry oxygen. Then major limitation associated with the artificial blood is that the hemoglobin present is not protected by the red blood cell membrane and hence undergoes rapid clearance from the body and also leads to the formation of many toxic metabolites. To combat all the limitations associated with the artificial blood, scientists have modified hemoglobin with PEG moiety of different molecular weights, such as 1900, 4000, and 5000. The modification of hemoglobin with PEG moiety results in the increased half-life of the artificial blood as well as decrease the generation of the toxic metabolic products (Arkosi et al. 2023).

#### 14.4.1.5 Oxygen Toxicity Diseases

The tissue damage is caused by the oxygen-free radicals. The various diseases caused by radicals include inflammation, thermal injury, and ischemia diseases. Superoxidase dismutase is an antioxidant enzyme was supposed to avoid the oxidative damage caused by the radicals. Also, the catalase enzyme was also employed for combating the harmful defects caused by the radicals. The disadvantage of administering the enzymes such as superoxide dismutase or catalase is that these enzymes possess very short half-life. When these enzymes are modified with the PEG moiety, the modified enzymes shown increased half-life and also increased uptake by the membrane. PEG moiety also shows the synergistic action with the enzymes by inactivating the hydroxyl radicals in vitro and also increase the function of perfused kidney (Filpula and Zhao 2008).

### 14.4.2 Pharmaceutical Applications

#### 14.4.2.1 PEGylated Liposomes

The PEGylated liposomes (Fig. 14.7) show prolonged circulation of the liposomes inside the body and also possess the clearance kinetics which is independent of the dosage amount. The PEGylation hence improves the pharmacokinetic properties of the drug substance and shows reduced toxicity to improve the therapeutic efficiency (Jain and Jain 2008).

By modification of the liposomal surface by PEG moiety, the efficacy of the liposome delivery can be improved by prolonging the circulation of the liposomes in the blood. The liposomes conjugated with PEG are popularly known as "stealth liposomes" and possess increased half-life, decreased clearance from the body and decreased volume of distribution. Also, the surface-modified liposomes are useful in targeting particular organ for the delivery of the therapeutic agent. When the PEG moiety is incorporated into the lipid bilayer of the liposome, it forms hydrated shell

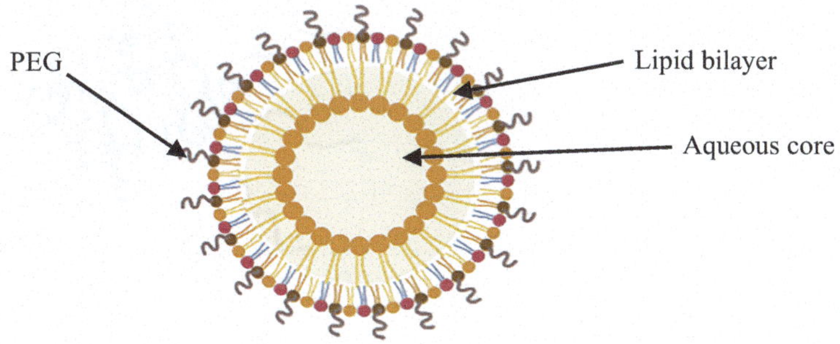


**Fig. 14.7** PEGylated liposome

that avoids the aggregation, hence improving the stability of the liposomal preparation. Also, inside the body, PEG helps to protect the liposomes from destruction by plasma proteins and avoids the premature release of the liposomes inside the body. PEGylation also avoids the interaction of the liposomes by the opsonin proteins and avoids the liposomal uptake by RES (Thomas et al. 2022). PEGylation of the liposomes also serve as the effective means for passive targeting of the drug substance to the desired action site. By virtue of the mechanism called enhanced permeation and retention (EPR) effect, the liposomes can be preferentially accumulated in the tumor areas and effectively deliver the drug there (Milla et al. 2012).

#### 14.4.2.2 PEGylated Proteins

PEGylated proteins show improved stability and lowered immunogenicity, compared to unmodified protein (Fig. 14.8). So, these modified proteins show sustained action inside the body and reduce the frequency of administration. The modified proteins are administered generally every week, but the unmodified proteins require more frequent administration, i.e., once every 1 or 2 days. The major limitation associated with the PEGylation of the proteins is that it may cause the chemical alteration of the protein molecule and sometimes shows the decreased therapeutic response. To avoid this, the care should be taken to avoid such chemical interactions. The scientists have studied the effect of the PEGylation of the proteins for the change in the pharmacokinetics, pharmacodynamics, and the biodistribution and to achieve higher clinical efficacy (Harris and Chess 2003). Table 14.2 contains the list of some PEGylated proteins.

The largest group of proteins, which has been modified with the PEGylation is of enzymes. Some enzymes, e.g., peroxidase, L-asparaginase, and alkaline phosphatase, upon conjugation with PEG moiety, showed excellent pharmacokinetic properties with increase stability (Milla et al. 2012). Some of the main enzymes which are modified with PEGylation are given in Table 14.3:

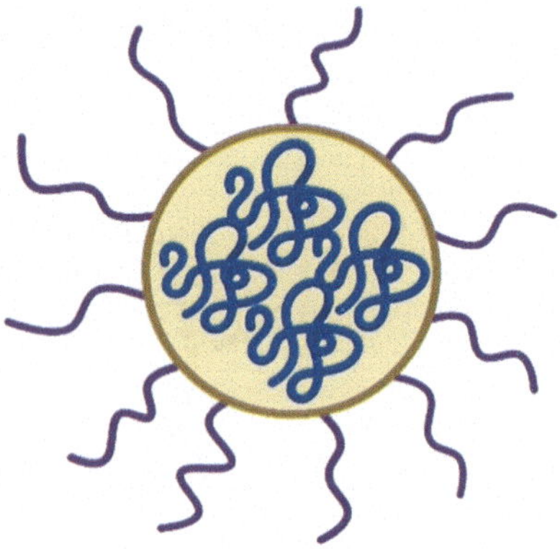

**Fig. 14.8** PEGylated protein

**Table 14.2** PEGylated proteins

| PEGylated proteins | Indication |
|---|---|
| Pegfilgrastim | Neutropenia |
| Certolizumab pegol | Rheumatoid arthritis |
| Pegloticase | Chronic gout |
| CERA | Anaemia |
| Pegademase | Severe combined immunodeficiency disease (SCID) |
| Pegaspargase | Acute lymphoblastic leukemia |
| Peginterferon Pegvisomant | Hepatitis C |
| Pegvisomant | Acromegaly |

**Table 14.3** PEGylated enzymes

| Protein | Application |
|---|---|
| Trypsin | Increased proteolytic activity |
| Chymotrypsin | Increased proteolytic activity |
| Ribonuclease | Increased activity inside body |
| Tissue plasmin activator | Increase circulation half life |
| Interferons | Antiviral applications |
| Interleukins | Immune response |
| Elastase | Increased hydrolyzing activity |
| Superoxide dismutase | Antiinflammatory activity |
| Alkaline phosphatase | Increase enzyme activity |
| L-asparaginase | Increased amidolytic activity |
| Peroxidase | Increased enzyme activity in vivo |
| Glucolactone oxidase | Increase stability of the conjugate |

#### 14.4.2.3 Targeted Delivery

When the nanoformulations are surface decorated with the PEG coatings, they specifically bind the desired targeted receptors and increase the effectiveness of the therapy, minimizing the drug loss by clearance and hence reduce the cost of the therapy. Various researchers have successfully surface-modified the nanoformulations for targeting the receptors at brain, airways, nose, gastrointestinal tract, etc. (Fig. 14.9) (Jain and Jain 2008).

##### Reversible PEGylation for Delivery to Airways

The nanoformulation delivery to the airways is essentially required to treat many ailments. However, the mucus layer present on the airways is rapidly cleared by the mucociliary clearance mechanisms and hence possess significant problems in the drug delivery. The scientists have examined the penetration of the PEG-coated nanoparticles through the sputum. Nanoparticles with 100 nm size range are found to effectively penetrate. Also, another barrier that hinders the drug action is the biofilm. The PEG-modified nanoparticles can effectively pass through biofilm. The studies shown that when the nanoparticles sized about 100 nm, they were unable to

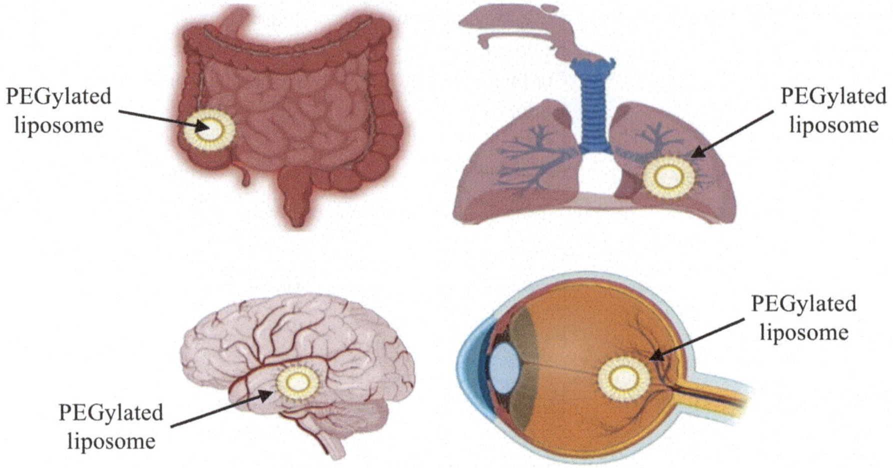


**Fig. 14.9** Reversible PEGylation for delivery to gastrointestinal tract, airways, brain and ocular delivery

penetrate; however, when the 200 nm nanoparticles well surface coated with the PEG, they efficiently penetrated the respiratory mucus (Wagner 2012).

### Reversible PEGylation for the Delivery to Gastrointestinal Tract

There are many barriers in the gastrointestinal delivery of the drugs. The formulation should possess good stability properties in order to effectively overcome the harsh environment. Researchers have studied the PEG coatings to improve the stability of the lipid based formulations. Various nonionic surfactants have been used to improve the stability of the drug substance across the gastrointestinal tract. The nanoparticles were further modified with the PEG coatings and found more effective than simple surfactant coatings. The nanoformulations also suffer because of the rapid clearance of the mucus in the gastrointestinal tract. The PEG coatings effectively increase the hydrophilicity of the lipid nanoparticles and hence minimize the mucoadhesion. The PEG coatings also shown the decreased cytotoxicity. For PEG coated nanoparticles who are targeting behavior, they show higher penetration to the inflamed tissue, but less penetration to the normal tissues (Sadekar and Ghandehari 2012).

### Reversible PEGylation for the Delivery to the Brain

As described in the delivery to the airways, the PEGylation increases the nanoparticle delivery to the brain through nose. There are mainly two barriers for the drug to reach the brain, first is the blood-brain barrier (BBB) and the second is the tissue extracellular matrix. Kannan group studied the penetration of the PEGylated nanoparticles through extracellular matrix, and the nanoparticles coated with the dense PEG coatings have been found to show the increased spread and the distribution, when studied in the rat model (Shechter et al. 2010).

**Reversible PEGylation for Ocular Delivery**

The main hurdles associated with the ocular drug delivery are poor absorption and the rapid clearance. Only less than 5% of the topically administered dose reaches the ocular tissues. To overcome these issues, the nanoformulations should be designed to possess the increased residence time and more penetration across the tissues. The inert coating produced by the PEG forms the interactions between the nanoparticle and the surface of an eye. PEG promotes mucoadhesion by formation of the hydrogen bonds between the nanoparticle and the mucus. Sanders studied the retinal drug delivery by using PEGylated nanoparticles and found that PEG coatings improve the mobility and penetration through the vitreous chamber and hence increase the drug delivery to the retina (Famili et al. 2014).

**Reversible PEGylation for the Vaccine Delivery**

Xua et al. have developed the hybrid nanoparticle-based nicotine vaccines by the PEGylation approach and found that when the nanoparticles are coated with 20% or lesser PEGylation, they possessed lesser stability. When the higher concentrations of PEGylation were used, specifically more than 30%, the nanoparticles shown the nanoparticles with the compromised core-shell structure and also showed the slow uptake by the dendritic cells. However, when the nanoparticles were coated by the PEG concentrations in the range of 20–30%, they possessed higher stability and enhanced uptake by the dendritic cells (Lv et al. 2022).

## 14.5 Challenges of Reversible PEGylation

Reversible PEGylation, a process involving the attachment of PEG chains to a molecule, has potential benefits in drug delivery and biotechnology. However, it presents several challenges, including controlled release, biocompatibility, stability, synthesis and scalability, specificity, and immunogenicity (Veronese 2001). Controlled release involves regulating the release of PEG chains to avoid premature release or insufficient stability. Biocompatibility concerns arise from the potential impact of PEG breakdown products or detached chains on biological systems (Shechter et al. 2008). Also, stability is crucial under various physiological conditions, such as pH, temperature, and enzyme presence. Developing efficient and scalable methods for synthesis is a significant challenge, as it must be economically viable and suitable for large-scale production. Specificity is another speed breaker in the path of PEGylation, as non-specific interactions or unintended PEGylation may compromise the effectiveness of the reversible PEGylation strategy. Immunogenicity is essential to avoid immune reactions that could impact therapeutic efficacy. Some of the key challenges associated with Reversible PEGylation are as follows (Fig. 14.10):

### 14.5.1 Design Complexity

Reversible PEGylation systems frequently entail the creation and synthesis of intricate compounds with the ability to release PEG under particular circumstances. It

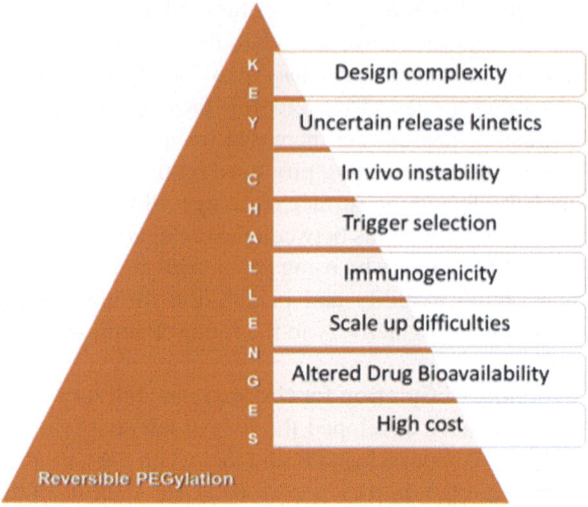


**Fig. 14.10** Key challenges involved in reversible PEGylation

can be difficult to make a balance between stability and controlled release; this calls for close examination of the conjugate's chemical makeup and physical characteristics.

### 14.5.2 Kinetics of Release

To maximize the therapeutic impact, PEG release kinetics and rate must be regulated. PEG can be released gradually and under control without releasing too much or too early because this is dependent on the particular stimuli or triggers used in the reversible PEGylation system (Thakur et al. 2015).

### 14.5.3 Biological Environment Stability

In order for reversible PEGylation systems to release PEG, they must first be stable in the biological environment. Premature release into circulation and stability against enzymatic degradation or other physiological circumstances are challenges (Shechter et al. 2008).

### 14.5.4 Trigger Selection

Selecting the right trigger to release PEG is essential. Changes in pH, temperature, or the presence of particular enzymes are examples of common triggers. However, because it depends on the particular biological context and requirements, choosing the best trigger for a certain application might be difficult (Zalipsky et al. 2007).

### 14.5.5 Immunogenicity

Despite PEG's well-known low immunogenicity, some researchers have reported experiencing immunological reactions. Reversible PEGylation may cause an immunological reaction when exposed to PEG repeatedly, which could eventually decrease the medicinal agent's effectiveness (Shiraishi and Yokoyama 2019; Veronese 2001).

### 14.5.6 Scale-up Difficulties

The shift from laboratory-scale synthesis to large-scale production may pose difficulties with cost-effectiveness, scalability, and reproducibility. For larger-scale applications, the synthesis and purification techniques developed on a small scale might need to be optimized.

### 14.5.7 Influence on Drug Bioavailability

PEG concentrations can affect a drug's bioavailability. The advantages of extended circulation times must be carefully weighed against possible changes in drug absorption, distribution, metabolism, and excretion (ADME) patterns in reversible PEGylation systems.

### 14.5.8 Cost

A major determinant of their uptake may be the expense of synthesizing and producing reversible PEGylation conjugates. One crucial factor to take into account is the expense of synthesis, purification, and regulatory compliance in relation to the possible advantages of better therapeutic outcomes (Tao et al. 2009).

Addressing these challenges requires interdisciplinary efforts, combining expertise in chemistry, biology, and materials science. Despite these obstacles, reversible PEGylation holds promise for enhancing the pharmacokinetics and therapeutic outcomes of various bioactive molecules (Pomroy and Deber 1998). Ongoing research aims to overcome these challenges and unlock the full potential of reversible PEGylation in medical and biotechnological applications.

## 14.6 Conclusion

To sum up, reversible PEGylation of nanocarriers offers a viable path towards drug delivery system optimization. Reversible PEGylation provides a careful balance among biocompatibility and effective payload delivery by addressing the drawbacks of irreversible PEGylation, including reduced cellular absorption and clinical efficacy. Stimuli-responsive polymers along with cleavable linkers provide controlled dissociation of PEG chains from nanocarrier, activated by specific physiological circumstances, thereby improving targeted release at a particular site of action. Reversible PEGylation techniques, such as pH-sensitive, temperature-triggered formulations, and enzyme-responsive, are highly versatile and can be used to achieve more precise drug release kinetics and better therapeutic results. Expanding up reversible PEGylation methods, maintaining stability in storage and transit, and negotiating regulatory processes for clinical approval are still difficult tasks. In order to fully realize the ability of reversible PEGylation to revolutionize drug delivery and ultimately result in less hazardous more efficient treatments for a variety of disorders, it will be imperative to address these obstacles.

## References


Abdelmohsen HA, Copeland NA, Hardy JG (2023) Light-responsive biomaterials for ocular drug delivery. Drug Deliv Transl Res 13:2159–2182

Alavi SE, Bakht U, Koohi Moftakhari Esfahani M, Adelnia H, Abdollahi SH, Ebrahimi Shahmabadi H, Raza A (2022) A PEGylated nanostructured lipid carrier for enhanced oral delivery of antibiotics. Pharmaceutics 14:1668

Alsawaftah NM, Awad NS, Pitt WG, Husseini GA (2022) pH-responsive nanocarriers in cancer therapy. Polymers 14:936

Alshawwa SZ, Kassem AA, Farid RM, Mostafa SK, Labib GS (2022) Nanocarrier drug delivery systems: characterization, limitations, future perspectives and implementation of artificial intelligence. Pharmaceutics 14:883

Amin M, Lammers T, Ten Hagen TL (2022) Temperature-sensitive polymers to promote heat-triggered drug release from liposomes: towards bypassing EPR. Adv Drug Deliv Rev 114503

Arkosi M-K, Mot AC, Lupan I, Tegla MGG, Silaghi-Dumitrescu R (2023) Selective attachment of polyethylene glycol to hemerythrin for potential use in blood substitutes. Protein J:1–9

Bista RK, Bruch RF (2008) Near-infrared spectroscopy of newly developed PEGylated lipids. Spectrochim Acta A Mol Biomol Spectrosc 71:410–416

Böttger R, Knappe D, Hoffmann R (2016) Readily adaptable release kinetics of prodrugs using protease-dependent reversible PEGylation. J Control Release 230:88–94

Cheng T-L, Chuang K-H, Chen B-M, Roffler SR (2012) Analytical measurement of PEGylated molecules. Bioconjug Chem 23:881–899

Famili A, Kahook MY, Park D (2014) A combined micelle and poly (serinol hexamethylene urea)-co-poly (N-isopropylacrylamide) reverse thermal gel as an injectable ocular drug delivery system. Macromol Biosci 14:1719–1729

Fella C, Walker GF, Ogris M, Wagner E (2008) Amine-reactive pyridylhydrazone-based PEG reagents for pH-reversible PEI polyplex shielding. Eur J Pharm Sci 34:309–320

Filpula D, Zhao H (2008) Releasable PEGylation of proteins with customized linkers. Adv Drug Deliv Rev 60:29–49

Gaberc-Porekar V, Zore I, Podobnik B, Menart V (2008) Obstacles and pitfalls in the PEGylation of therapeutic proteins. Curr Opin Drug Discov Devel 11:242

Ganji F, Abdekhodaie M (2008) Synthesis and characterization of a new thermosensitive chitosan–PEG diblock copolymer. Carbohydr Polym 74:435–441

Ghosal S, Walker JE, Alabi CA (2021) Predictive platforms of bond cleavage and drug release kinetics for macromolecule–drug conjugates. Annu Rev Chem Biomol Eng 12:241–261

Ghosh R, Dey J (2020) PH-responsive vesicle formation by PEGylated cholesterol derivatives: physicochemical characterization, stability, encapsulation, and release study. Langmuir 36:5829–5838

Gupta V, Bhavanasi S, Quadir M, Singh K, Ghosh G, Vasamreddy K, Ghosh A, Siahaan TJ, Banerjee S, Banerjee SK (2019) Protein PEGylation for cancer therapy: bench to bedside. J Cell Commun Signal 13:319–330
Hamidi M, Azadi A, Rafiei P (2006) Pharmacokinetic consequences of PEGylation. Drug Deliv 13:399–409
Harris JM, Chess RB (2003) Effect of PEGylation on pharmaceuticals. Nat Rev Drug Discov 2:214–221
Higashi T, Kogo T, Sato N, Hirotsu T, Misumi S, Nakamura H, Iohara D, Onodera R, Motoyama K, Arima H (2020) Efficient anticancer drug delivery for pancreatic cancer treatment utilizing supramolecular polyethylene-glycosylated bromelain. ACS Appl Bio Mater 3:3005–3014
Howard MD, Jay M, Dziubla TD, Lu X (2008) PEGylation of nanocarrier drug delivery systems: state of the art. J Biomed Nanotechnol 4:133–148
Hu F, Zhang R, Guo W, Yan T, He X, Hu F, Ren F, Ma X, Lei J, Zheng W (2020) PEGylated-PLGA nanoparticles coated with pH-responsive tannic acid–Fe (III) complexes for reduced premature doxorubicin release and enhanced targeting in breast cancer. Mol Pharm 18:2161–2173
Jain A, Jain SK (2008) PEGylation: an approach for drug delivery. A review. Crit Rev Ther Drug Carrier Syst 25:403
Karimi M, Eslami M, Sahandi-Zangabad P, Mirab F, Farajisafiloo N, Shafaei Z, Ghosh D, Bozorgomid M, Dashkhaneh F, Hamblin MR (2016) pH-sensitive stimulus-responsive nanocarriers for targeted delivery of therapeutic agents. Wiley Interdiscip Rev Nanomed Nanobiotechnol 8:696–716
Kodera Y, Matsushima A, Hiroto M, Nishimura H, Ishii A, Ueno T, Inada Y (1998) PEGylation of proteins and bioactive substances for medical and technical applications. Prog Polym Sci 23:1233–1271
Lehner R, Hunziker P (2012) Why not just switch on the light?: light and its versatile applications in the field of nanomedicine. Eur J Nanomed 4:73–80
Li D, Tang G, Yao H, Zhu Y, Shi C, Fu Q, Yang F, Wang X (2022) Formulation of pH-responsive PEGylated nanoparticles with high drug loading capacity and programmable drug release for enhanced antibacterial activity. Bioact Mater 16:47–56
Liu Y, Liu L, Yu N, Dai L, Stella C, Chang V, Kaur S, Xu K, Wakshull E (2020) Immunoaffinity LC–MS/MS is more suitable than ELISA to quantify a PEGylated molecule in cynomolgus monkey serum. Bioanalysis 12:1061–1069
Lv S, Song K, Yen A, Peeler DJ, Nguyen DC, Olshefsky A, Sylvestre M, Srinivasan S, Stayton PS, Pun SH (2022) Well-defined mannosylated polymer for peptide vaccine delivery with enhanced antitumor immunity. Adv Healthc Mater 11:2101651
Ma T, Walko M, Lepoitevin M, Janot JM, Balanzat E, Kocer A, Balme S (2018) Combining light-gated and pH-responsive nanopore based on PEG-Spiropyran functionalization. Adv Mater Interfaces 5:1701051
Massoumi B, Abbasian M, Jahanban-Esfahlan R, Motamedi S, Samadian H, Rezaei A, Derakhshankhah H, Farnudiyan-Habibi A, Jaymand M (2020) PEGylated hollow pH-responsive polymeric nanocapsules for controlled drug delivery. Polym Int 69:519–527
Mccombs JR, Owen SC (2015) Antibody drug conjugates: design and selection of linker, payload and conjugation chemistry. AAPS J 17:339–351
Milla P, Dosio F, Cattel L (2012) PEGylation of proteins and liposomes: a powerful and flexible strategy to improve the drug delivery. Curr Drug Metab 13:105–119
Mishra P, Nayak B, Dey R (2016) PEGylation in anti-cancer therapy: an overview. Asian J Pharm Sci 11:337–348
Moosmann A, Gerlach E, Lindner R, Böttinger H (2012) Purification of a PEGylated single chain Fv. J Chromatogr A 1236:90–96
Motlaq VF, Knudsen KD, Nyström B (2018) Effect of PEGylation on the stability of thermoresponsive nanogels. J Colloid Interface Sci 524:245–255
Nesher M, Vachutinsky Y, Fridkin G, Schwarz Y, Sasson K, Fridkin M, Shechter Y, Lichtstein D (2008) Reversible PEGylation prolongs the hypotensive effect of atrial natriuretic peptide. Bioconjug Chem 19:342–348

Nie Y, Günther M, Gu Z, Wagner E (2011) Pyridylhydrazone-based PEGylation for pH-reversible lipopolyplex shielding. Biomaterials 32:858–869

Ozer I, Pitoc GA, Layzer JM, Moreno A, Olson LB, Layzer KD, Hucknall AM, Sullenger BA, Chilkoti A (2022) PEG-like brush polymer conjugate of RNA aptamer that shows reversible anticoagulant activity and minimal immune response. Adv Mater 34:2107852

Paolino D, Accolla ML, Cilurzo F, Cristiano MC, Cosco D, Castelli F, Sarpietro MG, Fresta M, Celia C (2017) Interaction between PEG lipid and DSPE/DSPC phospholipids: an insight of PEGylation degree and kinetics of de-PEGylation. Colloids Surf B: Biointerfaces 155:266–275

Peleg-Shulman T, Tsubery H, Mironchik M, Fridkin M, Schreiber G, Shechter Y (2004) Reversible PEGylation: a novel technology to release native interferon α2 over a prolonged time period. J Med Chem 47:4897–4904

Pelosi C, Saitta F, Wurm FR, Fessas D, Tinè MR, Duce C (2019) Thermodynamic stability of myoglobin-poly (ethylene glycol) bioconjugates: a calorimetric study. Thermochim Acta 671:26–31

Pomroy NC, Deber CM (1998) Solubilization of hydrophobic peptides by reversible cysteine PEGylation. Biochem Biophys Res Commun 245:618–621

Rieger J, Grazon C, Charleux B, Alaimo D, Jérôme C (2009) Pegylated thermally responsive block copolymer micelles and nanogels via in situ RAFT aqueous dispersion polymerization. J Polym Sci A Polym Chem 47:2373–2390

Rustad EA, Von Hofsten S, Kumar R, Lænsman EA, Berge G, Škalko-Basnet N (2022) The pH-responsive liposomes—the effect of PEGylation on release kinetics and cellular uptake in glioblastoma cells. Pharmaceutics 14:1125

Sadekar S, Ghandehari H (2012) Transepithelial transport and toxicity of PAMAM dendrimers: implications for oral drug delivery. Adv Drug Deliv Rev 64:571–588

Shechter Y, Tsubery H, Mironchik M, Rubinstein M, Fridkin M (2005) Reversible PEGylation of peptide YY3-36 prolongs its inhibition of food intake in mice. FEBS Lett 579:2439–2444

Shechter Y, Mironchik M, Rubinraut S, Tsubery H, Sasson K, Marcus Y, Fridkin M (2008) Reversible PEGylation of insulin facilitates its prolonged action in vivo. Eur J Pharm Biopharm 70:19–28

Shechter Y, Heldman E, Sasson K, Bachar T, Popov M, Fridkin M (2010) Delivery of neuropeptides from the periphery to the brain: studies with enkephalin. ACS Chem Neurosci 1:399–406

Shete MB, Patil TS, Deshpande AS, Saraogi G, Vasdev N, Deshpande M, Rajpoot K, Tekade RK (2022) Current trends in theranostic nanomedicines. J Drug Deliv Sci Technol 71:103280

Shi D, Beasock D, Fessler A, Szebeni J, Ljubimova JY, Afonin KA, Dobrovolskaia MA (2022) To PEGylate or not to PEGylate: immunological properties of nanomedicine's most popular component, polyethylene glycol and its alternatives. Adv Drug Deliv Rev 180:114079

Shiraishi K, Yokoyama M (2019) Toxicity and immunogenicity concerns related to PEGylated-micelle carrier systems: a review. Sci Technol Adv Mater 20:324–336

Shiraishi Y, Shirakawa E, Tanaka K, Sakamoto H, Ichikawa S, Hirai T (2014) Spiropyran-modified gold nanoparticles: reversible size control of aggregates by UV and visible light irradiations. ACS Appl Mater Interfaces 6:7554–7562

Son J, Yi G, Yoo J, Park C, Koo H, Choi HS (2019) Light-responsive nanomedicine for biophotonic imaging and targeted therapy. Adv Drug Deliv Rev 138:133–147

Stigsnaes P, Frokjaer S, Bjerregaard S, Van De Weert M, Kingshott P, Moeller EH (2007) Characterisation and physical stability of PEGylated glucagon. Int J Pharm 330:89–98

Sun D, Ding J, Xiao C, Chen J, Zhuang X, Chen X (2015a) Drug delivery: pH-responsive reversible PEGylation improves performance of antineoplastic agent (Adv. Healthcare Mater. 6/2015). Adv Healthc Mater 4:786–786

Sun D, Ding J, Xiao C, Chen J, Zhuang X, Chen X (2015b) pH-responsive reversible PEGylation improves performance of antineoplastic agent. Adv Healthc Mater 4:844–855

Tao L, Liu J, Xu J, Davis TP (2009) Bio-reversible polyPEGylation. Chem Commun:6560–6562

Thakur S, Kesharwani P, Tekade RK, Jain NK (2015) Impact of PEGylation on biopharmaceutical properties of dendrimers. Polymer 59:67–92

Thomas OS, Rebmann B, Tonn M, Schirmeister IC, Wehrle S, Becker J, Zea Jimenez GJ, Hook S, Jäger S, Klenzendorf M (2022) Reversible shielding and immobilization of liposomes and viral vectors by tailored antibody-ligand interactions. Small 18:2105157

Tsubery H, Mironchik M, Fridkin M, Shechter Y (2004) Prolonging the action of protein and peptide drugs by a novel approach of reversible polyethylene glycol modification. J Biol Chem 279:38118–38124

Utatsu K, Kogo T, Taharabaru T, Onodera R, Motoyama K, Higashi T (2021) Supramolecular polymer-based transformable material for reversible PEGylation of protein drugs. Mater Today Bio 12:100160

Veronese FM (2001) Peptide and protein PEGylation: a review of problems and solutions. Biomaterials 22:405–417

Veronese FM, Pasut G (2005) PEGylation, successful approach to drug delivery. Drug Discov Today 10:1451–1458

Vllasaliu D, Fowler R, Stolnik S (2014) PEGylated nanomedicines: recent progress and remaining concerns. Expert Opin Drug Deliv 11:139–154

Wagner E (2012) Polymers for siRNA delivery: inspired by viruses to be targeted, dynamic, and precise. Acc Chem Res 45:1005–1013

Xing Y, Zeng B, Yang W (2022) Light responsive hydrogels for controlled drug delivery. Front Bioeng Biotechnol 10:1075670

Yadav D, Dewangan HK (2021) PEGYLATION: an important approach for novel drug delivery system. J Biomater Sci Polym Ed 32:266–280

Zalipsky S, Mullah N, Engbers C, Hutchins MU, Kiwan R (2007) Thiolytically cleavable dithiobenzyl urethane-linked polymer–protein conjugates as macromolecular prodrugs: reversible PEGylation of proteins. Bioconjug Chem 18:1869–1878

Zhang G, Jiang X (2019) Temperature responsive nanoparticles based on PEGylated polyaspartamide derivatives for drug delivery. Polymers 11:316

Zhang X, Fu C, Feng L, Ji Y, Tao L, Huang Q, Li S, Wei Y (2012) PEGylation and polyPEGylation of nanodiamond. Polymer 53:3178–3184

Zhang F, Liu M-R, Wan H-T (2014) Discussion about several potential drawbacks of PEGylated therapeutic proteins. Biol Pharm Bull 37:335–339

Zhang X, Feng L, Dong Z, Xin X, Yang Z, Deng D, Wagner E, Liu Z, Liu X (2020) Protein-drug conjugate programmed by pH-reversible linker for tumor hypoxia relief and enhanced cancer combination therapy. Int J Pharm 582:119321

Zhu S, Hong M, Tang G, Qian L, Lin J, Jiang Y, Pei Y (2010) Partly PEGylated polyamidoamine dendrimer for tumor-selective targeting of doxorubicin: the effects of PEGylation degree and drug conjugation style. Biomaterials 31:1360–1371

Zhu Y, Chen C, Cao Z, Shen S, Li L, Li D, Wang J, Yang X (2019) On-demand PEGylation and dePEGylation of PLA-based nanocarriers via amphiphilic mPEG-TK-Ce6 for nanoenabled cancer chemotherapy. Theranostics 9:8312

# Stimuli-Responsive PEGylated Nanocarriers

# 15

Samah Hamed Almurisi,
Prasanthi Sri Nagindera Rao,
and Thiagarajan Madheswaran

**Abstract**

Compared to traditional medication dosage forms, nanocarrier-based delivery systems confer several benefits, including drug protection, improved bioavailability, and tailored administration to disease sites. With increased availability in the body, better drug loading values, improved intracellular transport, and superior physiological outcomes, nanocarriers are reported to improve targeted drug delivery to cells and tissues. Stimuli-responsive nanocarriers with unique design characteristics can adapt to exogenous and endogenous stimuli; their structure and physicochemical properties change in response to various factors. This results in the release of drugs at a specific target site when triggered by certain physiological or external triggers. Numerous therapeutic and diagnostic processes can be made substantially more effective by creating pharmaceutical nanocarriers with multiresponsive systems. The most popular technique for giving drug nanocarriers stealthy characteristics is PEGylation. The PEG molecules have been altered to improve their uptake by particular targets and to allow controlled drug release. This chapter highlights the most recent developments in nanocarriers with multifunctional properties and stimuli sensitivity for therapeutic drug delivery applications. By delivering the drugs specifically to the disease cells, there is a pronounced enhancement in the efficacy of the treatment, improving the incidence of unwanted effects on nontarget cells, tissues, and organs.

S. H. Almurisi · P. S. Nagindera Rao · T. Madheswaran (✉)
Department of Pharmaceutical Technology, School of Pharmacy, International Medical University, Bukit Jalil, Kuala Lumpur, Malaysia
e-mail: Thiagarajan@imu.edu.my


R. K. Tekade, N. K. Jain (eds.), *PEGylated Nanocarriers in Medicine and Pharmacy*, https://doi.org/10.1007/978-981-97-7316-9_15

**Keywords**

Nanocarriers · Stimuli-responsive · Drug targeting · PEGylated · Theragnostic applications

## Abbreviations

| | |
|---|---|
| ABC | Ammonium bicarbonate |
| AMF | Magnetic field |
| CAFs | Cancer-associated fibroblasts |
| CAT | Catalase |
| CPs | Conductive polymers |
| CS | Chitosan |
| CST | Critical solution temperature |
| CUR | Curcumin |
| Cys | Cysteine |
| DOX | Doxorubicin |
| DTT | Dithiothreitol |
| ECs | Endothelial cells |
| EPR | Enhanced permeability and retention |
| Gox | Glucose oxidase |
| GPx | Glutathione peroxidase |
| GSH | Glutathione |
| HA | Hyaluronic acid |
| LCST | Lower critical solution temperature |
| MBs | Microbubbles |
| MC | Merocyanine |
| MHT | Magnetic hyperthermia |
| MMA | Methyl methacrylate |
| MMPs | Matrix metalloproteinases |
| MNPs | Magnetic nanoparticles |
| $MoS_2$ | Molydenum disulfide |
| NADPH oxidase | Nicotinamide adenine dinucleotide phosphate oxidase |
| NBs | Nanobubbles |
| NIR | Near-infrared |
| NPs | Nanoparticles |
| NR | Nile red |
| OG | Oil-soluble fluorescent green |
| PCL | Polycaprolactone |
| PEG | Polyethylene glycol |
| PEGylation | The process of polyethylene glycol (PEG)-based modification of molecules for biomedical application |
| PEI | Polyethyleneimine |
| pNIPAm | Poly(N-isopropylacrylamide) |

| | |
|---|---|
| PNIPAm | Poly(N-isopropyl acrylamide |
| PTX | Paclitaxel |
| RES | Reticuloendothelial system |
| ROS | Reactive oxygen species |
| Se | Selenium |
| Se-Se | Diselenide |
| SOD | Superoxide dismutases |
| SP | Spiropyran |
| S-S | Disulfide |
| US | Ultrasound |
| UV | Ultraviolet |
| γ-GCS | γ-glutamyl cysteine |
| γ-PGA | Poly-γ-glutamic acid (γ-PGA) |

## 15.1 Introduction

New technologies incorporating advanced nanomaterials that are applicable to theranostics have evolved to address the obstacles posed by conventional dosage forms and the adverse effects associated with pharmaceuticals. These new technologies have dual properties, including bioimaging for diagnostic purposes and targeted drug release for therapeutic function (Pham et al. 2020a). Engineers and researchers have been putting more effort into designing and producing stimuli-sensitive nanomaterials with combined therapy and diagnostic functions. This is being done to maximize the scope of nanotechnology applications for precise drug delivery and disease diagnosis in the human tissue while maintaining a higher level of safety and biocompatibility (Oladipo et al. 2023). These intelligent nanomaterials, utilized in controlled drug delivery strategies, are formulated to react primarily to external stimuli when required. Another way to implement this requirement is to trigger it when necessary (Ahmad and Packirisamy 2020).

In contrast to conventional nanocarriers, new-generation drug nanocarriers with stimuli-sensitive properties can recognize the environment in which bacteria are located and the site of the infection. Furthermore, these nanocarriers respond dynamically to specific triggers from outside factors (Bag et al. 2023). In addition to this, they can control drug delivery patterns by preventing drug leakage before reaching the target site to improve the delivery of loaded drugs at the target site while simultaneously reducing the likelihood of unwanted side effects. Therefore, nanocarriers created with nanomaterials with stimuli-sensitive properties present distinct, improved therapeutic action and more targeted delivery. Additionally, these nanocarriers have a superior capacity for cell penetration, which is necessary to treat various cancers effectively (Gessner and Neundorf 2020).

In the last 10 years, strategies for successfully delivering therapeutic drugs have been developed, including polymer-based nanomaterials sensitive to stimuli. The nanocarriers are designed to allow the release of drugs in the presence of selected stimuli across polymer, lipid, or metal nanoparticles. These stimuli can be internal or external or triggered by dual or multiple factors. The design of the material can be employed. The term "stimuli-responsive smart biomaterials" refers to materials triggered in response to the local microenvironment's pH, enzymes, physical force, or redox stimuli to release drugs loaded into the nanocarriers. A myriad of potential applications are associated with these biomaterials (Alves et al. 2021).

Redox potential, pH, enzymes, ions, and oxygen and glucose levels are all examples of endogenous stimuli, also known as internal or biological stimuli. These stimuli are mostly peculiar to sick tissue and have the potential to accumulate at the target site. They also have the potential to disrupt the structure of nanocarriers and change the functions they perform. These stimuli-responsive drug delivery systems observe various expression profiles in response to specific cells or tissues. The nanocarriers interact with antibodies and antigens and can detect host-guest products in a particular state to release the drug. An example of a highly specific potential trigger includes processes catalyzed by enzymes specific to particular substrates and catalytic properties (Sheshala et al. 2022). Enzymatically catalyzed processes possess high specificity and high catalytic properties, making them suitable as triggers for drug release. A critical challenge in drug release is an unspecific time and nontargeted release; to overcome this, careful design and selection of stimuli-sensitive biomaterials is necessary. This requirement must be met to fabricate nanocarriers (Joseph et al. 2023).

In many stimuli-based drug delivery strategies, exogenous or external stimuli are applied as chemical, biological, or physical techniques to allow the nanocarriers to release their therapeutic drug payloads in a regulated and targeted manner, as illustrated in Fig. 15.1. Exogenous triggers such as light, heat, ultrasound, and electrical and magnetic fields can all trigger controlled drug delivery. The benefits of applying external stimuli to disrupt the structure of stimuli-sensitive nanocarriers include the

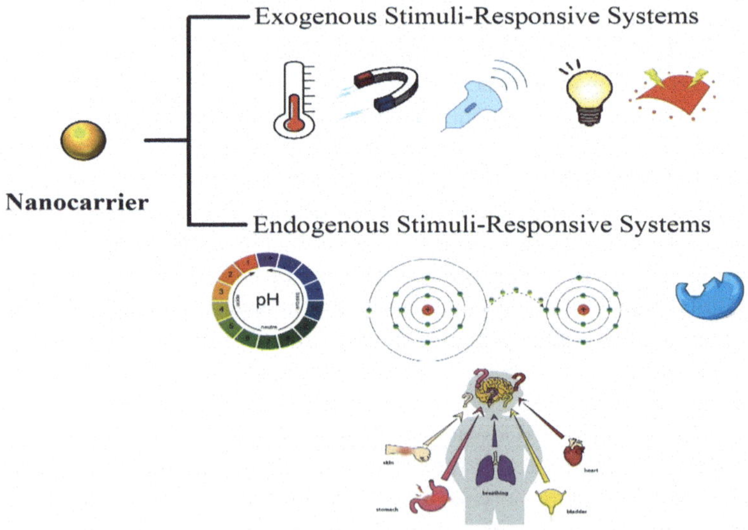


**Fig. 15.1** Illustration of exogenous and endogenous stimuli-responsive systems

ability to apply stimuli directly at the targeted drug release site to trigger the delivery of the drug and reduce off-target effects on healthy tissues. Therefore, the effectiveness of this method relies on being able to discriminate differentiate healthy from diseased tissue when applied (Yang et al. 2022).

Finding exogenous stimulus-responsive materials that can reach the desired tissue at the required depth and location for controlled, site-specific medication release presents a significant challenge. Two-photon methods or longer wavelengths of light (such as a near-infrared laser) are being developed to treat a wider variety of tissues to overcome this obstacle. Real-time monitoring of drug distribution, as well as diagnosis and management of disease, can be performed with the injection of a material activated by ultrasonic waves, cutting-edge light sources, or a powerful magnetic field; however, such treatments are not always feasible and/or cost-effective and thus require elaborate and stringent protocols (Huang et al. 2023).

## 15.2 External Stimuli-Responsive Systems

The most important physical triggers from external sources are light, heat, magnetic fields, and ultrasound waves. These stimuli signals rapidly trigger medication release when interacting with external stimuli-responsive nanocarriers (Li et al. 2019). Smart nanocarriers enable the controlled release of the amount of drugs and target timing and site for the release of encapsulated substances using external triggers as the remote, offering detectable regulation of nanocarrier-based drug delivery system (Li et al. 2020).

### 15.2.1 Thermoresponsive Systems

Targeted medication delivery often involves one of the three methods: active, passive, or stimuli-based targeting. Passive targeting relies on physiological processes; the combined application of thermoresponsive strategies for drug release and heat at the target tissue (such as the tumor site) improves the passive targeting potential. The basis for thermoresponsive drug delivery strategies is the use of custom polymers for nanocarrier fabrication that retain the drug at systemic circulation temperatures (37 °C) but can release the drug rapidly at increased temperatures, such as at tumor sites with increased local temperatures of 40–45 °C (Dastidar and Chakrabarti 2019).

The ability of heat-sensitive polymers to react to temperature changes makes them a valuable "smart material" with various applications; as a result, considerable research interest has been shown in this field (Gandhi et al. 2015). When the temperature changes slightly, the polymers in this innovative material transition reversibly from a hydrophilic to a hydrophobic state in aqueous solution. Thus, a sudden globule-to-coil transition brought on by a temperature shift produces the release of medicines integrated into the polymeric carrier (Gandhi et al. 2015).

The critical solution temperature (CST) is a crucial characteristic of thermoresponsive polymers. The CST represents the temperature at which solution phase separation of the solution happens (Sarwan et al. 2020). Thermoresponsive polymers with lower CST values (LCST) respond to increased temperatures by becoming insoluble, and upper CST (UCST) polymers become soluble (Dastidar and Chakrabarti 2019). Poly(N-isopropylacrylamide) (PNIPAm) is studied extensively as a thermoresponsive polymer because of its LCST in the water of around 32 °C (Bergueiro and Calderón 2015).

Additionally, polyethylene glycol (PEG) and its derivatives have been explored as a major class of polymers with thermoresponsive properties for synthesizing nanocarriers. In water, PEG displays an LCST greater than 90 °C. The LCST may be decreased by adding salt to the aqueous medium. As an example, its LCST approaches 35 °C when 450 mM of $K_2SO_4$ is present (Bordat et al. 2019). The LCST is significantly influenced by how many ethylene glycols repeating units are present: For $x = 2$, 3, and 8–9, the LCST is 28, 50, and 90 °C, respectively (Lutz and Hoth 2006). Additionally, as the molar mass rises, the LCST falls. This is explained by the carbon polymer's methyl-substituted backbone, which increases hydrophobicity and lowers LCST. The necessary LCST can be simply generated by selecting the molar mass and the number of repeating units (Bordat et al. 2019). It has been extensively researched how to lower the CSTs of "thermoresponsive" monomers using poly(ethylene glycol) methacrylates.

Hydrogels with thermoresponsive properties are being studied for drug release research, as they can change their structure with temperature change. During injection into the body, the change in temperature causes a phase change from sol to gel to allow controlled drug release. Delivery of the hydrogel via injection is critical to ensure minimally invasive applications and to ensure delivery of high drug loads at the target site while avoiding entry into the circulatory system (Lencina et al. 2018). Recently, studies on thermosensitive hydrogels using PEG and polycaprolactone (PCL) polymers, varying molecular weight (Mw), and ratios of the polymers were reported to design triblock copolymer sequences (Patel et al. 2019).

In the last 20 years, numerous thermoresponsive nanocarriers have been studied, but only ThermoDox® (Celsion) is currently undergoing a Phase III clinical trial (Bordat et al. 2019). A thermoresponsive liposomal formulation called ThermoDox® that contains doxorubicin (DOX) is used to treat primary liver cancer. The liposomes' combination of normal lipids and lysolipids allows them to be thermoresponsive between 40 and 45 °C.

### 15.2.2 Magnetically Responsive Systems

The first magnet field-based drug delivery system was developed in the 1980s. However, there has been gaining interest in magnet-based targeting recently with the availability of magnets with more extensive fields and sophisticated magnetic probes for use in various fields, such as theranostic probes (Gu et al. 2018). The probes enable the therapies to be delivered simultaneously, including precise drug delivery via magnetic field stimulation, hyperthermia, and diagnostic applications, such as magnetic particle or magnetic resonance imaging (MRI) (Price et al. 2018).

Typically, electromagnetic coils or permanent magnets can be used to direct magnetic drug targeting (Price et al. 2018). Nanoparticle magnets metal (such as iron, gadolinium, and manganese) or superparamagnetic oxides, such as superparamagnetic iron oxide ($Fe_3O_4$), are typically the main components of magnetic nanoparticles (MNPs) (Gu et al. 2018).

Our immune system combats pathogens by enlisting a variety of immune cells that discharge different inflammatory substances, producing heat (fever). Utilizing MNPs that have been externally energized by an alternating magnetic field (AMF) to generate heat has gained popularity in magnetic hyperthermia (MHT). The MHT can be used to activate MNPs that are deeply embedded in tissues or organs at frequencies (f) and field settings (H) that are safe for humans and provide direct damage, which can also trigger drug release (Mai et al. 2018). PEGylated liposomes activated magnetically by loading citrate-coated Fe3O4 magnetic NPs to carry DOX were produced in a recent study (Hardiansyah et al. 2019). Using a high-frequency magnetic field, DOX-loaded PEGylated magnetic liposomes could inductively heat surroundings to 56 °C from baseline physiological temperature. The results of in vitro studies of the liposomes indicate effectiveness against HeLa cell proliferation while not presenting cytotoxic effects in healthy fibroblast L-929 cells. The combination of MNP and DOX functionality in these liposomes has created a novel system for effective delivery and action of anticancer drugs.

Magnetic force has the potential to direct the site of a drug delivery system besides managing drug release. For instance, Li et al. produced platelets coated with L-arginine and $Fe_2O_3$ magnetic nanoparticles (PAMNs). The amount of PAMN at the lesion could increase by two times with the placement of a magnet that produces a static magnetic field above the lesion. The study of immunohistochemical slices demonstrated that PAMNs could more effectively access the region of the brain injury when the magnetic field was present (Ma et al. 2022). However, creating large gradients over enormous distances for applying this technology in the human body is still a challenge (Shamsi et al. 2018).

Therefore, micromotors and nanomotors are considered the latest innovations for personalized, precise drug delivery, with significant benefits compared to nonmotion, passive delivery strategies (Medina-Sánchez et al. 2018). Magnetic force for targeting drug delivery using micro−/nanomotors is very promising because the motor is adaptable to diverse conditions due to its amenability to be controlled (Choi et al. 2021). In contrast to superparamagnetic iron oxide nanoparticles, magnetically propelled helical nanomotors are typically activated by 5–10 mT rotating fields in a range of media. Additionally, it has been demonstrated that nanopropellers can spontaneously move through the tissues via biopolymeric networks (Kadiri et al. 2020). A widespread use for nano- and micromotors is the delivery of DNA or other pharmacoactive cargo (Li et al. 2017). A magnetic field is typically applied to deliver the magnetic targeted medication. However, using an external magnetic field for a prolonged duration is difficult and expensive. This issue may be resolved by implanting a magnet internally to produce the magnetic field, a technique that is reported to successfully deliver medications to deep tissues (Ge et al. 2017).

### 15.2.3 Ultrasound-Triggered Drug Delivery

Ultrasound (US) waves refer to mechanical waves with frequencies ranging from 20 kHz to 30 MHz and are employed for both diagnostic and treatment. Since the US waves don't ionize, they are a type of noninvasive radiation that delivers energy to biological tissue with a small chance of negative effects (Ahmadi et al. 2020). The various ways that the drug delivery system and ultrasonic waves interact encourage uptake and specificity in successful targeted drug delivery (Xia et al. 2016). Three distinct ways that improve in vivo medication distribution when administered by ultrasonic-responsible carriers: cell membrane permeabilization, vascular permeability expansion, and cellular endocytic uptake promotion (Fan et al. 2022).

Ultrasounds possess high spatial resolution and can target deep tissues to deliver stimuli. The use of ultrasound in medicine is not foreign. Its use in imaging, stone crushing, and thermal treatment has been reported (Kubota et al. 2021). The ultrasound functions by causing the particle to cavitation sonoporation, stable cavitation, or inertial cavitation (Kooiman et al. 2020).

Microbubbles (MBs) are frequently employed as ultrasonic contrast agents due to their high acoustic impedance mismatch with the surrounding tissue and their size of 1–10 μm, which facilitates simple circulation through the vasculature (Batchelor et al. 2020). There are several approaches to including therapeutics in MBs, including therapeutic gas (Fix et al. 2015), direct drug attachment to the lipid shell (Nesbitt et al. 2018), drug-filled liposome attachment (Malik et al. 2016), and release of the therapeutics by raising the US intensity. The delivery of targeted ultrasound pulses triggers the MB to disintegrate through a series of mechanical effects, starting with inertial cavitation to improve drug release and enhance cellular uptake of drugs (Browning et al. 2021). However, these MBs can only produce contrast-enhanced US imaging within the vasculature due to the limits of their large particle size in the range of several micrometers and their limited extravasation from the blood artery (Cui et al. 2019).

The nanobubbles (NBs), which are submicron bubbles typically 200–600 nm in diameter, are an appealing potential strategy to deliver drugs, showing increased tumor accumulation and retention compared to MBs. The leaky vasculature provides an environment for increased permeation and retention, improving tumor biodistribution (Wu et al. 2019). The development of nanosized contrast agents with the ability to effectively pass the leaky tumor vasculature and accumulate inside tumors (the gaps range in tumor vasculature is 600 to 800 nm) is crucial for anticancer drug delivery (Blanco et al. 2015). Also, the NB shells' surfaces can be functionalized to improve their stealth and molecular targeting capabilities. For instance, PEGylation could prolong the half-life of therapeutic agents in circulation and enable better targeting of the lesion tissues (Jin et al. 2020).

### 15.2.4 Light-Triggered Drug Delivery

Targeted drug delivery via the application of external light is an approach that promotes precise spatial resolution and temporal for the delivery of various physiologically active compounds, such as genetic material, proteins, and drug molecules. The nanocarriers can accumulate in the target cells and tissues, after which controlled light application irradiates the targeted diseased feature while leaving healthy tissues unaffected (Zhao et al. 2019).

Light responsiveness can be activated by a variety of wavelengths, including ultraviolet (UV, 200–400 nm), Vis (400–750 nm), or near-infrared (NIR, 750–2000 nm) light. UV and Vis light (short-wavelength light) are applied to attack photolabile moieties and trigger light-sensitive agents. However, light has poor penetration and reach due to its short wavelength, limiting its application in deep tissue targets for therapy. An additional concern is the risk of triggering cancer when tissues are exposed to these short wavelengths over extended time durations (Tao et al. 2020). NIR light, in contrast, is better for biological applications due to its higher penetration into deep tissues, better safety profile, and lower risk of signal interference (Karimi et al. 2017).

Many different types of nanomaterials are currently being used to create NIR-responsive delivery systems, including semiconductor nanomaterials (such as copper sulfides and bismuth sulfides), Au-based nanomaterials (such as Au NPs and Au nanorods (NRs)), carbon nanomaterials (such as GOs and carbon nanodots), and upconverting nanoparticles (UCNPs) (Karimi et al. 2017). These materials are used to fabricate NIR-activated nanoparticles due to their maximum optical absorbance specific to NIR wavelengths. Other than nanomaterials, outstanding NIR-responsive substances include conjugated polyelectrolytes and dye molecules ICG and IR780 that absorb light in the NIR region, resulting in an electron excitation and release of heat energy (Yan et al. 2016). These substances transform and produce heat and/or reactive oxygen species (ROS) with absorbed light, which then disrupts the nanoparticles to cause the site-specific release (Mohammed et al. 2019).

The matrices of thermosensitive hydrogels injections were typically incorporated with nanosized photothermal agents to facilitate change from light to heat and initiate the release of drugs on-demand by inducing phase transition from the gel to sol phase in traditional NIR light-responsive hydrogels (Qiu et al. 2018). The incorporated nanoscale NIR-absorbing components in some of these hydrogels may leach out, degrading the photothermal action and lowering the efficacy of medication delivery (Wu et al. 2018b). To achieve the desired outcomes from extended time in circulation and enable continuous treatment with just one administration of the hydrogel, the poly(ethylene glycol) chains incorporate hydrogel formulations so that their constituent parts can take up light in the NIR-II region, allowing for self-conversion of light into heat (Ruan et al. 2019).

### 15.2.5 Electroresponsive Systems

The electric fields, whether endogenous or exogenous, can be applied in several therapeutic methods (Kolosnjaj-Tabi et al. 2019). The type of stimuli-responsive materials determines what triggers are required to stimulate the release of drugs, whether extrinsic (external pulses) or intrinsic (changes in charge at the tissue microenvironment level). The release of drugs from their carriers is triggered by electric fields, one of the many applications. Systems that respond to electric fields, such as electroresponsive drug delivery nanosystems, allow for targeted drug release via spatial and temporal control (Uppalapati et al. 2016). Thus, the external application of a low-intensity electric field enables on-demand sustained or pulsed delivery of the active ingredient. Also, higher-intensity exogenous electric fields have been shown to directly affect cellular membrane permeability, stimulate medication delivery, and act as a tool for treatment to promote/improve (wound) healing and restore tissue integrity (Kolosnjaj-Tabi et al. 2019).

Electroresponsive systems could be administered as (thin) films or nanoparticles and employed as implants in the body or drug carriers. These devices might be implanted or injected, and upon electrical stimulation, they would (ideally) electrically activate surrounding tissue and cells over long periods (Raj et al. 2023). Concurrently, organic "conductive" polymers (CPs) with the ability to carry and allow the flow of electric charge have been developed (Lamprou 2023). The electronic structure of CPs, which alternates single and double bonds across its repeating structure, primarily determines their conductivity. Radical cations/anions (polarons) or dications/dianions (bipolarons) are made more accessible to form during the doping process, and counter-ions from the solution enter the polymeric material to balance the charge (Grancarić et al. 2018).

The CPs combine metallic and polymeric features, such as optical and electrical abilities, that can respond to electrical and electrochemical triggers to release the drug payload (Gupta 2022). Some examples of currently developed CPs with the highest potential for therapeutic applications are polypyrrole, polythiophene, polyaniline, and poly (3, 4-ethylenedioxythiophene) (Paramshetti et al. 2023). However, researchers are facing challenges in applying CPs for treatment due to poor biocompatibility, processability, and mechanical stability. This has prompted a study into integrating CP and non-CP materials via chemical modification to produce composites with combined properties of the CP and non-CP materials that exhibit superior drug delivery functionalities for biomedical applications. The conductivity of metals and flexibility of polymers are combined in these materials (Paramshetti et al. 2023).

The hydrogel is a polymer with hydrophilic, absorbing, and swelling capability for holding large amounts of liquid in its structure. Water-soluble polymers cross-linked to produce the 3D network structure include synthetic materials PEG, polyvinyl alcohol (PVA), polymethacrylate-2-hydroxyethyl ester (PHEMA), and natural biomaterials such as chitosan, starch, and cellulose (Rinoldi et al. 2021). The addition of an electric field-responsive element to hydrogels results in hydrogels with the added ability to produce mechanical energy from electricity, potentiating its application as a sensor, biomechanic agent, sound dampener, energy transductor, chemical separator, drug delivery carrier, and tissue engineering component (Liu et al. 2021).

## 15.3 Internal Stimuli-Responsive Systems

The advantage of using polymer nanoparticles for drug delivery lies in their ability to safely and stably carry the drug load to the targeted cells/tissue. To provide a therapeutic intervention effectively, there are numerous biological obstacles to overcome. These difficulties include the capacity to avoid immune detection, precisely reach the desired cells and tissues, and carry the drugs to specific intracellular sites. Applying stimuli-responsive nanoparticles is a crucial strategy for ensuring the successful carriage of a therapeutic payload to the target region. These nanoparticles are designed with the ability to alter their physical characteristics when triggered by certain internal factors (Deirram et al. 2019).

There are several advantages to using endogenous stimuli-sensitive drug delivery systems over conventional drug delivery systems, the most important being its capacity to react to particular physiological signals within the body, including enzyme action, pH levels, and redox (Abasian et al. 2021). The endogenous stimuli-responsive systems overcame the nonspecificity and toxicity linked to conventional medicine delivery techniques (Sethuraman et al. 2021). These endogenous stimuli-responsive systems are designed to be sensitive to biological signals that start the transport of medications by influencing the tissues in the microenvironment, upregulating certain enzymes, interacting with antibodies and antigens, and recognizing host-guest moieties in a particular state (Raza et al. 2019). The difference between pathological and healthy, normal tissue has contributed to the development of endostimuli-responsive nanocarriers for the transportation and precise targeting of therapeutic drugs (Das et al. 2020). Endogenous stimuli-responsive systems are gaining high interest in nanotechnology research because they offer the potential to develop multifunctional nanoplatforms for drug administration.

### 15.3.1 pH-Responsive Systems

The pH-triggered nanosystems take advantage of the distinct pH levels in healthy and diseased tissue to precisely deliver drugs to the local site for more effective treatment outcomes. This makes it feasible to regulate the delivery of medications to target organs, whether the liver or the pancreas, and intracellular compartments, including endosomes and lysosomes. The controlled release of drugs is linked to microenvironment indices present in disease, as is the case in infection or cancer (Goyal et al. 2019; Canaparo et al. 2019). For instance, tumor cells and their intracellular spaces (endosomes and lysosomes) have a slightly acidic pH, whereas normal pH is generally neutral (7.0–7.4) (Chen et al. 2018).

pH-responsive nanoparticles are designed to modify their surface chemistry or mechanical structure, disassemble, or release their load. It is possible to regulate drug release and cell absorption by adjusting the characteristics of nanoparticles. Therefore, pH-responsive nanoparticle is an effective method for designing therapeutic delivery systems using polymers with ionizable moieties (polymers with shifting charge), linkages susceptible to low/acidic pH, or crosslinkers that integrate polymers with non-cleavable linkages to produce particles that are able to swell or acid-labile linkages that disintegrate the nanoparticle when exposed to certain pH (Deirram et al. 2019).

Polymers containing ionizable moieties can be protonated or deprotonated, which changes the soluble properties of the nanoparticle in aqueous media. In contrast, acid-labile covalent connections in the polymer backbone are employed to cleave polymers and cause their breakdown in acidic conditions (Abasian et al. 2021). Acid-labile polymers frequently exhibit a slower internal structural transition than polymers with ionizable moieties, making them more suitable for drug-delivery systems (Tang et al. 2018).

Polyacids and polybases are the main categories used in polymers with ionizable moieties. Polyacids have acidic moieties, including carboxyl, phosphate, or boronic groups linked to the polymeric backbone, that become negatively charged when protons are released in an essential environment. In contrast, primary groups in polybasic polymers, such as amines or pyridines in an acidic pH, become positively charged (Huang et al. 2018). The ionization and deionization typically take place between a pH range of 4 and 8 for polyacids such as polyacrylic acid and polymethacrylic acid, where protons are taken at a low pH and released at a neutral or higher pH. In contrast, pH-responsive polymers from polybases such as methacrylic acid and N, N-dimethyl aminoethyl methacrylate, which have amino groups on their side chains, accept protons under acidic conditions as cations and then release them under alkaline conditions to reduce their solubility (Ofridam et al. 2021).

In addition, natural polymers like pectin, chitosan, or alginate, which are abundant, biodegradable, and biocompatible, have demonstrated distinct morphological alterations in response to environmental stimuli. Natural polymers often have high molecular weights; therefore, the appearance of many charges changes the conformation and assembly of the polymer chains intrinsically and has the benefit of being chemistry-modifiable, which enables the addition of new functions to the polymeric backbone (Schoeller et al. 2022).

Several studies have been conducted using pH-sensitive systems. For instance, a pH-sensitive polymeric-chitosan and hydroxypropyl methylcellulose crosslinked in the presence of tripolyphosphate nanoparticles was developed for the delivery of melatonin in breast cancer therapy (Jafari et al. 2021). Results show that melatonin encapsulated in polymeric chitosan and hydroxypropyl methylcellulose nanoparticles was more targeted at cancer cells than free melatonin, suggesting that the nanoparticles may effectively treat breast cancer.

Additionally, polymeric nanoparticles with acid-labile linkage have been extensively used in drug-delivery systems (Wu et al. 2018a). Imine, hydrazone, oxime, amide, acetals, and orthoester are examples of pH-sensitive chemical bonds that can be embedded within drug carriers. By cleaving the dynamic chemical bonds, the configuration of the carriers can be changed, allowing for the controlled release of the loaded pharmaceuticals (Zhuo et al. 2020).

Hydrazones were extensively investigated for their simple preparation, moderate stability, and favorable sensitivity among pH-sensitive linkages (Qi et al. 2018). Also, acetal is one of the most common pH-sensitive linkages used (Patil et al. 2012). Acetal bonds can be created as a connection between the carrier and the drug or can be found in the structure of the carrier. The internal structure of the carrier is damaged following the hydrolysis of the acetal link in the acidic environment, which results in the drug's release (Saadat et al. 2021). Other acid labile linkages such as orthoester, citraconic amide, and Schiff base bonds were reported. The chemistry of pH-sensitive bonds typically determines the effectiveness of drug delivery systems.

The highly targeted drugs can be applied to specific lesions by conjugating the drug molecules to delivery carriers via pH-responsive linkers. The polymer-drug conjugate is one of the pH-responsive drug delivery approaches with acid-sensitive links between the drug molecule and polymer. In a previous study, DOX was conjugated to a biodegradable, nontoxic, and nonimmunogenic nanoconjugate using PEG and pH-sensitive hydrazone linkage. These DOX-nanoconjugates were successfully demonstrated to inhibit the growth of invasive breast carcinoma cell lines in in vitro studies and were shown to be stable under physiological conditions (Patil 2012).

### 15.3.2 Redox-Responsive Systems

The fundamental processes of life, such as metabolism and respiration, depend on redox reactions. All parts of life that are mediated by or impacted by redox reactions are then referred to as redox biology (Franco and Vargas 2018). The availability of electrons is crucial for biological systems, particularly for the mitochondrial electron transport chain that provides energy to cells. The metabolic process produces ROS, which are highly reactive due to their existence as free radicals, ions, and molecules with a single unpaired electron (Mirhadi et al. 2020). ROS are elevated when there is insufficient electron supply, and unbalanced oxygen levels (hypoxia or hyperoxia) impede mitochondria from producing ATP (Kaur and New 2019).

Since ROS controls immunological function, autophagy, inflammation, and stress-related responses, it is produced in normal cells in a highly controlled manner. Free radicals are much more than unwanted side products that cause harm. Various cell types, such as cancer-associated fibroblasts (CAFs), endothelial cells (ECs), and inflammatory cells, along with processes like angiogenesis and hypoxia, contribute to the production of ROS (Mirhadi et al. 2020).

During immunological or inflammatory reactions, inflammatory cells, particularly macrophages and neutrophils, cause the formation of ROS; meanwhile, angiogenesis, which is crucial for tumor growth, development, and metastasis, serves as another significant source of ROS. Additionally, the activity of endogenous antioxidant enzymes, including thioredoxin, superoxide dismutases (SOD), and NADPH oxidase, may contribute to high ROS levels (Mirhadi et al. 2020). Cells have a buffer system to prevent oxidative damage, and a group of enzymes, including glutathione peroxidase (GPx) and catalase (CAT) are responsible for turning free radicals into stable and less harmful compounds (Kumari et al. 2018; Mirhadi et al. 2020). Cells experience oxidative stress when ROS production exceeds antioxidant defenses that scavenge the ROS molecules. High ROS levels in the body have been associated with various diseases such as diabetes, heart disease, obesity, neurological disease, inflammation, and cancer (Quader and Van Guyse 2022).

The antioxidant glutathione (GSH) is produced in the body from glycine and γ-glutamyl cysteine (γ-GCS) (Raj Rai et al. 2021). Many researchers are developing innovative redox-sensitive strategies for cancer diagnosis and therapy using GSH's antioxidant property, which detoxifies the elevated oxidative environment (Khan et al. 2022). The scientists can formulate and synthesize nanoparticles with the ability to respond to chemical triggers in the intracellular environment from elevated GSH accumulation representative of tumor pathophysiology.

To make drugs sensitive to ROS in tumor regions, the functional groups might be included in drug delivery systems. Diselenide bonds, succinimide-sulfide bonds, and the most extensively studies disulfide bonds are among the often utilized redox-sensitive chemical bonds (Fu et al. 2022). Glutathione readily breaks down disulfide bonds in the tumor microenvironment and transfers them to sulfhydryl groups (thiol), which causes the carrier to degrade. ROS thenxidises thiol groups to form sulfenic acid, sulfinic acid, and sulfonic acid (Gao and Dong 2018). Due to the disulfide bonds' intrinsic instability under many conditions, they could be considered as a great linker in a nanoparticle drug delivery system (Fu et al. 2022). For the prototype cancer drug paclitaxel, redox-sensitive poly(ethylene glycol) (PEG)-based prodrugs, chitosan (CS)-based micellar systems, or polyethyleneimine (PEI)-based amphiphilic micelles with disulfide bonds as redox-responsive linkages have all been successfully developed (Xu et al. 2019). The disulfide linkage is embedded in the nanoparticle structure and is triggered by changes in glutathione (GSH) concentrations, causing drug release (Gong et al. 2018).

Due to the attractive outcomes of disulfide bonds, selenium (Se) has been studied for its reactivity to reduction and high sensitivity to small amounts of oxidizing agents. Improved redox-responsive strategies could potentially from the use of a diselenide bond rather than the disulfide bond is foreseen due to the bond energies of the C-Se and Se-Se bond being less than compared to S-S bonds (Behroozi et al. 2018; Mirhadi et al. 2020). The redox sensitivity of the synthesized poly(ethylene oxide)b-poly(furfuryl methacrylate) using diselenide (Se-Se) and disulfide (S-S) redox-responsive core-crosslinked (CCL) micelles was studied. The particles prepared utilizing diselenide bonds were highly sensitive compared to disulfide bonds (Yadav et al. 2023). Moreover, Se partakes in many different kinds of physiological processes, including gene transcription, cell cycle, thyroid function, immunity, and antitumor activities (Behroozi et al. 2018).

### 15.3.3 Enzyme-Responsive Systems

Many biological reactions use enzymes as catalysts, and certain enzymes are dysregulated and elevated in pathological conditions, which are considered common features of several diseases, making them a potential biological trigger for treatments. Their incorporation into the design of targeted delivery is promising, as enzymes are highly specific for their substrate, allowing for accurate, intricate, biologically similar chemical reactions (Quader and Van Guyse 2022). Hydrolases play significant roles in breaking down numerous biomolecules, including sugars, lipids, proteins, DNA, RNA, and many more. The main hydrolase enzymes include protease, which hydrolyzes peptides; lipase, which hydrolyzes fat; glycosidase, which hydrolyzes complex sugar; esterase, which hydrolyzes ester; and elastase, which hydrolyzes elastin.

Numerous disorders, such as cancer, inflammation, atherosclerosis, Alzheimer's disease, and viral infections, have reported elevated levels of proteases (Smith et al. 2021). Polymeric nanoparticles can be modified with enzyme-labile links to trigger the release of their cargo when certain enzymes are present and reflectively cleaved by overexpressed enzymes (Shahriari et al. 2019). Several hydrolytic enzymes, including proteases, glycosidases, and lipases, as well as oxidoreductases like peroxidases, have been used in drug delivery because of their ability to cleave various natural and synthetic polymers, including those containing particular amino acid sequences and natural polysaccharides (Dou et al. 2020). The controlled release of drugs from enzyme-sensitive systems occurs under enzyme catalysis (Abasian et al. 2021).

Proteases can be grouped according to the peptide bond cleavage action and essential amino acids at the active site: cysteine, threonine, serine, aspartate, glutamic acid, and matrix metalloproteinases (MMPs) (Dudani et al. 2018). Another hydrolase, called an esterase, hydrolyzes ester bonds to produce acid and alcohol molecules. Numerous cancer cells overexpress the esterase enzymes, which can be used as triggers for delivering tumor-targeted drugs (Abbasi et al. 2023). Meanwhile, glycosidases catalyze hydrolysis reactions in complex sugars, breaking the glycosidic linkage in these molecules. The overexpression of this enzyme in pancreatic juice and saliva in diseased individuals is thought to be an enzymatic trigger for polysaccharide-based drug delivery systems (Kapalatiya et al. 2022). Additionally, oxidoreductase is vital in oxidative stress because of its significance in oxidative environments and its ability to detect glucose through the use of glucose oxidase (GOx); this has enabled this enzyme to be applied Alzheimer's and cancer treatments (Irshad et al. 2022).

Matrix metalloproteinases (MMP), hyaluronidases, and cathepsin are among the enzymes that directly contribute to the development of atherosclerosis. The proteolytic enzyme degrades the extracellular matrix proteins such as gelatin, collagen, fibrin, and elastin, while hyaluronidase is an enzyme family that can break down hyaluronic acid (HA) specifically and predominantly, it also plays a crucial part in the tissue's response to injury. Cathepsins are lysosomal cysteine proteases that could inactivate related enzymes, breakdown proteins, and increase atherosclerotic inflammation in response to microenvironment imbalance (Song et al. 2022).

Typically, enzyme-triggered drug-release nanocarriers have an enzyme-substrate linker unit linking the nanocarrier and drug, which serves as the cleavage site for the enzyme. The selection of an appropriate enzyme stimulus, informed by the disease pathophysiology and the desired enzyme activity, is the first design decision to be made when creating an enzyme-sensitive system (Duan et al. 2020). The mechanism of action behind enzyme-responsive systems is the physical and chemical modifications that the system undergoes in response to enzyme exposure, resulting in the release of drugs at the target site. Cleavage of the linker between the nanocarrier and the bioactive molecule via the nanocarrier shell or carrier degradation or cleavage of the functional groups leads to structural changes in the system that allow for the release of the drug (Kapalatiya et al. 2022) .

### 15.3.4 Self-Regulated Systems

Self-regulated nanodrug delivery systems comprise drug delivery platforms that can detect changes in physiological indices and adjust their characteristics or performance accordingly due to these external or internal biological signals. The drug delivery system can be designed for open-loop control or closed-loop (feedback). In open-loop control, the drug delivery system is established by a control system that governs drug release. In contrast, in closed-loop control, physiological data are received and fed back to the controller, which modifies the rate of drug release (Wang et al. 2021).

Feedback indicates a system that continuously triggers itself whenever necessary, meaning that the outputs of the system are fed back into it as inputs. The human body generates positive or negative feedback to maintain equilibrium (Arun et al. 2021). Positive feedback describes a system's output as enhancing the initial stimulus; examples include births, lactation, blood clotting, etc. When the body uses negative feedback, which is a typical form of feedback mechanism, the output of the system will suppress the initial stimulus, as in the case of thermoregulation, blood sugar regulation, osmoregulation, etc. Closed-loop drug delivery systems, often referred to as feedback-regulated systems, are a more advanced version of modified drug-releasing systems that respond to signals from the body to release drugs. Any biochemical factor (pH, temperature, pressure, and ion concentration) created in the body because of unexpected physiological conditions can serve as one of these triggers.

Traditional drug delivery is limited by brief circulation in the system, repeated administration, and adverse effects because of disorderly release, nonspecific biodistribution, rapid drug absorption, and the possibility that the drug release could reach highly toxic levels. Using self-regulated systems, drug delivery can be triggered, stopped, raised, decreased, or maintained at desirable levels in the body since the system will release the drug as long as it recognizes the abnormal biochemical change. This means that the drug will be released as and when required, and the system may be constructed so that the ratio of drug concentration to stimulus intensity is proportionate (Sawalwade et al. 2020). Thus, conventional drug delivery systems have been developed into new, smart systems that respond to specific stimuli, allowing the drug to bind to the target site and release the drug predictably over prolonged periods with enhanced bioactivity.

The basic design of self-regulating drug delivery systems consists of a drug reservoir, a rate-control membrane, and a sensor that responds to the biochemical triggering agent. The drug reservoir is encapsulated within a semipermeable polymeric membrane. The permeation of the triggering agent activates the system, and the drug is released depending on the reversible competitive binding of the triggering agent with the biosensor (Sawalwade et al. 2020). This sort of drug delivery is appropriate for molecules such as DNA, siRNA, nucleotides, proteins, peptides, and others that are specific in their action but are cytotoxic. Hence, fluctuations in their blood levels are undesirable. Recurring administration of these molecules can exacerbate their adverse effects and cytotoxicity, but closed-loop drug delivery can avoid this (Sawalwade et al. 2020). For instance, insulin injections are frequently used to treat diabetes. Even though regular insulin injections quickly lower blood sugar levels, they also always induce injection pain and a decrease in patients' quality of life.

Therefore, an alternative diabetic treatment that controls drug release continuously and automatically and is directly triggered by glucose is needed. The usage of glucose-sensitive materials, a type of "intelligent" polymer, has increased in self-regulating systems. These materials continually and automatically regulate insulin release, which is triggered by a raised blood glucose level. Drug delivery devices that respond to glucose levels can be helpful in treating diabetes and replacing repeated insulin injections by combining drug administration in response to glucose with minimum patient involvement and enhanced diabetic quality of life (Wang et al. 2019).

Recently, formulations have been developed that modify insulin release rates according to variations in local glucose levels. For instance, polymeric nanoparticles with the ability to alter their insulin release profiles have been developed using acid-degradable, acetylated-dextran polymers; nanoparticles made with high cyclic acetal content dextran (71% of residues) release insulin more rapidly than those produced from high acyclic acetal content dextran (94% of residues). The potential of this delivery system to release insulin in response to glucose following a therapeutic-specific profile represents a significant step forward in targeted insulin delivery (Volpatti et al. 2019).

## 15.4 Dual and Multiresponsive Systems

Besides single-responsive systems, dual or multistimuli-responsive systems are gaining interest in targeting different sites and disease conditions. Dual or multiple stimuli-responsive drug-delivery systems were first developed for targeted cancer treatment and respond to a combination of internal and external biological triggers for drug release (Zhao et al. 2021). Compared to a single responsive system, dual and/or multistimuli systems use various active triggers systematically incorporated into one nanoplatform, resulting in a multifaceted approach that has superior performance to simple nanosystems in facing physiological/pathological barriers. Dual/multiresponsive nanosystems respond to the stimuli simultaneously or subsequently to achieve targeted drug delivery (Zhao et al. 2021). Based on the specific targeting site and the response needed, a combination of exogenous triggers that use pH, redox potential, temperature, presence of enzymes, etc., can be considered triggers for dual or multiresponsive systems.

### 15.4.1 Temperature and pH Dual Stimuli-Responsive Systems

Temperature and pH-responsive polymers have been studied vastly in dual stimuli systems triggered by more precise, controlled, and targeted indices to stimulate the polymer responses. The trigger system is formulated by conjugating a pH-responsive polymer to a thermosensitive polymer. Several polymers fulfilling these criteria have been developed in the last 10 years for cancer treatment applications due to the increased temperature and acidic pH found in the tumor microenvironment when compared to healthy tissues (Pham et al. 2020b; Zhao et al. 2021; Zifar et al. 2023). The tumor cells' abnormal characteristics, including fast metabolism and abnormal proliferation of cells, cause accumulation of lactic acid, resulting in an acidic pH (pH 5.7–6.9) in the cells and microenvironment (Pham et al. 2020a, b).

The LCST is an essential characteristic of thermoresponsive polymers. At higher LCSTs, the bonds interacting between polymer and drug molecule weaken, increasing the hydrophobic interactions present in the molecule and causing the polymer to dehydrate and shrink. Poly (N-isopropyl acrylamide) (pNIPAAm) has been widely studied because of its LCST of approximately 32 °C; above this LCST temperature, its gel structure will collapse (Ghalehkhondabi et al. 2023; Pham et al. 2020b; Reza Soltani et al. 2023; Zhao et al. 2021). This applies to be used at body temperature (37 °C), where it will shrink. The LCST transition changes the polymer state from water-soluble to water-insoluble. The pH-responsive polymers commonly use weak acids, acrylic acids (AAs), poly(2-(diisopropylamine)) ethyl methacrylate (PDPA), and chitosan in their formulation (Zhao et al. 2021). Once both polymers are mixed, a dual responsiveness system in the form of micelles or nanoparticles will be created.

In the research of Reza Soltani et al., Molybdenum disulfide ($MoS_2$) nanosheets were mixed with pH and temperature-responsive monomers methyl methacrylate (MMA) and N-isopropyl acrylamide (pNIPAm) to allow controlled release and improved DOX delivery into breast tumor cells. The $MoS_2$@pNIPAM-MMA/GLN is a dual-triggered system, with the two-dimensional (2D) $MoS_2$ acting as a carrier for DOX loading (Reza Soltani et al. 2023). Characterization of the system in blood model fluid indicated slow drug release at pH 7.4, 37 °C, whereas 98% of the drug was released at pH 5.6, 50 °C after 6 h, an improvement in controlled DOX delivery across the tumor cells under certain pH and temperature (Reza Soltani et al. 2023).

Another report has also elaborated the fabrication and characterization of a potential, dual-responsive nanosystem consisting of pNIPAm-PAA spheres for controlled release of DOX in breast cancer treatment (Ghalehkhondabi et al. 2023). The poly(acrylic acid) (PAA) carriers were fabricated using polymerization via the precipitation method, followed by emulsion polymerization to add a thermoresponsive pNIPAm coating on the outside of the carrier spheres. The nanosphere cores were loaded with DOX as a model drug. The resulting nanospheres showed a pH- and temperature-responsive drug release profile in the release medium, with slight leaking at pH 7, and 37.0 °C, which improved significantly in more acidic environments (pH ~5.5), indicating the potential for drug release at the tumor site. The nanosphere release observed was approximately 2.3-fold higher than under neutral pH and temperature (Ghalehkhondabi et al. 2023). Lower temperatures decreased drug release profiles in both acidic and neutral pH.

Temperature-pH dual stimuli microcapsules system was also studied to improve systems for the controlled release of drugs. In a study by Chen et al. (2014), particles encapsulating Nile Red (NR) and fluorescent green (OG) dyes were prepared using temperature-sensitive pNIPAm particles. The authors reported significant pH- and temperature differences in the dye release. At low pH conditions, the microcapsule shell swelled and released the OG, but the NR was not released from the microparticles because of the electrostatic repulsion between the positively charged polymer chains of the shell. OG release was in response to pH change, but not NR. Variations in pH stimulus only triggered the release of OG but did not affect the NR. However, when the microcapsules were exposed to increasing temperature, it was noticed that the NR was released, but the OG was unaffected (Chen et al. 2020).

### 15.4.2 pH and Redox Potential Dual-Stimuli Responsive Systems

The redox-responsive DDs is a great system as well for targeting potential tumor cells; however, there are some challenges faced by this system, such as undesirable drug release behavior at non-targeted tissues with unresolved toxicities issues and unsatisfactory dose of chemotherapeutic agents reaching the targeted site (Jia et al. 2021; Pham et al. 2020b; Zifar et al. 2023). Therefore, modification of the initial redox system with the addition of pH-/ROS-sensitive groups and enzyme materials may provide great potential in addressing the challenges mentioned above. In this case, the pH-redox system is still predominant compared to the others. The synergistic actions from this dual system can help increase the drug release rate at a condition of low pH and high redox potential (Jia et al. 2021).

Dong et al. produced an oxidation-pH responsive hydrogel b PEG-*b*-P(METMA) with pendent thioether activity by performing reversible addition-fragmentation chain transfer polymerization of a tert-butyloxycarbonyl (Boc)-l-methionine-(2-methacryloylethyl)ester (Boc-METMA) monomer using a poly(ethylene glycol) (PEG)-based chain transfer agent. The b PEG-*b*-P(METMA) had self-assembly properties, forming micelles at pH over 6.0, and was hydrophilic at pH under 6.0, due to the protonable amine groups present in its structure. Curcumin (CUR) loaded into the micelles showed a pH-dependent release profile (Dong et al. 2020). In $H_2O_2$, swelling of the micelles was observed, followed by disintegration to release the drug. $H_2O_2$ triggered oxidation of the thioether to sulfoxide, changing the micelle solubility profile. The pH-dependent release and the $H_2O_2$ responsive behavior showed that PEG-*b*-P(METMA) micelles had exciting potential for delivering drugs that respond to ROS (Dong et al. 2020).

Ren et al. studied biocompatible, charged pH/redox dual-stimulus response nanoparticles using poly-γ-glutamic acid (γ-PGA), cysteine (Cys), and chitosan (CS) for the controlled release of DOX. The triggers acted on electrostatic interactions and amide bonds in the γ-PGA-CS particles crosslinked in situ using the disulfide bonds in cystine (Cys). The resulting γ-PGA-S-S-CS-DOX nanoparticles showed significantly higher DOX release in acidic and redox environments, triggers that are present in the tumor cell microenvironment for the release of DOX (Ren et al. 2019). Similarly, a pH/redox potential responsive dual stimuli system was developed by Ding et al. using polyacrylic acid (PAA-pH-responsive) and tocopherol succinate (TOS). The self-assembly constructed vesicle system consists of amphiphilic PAA-cystamine (cys)-TOS in the anticancer methotrexate drug aqueous solution. The dual stimuli polymers showed increased release of the drug in response to pH and GSH compared to pH alone, as GSH acts on the disulfide bond of cystamine. In contrast, pH acts on the amide bonds in PAA-cys-TOS to cause the vesicles to disintegrate, resulting in faster drug release (Ding et al. 2020).

### 15.4.3 Multistimuli Responsive Systems

Nanocarriers applying multistimuli responsive systems observe drug release in predetermined internal or external triggers as illustrated in Fig. 15.2 (Fu et al. 2018).

The drug delivery systems responsive to stimuli were developed for targeted cancer treatment and respond to various combinations of internal and external stimuli. Multiple stimuli-responsive drug-delivery systems (MSR-DDSs) can be designed with the advantages of increased drug encapsulation efficiency, increased time in circulation, improved stability, ability to recognize receptors, and drug release via controlled degradation at the target site (Chen et al. 2020; Jia et al. 2021).

A PAA-co-spiropyran (SP) methacrylate nanogel crosslinked by N,N-bis(acryloyl)cystamine containing disulfide was developed by Chen et al. with light, pH, and redox-sensitive properties. The nanogel was loaded with DOX-HCl as a model drug via electrostatic interaction with AA. The nanogel response to UV light irradiation and low pH, when the hydrophobic SP transformed to the hydrophilic merocyanine (MC), resulted in swelling (Chen et al. 2017). The MC was able to emit an intense green light upon uptake in the cancer cell's nucleus via endocytosis, facilitating cell imaging. Furthermore, the nanogel was also disrupted in the presence of a reducing agent, which reacts with the disulfide crosslinkers in the nanogel, resulting in oxidative breakage. The light activated the release of DOX-HCl, pH, and Dithiothreitol (DTT) reducing agent. In vitro assessments for cytotoxicity

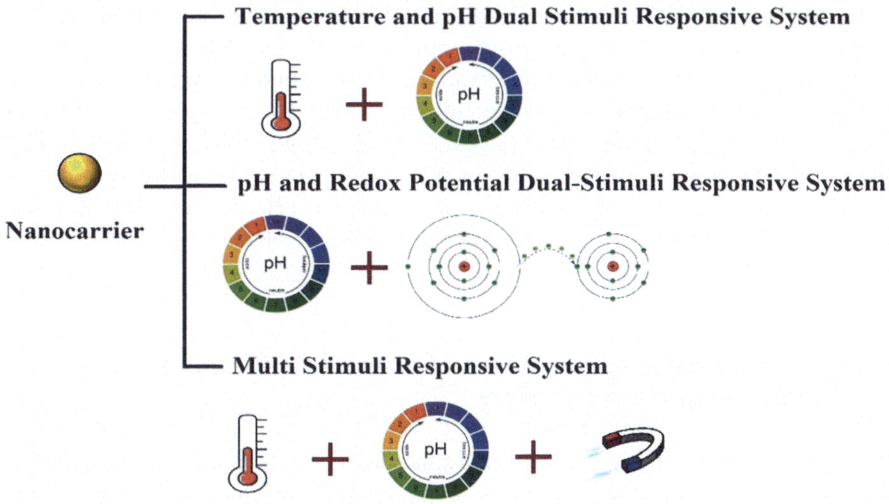


**Fig. 15.2** Illustration of multistimuli-responsive systems

showed that the nanogel encapsulating DOX was able to cause death in the cancer cells, an observation enhanced with UV light exposure (Chen et al. 2017).

Two different polymers were synthesized for sensitivity to pH, temperature, and redox to allow drug release under desired pH 5, 40 °C, and GSH ≥ 10 mM [(Poddar et al. 2020). The synthesized polymers were 2-(2-((4-(hexyloxy)benzyloxy)carbonyl)ethylthio)ethyl acrylate (HBCEEA), and N-isopropyl acrylamide (NIPA) and poly(ethylene glycol methyl ether acrylate) (PEGMA) copolymer, which are sensitive to pH, and temperature and redox potential, respectively. Combined, these two polymers result in polymer poly[HBCEEM-b-(NIPA-r-PEGMA)] (PHNP) with sensitivity to pH, temperature, and redox potential. Using multistimuli increases the drug release rate from the polymer when compared to dual-responsive stimuli systems (Poddar et al. 2020).

In some studies, dimers such as Au-$Fe_xO_y$ were used to fabricate advanced inorganic materials termed "heterostructures" (HSs). These HSs are produced using more than two inorganic domains to combine the functional moieties of NPs with various drugs onto the carrier, resulting in a structure with apparent geometric features (Jia et al. 2021). Kakwere et al. prepared the smart inorganic HSs that responded excellently to magnetic fields with heating. The pH- and thermoresponsive polymers were combined with Au-$Fe_xO_y$-HS moieties. The iron oxides present in the Au were used as magnetic resonance imaging (MRI) agents and thermal media for magnetism-induced hyperthermia or heat triggers (Kakwere et al. 2018).

Meanwhile, the Au-molecules allow for heat release of the $Fe_xO_y$ domain under an AMF. The heat-sensitive polymer pNIPAm was linked to the Au-domain via a gold mercaptan bond to allow Nile blue dye to be loaded. Finally, the surface of the $Fe_xO_y$ was attached with a pH-sensitive polymer, PDMAEA, to deliver non-green fluorescent protein-small interfering RNA (GFP-siRNA) (Kakwere et al. 2018). From the observation, HSs respond to low pH levels in the tumor microenvironment, triggering the PDMAEA to undergo hydrolysis to change the potential from positive to negative, releasing siRNA. It was also noted that at a lower temperature than the (LCST; 43 °C), pNIPAm in HSs dissolve in water and change its conformation to a globule to release the content while increasing the temperature (above LCST) by AMF (Kakwere et al. 2018; Ruiz et al. 2022; Zhao et al. 2021).

## 15.5 Theranostic Applications of Stimuli-Responsive PEGylated Nanocarriers

Theranostics refers to the application of a combination of therapeutic and diagnostic approaches to detect and treat diseases simultaneously or sequentially. The combination of drugs and techniques in this approach has seen major focus in cancer diagnosis and treatment. One such approach is the use of stimuli-sensitive PEGylated nanocarriers. The PEGylation technique involves the conjugation of PEG, the water-soluble polymer chains to any therapeutic molecule (Cheng et al. 2014; Mishra et al. 2016). The FDA has approved PEG as a safe, non-toxic, and poorly immunogenic material with poor uptake by the reticuloendothelial system (RES), allowing for longer drug circulation in the blood and better targeting due to enhanced permeability and retention (EPR) (Wang et al. 2016).

Further, the addition of an intelligent polymer in the PEGylated prodrugs increased the response of the nanocarrier to the tumor physiological microenvironment in vivo. It developed a novel system for site-specific targeted drug delivery. The thermoresponsive stimuli systems using smart materials are the most preferred systems to be studied for pharmaceutical and biomedical applications (Mishra et al. 2016).

A thermoresponsive PEGylated polyaspartamide derivative (mPEG-PAAHP) containing pendant phenyl moieties was developed with up to 99% paclitaxel (PTX) drug loading efficiency. The formulated mPEG-PAAHP NPs also had evident temperature responsiveness at 25 °C and anticancer activity against Hela cells, suggesting that temperature-responsive mPEG-PAAHP had good potential as a carrier for hydrophobic anticancer drugs (Zhang and Jiang 2019).

Alamoudi et al. recently reported a noninvasive thermoresponsive drug delivery strategy using a PEGylated liposome bubble carrier loaded with ammonium bicarbonate (ABC) salt. The authors also report the characterization of ABC–siRNA thermoresponsive lipoplexes to effectively deliver Bcl2-siRNA and/ or MRP1-siRNA to small carcinoma lung cancer cells via endosomal escape. Upon temperature elevation, the ABC load was degraded to produce water, ammonia, and carbon dioxide, creating a temporary lesion at the membrane of the nanocarrier and releasing the loaded drug (Chen et al. 2014). It was reported that the cancer cells took up the siRNA-liposomes in 5 minutes, followed by release of the siRNAs in the cytoplasm for successful gene silencing (Alamoudi et al. 2017; Chen et al. 2014).

A recently developed PEGylation polyphenol–cisplatin complexation-based core-shell structured nano-prodrug (PEG-GA*x*/Pt) has been studied using pH and ROS-responsive targeted system (Dong et al. 2022). First, the PEG-GA*x*/Pt, methoxyl-PEG is terminated with one (PEG-GA) or two (PEG-GA2) gallic acid moieties and was complexed with cisplatin (CDDP) into homogeneous NPs. The drug release of the Pt in vitro was studied at pH 5.0 and pH 7.4, respectively. The results indicate that at pH 5.0, in the span of 48 hours, the PEG-GA/Pt and PEG-GA2/Pt NPs had released 47% to 59% of Pt. These values were reduced to 20% less in both PEG-GA/Pt and PEG-GA2/Pt NPs at pH 7.4. Under acidic pH, Pt release acted on the polyphenols to weaken the bonds between them and Pt (II). To promote the ROS-responsive, $H_2O_2$ (reducing agent) was added, and an oxidation reaction with the galloyl group promote the release of the Pt (Dong et al. 2022). Therefore, the researchers suggested that PEG-GA*x*/Pt NPs stayed longer in the circulatory system and presented localized accumulation at the tumor site, thus lowering its toxicity profile and improving its action on the tumor.

In another study, Song et al. also fabricated pH-sensitive PEGylated DOX micelles with self-assembly ability in aqueous solutions via esterification and Schiff base reactions. The NPs had high drug encapsulation efficiencies and were highly sensitive to pH for controlled drug release. In MCF-7 cancer cells, the micelles showed fast prodrug internalization with good antitumor activity (Song et al. 2021). Under normal physiological pH, the bridged Schiff bases maintained the structure of the nanoparticle but were able to disassemble in more acidic environments. Furthermore, cell viability assays indicated that PEG-Schiff-DOX NPs were more potent against human breast cancer cells (MCF-7) when compared to free DOX. Further studies using the nanoparticle revealed a new DOX-based PEG-Schiff-DOX polymer with programmed DOX release behavior; the new polymer has the potential for higher drug loading, release over longer durations, and increased bioactivity (Song et al. 2021).

## 15.6 Clinical Status, Challenges, and Outlook

Polymeric nanocarriers have received interest from researchers in recent years due to their function as carriers for various molecules, including bioactive drugs, proteins, genes, and nucleic acids, for effective use in therapeutic and pharmacological applications (Bilal et al. 2021). The therapeutic nucleic acids cargo may be intercepted and/or destroyed before it reaches the target by the various physiological barriers in the blood and tissue via opsonization, endocytosis by the mononuclear phagocytic system, tissue pressure, endosomal escape, and intracellular transmission (Yu et al. 2021). Through years of study of nanomaterials, strategies to overcome these challenges have been developed. Therefore, acid- or enzyme-separable PEGylation is used to achieve surface charge shielding to increase steric stability. PEGylation is attaching PEG polymer chains, either covalently or noncovalently, to molecules and macrostructures (Osman et al. 2018). Interestingly, these polymeric nanocarriers can be designed and fabricated to respond to certain stimuli or triggers based on the mechanism of release; this includes endogenous or exogenous factors to ensure the drugs are delivered at the target site and following a controlled release pattern. Examples of triggers include enzymes, temperature, redox values, pH, glucose levels, or oxygen levels intracellularly, and light, ultrasound waves, magnetic fields, or light for outside triggers (Das et al. 2020).

Epirubicin-loaded polymeric micelles (NC6300) have begun phase I and II investigations (NCT03168061) to treat hepatocellular cancer in clinical trials as pH-stimulated nanoparticles (Das et al. 2020). The micelles are pH-responsive; in the patient, variables such as tumor pH or reducing agents present in the circulatory system have made them challenging to manage (Kamaly et al. 2016).

Therefore, most clinical trials are exogenous stimuli-responsive nanosystems (Abu-Thabit and Makhlouf 2018). Two magnetically sensitive iron-based nanocarriers, iron oxide magnetite (Phase IV, NCT00920023) and DOX-loaded iron and carbon (Phase I/II, NCT00041808), are currently undergoing clinical trials to treat cancer. Additionally, the following three clinical studies used the thermally sensitive DOX-incorporated liposomes: individuals suffering from recurring regional breast cancer are being studied in phases I and II to determine the safety of the treatment, including highest tolerated dose, pharmacokinetics, and effects of hyperthermia (NCT00826085); phase III of the ThermoDox with radiofrequency ablation for hepatocellular carcinoma treatment has become complete (NCT00617981). Also, targeted antibody-drug conjugate in a clinical trial for Trastuzumab Emtansine alongside Docetaxel and possibly Pertuzumab in patients with Metastatic breast cancer (Phase I/II, NCT00934856).

The design of stimuli-responsive nanocarriers still faces certain challenges. Although there has been progress, creating stable nanocarriers in a healthy medium is still difficult. The problem of overdesigning nanocarriers with various functionalities in one molecule complicates their ability to be used in clinical trials (Kaushik et al. 2022). Additionally, the clinical translation of nanotherapeutics depends on their large-scale, highly repeatable manufacture; many nanotherapeutics cannot reach the market due to their intricacy (Zhang et al. 2020).

More and more, researchers are focusing their attention on developing methods for early disease diagnosis and treatment using nanotechnology. Strategies for improved targeted drug delivery and diagnostics with high efficacy are explored to achieve this. The main objectives of such research are to shed insight into methods that can enhance targeted drug administration and imaging while preserving efficacy, and extensive research has been done. The engineered nanocarriers have helped overcome some of the problems associated with 'free' drug delivery (limited bioavailability, drug resistance, and drug toxicity), such as PEGylated nanosystems functionalized with biologically active moieties (Tapasya et al. 2022).

## 15.7 Conclusion

The subject of drug delivery in the past has paid interest to the creation of stimuli-responsive nanoparticles for the delivery of drugs to target sites following controlled release. However, in the past few years, much research and development has been focused on developing stimuli-responsive nanomaterials, starting from systems with a single stimuli-response to those responding to multiple stimuli. This chapter reviews developments in both single-stimulus-responsive nanomaterials and multistimulus-responsive nanomaterials. Stimuli-sensitive nanomaterials have exceptional stability and can rapidly respond to environmental stimuli. These nanomaterials can also be functionalized with immune modulators and cell-targeting ligands to deliver efficient and exact drug release.

## References

Abasian P, Shakibi S, Maniati MS, Nouri Khorasani S, Khalili S (2021) Targeted delivery, drug release strategies, and toxicity study of polymeric drug nanocarriers. Polym Adv Technol 32(3):931–944

Abbasi YF, Bera H, Cun D, Yang M (2023) Recent advances in pH/enzyme-responsive polysaccharide-small-molecule drug conjugates as nanotherapeutics. Carbohydr Polym 120797

Abu-Thabit NY, Makhlouf ASH (2018) Historical development of drug delivery systems: from conventional macroscale to controlled, targeted, and responsive nanoscale systems. In: Stimuli responsive polymeric nanocarriers for drug delivery applications, volume 1. Elsevier, pp 3–41

Ahmad N, Packirisamy G (2020) Intelligent nanomaterials for drug delivery applications. Elsevier

Ahmadi A, Hosseini-Nami S, Abed Z, Beik J, Aranda-Lara L, Samadian H, Morales-Avila E, Jaymand M, Shakeri-Zadeh A (2020) Recent advances in ultrasound-triggered drug delivery through lipid-based nanomaterials. Drug Discov Today 25(12):2182–2200

Alamoudi K, Martins P, Croissant JG, Patil S, Omar H, Khashab NM (2017) Thermoresponsive pegylated bubble liposome nanovectors for efficient siRNA delivery via endosomal escape. Nanomedicine 12(12):1421–1433

Alves PM, Barrias CC, Gomes P, Martins MCL (2021) Smart biomaterial-based systems for intrinsic stimuli-responsive chronic wound management. Mater Today Chem 22:100623

Arun A, Malrautu P, Laha A, Ramakrishna S (2021) Gelatin nanofibers in drug delivery systems and tissue engineering. Eng Sci 16:71–81

Bag N, Bardhan S, Roy S, Roy J, Mondal D, Guo B, Das S (2023) Nanoparticle-mediated stimulus-responsive antibacterial therapy. Biomater Sci 11(6):1994–2019

Batchelor DVB, Abou-Saleh RH, Coletta PL, McLaughlan JR, Peyman SA, Evans SD (2020) Nested nanobubbles for ultrasound-triggered drug release. ACS Appl Mater Interfaces 12(26):29085–29093

Behroozi F, Abdkhodaie M-J, Abandansari HS, Satarian L, Molazem M, Al-Jamal KT, Baharvand H (2018) Engineering folate-targeting diselenide-containing triblock copolymer as a redox-responsive shell-sheddable micelle for antitumor therapy in vivo. Acta Biomater 76:239–256

Bergueiro J, Calderón M (2015) Thermoresponsive nanodevices in biomedical applications. Macromol Biosci 15(2):183–199

Bilal M, Qindeel M, Raza A, Mehmood S, Rahdar A (2021) Stimuli-responsive nanoliposomes as prospective nanocarriers for targeted drug delivery. J Drug Deliv Sci Technol 66:102916

Blanco E, Shen H, Ferrari M (2015) Principles of nanoparticle design for overcoming biological barriers to drug delivery. Nat Biotechnol 33(9):941–951

Bordat A, Boissenot T, Nicolas J, Tsapis N (2019) Thermoresponsive polymer nanocarriers for biomedical applications. Adv Drug Deliv Rev 138:167–192

Browning R, Thomas N, Marsh LK, Tear LR, Owen J, Stride E, Farrer NJ (2021) Ultrasound-triggered delivery of iproplatin from microbubble-conjugated liposomes. ChemistryOpen 10(12):1170–1176

Canaparo R, Foglietta F, Giuntini F, Della Pepa C, Dosio F, Serpe L (2019) Recent developments in antibacterial therapy: focus on stimuli-responsive drug-delivery systems and therapeutic nanoparticles. Molecules 24(10):1991

Chen K-J, Chaung E-Y, Wey S-P, Lin K-J, Cheng F, Lin C-C, Liu H-L, Tseng H-W, Liu C-P, Wei M-C, Liu C-M, Sung H-W (2014) Hyperthermia-mediated local drug delivery by a bubble-generating liposomal system for tumor-specific chemotherapy. ACS Nano 8(5):5105–5115. https://doi.org/10.1021/nn501162x

Chen S, Bian Q, Wang P, Zheng X, Lv L, Dang Z, Wang G (2017) Photo, pH and redox multi-responsive nanogels for drug delivery and fluorescence cell imaging. Polym Chem 8(39):6150–6157. https://doi.org/10.1039/C7PY01424D

Chen H, Jin Y, Wang J, Wang Y, Jiang W, Dai H, Pang S, Lei L, Ji J, Wang B (2018) Design of smart targeted and responsive drug delivery systems with enhanced antibacterial properties. Nanoscale 10(45):20946–20962

Chen K, Liao S, Guo S, Zheng X, Wang B, Duan Z, Zhang H, Gong Q, Luo K (2020) Multistimuli-responsive PEGylated polymeric bioconjugate-based nano-aggregate for cancer therapy. Chem Eng J 391:123543. https://doi.org/10.1016/J.CEJ.2019.123543

Cheng L, Gong H, Zhu W, Liu J, Wang X, Liu G, Liu Z (2014) PEGylated Prussian blue nanocubes as a theranostic agent for simultaneous cancer imaging and photothermal therapy. Biomaterials 35(37):9844–9852. https://doi.org/10.1016/j.biomaterials.2014.09.004

Choi H, Yi J, Cho SH, Hahn SK (2021) Multifunctional micro/nanomotors as an emerging platform for smart healthcare applications. Biomaterials 279:121201

Cui X, Han X, Yu L, Zhang B, Chen Y (2019) Intrinsic chemistry and design principle of ultrasound-responsive nanomedicine. Nano Today 28:100773

Das SS, Bharadwaj P, Bilal M, Barani M, Rahdar A, Taboada P, Bungau S, Kyzas GZ (2020) Stimuli-responsive polymeric nanocarriers for drug delivery, imaging, and theragnosis. Polymers 12(6):1397

Dastidar DG, Chakrabarti G (2019) Thermoresponsive drug delivery systems, characterization and application. In: Applications of targeted nano drugs and delivery systems. Elsevier, pp 133–155

Deirram N, Zhang C, Kermaniyan SS, Johnston APR, Such GK (2019) pH-responsive polymer nanoparticles for drug delivery. Macromol Rapid Commun 40(10):1800917

Ding C, Wu H, Yin ZZ, Gao J, Wu D, Qin Y, Kong Y (2020) Disulfide-cleavage- and pH-triggered drug delivery based on a vesicle structured amphiphilic self-assembly. Mater Sci Eng C 107:110366. https://doi.org/10.1016/J.MSEC.2019.110366

Dong S, Jiang Y, Qin G, Liu L, Zhao H (2020) Methionine-based pH and oxidation dual-responsive block copolymer: synthesis and fabrication of protein nanogels. Biomacromolecules 21(10):4063–4075. https://doi.org/10.1021/ACS.BIOMAC.0C00879

Dong X, Brahma RK, Fang C, Yao SQ (2022) Stimulus-responsive self-assembled prodrugs in cancer therapy. Chem Sci 13(15):4239–4269). Royal Society of Chemistry. https://doi.org/10.1039/d2sc01003h

Dou Y, Li C, Li L, Guo J, Zhang J (2020) Bioresponsive drug delivery systems for the treatment of inflammatory diseases. J Control Release 327:641–666

Duan W, Ji S, Guan Y, Mu X, Fang S, Lu Y, Zhou X, Sun J, Li Z (2020) Esterase-responsive polypeptide vesicles as fast-response and sustained-release nanocompartments for fibroblast-exempt drug delivery. Biomacromolecules 21(12):5093–5103

Dudani JS, Warren AD, Bhatia SN (2018) Harnessing protease activity to improve cancer care. Annu Rev Cancer Biol 2:353–376

Fan C-H, Ho Y-J, Lin C-W, Wu N, Chiang P-H, Yeh C-K (2022) State-of-the-art of ultrasound-triggered drug delivery from ultrasound-responsive drug carriers. Expert Opin Drug Deliv 19(8):997–1009

Fix SM, Borden MA, Dayton PA (2015) Therapeutic gas delivery via microbubbles and liposomes. J Control Release 209:139–149

Franco R, Vargas MR (2018) Redox biology in neurological function, dysfunction, and aging. Antioxid Redox Signal 28(18):1583–1586. Mary Ann Liebert, Inc. New Rochelle, NY

Fu X, Hosta-Rigau L, Chandrawati R, Cui J (2018) Multi-stimuli-responsive polymer particles, films, and hydrogels for drug delivery. Chem 4(9):2084–2107

Fu S, Rempson CM, Puche V, Zhao B, Zhang F (2022) Construction of disulfide containing redox-responsive polymeric nanomedicine. Methods 199:67–79

Gandhi A, Paul A, Sen SO, Sen KK (2015) Studies on thermoresponsive polymers: phase behaviour, drug delivery and biomedical applications. Asian J Pharm Sci 10(2):99–107

Gao Y, Dong C (2018) Triple redox/temperature responsive diselenide-containing homopolypeptide micelles and supramolecular hydrogels thereof. J Polym Sci A Polym Chem 56(10):1067–1077

Ge J, Zhang Y, Dong Z, Jia J, Zhu J, Miao X, Yan B (2017) Initiation of targeted nanodrug delivery in vivo by a multifunctional magnetic implant. ACS Appl Mater Interfaces 9(24):20771–20778

Gessner I, Neundorf I (2020) Nanoparticles modified with cell-penetrating peptides: conjugation mechanisms, physicochemical properties, and application in cancer diagnosis and therapy. Int J Mol Sci 21(7):2536

Ghalehkhondabi V, Fazlali A, Soleymani M (2023) Temperature and pH-responsive PNIPAM@PAA Nanospheres with a core-shell structure for controlled release of doxorubicin in breast cancer treatment. J Pharm Sci. https://doi.org/10.1016/j.xphs.2023.04.009

Gong T, Liu T, Zhang L, Ye W, Guo X, Wang L, Quan L, Pan C (2018) Design redox-sensitive drug-loaded nanofibers for bone reconstruction. ACS Biomater Sci Eng 4(1):240–247

Goyal AK, Rath G, Faujdar C, Malik B (2019) Application and perspective of pH-responsive nano drug delivery systems. In: Applications of targeted Nano drugs and delivery systems. Elsevier, pp 15–33

Grancarić AM, Jerković I, Koncar V, Cochrane C, Kelly FM, Soulat D, Legrand X (2018) Conductive polymers for smart textile applications. J Ind Text 48(3):612–642

Gu M, Wang X, Toh TB, Chow EK-H (2018) Applications of stimuli-responsive nanoscale drug delivery systems in translational research. Drug Discov Today 23(5):1043–1052

Gupta RK (2022) Conducting polymers: chemistries, properties and biomedical applications. CRC Press

Hardiansyah A, Destyorini F, Irmawati Y, Yang M-C, Liu C-M, Chaldun ER, Yung M-C, Liu TY (2019) Characterizations of doxorubicin-loaded PEGylated magnetic liposomes for cancer cells therapy. J Polym Res 26:1–8

Huang B, Chen F, Shen Y, Qian K, Wang Y, Sun C, Zhao X, Cui B, Gao F, Zeng Z (2018) Advances in targeted pesticides with environmentally responsive controlled release by nanotechnology. Nano 8(2):102

Huang Q, Zhu W, Gao X, Liu X, Zhang Z, Xing B (2023) Nanoparticles-mediated ion channels manipulation: from their membrane interactions to bioapplications. Adv Drug Deliv Rev 195:114763

Irshad S, Siddiqui B, ur Rehman A, Farooq RK, Ahmed N (2022) Recent trends and development in targeted delivery of therapeutics through enzyme responsive intelligent nanoplatform. Int J Polym Mater Polym Biomater 71(6):403–413

Jafari H, Hassanpour M, Akbari A, Rezaie J, Gohari G, Mahdavinia GR, Jabbari E (2021) Characterization of pH-sensitive chitosan/hydroxypropyl methylcellulose composite nanoparticles for delivery of melatonin in cancer therapy. Mater Lett 282:128818

Jia R, Teng L, Gao L, Su T, Fu L, Qiu Z, Bi Y (2021) Advances in multiple stimuli-responsive drug-delivery systems for cancer therapy. Int J Nanomedicine 16:1525–1551). Dove Medical Press Ltd. https://doi.org/10.2147/IJN.S293427

Jin Z, Chang J, Dou P, Jin S, Jiao M, Tang H, Jiang W, Ren W, Zheng S (2020) Tumor targeted multifunctional magnetic nanobubbles for MR/US dual imaging and focused ultrasound triggered drug delivery. Front Bioeng Biotechnol 8:586874

Joseph TM, Kar Mahapatra D, Esmaeili A, Piszczyk Ł, Hasanin MS, Kattali M, Haponiuk J, Thomas S (2023) Nanoparticles: taking a unique position in medicine. Nano 13(3):574

Kadiri VM, Bussi C, Holle AW, Son K, Kwon H, Schütz G, Gutierrez MG, Fischer P (2020) Biocompatible magnetic micro-and nanodevices: fabrication of FePt nanopropellers and cell transfection. Adv Mater 32(25):2001114

Kakwere H, Materia ME, Curcio A, Prato M, Sathya A, Nitti S, Pellegrino T (2018) Dually responsive gold–iron oxide heterodimers: merging stimuli-responsive surface properties with intrinsic inorganic material features. Nanoscale 10(8):3930–3944

Kamaly N, Yameen B, Wu J, Farokhzad OC (2016) Degradable controlled-release polymers and polymeric nanoparticles: mechanisms of controlling drug release. Chem Rev 116(4):2602–2663

Kapalatiya H, Madav Y, Tambe VS, Wairkar S (2022) Enzyme-responsive smart nanocarriers for targeted chemotherapy: an overview. Drug Deliv Transl Res:1–13

Karimi M, Sahandi Zangabad P, Baghaee-Ravari S, Ghazadeh M, Mirshekari H, Hamblin MR (2017) Smart nanostructures for cargo delivery: uncaging and activating by light. J Am Chem Soc 139(13):4584–4610

Kaur A, New EJ (2019) Bioinspired small-molecule tools for the imaging of redox biology. Acc Chem Res 52(3):623–632

Kaushik N, Borkar SB, Nandanwar SK, Panda PK, Choi EH, Kaushik NK (2022) Nanocarrier cancer therapeutics with functional stimuli-responsive mechanisms. J Nanobiotechnol 20(1):1–23

Khan NH, Mir M, Qian L, Baloch M, Khan MFA, Ngowi EE, Wu D-D, Ji X-Y (2022) Skin cancer biology and barriers to treatment: recent applications of polymeric micro/nanostructures. J Adv Res 36:223–247

Kolosnjaj-Tabi J, Gibot L, Fourquaux I, Golzio M, Rols M-P (2019) Electric field-responsive nanoparticles and electric fields: physical, chemical, biological mechanisms and therapeutic prospects. Adv Drug Deliv Rev 138:56–67

Kooiman K, Roovers S, Langeveld SAG, Kleven RT, Dewitte H, O'Reilly MA, Escoffre J-M, Bouakaz A, Verweij MD, Hynynen K (2020) Ultrasound-responsive cavitation nuclei for therapy and drug delivery. Ultrasound Med Biol 46(6):1296–1325

Kubota T, Kurashina Y, Zhao J, Ando K, Onoe H (2021) Ultrasound-triggered on-demand drug delivery using hydrogel microbeads with release enhancer. Mater Des 203:109580

Kumari S, Badana AK, Malla R (2018) Reactive oxygen species: a key constituent in cancer survival. Biomark Insights 13:1177271918755391

Lamprou D (2023). Nano-and Microfabrication Techniques in Drug Delivery: Recent Developments and Future Prospects. Springer, pp 249–274.

Lencina MMS, Miconi EF, Leyes MDF, Domínguez C, Cuenca E, Ritacco HA (2018) Effect of surfactant concentration on the responsiveness of a thermoresponsive copolymer/surfactant mixture with potential application on "Smart" foams formulations. J Colloid Interface Sci 512:455–465

Li J, Esteban-Fernández de Ávila B, Gao W, Zhang L, Wang J (2017) Micro/nanorobots for biomedicine: delivery, surgery, sensing, and detoxification. Sci Robot 2(4):eaam6431

Li L, Yang W-W, Xu D-G (2019) Stimuli-responsive nanoscale drug delivery systems for cancer therapy. J Drug Target 27(4):423–433

Li F, Qin Y, Lee J, Liao H, Wang N, Davis TP, Qiao R, Ling D (2020) Stimuli-responsive nanoassemblies for remotely controlled drug delivery. J Control Release 322:566–592

Liu Z, Liu YD, Shi Q, Liang Y (2021) Electroactive dielectric polymer gels as new-generation soft actuators: a review. J Mater Sci 56(27):14943–14963

Lutz J-F, Hoth A (2006) Preparation of ideal PEG analogues with a tunable thermosensitivity by controlled radical copolymerization of 2-(2-methoxyethoxy) ethyl methacrylate and oligo (ethylene glycol) methacrylate. Macromolecules 39(2):893–896

Ma P, Lai X, Luo Z, Chen Y, Loh XJ, Ye E, Li Z, Wu C, Wu Y-L (2022) Recent advances in mechanical force-responsive drug delivery systems. Nanoscale Adv 4(17):3462–3478

Mai BT, Fernandes S, Balakrishnan PB, Pellegrino T (2018) Nanosystems based on magnetic nanoparticles and thermo-or pH-responsive polymers: An update and future perspectives. Acc Chem Res 51(5):999–1013

Malik R, Pancholi K, Melzer A (2016) Microbubble–liposome conjugate: payload evaluation of potential theranostic vehicle. Nano 3:1849543516670806

Medina-Sánchez M, Xu H, Schmidt OG (2018) Micro-and nano-motors: the new generation of drug carriers. Ther Deliv 9(4):303–316

Mirhadi E, Mashreghi M, Maleki MF, Alavizadeh SH, Arabi L, Badiee A, Jaafari MR (2020) Redox-sensitive nanoscale drug delivery systems for cancer treatment. Int J Pharm 589:119882

Mishra P, Nayak B, Dey RK (2016) PEGylation in anti-cancer therapy: an overview. Asian J Pharm Sci 11(3):337–348. https://doi.org/10.1016/J.AJPS.2015.08.011

Mohammed F, Ke W, Mukerabigwi JF, Japir AA-WMM, Ibrahim A, Wang Y, Zha Z, Lu N, Zhou M, Ge Z (2019) ROS-responsive polymeric nanocarriers with photoinduced exposure of cell-penetrating moieties for specific intracellular drug delivery. ACS Appl Mater Interfaces 11(35):31681–31692

Patil R, Portilla-Arias J, Ding H, Konda B, Rekechenetskiy A, Inoue S, Black KL, Holler E, Ljubimova JY (2012) Cellular delivery of doxorubicin via pH-controlled hydrazone linkage using multifunctional nanovehicle based on poly (β-L-malic acid). Int J Mol Sci 13(9):11681–11693.

Nesbitt H, Sheng Y, Kamila S, Logan K, Thomas K, Callan B, Taylor MA, Love M, O'Rourke D, Kelly P (2018) Gemcitabine loaded microbubbles for targeted chemo-sonodynamic therapy of pancreatic cancer. J Control Release 279:8–16

Ofridam F, Tarhini M, Lebaz N, Gagnière É, Mangin D, Elaissari A (2021) pH-sensitive polymers: classification and some fine potential applications. Polym Adv Technol 32(4):1455–1484

Oladipo AO, Lebelo SL, Msagati TAM (2023) Nanocarrier design–function relationship: the prodigious role of properties in regulating biocompatibility for drug delivery applications. Chem Biol Interact 110466

Osman G, Rodriguez J, Chan SY, Chisholm J, Duncan G, Kim N, Tatler AL, Shakesheff KM, Hanes J, Suk JS (2018) PEGylated enhanced cell penetrating peptide nanoparticles for lung gene therapy. J Control Release 285:35–45

Paramshetti S, Angolkar M, Al Fatease A, Alshahrani SM, Hani U, Garg A, Ravi G, Osmani RAM (2023) Revolutionizing drug delivery and therapeutics: the biomedical applications of conductive polymers and composites-based systems. Pharmaceutics 15(4):1204

Patel P, Mandal A, Gote V, Pal D, Mitra AK (2019) Thermosensitive hydrogel-based drug delivery system for sustained drug release. J Polym Res 26:1–11

Patil R, Portilla-Arias J, Ding H, Konda B, Rekechenetskiy A, Inoue S, Black KL, Holler E, Ljubimova JY (2012) Cellular delivery of doxorubicin via pH-controlled hydrazone linkage using multifunctional nano vehicle based on poly (β-L-malic acid). Int J Mol Sci 13(9):11681–11693

Pham SH, Choi Y, Choi J (2020a) Stimuli-responsive nanomaterials for application in antitumor therapy and drug delivery. Pharmaceutics 12(7):630

Pham SH, Choi Y, Choi J (2020b) Stimuli-responsive nanomaterials for application in antitumor therapy and drug delivery. Pharmaceutics 12(7):1–19). MDPI AG. https://doi.org/10.3390/pharmaceutics12070630

Poddar P, Maity P, Maiti S, Sahoo S, Dhara S, Dhara D (2020) Synthesis of a new triple-responsive biocompatible block copolymer: self-assembled nanoparticles as potent anti-cancer drug delivery vehicle. React Funct Polym 154:104679. https://doi.org/10.1016/J.REACTFUNCTPOLYM.2020.104679

Price PM, Mahmoud WE, Al-Ghamdi AA, Bronstein LM (2018) Magnetic drug delivery: where the field is going. Front Chem 6:619

Qi P, Wu X, Liu L, Yu H, Song S (2018) Hydrazone-containing triblock copolymeric micelles for pH-controlled drug delivery. Front Pharmacol 9:12

Qiu M, Wang D, Liang W, Liu L, Zhang Y, Chen X, Sang DK, Xing C, Li Z, Dong B (2018) Novel concept of the smart NIR-light–controlled drug release of black phosphorus nanostructure for cancer therapy. Proc Natl Acad Sci 115(3):501–506

Quader S, Van Guyse JFR (2022) Bioresponsive polymers for nanomedicine—expectations and reality! Polymers 14(17):3659

Raj Rai S, Bhattacharyya C, Sarkar A, Chakraborty S, Sircar E, Dutta S, Sengupta R (2021) Glutathione: role in oxidative/nitrosative stress, antioxidant defense, and treatments. ChemistrySelect 6(18):4566–4590

Raj GK, Singh E, Hani U, Ramesh K, Talath S, Garg A, Savadatti K, Bhatt T, Madhuchandra K, Osmani RAM (2023) Conductive polymers and composites-based systems: An incipient stride in drug delivery and therapeutics realm. J Control Release 355:709–729

Raza A, Rasheed T, Nabeel F, Hayat U, Bilal M, Iqbal HMN (2019) Endogenous and exogenous stimuli-responsive drug delivery systems for programmed site-specific release. Molecules 24:6). MDPI AG. https://doi.org/10.3390/molecules24061117

Ren DX, Chen PC, Zheng P, Xu ZN (2019) pH/redox dual response nanoparticles with poly-γ-glutamic acid for enhanced intracellular drug delivery. Colloids Surf A Physicochem Eng Asp 577:412–420. https://doi.org/10.1016/J.COLSURFA.2019.06.002

Reza Soltani E, Ahmad Panahi H, Moniri E, Torabi Fard N, Raeisi I, Beik J, Yousefi Siavoshani A (2023) Construction of a pH/temperature dual-responsive drug delivery platform based on exfoliated MoS2 nanosheets for effective delivery of doxorubicin: parametric optimization via central composite design. Mater Chem Phys 295:127159. https://doi.org/10.1016/J.MATCHEMPHYS.2022.127159

Rinoldi C, Lanzi M, Fiorelli R, Nakielski P, Zembrzycki K, Kowalewski T, Urbanek O, Grippo V, Jezierska-Woźniak K, Maksymowicz W (2021) Three-dimensional printable conductive semi-interpenetrating polymer network hydrogel for neural tissue applications. Biomacromolecules 22(7):3084–3098

Ruan C, Liu C, Hu H, Guo X-L, Jiang B-P, Liang H, Shen X-C (2019) NIR-II light-modulated thermosensitive hydrogel for light-triggered cisplatin release and repeatable chemo-photothermal therapy. Chem Sci 10(17):4699–4706

Ruiz AL, Ramirez A, McEnnis K (2022) Single and multiple stimuli-responsive polymer particles for controlled drug delivery. Pharmaceutics 14:2. https://doi.org/10.3390/PHARMACEUTICS14020421

Saadat M, Mostafaei F, Mahdinloo S, Abdi M, Zahednezhad F, Zakeri-Milani P, Valizadeh H (2021) Drug delivery of pH-sensitive nanoparticles into the liver cancer cells. J Drug Deliv Sci Technol 63:102557

Sarwan T, Kumar P, Choonara YE, Pillay V (2020) Hybrid thermo-responsive polymer systems and their biomedical applications. Front Mater 7:73

Sawalwade M, Naha A, Zende S (2020) A review on novel biochemical mediator triggered on demand drug delivery system. J Glob Pharma Technol 12:09

Schoeller J, Itel F, Wuertz-Kozak K, Fortunato G, Rossi RM (2022) pH-responsive electrospun nanofibers and their applications. Polym Rev 62(2):351–399

Sethuraman V, Janakiraman K, Krishnaswami V, Kandasamy R (2021) Recent progress in stimuli-responsive intelligent nano scale drug delivery systems: a special focus towards pH-sensitive systems. Curr Drug Targets 22(8):947–966

Shahriari M, Zahiri M, Abnous K, Taghdisi SM, Ramezani M, Alibolandi M (2019) Enzyme responsive drug delivery systems in cancer treatment. J Control Release 308:172–189

Shamsi M, Sedaghatkish A, Dejam M, Saghafian M, Mohammadi M, Sanati-Nezhad A (2018) Magnetically assisted intraperitoneal drug delivery for cancer chemotherapy. Drug Deliv 25(1):846–861

Sheshala R, Madheswaran T, Panneerselvam J, Vora L, Thakur RRS (2022) Stimuli-responsive nanomaterials in infectious diseases. In: Nanotheranostics for treatment and diagnosis of infectious diseases. Elsevier, pp 151–198

Smith J, Petrovic P, Rose M, De Souz C, Muller L, Nowak B, Martinez J (2021) Placeholder text: a study. J Citation Styles 3

Song J, Xu B, Yao H, Lu X, Tan Y, Wang B, Wang X, Yang Z (2021) Schiff-linked PEGylated doxorubicin prodrug forming pH-responsive nanoparticles with high drug loading and effective anticancer therapy. Front Oncol 11:656717

Song Y, Jing H, Vong LB, Wang J, Li N (2022) Recent advances in targeted stimuli-responsive nano-based drug delivery systems combating atherosclerosis. Chin Chem Lett 33(4):1705–1717

Tang H, Zhao W, Yu J, Li Y, Zhao C (2018) Recent development of pH-responsive polymers for cancer nanomedicine. Molecules 24(1):4

Tao Y, Chan HF, Shi B, Li M, Leong KW (2020) Light: a magical tool for controlled drug delivery. Adv Funct Mater 30(49):2005029

Tapasya K, Kumar AS, Dharmarajan A, Parvathi VD (2022) Nanocarriers: the promising future to cancer diagnostics and treatment. Biomed Pharmacol J 15(2):785–802

Uppalapati D, Boyd BJ, Garg S, Travas-Sejdic J, Svirskis D (2016) Conducting polymers with defined micro-or nanostructures for drug delivery. Biomaterials 111:149–162

Volpatti LR, Matranga MA, Cortinas AB, Delcassian D, Daniel KB, Langer R, Anderson DG (2019) Glucose-responsive nanoparticles for rapid and extended self-regulated insulin delivery. ACS Nano 14(1):488–497

Wang X, Yang Y, Zhuang Y, Gao P, Yang F, Shen H, Guo H, Wu D (2016) Fabrication of pH-responsive nanoparticles with an AIE feature for imaging intracellular drug delivery. Biomacromolecules 17(9):2920–2929

Wang C, Lin B, Zhu H, Bi F, Xiao S, Wang L, Gai G, Zhao L (2019) Recent advances in phenylboronic acid-based gels with potential for self-regulated drug delivery. Molecules 24(6):1089

Wang Y, Li Z, Hu Q (2021) Emerging self-regulated micro/nano drug delivery devices: a step forward towards intelligent diagnosis and therapy. Nano Today 38:101127

Wu W, Luo L, Wang Y, Wu Q, Dai H-B, Li J-S, Durkan C, Wang N, Wang G-X (2018a) Endogenous pH-responsive nanoparticles with programmable size changes for targeted tumor therapy and imaging applications. Theranostics 8(11):3038

Wu Y, Wang H, Gao F, Xu Z, Dai F, Liu W (2018b) An injectable supramolecular polymer nanocomposite hydrogel for prevention of breast cancer recurrence with theranostic and mammoplastic functions. Adv Funct Mater 28(21):1801000

Wu H, Abenojar EC, Perera R, An T, Exner AA (2019) Time-intensity-curve analysis and tumor extravasation of nanobubble ultrasound contrast agents. Ultrasound Med Biol 45(9):2502–2514

Xia H, Zhao Y, Tong R (2016) Ultrasound-mediated polymeric micelle drug delivery. Therapeutic Ultrasound:365–384

Xu W, Wang H, Dong L, Zhang P, Mu Y, Cui X, Zhou J, Huo M, Yin T (2019) Hyaluronic acid-decorated redox-sensitive chitosan micelles for tumor-specific intracellular delivery of gambogic acid. Int J Nanomedicine 14:4649–4666

Yadav S, Ramesh K, Reddy OS, Karthika V, Kumar P, Jo S-H, Yoo SII, Park S-H, Lim KT (2023) Redox-responsive comparison of diselenide and disulfide core-cross-linked micelles for drug delivery application. Pharmaceutics 15(4):1159

Yan F, Duan W, Li Y, Wu H, Zhou Y, Pan M, Liu H, Liu X, Zheng H (2016) NIR-laser-controlled drug release from DOX/IR-780-loaded temperature-sensitive-liposomes for chemo-photothermal synergistic tumor therapy. Theranostics 6(13):2337

Yang G, Liu Y, Chen J, Ding J, Chen X (2022) Self-adaptive nanomaterials for rational drug delivery in cancer therapy. Acc Mater Res 3(12):1232–1247

Yu C, Li L, Hu P, Yang Y, Wei W, Deng X, Wang L, Tay FR, Ma J (2021) Recent advances in stimulus-responsive nanocarriers for gene therapy. Adv Sci 8(14):2100540

Zhang G, Jiang X (2019) Temperature responsive nanoparticles based on PEGylated polyaspartamide derivatives for drug delivery. Polymers 11(2):316. https://doi.org/10.3390/polym11020316

Zhang C, Yan L, Wang X, Zhu S, Chen C, Gu Z, Zhao Y (2020) Progress, challenges, and future of nanomedicine. Nano Today 35:101008

Zhao W, Zhao Y, Wang Q, Liu T, Sun J, Zhang R (2019) Remote light-responsive nanocarriers for controlled drug delivery: advances and perspectives. Small 15(45):1903060

Zhao X, Bai J, Yang W (2021) Stimuli-responsive nanocarriers for therapeutic applications in cancer. Cancer Biol Med 18(2):319–335. https://doi.org/10.20892/j.issn.2095-3941.2020.0496

Zhuo S, Zhang F, Yu J, Zhang X, Yang G, Liu X (2020) pH-sensitive biomaterials for drug delivery. Molecules 25(23):5649

Zifar M, Ahmad Panahi H, Daghighi Asli M, Eftekhari-Sis B, Rafiee A (2023) Fabrication of pH and thermosensitive polymer conjugated tungsten disulfide for photo-thermal prostate cancer therapy under near-infrared laser irradiation: in vitro cytotoxicity study. Eur Polym J 192. https://doi.org/10.1016/j.eurpolymj.2023.112066

# Beyond PEGylation "PEGylation and its Alternatives"

# 16

Sehasree Mohanta, Anuja Muley, Mansi Upadhyay,
Dwiptesha Dahake, Muktika Tekade,
Aprameya Ganesh Prasad, and Rakesh Kumar Tekade

**Abstract**

In the realm of pharmaceuticals and biotechnology, the science of drug delivery and therapeutic efficacy has undergone a remarkable evolution over the past few decades. Central to this progress has been the advent of PEGylation, a groundbreaking technology that has revolutionized how drugs are administered and their pharmacokinetics optimized. However, as with any revolutionary concept, the limitations and drawbacks of PEGylation have become increasingly apparent. This has prompted a new wave of innovation and exploration in the field, seeking alternative strategies and technologies to go "Beyond PEGylation". In

S. Mohanta
School of Biological Sciences, Indian Association for the Cultivation of Science, Kolkata, West Bengal, India

A. Muley · M. Upadhyay · D. Dahake
Department of Pharmaceuticals, National Institute of Pharmaceutical Education and Research (NIPER), Ahmedabad, An Institute of National Importance, Government of India, Ministry of Chemicals and Fertilizers, Palaj, Opposite Air Force Station, Gandhinagar, Gujarat, India

M. Tekade
School of Pharmacy, Devi Ahilya Vishwavidyalaya, Takshila, Indore, Madhya Pradesh, India

School of Pharmaceutical Sciences, Lovely Professional University, Phagwara, Punjab, India

A. G. Prasad
Department of Chemical and Biomolecular Engineering, Johns Hopkins University, Baltimore, MD, USA

R. K. Tekade (✉)
National Institute of Pharmaceutical Education and Research (NIPER) Ahmedabad, An Institute of National Importance, Government of India, Department of Pharmaceuticals, Ministry of Chemicals and Fertilizers, Palaj, Opp. Air force station, Gandhinagar, Gujarat, India
e-mail: rakeshtekade@niperahm.res.in



R. K. Tekade, N. K. Jain (eds.), *PEGylated Nanocarriers in Medicine and Pharmacy*, https://doi.org/10.1007/978-981-97-7316-9_16

this discussion, we delve into the world of PEGylation and its alternatives, exploring the current state of the art, the challenges it presents, and the novel approaches that promise to reshape drug delivery and therapeutic strategies in the future.

**Keywords**

PEGylation · Pharmacokinetics · Drug delivery

## 16.1 Introduction

The pharmaceutical and biotechnology industries are in perpetual upheaval, defined by the unrelenting quest for creative techniques to improve treatment efficacy, reduce side effects, and improve patient outcomes. One such breakthrough has emerged as a game-changer in this pursuit throughout the years: PEGylation. This method, which includes the covalent attachment of polyethylene glycol (PEG) chains to various medicinal molecules, has proven to be an excellent tool for improving the therapeutic potential of a wide range of medications. Its effectiveness is based on its capacity to change pharmacological properties such as solubility, stability, and circulation half-life while lowering immunogenicity and side effects. However, as the pharmaceutical and biotech sectors embraced PEGylation, a rising knowledge of its limitations and downsides emerged. The technology that was previously marked as the pinnacle of drug delivery tactics now faces a fundamental challenge: moving "Beyond PEGylation". This shift in attitude has ushered into a new era of drug delivery and therapeutic improvement innovation, inquiry, and reevaluation.

Our investigation begins with thoroughly examining the principles and mechanisms underlying PEGylation. We investigate the science underpinning the covalent attachment of PEG chains to medicinal molecules and the numerous advantages it has brought to the sector. PEGylation is helpful in a wide range of medicinal applications, from prolonging circulation times to lowering immunogenicity and enhancing bioavailability. However, as we investigate these achievements, we will also shed light on the emerging concerns about PEG-associated toxicities and their potential to reduce the efficacy of some medicines. As we progress through the topic, we will examine the issues that have emerged as PEGylation has become the gold standard in drug delivery. These challenges include the potential for PEGylated drug resistance, restricted targeting capabilities, and the immune system's ability to recognize and neutralize PEGylated entities. By examining these difficulties, we will highlight the need for novel solutions to overcome these limitations and reinvent drug delivery systems (Veronese and Pasut 2005).

The investigation of these diverse techniques is at the center of our debate. We will examine a wide range of novel tactics and technologies with the goal of outperforming PEGylation. Each option offers distinct benefits, from lipid-based drug delivery techniques to nanotechnology and cell-penetrating peptides to antibody-drug conjugates (Howard et al. 2008). The goal is to understand how these innovative methods shape the future of drug delivery and therapeutic strategies, open doors to customized treatments, reduce side effects, and improve patient outcomes. PEGylation has undeniably played a crucial role in enhancing drug delivery in the dynamic world of pharmaceuticals and biotechnology. However, it is no longer enough to rely on the achievements of this pioneering technology. Precision medicine, personalized therapy, and improved therapeutic efficacy necessitate a new paradigm that takes us "Beyond PEGylation". As we embark on this path, we should expect to see a shift in how pharmaceuticals are developed, administered, and adjusted to meet the changing requirements of patients. This conversation is a vital step in the process, providing insight, analysis, and exploration of the potential beyond PEGylation in the modern era of drug delivery (Yadav and Dewangan 2021). In this in-depth review, we will dissect the concepts and applications of PEGylation and critically explore the novel alternatives that can potentially reinvent drug delivery in the modern day. The key goal is to comprehend how these alternatives are transforming the landscape of drug research by providing improved solutions and addressing the limitations of PEGylation.

## 16.2 PEGylation and Drug Delivery Systems

Drug delivery systems (DDSs) are rationally designed to administer medications in a controlled manner to achieve superior efficacy. Even though DDS has been extensively studied over the last couple of decades, they frequently suffer from rapid clearance from the blood and kidneys, undesired nonspecific interactions, enzymatic degradation, poor aqueous solubility, and a proclivity to create neutralizing antibodies (Allen and Cullis 2004; Tiwari et al. 2012). Over the years, various approaches have been tried to bypass these concerns by making the DDS stealthy and less susceptible to the aforementioned challenges. The surface of DDS has been successfully decorated with neutral and hydrophilic molecules since foreign materials and opsonins primarily interact via hydrophobic, ionic, van der Waals, and electrostatic forces (Leucuta 2012). PEGylation is the process of covalently or noncovalently attaching PEG (polyethylene glycol) molecules to different classes of drugs or macromolecular structures such as proteins, oligonucleotides, or vesicles. This modification results in improved pharmacokinetic and pharmacodynamic properties of the DDS (Allen and Cullis 2004; Nakamura et al. 2012; Jevševar et al. 2010) (Fig. 16.1).

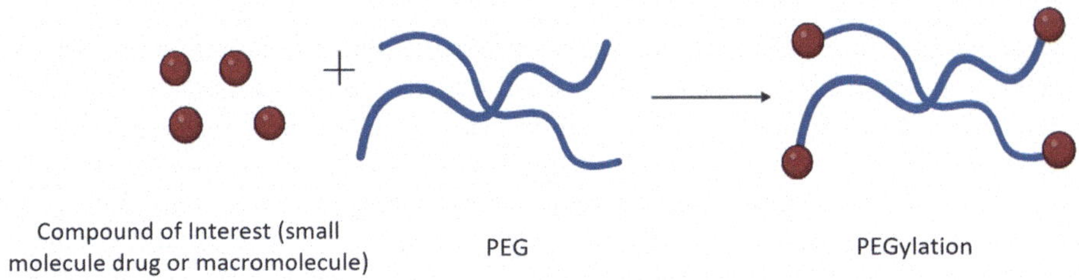


**Fig. 16.1** PEGylation of compound of interest

Frank Davis and colleagues pioneered the PEGylation technique in the 1970s to achieve enhanced efficacy of polypeptide-based medications (Hoffman 2016). Due to their neutral charge, flexibility, and hydrophilicity, PEG chains can form a steric barrier that reduces contact between the reticuloendothelial system and the DDS, reducing immunogenicity and masking them from phagocytes. This is referred to as PEG-chain-imparted stealthiness (Hamidi et al. 2006).

Furthermore, the strong hydration shield that envelopes the PEG molecules improves the serum solubility of PEGylated DDS and protects the drug and formulation from enzyme-mediated degradation and quick elimination from kidneys (Liechty et al. 2010).

The notion of PEGylation has since been generalized to several disciplines and various chemical, enzymatic, and physical entrapment methods for functionalizing the surface of DDS. PEGylation can play a critical role in biomolecules as it can affect protein function in vitro and lacks site specificity. However, these disadvantages are often outweighed in biological systems by other essential benefits (Francis et al. 1998; Dozier and Distefano 2015; Padín-González et al. 2022).

On the other hand, PEGylation has been shown in several studies over the last decade to produce significant reductions in drug delivery, including improved serum protein binding, lower uptake by target cells, and elicitation of an immune response that promotes clearance in vivo. Some of the literature indicates the adverse effects of PEGylation, calling into question the wisdom of using this technique in drug development.

## 16.3 PEGylated Products

PEG stands for polyethylene glycol, a synthetic polymer composed of repeating ethylene glycol units. PEGylated products refer to drugs, therapeutic agents, or other biologically active molecules that have been modified through a process called PEGylation. PEGylation involves the covalent attachment of PEG chains to the target molecule (lipid, protein, drug), typically through chemical reactions with functional groups on the molecule's surface. This chapter discusses PEGylated proteins, PEGylated liposomes, PEGylated G-CSF, and PEGylated nanoparticles.

### 16.3.1 PEGylated Liposomes

Liposomes are spherical nanoparticles composed of amphiphilic lipids with charged or neutral headgroups. They can entrap small molecule drugs or diagnostic agents, oligonucleotides, and other biomacromolecules such as proteins, vaccines, or enzymes (Nakamura et al. 2012). While liposomes improve the therapeutic index of drugs, they are susceptible to opsonization and subsequent destruction. Their biophysical properties and in vivo efficacy can be adjusted by varying variables such as size, surface charge, lipid composition, hydrophobic chain length, and degree of unsaturation (Allen and Cullis 2013; Dzieciuch et al. 2015). In comparison to

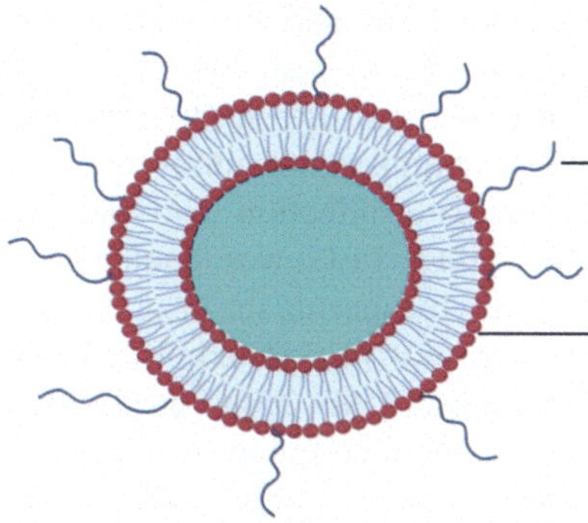


**Fig. 16.2** Surface modification of liposomes through PEGylation

unfunctionalized liposomes, PEGylated liposomes (Fig. 16.2) demonstrate prolonged blood circulation, enhanced drug bioavailability by bypassing the digestive tract, improved safety profile, and passive targeting owing to the enhanced permeability and retention effect (Dzieciuch et al. 2015; Moghimi and Szebeni 2003; Immordino et al. 2006). Several liposomal products like AmBisome® and Doxil®are available in market and currently many others are being investigated in clinical trials.

The field of PEGylated liposomes is still evolving. Ongoing research focuses on improving liposome characteristics, discovering novel ligands for targeted delivery, and expanding the technology's versatility and safety. PEGylated liposomes are projected to play an increasingly important role in tailoring medicines to specific patients, maximizing therapeutic outcomes, and minimizing side effects as precision medicine gets a grip. With ongoing research and innovation, PEGylated liposomes are poised to revolutionize medication delivery and tailored therapy in the coming years.

### 16.3.2 PEGylated G-CSF

PEGylated Granulocyte Colony-Stimulating Factor (PEG-G-CSF), a modified and prolonged-release version of human granulocyte colony-stimulating factor, has emerged as a transformational agent in hematology and oncology. This pharmacological breakthrough has been critical in the treatment of neutropenia, a disorder that is characterized by unusually low neutrophil count, a type of white blood cell essential for immunological function (De Volder et al. 2020). PEGylated G-CSF is a big step forward in the treatment of neutropenia, particularly in patients undergoing chemotherapy or stem cell transplantation.

G-CSF is a cytokine that mimics the proliferation and differentiation of neutrophil precursors. They are also known to improve the function of mature neutrophils (Marsh et al. 2007). Filgrastim, the first approved G-CSF, has been indicated for use in cancer patients receiving chemotherapy (Welte et al. 1996). G-CSF-type drugs have been extensively used in the clinic to minimize chemotherapy-induced neutropenia (CIN) and are recommended for patients undergoing high febrile neutropenia(FN) chemotherapy regimens (Aapro et al. 2011; Crawford et al. 2010). Human G-CSF, on the other hand, has a short half-life due to its primary elimination through the kidney, necessitating daily administration intravenously or subcutaneously, resulting in frequent injection site infection and decreased tolerance (Fernandes et al. 2017; Tan et al. 2011). PEG was conjugated with G-CSF (PEGylated G-CSF) to extend the half-life by altering the clearance site, resulting in lowered systemic clearance. A single dose of PEGylated G-CSF showed efficacy comparable to daily injections of normal G-CSF (Bond et al. 2018; Kuan et al. 2017). Fewer injections have improved patient compliance and reduced the burden on the healthcare system in the case of various tumors and non-Hodgkin's lymphoma (NHL) (Botteri et al. 2018). Hendler et al. (2011) discovered that both types of G-CSF reduced the incidence of FN, enhanced safety, and lowered the cost of chemotherapy in breast cancer patients. Many studies have compared data between PEGylated and non-PEGylated G-CSF in breast cancer patients, and the superiority of the former has not been established in terms of safety or effectiveness (Bond et al. 2018; Botteri et al. 2018; Kuan et al. 2017; Cornes et al. 2018; Schwartzberg et al. 2018).

As research continues to unveil its broader potential in other clinical contexts, PEGylated G-CSF stands as a shining example of how biopharmaceutical innovation can profoundly impact patient care and outcomes in the fields of oncology and hematology (Li et al. 2020).

### 16.3.3 PEGylated Proteins

Because of their great biological activity and selectivity, protein therapeutics have become an essential modern pharmaceutical product. Due to its nontoxic and highly hydrophilic properties, PEG can be combined with the protein's inactive regions to increase the protein drug's molecular weight. Bound PEG can enhance protein stability, extend its circulation half-life, and lower undesired immune responses (Sun et al. 2023). The use of PEGylation in protein therapeutics has, however, sparked widespread criticism around the globe (Fig. 16.3).

Since 1982, the US Food and Drug Administration (FDA) has licensed approximately 200 protein or peptide-based drugs for clinical use in the treatment of various human disorders (Cattani et al. 2015; Jevševar et al. 2010). Although these modalities have been known for some time, some limitations still need to be addressed. Proteins frequently suffer rapid degradation, aggregation, breakdown by various proteases, solubility concerns, short circulation times, and immunogenicity issues (Dozier and Distefano 2015). Davis and Abuchowski demonstrated for the first time in 1977 that PEGylated bovine serum albumin demonstrated an extended serum half-life and reduced immunogenicity compared to native proteins (Alconcel et al. 2011). Several studies have reported improved protein shell life and stability, enhanced solubility, decreased aggregation, and proteolysis (Zuma et al. 2022).

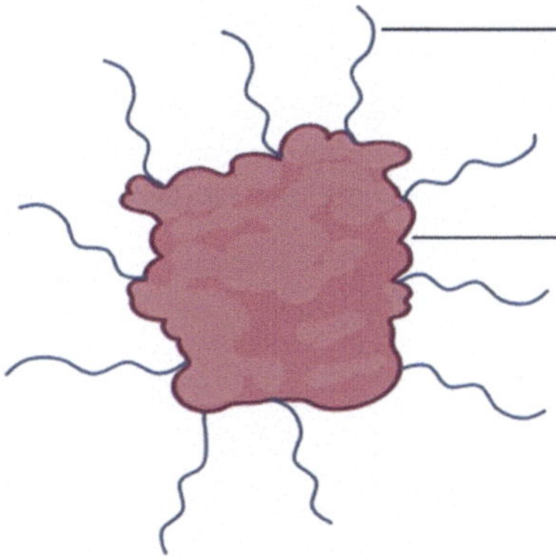


**Fig. 16.3** PEGylation of protein

As a result, PEGylation has been explored and has evolved into a diverse method for improving the pharmacokinetics of proteins and peptide-based therapeutics. The first PEGylated protein was introduced to the market in 1990, and nine distinct PEGylated protein therapeutics are now in the clinic for various indications. A few additional PEGylated biotherapeutics are also in late-stage clinical studies (Carter 2011; Alconcel et al. 2011; Milla et al. 2012). This technology has expanded the possibilities in drug development and enhanced the quality of life for patients needing protein-based therapies. As research advances and new applications emerge, PEGylated proteins will continue to play a crucial role in modern medicine, offering innovative solutions for various medical conditions.

### 16.3.4 PEGylated Nanoparticles

Nanotechnology has opened up a new frontier in drug delivery, diagnostics, and various other sectors. PEGylated nanoparticles have emerged as a spectacular innovation in this field, potentially revolutionizing several parts of medical research and technology. PEGylated nanoparticles have become a flexible tool with a wide range of uses, particularly in the field of drug administration, by combining the benefits of nanoparticles with the benefits of polyethylene glycol (PEG) coatings. Nanoparticles are microscopic structures of nanoscale dimensions (usually fewer than 100 nanometers). These nanoparticles can be programmed to deliver medicines, genetic material, or other bioactive molecules. PEGylation, on the other hand, is the process by which PEG chains are attached to the surface of these nanoparticles (Fig. 16.4). PEG is a hydrophilic polymer known for its biocompatibility and stealth-like qualities (Otsuka et al. 2003).

PEGylated nanoparticles have diverse applications in the fields of medicine and biotechnology. PEGylated nanoparticles are widely used to deliver chemotherapeutic agents, vaccines, and various drugs. Their targeted drug delivery capabilities reduce systemic side effects while improving therapeutic efficacy. Nanoparticles

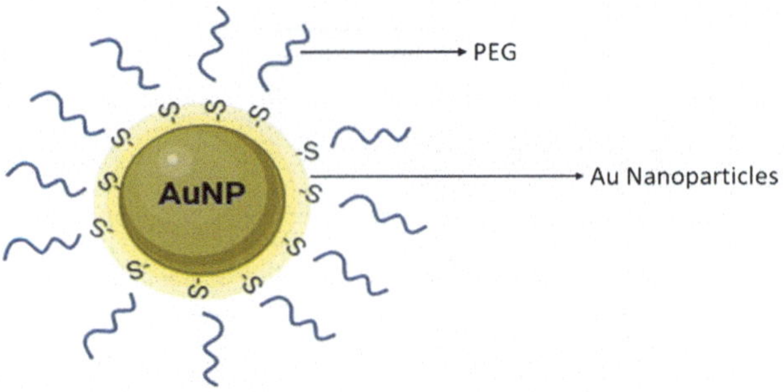


**Fig. 16.4** PEGylation of gold nanoparticle (AuNP)

can be loaded with contrast agents for diagnosis, such as computed tomography (CT) or magnetic resonance imaging (MRI). PEGylation enhances their circulation time and biocompatibility in vivo (Park et al. 2017). PEGylated nanoparticles can efficiently deliver genetic material, such as plasmid DNA or RNA, to target cells. This is a critical aspect of gene therapy and RNA-based treatments. PEGylated nanoparticles have shown promise as vaccine carriers, enhancing the immune response to antigens and improving the duration of protection (Vila et al. 2004). PEGylated lipid nanoparticles (LNPs) were used to formulate mRNA-based COVID-19 vaccines like Comirnaty™ (McCrudden et al. 2023).

## 16.4 Limitations of PEGylation

Over the past 20 years, PEGylated products have become an important element in medication, surfactants, and dispersion agents, for both clinical and industrial applications. While the notion of PEGylation is straightforward and effective in most circumstances, numerous possible safety issues have lately been suggested due to the long-term use of PEG-related products. Several studies have shown PEGylated products to cause immunological responses (with both oral and intravenous administration), cytoplasmic vacuolation, hypersensitivity, and antibody induction under certain conditions. Furthermore, PEG is nonbiodegradable, and larger molecular weight PEG tends to accumulate in tissues (Engler et al. 2015; Hatakeyama et al. 2013; Knop et al. 2010; Zhang et al. 2014; Han et al. 1997; Pisal et al. 2010).

While PEGs of lower molecular weight are preferred for use in several biomedical applications, they are known to generate toxic oxidative side products. PEG is also known to degrade under thermal and mechanical stresses (Han et al. 1997). This could have an impact on PEG medication formulation and storage. A few reports have claimed that protein PEGylation decreases biologic activity and binding affinity as concentrated solutions tend to be highly waxy. Product analysis becomes more complicated due to the inherent polydispersity of commercialized activated PEG and its derivatives (Hatakeyama et al. 2013; Zhang et al. 2014; Han et al. 1997). Although PEG conjugation with biologics is conceptually straightforward, it frequently necessitates extensive optimization; specifically, the paucity of site-specific conjugation techniques results in poorly defined structures, producing unsatisfactory outcomes. Synthetic contaminants, such as formaldehyde, 1,4-dioxane, and cyclic dimers of ethylene oxide, highlight the need for highly purified PEG for biomedical applications. It's important to note that the drawbacks of PEGylation vary depending on the specific drug or nanoparticle, the PEGylation method employed, and individual patient factors. Ongoing research aims to address these limitations and develop innovative strategies to optimize PEGylation and mitigate its potential downsides while harnessing its advantages in various biomedical applications.

## 16.5 Potential Alternatives of PEG

As PEG-mediated immune response limits its utility, many alternative polymers that mirror the physicochemical features of PEG without affecting its pharmacokinetic behavior have been developed. The primary requirements for developing PEG-alternative materials are outstanding stealth qualities, facile synthesis, degradability, biocompatibility, and a delicate balance of hydrophilicity and hydrophobicity. Recent advances in material chemistry and polymer science have played a critical role in developing PEG alternatives (Fig. 16.5) and exploring their potential biological applications (Hoang Thi et al. 2020).

### 16.5.1 Poly(Zwitterions)

Poly(zwitterions) are amphiphilic polymers that have found use as a potential alternative to nonionic PEG polymers in recent years (Wang et al. 2016). Poly(zwitterions), particularly poly(sulfobetaine), poly(phosphobetaine), and poly(carboxybetaine) (Fig. 16.6) or other materials coated with these polymers, demonstrated enhanced protein immobilization and non-fouling characteristics, as well as high resistance to biofilm formation and bacterial adhesion (Wu et al. 2016; Keefe and Jiang 2012).

Due to their unique chemical structure and characteristics, poly(zwitterions) have a PEG-like effect. As zwitterions are molecules with both positive and negative charges in the same structure, resulting in no net charge, this feature, together with zwitterions' hydrophilic nature, contributes to their PEG-like actions via numerous mechanisms: poly(zwitterions), like PEG, can provide a hydrophilic surface that repels protein adsorption. This avoids opsonization, a process in which proteins in the blood bind to the surface of nanoparticles or other drug delivery systems, potentially identifying them for elimination by the immune system. Poly(zwitterions) assist nanoparticles in evading rapid clearance from circulation by avoiding protein adsorption. Opsonization resistance and decreased absorption by macrophages and the reticuloendothelial system (RES) result in longer circulation times for drug delivery systems, including poly(zwitterions). This increased circulation time increases the possibility that the drug will reach its intended place within the body.

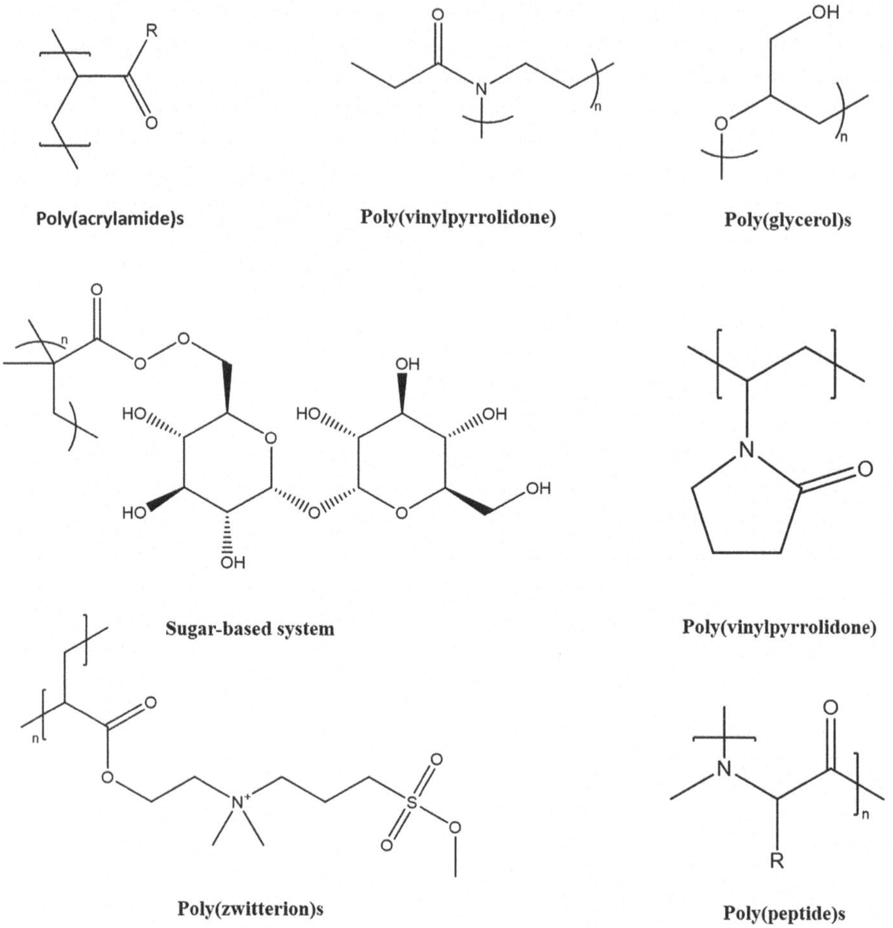


**Fig. 16.5** Alternatives of PEG

Because, poly(zwitterions) lack a net charge and are hydrophilic, they are less likely to elicit immunological responses (Leng et al. 2015; Xing et al. 2017) (Fig. 16.7).

Several chemistries can be used to conjugate poly(zwitterions) to various drug carriers or nanoparticles. The functional groups on the polymer and the surface of the drug carrier or nanoparticle determine the chemistry utilized. The following is a broad overview of how poly(zwitterions) can be conjugated; reactive groups must be added to both the poly(zwitterion) and the drug carrier or nanoparticle. Amino (-NH2), carboxyl (-COOH), and thiol (-SH) groups are common functional groups that have been used for amido bond formation, click chemistry, and thiol-ene reactions. If both the poly(zwitterion) and the carrier contain amine groups, they can be conjugated by first activating one of them with a coupling reagent like N-hydroxysuccinimide (NHS). This activated compound then can react with the other, which contains carboxyl groups (Keefe and Jiang 2012). In some

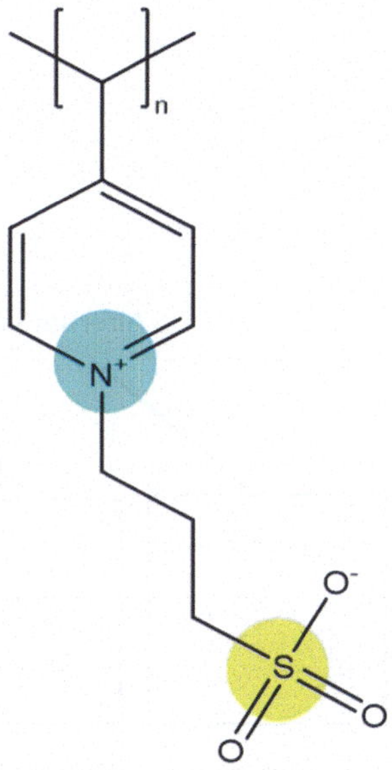


**Fig. 16.6** Chemical structure of poly(sulfobetaine), a zwitterionic polymer

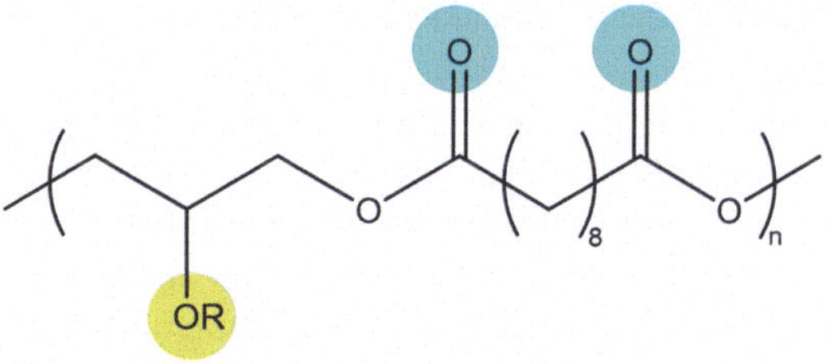


**Fig. 16.7** Chemical structure of poly(glycerol sebacate), a poly(glycerol)

circumstances, poly(zwitterions) may be employed for surface modification rather than direct conjugation. The polymer forms a protective layer around the drug carrier or nanoparticle; in this case, creating a stealthy, biocompatible surface. It is vital to note that the individual application and the attributes of the intended drug delivery system will determine the poly(zwitterion) and conjugation chemistry used. Poly(zwitterions) with different zwitterionic moieties and side chains may function differently in a given application. The functional group compatibility determines not only the specific chemistry utilized for conjugation, but also the stability of the resultant bond (Liu et al. 2016b) (Fig. 16.8).

To ensure the safe and efficient use of poly(zwitterions) in drug delivery systems or biomaterial applications, many precautions should be taken. This primarily includes biocompatibility testing to evaluate the substances' immunogenicity, cytotoxicity, and possible allergic reactions. Testing the safety and effectiveness of poly(zwitterion)-based systems can be done in vitro and in vivo, and the chemical stability should be analyzed in the environment where it will be used. Some polymers may deteriorate or lose their useful properties under particular pH, temperature, or enzymatic conditions. To ensure the polymer's performance for the necessary

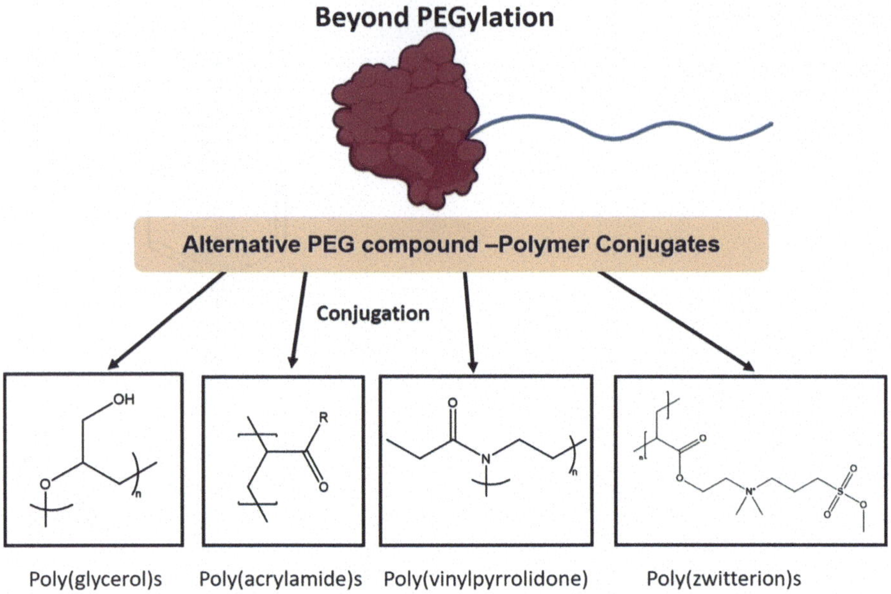


**Fig. 16.8** Conjugation of PEG alternatives

amount of time, it is crucial to understand its stability profile. Proper dosage and method of administration for the poly(zwitterion)-based system should be determined, and to avoid any potential negative responses or side effects, it is essential to comprehend the ideal concentration and frequency of administration. It is recommended to investigate possible interactions between the poly(zwitterion) and other elements of the drug delivery system or the biological setting in material interaction studies (Jin et al. 2014). Knowing how the polymer interacts with other substances might assist in avoiding unanticipated reactions or changes in the system's intended function. The effect of employing poly(zwitterions) in the intended application on the long-term outcomes should also be examined. To guarantee the polymer's safety and efficacy in prolonged use, it is essential to understand the potential accumulation, degradation, or removal of the polymer over a protracted period.

As for the safety profile, one example involves using poly(zwitterions) to develop stealth nanoparticles for targeted drug delivery. In this case, a specific poly(zwitterion) is utilized to coat the surface of nanoparticles, preventing opsonization and improving their circulation time in the bloodstream. This leads to reduced recognition by the immune system, enhancing their capacity to reach the intended target site. Poly(carboxybetaine methacrylate) is a poly(zwitterion) polymer that has shown promise in drug delivery applications. It is known for its PEG-like effect and biocompatibility. Here's a simplified safety profile example for pCBMA:

(a) The biocompatibility of pCBMA has been thoroughly investigated.
(b) In vitro studies have revealed negligible cytotoxicity and low immunogenicity.

(c) Reduced protein adsorption on surfaces coated with pCBMA has been shown in in vivo investigations using animal models, leading to extended circulation durations for drug carriers. This shows that the body tolerates pCBMA well (Taylor et al. 2021).

According to studies on its toxicity, when manufactured correctly and delivered in the proper doses, pCBMA does not demonstrate acute toxicity or adverse effects on essential organs. However, like with any substance, negative consequences could result from prolonged exposure or high doses. To avoid contamination, pCBMA should be synthesized and handled in sterile circumstances. Any impurities added during synthesis or handling can jeopardize the product's safety profile. For each unique drug delivery application, the pCBMA dosage and concentration should be calculated. Appropriate dosage is essential to get the desired PEG-like effects while lowering the risk of overexposure. It's important to do thorough safety assessments and research for each situation because the safety profile can vary depending on the specific poly(zwitterion) and its application (Marcelino 2016).

Since poly(zwitterions) exhibit much stronger intramolecular interactions via ionic binding than PEG, this could explain their high nonspecific protein resistance capability mediated by higher charge density (Cao and Jiang 2012). Since synthetic zwitterionic materials like poly(carboxybetaine) (pCB) and poly(sulfobetaine) (pSB) (Fig. 16.6) have strong hydration shells, they have been suggested as PEG substitutes. This property imparts them with low immunogenicity while also making them more resistant to nonspecific protein fouling (Yang et al. 2009; Jiang and Cao 2010). These polyelectrolytes' excellent non-fouling properties are due to their intense hydration, which is caused by electrostatic interactions between the ionic functional groups and water molecules. This property is essential in the construction of polymer protein conjugates, self-assembled nanoparticles, and vesicles. Poly(zwitterions) are very hydrophilic and poorly soluble in most nonpolar solvents, in contrast to PEG (Cao et al. 2011; Keefe and Jiang 2012; Jiang and Cao 2010).

Here's how poly(zwitterions) are used for PEGylation and their applications in drug delivery systems. Like PEGylation, polyzwitterion conjugation can improve the pharmacokinetics of drugs and therapeutic molecules. They can, therefore, extend the drug's circulation time in the bloodstream, increasing its bioavailability. Poly(zwitterions), due to their neutral charge and biocompatible properties, can reduce the immunogenic response of the drug. This is important for enhancing the safety profile of biopharmaceuticals. It can create a hydrophilic surface that resists protein adsorption, which helps prevent opsonization. Opsonization can mark drug delivery systems for removal by the immune system, so minimizing this process improves drug delivery efficiency (Laschewsky 2014). Poly(zwitterions) can be used as biocompatible coatings on medical implants and devices to reduce the risk of inflammation and infection. Their biocompatibility and resistance to protein adsorption make them ideal for improving the safety and performance of medical implants. Additionally, poly(zwitterions) can be used to functionalize nanoparticles or liposomes for drug delivery to tumor sites. The stealthy nature of the polymer can help evade the immune system and enhance tumor-specific drug delivery (Zheng et al. 2017), potentially improving the efficacy of cancer treatments. The use of polyzwitterions for PEGylation offers a versatile and biocompatible alternative to traditional PEGylation, and it continues to be an area of active research in the development of advanced drug delivery systems and biomaterials. Although poly(zwitterions) have substantial advantages, more research may be required to establish them as an alternative to PEG polymers.

### 16.5.2 Poly(Glycerols)

Poly(glycerols) (PGs) have recently emerged as a potential alternative to PEG in a variety of pharmaceutical applications. Frey and colleagues first synthesized PGs, and their applications were further explored and expanded by the research groups of Haag and Brooks (Imran Ul-Haq et al. 2012; Kainthan et al. 2006; Kurniasih et al. 2015). Poly(glycerols) of various molecular weights have been synthesized in a highly regulated manner using anionic ring-opening polymerization of glycidol (Kainthan et al. 2006; Kurniasih et al. 2015; Sunder et al. 1999). Poly(glycerols) are hyperbranched, compact structures with high hydrophilicity and biocompatibility. Their low intrinsic viscosity in water and a structure amenable to multifunctionalization make them attractive alternatives to PEG.

Poly(glycerols) are extremely hydrophilic polymers with a high attraction for water. This characteristic avoids protein and other biomolecule adsorption on the surface of drug carriers or biomaterials, resulting in a hydrophilic and "stealthy" layer. This, like PEG, decreases opsonization (the immune system's identification and clearance of foreign elements) and increases circulation times. Poly(glycerols) are biocompatible and are less prone to elicit an immunological response. They lack the immunogenicity that certain other materials may have, which is critical for improving the safety of drug delivery systems and biomaterials. Poly(glycerol) characteristics, such as molecular weight and architecture, can be tuned to tailor their performance to specific applications. This adaptability enables researchers to optimize drug delivery systems for various therapeutic objectives (Sunder et al. 1999).

A variety of chemical processes can be used to conjugate poly(glycerols) to drug carriers or nanoparticles. The specific chemistry used is determined by the functional groups present on the polymer and the drug carrier's surface. The following is a general overview of how poly(glycerols) can be conjugated: both poly(glycerol) and the drug carrier must be functionalized with reactive groups. Hydroxyl (-OH), amino (-NH2), and carboxyl (-COOH) groups are the typical functional groups required for PG conjugation (Gheybi et al. 2018). Esterification, amide bond formation, and click chemistry are examples of common chemistries involved. Suppose both the poly(glycerol) and the carrier have hydroxyl groups. In that case, they can be conjugated by activating one with a coupling reagent (e.g., N-hydroxysuccinimide or NHS) and reacting it with the other, which has compatible functional groups (Cao et al. 2016). In some circumstances, poly(glycerols) may be utilized for surface modification rather than direct conjugation. In this case, the polymer forms a protective layer over the drug carrier or nanoparticle, resulting in a biocompatible, stealthy surface.

The poly(glycerols) employed and the chemistry used for conjugation will be determined by the application and desired attributes of the drug delivery system. Due to their high purity and quality, poly(glycerols) are essential in drug delivery systems and biomaterial applications. They must undergo biocompatibility testing to assess their safety and interactions with biological systems. Chemical compatibility with other drug delivery system components is crucial, and contamination control is essential to assure sterility. The optimal dosage and concentration of poly(glycerols) should be determined, as excessive amounts may not improve the PEG-like effect and may result in unnecessary exposure (Abbina et al. 2017).

Like poly(glycerol), PEGylated liposomes are a commonly studied and utilized drug delivery system. These liposomes have a poly(glycerol) PEG layer on their surface, which provides improved biocompatibility and stability. Poly(glycerol) PEGylated liposomes have shown high biocompatibility, low cytotoxicity, minimal immune response, and extended circulation times in the bloodstream. They do not exhibit acute toxicity, but exposure to high concentrations may cause adverse effects (Liang et al. 2010). They should be prepared and handled under sterile conditions to prevent contamination. The optimal dosage and concentration for each drug delivery application should be determined, achieving desired PEG-like effects while minimizing overexposure risk.

Kizhakkedathu's research group has created biodegradable PGs by including acid-sensitive ketal moieties. They were able to establish high degradation profiles, which aided in the reduction of hepatic and renal biodistributions. They found no appreciable PG buildup during the 1–7 day injection period. Because of these outstanding features, PGs may be promising candidates for developing long-circulating multifunctional drug delivery systems. There are a few PG-based investigational therapeutics that are undergoing clinical testing at various stages (Shenoi et al. 2016, 2012; Yu et al. 2012).

### 16.5.3 Poly(Amino Acids)

Poly(amino acids) have been extensively researched as structural and biofunctional structural materials for applications in drug delivery, and they are typically synthesized via condensation polymerization of amino acid monomers (Romberg et al. 2007; Obst and Steinbüchel 2004; Sun et al. 2011). Poly(amino acids) are a potential alternative to PEG due to their charged side chains, biodegradability, biocompatibility, and minimal toxicity. Though they are hydrophilic in nature, they can be easily functionalized into the polymer backbone with groups like alcohol, amine, and thiol (Nishikawa and Ogawa 2004) (Fig. 16.9).

Poly(amino acids) exhibit a "PEG-like effect" in drug delivery and biomaterial applications, improving pharmacokinetics and biocompatibility. They are hydrophilic, preventing protein adsorption through a "stealthy" layer, and biocompatible,

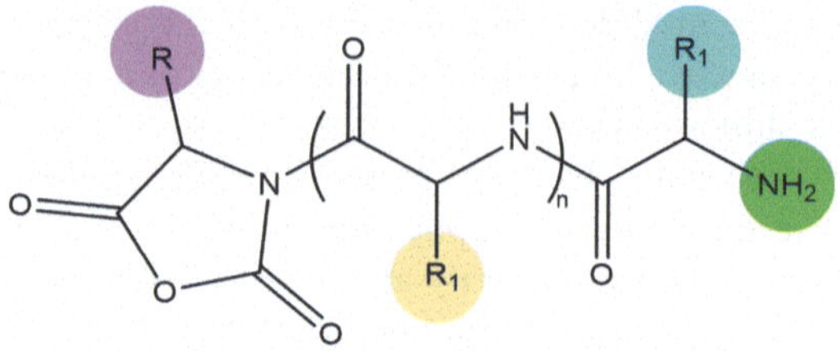


**Fig. 16.9** Chemical structure of poly(amino acid)s

reducing immune responses. They are also customizable, allowing researchers to optimize drug delivery systems for various therapeutic goals. This effect is attributed to their strong affinity for water, biocompatibility, and ability to be tailored to specific applications.

Poly(amino acids) can be conjugated to drug carriers or nanoparticles through various chemical reactions, depending on the functional groups on the polymer and the drug carrier's surface. Typical functional groups include amino (-NH2), carboxyl (-COOH), and thiol (-SH). The specific chemistry used for conjugation depends on the functional groups involved, such as amide bond formation, click chemistry, and thiol-disulfide exchange reactions. For example, suppose both the poly(amino acid) and carrier have carboxyl groups. In that case, they can be conjugated by activating one with a coupling reagent like N-hydroxysuccinimide (NHS) and reacting with the other with amino groups. Poly(amino acids) can also be used for surface modification, creating a protective layer around the drug carrier or nanoparticle. The choice of poly(amino acids) and the specific chemistry used for conjugation depends on the application and desired properties of the drug delivery system (Yang et al. 2006).

PEGylated nanoparticles are a widely used drug delivery system with a poly(glutamic acid) PEG layer on their surface, providing enhanced biocompatibility and stability. Studies have shown that these nanoparticles exhibit low cytotoxicity and minimal immune response, and in vivo studies in animal models show extended circulation times in the bloodstream. Toxicity studies show that these nanoparticles do not exhibit acute toxicity, but extensive exposure to high concentrations may lead to adverse effects. They should be prepared and handled under sterile conditions to prevent contamination. The optimal dosage and concentration should be determined for each specific drug delivery application, achieving desired PEG-like effects while minimizing overexposure risk. Safety profiles can vary depending on the particular poly(amino acid) used and its application, and detailed safety assessments and studies are essential (Mccormick-Thomson et al. 1989).

Poly(amino acids) are used in a variety of biopharmaceutical and cosmetic applications, as well as for environmental and agricultural purposes (Sun et al. 2011; Li and Wallace 2008; Shima and Sakai 1977). There are many natural poly(amino acids) such as cyanophycin, poly(alanine), poly(-glutamic acid), and poly(lysine), as well as synthetic poly(amino acids) like poly(hydroxyethyl-L-glutamine) and poly(hydroxyethyl-L-asparagine) have been investigated for various applications. Phase III clinical studies have also investigated poly(L-glutamic acid) (PGs). PGs find other applications as well, such as their use as a food thickening agent, in fertilizers, and in cosmetics as a wetting agent. Even though poly(amino acids) have made great advances in recent years, they still have flaws, particularly synthetic restrictions (Romberg et al. 2007; Yang et al. 2010; Shih et al. 2004; Melancon and Li 2011).

### 16.5.4 Poly(Oxazolines)

Poly(oxazolines) (POX) are another intriguing class of bioinspired polymers that have been investigated for various pharmaceutical applications. POX is biocompatible, can prevent rapid clearance, and control protein adsorption. They are neutrally charged and exhibit high stability and solubility in both polar and nonpolar solvents. POX also comes in a variety of architectural and chemical configurations. They are easily synthesized using the controlled cationic ring-opening polymerization of cyclic oxazolines (Guillerm et al. 2012). In many cases, the physicochemical features of POX have been reported to be equivalent to those of PEG, and they may be easily modified by inserting various functional groups on the monomer's oxazoline ring and altering the monomer concentration (Guillerm et al. 2012; Lava et al. 2015; Hoogenboom 2009) (Fig. 16.10).

Poly(oxazolines) are polymers that mimic the characteristics of poly(ethylene glycol) (PEG), exhibiting a PEG-like effect in drug delivery and biomaterials. They have hydrophilicity, creating a protective layer on drug delivery systems or biomaterials, minimizing protein adsorption, and immune system recognition. They also produce a stealth effect by reducing interactions with serum proteins and cells, ensuring the stability of the drug delivery system. Their biocompatibility reduces adverse reactions and improves safety.

Poly(oxazolines) can be conjugated to various molecules, surfaces, or particles through various chemical reactions, depending on the functional groups of the polymer and the substrate. These reactions can be initiated during synthesis or post-polymerization. Poly(oxazolines) can be conjugated to a substrate using amidation, thiol-ene click reactions, or carbodiimide-mediated coupling. If both the polymer and substrate have reactive functional groups, they can be linked to form a stable covalent bond. Poly(oxazolines) can also be used to modify nanoparticles, micelles, or other drug delivery systems, providing a stealthy and biocompatible surface

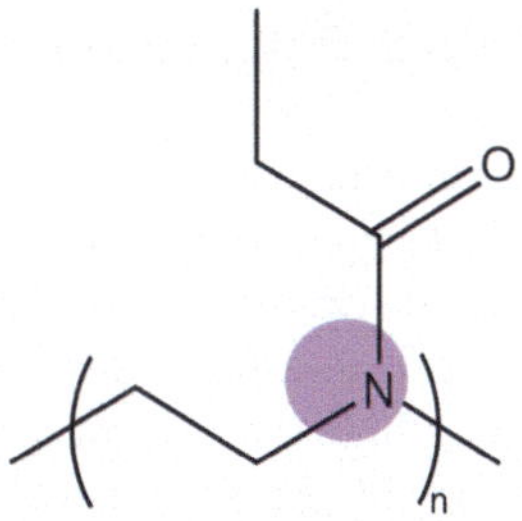


**Fig. 16.10** Chemical structure of poly(2-ethyl-2-oxazoline), PEOXA, a poly(oxazoline)

(Harris et al. 2019). The choice of poly(oxazolines) and chemistry for conjugation depends on the application and desired properties of the drug delivery system. Poly(oxazolines) are promising alternatives to PEG, offering similar benefits in drug delivery and biomaterial applications while providing opportunities for fine-tuning their properties to meet specific biomedical needs (Sedlacek and Hoogenboom 2020).

Due to their high purity and quality, poly(oxazolines) are used in drug delivery systems and biomaterial applications. They must undergo biocompatibility testing to assess their safety and interactions with biological systems. Chemical compatibility is also crucial, as incompatibilities can lead to degradation, reduced drug efficacy, or undesired by-products. Sterility and contamination control are essential to prevent contamination during the synthesis, handling, and storage of poly(oxazolines). The optimal dosage and concentration of poly(oxazolines) in drug delivery systems should be determined, as excessive amounts may not improve the PEG-like effect and may result in unnecessary exposure. These precautions ensure the safety and performance of the materials used in these applications (Kronek et al. 2013).

Poly(2-methyl-2-oxazoline) (PMeOx) is a poly(oxazoline) polymer used to modify micelles for drug delivery applications. These micelles have shown high biocompatibility, minimal cytotoxicity, and low immunogenicity in in vitro studies. In vivo studies in animal models indicate that they extend circulation in the blood, reducing protein adsorption and opsonization, enhancing safety. Toxicity studies show that PMeOx-modified micelles do not exhibit acute toxicity, but excessive exposure may lead to adverse effects. Sterility is crucial for handling PMeOx-modified micelles (Drago et al. 2021). The optimal dosage and concentration should be determined for each specific drug delivery application, achieving PEG-like effects while minimizing overexposure risk. Safety profiles can vary depending on the poly(oxazolines) used and the application, thereby necessitating detailed safety assessments and studies.

As an alternative to PEGylation, the concept of POXylation has been warmly embraced. Various POX polymers, particularly poly(2-ethyl-2-oxazoline) (PEtOx) and poly(2-methyl-2-oxazoline) (PMeOx), have been studied for self-assembly applications mediated by physical and chemical coupling techniques. They have also been explored as a stealth polymer for therapeutics and antimicrobial agent delivery. As they are thermo-sensitive, they have also been applied in responsive polymers and hydrogels (Viegas et al. 2011; Hoogenboom 2009; Seeliger et al. 1966).

### 16.5.5 Poly(Acrylamides) and Allied Other Systems

Several research groups have made substantial contributions towards developing vinyl-based polymers as a viable substitute for PEG over the last 50 years. Polyacrylamide, a derivative of acrylamide, has been utilized for various applications, primarily as a support matrix for electrophoresis (Milla et al. 2012; Nag and Awasthi 2013). Nonionic polyacrylamides can be linear, branched, or substituted. Some of the properties that make polyacrylamide a promising candidate are its biocompatibility, nontoxicity, cost-effectiveness, and stability over a wide pH range (pH 3–11), (Amoozgar and Yeo 2012; Kadajji and Betageri 2011; Nag and Awasthi 2013) (Fig. 16.11).

Poly(acrylamides) exhibit a PEG-like effect due to their hydrophilicity and steric hindrance, creating a protective layer on drug delivery systems or biomaterials. This prevents protein adsorption, thereby enhancing drug bioavailability. Poly(acrylamides) are biocompatible, reducing immune responses and adverse reactions. Their properties can be adjusted to optimize performance for specific applications, allowing researchers to fine-tune their performance.

Poly(acrylamides) can be conjugated to drug carriers or nanoparticles through various chemical reactions, depending on the functional groups present on the polymer and the drug carrier's surface. Functional groups like amino ($-NH_2$), carboxyl (-COOH), or thiol (-SH) can be introduced during synthesis or post-polymerization. Chemical conjugation can occur through amidation, thiol-ene click reactions, and carbodiimide-mediated coupling. If both polymers have reactive functional groups, they can be linked to form a stable covalent bond. Poly(acrylamides) can also be used to modify the surface of nanoparticles, micelles, or other drug delivery systems, providing a hydrophilic and biocompatible layer for enhanced stability and circulation times (Liu et al. 2016a).

Poly(acrylamides) are essential materials in drug delivery systems and biomaterial applications, but they must be of high purity and quality to ensure safety and performance. Biocompatibility testing is crucial to assess their safety and interactions with biological systems, including in vitro and in vivo studies. Chemical compatibility is essential to prevent degradation, reduced drug efficacy, or unwanted by-products. Sterility and contamination control are imperative to prevent contamination during the synthesis, handling, and storage of poly(acrylamides). Dosage optimization is crucial to determine the optimal dosage and concentration of poly(acrylamides) in drug delivery systems, as excessive amounts may not improve the PEG-like effect and may result in unnecessary exposure (Pathania et al. 2016).

Poly(acrylamide) hydrogels are used in tissue engineering due to their biocompatibility, low cytotoxicity, and compatibility with surrounding tissues. They do not exhibit acute toxicity, but cross-linking and polymerization are crucial. Sterile handling is essential to prevent contamination. The optimal dosage and concentration should be determined for each tissue engineering application to achieve desired therapeutic effects without unnecessary exposure. The safety profile can vary depending on the specific poly(acrylamides) used and the application (King and Noss 1989).

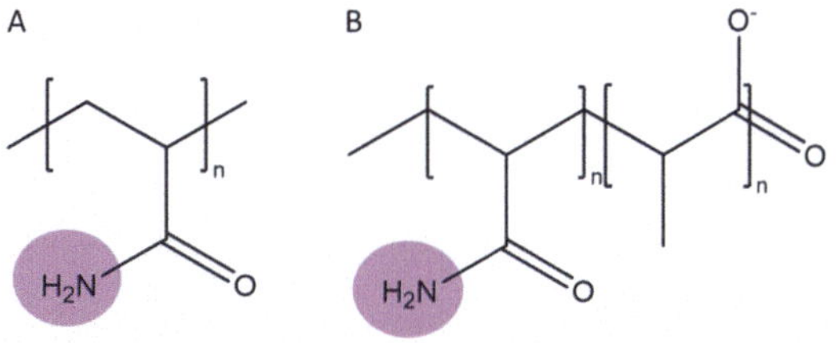


**Fig. 16.11** Chemical structures of poly(acrylamide)s: (**a**) Polyacrylamide (PAM) and (**b**) partially hydrolyzed polyacrylamide (PHPA)

Poly(acrylamides) are generally employed in hydrogels for applications in implantable devices, protein separation, and delivery of protein and oligonucleotides. While polyacrylamide coating tends not to elicit an immune response, with negligible cytotoxicity, there are reports (Smith and Oehme 1991; Caló and Khutoryanskiy 2015) of its implants inducing an inflammatory response. Another significant disadvantage of these materials is the monomer's highly hazardous side effects. These disadvantages limit polyacrylamides' extensive use in biological applications. Poly(N-(2-hydroxypropyl) methacrylamide) (PHPMA) is a fascinating acrylamide-based polymer, and its conjugates have been investigated in clinical studies, particularly with chemotherapeutics (Kopeček and Kopečková 2010; Ulbrich and Šubr 2010).

### 16.5.6 Poly(Vinylpyrrolidones)

Solubilizing chemicals tend to solubilize and enhance the bioavailability of several formulations. Poly(vinylpyrrolidone) (PVP), povidone, is a commonly used excipient and binder in the pharmaceutical industry. Over the previous four decades, significant advances have been made to commercialize PVP-based materials for a wide range of applications (Francis et al. 1998; Hsiao and Huang 2005; Liu et al. 2013) (Fig. 16.12).

Poly(vinylpyrrolidones) are a class of polymers that can create a "PEG-like effect" in drug delivery and biomaterial applications. This effect is similar to poly(ethylene glycol) (PEG) and involves the hydrophilicity and steric hindrance of poly(vinylpyrrolidones) to enhance the pharmacokinetics and biocompatibility of drug delivery systems. This hydrophilicity prevents protein adsorption, opsonization, and immune recognition, allowing the drug delivery system to circulate in the bloodstream for more extended periods, enhancing drug bioavailability. Poly(vinylpyrrolidones) are generally biocompatible, reducing the risk of adverse reactions. Additionally, the properties of poly(vinylpyrrolidones), such as molecular weight, architecture, and functional groups, can be adjusted to optimize

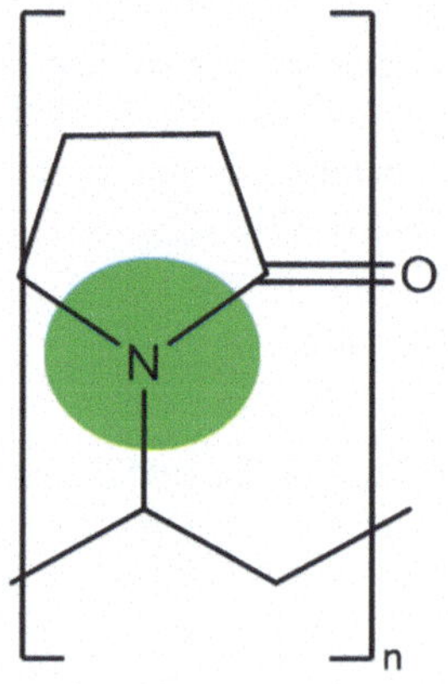


**Fig. 16.12** Chemical structure of poly(vinylpyrrolidone)

performance for specific applications, allowing researchers to fine-tune these properties to meet the requirements of the drug delivery system (Koczkur et al. 2015).

Poly(vinylpyrrolidones) can be conjugated to drug carriers or nanoparticles through various chemical reactions, depending on the functional groups on the polymer and the drug carrier's surface. Functional groups like amino (-NH2), carboxyl (-COOH), or thiol (-SH) can be introduced during synthesis or post-polymerization. Chemical conjugation can occur through amidation, thiol-ene click reactions, and carbodiimide-mediated coupling. If both polymers have reactive functional groups, they can be linked to form a stable covalent bond. Poly(vinylpyrrolidones) can also be used to modify the surface of nanoparticles, micelles, or other drug delivery systems, providing a hydrophilic and biocompatible layer that enhances stability and circulation times. The choice of poly(acrylamides) and chemistry depends on the application and desired drug delivery system properties (Zelikin et al. 2007).

Poly(vinylpyrrolidone) (PVP) is a drug delivery system that requires strict quality control and biocompatibility testing. Its chemical compatibility with other drug delivery components is crucial to prevent degradation or reduced efficacy. Sterility and contamination control are essential to prevent contamination. The optimal dosage and concentration of PVP should be determined to ensure the best PEG-like effect and avoid unnecessary exposure (Yu et al. 2009).

PVP is a widely used material for drug delivery. Studies have shown that PVP-modified liposomes have low cytotoxicity and extended circulation times in the bloodstream, enhancing drug delivery safety and efficacy. Toxicity studies reveal that PVP-modified liposomes do not exhibit acute toxicity, but thorough preparation and purification are crucial. Sterility is also essential to prevent contamination. The optimal dosage and concentration of PVP-modified liposomes should be determined for each specific drug delivery application, achieving PEG-like effects while minimizing overexposure risks (Schwarz 2018). The safety profile varies depending on the PVP used and the application, and detailed safety assessments and studies are essential.

PVP's strong hydration shell aids in inhibiting immune system interaction and enabling enhanced blood circulation and minimal protein adsorption. PVP degrades slower than PEG when exposed to ultraviolet light. PVP-based compounds are used in several industries with applications in adhesives, dyes, coatings, photoresists, photography, textiles, fibers and textiles, disinfectants, and food additives (Liu et al. 2013). PVP can be synthesized in various molecular weights by free and controlled radical polymerization methods. As with polyacrylamides, the free monomer in PVP should be entirely eliminated because it is highly carcinogenic. Although PVP's biocompatibility is promising, it suffers from issues such as uncertain immunological behavior and organ accumulations above excretion limitations (Le Garrec et al. 2002; Torchilin 1998; Zelikin et al. 2007).

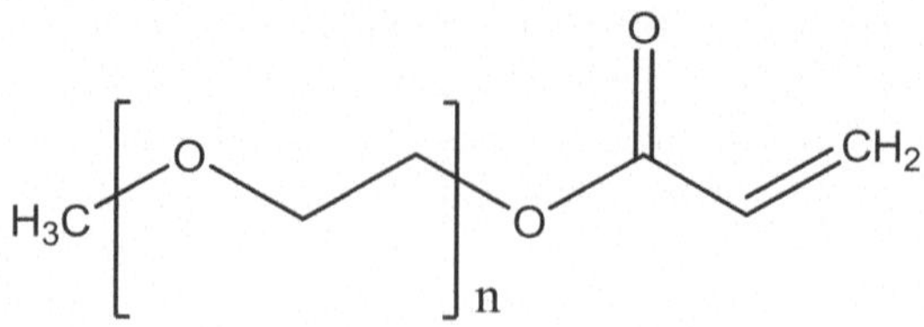


**Fig. 16.13** Chemical structure of Poly [oligo(ethylene glycol) methyl ether methacrylate] (POEGMA)

### 16.5.7 Poly[Oligo(Ethylene Glycol) Methyl Ether Methacrylate] (POEGMA)

POEGMA is a water-soluble polymer composed of numerous ethylene glycol (EG) chains grafted over a hydrophobic methyl methacrylate framework (Lutz 2008). The hydrophilicity and biocompatibility of methacrylate polymers are linked to oligoethylene glycol (EG) chain grafting, which makes them appropriate for application in biomedicine. Lutz's and coworkers further demonstrated thermo-responsive behavior and antifouling properties of POEGMA (Lutz et al. 2007). POEGMA was developed as an alternative to linear PEG by modifying PEG into a bottlebrush design to address PEG-associated ABC (Joh et al. 2019). POEGMA, which was synthesized, has a 3D hyperbranched structure with numerous side chains of EG moieties and effective stealth characteristics (Pires-Oliveira et al. 2020; Hucknall et al. 2009) (Fig. 16.13).

POEGMA is often conjugated with various therapeutic agents, such as peptides (Qi et al. 2016), proteins (Liu et al. 2014), nanoparticles (Nastyshyn et al. 2020), nanocrystals (Roberts et al. 2020), and micelles (Skandalis and Pispas 2017; Zhao et al. 2016). This can prevent immunogenicity and improve the circulation half-life of these therapeutics, resulting in better efficacy. Moreover, this hydrophilic polymer also has wider applications in the field of gene delivery (Üzgün et al. 2010), tumor targeting (Sano et al. 2022), and tissue engineering (Harrison et al. 2015).

### 16.5.8 Polypeptides

Polypeptides have various advantages over PEG (polyethylene glycol) and may be a better choice in some cases. Here are some of the reasons why polypeptides are employed as PEG substitutes:

(a) Polypeptides, made up of amino acids, are biodegradable and can be broken down by the body's enzymatic processes, making them suitable for applications where long-term retention is not desired.
(b) They are also minimally immunogenic, as they are based on naturally occurring sequences in the human body, making them suitable for immunogenicity concerns. Polypeptides can be designed with specific targeting sequences, allowing for more precise drug delivery (González-Aramundiz et al. 2012).

(c) Additionally, polypeptides can be customized with specific sequences and lengths, allowing for precise control over their properties and interactions with other molecules, making them valuable for tailoring drug delivery systems to specific therapeutic needs.
(d) Polypeptides offer diverse functionalities, including amino acids with side chains, which can be used for drug conjugation and targeting ligands.
(e) They are naturally compatible with biological systems, allowing them to interact more effectively with cells, tissues, and biological molecules than synthetic polymers.
(f) Some polypeptides, like β-sheet-forming peptides, can self-assemble into stable secondary structures, enhancing drug delivery and controlled release. Polypeptides are particularly useful in applications like tissue engineering, wound healing, and regenerative medicine, as they mimic extracellular matrix components and are biocompatible with biological structures (Hosseinkhani et al. 2013).

Excluding hydrophobic and highly polar amino acids from a polypeptide sequence increases drug delivery or the system's solubility. Taking advantage of this benefit, the circulation half-life of the drug can be enhanced by using homopolymers of genetically encoded amino acids (Bolze et al. 2016; Falvo et al. 2016). PAS, a polypeptide composed of proline, alanine, and serine, was biodegradable and hydrophilic, with a large hydrodynamic volume and surface charge. The highly soluble formulation not only showed antihemolytic activity but was also distributed and cleared similarly to PEG. PASylation is defined as "the genetic infusion of a conformationally unfolded homopolymer of PAS with therapeutic proteins or drugs" (Harari et al. 2014; Kuhn et al. 2016). In mouse models, the stability of PASylated products was demonstrated in serum, with no organ distribution and minimal immunogenicity and toxicity. Though this class of products is yet to be investigated in the clinic, they have been widely employed to extend the systemic circulation of leptins, antibody fragments, chemotherapeutics, interferons, and growth hormones and for various therapeutic as well as diagnostic purposes (Mendler et al. 2015; Morath et al. 2015; Schlapschy et al. 2013). Another study reported an intriguing genetic fusion of XTEN, a polypeptide sequence, with a protein that resulted in a 60-fold increase in plasma half-life. XTEN is a hydrophilic, stable homopolymer with an indefinite structure. It is non-immunogenic and consists of certain amino acid sequences (alanine, glutamic acid, glycine, proline, serine, threonine). XTEN fusions have been further extended to glucagon, factor VII, green fluorescent protein, and human growth hormone to improve the efficacy of the respective systems in diverse in vivo models (Podust et al. 2016; Schellenberger et al. 2009; Strohl 2015).

Polypeptides offer advantages like enzymatic degradation and structural stability, but may also face challenges in large-scale production. The choice between PEG and polypeptides depends on the drug delivery system's requirements and desired properties, and researchers carefully consider these factors when selecting the most suitable polymer (Fig. 16.14).

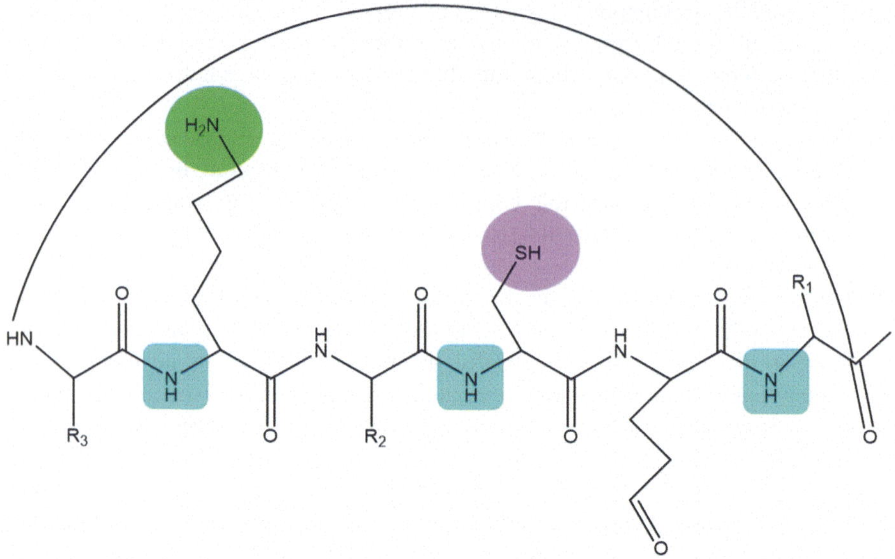


**Fig. 16.14** Chemical structure of cyclized polypeptide as an alternative to PEG

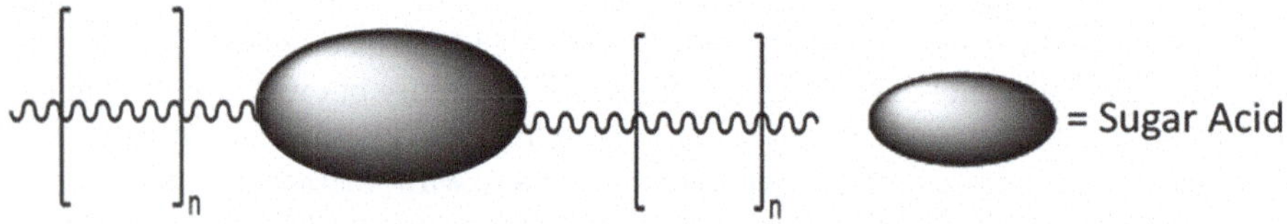


**Fig. 16.15** Structure of carbohydrate-based polymers

### 16.5.9 Carbohydrate-Based Systems

Carbohydrates, nucleic acids, and proteins are important natural macromolecules for life. Synthetic carbohydrates can be good candidates for applications in polymer and pharmaceutical research. They can have a linear, comb-type, branched, or dendrimer-like structure (Tolstyka et al. 2016) (Fig. 16.15).

Carbohydrate-based systems exhibit a PEG-like effect due to their hydrophilicity and steric hindrance, which form a protective layer on drug delivery systems or biomaterials. This layer minimizes protein adsorption, extending the circulation time and enhancing bioavailability. Carbohydrate-based systems can also mimic glycoproteins and glycolipids found on cell surfaces, effectively "camouflaging" themselves from the immune system. This molecular mimicry can reduce immune responses and prevent immune-mediated clearance of the drug delivery system. Carbohydrates are biocompatible, making them suitable for drug delivery reducing the likelihood of adverse reactions or immunogenicity. The chemistry in their conjugation is crucial in understanding these mechanisms (Moros et al. 2012).

Carbohydrate-based systems can be conjugated to drug carriers, nanoparticles, or other substrates through various chemical reactions. The specific chemistry used depends on the functional groups on the carbohydrate and the substrate's surface. Carbohydrates can be functionalized with specific reactive groups, such as amino (-NH2), carboxyl (-COOH), or thiol (-SH) groups, during synthesis or chemical modifications. Chemical conjugation can occur through amidation, thiol-disulfide exchange, or carbodiimide-mediated coupling. If both carbohydrate and substrate have compatible functional groups, they can be covalently linked to form a stable bond. Some carbohydrate-based systems can be conjugated using enzymatic or glycosylation reactions, mimicking natural glycosylation processes in living organisms. The carbohydrate-based system and chemistry choice depend on the application and desired drug delivery properties. Carbohydrate-based systems offer an attractive alternative to PEG, providing similar benefits in drug delivery and biomaterials while aligning with the biocompatibility of natural carbohydrates (Su et al. 2021).

Aside from their biological activities, they are suitable candidates for PEG substitutes due to intrinsic qualities that render them chemically well-defined, compact, biocompatible, highly hydrated, synthetically flexible, and less susceptible to aggregation and protein adsorption. Both polysaccharides (dextran, starch, hydroxyethyl starch (HES), and hyaluronic acid) and oligosaccharides (from one to twelve sugar units) have been widely investigated. Dextran-coated nanocarriers, like PEG and HES, have demonstrated low protein adsorption and prolonged circulation durations (Besheer et al. 2007). Carbohydrates, combined with their active biological role of interacting with cell surfaces or specific proteins, have emerged as desirable candidates for biotherapeutic applications (Sizovs et al. 2013; Luyckx and Baudouin 2011). Trehalose is an exciting example of a naturally occurring disaccharide. It is a nonreducing sugar with glucose units joined by α,α'-1,1-glucosidic bond, another compelling example of a carbohydrate. It has shown fascinating biochemical properties, rendering them appealing for research in academia and industry (Gu et al. 2014).

### 16.5.10 Hydrophilic Polymers

Hydrophilic polymers have numerous applications, including drug delivery (Knop et al. 2010; Hildebrand et al. 2017), self-assembly (Willersinn et al. 2017), and catalysis (Ge et al. 2007). As a result, research into hydrophilic polymers is an essential aspect of polymer science. Significant areas of investigation have been structure-property relationships and their interactions with biological entities (Rudolph et al. 2013; Takahashi et al. 2018; Cuomo et al. 2012). Synthetic polymer chemistry offers new possibilities for advanced characteristics and prospects using innovative hydrophilic polymers and structures (Liu et al. 2018; Mccormick and Lowe 2004). Hydrophilic polymers have essential features not just when dissolved but also when crosslinked, as in the case of hydrogels. This class of materials has various intriguing features as well as customizable size and morphology, which are of great importance for pharmaceutical applications (Thiele et al. 2014). Hydrophilic polymers include poly(acrylamide) and poly(ethylene glycol) (PEG). It has to be noted that some polymers, such as poly(2-hydroxyethyl methacrylate), are hydrophilic and water-swellable albeit with poor aqueous solubility. Even for hydrophilic polymers, water solubility varies according to polymer concentration, type, and molecular weight. As a result, they may not be distinguished based just on solubility (Schmidt 2019) (Fig. 16.16).

Hydrogels are hydrophilic polymer materials known for their biocompatibility, swelling, and soft texture. They find applications in areas such as tissue engineering and implants. Reports have described polymer self-assembly combined with hydrogel formation has been described, with the double hydrophilic block copolymer poly(N-vinylpyrrolidone)-b-poly (oligo ethylene glycol methacrylate) coupled with α-cyclodextrin (α-CD). This results in thermo-responsive hydrogels that exhibit distinct mechanical properties upon cooling to room temperature (Jiang et al. 2016). Cyclic cyclodextrin (CD) can not only be applied for gelation but can also be utilized in stimulus-responsive hydrogels for drug entrapment. Bian et al. synthesized an amphiphilic polymer by combining hydrophilic acrylamide, hydrophobic N-dodecyl acrylamide, and sodium acrylate, which delayed the formation of phenolic network crosslinking points (Xu et al. 2019). Cuomo and coworkers described a

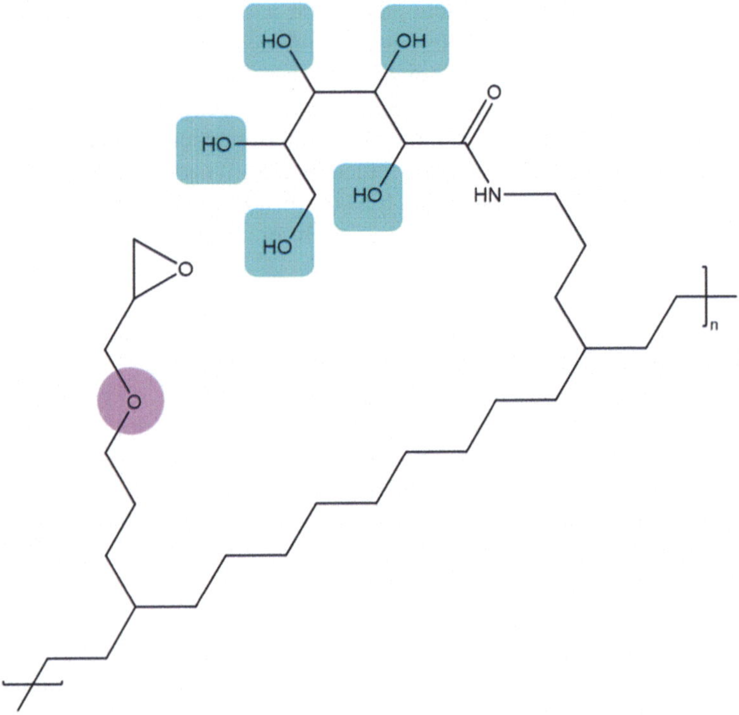


**Fig. 16.16** Chemical structure of epoxy poly(AAG-AA), a hydrophilic polymer

hydrogel based on an emulsion of alginate and lemon grass essential oil, which was studied for mechanical properties.

Thiele and coworkers reported a polysaccharide hydrogel targeting microgels via microfluidics (Hauck et al. 2018). Combining chitosan or hyaluronic acid with tert-butyl isocyanide and a PEG crosslinker allowed for the fabrication of mono-disperse microgels with sizes in the 70–100 μm range. Functional microgels were formed by adding functional carboxylic acids in the Ugi crosslinking reaction, potentially valuable for drug delivery. Lensen and coworkers described the microstructure of PEG-based hydrogels, revealing the strong effect of the polymerization process on the size of crystalline domains. With this understanding, crystallization behavior in PEG-based hydrogels can be fine-tuned to improve material properties (Zhang et al. 2018).

Reinforcement via nanoparticles has also been explored to investigate the formation and mechanical properties of Laponite/cellulose-based hydrogels (Xie et al. 2018). When Laponite was combined with silanized hydroxypropyl methylcellulose, a significant reinforcement effect was observed, increasing the storage modulus by an order of magnitude. Fluorescent microsphere tracking analysis was used to investigate the hydrogel architecture, revealing two domains: loose aggregates in the periphery and dense Laponite structures in the core.

Hydrophilic polymers have been applied in the biomedical field with advances in drugs and vaccines in the form of polymeric nanomaterials. These nanoparticles exhibit targeted delivery and sustained release (Secker et al. 2015). Poly(peptoids) have gained attention, with Barz et al. developing siRNA polyplexes with poly(sarcosine) as a shielding agent. They reported the formation of lipo-oligomer structure and siRNA polyplexes, with the oligomer consisting of cholanic acids coupled through a bioreducible disulfide linker. The ability to shield was tested using in vivo imaging and biophysical assays, revealing prolonged systemic circulation time on par with the gold standard PEG (Klein et al. 2018) (Table 16.1).

### 16.5.11 Non-PEGylated Nanoparticles

Non-PEGylated microparticles and nanoparticles (Fig. 16.17), mainly inorganic nanoparticles such as iron, gold, silver, hafnium, ceramic silica, and carbon nanotubes, have sparked considerable interest in tissue engineering, drug delivery, and diagnostic formulations (Anselmo and Mitragotri 2015, 2016). Metal nanoparticles with about 1–100 nm diameters are appealing owing to their distinct physicochemical characteristics, such as enhanced surface-area-to-volume ratio, small size, and conductivity. They are also amenable to surface modification and can be designed to respond to external stimuli with pharmacokinetics suitable for systemic delivery (Huang et al. 2011). Because of their increased permeability and retention, nanoparticles preferentially concentrate at malignant tumor sites.

Non-PEGylated nanoparticles can be modified to achieve a PEG-like effect through chemical reactions and surface functionalization. These methods include hydrophilic coating, ligand attachment, surface modification, and stealth coatings.

**Table 16.1** Comparison between different alternative polymers for PEG

| Name of Polymer | Advantages | Disadvantages | Reference |
|---|---|---|---|
| Polyzwitterions | Chemical flexibility, zwitterionic nano-vectors, biocompatibility, non-fouling, prolonged blood circulation, and multifunctionality | Nondegradable, high synthesis and modification cost | Qian et al. (2022), Zheng et al. (2017) |
| Polyglycerols (PGs) | Biocompatibility, easy synthesis, immune invasion, and prolonged circulation due to hyperbranched polyglycerol (HPG)s | Nonbiodegradability, accumulation in tissues | Hoang Thi et al. (2020) |
| Polyvinyl pyrolidones (PVP) | Biocompatible and stabilizer | Enhanced protein adsorption, slower biodegradation | Hadjesfandiari and Parambath (2018), Knop et al. (2010) |
| Polyoxazolines (Pox) | Low PDI values, flexible end-group chemistry, biodegradability, and biocompatibility | Higher synthesis cost and no FDA approval | Hoang Thi et al. (2020) |
| Polyacrylamides | Easy functionalization due to hydroxyl groups, Biocompatible and fouling | Nonbiodegradability, degradation by thermal and photolytic effect, resultant monomer gives neurotoxic effect | Hadjesfandiari and Parambath (2018), Knop et al. (2010) |
| Poly[oligo(ethylene glycol) methyl ether methacrylate] (POEGMA) | Biocompatible, Easy chemical modification, Thermoresponsive, Biodegradable | | Lutz (2008) |
| Polyamino acids | Biodegradable, Decreased ABC phenomenon, Biocompatible | Microbial contamination, excessive hydration, and reduced viscosity on storage | Boddu et al. (2021) |
| Carbohydrate-based | Biodegradable, low immunogenicity, nontoxic, easy chemical modification, biocompatible | Variable chemistry, broad or mixed molecular weights | Barclay et al. (2019), Hoang Thi et al. (2020) |

Hydrophilic polymers like dextran, chitosan, or albumin can be conjugated to the nanoparticle's surface through covalent bonding (Salatin and Yari Khosroushahi 2017). Ligands, peptides, or biomimetic materials can enhance biocompatibility and enable targeted drug delivery. Surface modifications, such as adsorption or encapsulation of hydrophilic materials, create a hydrophilic shell around the nanoparticles. Stealth coatings can be formed using materials like zwitterions or synthetic polymers.

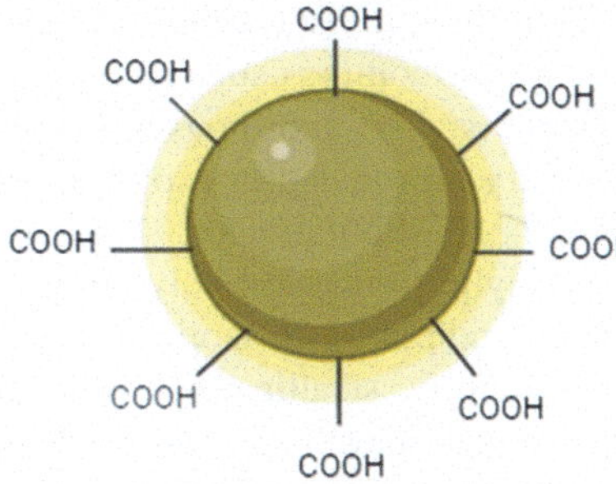


**Fig. 16.17** Structure of a non-PEGylated nanoparticle

Non-PEGylated nanoparticles should be synthesized and characterized to meet high standards of quality, purity, and consistency. Biocompatibility testing should be conducted to evaluate safety and interactions with biological systems. Sterility should be maintained during preparation and handling, especially in medical and biological applications. Chemical compatibility should be verified to avoid degradation, reduced efficacy, or undesired by-products. Dosage optimization should be determined for each specific application, as excessive amounts may not improve effectiveness or lead to undesirable side effects.

Non-PEGylated gold nanoparticles are used in cancer therapeutic applications like drug delivery and photothermal therapy due to their high biocompatibility, minimal cytotoxicity, and low immunogenicity. They are compatible with biological systems and have limited inflammation and foreign body responses. Toxicity studies show no acute toxicity, but thorough characterization and purification are crucial. Sterile handling is essential to prevent contamination. The optimal dosage and concentration should be determined for each cancer therapeutic application. The safety profile varies depending on the nanoparticles used and their application, and comprehensive safety assessments and studies are essential. Researchers should reference peer-reviewed sources for safety assessments and research (Gamucci et al. 2014).

Metal nanoparticle-based therapeutic systems are being investigated in various stages of preclinical and clinical development for the therapy as well as diagnosis of diseases such as cancer, liver cirrhosis, and anemia. While more research is warranted to understand the long-term effects of nanoparticles, current treatment options can be improved by enhancing bioavailability and efficacy via an approach that combines inorganic metal nanoparticles conjugated to small molecule entities (Na et al. 2009).

## 16.6 Future Prospects

Although the FDA has approved natural polymers such as polypeptides, polysaccharides, and nucleotides for various indications, scientists have tried to develop synthetic polymers that can perform better. Among other synthetic polymers, linear PEG has been demonstrated to be superior for multiple applications.

Even though PEGylation provides numerous benefits, new research has shown critical shortcomings such as immune activation, poor degradability, and potential aggregation. The enormous advances in macromolecular engineering and polymer chemistry have aided in exploring substitutes for PEG that can address the above-mentioned limitations. Polymeric systems such as poly(glycerol)s, poly(vinylpyrrolidone), poly(amino acid)s, poly(N-(2-hydro-xypropyl) methacrylamide), and poly(2-oxazoline)s have shown potential as far as hydrophilicity, aqueous solubility, stealth properties, and biocompatibility are concerned. However, none of them have been approved by the FDA yet, and further research is required to properly evaluate and compare them with PEG through a comprehensive analysis of polymer structure as well as cellular interactions.

## 16.7 Conclusion

PEGylation is, without a doubt, a very versatile and effective bioconjugation method. PEGylation of nanoparticles allows for effective transport of hydrophobic medicines, prolongs systemic circulation, enhances mucus penetration, and improves the stability of biotherapeutics. As a result, PEGylation has emerged as an invaluable technique for nanomedicine modification to enable the development of safe and efficacious drug delivery systems. Nonetheless, the anti-PEG immune response is a vexing problem that compromises the safety of drugs. So far, potential PEG alternatives are yet to be successfully translated to the clinic. Studies investigating the immunogenic potential of PEG have demonstrated that the clearance of subsequent doses of the drug can be primarily attributed to the anti-PEG antibodies produced after the first dose. If the patient has preexisting antibodies due to prior exposure to PEG in domestic or cosmetic goods, the first dose can elicit an immune response. It is, therefore, critical to understand the parameters that influence the development of anti-PEG antibodies to design innovative drug vectors or change the injection schedule and/or administration route and achieve therapeutic efficacy. While PEG alternatives will be aggressively investigated in the near future, we believe that PEG will continue to be extensively used for bioconjugation to enable effective drug delivery.

**Acknowledgments** The author RKT acknowledges the Department of Pharmaceuticals, Ministry of Chemicals and Fertilizers, India, for supporting the drug discovery and formulation research at NIPER Ahmedabad. R.K.T also acknowledges the Department of Science and Technology, Government of India, for a Core Research Grant funding (File No. CRG/2021/005402) and also acknowledges the Indian Council of Medical Research (ICMR), New Delhi, for the grant File Id: 2021-14161 and grant File Id: IIRP-2023-4849/F1 for supporting research in RKT lab.

## References

Aapro M, Bohlius J, Cameron DA, Dal Lago L, Donnelly JP, Kearney N, Lyman G, Pettengell R, Tjan-Heijnen V, Walewski (2011) 2010 update of EORTC guidelines for the use of granulocyte-colony stimulating factor to reduce the incidence of chemotherapy-induced febrile neutropenia in adult patients with lymphoproliferative disorders and solid tumours. European J Cancer 47:8–32

Abbina S, Vappala S, Kumar P, Siren EM, La CC, Abbasi U, Brooks DE, Kizhakkedathu JN (2017) Hyperbranched polyglycerols: recent advances in synthesis, biocompatibility and biomedical applications. J Mater Chem B 5:9249–9277

Alconcel SN, Baas AS, Maynard HD (2011) FDA-approved poly (ethylene glycol)–protein conjugate drugs. Polym Chem 2:1442–1448

Allen TM, Cullis PR (2004) Drug delivery systems: entering the mainstream. Science 303:1818–1822

Allen TM, Cullis PR (2013) Liposomal drug delivery systems: from concept to clinical applications. Adv Drug Deliv Rev 65:36–48

Amoozgar Z, Yeo Y (2012) Recent advances in stealth coating of nanoparticle drug delivery systems. Wiley Interdiscip Rev Nanomed Nanobiotechnol 4:219–233

Anselmo AC, Mitragotri S (2015) A review of clinical translation of inorganic nanoparticles. AAPS J 17:1041–1054

Anselmo A, Mitragotri S (2016) Nanoparticles in the clinic. Bioeng Transl Med 1:10–29

Barclay TG, Day CM, Petrovsky N, Garg S (2019) Review of polysaccharide particle-based functional drug delivery. Carbohydr Polym 221:94–112

Besheer A, Hause G, Kressler J, Mäder K (2007) Hydrophobically modified hydroxyethyl starch: synthesis, characterization, and aqueous self-assembly into nano-sized polymeric micelles and vesicles. Biomacromolecules 8:359–367

Boddu SH, Bhagav P, Karla PK, Jacob S, Adatiya MD, Dhameliya TM, Ranch KM, Tiwari AK (2021) Polyamide/poly (amino acid) polymers for drug delivery. J Funct Biomater 12:58

Bolze F, Morath V, Bast A, Rink N, Schlapschy M, Mocek S, Skerra A, Klingenspor M (2016) Long-acting PASylated leptin ameliorates obesity by promoting satiety and preventing hypometabolism in leptin-deficient Lepob/ob mice. Endocrinology 157:233–244

Bond TC, Szabo E, Gabriel S, Klastersky J, Tomey O, Mueller U, Schwartzberg L, Tang B (2018) Meta-analysis and indirect treatment comparison of lipegfilgrastim with pegfilgrastim and filgrastim for the reduction of chemotherapy-induced neutropenia-related events. J Oncol Pharm Pract 24:412–423

Botteri E, Krendyukov A, Curigliano G (2018) Comparing granulocyte colony–stimulating factor filgrastim and pegfilgrastim to its biosimilars in terms of efficacy and safety: a meta-analysis of randomised clinical trials in breast cancer patients. Eur J Cancer 89:49–55

Caló E, Khutoryanskiy VV (2015) Biomedical applications of hydrogels: a review of patents and commercial products. Eur Polym J 65:252–267

Cao Z, Jiang S (2012) Super-hydrophilic zwitterionic poly (carboxybetaine) and amphiphilic non-ionic poly (ethylene glycol) for stealth nanoparticles. Nano Today 7:404–413

Cao Z, Brault N, Xue H, Keefe A, Jiang S (2011) Manipulating sticky and non-sticky properties in a single material. Angew Chem Int Ed 50:6102–6104

Cao Y, Liu M, Zhang K, Dong J, Zu G, Chen Y, Zhang T, Xiong D, Pei R (2016) Preparation of linear poly (glycerol) as a T 1 contrast agent for tumor-targeted magnetic resonance imaging. J Mater Chem B 4:6716–6725

Carter PJ (2011) Introduction to current and future protein therapeutics: a protein engineering perspective. Exp Cell Res 317:1261–1269

Cattani G, Vogeley L, Crowley PB (2015) Structure of a PEGylated protein reveals a highly porous double-helical assembly. Nat Chem 7:823–828

Cornes P, Gascon P, Chan S, Hameed K, Mitchell CR, Field P, Latymer M, Arantes LH (2018) Systematic review and meta-analysis of short-versus long-acting granulocyte colony-

stimulating factors for reduction of chemotherapy-induced febrile neutropenia. Adv Ther 35:1816–1829
Crawford J, Caserta C, Roila F (2010) Hematopoietic growth factors: ESMO clinical practice guidelines for the applications. Ann Oncol 21:v248–v251
Cuomo F, Lopez F, Ceglie A, Maiuro L, Miguel MG, Lindman B (2012) pH-responsive liposome-templated polyelectrolyte nanocapsules. Soft Matter 8:4415–4420
De Volder J, Vereecke L, Joos G, Maes T (2020) Targeting neutrophils in asthma: a therapeutic opportunity? Biochem Pharmacol 182:114292
Dozier JK, Distefano MD (2015) Site-specific PEGylation of therapeutic proteins. Int J Mol Sci 16:25831–25864
Drago SE, Craparo EF, Luxenhofer R, Cavallaro G (2021) Development of polymer-based nanoparticles for zileuton delivery to the lung: PMeOx and PMeOzi surface chemistry reduces interactions with mucins. Nanomedicine 37:102451
Dzieciuch M, Rissanen S, Szydłowska N, Bunker A, Kumorek M, Jamroz D, Vattulainen I, Nowakowska M, Rog T, Kepczynski M (2015) PEGylated liposomes as carriers of hydrophobic porphyrins. J Phys Chem B 119:6646–6657
Engler AC, Ke X, Gao S, Chan JM, Coady DJ, Ono RJ, Lubbers R, Nelson A, Yang YY, Hedrick JL (2015) Hydrophilic polycarbonates: promising degradable alternatives to poly (ethylene glycol)-based stealth materials. Macromolecules 48:1673–1678
Falvo E, Tremante E, Arcovito A, Papi M, Elad N, Boffi A, Morea V, Conti G, Toffoli G, Fracasso G (2016) Improved doxorubicin encapsulation and pharmacokinetics of ferritin–fusion protein nanocarriers bearing proline, serine, and alanine elements. Biomacromolecules 17:514–522
Fernandes R, Mazzarello S, Stober C, Vandermeer L, Dudani S, Ibrahim MF, Majeed H, Perdrizet K, Shorr R, Hutton BJBCR (2017) Optimal primary febrile neutropenia prophylaxis for patients receiving docetaxel–cyclophosphamide chemotherapy for breast cancer: a systematic review. Breast Cancer Res Treat 161:1–10
Francis G, Fisher D, Delgado C, Malik F, Gardiner A, Neale D (1998) PEGylation of cytokines and other therapeutic proteins and peptides: the importance of biological optimisation of coupling techniques. Int J Hematol 68:1–18
Gamucci O, Bertero A, Gagliardi M, Bardi G (2014) Biomedical nanoparticles: overview of their surface immune-compatibility. Coatings 4:139–159
Ge Z, Xie D, Chen D, Jiang X, Zhang Y, Liu H, Liu S (2007) Stimuli-responsive double hydrophilic block copolymer micelles with switchable catalytic activity. Macromolecules 40:3538–3546
Gheybi H, Sattari S, Bodaghi A, Soleimani K, Dadkhah A, Adeli M (2018) Polyglycerols. Engineering of biomaterials for drug delivery systems. Elsevier
González-Aramundiz JV, Lozano MV, Sousa-Herves A, Fernandez-Megia E, Csaba N (2012) Polypeptides and polyaminoacids in drug delivery. Expert Opin Drug Deliv 9:183–201
Gu L, Faig A, Abdelhamid D, Uhrich K (2014) Sugar-based amphiphilic polymers for biomedical applications: from nanocarriers to therapeutics. Acc Chem Res 47:2867–2877
Guillerm B, Monge S, Lapinte V, Robin JJ (2012) How to modulate the chemical structure of polyoxazolines by appropriate functionalization. Macromol Rapid Commun 33:1600–1612
Hadjesfandiari N, Parambath A (2018) Stealth coatings for nanoparticles: polyethylene glycol alternatives. Engineering of biomaterials for drug delivery systems. Elsevier
Hamidi M, Azadi A, Rafiei P (2006) Pharmacokinetic consequences of pegylation. Drug Deliv 13:399–409
Han S, Kim C, Kwon D (1997) Thermal/oxidative degradation and stabilization of polyethylene glycol. Polymer 38:317–323
Harari D, Kuhn N, Abramovich R, Sasson K, Zozulya AL, Smith P, Schlapschy M, Aharoni R, Köster M, Eilam R (2014) Enhanced in vivo efficacy of a type I interferon superagonist with extended plasma half-life in a mouse model of multiple sclerosis. J Biol Chem 289:29014–29029
Harris JM, Bentley MD, Moreadith RW, Viegas TX, Fang Z, Yoon K, Weimer R, Dizman B, Nordstierna L (2019) Tuning drug release from polyoxazoline-drug conjugates. Eur Polym J 120:109241

Harrison RH, Steele JA, Chapman R, Gormley AJ, Chow LW, Mahat MM, Podhorska L, Palgrave RG, Payne DJ, Hettiaratchy SP (2015) Modular and versatile spatial functionalization of tissue engineering scaffolds through fiber-initiated controlled radical polymerization. Adv Funct Mater 25:5748–5757
Hatakeyama H, Akita H, Harashima H (2013) The polyethyleneglycol dilemma: advantage and disadvantage of PEGylation of liposomes for systemic genes and nucleic acids delivery to tumors. Biol Pharm Bull 36:892–899
Hauck N, Seixas N, Centeno SP, Schlüßler R, Cojoc G, Müller P, Guck J, Wöll D, Wessjohann LA, Thiele J (2018) Droplet-assisted microfluidic fabrication and characterization of multifunctional polysaccharide microgels formed by multicomponent reactions. Polymers 10:1055
Hendler D, Rizel S, Yerushalmi R, Neiman V, Bonilla L, Braunstein R, Sulkes A, Stemmer SM (2011) Different schedules of granulocyte growth factor support for patients with breast cancer receiving adjuvant dose-dense chemotherapy: a prospective nonrandomized study. Am J Clin Oncol 34:619–624
Hildebrand V, Heydenreich M, Laschewsky A, Möller HM, Müller-Buschbaum P, Papadakis CM, Schanzenbach D, Wischerhoff E (2017) "Schizophrenic" self-assembly of dual thermoresponsive block copolymers bearing a zwitterionic and a non-ionic hydrophilic block. Polymer 122:347–357
Hoang Thi TT, Pilkington EH, Nguyen DH, Lee JS, Park KD, Truong NP (2020) The importance of poly (ethylene glycol) alternatives for overcoming PEG immunogenicity in drug delivery and bioconjugation. Polymers 12:298
Hoffman AS (2016) The early days of PEG and PEGylation (1970s–1990s). Acta Biomater 40:1–5
Hoogenboom R (2009) Poly (2-oxazoline) s: a polymer class with numerous potential applications. Angew Chem Int Ed 48:7978–7994
Hosseinkhani H, Hong P-D, Yu D-S (2013) Self-assembled proteins and peptides for regenerative medicine. Chem Rev 113:4837–4861
Howard MD, Jay M, Dziubla TD, Lu X (2008) PEGylation of nanocarrier drug delivery systems: state of the art. J Biomed Nanotechnol 4:133–148
Hsiao C, Huang K (2005) Synthesis, characterization, and applications of polyvinylpyrrolidone/SiO2 hybrid materials. J Appl Polym Sci 96:1936–1942
Huang H-C, Barua S, Sharma G, Dey SK, Rege K (2011) Inorganic nanoparticles for cancer imaging and therapy. J Control Release 155:344–357
Hucknall A, Rangarajan S, Chilkoti A (2009) In pursuit of zero: polymer brushes that resist the adsorption of proteins. Adv Mater 21:2441–2446
Immordino ML, Dosio F, Cattel L (2006) Stealth liposomes: review of the basic science, rationale, and clinical applications, existing and potential. Int J Nanomedicine 1:297
Imran Ul-Haq M, Lai BF, Chapanian R, Kizhakkedathu JN (2012) Influence of architecture of high molecular weight linear and branched polyglycerols on their biocompatibility and biodistribution. Biomaterials 33:9135–9147
Jevševar S, Kunstelj M, Porekar VG (2010) PEGylation of therapeutic proteins. Biotechnology J 5:113–128
Jiang S, Cao Z (2010) Ultralow-fouling, functionalizable, and hydrolyzable zwitterionic materials and their derivatives for biological applications. Adv Mater 22:920–932
Jiang Y, Lu H, Chen F, Callari M, Pourgholami M, Morris DL, Stenzel MH (2016) PEGylated albumin-based polyion complex micelles for protein delivery. Biomacromolecules 17:808–817
Jin Q, Chen Y, Wang Y, Ji J (2014) Zwitterionic drug nanocarriers: a biomimetic strategy for drug delivery. Colloids Surf B: Biointerfaces 124:80–86
Joh DY, Zimmers Z, Avlani M, Heggestad JT, Aydin HB, Ganson N, Kumar S, Fontes CM, Achar RK, Hershfield MS (2019) Architectural modification of conformal PEG-bottlebrush coatings minimizes anti-PEG antigenicity while preserving stealth properties. Adv Healthc Mater 8:1801177
Kadajji VG, Betageri GV (2011) Water soluble polymers for pharmaceutical applications. Polymers 3:1972–2009

Kainthan RK, Janzen J, Levin E, Devine DV, Brooks DE (2006) Biocompatibility testing of branched and linear polyglycidol. Biomacromolecules 7:703–709
Keefe AJ, Jiang S (2012) Poly (zwitterionic) protein conjugates offer increased stability without sacrificing binding affinity or bioactivity. Nat Chem 4:59–63
King DJ, Noss RR (1989) Toxicity of polyacrylamide and acrylamide monome. Rev Environ Health 8:3–16
Klein PM, Klinker K, Zhang W, Kern S, Kessel E, Wagner E, Barz M (2018) Efficient shielding of polyplexes using heterotelechelic polysarcosines. Polymers 10:689
Knop K, Hoogenboom R, Fischer D, Schubert US (2010) Poly (ethylene glycol) in drug delivery: pros and cons as well as potential alternatives. Angew Chem Int Ed 49:6288–6308
Koczkur KM, Mourdikoudis S, Polavarapu L, Skrabalak SE (2015) Polyvinylpyrrolidone (PVP) in nanoparticle synthesis. Dalton Trans 44:17883–17905
Kopeček J, Kopečková P (2010) HPMA copolymers: origins, early developments, present, and future. Adv Drug Deliv Rev 62:122–149
Kronek J, Paulovičová E, Paulovičová L, Kroneková Z, Lustoň J (2013) Biocompatibility and immunocompatibility assessment of poly (2-oxazolines). INTECH
Kuan JW, Su AT, Leong CF (2017) Pegylated granulocyte-colony stimulating factor versus non-pegylated granulocyte-colony stimulating factor for peripheral blood stem cell mobilization: a systematic review and meta-analysis. J Clin Apher 32:517–542
Kuhn N, Schmidt CQ, Schlapschy M, Skerra A (2016) PASylated coversin, a C5-specific complement inhibitor with extended pharmacokinetics, shows enhanced anti-hemolytic activity in vitro. Bioconjug Chem 27:2359–2371
Kurniasih IN, Keilitz J, Haag R (2015) Dendritic nanocarriers based on hyperbranched polymers. Chem Soc Rev 44:4145–4164
Laschewsky A (2014) Structures and synthesis of zwitterionic polymers. Polymers 6:1544–1601
Lava K, Verbraeken B, Hoogenboom R (2015) Poly (2-oxazoline) s and click chemistry: a versatile toolbox toward multi-functional polymers. Eur Polym J 65:98–111
Le Garrec D, Taillefer J, Van Lier J, Lenaerts V, Leroux J-C (2002) Optimizing pH-responsive polymeric micelles for drug delivery in a cancer photodynamic therapy model. J Drug Target 10:429–437
Leng C, Hung H-C, Sun S, Wang D, Li Y, Jiang S, Chen Z (2015) Probing the surface hydration of nonfouling zwitterionic and PEG materials in contact with proteins. ACS Appl Mater Interfaces 7:16881–16888
Leucuta SE (2012) Drug delivery systems with modified release for systemic and biophase bioavailability. Curr Clin Pharmacol 7:282–317
Li C, Wallace S (2008) Polymer-drug conjugates: recent development in clinical oncology. Adv Drug Deliv Rev 60:886–898
Li X, Zheng H, Yu M-C, Wang W, Wu X-H, Yang D-M, Xu J (2020) Is PEGylated G-CSF superior to G-CSF in patients with breast cancer receiving chemotherapy? A systematic review and meta-analysis. Support Care Cancer 28:5085–5097
Liang S-L, Cook WD, Thouas GA, Chen Q-Z (2010) The mechanical characteristics and in vitro biocompatibility of poly (glycerol sebacate)-Bioglass® elastomeric composites. Biomaterials 31:8516–8529
Liechty WB, Kryscio DR, Slaughter BV, Peppas NA (2010) Polymers for drug delivery systems. Annu Rev Chem Biomol Eng 1:149–173
Liu X, Xu Y, Wu Z, Chen H (2013) Poly (N-vinylpyrrolidone)-modified surfaces for biomedical applications. Macromol Biosci 13:147–154
Liu M, Johansen P, Zabel F, Leroux J-C, Gauthier MA (2014) Semi-permeable coatings fabricated from comb-polymers efficiently protect proteins in vivo. Nat Commun 5:5526
Liu EY, Jung S, Yi H (2016a) Improved protein conjugation with uniform, macroporous poly (acrylamide-co-acrylic acid) hydrogel microspheres via EDC/NHS chemistry. Langmuir 32:11043–11054
Liu Y, Duzhko VV, Page ZA, Emrick T, Russell TP (2016b) Conjugated polymer zwitterions: efficient interlayer materials in organic electronics. Acc Chem Res 49:2478–2488

Liu F, Wu Y, Bai L, Peng X, Zhang H, Zhang Y, An P, Wang S, Ma G, Ba X (2018) Facile preparation of hyperbranched glycopolymers via an AB 3* inimer promoted by a hydroxy/cerium (IV) redox process. Polym Chem 9:5024–5031

Lutz JF (2008) Polymerization of oligo (ethylene glycol)(meth) acrylates: toward new generations of smart biocompatible materials. J Polym Sci A Polym Chem 46:3459–3470

Lutz J-F, Andrieu J, Üzgün S, Rudolph C, Agarwal S (2007) Biocompatible, thermoresponsive, and biodegradable: simple preparation of "all-in-one" biorelevant polymers. Macromolecules 40:8540–8543

Luyckx J, Baudouin C (2011) Trehalose: an intriguing disaccharide with potential for medical application in ophthalmology. Clin Ophthalmol:577–581

Marcelino EMBF (2016) Protein disulfide isomerase as a target for besnoitiosis therapy: molecular characterization and studies of its role in infection and host immune response. Universidade de Lisboa (Portugal)

Marsh JC, Ganser A, Stadler M (2007) Hematopoietic growth factors in the treatment of acquired bone marrow failure states. Semin Hematol 44:138–147. Elsevier

Mccormick CL, Lowe AB (2004) Aqueous RAFT polymerization: recent developments in synthesis of functional water-soluble (co) polymers with controlled structures. Acc Chem Res 37:312–325

Mccormick-Thomson LA, Sgouras D, Duncan R (1989) Poly (amino acid) copolymers as a potential soluble drug delivery system. 2. Body distribution and preliminary biocompatibility testing in vitro and in vivo. J Bioact Compat Polym 4:252–268

McCrudden CM, Bennie L, Chambers P, Wilson J, Kerr M, Ziminska M, Douglas H, Kuhn S, Carroll E, O'brien, G. (2023) Peptide delivery of a multivalent mRNA SARS-CoV-2 vaccine. J Ophthalmol Clin Res 362:536–547

Melancon MP, Li C (2011) Multifunctional synthetic poly (L-glutamic acid)–based cancer therapeutic and imaging agents. Mol Imaging 10:7290.2011.00007

Mendler CT, Friedrich L, Laitinen I, Schlapschy M, Schwaiger M, Wester H-J, Skerra A (2015) High contrast tumor imaging with radio-labeled antibody fab fragments tailored for optimized pharmacokinetics via PASylation. MAbs 7:96–109. Taylor & Francis

Milla P, Dosio F, Cattel L (2012) PEGylation of proteins and liposomes: a powerful and flexible strategy to improve the drug delivery. Curr Drug Metab 13:105–119

Moghimi SM, Szebeni J (2003) Stealth liposomes and long circulating nanoparticles: critical issues in pharmacokinetics, opsonization and protein-binding properties. Prog Lipid Res 42:463–478

Morath V, Bolze F, Schlapschy M, Schneider S, Sedlmayer F, Seyfarth K, Klingenspor M, Skerra A (2015) PASylation of murine leptin leads to extended plasma half-life and enhanced in vivo efficacy. Mol Pharm 12:1431–1442

Moros M, Hernaez B, Garet E, Dias JT, Saez B, Grazu V, Gonzalez-Fernandez A, Alonso C, De La Fuente JM (2012) Monosaccharides versus PEG-functionalized NPs: influence in the cellular uptake. ACS Nano 6:1565–1577

Na HB, Song IC, Hyeon T (2009) Inorganic nanoparticles for MRI contrast agents. Adv Mater 21:2133–2148

Nag OK, Awasthi V (2013) Surface engineering of liposomes for stealth behavior. Pharmaceutics 5:542–569

Nakamura K, Yamashita K, Itoh Y, Yoshino K, Nozawa S, Kasukawa H (2012) Comparative studies of polyethylene glycol-modified liposomes prepared using different PEG-modification methods. Biochim Biophys Acta Biomembr 1818:2801–2807

Nastyshyn S, Raczkowska J, Stetsyshyn Y, Orzechowska B, Bernasik A, Shymborska Y, Brzychczy-Włoch M, Gosiewski T, Lishchynskyi O, Ohar H (2020) Non-cytotoxic, temperature-responsive and antibacterial POEGMA based nanocomposite coatings with silver nanoparticles. RSC Adv 10:10155–10166

Nishikawa M, Ogawa K (2004) Occurrence of d-histidine residues in antimicrobial poly (arginyl–histidine), conferring resistance to enzymatic hydrolysis. FEMS Microbiol Lett 239:255–259

Obst M, Steinbüchel A (2004) Microbial degradation of poly (amino acid)s. Biomacromolecules 5:1166–1176

Otsuka H, Nagasaki Y, Kataoka K (2003) PEGylated nanoparticles for biological and pharmaceutical applications. Adv Drug Deliv Rev 55:403–419

Padín-González E, Lancaster P, Bottini M, Gasco P, Tran L, Fadeel B, Wilkins T, Monopoli MP (2022) Understanding the role and impact of poly (ethylene glycol)(PEG) on nanoparticle formulation: implications for COVID-19 vaccines. Front Bioeng Biotechnol 10:882363

Park J, Fong PM, Lu J, Russell KS, Booth CJ, Saltzman WM, Fahmy TM (2017) PEGylated PLGA nanoparticles for the improved delivery of doxorubicin. Nanomed Cancer. Jenny Stanford Publishing

Pathania D, Gupta D, Kothiyal N, Eldesoky G, Naushad M (2016) Preparation of a novel chitosan-g-poly (acrylamide)/Zn nanocomposite hydrogel and its applications for controlled drug delivery of ofloxacin. Int J Biol Macromol 84:340–348

Pires-Oliveira R, Tang J, Percebom AM, Petzhold CL, Tam KC, Loh W (2020) Effect of molecular architecture and composition on the aggregation pathways of POEGMA random copolymers in water. Langmuir 36:15018–15029

Pisal DS, Kosloski MP, Balu-Iyer SV (2010) Delivery of therapeutic proteins. J Pharm Sci 99:2557–2575

Podust VN, Balan S, Sim B-C, Coyle MP, Ernst U, Peters RT, Schellenberger V (2016) Extension of in vivo half-life of biologically active molecules by XTEN protein polymers. J Control Release 240:52–66

Qi Y, Simakova A, Ganson NJ, Li X, Luginbuhl KM, Ozer I, Liu W, Hershfield MS, Matyjaszewski K, Chilkoti A (2016) A brush-polymer/exendin-4 conjugate reduces blood glucose levels for up to five days and eliminates poly (ethylene glycol) antigenicity. Nat Biomed Eng 1:0002

Qian H, Wang K, Lv M, Zhao C, Wang H, Wen S, Huang D, Chen W, Zhong Y (2022) Recent advances on next generation of polyzwitterion-based nano-vectors for targeted drug delivery. J Control Release 343:492–505

Roberts MG, Yu Q, Keunen R, Liu J, Ngae Wong EC, Rastogi CK, Reilly RM, Allen C, Winnik MA (2020) Functionalization of cellulose nanocrystals with POEGMA copolymers via copper-catalyzed azide–alkyne cycloaddition for potential drug-delivery applications. Biomacromolecules 21:2014–2023

Romberg B, Metselaar JM, Baranyi L, Snel CJ, Bünger R, Hennink WE, Szebeni J, Storm G (2007) Poly (amino acid)s: promising enzymatically degradable stealth coatings for liposomes. Int J Pharm 331:186–189

Rudolph T, Crotty S, Von Der Lühe M, Pretzel D, Schubert US, Schacher FH (2013) Synthesis and solution properties of double hydrophilic poly (ethylene oxide)-block-poly (2-ethyl-2-oxazoline)(PEO-b-PEtOx) star block copolymers. Polymers 5:1081–1101

Salatin S, Yari Khosroushahi A (2017) Overviews on the cellular uptake mechanism of polysaccharide colloidal nanoparticles. J Cell Mol Med 21:1668–1686

Sano K, Umemoto K, Miura H, Ohno S, Iwata K, Kawakami R, Munekane M, Yamasaki T, Citterio D, Hiruta Y (2022) Feasibility of using poly [oligo (ethylene glycol) methyl ether methacrylate] as tumor-targeted carriers of diagnostic drugs. ACS Appl Polym Mater 4:4734–4740

Schellenberger V, Wang C-W, Geething NC, Spink BJ, Campbell A, To, W, Scholle MD, Yin Y, Yao Y, Bogin O (2009) A recombinant polypeptide extends the in vivo half-life of peptides and proteins in a tunable manner. Nat Biotechnol 27:1186–1190

Schlapschy M, Binder U, Börger C, Theobald I, Wachinger K, Kisling S, Haller D, Skerra A (2013) PASylation: a biological alternative to PEGylation for extending the plasma half-life of pharmaceutically active proteins. Protein Eng Des Sel 26:489–501

Schmidt BV (2019) Hydrophilic polymers. Polymers 11(4):693

Schwartzberg LS, Yang Z, Peguero JA, Agajanian R, Bharadwaj JS, Restrepo A, Hlalah O, Mehmi I, Bhat G, Cobb PW (2018) Safety and efficacy of eflapegrastim in reducing severe neutropenia in patients treated with myelosuppressive chemotherapy in a phase 3 randomized controlled trial compared to pegfilgrastim (ADVANCE trial). American Society of Clinical Oncology

Schwarz W (2018) PVP: a critical review of the kinetics and toxicology of polyvinylpyrrolidone (povidone). CRC Press

Secker C, Brosnan SM, Luxenhofer R, Schlaad H (2015) Poly (α-Peptoid) s revisited: synthesis, properties, and use as biomaterial. Macromol Biosci 15:881–891
Sedlacek O, Hoogenboom R (2020) Drug delivery systems based on poly (2-oxazoline) s and poly (2-oxazine) s. Adv Ther 3:1900168
Seeliger W, Aufderhaar E, Diepers W, Feinauer R, Nehring R, Thier W, Hellmann H (1966) Recent syntheses and reactions of cyclic imidic esters. Angew Chem Int Ed Engl 5:875–888
Shenoi RA, Narayanannair JK, Hamilton JL, Lai BF, Horte S, Kainthan RK, Varghese JP, Rajeev KG, Manoharan M, Kizhakkedathu JN (2012) Branched multifunctional polyether polyketals: variation of ketal group structure enables unprecedented control over polymer degradation in solution and within cells. J Am Chem Soc 134:14945–14957
Shenoi RA, Abbina S, Kizhakkedathu JN (2016) In vivo biological evaluation of high molecular weight multifunctional acid-degradable polymeric drug carriers with structurally different ketals. Biomacromolecules 17:3683–3693
Shih I-L, Van Y-T, Shen M-H (2004) Biomedical applications of chemically and microbiologically synthesized poly (glutamic acid) and poly (lysine). Mini Rev Med Chem 4:179–188
Shima S, Sakai H (1977) Polylysine produced by Streptomyces. Agric Biol Chem 41:1807–1809
Sizovs A, Xue L, Tolstyka ZP, Ingle NP, Wu Y, Cortez M, Reineke TM (2013) Poly (trehalose): sugar-coated nanocomplexes promote stabilization and effective polyplex-mediated siRNA delivery. J Am Chem Soc 135:15417–15424
Skandalis A, Pispas S (2017) PLMA-b-POEGMA amphiphilic block copolymers: synthesis and self-assembly in aqueous media. J Polym Sci A Polym Chem 55:155–163
Smith EA, Oehme FW (1991) Acrylamide and polyacrylamide: a review of production, use, environmental fate and neurotoxicity. Rev Environ Health 9:215–228
Strohl WR (2015) Fusion proteins for half-life extension of biologics as a strategy to make biobetters. BioDrugs 29:215–239
Su L, Feng Y, Wei K, Xu X, Liu R, Chen G (2021) Carbohydrate-based macromolecular biomaterials. Chem Rev 121:10950–11029
Sun H, Meng F, Dias AA, Hendriks M, Feijen J, Zhong Z (2011) α-Amino acid containing degradable polymers as functional biomaterials: rational design, synthetic pathway, and biomedical applications. Biomacromolecules 12:1937–1955
Sun Q, Yang Z, Qi X (2023) Design and application of hybrid polymer-protein systems in cancer therapy. Polymers 15:2219
Sunder A, Hanselmann R, Frey H, Mülhaupt R (1999) Controlled synthesis of hyperbranched polyglycerols by ring-opening multibranching polymerization. Macromolecules 32:4240–4246
Takahashi R, Narayanan T, Yusa S-I, Sato T (2018) Kinetics of morphological transition between cylindrical and spherical micelles in a mixture of anionic–neutral and cationic–neutral block copolymers studied by time-resolved SAXS and USAXS. Macromolecules 51:3654–3662
Tan H, Tomic K, Hurley D, Daniel G, Barron R, Malin J (2011) Comparative effectiveness of colony-stimulating factors for febrile neutropenia: a retrospective study. Curr Med Res Opin 27:79–86
Taylor ME, Clarkson D, Greenbaum SG, Panzer MJ (2021) Examining the impact of polyzwitterion chemistry on lithium ion transport in ionogel electrolytes. ACS Appl Polym Mater 3:2635–2645
Thiele J, Ma Y, Bruekers SM, Ma S, Huck WT (2014) 25th anniversary article: designer hydrogels for cell cultures: a materials selection guide. Adv Mater 26:125–148
Tiwari G, Tiwari R, Sriwastawa B, Bhati L, Pandey S, Pandey P, Bannerjee SK (2012) Drug delivery systems: an updated review. Int J Pharm Investig 2:2
Tolstyka ZP, Phillips H, Cortez M, Wu Y, Ingle N, Bell JB, Hackett PB, Reineke TM (2016) Trehalose-based block copolycations promote polyplex stabilization for lyophilization and in vivo pDNA delivery. ACS Biomater Sci Eng 2:43–55
Torchilin V (1998) Polymer-coated long-circulating microparticulate pharmaceuticals. J Microencapsul 15:1–19
Ulbrich K, Šubr V (2010) Structural and chemical aspects of HPMA copolymers as drug carriers. Adv Drug Deliv Rev 62:150–166

Üzgün S, Akdemir OZR, Hasenpusch GN, Maucksch C, Golas MM, Sander B, Stark H, Imker R, Lutz J-F, Rudolph C (2010) Characterization of tailor-made copolymers of oligo (ethylene glycol) methyl ether methacrylate and N, N-dimethylaminoethyl methacrylate as nonviral gene transfer agents: influence of macromolecular structure on gene vector particle properties and transfection efficiency. Biomacromolecules 11:39–50

Veronese FM, Pasut G (2005) PEGylation, successful approach to drug delivery. Drug Discov Today 10:1451–1458

Viegas TX, Bentley MD, Harris JM, Fang Z, Yoon K, Dizman B, Weimer R, Mero A, Pasut G, Veronese FM (2011) Polyoxazoline: chemistry, properties, and applications in drug delivery. Bioconjug Chem 22:976–986

Vila A, Sanchez A, Evora C, Soriano I, Vila Jato J, Alonso M (2004) PEG-PLA nanoparticles as carriers for nasal vaccine delivery. J Aerosol Med 17:174–185

Wang J, Yuan S, Zhang Y, Wu W, Hu Y, Jiang X (2016) The effects of poly (zwitterions) s versus poly (ethylene glycol) surface coatings on the biodistribution of protein nanoparticles. Biomater Sci 4:1351–1360

Welte K, Gabrilove J, Bronchud MH, Platzer E, Morstyn G (1996) Filgrastim (r-metHuG-CSF): the first 10 years. 1907–1929

Willersinn J, Bogomolova A, Cabré MB, Schmidt BV (2017) Vesicles of double hydrophilic pullulan and poly (acrylamide) block copolymers: a combination of synthetic-and bio-derived blocks. Polym Chem 8:1244–1254

Wu C-J, Huang C-J, Jiang S, Sheng Y-J, Tsao H-K (2016) Superhydrophilicity and spontaneous spreading on zwitterionic surfaces: carboxybetaine and sulfobetaine. RSC Adv 6:24827–24834

Xie F, Boyer C, Gaborit V, Rouillon T, Guicheux J, Tassin J-F, Geoffroy V, Réthoré G, Weiss P (2018) A cellulose/laponite interpenetrated polymer network (IPN) hydrogel: controllable double-network structure with high modulus. Polymers 10:634

Xing C-M, Meng F-N, Quan M, Ding K, Dang Y, Gong Y-K (2017) Quantitative fabrication, performance optimization and comparison of PEG and zwitterionic polymer antifouling coatings. Acta Biomater 59:129–138

Xu B, Zhang H, Bian H (2019) Delayed crosslinking amphiphilic polymer gel system with adjustable gelation time based on competitive inclusion method. Polymers 11:381

Yadav D, Dewangan HK (2021) PEGYLATION: an important approach for novel drug delivery system. J Biomater Sci Polym Ed 32:266–280

Yang SR, Lee HJ, Kim J-D (2006) Histidine-conjugated poly (amino acid) derivatives for the novel endosomolytic delivery carrier of doxorubicin. J Control Release 114:60–68

Yang W, Zhang L, Wang S, White AD, Jiang S (2009) Functionalizable and ultra stable nanoparticles coated with zwitterionic poly (carboxybetaine) in undiluted blood serum. Biomaterials 30:5617–5621

Yang D, Yu L, Van S (2010) Clinically relevant anticancer polymer paclitaxel therapeutics. Cancers 3:17–42

Yu D-G, Shen X-X, Branford-White C, White K, Zhu L-M, Bligh SA (2009) Oral fast-dissolving drug delivery membranes prepared from electrospun polyvinylpyrrolidone ultrafine fibers. Nanotechnology 20:055104

Yu X, Liu Z, Janzen J, Chafeeva I, Horte S, Chen W, Kainthan RK, Kizhakkedathu JN, Brooks DE (2012) Polyvalent choline phosphate as a universal biomembrane adhesive. Nat Mater 11:468–476

Zelikin AN, Such GK, Postma A, Caruso F (2007) Poly (vinylpyrrolidone) for bioconjugation and surface ligand immobilization. Biomacromolecules 8:2950–2953

Zhang F, Liu M-R, Wan H-T (2014) Discussion about several potential drawbacks of PEGylated therapeutic proteins. Biol Pharm Bull 37:335–339

Zhang Z, Li Q, Yesildag C, Bartsch C, Zhang X, Liu W, Loebus A, Su Z, Lensen MC (2018) Influence of network structure on the crystallization behavior in chemically crosslinked hydrogels. Polymers 10:970

Zhao S, Yang H, Zuo C, Sun L, Ma L, Wei H (2016) pH-sensitive drug release of star-shaped micelles with OEG brush corona. RSC Adv 6:111217–111225
Zheng L, Sundaram HS, Wei Z, Li C, Yuan Z (2017) Applications of zwitterionic polymers. React Funct Polym 118:51–61
Zuma LK, Gasa NL, Makhoba XH, Pooe OJ (2022) Protein PEGylation: navigating recombinant protein stability, aggregation, and bioactivity. Biomed Res Int 2022:8929715

# Current Clinical, Regulatory, and Patent Aspects of PEGylated Nanoproducts

# 17

Sombir Saharan, Sajidul Hoque, Piyush Neware, V. Ravichandiran, and Vinod Gaikwad

**Abstract**

PEGylated nanoproducts are a type of nanomedicine that consists of a drug molecule that is conjugated to a PEG molecule. The PEG molecule has several properties that make it well-suited for drug delivery such as its biodegradability and safe to consume, aqueous-solubility (easy uptake by the body), protecting drug from degradation with good longevity, and facilitating the drug to be administered to the target tissues for an extended period. PEGylated nanoproducts are an exciting new class of pharmaceutical drugs for several disorders like cancer, cardiovascular disease, and infectious diseases because of these characteristics. However, several regulatory and patent issues need to be addressed before these products can be widely adopted. The FDA has not yet developed specific guidelines for the clinical development of PEGylated nanoproducts. This means that companies developing these products need to work closely with the FDA to ensure that their products meet all of the necessary safety and efficacy requirements. Several patent issues need to be considered. The patent landscape for PEGylated nanoproducts is complex and there is a risk of patent infringement. Companies developing these products need to carefully assess the patent landscape before they begin clinical development. Despite the challenges, the potential benefits of PEGylated nanoproducts are significant. These products have the potential to enhance the efficacy and safety of drugs and they could revolutionize the treatment of a variety of diseases.

S. Saharan · S. Hoque · P. Neware · V. Gaikwad (✉)
Department of Pharmaceutics, National Institute of Pharmaceutical Education and Research (NIPER), Hajipur, Bihar, India

V. Ravichandiran
National Institute of Pharmaceutical Education and Research (NIPER), Hajipur, Bihar, India





R. K. Tekade, N. K. Jain (eds.), *PEGylated Nanocarriers in Medicine and Pharmacy*, https://doi.org/10.1007/978-981-97-7316-9_17

**Keywords**

PEGylated nanocarriers · PEGylation · Targeted drug delivery · Clinical aspects Regulatory consideration · Patent status

## Abbreviations

| | |
|---|---|
| CNS | Central nervous system |
| EDQM | European Directorate for the Quality of Medicines |
| EMA | European Medicines Agency |
| EPR | Enhanced permeability and retention |
| FD&C Act | Federal Food, Drug, and Cosmetic Act |
| FDA | Food and Drug Administration |
| IBCA | Isobutyl cyanoacrylate |
| MHRA | Medicines and Healthcare Products Regulatory Agency |
| MPS | Mononuclear phagocyte system |
| NPs | Nanoparticles |
| PEG | Polyethylene glycol |
| PEO | Polyethylene oxide |
| PGA | Polyglycolide |
| PLA | Polylactide |
| PLGA | Poly-(D,L-lactide-co-glycolide) |
| PPO | Polypropylene oxide |
| PTX | Paclitaxel |
| SLN | Solid lipid nanoparticle |
| WHO | World Health Organization |

## 17.1 Introduction

One of the most exciting fields of research at present is nanomedicine. One thousand five hundred patents have already been filed in this subject over the past 20 years, and numerous clinical trials have also been completed. The usage of nanomedicine and nano-drug delivery devices is unquestionably a trend that will be the future field of study and development for decades to come. Various types of nanoparticles are used to efficiently transport drugs to cancer/tumor cells without disrupting normal cell physiology (Patra et al. 2018). Different kinds of nanocarrier systems provide pockets that could be used for maintaining bioactive substances that lack pharmaceutical qualities such as poor solubility, permeability, and stability-free crucial factors affecting therapeutic efficacy. These nano systems' surface-specific architectures can be altered to achieve particular features, such as for selective distribution and controlled release, making them appropriate carriers in the delivery of bioactives. An appropriate nanosystem can be created by considering the target of therapy and using novel material-forming nanosystems and manufacturing technology (Rachmawati et al. 2020).

To improve the safety and effectiveness of active substances in medical, veterinary, or agricultural fields, nanocarrier-based delivery techniques can be used. Even while the bulk of nanocarriers have been developed to combat cancer, several formulations being made for diverse applications, notably those used in the cure of infectious diseases and parasitic infections that impact people, livestock, and pet animals, are rapidly rising. The study of designed agrochemical delivery methods that enhance efficacy while minimizing their negative effects on the environment has also been prompted by the restriction and prohibition of numerous pesticides (Chariou et al. 2020).

### 17.1.1 Surface-Decorated Nanocarriers for Targeted Drug Delivery

NPs were recognized for their reliable availability, simple functionalization, biocompatibility, stability, and effective drug targeting. Several NP features need to be managed to deliver medications effectively to target areas with the fewest negative effects. These qualities are the optimum size, surface charge, and geometry to prevent any potentially harmful side effects. NPs typically exhibit a variety of magnetic, optical, thermal, electrical, and other properties because they have a large surface area and only a limited number of quantum mechanical effects (Abed et al. 2019). To increase their affinity and selectivity towards cancer cells, targeting moieties are covalently added to the surface of NPs. However, selecting the ideal one for specific applications might be challenging due to the wide variety of surface decorating molecules that are available. In this article, we'll examine several illustrative experimental and computational experiments to choose the best surface functional groups. With adjustable stiffness and the same surface chemistry, the NPs with surface-decorated polymers that are amphiphilic are capable of entering the cell interior through penetrating processes, they can be selectively accepted by diseased cells based on stiffness, and they can be combined with pH-responsive polymers to be embraced or turned down by the cells based on the pH of the surrounding environment. Different types of surface-decorated nanocarrier are listed in Table 17.1 (Baek and Cho 2017).

### 17.1.2 In Vivo Performance of PEGylated Nanocarriers

The medications that already have FDA approval serve as a convincing example of how PEGylation might enhance the therapeutic value of pharmaceuticals. The most significant benefits include increased stability against nucleases and proteases, the longer residence time in the body, which enables less routinely administrations, and the decreased immunogenicity, which is crucial for nonhomologous proteins like asparaginase and adenosine deaminase. PEGylation's characteristics enabled the development of successful products like PEG-Intron® and PEGasys®. Particularly, the accessibility of PEGs with a wide range of molecular weights and active forms needed to fulfill the various drug-modification requirements (mostly from Nektar

**Table 17.1** List of different surface-decorated nanocarriers

| Type of drug delivery system | Type of modification | Drug used | Advancements | References |
|---|---|---|---|---|
| SLN | Coated with N-carboxymethyl chitosan (NCC) | Curcumin | Improved bioavailability | Baek and Cho (2017) |
| SLN | Coated with chitosan | Ferulic acid, caffeic acid, and resveratrol | Enhanced drug delivery | Ramalingam and Ko (2016a) |
| SLN | Coated with trimethyl chitosan | Curcumin and resveratrol | Improved bioavailability | Ramalingam and Ko (2015) |
| SLN | Coating of N-trimethyl chitosan-palmitic acid | Resveratrol drug | Improved bioavailability | Ramalingam and Ko (2016b) |
| Liposomal drug delivery | Coating of PEG | Coenzyme Q10 | Controlled drug delivery | Yang (2018) |

Therapeutics) made it possible for PEGylation to be used more frequently (Plosker 2008). When PEG is applied to the surface of NPs, electrostatic and hydrophobic interactions are blocked, which lengthens the half-life time of the NPs and increases their selective binding and internalization into target tumor cells or organs. The blood circulation time can be increased 1.5–5 times by incorporating PEG into nanosystems, which also increases plasma half-life (Karra and Benita 2011). Due to their excellent binding capacity and tolerability without producing negative side effects, albumin NPs have attracted a lot of attention. Some benefits of albumin NPs include biodegradability, simplicity in preparation, and reproducibility. In a phase II trial, PTX as albumin-stabilized nanocarrier was one form of nanocarrier used to treat elderly people with locally progressed and metastatic carcinoma of the breast (Abed et al. 2019). PEGylation may also be utilized to enhance the properties of frequently administered anticancer medications with poor solubilities and significant toxicities, such as docetaxel, PTX, and camptothecin. Additionally, the heterogeneous endothelial linings of the cancerous tissues' vasculatures might lead to the buildup of supramolecular substances including polymers and liposomes. To achieve prolonged drug release, PEGylated molecules with elevated molecular weight anticancer drugs are absorbed selectively in cancer tissues (Duncan and Gaspar 2011).

### 17.1.3 Significance of PEGylation

PEG has been widely employed to increase the drug delivery efficacy of therapeutic micro- and nanocarriers. PEG has been incorporated into drug delivery systems to enhance the pharmacokinetics profile of macromolecular drugs as well as particle formulations. PEGylation of therapeutics drug delivery systems has been linked to several kinds of additional beneficial properties, including an increase in the durability of NPs as systems with elevated surface energies tend to gather, an increase in the movement of NPs in biological tissues, such as the mucus and cerebral extracellular space, and an increase in the physicochemical properties (e.g., solubility) and the biocompatibility profiles (reduction). The polymer brushes prevent serum proteins from adhering to them because of the hydrophilic nature of PEG polymers. This artificial nanosphere has been shown to have a lengthy blood circulation period in mice and reduced liver buildup. PEG polymers on the surface of NP will create a substantial and effective hydration shell because they are adaptable and hydrophilic, preventing serum proteins from adhering to the surface. By decreasing opsonization and clearance by the MPS, the grafted PEG polymers on an NP surface can prolong blood circulation time (Carvalho et al. 2020). PEG conjugation is a fast-developing approach that is used as an adjunct to conventional drug delivery techniques to overcome obstacles in therapeutic delivery. In chemical terms, the word "PEGylation" refers to the conjugation of PEG to medicinal compounds to modify them. To expand PEG's current usage, efforts are continuously conducted to create new methods for conjugating it with these compounds. These methods depend on the availability of active PEGylating agents and are particular to the treatment system in use. Therefore, choosing the best procedure for PEGylation requires prior knowledge. An effective PEGylation can change the pharmacokinetic and pharmacodynamic effects of medications after it is accomplished. The main focus is on their capacity to lessen reticuloendothelial system absorption, extend blood residence, lessen metabolic enzyme breakdown, and lessen protein immunogenicity. Numerous clinical studies have been produced as a result of the substantial research in this area (Peracchia et al. 1997).

## 17.1.4 PEGylated Nanomedicine Products

The majority of PEGylated nanomedicine products available on the market occupy the form of therapeutic PEGylated proteins without a nanocarrier. Doxil and Abraxane (PTX-based NPs stabilized with albumin; approved in 2005) stand out as prominent exceptions. These drugs made $402 million and $386 million in revenue, respectively, in 2011, according to Reuters. Celgene Corp. expects that in 2015, Abraxane sales ranged from $1 billion to $1.25 billion.

### 17.1.4.1 Lipid-Based Products

In comparison to conventional doxorubicin, PEG-coated liposomes of doxorubicin (Doxil®, Caelyx™) have a better pharmacokinetics profile and can be administered intravenously over an hour. It is also less frequently linked with neutropenia, alopecia, and cardiotoxicity. In the European Union, the United States, and a few other countries, PEG-coated liposomes of doxorubicin have been approved in combination with the proteasome antagonist bortezomib for the therapy of patients with recurrent or refractory multiple myeloma (Veronese and Pasut 2005). PEGylated liposomal doxorubicin + bortezomib was associated with a greater frequency of moderate to severe adverse events, which was mostly due to the combination's enhanced myelosuppression and gastrointestinal issues. PEGylated liposomal doxorubicin with bortezomib is a new standard of care for patients with recurrent or refractory multiple myeloma who have undergone at least one prior drug (Plosker 2008).

#### 17.1.4.2 Polymer-Based Products

A variety of synthetic and natural polymers, such as PLA, PGA, polyacrylates, polycaprolactone, poly(alkyl cyanoacrylates), polyethyleneimine, and their analogs, have been thoroughly studied for potential applications in the development of nanomedicine. PEG, PLA, and PLGA, as well as their copolymers such as PEG-PGA, PEG-PLA, PLGA, and PEG-PLGA, are potentially the most thoroughly explored synthetic polymers (Karra and Benita 2011).

#### 17.1.4.3 Other PEGylated Platforms

Aurimune (Cytimmune Sciences Inc.), an anticancer nanomedicine, is made up of recombinant human TNF-alpha having dispersed gold NPs with surface-grafted PEG having a size of 27 nm. According to Aurimune's producer, the drug acts preferentially and largely travels in the bloodstream through leaky, freshly formed microvascular at tumor locations, passing through holes in the walls of blood vessels. It is believed that the TNF-a-induced increase in vascular permeability and the modification of penetration in cancer tissue by TNF-alpha-gold NPs offer opportunities to improve the EPR targeting approach The nanomedicine alone may be capable of initiating the treatment process immediately as it enters the region whether the biological target is a plasma membrane-localized receptor or an extracellular targeting (such as an enzyme substrate, such as the situation with some PEGylated enzymes). The carrier must normally be broken down to make the chemical to be biologically active. To achieve the cytosolic access required for their pharmacological effectiveness, the majority of new macromolecular medications (such as siRNA, genes, and proteins) require extra help to cross the plasma or endo/lysosomal wall barriers. On the other hand, if a low molecular weight drug is released extracellularly, it can commonly cross the plasma membrane and reach a therapeutic target inside the cell (Duncan and Gaspar 2011).

### 17.1.5 Current Status and Challenges for PEGylation of Nanocarriers

PEGylation is an efficient way to prolong plasma half-life and decrease early clearance from the body by lowering the opsonization of nanocarriers. Longer circulation times, reduced dosages and frequency, and more site-specific drug delivery are the results of PEGylation, which affords nanocarrier surfaces stealth properties that inhibit immune system cells detection or absorption (Peracchia et al. 1997). To improve tumor-specific targetability, overcome treatment resistance, improve drug pharmacokinetics profiles, and increase anticancer effectiveness, nanocarriers can be therapeutically PEGylated. PEGylated cancer nanomedicines face difficulties, and alternative solutions have been considered (Duncan and Gaspar 2011). Interdisciplinary research is crucial for the development of a logical anticancer regimen. Every vascularized organ and tissue in our body may be accessible to and receive therapeutic payloads from NPs given systemically. Since increasing the retention duration in the blood increases the likelihood that circulating NPs will come into contact with and partition into the desired targets, this method has been acknowledged as the first line of defense. However, the MPS's presence has made this effort difficult, especially in the beginning. The MPS, which is in charge of removing, digesting, and destroying foreign substances in the bloodstream, is made up of dendritic cells, granulocytes, blood monocytes, and tissue-resident macrophages in the liver, spleen, and lymph nodes (Kwilas et al. 2017). Additionally, the creation of nanomedicines may strengthen pharmaceutical companies' market share globally and increase their bottom line in an economically competitive environment. The primary nano transporter liposomes, nanoemulsions, micelles, lipid NPs, polymeric NPs, metallic NPs, and dendrimers, as well as the most recent developments in their application in drug delivery systems for animal use, contribute to this situation (Carvalho et al. 2020).

## 17.2 Methods of PEGylation

There is growing interest in creating injectable drug polymeric carriers that are quickly removed from the bloodstream and undetectable as foreign particles. According to the PEG density, a PEG coating on particles reduced both protein adsorption and complement consumption. When IBCA was emulsified or polymerized, PEG chemical coupling resulted in the formation of NPs (Peracchia et al. 1997). The assessment of NPs includes an examination of their surface characteristics, including the overall quantity of PEG, the surface-grafted PEG chains' hydrophilic nature, and structural flexibility. The rate of IBCA polymer formation in a mixture of PEG or the methyl group containing PEG was further studied. The number of particles in contact with human serum and the PEG conformation were found to have a significant impact on complement uptake (Park et al. 2019).

### 17.2.1 PEGylation Through Self-Assembly

The two primary processes employed by self-assembly to PEGylate nanocarriers are nanoprecipitation (solvent diffusion) and emulsification (solvent evaporation or nanoemulsion). Using nanoprecipitation, PEGylated nanocarriers for specific administration of therapeutic drugs after intravenous injections have regularly been generated. The organic solvent makeup and volume of these PEGylated nanocarriers can vary their physicochemical attributes (particle size, the zeta potential), releasing kinetics, stealthy effects, and targeting efficacy (Cabello 2008).

### 17.2.2 PEGylation by Physical Adsorption

As a component of a physical adsorption procedure, PEG-containing molecules are dissolved in an aqueous solution and contacts with the outermost layer of the nanocarrier are achieved by electrostatic, hydrophobic, or ligand-binding interactions. Pluronic is a form of triblock copolymer composed of PEO-PPO-PEO, where the hydrophobic PPO section is enclosed by two hydrophilic PEO (in this context, PEO is equivalent to PEG) chains. Pluronic molecules can be adsorbed onto hydrophobic nanocarrier surfaces by interactions between the PPO segment and the hydrophobic surface. Electrostatic interactions allow changes in the surface of nanocarriers by coating them with PEG molecules that have opposite charges. For instance, PEG polymers are cation-containing polymers like PEG-PEI and PEG-PLL can be used to bond with negatively charged PLGA nanocarriers and modify them. The customized co-delivery of medication and gene was achieved by PEGylation-aided manipulating liposomal NPs (Wang et al. 2010).

### 17.2.3 PEGylation by Polymeric Conjugation

It is also possible to PEGylate nanocarriers by chemically coupling PEG to the surfaces of manufactured nanocarriers. This method has been used to create a variety of PEGylated nanocarriers, including gold nanocarriers, dendrimers, and polymeric NPs. PEGylated polymeric nanocarriers were created by Nance et al. (2012) to carry the anticancer medication PTX to the tissues of the brain. In comparison to polymeric nanocarriers that are not coated, they found that PEGylation of nanocarriers significantly increases the trafficking of PTX to brain regions. The injection of PEG-lipids must be under the critical concentration of micelle to prevent the development of micelle. Furthermore, PEG-lipids must be added gently at temperatures that are closer to their melting point (Iqbal et al. 2019).

## 17.3 Preclinical and Clinical Status of PEGylated Nanocarriers

Preclinical and clinical trials have been conducted on different PEGylated anticancer nanomedicines, with some of them being authorized for clinical use. Several varieties of PEGylated antitumor nanomedicines are currently in clinical development or have been authorized for clinical use (Hussain et al. 2019). A prospective, multicenter, large-scale clinical trial found that the L-DEP regimen provides an efficient and secure treatment option for refractory EBV-HLH before allo-hematopoietic stem cell transplantation. The overall response rate (ORR) for this trial was 85.7%. L-asparagine is hydrolyzed by PEG-asparaginase, which prevents target cells from synthesizing proteins, DNA, and RNA. This eventually prevents the growth of the target cells. Insufficient asparagine results in cell death in lymphoblasts (Chen et al. 2022; Wang et al. 2016; Schore et al. 2019). By giving PEG-rhG-CSF after each chemotherapy cycle, the incidence of grade 1–4 neutropenia and treatment delay in patients with cervical cancer receiving cisplatin and docetaxel were considerably reduced. High neutrophil counts, rapid recovery, and a low incidence of grade 1–4 neutropenia were all results of using PEG-rhG-CSF. There were no significant side effects of PEG-rhG-CSF treatment (Li et al. 2023). Balugrastim (non-PEGylated HSA-GCSF fusion protein) and PEGfilgrastim (PEGylated GCSF) were evaluated in three clinical studies for effectiveness, safety, and immunogenicity. In addition, two clinical trials revealed that balugrastim was superior to PEGfilgrastim in reducing neutropenia, while a third experiment was still underway and yielding no results (Tao et al. 2021; Gladkov et al. 2016). Even though PEG has no medicinal value, it is no longer considered a basic component. This fact suggests that the associated molecule must be regarded as a new drug after coupling with chemical moiety. It can change how conjugated chemical moiety pharmacokinetics and pharmacodynamics work. PEGylated products consequently form a part of the procedure that will carry out all the requirements for new medication approval. Before being commercialized, the PEGylated product's stability, toxicity, and biological destiny must be understood. A rapidly growing body of research and studies in this area will raise consumer expectations for PEGylated products. EMEA in Europe, FDA in the USA, and MHRA in the UK are the primary sources of guidelines and recommendations for nanomedicines. No drug product is guaranteed to be risk-free; therefore, it is the drug sponsor's job to ensure an appropriate risk/benefit ratio. In addition, other PEGylated products are undergoing clinical testing. To create prospective therapies for clinical usage, PEGylation is a promising method to address a variety of physicochemical and biological issues related to pharmaceuticals. New PEGylated product preparation techniques are a result of technological advancements (Kolate et al. 2014). Several PEGylated drugs for cancer treatment are currently being developed. For instance, topoisomerase I antagonist camptothecin-based drugs have recently been found to be helpful in the management of a range of solid tumors. The hydrophobic properties of such materials continue to have an impact on their therapeutic effectiveness. It was discovered that PEGylation processes of arginine deiminase were a more successful therapeutic approach for the treatment of cancer. The results of an in vivo study demonstrated the superiority of the ADI containing PEG 20,000 formulation, prepared by combining 10–12 chains of 20 kDa PEG with ADI using the succinimidyl succinate linker. Clinical research has demonstrated that ADI PEG 200,000 has greater anticancer activity and tolerability. A phase III clinical trial comparing ADI containing PEG 200,000 to placebo is currently being conducted for advanced hepatocellular cancer. Phase I (for metastatic melanoma in conjunction with cis-platin; for solid tumors in combination with docetaxel) and phase II (for acute myeloid leukemia/non-Hodgkin's lymphoma) are also being tested (Mishra et al. 2016; Holtsberg et al. 2002; Glazer et al. 2010; Ascierto et al. 2005). Nektar Therapeutics created NKTR-102 as a PEGylated formulation employing the multi-arm PEG architecture for the therapy of solid malignancies like cancer of the colon. The topoisomerase I enzyme is the first long-lasting antagonist. During the process of synthesis, irinotecan is attached covalently to a PEG containing four arms. The active breakdown product of NKTR-102 is called SN38 (10-hydroxy-7-ethyl-camptothecin, a camptothecin derivative). In several phase I and phase II studies for NKTR-102, extended exposure to the active ingredient was linked with beneficial antitumor efficacy. The phase III clinical study for NKTR-102 is now being conducted for individuals with metastatic or locally frequent breast cancer. For individuals with solid tumor carcinomas, such as colorectal, lung cancers, ovarian cancer, and small cell lung cancer, a phase II clinical study for NKTR-102 is now underway (Mishra et al. 2016; Hoch et al. 2014; Jameson et al. 2013; Vergote et al. 2013; Awada et al. 2013). The first phase I clinical study for solid tumors has begun with Calando Pharmaceuticals' CALLA 01 nano product (cyclodextrin NPs coupled to transferrin and encapsulated with PEG) (Mishra et al. 2016; Lee et al. 2013). VEGFR-2, or Vascular Endothelial Growth Factor Receptor 2, is a renowned angiogenesis stimulatory protein that is involved in tumor formation and is inhibited by the PEGylated diFab antibody fragment CDP791. The unaltered CDP791 antibody fragment has a short plasma half-life, a low molecular weight, and a poor therapeutic index because it lacks the Fc domain. This problem could be solved by lowering the renal clearance of the natural antibody by PEGylating the cysteine amino acid located at its C-terminus (Mishra et al. 2016; Ton et al. 2007). The MPS cells remove liposomes by identifying opsonins and complement proteins (such as C5b-9 conjugates), attached to the liposome's surface. The topoisomerase I inhibitor Onivyde (liposomal irinotecan), which is based on passive targeting, is the most recent liposomal drug carrier to receive approval. In therapy for metastatic pancreatic cancer, onivyde has received approval. Patients who received onivyde together with fluorouracil and folinic acid had a median survival of 6.1 months as opposed to 4.2 months when irinotecan containing liposomal formulation was added. PEGylated liposomes are also used to transport other medications, for example, Visudyne delivers verteporfin to cure macular degeneration and Ambisome helps to deliver amphotericin-B to cure fungal infections. A few liposome systems are now undergoing clinical trials. Many different medications' pharmacokinetics are changed and enhanced by PEGylation. In order to adapt to the physiological milieu of the targeted location with reduced harmful effects and greater efficacy, recent studies have focused on the production of stimuli-sensitive polymers/smart polymers and polymeric micelles. By shielding them from different cellular or tissue-based deteriorating mechanisms, PEGylation gives therapeutic agents like proteins, enzymes, small-molecule drugs, and liposomes the capacity to increase their retention period, which in turn increases their therapeutic efficacy. PEGylated NPs for systemic applications were first studied in the early 1980s and 1990s as a result of the success of PEGylation of particles for extending circulation periods. In 1995, the first PEGylated nanoparticle product (Doxil) was authorized by the FDA. Doxil is the brand name for PEGylated Liposomal Doxorubicin, a version of the medicine designed to improve the effectiveness of cancer treatment. It has doxorubicin (Adriamycin) inside a PEG-coated, unilamellar liposome that is 80–90 nm in size. Doxorubicin which is PEGylated Liposomal is less harmful to healthy cells than doxorubicin which is not. The chemotherapeutic agent camptothecin has been enclosed in cyclodextrin-PEG copolymers. To encourage the EPR effect and improve targeting, the resulting NPs (CRLX101) are injected intravenously. In phase I and phase II clinical studies, CRLX101 showed good outcomes in patients with peritoneal, ovarian, tubal, and rectal malignancies (Abed et al. 2019; Mishra et al. 2016; Bobo et al. 2016; Clark et al. 2016; Svenson et al. 2011).

### 17.3.1 Preclinical Status

A PEG-coated PTX liposome product demonstrated extended half-life and improved anticancer activity due to excessive accumulation in MDAMB-231 breast carcinoma in nude mice in a preclinical investigation. PEGylation dramatically reduces drug absorption in RES-containing organs while boosting it in tumor tissues, according to biodistribution studies in a breast tumor xenografted nude mice model. Another preclinical investigation found that PEGylated micelles composed of polymers, and even at low doses, the PTX drug increased drug toxicity against the MDA-MB-435 cancer cell line. The MDA-MB-435 carcinoma-bearing nude mice treated with these therapeutic-loaded PEGylated micelles saw enhanced medicine absorption, which significantly slowed the development of the tumor. A preclinical study using docetaxel-loaded PEGylated lipid NPs in mice with subcutaneously implanted C26 colon carcinoma showed a boost in the drug's pharmacological half-life and a considerable deposition at the tumor site. A preclinical study of cisplatin encapsulated in PEGylated NPs of polymer showed a considerable development of tumor inhibition in SCID mice with the HT 29 tumor. Similar to this, PEGylated polymeric nanoparticle-treated cisplatin-treated mice had longer lifespans. Mice's solid tumors (NK911), an additional DOX-encapsulated PEGylated Polymer Micelle System, accumulated effectively (Hussain et al. 2019; Yang et al. 2007; Hu et al. 2008; Khalid et al. 2006; Mattheolabakis et al. 2009). The residency period of high molecular weight (HMW) PEG40 in alveolar macrophages was investigated in the preclinical study, as well as its impact on macrophage activities in vitro. Furthermore, the pulmonary toxicity of a mono-PEGylated Fab′ anti-IL17A (PEG40-Fab′ anti-IL17A or PEG40-Fab′) delivered by intratracheal route in mice was studied in acute and chronic tests. During 12 weeks, mice were intratracheally treated to one treatment per week with PEG40-Fab′ anti-IL17A, Fab′ anti-IL17A, PEG40 alone, or PBS to assess possible chronic toxicity. Mice showed no significant difference in weight increase after 12 weeks of therapy. Results show that intratracheal administration of PEG40 or PEG40-Fab′ anti-IL17A has a favorable safety profile, indicating that no evident toxicity is anticipated when these compounds are supplied orally. Furthermore, we demonstrate that pulmonary administration of PEG40 does not harm macrophages and that alveolar macrophages are gradually removed from PEG40, indicating that accumulation in the lungs should be watched in cases of chronic administration, as intended for PEGylated antibody responses treatment, regarding the interval between consecutive doses. Before concluding that HMW PEG inhalation is safe, more research employing GLP-produced PEGylated antibodies in rodents and non-rodents is required (Freches et al. 2019). High molecular weight PEG compounds utilized in the mono-PEGylation of therapeutic proteins do not exhibit preclinical toxicity. Given that BAY 94-9027's expected clinical dose only contains a very small quantity of PEG, data suggest that long-term chronic treatment with a 60 kDa PEGylated rFVIII would appear to be safe and shouldn't cause PEG-related side effects. To substantiate these findings, additional clinical trial results

**Table 17.2** Preclinical study of PEGylated products

| S. No. | Nanocarrier type | Status | References |
|---|---|---|---|
| 1 | Liposome | Preclinical phase | Yang et al. (2007) |
| 2 | Polymeric micelle | Preclinical phase | Hu et al. (2008) |
| 3 | Lipid nanoparticle | Preclinical phase | Khalid et al. (2006) |
| 4 | Polymeric nanoparticle | Preclinical phase | Mattheolabakis et al. (2009) |
| 5 | Antibody fragment | Preclinical phase | Freches et al. (2019) |
| 6 | Recombinant F-VIII | Preclinical phase | Ivens et al. (2013) |

must be obtained. Additionally, more study is required to determine the comparative safety of various biologics and the long-term safety of modified coagulation products. There are some PEGylated products enlisted in Table 17.2 which are under preclinical studies (Ivens et al. 2013).

### 17.3.2 Clinical Status

Regarding the fact that PEG lacks medicinal value, it is no longer considered a simple component. This means that after attaching to a chemical moiety, the resulting conjugate must be treated as a new chemical entity. It has the potential to change both the pharmacodynamics and pharmacokinetics of linked chemical moiety. As a result, PEGylated product turns into a component of the process that will carry out all of the approval stages required for novel drugs. Before a PEGylated product can be commercialized, its stability, toxicity, and biological fate must be known. A rapidly expanding variety of studies as well as research in this particular area will raise market expectations for PEGylated products. The EMEA in Europe, the FDA in the United States, and the MHRA in the United Kingdom provide the majority of the guidelines and recommendations in the field of nanomedicines. It is widely acknowledged that no pharmacological product is without risk, and it is the medication sponsor's job to ensure an appropriate risk/benefit balance. Furthermore, a huge variety of PEGylated compounds are being tested in clinical trials (Kolate et al. 2014).

#### 17.3.2.1 Clinical Phase I and Phase II Trials

PEGylated formulations (PEG-liposomal Doxorubicin, CaelyxTM, PEG Dox) enhanced blood-brain barrier penetration. One of the most widely used effective drugs for glioblastoma cells is doxorubicin. However, due to inadequate blood-brain barrier penetration, it has little to no effect in vivo. When effectively stabilized liposomes were employed in a tumor model, the concentrations of doxorubicin in the tissue and CSF were significantly raised. This led to a comparable clinical response with only about half the dose of stabilized liposomes needed in comparison with traditional doxorubicin (Beier et al. 2009). Phase II study was done using a fixed dose of PEG-Filgrastim, and an increasing dose of GM-CSF (Leukine) up to 250 mcg, the optimal dose of CHOP-R, and PEG-Filgrastim should be determined. If it is safe to do so, the dose of Neulasta will be gradually increased once that level is reached. Every 14 days, CHOP-R will be administered at a fixed standard dose, with PEG-Filgrastim and GM-CSF doses being adjusted upwards and downwards by the minimum and maximum blood counts (Vassilakopoulos et al. 2012). Phase II clinical trial of Carboplatin-PEGylated Liposomal Doxorubicin (PLD) was done in ovarian cancer patients in which high IL-6 levels are associated with a poor prognosis and chemoresistance. To overcome drug resistance and boost host immunity in patients with recurrent ovarian cancer, modulating the IL-6 pathway by inhibiting the IL-6 receptor may be a promising approach in cases of chemo-resistant ovarian cancer. The local normal DC maturation may be restored and immunogenic cell death may be enhanced by blocking the IL-6/IL-6R pathway. The full maturation of DC is also made possible by the use of interferon-alpha (PEG-Intron), which improves the antitumor response (Dijkgraaf et al. 2015). The FDA IND approved the use of newer generation cancer-selective C dots as a single-(optical) or dual-modality (optical-PET) platform. These C dots have prolonged blood residency half-times (hours) and show significant renal clearance. Their special physicochemical and photophysical properties have been used in several minimally invasive and/or surgical applications. A clinically authorized integrated technological platform for viewing and molecularly characterizing regional tumor-draining lymph nodes inside the nodal basin of melanoma models has been created by integrating with a real-time handheld fluorescence camera system (Bradbury et al. 2016).

#### 17.3.2.2 Clinical Phase III Trials

Phase III clinical study of newly diagnosed patients with acute lymphoblastic leukemia was done to compare the clinical benefit, the pharmacokinetics, and the pharmacodynamics of PEG-conjugated asparaginase given in higher dose versus conventional dose during the continuation phase (Omar et al. 2022). In phase I, II, and III clinical investigations, lipopolytin significantly decreased the renal toxicity, ototoxicity, myelotoxicity, peripheral neuropathy, nausea/vomiting, and asthenia of cisplatin with improved or comparable effectiveness to cisplatin. Patients' solid tumors accumulated Lipoplatin 10- to 200-fold more in clinical trials than the surrounding normal tissue did. Lipoplatin was postulated to serve as a double-edged sword: as chemotherapy and an antiangiogenesis medicine. Lipoplatin's targeting of tumor vasculature in animals revealed its antiangiogenesis potential. When used in conjunction with PTX as first-line therapy for non-small cell lung cancer, lipopolytin has successfully completed one phase III non-inferiority clinical research, demonstrating a statistically significant decrease in nephrotoxicity (Boulikas 2009). The PEGylated products that are under clinical investigation along with their current status are summarized in Table 17.3.

### 17.3.3 Challenges Involved in Clinical Testing of PEGylated Nanoproducts

The PEGylation method has shown promise in enhancing the PK/PD profiles of chemotherapeutic drugs and reducing their toxicity, according to numerous studies

**Table 17.3** PEGylated products under Clinical trials

| S. No. | Study title | NCT number | Phase | Interventions | Status | Study completion | References |
|---|---|---|---|---|---|---|---|
| 1. | PEGylated liposomal doxorubicin and prolonged temozolomide in addition to radiotherapy in newly diagnosed glioblastoma | NCT00944801 | Phase 1<br>Phase 2 | Drug: PEGylated liposomal doxorubicin | Completed | 2009 | Beier et al. (2009) |
| 2. | Cyclophosphamide, doxorubicin, vincristine, prednisone, rituximab patients with aggressive NHL | NCT01527422 | Phase 1<br>Phase 2 | Drug: cyclophosphamide, doxorubicin, vincristine, and prednisone | Completed | 2011 | Vassilakopoulos et al. (2012) |
| 3. | Combination Study of SB-485232 (interleukin 18) and doxil for advanced stage epithelial ovarian cancer | NCT00659178 | Phase 1 | Drug: SB-485232 (interleukin 18), PEGylated liposomal doxorubicin | Completed | 2011 | Simpkins et al. (2013) |
| 4. | EGEN-001 and PEGylated liposomal doxorubicin hydrochloride in treating patients with recurrent or persistent ovarian epithelial cancer, fallopian tube cancer, or primary peritoneal cancer | NCT01489371 | Phase 1 | Other: Laboratory biomarker analysis<br>Biological: PEG-PEI-cholesterol lipopolymer-encased IL-12 DNA plasmid vector GEN-1<br>Drug: PEGylated liposomal doxorubicin hydrochloride | Completed | 2018 | Thaker et al. (2017) |
| 5. | PEG-aspargase and combination chemotherapy in treating younger patients with newly diagnosed high-risk acute lymphoblastic leukemia (closed to accrual 4-22-2011) | NCT00866307 | Phase 1 | Drug: cyclophosphamide<br>Drug: cytarabine<br>Drug: daunorubicin Hydrochloride<br>10 more | Completed | 2021 | Rodriguez et al. (2016) |

(continued)

| | | | | | | | |
|---|---|---|---|---|---|---|---|
| 6. | Feasibility of the combination of chemotherapy (carbo/caelyx or carbo/doxorubicin) with tocilizumab (mAb IL-6R) and PEG-Intron in patients with recurrent ovarian cancer | NCT01637532 | Phase 1<br>Phase 2 | Drug: tocilizumab and interferon alpha 2-b<br>Drug: carboplatin and caelyx or doxorubicin | Completed | 2013 | Dijkgraaf et al. (2015) |
| 7. | Targeted silica NPs for real-time image-guided intraoperative mapping of nodal metastases | NCT02106598 | Phase 1<br>Phase 2 | Drug: fluorescent cRGDY-PEG-Cy5.5-C dots | Recruiting | 2024 | Bradbury et al. (2016) |
| 8. | Total therapy study XVI for newly diagnosed patients with acute lymphoblastic leukemia | NCT00549848 | Phase 3 | Drug: prednisone, vincristine, daunorubicin, PEG-L-asparaginase, erwinia L-asparaginase, doxorubicin, cyclophosphamide, cytarabine, thioguanine | Complete | 2022 | Omar et al. (2022) |

(Hussain et al. 2019; Bocca et al. 1998). The PEGylation method led to a notable increase in half-life, selective entry into cancer cells, and lower plasma clearance because RES and MPS systems were less effective at recognizing and clearing the substance from the body (Howard et al. 2008). Despite these accomplishments, the PEGylation method is still hampered by several obstacles and constraints that prevent its widespread use in the design and production of anticancer nanomedicines. By preventing drug recognition and eventual clearance by RES or MPS systems, PEGylation of nanocarriers lengthens plasma half-life. It also enhances medication penetration into cancer cells through leaky vasculature via EPR effects. However, other research has shown that the hydrophilic layer of the outer PEG coating also tends to greatly reduce the interaction of PEGylated nanocarriers with tumor cells and other biological tissues, resulting in inadequate cellular absorption and anticancer effectiveness (Hatakeyama et al. 2007; Chen et al. 2017). Actively targeted nanocarriers that depend on ligand-receptor interaction to produce clinical advantages may find this issue to be more difficult. PEG on the surfaces of nanocarriers prevents surface-grafted ligands from interacting with biological targets (receptors, biomarkers) that are overexpressed on the surfaces of tumor cells and tissues (Vllasaliu et al. 2014). Newer design nanocarriers, also known as sheddable nanoparticle systems, have been created on the theory of microenvironment-triggered PEG cleavage from the nanocarrier's surface to solve these issues. The outer PEG layer protects the hydrophobicity and positive charge of the core during their homing into plasma circulation, keeping the nanocarriers stable and extending the duration of blood circulation. Upon reaching the tumor site, the stimuli in the tumor microenvironment cause the linker between PEG and the nanoparticle core to cleave, which causes PEG to detach/desorb from the nanocarrier's surface and expose them to cancer tissues for the best possible interaction and enhanced antitumor effects (Perche et al. 2014; Dong et al. 2015; Xu et al. 2016; Sun et al. 2015). Another problem with PEGylated nanomedicines is their immunogenicity. It has been determined that complement activation-related pseudo allergy (CARPA) is brought on by liposomes made of anionic phospholipid-PEG conjugates. While complement activation in this case was at least partially attributable to the highly cationic region of the globular C1q protein attaching to the anionic charge on the phosphate oxygen of the lipid-PEG conjugate through electrostatic contact (Moein Moghimi et al. 2006), another study described PEG-induced complement activation that happens via traditional (C1q-dependent), lectin, or other routes and is concentration- and molecular weight-dependent. Anti-PEG antibody reactions are also capable of being induced by PEG (Hamad et al. 2008). PEGylated liposomes have been demonstrated to produce IgM antibodies, which result in the post-repeat ABC phenomenon. Numerous studies have shown that PEGylated liposomes quickly leave the blood after repeated injections. It was discovered that the ABC of the second dosage of PEGylated liposomes is the reason for the binding of anti-PEG IgM, which was created by the first dose of liposomes and led to the complement system's subsequent activation (Ishida et al. 2006, 2003; Pacheco et al. 2023; Kamaly et al. 2012). The amount of PEG present on the surface of a nanoparticle affects both the complement activation and anti-PEG IgM production. To eliminate unwanted immunological responses, PEG density and molecular weight optimization on the surface are essential. Additionally, it appears that other properties of nanocarriers, such as surface charge and the type of medication contained, are related to the development of anti-PEG IgM. Therefore, adjusting the dosage, PEG molecular weight and density, and surface charge of nanocarriers are all effective ways to reduce the immune response against PEG. The addition of a substance that reduces macrophage activity would also be beneficial. The biodegradability of PEGylated nanomedicines and the possibility for nanocarrier buildup in the lysosomes after prolonged treatment are additional concerns about their safety. The adjustment of dosage, PEG molecular weight, and density, as well as a surface charge of nanocarriers, are thus techniques to reduce the anti-PEG immune response. It would also be advantageous to include a substance that reduces macrophage activity (Duncan and Gaspar 2011). According to certain preclinical research, the kidneys generate organ-specific intracellular vesicles containing PEG (Bvacuolation), which is related to the way PEGylated proteins are cleared from the body (Webster et al. 2007). Furthermore, prolonged circulation of PEGylated nanomedicines, made possible by PEG on the surface of nanocarriers (Markovsky et al. 2012), may increase exposure to normal tissue, which could result in unforeseen adverse effects (as is thought to have happened with the Bhand-and-foot syndrome after DOX delivery) (Robert et al. 2004).

### 17.3.4 Reasons Behind Failure in Clinical Testing of PEGylated Nanocarriers

Lack of clinical efficacy, uncontrollable toxicity, substandard drug-like qualities, a lack of commercial demands, and inadequate strategic planning are four potential causes of clinical research failure. Polydispersity is exhibited by PEGylated medicines. Therefore, proper synthesis and refinement control must be established to ensure product uniformity, which in turn ensures reproducibility. It is important to thoroughly analyze the raw materials that are used (PEG and protein), ensure the consistency of PEGylated pharmaceuticals, and evaluate the production and derivatization methods. When a business creates a PEGylated variant of an existing medication, it must show that the PEGylated version is more affordable than the original medication (Park et al. 2019). Pharmacoeconomic studies showed that PEGaspargase, PEGinterferon, and PEGfilgrastim were inexpensive. PEGaspargase is a treatment for ALL that works just as well as L-asparaginase but costs much more. However, it has a 357-h half-life and causes less hypersensitivity. According to a cost-minimization analysis using a standard adult ALL procedure, PEGaspargase is more economical than L-asparaginase. PEGamotecan (PEG-camptothecin) was discontinued from a Phase IIb trial in 2005 due to a lack of efficacy (Graham 2003). Cytarabine liposomal terminated due to the resignation of the trial's original investigator. With aggressive chemotherapy and cytarabine liposome (Depocyt®) CNS prophylaxis, the goal of this treatment is to prolong survival for individuals with chronic lymphoblastic leukemia while minimizing systemic and CNS recurrence with tolerable toxicity. Due to an upsurge in bacterial infections, a clinical trial using PEGaspargase and combination chemotherapy to treat those suffering from newly identified cases of acute lymphoblastic leukemia was stopped.

### 17.3.5 Commercially Available PEGylated Nanocarriers

PEGylated products have a well-established effectiveness in the pharmaceutical and biomedical sectors. The first PEGylated protein for the treatment of immunogenicity syndrome to get USFDA approval was bovine adenosine deaminase (marketed under the trade name Adagen® by Enzon Pharmaceuticals). Following Adagen®, a sizable number of pharmaceutical products using proteins and peptides have been introduced. Many PEGylated proteins and delivery technologies are in clinical trials or, as previously said, are still in the developing phases (Kolate et al. 2014).

## 17.4 Regulatory Considerations for PEGylated Nanocarriers

PEGylated nanocarriers are suitable for a wide range of FDA-regulated products, including human drug products and biological products (FDA 2022; USCODE 1970). Usually, the FDA's guidance statements do not define obligations that are enforceable by regulations. Instead, unless specific regulatory or statutory requirements are stated, guidance should only be considered as recommendations. It reflects the Agency's current perspective on an issue (FDA 2022). In Agency guidelines, the word should denote anything that is advised or recommended but not mandated (FDA 2022). To overcome the therapy's limitations, such as potential low solubility, high toxicity, undesirable pharmaceutical characteristics, and a nonspecific biodistribution profile, nanocarrier technology may be employed to manufacture PEGylated pharmaceuticals (Li et al. 2013). The FDA has approved eleven PEGylated biomolecules for use in the treatment of associated diseases (Li et al. 2013). PEGylation technology is typically used with macromolecular drugs, such as proteins and enzymes (Li et al. 2013). The most effective method of chemically modifying these biopharmaceuticals to date has been PEGylation (Ivens et al. 2015). PEG is a molecule that can be added to proteins to increase stability, prolong half-lives, and, in some cases, lessen the immunogenicity of highly immunogenic proteins like enzymes (Ivens et al. 2015). As of 2015, 10 peptides/proteins and 1 aptamer were among the 11 PEGylated biopharmaceuticals that were approved in the United States and Europe (Ivens et al. 2015). Many more are currently undergoing nonclinical and clinical testing. We'll focus on the following issues in particular (Ivens et al. 2015):

- Regulations that support clinical development, PEG disposition and metabolism, and excretion are advised by health authorities.
- Immunogenicity of PEG.
- A concise examination of the toxicological information for 11 PEGylated biopharmaceuticals that have obtained regulatory approval.
- The potential risk for the patient population posed by cellular vacuolation seen in nonclinical toxicology research, as well as its toxicological relevance.
- 2013 Biosafe survey of the pharmaceutical sector, which included nonclinical case studies for particular 17 items from 10 businesses before approval.

- Factors to consider when evaluating the nonclinical safety of PEGylated biopharmaceuticals.

The first liposomal formulation for the treatment of cancer was approved, and it contained doxorubicin-loaded PEGylated liposomes (Gaspar et al. 2014). Additionally, Doxil® and Caelyx®, the first PEGylated formulations, received official clearance from the USA and the European Union in 1994/1995 (Gaspar et al. 2014). These formulations included a long-circulating liposomal formulation coated with PEG2000 (hydrogenated soy phosphatidylcholine, cholesterol, MPEG2000-distearoyl phosphatidylethanolamine, 56:39:5) and containing Doxorubicin hydrochloride that was enclosed/entrapped in the internal aqueous space (Gaspar et al. 2014). The treatment indications for this PEGylated Liposomal Doxorubicin include Kaposi sarcoma, breast and ovarian cancer, and included Distearoyl Phosphatidylcholine, Cholesterol, and MPEG2000-Distearoyl Phosphatidylethanolamine (56:39:5) (Gaspar et al. 2014). According to reports, medicinal products with PEGylated nanocarriers could help with the following:

1. Formulation difficulties for poor solubility drugs
2. In vivo and in vivo drug solubility issues
3. Pharmacokinetic and biodistribution-related issues
4. Efficacy

Every one of which could potentially minimize toxicity (Zolnik and Sadrieh 2009; Couvreur and Vauthier 2006).

### 17.4.1 Current Regulatory Status of PEGylated Nanocarriers

The book chapter examines regulatory actor's perspectives on challenges posed by PEGylated nanocarriers and their corresponding governance responses (Dorbeck-Jung and Chowdhury 2011). Key stakeholders, including authorities, pharmaceutical companies, scientists, and organizations like ISO, SCENIHR, and EGE, are actively addressing regulatory issues (Dorbeck-Jung and Chowdhury 2011). The analysis questions the appropriateness of current governance responses and their effectiveness in addressing identified challenges (Dorbeck-Jung and Chowdhury 2011). As the European medicinal product regulation regime influences another regulatory framework for medical technology, responses to nano pharmaceuticals may indicate the future direction of PEGylated nanocarrier product regulation (Dorbeck-Jung and Chowdhury 2011; Parvizi and Woods 2014). The study relies on literature, legal, and policy documents (Dorbeck-Jung and Chowdhury 2011).

#### 17.4.1.1 Development of PEGylated Nanoproducts

Nanopharmaceuticals are a subset of the PEGylated nanocarriers therapeutic field, which encompasses both "nano pharmaceuticals" and "nano-enables devices" (Dorbeck-Jung and Chowdhury 2011; Farjadian et al. 2019). PEGylated nanocarriers entail advancements in drug delivery and medication utilizing NPs containing the active ingredient (Dorbeck-Jung and Chowdhury 2011). These PEGylated nanocarriers can also serve as carriers or porous materials for the controlled release of the active component (Dorbeck-Jung and Chowdhury 2011). Ongoing developments are focused on addressing conditions such as cancer, Parkinson's disease, and Alzheimer's disease (Dorbeck-Jung and Chowdhury 2011; Franzè et al. 2020).

#### 17.4.1.2 Potential Benefits

Nanopharmaceuticals are highly promising in PEGylated nanomedicine (Dorbeck-Jung and Chowdhury 2011). The controlled release system is a key advantage, with surface modifications enhancing drug effectiveness (Dorbeck-Jung and Chowdhury 2011). Cancer, infections, autoimmune illnesses, and inflammation are some of the diseases that nano drugs may be able to prevent, diagnose, and cure (Dorbeck-Jung and Chowdhury 2011). They might be used to address conditions that are currently treated with high side effects of drugs (Dorbeck-Jung and Chowdhury 2011). Nanomedicines are expected to minimize adverse effects by precise targeting and reduced drug dosage requirements (Dorbeck-Jung and Chowdhury 2011; Farjadian et al. 2019).

#### 17.4.1.3 Regulatory Challenges

PEGylated nanotechnologies present a challenge to existing regulatory frameworks, as they require tailored rule implementation for new technological advancements (Dorbeck-Jung and Chowdhury 2011). The rapid advancements in nanomedical science, applications, and risk assessment make it difficult for regulators to keep up (Bobo et al. 2016; Dorbeck-Jung and Chowdhury 2011; Forloni 2012). The skill of regulators in balancing the advantages and risks of PEGylated nanotechnologies is put to the test, as is their capacity to win over stakeholders' trust, including the general public (Dorbeck-Jung and Chowdhury 2011). The uncertainties surrounding nanotechnologies encompass technological, scientific, normative, conceptual, and institutional aspects (Dorbeck-Jung and Chowdhury 2011).

Insufficient knowledge exists regarding the trajectory of technological advancements, the behavior of nanomaterials, and data on exposure and hazards (Dorbeck-Jung and Chowdhury 2011). Knowledge gaps extend to product formulations and nanoparticle concentrations, questioning the applicability of European regulations such as REACH (Dorbeck-Jung and Chowdhury 2011). Uncertainty surrounding the risks of nanotechnologies raises the question of whether regulators should wait for more knowledge or respond promptly (Dorbeck-Jung and Chowdhury 2011). According to the European Commission's established precautionary principle, action should be taken even in the face of scientific uncertainty where there is a chance that it could have severe negative impacts (Dorbeck-Jung and Chowdhury 2011).

The perception and interpretation of adverse results, risk-benefit evaluations, public trust, and acceptance of regulation have all been challenged by conflicting standards and ideals (Dorbeck-Jung and Chowdhury 2011). Institutional uncertainties result from ignorance of the costs and efficacy of dynamic, fragmented, multilevel, and multi-actor regulation (Dorbeck-Jung and Chowdhury 2011). There are major knowledge gaps about the potential risks of nanomaterials to human health and the environment, according to scientific bodies of the European Union, member states, the US, and the OECD (Dorbeck-Jung and Chowdhury 2011).

Nanomaterials' potentials to cross blood-brain barriers, cling to capillaries, affect the immune system, and be consumed by macrophages are all toxicological unknowns (Dorbeck-Jung and Chowdhury 2011). Biocompatibility concerns arise when PEGylated NPs are used in medical products or materials (Dorbeck-Jung and Chowdhury 2011). Toxicologically, it is unknown whether nanomaterials can penetrate the blood-brain barrier, stick to capillaries, impact the immune system, or be ingested by macrophages (Dorbeck-Jung and Chowdhury 2011).

In summary, the regulatory challenges posed by PEGylated nanotechnologies stem from the need to address uncertainties in various domains and bridge knowledge gaps related to toxicological risks, biocompatibility, and the implications for human health and the environment (Dorbeck-Jung and Chowdhury 2011).

#### 17.4.1.4 European Regulatory Framework for Approval of PEGylated Nanoproducts

The pharmaceuticals regime in the EU, which covers PEGylated Nanomedicines, is a thorough and complex system of multilayer control. It comprises both European and domestic law, as well as a sizable body of nonbinding non-statutory law (Dorbeck-Jung and Chowdhury 2011; EMA 2010a). With business and national governments playing a prominent role in its execution, the EU medical devices legislation allows for broader engagement of private authorization bodies and private standards as compared to drug regulation (Dorbeck-Jung and Chowdhury 2011).

Drug catastrophes like the thalidomide crisis led to the creation of European rules in the 1960s to ensure a high level of public health protection in terms of safety, efficacy, and quality (Dorbeck-Jung and Chowdhury 2011). Two new licensing methods, the "centralized procedure" and the "mutual recognition procedure," were introduced in 1993 as a result of institutional changes intended to improve scientific evaluation for high-technology pharmaceutical items, particularly those produced through biotechnological processes (Dorbeck-Jung and Chowdhury 2011; Communities 1993).

Regulation 726/2004, which broadened the centralized procedure's application to new active chemicals for particular medicinal reasons, brought forth more advancements in 2004. The examination of potential environmental impact for pharmaceutical products was introduced by Directive 2004/27/EC, requiring applicants to evaluate and resolve environmental hazards associated with their medicines (Dorbeck-Jung and Chowdhury 2011; European Medicines Agency 1994, 2004a, b; CHMP 2006).

Regulation 1394/2007/EC on advanced treatments medical products (ATMPs) was implemented in 2007 in response to developments in biomedicine, and it created the Committee on Advanced treatments inside the EMA to give scientific expertise during the authorization process (Dorbeck-Jung and Chowdhury 2011; European Medicines Agency 2022a). Specific regulatory provisions also exist for Medicinal Products for Pediatric Use and Orphan Medicinal Products (Dorbeck-Jung and Chowdhury 2011; European Union 2006).

Overall, the EU pharmaceuticals regime is a dynamic framework with ongoing amendments and adaptations to accommodate technological advancements and ensure the safety, efficacy, and quality of medicinal products, including PEGylated nanomedicines (Dorbeck-Jung and Chowdhury 2011; European Parliament 2019).

#### 17.4.1.5 Rules Governing Nanomedicines

In the European Union, PEGylated nanomedicines are currently regulated under the existing regime applicable to medicinal products. Specific rules for nanomedicines have not been established (Dorbeck-Jung and Chowdhury 2011). The centralized procedure and the mutual recognition processes specified in the medical product regulation are both used in the market authorization process for nanodrugs (Dorbeck-Jung and Chowdhury 2011).

For nano pharmaceuticals classified as advanced therapy medicinal products, ATMPs or medical device approval can only be obtained through the centralized procedure (Dorbeck-Jung and Chowdhury 2011; MHRA 2020; EMA, CAT, CHMP 2019; FDA 2020). Depending on how the features are incorporated into the product, the regulatory regimes for medical devices and pharmaceuticals may concurrently apply when PEGylated Nano Medicines, also known as "Borderline Products," exhibit characteristics of both Medical Devices and Medicines (Dorbeck-Jung and Chowdhury 2011).

The major mechanism of action of the product will determine the regulatory framework that applies. The regulatory framework for medical devices covers nanomedicines having a primary mechanical action and a secondary pharmacological activity, and authorization is given by a notified body designated by a member state based on conformity assessment and certification (Dorbeck-Jung and Chowdhury 2011; Contardi 2019). The integrated method, however, becomes necessary when advanced treatments and pharmaceutical products are combined with devices, and the EMA takes into account the notified body's evaluation regarding the safety standards for the device part (Dorbeck-Jung and Chowdhury 2011; European Parliament 2007).

To obtain market authorization, a product must be proven to be safe, high-quality, and effective using extensive scientific evidence (Dorbeck-Jung and Chowdhury 2011; European Parliament 2007). Regarding investigational medicinal products, it is crucial to adhere to the rules for good manufacturing application, good clinical procedure, and good manufacturing execution (Dorbeck-Jung and Chowdhury 2011). Further limitations apply to products containing genetically modified organisms, human blood and plasma, human tissues, and orphan medicinal products, as well as to ATMPs, pediatric, and orphan medical products (Dorbeck-Jung and Chowdhury 2011; EMA 2006, 2003, 2022).

In summary, the applicable rules for marketing approval of PEGylated nanomedicines depend on the type of nanomedicine and its classification as well as the combination of medical devices and medicinal products (Dorbeck-Jung and Chowdhury 2011).

#### 17.4.1.6 Significant Regulatory Factors

The regulatory process of PEGylated nanomedicines in the European Union involves various regulatory factors operating at different levels and with different degrees of influence (Dorbeck-Jung and Chowdhury 2011). Based on their level of activity, these factors can be divided into four groups: highly active, moderately active, slightly active, and dormant (Dorbeck-Jung and Chowdhury 2011).

EMA, member state competent authorities, notified bodies, the European Commission, the European Parliament and Council, and the EDQM are among the influential organizations that make up the first group, regulatory bodies (Dorbeck-Jung and Chowdhury 2011). The EMA has been moderately active, organizing regulatory workshops and increasing its role in the regulation of PEGylated Nanomedicine. Competent authorities of member states have taken few regulatory actions, and notified organizations have remained inactive up to this point despite being expected to play a significant role in the regulation of combination products (Dorbeck-Jung and Chowdhury 2011).

The DG SANCO (Director General for Health and Consumers) within the European Commission has been moderately active, organizing workshops and funding research activities related to nanomedicine. On the other hand, the European Parliament, Council, and EDQM have not been involved in nanomedicine regulation, thus classified as dormant (Dorbeck-Jung and Chowdhury 2011; EMA 2008).

Large, medium-sized, and small pharmaceutical enterprises are included in the second group, which consists of manufacturers of nanomedicines and their organizations. While medium-sized companies have demonstrated low regulatory activity, large organizations have been relatively active in contacting regulatory bodies (Dorbeck-Jung and Chowdhury 2011). The European Technology Platform on Nanomedicine plays a role in coordinating nanomedicine development and regulatory policy discussions, making it marginally active at present (Dorbeck-Jung and Chowdhury 2011).

Users who might have an impact on regulatory actions but have not yet engaged in discussions on PEGylated nanomedicine regulation include associations for doctors and patients (Dorbeck-Jung and Chowdhury 2011). Similar to inform bodies, it is anticipated that this group will play a bigger role in the future (Dorbeck-Jung and Chowdhury 2011).

The European Group on Ethics in Science and New Technologies, the Scientific Committee on Emerging and Newly Identified Health Risks, and the N&ET Working Group make up the fourth group, consultative bodies (Dorbeck-Jung and Chowdhury 2011). The N&ET Working Group and SCENIHR have been quite active, contributing reports and viewpoints on the health hazards and legal issues associated with nanotechnologies. The EGE's influence has been marginal, as it focuses on providing ethical advice rather than long-term engagement in specific issues (Dorbeck-Jung and Chowdhury 2011).

Other regulatory organizations, including the WHO, the International Standards Organization (ISO), the Organization for Economic Co-operation and Development

(OECD), and the International Conference on Harmonization of Technical Requirements for Registration of Pharmaceuticals for Human Use (ICH), have made significant contributions to setting standards and allocating resources to nanotechnology. These organizations have provided leadership that has been accepted and recognized by other regulations (Dorbeck-Jung and Chowdhury 2011).

In conclusion, the functional relationship between regulatory factors and regulation determines how influential they are in the regulation of PEGylated nanomedicines, not the level of regulation. Different regulatory factor groups have varying levels of activity, with some being very active, moderately active, barely active, or inert.

### 17.4.1.7 Issues in Regulation and Government Response

#### Issues in Regulation

The interviewees brought out several regulatory issues with PEGylated nanomedicines. These problems include the absence of a precise definition of nanomedicines, pointless regulatory hype, deficient safety, quality, and efficacy criteria, and a lack of suitable techniques to evaluate potential environmental implications (Dorbeck-Jung and Chowdhury 2011). Additional issues, such as the regulatory regime's applicability and the absence of criteria specifically designed for medical device assessment, have been observed in the case of combination items that contain both medical devices and pharmaceuticals (Dorbeck-Jung and Chowdhury 2011).

However, only a small number of distinct regulatory issues have been faced by the majority of prominent regulatory players active in nanomedicine regulation (Dorbeck-Jung and Chowdhury 2011). Because assessment companies considered nanomedicines as conventional pharmaceuticals, business representatives did not report any nano-related issues with the market authorization process for such products. The breadth and nature of these regulatory issues are yet unknown, despite rising awareness of questions and challenges connected to nanodrugs (Dorbeck-Jung and Chowdhury 2011). Some people who participated in the interview voiced concern that the theory of regulatory difficulties can fuel unwarranted regulatory hype, leading to the construction of a particular regulatory framework that might not be required (Dorbeck-Jung and Chowdhury 2011).

Most interviewees felt that the current European criteria for the authorization of medicinal products are strong enough to address nanomedicines about the evaluation of nanomedicines (Dorbeck-Jung and Chowdhury 2011). The existing risk-benefit balance approach and the evaluation criteria were considered sufficient by representatives of authorization bodies. Clinical testing procedures using in vitro and in vivo animals were deemed to be reliable for verifying the safety of products (Dorbeck-Jung and Chowdhury 2011; Park et al. 2009). Since none of the approved nanomedicines experienced substantial risk, quality, or efficacy issues attributable to their "nano" qualities, no specific review criteria were judged essential for them (Dorbeck-Jung and Chowdhury 2011).

However, some challenges were identified, such as the need for improved characterization of nanomedicines, particularly liposomal particles. The term "nanomedicines" was used to describe the activities of ISO Technical Committee 229

(Dorbeck-Jung and Chowdhury 2011). Medium- and large-sized companies faced difficulties with environmental impact analysis since it showed a lack of capacity to evaluate environmental impact. Some evaluation agencies mentioned that major pharmaceutical corporations have conducted environmental risk assessments in their arguments. Instead of rejecting authorization requests due to risk to the environment, the focus was on demonstrating strategies to reduce potential environmental damage.

For combined products that have both pharmacological, immunological, or metabolic action and mechanical action, a classification problem was identified (Dorbeck-Jung and Chowdhury 2011; Novelli et al. 2008). The current principle of applying the regulatory regime based on the primary mode of action was considered too simplistic for complex multifunctional nano pharmaceuticals (Dorbeck-Jung and Chowdhury 2011; European Commission 2011). Companies sought clarity from competent bodies regarding the applicable regulatory regime for such borderline products, as uncertainty could harm the sector's reputation (Dorbeck-Jung and Chowdhury 2011). Furthermore, it was noted that the current guidelines for medical device assessment lack specific standards validated for NPs, and the voluntary nature of scientific advice requested from drug evaluation bodies by notified bodies has led to instances where it was ignored without sanctions (Dorbeck-Jung and Chowdhury 2011).

In summary, regulatory problems identified in the interviews revolved around issues such as definition, hype, standards, environmental impact analysis, and the regulatory regime for combined products. While some challenges were acknowledged, the current regulatory framework was generally deemed sufficient by evaluation bodies, with the need for improved characterization standards and specific guidelines for medical device assessment highlighted as areas for improvement.

#### Government Responses

Data gathering, the creation of specialized forums, ad hoc reflection, and funding of particular initiatives have been the main EU governance responses to the regulatory issues faced by nanomedicines (Dorbeck-Jung and Chowdhury 2011; EMA 2010b). In this domain, regulatory action is still in the planning and consideration stages. The EMA acknowledged that nanotechnologies, including combined products, presented challenges to existing regulatory classification rules and expertise in 2006 by publishing the Scientific Committee's Reflection Paper on nanotechnology-based medicinal products (Dorbeck-Jung and Chowdhury 2011).

The EMA established the Innovation Task Force to address these issues; it is in charge of coordinating scientific and regulatory expertise and promoting early communication with applicants. An ad hoc Expert Group of the Scientific Committee was established, and the Small and Medium Size Enterprises Office of the EMA provides regulatory support. Regulatory conversations take place informally between the EMA and the U.S. FDA, as well as national authorities. Response times from the evaluation bodies of the member states have been rather slow, with some of them relying on established standards and processes while collaborating with the EMA (Dorbeck-Jung and Chowdhury 2011).

Annual workshops on nanotechnology have been organized by the Directorate-General for Health and Food Safety (DG SANTE) since 2007. Interviews highlighted the importance of mutual learning and partnership between evaluation bodies and companies in addressing regulatory questions. Companies aim to establish early contact with evaluation authorities and build collaborative relationships. Briefing meetings offered by evaluation bodies to companies and academic centers developing innovative nano-medicinal products are considered crucial for understanding the direction of product development (Dorbeck-Jung and Chowdhury 2011).

The EMA's proposed Road Map to 2015 emphasizes greater involvement of patients and healthcare professionals, as well as increased transparency. Regarding definitions and standardization, the EMA participates in ISO initiatives as observers or invited experts, with a focus on monitoring the ISO's definition modifications and improvements. The EMA closely follows the sponsorship programs for manufactured nanomaterials undertaken by the OECD's Working Party on Manufactured PEGylated Nanomaterials (Dorbeck-Jung and Chowdhury 2011).

The interviews revealed no specific regulatory activities by the International Conference on Pharmaceuticals, the WHO, and EDQM. It has been suggested that lobbying efforts be used to advocate for the harmonization of medical device regulations to impose stronger safety, quality, and efficacy requirements to address the issues with borderline items. Harmonization would discourage manufacturers from seeking lenient operating standards through forum shopping. Another response has been to apply drug regulation in the authorization process, even for products initially categorized based on their apparent physical mechanism of action, such as certain contrast media (Dorbeck-Jung and Chowdhury 2011).

In summary, EU governance responses to PEGylated nanomedicine regulatory challenges involve data collection, specialized forums, ad hoc reflection, and funding initiatives. The EMA plays a central role in coordinating scientific and regulatory expertise, while informal exchanges serve as platforms for regulatory discussions. Harmonization efforts and the application of drug regulation to borderline products are among the responses to address regulatory uncertainties.

### 17.4.2 Pharmacokinetic Assessment of PEGylated Nanocarriers by Regulatory Agencies

Aspects of PEG biodistribution, metabolism, and excretion are covered in the book chapter in the context of nonclinical toxicological evaluation of PEGylated biopharmaceuticals (Ivens et al. 2015; Pasut and Zalipsky 2020; Mero et al. 2009). PEG's distribution, metabolism, and excretion are all poorly understood, especially for PEGs with molecular weights of more than 20 Koda. Sensitive techniques are needed to measure free and conjugated PEG in tissues and bodily fluids; albeit recently improved, these techniques are still difficult (Ivens et al. 2015; Baumann et al. 2014).

Studies have used a variety of methods to examine the distribution of PEG, such as injecting mice with labeled PEG of various molecular weights (Ivens et al. 2015;

Pasut and Zalipsky 2020). These investigations have demonstrated that the distribution of PEG throughout the body may be explained by a two-compartment model, with larger PEG molecules slowly translocating from the circulation (Ivens et al. 2015; Pasut and Zalipsky 2020). Liver clearance, potentially due to Kupffer cell uptake, is more pronounced for higher molecular weights (Ivens et al. 2015; Caliceti and Veronese 2003; Cheng et al. 2012; European Medicines Agency 2012). In earlier research that used radioactive tagging, the brain was shown to contain the least amount of PEG-related radioactivity compared to other tissues when examining the tissue distribution of the PEG 40 kDa protein in mice (Ivens et al. 2015; European Medicines Agency 2012).

Immunohistochemistry (IHC) has been used to identify PEG in tissues, but it is not quantitative and cannot provide a complete understanding of the catabolism of PEGylated biopharmaceuticals (Ivens et al. 2015). The cellular uptake mechanisms of PEG, whether nonspecific or drug target-mediated, likely depend on various factors such as dose, duration of treatment, pharmacological activity, receptor distribution, immunogenicity, overall molecular weight, and PEG molecular weight (Ivens et al. 2015; Pasut and Zalipsky 2020; Bendele et al. 1998). PEG and PEG-associated vacuoles are present in phagocytic cells, which indicates that phagocytic cell-specific absorption processes are important for tissue responses to high molecular weight PEG.

The main method of elimination for both small and large unconjugated PEG molecules appears to be renal clearance (Ivens et al. 2015). Despite their molecular weights above the glomerular filtration rate cutoff for globular proteins, large PEG molecules can pass past the glomerulus and be eliminated by the kidney (Ivens et al. 2015; Pasut and Zalipsky 2020). PEG is localized in the kidney's proximal convoluted tubules, indicating renal excretion followed by reuptake mechanisms frequently seen for other freely filtered solutes (Ivens et al. 2015; Ganson et al. 2005).

In summary, the biodistribution, metabolism, and excretion of PEG in PEGylated biopharmaceuticals involve complex processes that are still not fully understood. To improve our comprehension of these processes and their consequences for the safety and effectiveness of PEGylated biopharmaceuticals, more investigation and the development of sensitive analytical tools are required.

#### 17.4.2.1 Food and Drug Administration

Regarding the evaluation of substantial modifications in the production process of food ingredients currently on the market, the FDA has provided recommendations. This recommendation describes elements to consider when assessing how such changes would affect the identification, safety, and regulatory standing of the food substance. It emphasizes how crucial FDA consultation is, especially when improvements incorporate cutting-edge technologies like nanotechnology. The recommendations also underline the possible repercussions of disseminating a food ingredient that later turns out to be unsafe.

The FD&C Act establishes the rules that the FDA must follow when regulating food substances. While certain food additives need FDA premarket evaluation and clearance, others could merely need premarket notification. It's possible that some food ingredients don't need premarket notification or assessment. When evaluating the effects of significant modifications on the manufacturing process, the FDA's guidance takes these varied needs into account.

The FD&C Act places responsibility for ensuring the safe and legal use of food ingredients on both manufacturers and end users. The FD&C Act has several remedies that can be used, including seizure, injunctions, and criminal prosecution. According to the Act, adulterated food is any food that has been added with harmful or deadly ingredients, dangerous food additives, or unsafe color additives.

Both food additives and color additives are covered by the FD&C Act. For food additives and food contact compounds, a premarket approval procedure is established under the Food Additives Amendment and the FDA Modernization Act. These laws identify food additives and set guidelines for their secure application. The Color Additive Amendments, which provide a listing and certification process for color additives, regulate color additives.

Analyzing data on a food substance's functionality and safety is part of the safety assessment process. investigations and testing for identification, stability, purity, potency, performance, and usefulness are all included in this assessment, as are investigations and tests carried out on humans and animals. Name, chemical formula, source, composition, impurities, contaminants, and physical and chemical qualities all help to identify a food ingredient.

Overall, the FDA's advice strives to make sure that significant modifications to the way food compounds are made are thoroughly assessed for how they may affect the substances' identification, safety, and regulatory status. It emphasizes how crucial industry and FDA cooperation is to ensuring the safe and legal use of food ingredients (FDA 2014).

#### Status of Approved Nanoproducts

Some FDA-approved PEGylated nanoproducts are enlisted in Table 17.4 (FDA 2014).

#### Suggested Considerations for the Evaluation of Nanoproducts

According to the FDA, any special qualities and behaviors that the application of PEGylated nanotechnology may impart should be considered when evaluating the safety, efficacy, public health impact, or regulatory status of PEGylated nanotechnology products. This belief is based on the current technical and scientific understanding of nanomaterials along with their characteristics. In this advice, two points to consider are listed that should be utilized to determine if FDA-regulated devices use PEGylated nanotechnology. Both particle dimensions and dimension-dependent characteristics or phenomena are included in these areas. When necessary, product-specific premarket reviews provide the FDA with an opportunity to apply these guidelines, and when products are not subject to premarket reviews, the industry should take these guidelines into account. If either of these statements is true for a particular product, the industry and FDA should take into account whether any special characteristics or behaviors of the product have been discovered and effectively addressed during the PEGylated product's safety, efficacy, public health impact, or regulatory status studies (FDA 2023c).

**Table 17.4** List of FDA-approved PEGylated nanoproducts

| Name of product | Material description | Liposomal advantage | Indication | References |
|---|---|---|---|---|
| Doxil® | Liposomal doxorubicin | Enhanced disease-specific delivery; reduced systemic toxicity of free medication | Ovarian cancer, multiple myeloma, Karposi's sarcoma | FDA (2023a) |
| Onivyde® | PEGylated liposomal irinotecan | Enhanced delivery to the cancer site; less systemic toxicity from side effects | Metastatic pancreatic cancer | EMA (2016) |
| PEG-intron® | PEG-interferonalfa-2b | Enhanced protein stability with PEGylation | Chronic hepatitis C infection | Zheng (2004) |
| Adagen® | PEG-ademase bovine | Enhanced circulation and reduced immunogenicity | Adenosine deaminase (ADA) replacement therapy for severe combined immunodeficiency (SCID) patients | FDA (2023b) |
| Kadcyla® | Ado-trastuzumab emtansine | Help to delay or overcome resistance, reduce systemic toxicity | Treat HER2-positive breast cancer | CDER (2013) |
| Mircera® | Methoxy PEG-epoetin beta | To improve drug delivery, stability, and bioavailability. | Treatment for chronic renal disease-related anemia. Management of chronic renal disease-related anemia | Schwarsin (2018) |
| Krystexxa® | PEGloticase | To improve drug delivery, enhance drug stability, and reduce side effects | Treatment of chronic gout | FDA (2004) |

#### 17.4.2.2 European Medicines Agency

The precise nature of the interface between a nanomedicine product's "engineered" surface and the biological environment depends on the therapeutic use being considered as well as the administration method, such as blood plasma in intravenous delivery. A non-covalent or covalently bonded coating is frequently a key element of the design of nanomedicines that have been approved as products and/or are currently undergoing development. In general, the coating is used to reduce aggregation and improve stability (e.g., iron solutions for treating anemia), or in some circumstances, to reduce reticuloendothelial system clearance after intravenous administration to extend plasma circulation time and provide a chance for improved disease-specific targeting (e.g., PEGylation as applied to PEGylated liposomes) (European Medicine Agency 2013; CHMP 2013). Additionally, coatings have been employed to reduce antigenicity and enhance hemato-compatibility. These occurrences may occur as a result of the PEGylated nanomedicine's inherent physicochemical makeup or as a result of biomolecules adhering to the surface from the biological milieu to which they are exposed, such as when plasma proteins interact to form the so-called "protein corona."

The existence of a coating could affect the crucial aspects of the PEGylated nanocarrier safety and effectiveness. The product's pharmacokinetics, bio-distribution, and intracellular fate will be determined by the coating's physicochemical makeup, uniformity of surface coverage, and coating stability (both in terms of adhesion and susceptibility to degradation). Additionally, the clinically observed infusion-related reactions for some coated nanomedicines (such as iron solutions and PEGylated liposomes) may be caused by the coating material's physicochemical characteristics, a particular biomolecular interaction with the coated nanomedicines (such as complement activation), and/or cell interaction (European Medicine Agency 2013).

## Status of Approved Nanoproducts

Table 17.5 summarizes some PEGylated nanocarrier products that are under additional monitoring (European Medicine Agency 2013).

**Table 17.5** List of PEGylated nanocarrier products under additional monitoring

| Name of product | Liposomal advantage | Indication | References |
|---|---|---|---|
| Zolsketil PEGylated liposomal | In Zolsketil PEGylated liposomal, it is enclosed in "PEGylated liposomes". Because of this, the medication takes longer to leave the body and clears more slowly. It also decreases the effect on healthy tissues and cells, reducing the likelihood | Treat the ovarian neoplasm, sarcoma, kaposi, multiple myeloma | CHMP (2022) |
| Exparel liposomal | Exparel is designed to provide prolonged analgesia by releasing bupivacaine over an extended period, thereby reducing the need for additional pain medications after surgery | Brachial plexus block or femoral nerve block for treatment of postoperative pain in adults, and as a field block for treatment of somatic postoperative pain from small- to medium-sized surgical wounds in adults | European Medicines Agency (2021) |
| Arikayce liposomal | Treatment of Mycobacterium avium complex lung disease in adults who have not responded to conventional therapy | Treatment for a mycobacterial lung infection that is not tuberculous. Patients with cystic fibrosis receiving treatment for Pseudomonas aeruginosa lung infection/colonization | European Medicines Agency (2022b) |
| Myocet liposomal | Reduced cardiotoxicity, enhanced tumor targeting | Metastatic breast cancer | European Medicines Agency (2023) |

### Suggested Considerations for the Evaluation of Nanoproducts

The scientific difficulties involved assessing the effectiveness, safety, and quality of newly developed PEGylated nanomedicines. Due to the peculiar behavior and nano-sized nature of these systems in biological structures, current testing methods have limits, and the dependability of novel approaches is uncertain. To create a strong scientific foundation for an accurate assessment of nanomedicines, more research is required.

Determining the risk, which combines the possibility of harm occurring with the seriousness of that harm, is a step in the safety review process. Hazards are evaluated using both in vitro and in vivo assays, with in vivo assays typically being more pertinent for risk assessment. However, NPs add confusing elements. Nanoparticle identification and characterization are crucial but frequently difficult tasks. It is unclear how the physicochemical characteristics of hazardous responses work. Because of their poor dispersibility and tendency to aggregate, nanostructures may be challenging to use in tests. Due to inconsistent manufacture and the existence of hazardous solvents or processing residues, the quality of nanoparticle preparations can vary. Additionally, as size measurements can differ depending on the method utilized, it is important to carefully evaluate how to precisely determine the size of nanomaterials, which is essential for their characterization.

On an individual basis, the regulatory assessment of biosimilars, which takes a multidisciplinary approach, is offered as a potential framework for contrasting liposomal formulations of nanomedicines. Although it is acknowledged that nanomedicines are complicated, the quality, safety, and efficacy data that have been provided are sufficient to support the assessment of benefit-risk for nanomedicines that are already on the market. Instead of creating new guidelines, it is necessary to improve the current ones to take into account particular features of emerging nanotechnologies. The difficulties of setting acceptable difference criteria and of comparing products are also highlighted.

In summary, the evaluation of PEGylated nanomedicines poses scientific challenges due to limitations in testing methods and uncertainties regarding the reliability of novel approaches. Extensive research is required to establish a robust scientific basis for evaluating these products (European Medicines Agency 2010).

## 17.5 Patent Status of PEGylated Nanocarriers

A patent is an intellectual property that protects the legal rights of its owner by excluding others from unauthorized making, using, selling, and importing an invention for a limited period (Shiva 2004; Raj et al. 2015). As per the Trade-Related Aspects of Intellectual Property Rights agreement of the World Trade Organization, an invention having novelty, usefulness, and nonobviousness, limited legal monopoly will be granted for 20 years in the form of a patent. It is an exclusive right granted by each country to the inventor according to their laws and international agreements (Correa 2000). According to different areas of utility in technology, patents are classified by a hierarchical language-independent symbol in the International Patent Classification system, which was constituted by the Strasbourg Agreement 1971 (Gao et al. 2018). PEGylated nanocarriers fall under the IPC class of A61K- corresponds to preparations for medical, dental, or toiletry purposes with different subclassifications depending on their areas of application (USPTO, EPO 2023; Leitch et al. 2012).

### 17.5.1 Criteria for Patentability

A patent may have one or more claims that define the invention and the specific property right. These claims should also fulfill the patentability criteria like novelty, usefulness, and nonobviousness (Jennings Taylor and Inman 2017; Mehta et al. 2017). As per the Indian Patent Act, a patentable invention means, "a new product or process involving an inventive step and capable of industrial application" (Jennings Taylor and Inman 2017). The criteria of patentability are newness, inventive step, and industrial applicability. The Patents (Amendment) Act, of 2005, reports that the patentable inventions must not have been known earlier. It will be considered as new, if it has not been published or used anywhere in the world (Basheer 2005; Mishra 2014). The invention should have technical advancements better than existing ones or should have economic significance, or both, which should not be obvious. It should have a property like industrial applicability (Rosenberg 1998; Schmookler 1966; Kitch 2013).

### 17.5.2 Difficulties in Fulfilling the Patentability Criteria

Fulfilling the patentability criteria can be a challenging process, and there are several difficulties that inventors and applicants may encounter. The patentability criteria vary from country to country, but some common difficulties include:

- Prior art search and meeting the novelty criteria of a patent: The key requirement for a patent filing is that the invention must be novel, but difficulty lies in conducting comprehensive prior art searches for ensuring truly novelty of the invention and it shouldn't be part of existing literature.
- Demonstration of nonobviousness: The invention must be nonobvious to a person skilled in the relevant field and requires strong arguments and evidence to convince patent examiners or authorities.
- Defining the practical utility of an invention: The invention should have industrial applicability or can have tangible benefit or use. Defining this practical use of emerging or speculative technologies is difficult.
- Complexation in selecting the subject matter eligibility: In some jurisdictions, there are restrictions on the types of subject matter that can be patented. For example, laws may prohibit patenting abstract ideas, natural phenomena, or laws of nature. Determining whether the invention falls within the eligible subject matter can be complex.

- Inadequate description: The patent application must contain a clear and complete description of the invention, enabling someone skilled in the field to replicate and use the invention.
- Public disclosure: Timing of disclosure of a patent and its understanding of the implications to the public can be difficult. Publicly disclosing the invention before filing a patent application can jeopardize its novelty.
- Legal and technical complexity: The patent application process involves both legal and technical aspects. Understanding patent law and drafting a strong application requires expertise and can be overwhelming for inventors without prior experience.
- Cost and time: The patent application process can be expensive and time-consuming, particularly if multiple rounds of examination and revisions are necessary.

### 17.5.3 Preliminary Steps in Adapting Patent Regime for Nanocarriers

Adapting a patent regime involves a complex process that requires careful consideration of legal, scientific, and policy aspects. Some general steps could be involved in adapting the patent regime for nanocarriers; it's important to note that these steps may vary depending on the jurisdiction and specific requirements (Patel et al. 2023; Oldham et al. 2013; Coriat and Weinstein 2012). Here are some preliminary steps that could be considered:

1. Study the existing patent laws and analyze the international practices: Understand the current patent laws and regulations in the relevant jurisdiction. Study the patent regimes of other jurisdictions that have experience or expertise in nanotechnology or similar fields. Identify best practices and lessons learned from these jurisdictions that could be relevant for adapting the patent regime for nanocarriers (Acharya and Pal 2020; Lombi et al. 2019; Tenchov et al. 2022). This includes reviewing the definitions, requirements, and criteria for patentability, as well as any specific provisions that may apply to nanocarriers or similar technologies (Gürkaynak et al. 2018; Wagner 2009).
2. Assessing scientific advancement and identifying specific challenges: Evaluate the challenges, updated from the latest scientific advancements and developments in nanocarriers and unique aspects associated with nanocarriers that may require adaptation in the patent regime. This could include issues related to patentability criteria, examination procedures, or the scope of protection aligned with the state of the art and evolving technological landscape (Patel et al. 2023; Murthy 2012; Van Calster and Law 2009).
3. Engage stakeholders: Consult with stakeholders, including researchers, industry experts, patent attorneys, and policymakers. Gather their insights and perspectives on the existing patent regime and potential areas that need to be addressed or improved to accommodate nanocarriers (Susana et al. 2007; Lamprinopoulou et al. 2014).

4. Evaluate patentability criteria: Consider whether the existing patentability criteria, such as novelty, inventive step, industrial applicability, and enablement, adequately address the unique aspects of nanocarriers. Determine if any specific modifications or additions are required to accommodate particular technology (Cisneros 2014; Chowdhury et al. 2016).
5. Consider examination procedures: Examine the patent examination procedures and determine if they need to be tailored or streamlined for nanocarrier-related inventions. This may involve specialized training for patent examiners and establishing guidelines specific to nanocarriers.
6. Explore patent scope and claims: Analyze the scope of protection provided by patents for nanocarriers. Assess whether the existing patent claim structures effectively capture the features and variations of nanocarriers or if specific claim formats are required (Bhatia 2016; O'Neill et al. 2019; Paradise 2012).
7. Review patent term and enforcement: Evaluate the patent term duration and potential extensions to ensure that it is adequate for nanocarriers, considering their unique development cycles and regulatory requirements. Also, review the enforcement mechanisms to address any potential infringement or validity challenges specific to nanocarriers (Bawa 2020, 2007; Chin et al. 2014).
8. Draft and propose amendments: Based on the findings and considerations from the previous steps, draft proposed amendments to the existing patent laws or regulations to adapt the patent regime for nanocarriers. Work with relevant authorities, policymakers, and legal experts to refine and implement these amendments. Patent regime is a complex and ongoing process that requires a deep understanding of the technology, legal expertise, and stakeholder involvement. Consulting with experts and engaging in a comprehensive policy-making approach is crucial to ensure that the adapted patent regime effectively addresses the unique aspects of nanocarriers while promoting innovation and protecting intellectual property rights.

### 17.5.4 Patenting Issues in Developing PEGylated Nanocarrier

Despite being advantageous of being PEGylation-like solubility, stability, and bioavailability of nanocarrier system, there are several issues involved in patenting:

- **Prior art issues:** It defines the existing knowledge, methods, or technologies that are in the public domain before filing a patent. A PEGylated nanocarrier may already exist in publication or in a patent of similar methods or technologies, which could potentially hinder the novelty and nonobviousness requirements of filling a patent.
- **Obviousness:** To grant a patent, the filed patent must be nonobvious to a person skilled in the field. If it is considered obvious based on current knowledge literature or practices, it's difficult to secure a patent.
- **Lack of novelty:** Novelty is a fundamental requirement for patentability. If the PEGylation method or nanocarrier formulation is already disclosed in the prior art, even if it is not identical, it may lack novelty and be ineligible for patent protection.

- **Insufficient disclosure:** A patent application must provide a clear and complete disclosure of the invention to enable a person skilled in the field to reproduce it. If the description of the PEGylated nanocarrier is insufficient or lacks the necessary details, it may lead to patent rejection or a limited scope of protection.
- **Patentability of natural substances:** PEGylation often involves modifying natural substances such as proteins or peptides. In some jurisdictions, the patentability of naturally occurring substances or their modifications can be limited or subject to specific requirements, potentially affecting the patentability of PEGylated nanocarriers.
- **Patent term exclusivity:** Patents have a limited term of protection, typically 20 years from the filing date. However, the development and regulatory approval process for nanocarriers can be lengthy, which may result in a shorter effective exclusivity period for the commercialization of PEGylated nanocarriers.
- **Patent infringement risks:** Even if a patent is granted, enforcing and protecting the patent rights can be challenging. Competitors may attempt to develop similar PEGylation methods or formulations that could potentially infringe on the granted patent, leading to legal disputes and potential loss of market exclusivity.

#### 17.5.4.1 Patented PEGylated Products and their Regulatory Evolution

There are some patented PEGylated products as enlisted in Table 17.6.

### 17.5.5 Regulatory Evolution of PEGylated Nanocarrier

Regulatory consideration is an ongoing process. As the field continues to advance, regulatory agencies are likely to refine their guidelines and requirements to address emerging challenges and ensure the safe and effective use of these technologies in drug delivery.

Early regulatory Consideration: Earlier regulatory agencies like the FDA and EMA evaluated these products under the existing regulatory framework for drug delivery systems. The primary focus was accessing the safety and efficacy of the drug product utilizing PEGylated nanocarriers.

Specific guidelines: As the use of the nanocarriers became more prevalent, the agencies recognized to address the unique characteristics and considerations associated with these systems. The guidelines were developed and recommended individual characterization, formulation, and preclinical and clinical evaluations of these carriers.

Safety concerns and regulations: Over time, safety concerns related to PEGylated nanocarriers emerged, particularly regarding the potential for immune reactions, hypersensitivity, and accumulation in specific organs. Regulatory agencies started to pay closer attention to these issues and introduced stricter regulations for the evaluation of safety profiles, including immunogenicity and toxicity assessments.

Quality and manufacturing considerations: Regulatory agencies have also emphasized the importance of quality and manufacturing controls for PEGylated nanocarriers. Good Manufacturing Practices regulations have been applied to ensure the

**Table 17.6** List of patented PEGylated products

| S. No. | Patent number | Title | Description | Year | References |
|---|---|---|---|---|---|
| 1. | IN202321006175 | Trastuzumab-conjugated drug-loaded PEGylated pH-sensitive nanocarriers composition for breast cancer treatment and its synthesizing process | Antibody-conjugated PEGylated pH-responsive lipid nanocarrier provides composition and synthesizing process of nanocarriers loaded with anticancer drug for breast cancer. The vesicular drug delivery system bearing anticancer drug (PTX) and P-glycoprotein efflux pump inhibitor (Elacridar) is fabricated for effective treatment of breast cancer | 2023 | Raikwar and Jain (2023) |
| 2. | US20170258927, 20170258927(US), 2017230891(AU), 3017365(CA), 109152819 (CN), 3426285 (EP), 201817038590 (EP), 20210187081 (US), 4054531 (US), 2020380944(AU), 3160642 (CA), WO/2017/156513 (WO), 2022068180(JP) | Formulations and doses of PEGylated uricase | Polymeric synthetic nanocarrier comprising composition preparation method of immunosuppressant, uricase, and nanocarrier compositions like PLA, and PLA-PEG for gout flare treatment | 2017, 2019, 2021, 2022 | Johnston (2022) |
| 3. | CN110996963 | Conjugates of active pharmaceutical ingredients | Nanocarrier can be prepared by hydrophilic and photosensitive molecules by drug-drug conjugates, drug-porphyrin-conjugated NPs were PEGylates on the surface. The conjugates of NPs were claimed in the applications for imaging in various diseases | 2020 | Li et al. (2022) |

(continued)

**Table 17.6** (continued)

| S. No. | Patent number | Title | Description | Year | References |
|---|---|---|---|---|---|
| 4. | RO133819 | Process for preparing nanocarriers to be used for vectorization of interference ribonucleic acid (RNA) and targeted delivery to aortic valve cells | PEGylated cationic liposomes (50M % DOTAP, 35M% DOPE, 1-14M% PHC, 1% DSPE-PEG-Maleimide) preparation process for targeted delivery of nucleic acids (siRNA) to the cells of a malfunctioning aortic valve | 2020 | Manuela et al. (2020) |
| 5. | CN109260176B | Tumor-specific cleavable PEG NPs as well as preparation method and application thereof | The invention discloses methods of preparation and application of tumor-specific cleavable PEG mesoporous silica NPs with high drug loading. The carrier has the characteristics that PEG molecules in a tumor microenvironment can be dropped and can respond to oxidation-reduction of glutathione in cells | 2020 | Guanhai et al. (2020) |
| 6. | EP2707419B1 | Reversibly crosslinked micelle systems | Nanocarrier micelles formed from the telodendrimeric cross-linking of dendric polymer and PEG group for improving the stability of the nanocarrier micelle. The above-mentioned nanocarrier claimed to have therapeutic applications in cancer and asthma | 2019 | Lam et al. (2019) |
| 7. | US20200046841A1 | Porphyrin modified telo dendrimers | Amphiphilic telo dendrimers core containing cholic acid or cholic acid derivatives and with exterior branched or linear poly (ethylene glycol) segments that aggregate to form nanocarriers. Hydrophobic drug molecules sequester inside the core via a covalent bond with a telodendrimer unit. The altered structure leads to an effect on drug loading and interaction with biological membranes | 2022 | Lam et al. (2020) |

(continued)

**Table 17.6** (continued)

| S. No. | Patent number | Title | Description | Year | References |
|---|---|---|---|---|---|
| 8. | US20190388348A1 | Compositions and methods for treating cardiovascular and pulmonary diseases with apelin | The invention contains a liposomal formulation encapsulating apelin. PEGylated lipid nanocarrier shows improvements in efficacy, cellular activity, and controlled release of apelin under cardiovascular and pulmonary physiology conditions | 2019 | Rajadas et al. (2019) |
| 9. | TWI653244B | Use of bi-specific antibodies for modifying PEGylated substance | The Invention relates to the modification of PEGylated substances and is characterized in that bi-specific antibodies (bsAb) having the ability to bind to PEGylation are used to modify PEGylated substances. The humanized bifunctional antibody has the first antigen-binding site for binding a terminal methoxy or hydroxyl group of the PEG | 2019 | Cheng et al. (2019) |
| 10. | JP2019176872A | Linear DNA molecule delivery using PEGylated quantum dots for stable transformation in plants | Methods for noninvasively delivering nucleic acid by producing a PEGylated nanoparticle | 2019 | Samboju et al. (2019) |
| 11. | CN112402623 | Preparation method and application of PEGylated msns nanoparticle and mesoporous silica modified sustained-release hydrogel composite material | PEGylated block copolymer (PF 127) self-assembled hydrophobic msns are prepared by capping and loading, the prepared PEGylated system shows good dispersity, high affinity with dental pulp stem cells, and release of drug is prolonged by preventing prores of the diffusion layer and maintaining in vivo to provide more effective treatment | 2021 | Qingsong et al. (2021) |
| 12. | WO2022092396 | Method of preparing PEGylated albumin-bound steroid ginsenoside derivative NPs | The method comprises preparing a PEGylated bovine serum albumin by mixing binding and loading of ganglioside by dissolving and then filtering the suspension. The invention may contribute to improving the morbidity rate and death rate during the SARS-cov-2 pandemic until antivirus vaccines are developed | 2020 | Lee et al. (2022) |

(continued)

**Table 17.6** (continued)

| S. No. | Patent number | Title | Description | Year | References |
|---|---|---|---|---|---|
| 13. | US20140044791, US20200054753, WO/2012/106713 | Targeted nanoparticle conjugates | The composition comprises a PEGylated nanoparticle where a hydrophobic therapeutic agent (has amine or amino functionality) is non-covalently coupled and the surface of the nanoparticle has targeting moiety (polypeptide EGF) coupled to PEG of the nanoparticle for targeting a cell-associated disorder like cancer | 2019 | Basilion and Burda (2020) |
| 14. | WO2023056872 | Lipid nanoparticle composition and drug delivery system prepared thereby | Lipid NPs composed of the different types of lipids such as ionizable cationic, neutral, cholesterol, and a PEGylated lipid molecule. The lipid nanoparticle is used as a delivery carrier for nucleic acid molecules (such as mRNA) and has the advantages of small and uniform particle size, high encapsulation efficiency, and high cell transfection efficiency | 2023 | Lin et al. (2023) |
| 15. | WO2023044034 | Nanoparticle surface topography modified with bottlebrush block copolymers | It is the method for tailoring the surface topography of PEGylated NPs (having PEO or PEG covalently bound). It was demonstrated that nanoparticle formation and surface topography can be controlled by systematically changing the structural parameters of BBCP architecture and their significance on biological performance | 2023 | Saltzman et al. (2023) |
| 16. | IN202111026752 | Temozolomide lyotropic lipid nanoparticle composition | The invention claims the composition of lyotropic lipid NPs for sustained release of temozolomide comprising 1–20% (w/w) of lipid, 0.1–10% (w/w) of surfactant, 0.05–5% (w/w) of PEGylated lipid, 0.05–5% (w/w) of temozolomide and a buffer (pH 4–6). The cryoprotectant used in the composition includes mannitol | 2023 | Tejashree et al. (2023) |

(continued)

**Table 17.6** (continued)

| S. No. | Patent number | Title | Description | Year | References |
|---|---|---|---|---|---|
| 17. | CN115708809 | Novel lipid nanoparticle for delivering nucleic acid medicine and preparation method of novel lipid nanoparticle | The invention claims lipid nanoparticle preparation methods and compositions for delivery of nucleic acid drug (siRNA). The lipid NPs are composed of surfactant, natural, cationic, PEGylated phospholipid, cholesterol, buffering agent, and cane sugar as cryoprotectant. According to the claim, transfection efficiency, blood circulation efficiency, half-life, and immunogenicity are improved by adjusting the different lipid compositions | 2023 | Ping et al. (2023) |
| 18. | WO2023019310 | Lipid nanoparticle formulations | Lipid NPs composed of an active agent; a cationic and/or ionizable lipid (40–60 mol %), a phospholipid (5–20 mol %), a structural lipid (30–50 mol %), a PEGylated lipid (0.05–0.5 mol %) of total lipid composition. The invention claimed to have targeted splenic tissue delivery | 2023 | Al-Wasiti et al. (2023) |
| 19. | WO2023023410 | Ionizable lipids, lipid NPs for mRNA delivery, and methods of making the same | The invention provides the composition and methods of preparation of ionizable lipids and their NPs. The ionizable lipid includes 2AEOAP2, 2AEOAP4, 2AELAP2, 2AELAP4, Lipid 16A, Lipid 16B, Lipid 16C, Lipid 16B, and Lipid 20B. The nanoparticle is composed of ionizable, a helper lipid, sterol, and PEG-coated lipid conjugate | 2023 | Lee and Sriram (2023) |
| 20. | GB2606038, GB2600800 | Compositions and methods for organ-specific delivery of nucleic acids | The present invention provides compositions of NPs, a method of preferential delivery of a therapeutic agent to a non-liver organ in a subject, comprising a therapeutic agent. Lipid nanoparticle comprising selective organ targeting (SORT) compound (ionizable cationic lipid), a phospholipid, a PEGylated lipid (e.g., DMG-PEG), and a steroid, preferably cholesterol | 2022 | Cheng et al. (2022) |

(continued)

**Table 17.6** (continued)

| S. No. | Patent number | Title | Description | Year | References |
|---|---|---|---|---|---|
| 21. | RU0002772520 | Method for treating proliferative vitreoretinopathy using liposome-associated melphalan in an experiment | The present invention finds its application in ophthalmology and is intended for treatment in proliferative vitreoretinopathy using PEGylated liposomes of melphalan. Melphalan is used as a lipophilic prodrug, diglyceride ester conjugate in the composition of a stealth liposomal form with a particle size of 500 nm. Melphalan is used in low doses and stealth liposomal formulation reduces the retino-toxic effect of the compound | 2022 | Leontinovna et al. (2022) |
| 22. | CN114222802, CN114222803 | Narrow emission dyes, compositions comprising the same, and methods of making and using the same | The invention provides a bacteriochlorin derivatives synthesis procedure, having narrow-band emission. The derivatives contain a bio-conjugatable group (with an antibody or nanoparticle), PEGylated, and having high water solubility. Thus, it finds application in imaging and therapeutic | 2022 | Macnevin et al. (2022) |
| 23. | CN114177305 | Prodrug nanoparticle for inducing multi-mechanism death of tumor cells as well as preparation method and application of prodrug nanoparticle | The invention gives preparation method and application of prodrug nanoparticle (self-assembling a prodrug with a ROS sensitive bond and PEGylated ferrocene) for inducing multi-mechanism death of tumor cells (EPR effect and circulation half-life) | 2023 | Lifang et al. (2022) |
| 24. | MYPI 2020004701 | A NPs conjugate drug delivery system targeted for cancer cells and cancer stem cells | The invention claims nanoparticle conjugates of PAMAM-PEG-B19 aptamer preparation method, purification, and its characterization. The dendrimer conjugates effectively target and inhibit CD133 protein, and subsequently penetrate cancer stem cells to co-deliver drugs in brain cancer | 2022 | Behrooz et al. (2022) |

(continued)

**Table 17.6** (continued)

| S. No. | Patent number | Title | Description | Year | References |
|---|---|---|---|---|---|
| 25. | US20220031632, WO2021231898 | Nanoparticle compositions and uses thereof | Present developments provide methods and compositions of short-peptides loaded or peptide-metal conjugate of nanocapsules, or PEGylated NPs for treating and/or preventing autoimmune diseases like type 1 diabetes, and sclerosis by interacting with specific receptor proteins | 2022 | David et al. (2022) |
| 26. | CN113855788 | Antibody NPs as well as preparation method and application thereof | The antibody NPs comprise a PEGylated anti-HER-2 nano antibody chemically bonded with HCC NPs (chlorine and catalase encapsulated by HSA). The application of the invention is photosensitizer-induced deep tumor-targeted delivery, which induce catalytic oxygen supply, tumor metastasis, and reduced expression of HIF-1alpha in ovarian cancer | 2021 | Qing et al. (2021) |
| 27. | GB2596224 | Lipid nanoparticle compositions for delivery of mRNA and long nucleic acids | According to the invention, lipid nanoparticle compositions are useful for delivering mRNA and nucleic acids (each unit contains 80 nucleotides). These formulations use less cationic ionizable lipids than formulations with shorter nucleic acids and also contain phospholipids, PEGylated lipids, and steroids | 2021 | Siegwart and Cheng (2021) |
| 28. | US20210369867, EP3856146 | Lipid nanoparticle formulations comprising lipidated cationic peptide compounds for nucleic acid delivery | The present disclosure relates to the preparation and use of lipid NPs of lipid mixtures of lipidated cationic peptide compounds (tertiary amino lipidated and/or PEGylated cationic peptide compounds or lipitoids) and other structural lipid, phospholipid, and shielding lipids for nucleic acid delivery | 2021 | McKinlay (2021) |

(continued)

**Table 17.6** (continued)

| S. No. | Patent number | Title | Description | Year | References |
|---|---|---|---|---|---|
| 29. | US11173152 | Glutathione-coated NPs for delivery of MKT-077 across the blood-brain barrier | The invention provides the PEG-PLGA conjugation method, encapsulating MKT and further nanoparticle coated with glutathione for BBB permeability for the treatment of Alzheimer's in an experimental model | 2021 | Sutariya and Jinwal (2021) |
| 30. | JP2021172597 | Lipid nanoparticle | PEGylated lipid nanoparticle comprising an anticancer drug(sorafenib), a small molecule nucleic acid that suppresses the expression of midkine gene by RNA interference, and selectively targets hepatocellular carcinoma cells | 2021 | Hideyoshi et al. (2021) |

consistent production of safe and effective PEGylated nanocarrier-based drug products. Manufacturers are required to demonstrate appropriate control over critical quality attributes, such as particle size, drug loading, and PEGylation efficiency. Therefore, developers and manufacturers must stay updated with the latest regulatory developments and engage with regulatory authorities early in the development process.

Biocompatibility and biodistribution assessments: As the understanding of PEGylated nanocarriers' biocompatibility and biodistribution improved, regulatory agencies began to emphasize the need for thorough characterization and assessment of these parameters. Manufacturers are required to provide detailed data on the nanocarriers' behavior in biological systems, including their interactions with cells, tissues, and the immune system.

Specific regulatory pathways: In some cases, regulatory agencies have introduced specific regulatory pathways or designations for PEGylated nanocarrier-based drug products (Table 17.7). For example, in the United States, the FDA's Nanotechnology Task Force has provided recommendations for the evaluation of nanotechnology-based products, including nanocarriers, to ensure their safety and effectiveness.

**Table 17.7** List of approved PEGylated nanocarriers by regulatory agencies (Patra et al. 2018; Mishra et al. 2016)

| S. No. | Product name | Regulatory body | Year of approval | Details |
|---|---|---|---|---|
| 1. | Doxorubicin (doxil)/ calyx-liposomal | FDA | 1995 | Metastatic ovarian cancer treatment<br>Liposomes coated with PEG. DDS improves half-life of drug |
| 2. | Irinotecan (onivyde)-liposomal | FDA and EMA, AUSPAR | 2015 | PEG-coated liposome of irinoteca used in pancreatic adenocarcinoma |
| 3. | Certolizumab PEGol (cimzia)-polymer NPs | FDA | 2008 | Cimzia is a PEGylated fab fragment that binds to TNF-A, thus neutralizing its activity |
| 4. | PEGfilgrastim (neulasta) | FDA | 2002 | Neulasta is prepared by conjugation of a monomethoxy-PEG aldehyde chain (20 kDa) to the n-terminal methionine residue of filgrastim |
| 5. | PEGaspargase (oncaspar) | FDA | 1994 | PEG-coated L-asparaginase is used in the treatment of leukemia |
| 6. | PEGinterferon alfa-2a (PEGasys) | FDA | 2002 | Alfa-2a interferon conjugated to branched PEG for the treatment of hepatitis C |
| 7. | PEGvisomant (somavert) | FDA | 2003 | It is PEG-coated analog of human growth hormone approved for acromegaly treatment |

(continued)

**Table 17.7** (continued)

| S. No. | Product name | Regulatory body | Year of approval | Details |
|---|---|---|---|---|
| 8. | Methoxy PEG glycol-epoetin B (mircera) | FDA/EMA | 2007 | Epoetin β (epo) conjugated to methoxy-PEG approved to treat anemia. The formulation provides a controlled release system with a significantly improve in half-life |
| 9. | PEGinterferon Alfa-2b (PEG-INTRON) | FDA | 2001 | Alpha interferon (inf) molecule is a conjugation of mono PEG chain with succinimidyl carbonate |
| 10. | PEGloticase (krystexxa) | FDA/EMA | 2010/2013 | PEGloticase is efficacious, improving the solubility of the compound, and is approved for the treatment of chronic gouty arthritis |
| 11. | PEGinterferon beta-1a (plegridy) | FDA | 2014 | Polymer–protein conjugate (PEGylated Ifnbeta-1a) approved for multiple sclerosis |
| 12. | PEGylated factor VIII (adynovate) | FDA | 2015 | Polymer–protein conjugate (PEGylated factor viii) is approved for hemophilia |
| 13. | Factor IX (Rebinyn) | FDA | 2017 | Glyco-PEGylated coagulation factor IX approved for hemophilia |
| 14. | Jivi (damoctocog alfa PEGol) | FDA | 2017 | Recombinant antihemophilic factor for hemophilia A |
| 15. | Asparlas (calaspargase PEGol) | FDA | 2018 | L-asparaginase for leukemia |
| 16. | Fulphila (PEGfilgrastim-jmdb) | FDA | 2018 | G-CSF for infection during chemotherapy |
| 17. | Revcovi (elapegademase-lvlr) | FDA | 2018 | Recombinant adenosine deaminase for ADA-SCID |
| 18. | Palynziq (PEGvaliase) | FDA | 2018 | Recombinant phenylalanine ammonia lyase for phenylketonuria |
| 19. | Udenyca (PEGfilgrastim-cbqv injection) | FDA | 2018 | G-CSF for infection during chemotherapy |
| 20. | Ziextenzo (PEGfilgrastim-bmez) | FDA | 2019 | G-CSF for infection during chemotherapy |
| 21. | Esperoct (turoctocog alfa PEGol) | FDA | 2019 | Recombinant antihemophilic factor for hemophilia A |
| 22. | Nyvepria (PEGfilgrastim-apgf) | FDA | 2020 | G-CSF for G-CSF |
| 23. | Empaveli (PEGcetacoplan) | FDA | 2021 | Pentadecapeptide paroxysmal nocturnal hemoglobinuria (PNH) |

(continued)

**Table 17.7** (continued)

| S. No. | Product name | Regulatory body | Year of approval | Details |
|---|---|---|---|---|
| 24. | Skytrofa (lonapegsomatropin) | FDA | 2021 | Human growth hormone for growth hormone deficiency |
| 25. | BESREMi (ropeginterferon alfa-2b) | FDA | 2021 | Interferon for polycythemia vera |
| 26. | Fylnetra (PEGfilgrastim-pbbk) | FDA | 2022 | G-CSF for neutropenia |
| 27. | Stimufend (PEGfilgrastim-fpgk) | FDA | 2022 | G-CSF for neutropenia |
| 28. | Rolvedon (eflapegrastim-xnst) | FDA | 2022 | G-CSF for febrile neutropenia |
| 29. | SYFOVRE (PEGcetacoplan injection) | FDA | 2023 | Complement C3 inhibitor peptide for geographic atrophy (GA) |
| 30. | Elfabrio (PEGunigalsidase alfa-iwxj) | FDA | 2023 | Recombinant human GLA enzyme for fabry disease |

## 17.6 Potential Alternatives to PEG for Surface Decoration

A few potential alternatives to PEG for the surface decoration of nanoproducts are discussed below. However, the best choice will depend on the specific application and the desired properties of the nanoproduct.

- **Polysorbate 80 (Tween® 80)** is a nonionic surfactant that is also water-soluble and biocompatible (Fig. 17.1). It has a shorter half-life in the bloodstream than PEG, which could make it a better option for targeted drug delivery (Hoffmann et al. 2009).
- **Polyethylene oxide** is a synthetic polymer that is similar to PEG as shown in Fig. 17.2, but it is more hydrophilic and has a lower molecular weight. This could make it a better option for nanoproducts that need to be taken up by cells more easily (Wikipedia 2023a).
- **Polylactic acid** is a biodegradable polymer that is used in a variety of medical applications (Fig. 17.3). It is biocompatible and has a long half-life in the bloodstream, which could make it a good option for sustained drug release (Wikipedia 2023b).
- **Polyglycolic acid** is another biodegradable polymer that is similar to PLA (Fig. 17.4). It is also biocompatible and has a long half-life in the bloodstream (Wikipedia 2023c).

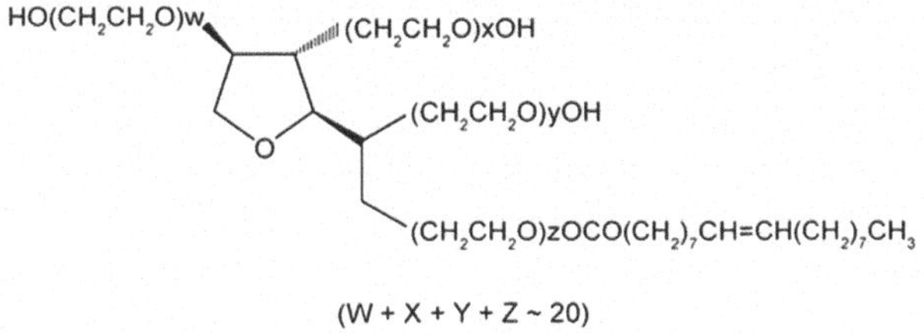


**Fig. 17.1** Chemical structure of Tween® 80 (Schwartzberg and Navari 2018)

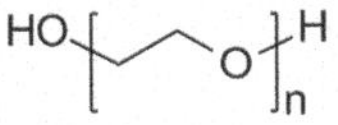

**Fig. 17.2** Molecular structure of polyethylene oxide (Wikipedia 2023a)

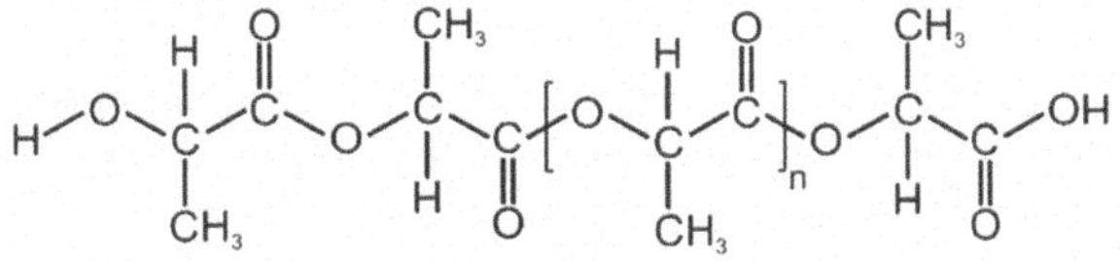

**Fig. 17.3** Chemical structure of polylactic acid (Rong and Keif 2007)

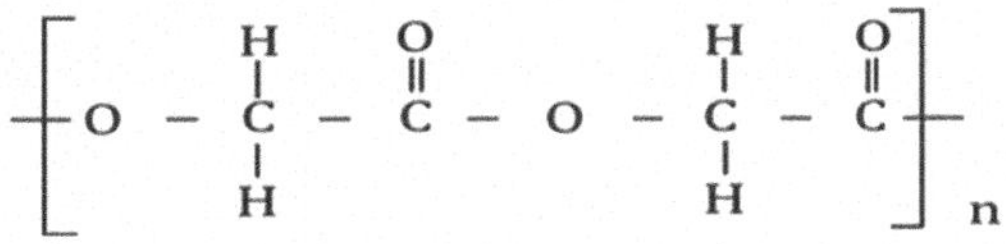

**Fig. 17.4** Chemical structure of polyglycolic acid (Wikipedia 2023c)

It is important to note that the use of PEGylated nanoproducts is still in its early stages of development, and there are several challenges that need to be addressed before these products can be widely adopted. However, the potential benefits of PEGylated nanoproducts are significant, and they could revolutionize the treatment of a variety of diseases.

## 17.7 A Vision for the Future

The future of PEGylated nanoproducts is very bright. These products have the potential to revolutionize the treatment of a variety of diseases, including cancer, cardiovascular disease, and infectious diseases.

Shortly, we can expect to see more PEGylated nanoproducts entering clinical trials. These trials will help to demonstrate the safety and efficacy of these products, and they will pave the way for their wider adoption.

In the long term, PEGylated nanoproducts could become the standard of care for a variety of diseases. These products could improve the efficacy and safety of drug delivery, and they could help to reduce the side effects of treatment.

Future scope for use of PEGylated nanoproducts in different fields is summarized below:

- **Cancer treatment**: PEGylated nanoproducts could be used to deliver chemotherapy drugs directly to cancer cells (Fig. 17.5). This would allow for higher doses of drugs to be used while minimizing the side effects on healthy cells (Sharma et al. 2022).
- **Cardiovascular disease**: PEGylated nanoproducts could be used to deliver drugs that target plaque buildup in arteries (Fig. 17.6). This could help to prevent heart attacks and strokes.
- **Infectious diseases**: PEGylated nanoproducts could be used to deliver vaccines and antibiotics directly to infected cells. This could help to prevent the spread of disease and improve the efficacy of treatment (Fig. 17.7).

The possibilities are endless. PEGylated nanoproducts have the potential to revolutionize the way we treat disease. I am excited to see what the future holds for this promising technology.

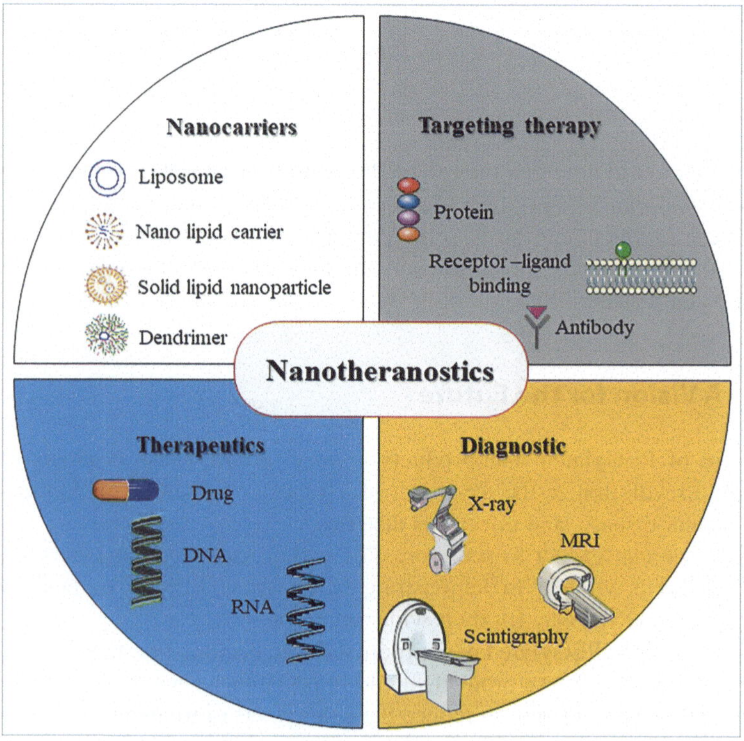


**Fig. 17.5** Theranostic approaches of nanoparticles [inspired by (Gupta et al. 2024)]. Some of the figure components are adapted from "Servier Medical Art" (https://smart.servier.com/)]

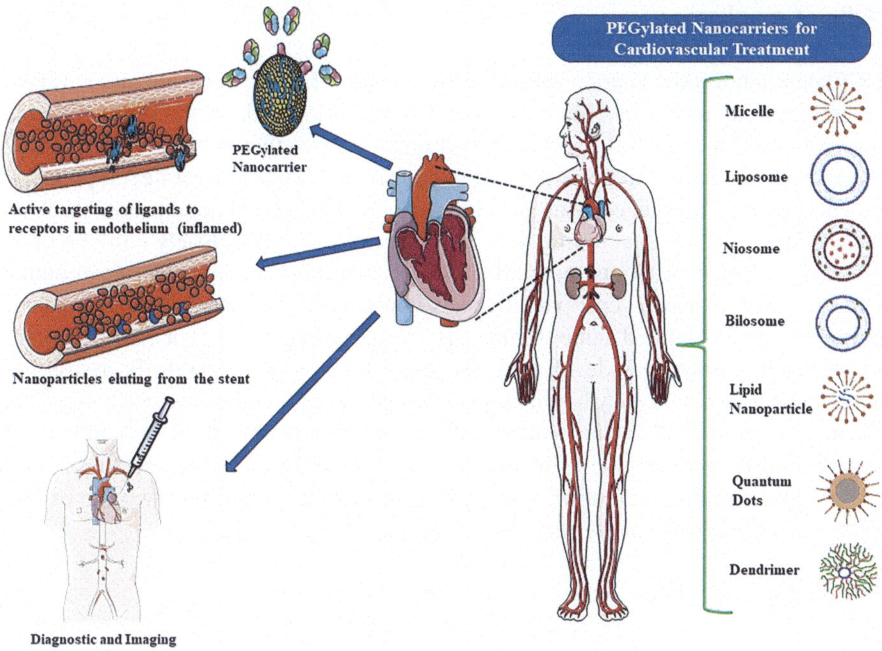


**Fig. 17.6** PEGylated nanocarriers in cardiovascular disease treatment [inspired by (Sabir et al. 2021)]. Some of the figure components are adapted from "Servier Medical Art" (https://smart.servier.com/)]

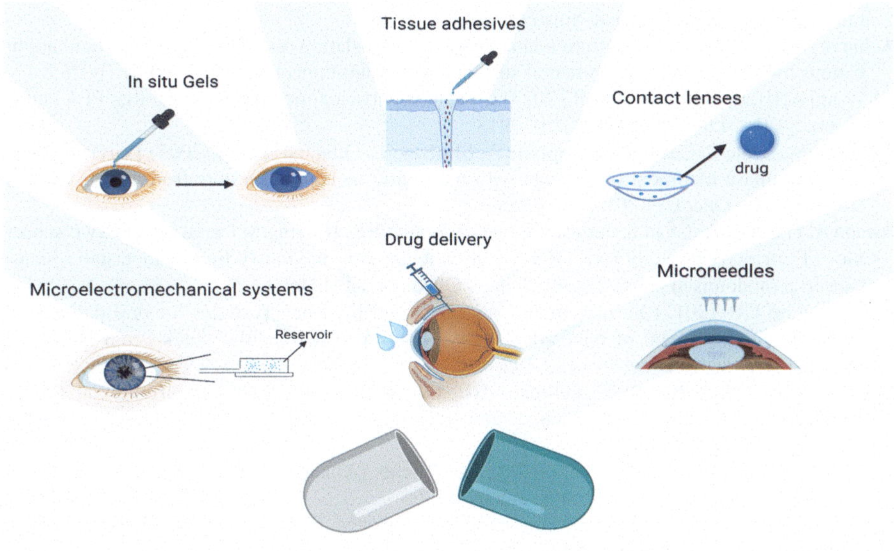


**Fig. 17.7** Ocular disease treatment (Bal-Öztürk et al. 2023). "Reproduced with permission from Bal-Öztürk A, Özcan-Bülbül E, Gültekin HE, Cecen B, Demir E, Zarepour A et al (2023) Application of convergent science and technology toward ocular disease treatment. Pharmaceuticals 16(3):445 published by MDPI, Basel, Switzerland under the terms and conditions of CC BY license (https://creativecommons.org/licenses/by/4.0/)"

## 17.8 Conclusion

PEGylated nanoproducts are a promising new class of therapeutic agents that have the potential to overcome the limitations of traditional drug delivery systems. However, there are several regulatory and patent issues that need to be addressed before these products can be widely adopted. The FDA has not yet developed specific guidelines for the clinical development of PEGylated nanoproducts. This means that companies developing these products need to work closely with the FDA to ensure that their products meet all of the necessary safety and efficacy requirements. There are also several patent issues that need to be considered. The patent landscape for PEGylated nanoproducts is complex and there is a risk of patent infringement. Companies developing these products need to carefully assess the patent landscape before they begin clinical development. Despite the challenges, the potential benefits of PEGylated nanoproducts are significant. These products have the potential to improve the efficacy and safety of drug delivery, and they could revolutionize the treatment of a variety of diseases. In the near future, we can expect to see more PEGylated nanoproducts entering clinical trials.

**Conflicts of Interest** All the authors declare no conflicts of interest.

## References

Abed SN, Deb PK, Surchi HS, Kokaz SF, Jamal SM, Bandopadhyay S et al (2019) Chapter 17: Nanocarriers in different preclinical and clinical stages. In: Tekade RK (ed) Basic fundamentals of drug delivery. Elsevier, Amsterdam, pp 685–731

Acharya A, Pal P (2020) Agriculture nanotechnology: translating research outcome to field applications by influencing environmental sustainability. Nanoimpact Sci Direct 19(4):100232

Al-Wasiti AHB, Pouton CW, Ho KT (2023) Lipid nanoparticle formulations. Australia: FPA Patent Attorneys Pvt. Ltd.; PCT/AU2022/050913

Ascierto PA, Scala S, Castello G, Daponte A, Simeone E, Ottaiano A et al (2005) Pegylated arginine deiminase treatment of patients with metastatic melanoma: results from phase I and II studies. J Clin Oncol 23(30):7660–7668

Awada A, Garcia AA, Chan S, Jerusalem GH, Coleman RE, Huizing MT et al (2013) Two schedules of etirinotecan pegol (NKTR-102) in patients with previously treated metastatic breast cancer: a randomised phase 2 study. Lancet Oncol 14(12):1216–1225

Baek JS, Cho CW (2017) Surface modification of solid lipid nanoparticles for oral delivery of curcumin: improvement of bioavailability through enhanced cellular uptake, and lymphatic uptake. Eur J Pharm Biopharm 117:132–140

Bal-Öztürk A, Özcan-Bülbül E, Gültekin HE, Cecen B, Demir E, Zarepour A et al (2023) Application of convergent science and technology toward ocular disease treatment. Pharmaceuticals 16(3):445

Basheer S (2005) India's tryst with TRIPS: The Patents (Amendment) Act, 2005. Indian J Law Technol 1(1):32. https://ssrn.com/abstract=764066

Basilion J, Burda C (2020) Targeted nanoparticle conjugates. USA: Tarolli, Sundheim, Covell & Tummino LLP; US20200054753

Baumann A, Tuerck D, Prabhu S, Dickmann L, Sims J (2014) Pharmacokinetics, metabolism and distribution of PEGs and PEGylated proteins: Quo vadis? Drug Discov Today 19(10):1623–1631

Bawa R (2007) Patents and nanomedicine. Nanomedicine 2(3):351–374

Bawa R (2020) Drug delivery at the nanoscale: a guide for scientists, physicians and lawyers. In: Mousa SA, Bawa R, Audette GF (eds) The road from nanomedicine to precision medicine, 1st edn. Taylor and Francis, Boca Raton, p 134

Behrooz AB, Vazifehmand R, Tajudin AA, Masarudin MJ, Sekawi Z, Hamzah ASA (2022) A nanoparticles conjugate drug delivery system targeted for cancer cells and cancer stem cells. Malaysia: Norunnuha Binti Datuk Hj. Nawawi; MYPI 2020004701

Beier CP, Schmid C, Gorlia T, Kleinletzenberger C, Beier D, Grauer O et al (2009) RNOP-09: pegylated liposomal doxorubicine and prolonged temozolomide in addition to radiotherapy in newly diagnosed glioblastoma - a phase II study. BMC Cancer 9:308

Bendele A, Seely J, Richey C, Sennello G, Shopp G (1998) Short communication: renal tubular vacuolation in animals treated with polyethylene-glycol-conjugated proteins. Toxicol Sci 42(2):152

Bhatia S (2016) Nanoparticles types, classification, characterization, fabrication methods and drug delivery applications. In: Bhatia S (ed) Natural polymer drug delivery systems, 1st edn. Cham, Springer, pp 33–93

Bobo D, Robinson KJ, Islam J, Thurecht KJ, Corrie SR (2016) Nanoparticle-based medicines: a review of FDA-approved materials and clinical trials to date. Pharm Res 33(10):2373–2387

Bocca C, Caputo O, Cavalli R, Gabriel L, Miglietta A, Gasco MR (1998) Phagocytic uptake of fluorescent stealth and non-stealth solid lipid nanoparticles. Int J Pharm 175(2):185–193

Boulikas T (2009) Clinical overview on lipoplatin: a successful liposomal formulation of cisplatin. Expert Opin Investig Drugs 18(8):1197–1218

Bradbury MS, Pauliah M, Zanzonico P, Wiesner U, Patel S (2016) Intraoperative mapping of sentinel lymph node metastases using a clinically translated ultrasmall silica nanoparticle. Wiley Interdiscip Rev Nanomed Nanobiotechnol 8(4):535–553

Cabello SF (2008) Aspectos psicológicos de la disfunción eréctil. Rev Int Androl 6(3):210–213

Caliceti P, Veronese FM (2003) Pharmacokinetic and biodistribution properties of poly(ethylene glycol)-protein conjugates. Adv Drug Deliv Rev 55(10):1261–1277

Carvalho SG, Araujo VHS, dos Santos AM, Duarte JL, Silvestre ALP, Fonseca-Santos B et al (2020) Advances and challenges in nanocarriers and nanomedicines for veterinary application. Int J Pharm 580:119214

CDER (2013) 2013 CDER fast track calendar year approvals* data as of 12-31-2013/revised 12-23-21* total of 15 approvals. Ammendale Road Beltsville, p 2. Available from: https://www.fda.gov/media/131200/download

Chariou PL, Ortega-Rivera OA, Steinmetz NF (2020) Nanocarriers for the delivery of medical, veterinary, and agricultural active ingredients. ACS Nano 14(3):2678–2701

Chen B, Dai W, He B, Zhang H, Wang X, Wang Y et al (2017) Current multistage drug delivery systems based on the tumor microenvironment. Theranostics 7(3):538–558

Chen L, Wang J, Wang Z (2022) L-DEP regimen is effective as an initial therapy for adult EBV-HLH. Ann Hematol 101(11):2461–2470

Cheng TL, Chuang KH, Chen BM, Roffler SR (2012) Analytical measurement of PEGylated molecules. Bioconjug Chem 23(5):881–899

Cheng TL, Kao CH, Chuang, Roffler SR, Su YC, Tung HY, et al (2019) Use of bi-specific antibodies for modifying pegylated substance - Google Patents. Google Patents. China, Taiwan; TWI653244B

Cheng Q, Siegwart DJ, Wei T (2022) Compositions and methods for organ specific delivery of nucleic acids. united kingdom: United Kingdom; GB2600800

Chin WWL, Parmentier J, Widzinski M, Tan EH, Gokhal R (2014) A brief literature and patent review of nanosuspensions to a final drug product. J Pharm Sci 103(10):80–99

CHMP (2006) Guideline on the environmental risk assessment of medicinal products for human use, p 12. https://www.ema.europa.eu/en/documents/scientific-guideline/guideline-environmental-risk-assessment-medicinal-products-human-use-first-version_en.pdf

CHMP (2013) Reflection paper on the data requirements for intravenous liposomal products developed with reference to an innovator liposomal product Final, p 13. https://www.ema.europa.eu/

en/documents/scientific-guideline/reflection-paper-data-requirements-intravenous-liposomal-products-developed-reference-innovator_en.pdf
CHMP (2022) Zolsketil pegylated liposomal, INN-doxorubicin. Domenico Scarlattilaan, p 2. https://www.ema.europa.eu/en/documents/smop-initial/chmp-summary-opinion-zolsketil-pegylated-liposomal_en.pdf
Chowdhury P, Gogoi M, Das S, Zaman A, Hazarika P, Borchetia S et al (2016) Intellectual property rights for nanotechnology in agriculture. In: Ranjan S, Dasgupta N, Lichtfouse E (eds) Nanoscience in food and agriculture, vol 2. Cham, Springer, pp 1–25
Cisneros M (2014) Patentability requirements for nanotechnological inventions: an approach form the European Patent convention perspective. Soc Sci Res Netw 5:1–51
Clark AJ, Wiley DT, Zuckerman JE, Webster P, Chao J, Lin J et al (2016) CRLX101 nanoparticles localize in human tumors and not in adjacent, nonneoplastic tissue after intravenous dosing. Proc Natl Acad Sci USA 113(14):3850–3854
Communities (1993) EUR-Lex & Legal Information Unit, p 159 Council Regulation (EEC) No 2309/93 of 22 July 1993 laying down Community procedures for the authorization and supervision of medicinal products for human and veterinary use and establishing a European Agency for the Evaluation of Medicinal Products. Available from: https://eur-lex.europa.eu/legal-content/EN/TXT/?uri=celex%3A31993R2309
Contardi M (2019) Changes in the medical device's regulatory framework and its impact on the medical device's industry: from the medical device directives to the medical device regulations. Erasmus Law Rev 12(2):166–177
Coriat B, Weinstein O (2012) Patent regimes, firms and the commodification of knowledge. Socio-Econ Rev 10(2):267–292
Correa C (2000) Intellectual property rights, the WTO and developing countries: the TRIPS agreement and policy options. Zed Books, Penang, pp 1–254
Couvreur P, Vauthier C (2006) Nanotechnology: intelligent design to treat complex disease. Pharm Res 23(7):1417–1450
David H, Wagner J, Yussman MG, Vaitaitis GM, Henry CW (2022) Nanoparticle compositions and uses thereof. United States of America; US20220031632
Dijkgraaf EM, Santegoets SJAM, Reyners AKL, Goedemans R, Wouters MCA, Kenter GG et al (2015) A phase I trial combining carboplatin/doxorubicin with tocilizumab, an anti-IL-6R monoclonal antibody, and interferon-α2b in patients with recurrent epithelial ovarian cancer. Ann Oncol 26(10):2141–2149
Dong H, Tang M, Li Y, Li Y, Qian D, Shi D (2015) Disulfide-bridged cleavable PEGylation in polymeric nanomedicine for controlled therapeutic delivery. Nanomedicine 10(12):1941–1958
Dorbeck-Jung BR, Chowdhury N (2011) Is the European medical products authorisation regulation equipped to cope with the challenges of nanomedicines? Law Policy 33(2):276–303
Duncan R, Gaspar R (2011) Nanomedicine(s) under the microscope. Mol Pharm 8(6):2101–2141
EMA (2003) Commission directive 2003/94/EC of 8 October 2003 laying down the principles and guidelines of good manufacturing practice in respect of medicinal products for human use and investigational medicinal products for human use (text with EEA relevance), pp 73–427. https://eur-lex.europa.eu/legal-content/EN/TXT/?uri=CELEX%3A32003L0094
EMA (2006) B Directive 2001/20/EC of the European Parliament and of the Council, pp 1–19. https://health.ec.europa.eu/system/files/2016-11/dir_2001_20_en_0.pdf
EMA (2008) Status of EMEA scientific guidelines and european pharmacopoeia monographs and chapters in the regulatory framework applicable to medicinal products, p 3. https://www.ema.europa.eu/en/documents/scientific-guideline/status-emea-scientific-guidelines-european-pharmacopoeia-monographs-chapters-regulatory-framework_en.pdf
EMA (2010a) Guideline on the minimum quality and non-clinical data for certification of advanced therapy medicinal products, p 22. https://www.gmp-compliance.org/guidelines/advanced-therapy-guidelines?
EMA (2010b) Evaluation HMD and 1st International Workshop on nanomedicines 2010 summary report, p 34. https://www.ema.europa.eu/en/documents/report/summary-report-workshop-nanomedicines-2-3-september-2010_en.pdf

EMA (2016) Onivyde pegylated liposomal (previously known as Onivyde). https://www.ema.europa.eu/en/medicines/human/EPAR/onivyde-pegylated-liposomal

EMA (2022) Commission directive 2005/28/EC of 8 April 2005 laying down principles and detailed guidelines for good clinical practice as regards investigational medicinal products for human use, as well as the requirements for authorisation of the manufacturing or im. https://www.ema.europa.eu/en/human-regulatory/research-development/compliance/good-clinical-practice/qa-good-clinical-practice-gcp

EMA, CAT, CHMP (2019) European Commission DG Health and Food Safety and European Medicines Agency Action Plan on ATMPs, pp 1–6. https://www.ema.europa.eu/en/documents/other/european-commission-dg-health-food-safety-european-medicines-agency-action-plan-advanced-therapy_en.pdf

European Commission (2011) Final report summary - nanomed round table (nanomedicine ethical, regulatory, social and economic environment), p 3. https://cordis.europa.eu/project/id/218732/reporting/fr

European Medicine Agency (2013) Reflection paper on surface coatings: general issues for consideration regarding parenteral administration of coated nanomedicine products, p 5. https://etp-nanomedicine.eu/wp-content/uploads/2018/10/reflection-paper-surface-coatings-general-issues-consideration-regarding-parenteral-administration_en.pdf

European Medicines Agency (1994) Commission communication on the Community marketing authorisation procedures for medicinal products, p 14. https://eur-lex.europa.eu/LexUriServ/LexUriServ.do?uri=OJ:C:1998:229:0004:0017:EN:PDF

European Medicines Agency (2004a) Laying down Community procedures for the authorisation and supervision of medicinal products for human and veterinary use and establishing a European Medicines Agency (text with EEA relevance), p 33. Available from: https://eur-lex.europa.eu/legal-content/EN/TXT/?uri=celex%3A32004R0726

European Medicines Agency (2004b) Laying down community procedures for the authorisation and supervision of medicinal products for human and veterinary use and establishing a European Medicines Agency, p 1–70. https://health.ec.europa.eu/system/files/2016-11/reg_2004_726_en_0.pdf

European Medicines Agency (2010) 1st international workshop on nanomedicines 2010 summary report. vol. 44. London, p. 1–34. https://www.ema.europa.eu/en/documents/report/summary-report-workshop-nanomedicines-2-3-september-2010_en.pdf

European Medicines Agency (2012) CHMP safety working party's response to the PDCO regarding the use of PEGylated drug products in the paediatric population. https://www.ema.europa.eu/en/documents/scientific-guideline/chmp-safety-working-partys-response-pdco-regarding-use-pegylated-drug-products-paediatric-population_en.pdf

European Medicines Agency (2021) European Medicines Agency decision P/0342/2021 of 9 August 2021 on the acceptance of a modification of an agreed paediatric investigation plan for bupivacaine (Exparel), (EMEA-000877-PIP03-17-M03), p 10. https://www.ema.europa.eu/en/documents/pip-decision/p/0342/2021-ema-decision-9-august-2021-acceptance-modification-agreed-paediatric-investigation-plan_en.pdf

European Medicines Agency (2022a) EMA/911829. Committee for medicinal product for human use rules of procedure, p 12. https://www.ema.europa.eu/en/documents/work-programme/chmp-work-plan-2023_en.pdf

European Medicines Agency (2022b) European Medicines Agency decision P/0107/2022 of 13 April 2022 on the acceptance of a modification of an agreed paediatric investigation plan for amikacin (sulfate) (Arikayce liposomal), (EMEA-000525-PIP01-08-M08), pp 1–10. https://www.ema.europa.eu/en/documents/pip-decision/p/0107/2022-ema-decision-13-april-2022-acceptance-modification-agreed-paediatric-investigation-plan_en.pdf

European Medicines Agency (2023) Myocet liposomal, INN-doxorubicin, p 37. https://www.ema.europa.eu/en/documents/product-information/myocet-liposomal-previously-myocet-epar-product-information_en.pdf

European Parliament (2007) Regulation (EC) No 1394/2007 of the European Parliament and of the Council of 13 November 2007 on advanced therapy medicinal products and amending

Directive 2001/83/EC and Regulation (EC) No 726/2004 (text with EEA relevance). https://eur-lex.europa.eu/LexUriServ/LexUriServ.do?uri=OJ:L:2007:324:0121:0137:en:PDF
European Parliament (2019) Regulation (EC) No 141/2000 of the European Parliament and of the Council of 16 December 1999 on orphan medicinal products. https://eur-lex.europa.eu/legal-content/EN/TXT/PDF/?uri=CELEX:02000R0141-20190726
European Union (2006) B Regulation (Ec) No 1901/2006 of The European Parliament and of the Council. https://eur-lex.europa.eu/LexUriServ/LexUriServ.do?uri=CONSLEG:2006R1901:20070126:EN:PDF
Farjadian F, Ghasemi A, Gohari O, Roointan A, Karimi M, Hamblin MR (2019) Nanopharmaceuticals and nanomedicines currently on the market: challenges and opportunities. Nanomedicine 14(1):93–126
FDA (2004) Changes to an approved NDA or ANDA, Rockville, MD, p 40. https://www.fda.gov/files/drugs/published/Changes-to-an-Approved-NDA-or-ANDA.pdf
FDA (2014) Guidance for industry: assessing the effects of significant manufacturing process changes, including emerging technologies, on the safety and regulatory status of food ingredients and food contact substances, including food ingredients that are color addi. https://www.fda.gov/regulatory-information/search-fda-guidance-documents/guidance-industry-assessing-effects-significant-manufacturing-process-changes-including-emerging
FDA (2020) Medical device reporting - 21 CFR Part 803. https://www.fda.gov/medical-devices/device-advice-comprehensive-regulatory-assistance/overview-device-regulation
FDA (2022) Drug products, including biological products, that contain nanomaterials - guidance for industry. U.S. Food and Drug Administration, Silver Spring, p 33
FDA (2023a) FDA listing of authorized generics, p 98. https://www.fda.gov/media/77725/download?attachment
FDA (2023b) Transfer of therapeutic biological products to the center for drug evaluation and research. https://www.fda.gov/combination-products/jurisdictional-information/transfer-therapeutic-biological-products-center-drug-evaluation-and-research
FDA (2023c) Drug products, including biological products, that contain nanomaterials guidance for industry. U.S. Food and Drug Administration, Silver Spring, p 29. https://www.fda.gov/files/drugs/published/Drug-Products-Including-Biological-Products-that-Contain-Nanomaterials-Guidance-for-Industry.pdf
Forloni G (2012) Responsible nanotechnology development. J Nanopart Res 14:8
Franzè S, Selmin F, Rocco P, Colombo G, Casiraghi A, Cilurzo F (2020) Preserving the integrity of liposomes prepared by ethanol injection upon freeze-drying: insights from combined molecular dynamics simulations and experimental data. Pharmaceutics 12(6):1–15
Freches D, Rocks N, Patil HP, Perin F, Van Snick J, Vanbever R et al (2019) Preclinical evaluation of topically-administered PEGylated Fab' lung toxicity. Int J Pharm 1:100019
Ganson NJ, Kelly SJ, Scarlett E, Sundy JS, Hershfield MS (2005) Control of hyperuricemia in subjects with refractory gout, and induction of antibody against poly(ethylene glycol) (PEG), in a phase I trial of subcutaneous PEGylated urate oxidase. Arthritis Res Ther 8(1):12
Gao Y, Zhu Z, Riccaboni M (2018) Consistency and trends of technological innovations: a network approach to the international patent classification data. Stud Comput Intell 689:744–756
Gaspar RS, Florindo HF, Silva LC, Videira MA, Corvo ML, Martins BF et al (2014) Regulatory aspects of oncologicals: nanosystems main challenges. In: Alonso MJ, Garcia-Fuentes M (eds) Nano-oncologicals advances in delivery science and technology, 1st edn. Springer, Cham, pp 425–452
Gladkov O, Moiseyenko V, Bondarenko IN, Shparyk Y, Barash S, Adar L et al (2016) A phase III study of balugrastim versus pegfilgrastim in breast cancer patients receiving chemotherapy with doxorubicin and docetaxel. Oncologist 21(1):7–15
Glazer ES, Piccirillo M, Albino V, Di Giacomo R, Palaia R, Mastro AA et al (2010) Phase II study of pegylated arginine deiminase for nonresectable and metastatic hepatocellular carcinoma. J Clin Oncol 28(13):2220–2226
Graham ML (2003) Pegaspargase: a review of clinical studies. Adv Drug Deliv Rev 55(10):1293–1302

Guanhai W, Jiantao L, Dawei Z, Haibing Y (2020) Tumor-specific cleavable PEG (polyethylene glycol) nanoparticle as well as preparation method and application thereof - Google Patents. Google Patents

Gupta D, Roy P, Sharma R, Kasana R, Rathore P, Gupta TK (2024) Recent nanotheranostic approaches in cancer research. Clin Exp Med 24(1):8. https://doi.org/10.1007/s10238-023-01262-3

Hamad I, Hunter AC, Szebeni J, Moghimi SM (2008) Poly(ethylene glycol)s generate complement activation products in human serum through increased alternative pathway turnover and a MASP-2-dependent process. Mol Immunol 46(2):225–232

Hatakeyama H, Akita H, Kogure K, Oishi M, Nagasaki Y, Kihira Y et al (2007) Development of a novel systemic gene delivery system for cancer therapy with a tumor-specific cleavable PEG-lipid. Gene Ther 14(1):68–77

Hideyoshi, Harashima Yonis MA, Khalil Abdelraheem I (2021) Lipid nanoparticle. Japan; JP2021172597

Hoch U, Staschen CM, Johnson RK, Eldon MA (2014) Nonclinical pharmacokinetics and activity of etirinotecan pegol (NKTR-102), a long-acting topoisomerase 1 inhibitor, in multiple cancer models. Cancer Chemother Pharmacol 74(6):1125–1137

Hoffmann C, Blume A, Miller I, Garidel P (2009) Insights into protein-polysorbate interactions analysed by means of isothermal titration and differential scanning calorimetry. Eur Biophys J 38(5):557–568

Holtsberg FW, Ensor CM, Steiner MR, Bomalaski JS, Clark MA (2002) Poly(ethylene glycol) (PEG) conjugated arginine deiminase: effects of PEG formulations on its pharmacological properties. J Control Release 80(1-3):259–271

Howard MD, Jay M, Dziubla TD, Lu X (2008) PEGylation of nanocarrier drug delivery systems: state of the art. J Biomed Nanotechnol 4(2):133–148

Hu Z, Luo F, Pan Y, Hou C, Ren L, Chen J et al (2008) Arg-Gly-Asp (RGD) peptide conjugated poly(lactic acid)-poly(ethylene oxide) micelle for targeted drug delivery. J Biomed Mater Res 85(3):797–807

Hussain Z, Khan S, Imran M, Sohail M, Shah SWA, de Matas M (2019) PEGylation: a promising strategy to overcome challenges to cancer-targeted nanomedicines: a review of challenges to clinical transition and promising resolution. Drug Deliv Transl Res 9(3):721–734

Iqbal S, Naveed Yasin M, Sheardown H (2019) Engineering of targeted nanoparticles by using self-assembled biointegrated block copolymers. In: Pathak YV (ed) Surface modification of nanoparticles for targeted drug delivery. Cham, Springer, pp 451–466

Ishida T, Masuda K, Ichikawa T, Ichihara M, Irimura K, Kiwada H (2003) Accelerated clearance of a second injection of PEGylated liposomes in mice. Int J Pharm 255(1-2):167–174

Ishida T, Ichihara M, Wang XY, Yamamoto K, Kimura J, Majima E et al (2006) Injection of PEGylated liposomes in rats elicits PEG-specific IgM, which is responsible for rapid elimination of a second dose of PEGylated liposomes. J Control Release 112(1):15–25

Ivens IA, Baumann A, McDonald TA, Humphries TJ, Michaels LA, Mathew P (2013) PEGylated therapeutic proteins for haemophilia treatment: a review for haemophilia caregivers. Haemophilia 19(1):11–20

Ivens IA, Achanzar W, Baumann A, Brändli-Baiocco A, Cavagnaro J, Dempster M et al (2015) PEGylated biopharmaceuticals: current experience and considerations for nonclinical development. Toxicol Pathol 43(7):959–983

Jameson GS, Hamm JT, Weiss GJ, Alemany C, Anthony S, Basche M et al (2013) A multicenter, phase I, dose-escalation study to assess the safety, tolerability, and pharmacokinetics of etirinotecan pegol in patients with refractory solid tumors. Clin Cancer Res 19(1):268–278

Jennings Taylor E, Inman M (2017) Looking at patent law:patentable inventions, conditions for receiving a patent, and claims. Electrochem Soc Interface 26(3):39–43

Johnston L (2022) Formulations and doses of pegylated uricase. Google Patents. US Patent App. 15/456, 520-2017, pp 1–38. https://patentimages.storage.googleapis.com/cb/51/4a/aceb17eace12bb/US20170258927A1.pdf

Kamaly N, Xiao Z, Valencia PM, Radovic-Moreno AF, Farokhzad OC (2012) Targeted polymeric therapeutic nanoparticles: Design, development and clinical translation. Chem Soc Rev 41(7):2971–3010

Karra N, Benita S (2011) The ligand nanoparticle conjugation approach for targeted cancer therapy. Curr Drug Metab 13(1):22–41

Khalid MN, Simard P, Hoarau D, Dragomir A, Leroux JC (2006) Long circulating poly(ethylene glycol)-decorated lipid nanocapsules deliver docetaxel to solid tumors. Pharm Res 23(4):752–758

Kitch EW (2013) The nature and function of the patent system. Priv Common Prop Lib Prop Law, pp 129–154

Kolate A, Baradia D, Patil S, Vhora I, Kore G, Misra A (2014) PEG - a versatile conjugating ligand for drugs and drug delivery systems. J Control Release 192:67–81

Kwilas AR, Donahue RN, Tsang KY, Hodge JW (2017) PEGylation as a strategy for improving nanoparticle-based drug and gene delivery. Adv Drug Deliv Rev 99:28–51

Lam KS, Li Y, Luo J, Xiao K (2019) Reversibly crosslinked micelle systems. US: Mintz Levin Cohn Ferris Glovsky and Popeo, P.C.; US20190112423

Lam KS, Li Y, Pan CX (2020) Porphyrin modified telodendrimers. USA: Mintz, Levin, Cohn, Ferris, Glovsky and Popeo, P.C.; US20200046841

Lamprinopoulou C, Renwick A, Klerkx L, Hermans F (2014) Application of an integrated systemic framework for analysing agricultural innovation systems and informing innovation policies: comparing the Dutch and Scottish. Agric Syst Sci Direct 129:40–54

Lee JY, Sriram V (2023) Ionizable lipids, lipid nanoparticles for mRNA delivery and methods of making the same. USA: Hedrick, Megan D. Livingston, Jennifer L. Oberhaus, Geoffrey L. Tent, Brian A. Rericha, Nicholaus R. Elswick, Christopher W.; PCT/US2022/042761

Lee JM, Yoon TJ, Cho YS (2013) Recent developments in nanoparticle-based siRNA delivery for cancer therapy. Biomed Res Int 2013:782041

Lee JY, Song GY, Lee HS, Mun YH, Ahn JS, Yoo SY et al (2022) Method of preparing Pegylated Albumin-Bound Steroid Ginsenoside derivative nanoparticles. Korea: Ghong-Gan International Patent Law Firm; WO2022092396

Leitch ME, Casman E, Lowry GV (2012) Nanotechnology patenting trends through an environmental lens: analysis of materials and applications. J Nanopart Res 14:12

Leontinovna LN, Petrovna KI, Lvovna, Elena V, Kaplun Aleksandr Petrovich AA, Sergeevna TD (2022) Method for treating proliferative vitreoretinopathy using liposome-associated melphalan in an experiment. Russia: Russian Fedaration; RU0002772520

Li W, Zhan P, De Clercq E, Lou H, Liu X (2013) Current drug research on PEGylation with small molecular agents. Prog Polym Sci 38(3-4):421–444

Li Y, Xue X, Huang Y, Ma Z (2022) Conjugates of active pharmaceutical ingredients. US, Australia, New Zealand: Trimble, Alexander R. Hom, Roy K. Kezer, William B. Dolan, Gregory F.; WO2018232334, US20210138084, AU2018283383, NZ760876, CN110996963,EP3638248

Li W, Dong M, Huang S, Shi L, Yang H, Zhang Y et al (2023) Efficacy and safety of PEG-rhG-CSF in preventing chemoradiotherapy-induced neutropenia in patients with locally advanced cervical cancer. Biomol Biomed 23(1):310–316

Lifang Y, Xiaopeng H, Yifei Z (2022) Prodrug nanoparticle for inducing multi-mechanism death of tumor cells as well as preparation method and application of prodrug nanoparticle. China; CN114177305

Lin Y et al (2023) Lipid Nanoparticle composition and drug delivery system prepared thereby. China: Beijing Tricisionbio Therapeutics Inc.; PCT/CN2022/122221

Lombi E, Donner E, Dusinska M (2019) A one health approach to managing the applications and implications of nanotechnologies in agriculture. Nat Nanotechnol 14(6):523–531

Macnevin CJ, Pitner JB, Chen C-Y (2022) Narrow emission dyes, compositions comprising same, and methods of making and using same. China; CN114222802

Manuela DC, Cristina R, Constantinescu AG, Mariana V, Ileana D et al (2020) Process for preparing nanocarriers to be used for vectorization of interference ribonucleic acid (RNA) and

targeted delivery to aortic valve cells. Romania: Cabinet De Proprietate Industrială "Lazăr Elena"; RO133819
Markovsky E, Baabur-Cohen H, Eldar-Boock A, Omer L, Tiram G, Ferber S et al (2012) Administration, distribution, metabolism and elimination of polymer therapeutics. J Control Release 161(2):446–460
Mattheolabakis G, Taoufik E, Haralambous S, Roberts ML, Avgoustakis K (2009) In vivo investigation of tolerance and antitumor activity of cisplatin-loaded PLGA-mPEG nanoparticles. Eur J Pharm Biopharm 71(2):190–195
McKinlay CJ (2021) Lipid nanoparticle formulations comprising lipidated cationic peptide compounds for nucleic acid delivery. United States of America: Marshall, Gerstein & Borun LLP; US20210369867
Mehta H, Tidwell L, Liotta LA (2017) Inventions and patents: a practical tutorial. Methods Mol Biol 1606:379–397
Mero A, Pasut G, Veronese FM (2009) PEG: a useful technology in anticancer therapy. In: Veronese FM (ed) PEGylated protein drugs: basic science and clinical applications. Basel, Birkhäuser Verlag, pp 255–271
MHRA (2020) A guide to what is a medicinal product. South Colonnade Canary Wharf, London, p 71. https://assets.publishing.service.gov.uk/government/uploads/system/uploads/attachment_data/file/872742/GN8_FINAL_10_03_2020__combined_pdf
Mishra U (2014) Patentability criteria in different countries. SSRN electronic. https://papers.ssrn.com/sol3/papers.cfm?abstract_id=2430768
Mishra P, Nayak B, Dey RK (2016) PEGylation in anti-cancer therapy: an overview. Asian J Pharm Sci 11(3):337–348
Moein Moghimi S, Hamad I, Andresen TL, Jørgensen K, Szebeni J, Moein Moghimi S et al (2006) Methylation of the phosphate oxygen moiety of phospholipid-methoxy(polyethylene glycol) conjugate prevents PEGylated liposome-mediated complement activation and anaphylatoxin production. FASEB J 20(14):2591–2593
Murthy RSR (2012) Challenges and emerging issues in patenting nanomedicines. In: Souto EB (ed) Patenting nanomedicines. Berlin, Springer, pp 25–48
Nance EA, Woodworth GF, Sailor KA, Shih TY, Xu Q, Swaminathan G et al (2012) A dense poly(ethylene glycol) coating improves penetration of large polymeric nanoparticles within brain tissue. Sci Transl Med 4(149):119
Novelli G, Ciccacci C, Borgiani P, Amati MP, Abadiec E (2008) Genetic tests and genomic biomarkers: regulation, qualification and validation. Clin Cases Miner Bone Metab 5(2):149–154
O'Neill S, Hermann K, Klein M, Bus JL (2019) Broad claiming in nanotechnology patents: is litigation inevitable. Am Bar Assoc 16(1):1–32. http://nanolabweb.com/index.cfm/action/main.default.viewArticle/articleID/175/CFID/549403/CFTOKEN/88339636/index.html
Oldham P, Hall S, Forero O (2013) Biological diversity in the patent system. PLoS One 8(11):e78737
Omar AA, Basiouny L, Elnoby AS, Zaki A, Abouzid M (2022) Jude total therapy studies from I to XVII for childhood acute lymphoblastic leukemia: a brief review. J Egypt Natl Canc Inst 34(1):25
Pacheco C, Baião A, Ding T, Cui W, Sarmento B (2023) Recent advances in long-acting drug delivery systems for anticancer drug. Adv Drug Deliv Rev 194:114724
Paradise J (2012) Claiming nanotechnology: improving USPTO efforts at classification of emerging nano-enabled pharmaceutical technologies. Northwest J Technol Intellect Prop 10(3):169
Park MVDZ, Lankveld DPK, van Loveren H, de Jong WH (2009) The status of in vitro toxicity studies in the risk assessment of nanomaterials. Nanomedicine 4(6):669–685
Park EJ, Choi J, Lee KC, Na DH (2019) Emerging PEGylated non-biologic drugs. Expert Opin Emerg Drugs 24(2):107–119
Parvizi N, Woods K (2014) Regulation of medicines and medical devices: contrasts and similarities. Clin Med J R Coll Phys 14(1):6–12
Pasut G, Zalipsky S (2020) Polymer-protein conjugates: from pegylation and beyond. Elsevier, Amsterdam, p 514

Patel D, Talele D, Khunt D, Pardeshi S (2023) Patenting nanomaterials: regulatory aspects. In: Nanomaterial-based drug delivery systems. Springer, Cham, pp 369–380

Patra JK, Das G, Fraceto LF, Campos EVR, Rodriguez-Torres MDP, Acosta-Torres LS et al (2018) Nano based drug delivery systems: recent developments and future prospects. J Nanobiotechnol 16(1):1–33

Peracchia MT, Vauthier C, Passirani C, Couvreur P, Labarre D (1997) Complement consumption by poly(ethylene glycol) in different conformations chemically coupled to poly(isobutyl 2-cyanoacrylate) nanoparticles. Life Sci 61(7):749–761

Perche DF, Biswas DS, Wang DT, Zhu DL, Torchilin PV (2014) Hypoxia-targeted siRNA delivery. Angew Chem Int Ed Engl 53(13):3362–3366

Ping W, Yong L, Yangyang M, Zhao L, Mengyan Z, Jiaqi W (2023) Novel lipid nanoparticle for delivering nucleic acid medicine and preparation method of novel lipid nanoparticle. China: Li Yong; CN115708809

Plosker GL (2008) Pegylated liposomal doxorubicin in multiple myeloma. ADIS Drug Eval 68(17):2535–2551

Qing Z, Lian W, Xiangmin L, Qing Z, Shaozheng L, Qingjie C (2021) Antibody nanoparticles as well as preparation method and application thereof. China; CN113855788

Qingsong Y, Ghalib AA, Yan H, Lina Z (2021) Preparation method and application of PEGylated MSNs nanoparticle and mesoporous silica modified sustained-release hydrogel composite material. China; CN112402623

Rachmawati H, Larasati A, Adi AC, Shegokar R (2020) Chapter 2: Role of nanocarriers and their surface modification in targeting delivery of bioactive compounds. In: Shegokar R (ed) Nanopharmaceuticals: expectations and realities of multifunctional drug delivery systems, vol 1. Elsevier, Amsterdam, pp 17–43

Raikwar S, Jain SK (2023) Trastuzumab conjugated drug-loaded PEGylated pH-sensitive nanocarriers composition for breast cancer treatment and its synthesizing process. India; IN202321006175

Raj GM, Priyadarshini R, Mathaiyan J (2015) Drug patents and intellectual property rights. Eur J Clin Pharmacol 71(4):403–409

Rajadas J, Ruiz-Lozano P, Serpoosan V (2019) Compositions and methods for treating cardiovascular and pulmonary diseases with Apelin. US, CN; US20190388348

Ramalingam P, Ko YT (2015) Enhanced oral delivery of curcumin from N-trimethyl chitosan surface-modified solid lipid nanoparticles: pharmacokinetic and brain distribution evaluations. Pharm Res 32(2):389–402

Ramalingam P, Ko YT (2016a) Validated LC-MS/MS method for simultaneous quantification of resveratrol levels in mouse plasma and brain and its application to pharmacokinetic and brain distribution studies. J Pharm Biomed Anal 119:71–75

Ramalingam P, Ko YT (2016b) Improved oral delivery of resveratrol from N-trimethyl chitosan-g-palmitic acid surface-modified solid lipid nanoparticles. Colloids Surf B Biointerfaces 139:52–61

Robert NJ, Vogel CL, Henderson IC, Sparano JA, Moore MR, Silverman P et al (2004) The role of the liposomal anthracyclines and other systemic therapies in the management of advanced breast cancer. Semin Oncol 31(13):106–146

Rodriguez V, Kairalla J, Salzer WL, Raetz EA, Loh MLC, Carroll AJ et al (2016) A pilot study of intensified PEG-asparaginase in high-risk acute lymphoblastic leukemia: Children's Oncology Group Study AALL08P1. J Pediatr Hematol Oncol 38(6):409–412

Rong X, Keif M (2007) A study of PLA printability with flexography. Proc Tech Assoc Graph Arts, TAGA, pp 605–613. https://digitalcommons.calpoly.edu/cgi/viewcontent.cgi?article=1017&context=grc_fac

Rosenberg N (1998) Uncertainty and technological change. In: Siesfeld T, Cefola J, Neef D (eds) The economic impact of knowledge, 1st edn. Taylor and Francis, London, p 18

Sabir F, Barani M, Mukhtar M, Rahdar A, Cucchiarini M, Zafar MN et al (2021) Nanodiagnosis and nanotreatment of cardiovascular diseases: an overview. Chemosensors 9(4):67. https://doi.org/10.3390/chemosensors9040067

Gürkaynak G, Yılmaz I, Yeşilaltay B (2018) Intellectual property law and practice in the blockchain realm. Comput Law Secur Rev 34(4):847–862
Saltzman G, Mark W, Zhong JM (2023) Nanoparticle surface topography modified with bottlebrush block copolymers. United States of America: Yale University; PCT/US2022/043842
Samboju NC, Yau KY, Burroughs FG, Samuel JP Webb SR (2019) Linear DNA molecule delivery using pegylated quantum dots for stable transformation in plants. Japan: Narasimha Chary Samboju, Kerrm Y Yau, Frank G Burroughs, Jayakumar Pon Samuel, Steven R Webb; JP2019176872
Schmookler J (1966) Invention and economic growth. Bus Hist Rev 41(1):110–112
Schore RJ, Devidas M, Bleyer A, Reaman GH, Winick N, Loh ML et al (2019) Plasma asparaginase activity and asparagine depletion in acute lymphoblastic leukemia patients treated with pegaspargase on Children's Oncology Group AALL07P4. Leuk Lymphoma 60(7):1740–1748
Schwarsin A (2018) Clinical review Alexandria Schwarsin, MD BLA 125164 Supplement-078 Mircera® (methoxy polyethylene glycol-epoetin beta) [Internet]. Ammendale Road Beltsville, p 34. https://www.fda.gov/media/114558/download
Schwartzberg LS, Navari RM (2018) Safety of polysorbate 80 in the oncology setting. Adv Ther 35(6):754–767
Sharma A, Shambhwani D, Pandey S, Singh J, Lalhlenmawia H, Kumarasamy M et al (2022) Advances in lung cancer treatment using nanomedicines. ACS Omega 8(1):10–41
Shiva V (2004) Protect or plunder? Understanding intellectual property rights. JSTOR 17(4):547–549. https://www.jstor.org/stable/26162411
Siegwart DJ, Cheng Q (2021) Lipid nanoparticle compositions for delivery of mRNA and long nucleic acids. United Kingdom, United States of America: United kingdom, US; GB2596224, US20210121411
Simpkins F, Flores A, Chu C, Berek JS, Lucci J, Murray S et al (2013) Chemoimmunotherapy using pegylated liposomal Doxorubicin and interleukin-18 in recurrent ovarian cancer: a phase I dose-escalation study. Cancer Immunol Res 1(3):168–178
Sun CY, Shen S, Xu CF, Li HJ, Liu Y, Cao ZT et al (2015) Tumor acidity-sensitive polymeric vector for active targeted siRNA delivery. J Am Chem Soc 137(48):15217–15224
Susana B, Charalampos K, Frank W (2007) European agencies and input legitimacy: EFSA, EMeA and EPO in the post-delegation phase. J Eur Integr 29(5):583–600
Sutariya VB, Jinwal UK (2021) Glutathione-coated nanoparticles for delivery of MKT-077 across the blood-brain barrier. United States of America: Quarles & Brady LLP; US11173152
Svenson S, Wolfgang M, Hwang J, Ryan J, Eliasof S (2011) Preclinical to clinical development of the novel camptothecin nanopharmaceutical CRLX101. J Control Release 153(1):49–55
Tao H, Wang R, Sheng W, Zhen Y (2021) The development of human serum albumin-based drugs and relevant fusion proteins for cancer therapy. Int J Biol Macromol 187:24–34
Tejashree W, Singhvi G, Narayan SR (2023) Temozolomide lyotropic lipid nanoparticle composition. India: Birla Institute of Technology & Science (BITS), Pilani; IN202111026752
Tenchov R, Sasso JM, Wang X, Liaw WS, Chen CA, Zhou QA (2022) Exosomes nature's lipid nanoparticles, a rising star in drug delivery and diagnostics. ACS Nano 16(11):17802–17846
Thaker PH, Brady WE, Lankes HA, Odunsi K, Bradley WH, Moore KN et al (2017) A phase I trial of intraperitoneal GEN-1, an IL-12 plasmid formulated with PEG-PEI-cholesterol lipopolymer, administered with pegylated liposomal doxorubicin in patients with recurrent or persistent epithelial ovarian, fallopian tube or primary peritoneal. Gynecol Oncol 147(2):283–290
Ton NC, Parker GJM, Jackson A, Mullamitha S, Buonaccorsi GA, Roberts C et al (2007) Phase I evaluation of CDP791, a PEGylated Di-Fab′ conjugate that binds vascular endothelial growth factor receptor 2. Clin Cancer Res 13(23):7113–7118
USCODE (1970) Title 42-the public health and welfare, vol. 110, p 372–382. https://www.govinfo.gov/content/pkg/USCODE-2019-title42/pdf
USPTO, EPO (2023) United States patent classification. CPC Cooperative patent classification, pp 1–26. https://www.cooperativepatentclassification.org/home
Van Calster G, Law JD (2009) Taking temperature-A review of European Union regulation in nanomedicine. Eur J Health Law 16(3):69–249

Vassilakopoulos TP, Pangalis GA, Katsigiannis A, Papageorgiou SG, Constantinou N, Terpos E et al (2012) Rituximab, cyclophosphamide, doxorubicin, vincristine, and prednisone with or without radiotherapy in primary mediastinal large B-cell lymphoma: the emerging standard of care. Oncologist 17(2):239–249

Vergote IB, Garcia A, Micha J, Pippitt C, Bendell J, Spitz D et al (2013) Randomized multicenter phase II trial comparing two schedules of etirinotecan pegol (NKTR-102) in women with recurrent platinum-resistant/refractory epithelial ovarian cancer. J Clin Oncol 31(32):4060–4066

Veronese FM, Pasut G (2005) PEGylation, successful approach to drug delivery. Drug Discov Today 10(21):1451–1458

Vllasaliu D, Fowler R, Stolnik S (2014) PEGylated nanomedicines: recent progress and remaining concerns. Expert Opin Drug Deliv 11(1):139–154

Wagner R (2009) Understanding patent-quality mechanisms. Univ Pa Law Rev 157:9–29. https://ssrn.com/abstract=1452862

Wang H, Zhao P, Su W, Wang S, Liao Z, Niu R et al (2010) PLGA/polymeric liposome for targeted drug and gene co-delivery. Biomaterials 31(33):8741–8748

Wang J, Wang Y, Wu L, Zhang J, Lai W, Wang Z (2016) PEG-aspargase and DEP regimen combination therapy for refractory Epstein-Barr virus-associated hemophagocytic lymphohistiocytosis. J Hematol Oncol 9(1):84

Webster R, Didier E, Harris P, Siegel N, Stadler J, Tilbury L et al (2007) PEGylated proteins: Evaluation of their safety in the absence of definitive metabolism studies. Drug Metab Dispos 35(1):9–16

Wikipedia (2023a) Structure of polyethylene oxide. https://en.wikipedia.org/wiki/Polyethylene_glycol

Wikipedia (2023b). Structure of polylactic acid. https://en.wikipedia.org/wiki/Polylactic_acid

Wikipedia (2023c). Structure of polyglycolide. https://en.wikipedia.org/wiki/Polyglycolide

Xu CF, Zhang HB, Sun CY, Liu Y, Shen S, Yang XZ et al (2016) Tumor acidity-sensitive linkage-bridged block copolymer for therapeutic siRNA delivery. Biomaterials 88:48–59

Yang S (2018) Preparation, in vitro characterization and pharmacokinetic study of coenzyme Q10 long-circulating liposomes. Drug Res 68(5):270–279

Yang T, De Cui F, Choi MK, Cho JW, Chung SJ, Shim CK et al (2007) Enhanced solubility and stability of PEGylated liposomal paclitaxel: in vitro and in vivo evaluation. Int J Pharm 338(1–2):317–326

Zheng JH (2004) Clinical pharmacology and biopharmaceutics review. Ammendale Road Beltsville, p 223. https://www.accessdata.fda.gov/drugsatfda_docs/nda/2005/21797_BARACLUDE_biopharmr.PDF

Zolnik BS, Sadrieh N (2009) Regulatory perspective on the importance of ADME assessment of nanoscale material containing drugs. Adv Drug Deliv Rev 61(6):422–427

# Therapeutic and Pharmaceutical Applications of PEGylated Nano-Carriers

# 18

Raheleh Shakeri, Seyedeh Zohreh Mirjalili,
Ceyda Oksel Karakus, and Maliheh Safavi

**Abstract**

The coating of the exterior of nanocarriers with polyethylene glycol (PEG; known as "PEGylation"), has emerged as one of the most widely adopted approaches to impart stealth properties to pharmaceutical nanocarrier systems. The PEGylation tactic offers a stealth barrier towards the adsorption of circulatory proteins owing to its hydrophilicity, elasticity, neutralization, and hydration properties. PEGylation dramatically provides sustained drug release benefits, reduces immune sensitivity, as well as improves the systemic circulation period, Overall, PEGylation effectively improves the therapeutic efficacy of the loaded therapeutic moiety in treating target cells and tissues. This chapter offers discussion on drug delivery nanosystems that are developed based on the principle of PEGylation strategy. Furthermore, a detailed account of the pharmaceutical and biomedical applications of PEGylated nanocarriers has also been presented in detail.

**Keywords**

Therapeutic benefits · Stealth properties · Immunosensitivity · Pharmaceutical applications

---

R. Shakeri
Faculty of Science, Department of Biological Science, University of Kurdistan, Sanandaj, Iran

S. Z. Mirjalili
Faculty of Pharmacy, Drug and Food Control Department, Tehran University of Medical Sciences, Tehran, Iran

C. O. Karakus
Department of Bioengineering, Izmir Institute of Technology, Izmir, Turkey

M. Safavi (✉)
Department of Biotechnology, Iranian Research Organization for Science and Technology, Tehran, Iran



R. K. Tekade, N. K. Jain (eds.), *PEGylated Nanocarriers in Medicine and Pharmacy*, https://doi.org/10.1007/978-981-97-7316-9_18

## Abbreviations

| | |
|---|---|
| ABC | Accelerated blood clearance |
| ACSF | Artificial cerebrospinal fluid |
| AL | Alendronate |
| ALS | Amyotrophic lateral sclerosis |
| AMPs | Antimicrobial peptides |
| Au DENPs | Dendrimer-entrapped AuNPs |
| Au PENPs | PEI-entrapped AuNPs |
| AuNPs | Gold NPs |
| BBB | Blood-brain barrier |
| CAs | Contrast agents |
| $Ce_6$ | $Chlorine_6$ |
| $CeO_2$ | Cerium oxide |
| CF | Cystic fibrosis |
| $CK_{30}PEG_{10k}$ | Copolymers of poly-l-lysine and 10 kDa polyethylene glycol |
| CMC | Critical micelle concentration |
| CNPs | Ceria NPs |
| CNS | Central nervous system |
| COPD | Chronic obstructive pulmonary disease |
| CpG | Cytosine-phosphate-guanine |
| CRS | Chronic rhinosinusitis |
| CRSM | CRS mucus |
| CT | Computerized tomography |
| CTLs | Tumor-specific killer T cells |
| Cur | Curcumin |
| CVM | Cervicovaginal mucus |
| DCs | Dendritic cells |
| DDS | Drug-delivery system |
| DES | Diethylstilbestrol |
| DEXA | Dexamethasone |
| DSPE-PEG2000 | 1,2-distearoyl-sn-glycero-3-phospho-ethanolamine-PEG |
| EBRT | External beam radiotherapy |
| ECM | Extracellular matrix |
| EPR | Permeability and retention effect |
| FA | Folic acid |
| FDA | Food and Drug Administration |
| FNPs | Nanostructured iron oxide |
| GA | Gambogic acid |
| GCs | Glucocorticoids |
| $Gd^{3+}$ | Gadolinium ions |
| GFP | Green fluorescent protein |
| GI | Gastrointestinal |
| GO | Graphene oxide |
| HSV | Herpes simplex virus |
| IBD | Inflammatory bowel disease |

| | |
|---|---|
| ICIs | Immune checkpoint inhibitors |
| IgM | Immunoglobulin M |
| IR | Ionizing radiation |
| IRT | Internal radioisotope therapy |
| LBFs | Lipid-based formulations |
| LDNPs | PEGylated lipid NPs |
| LE | Loteprednol etabonate |
| LH | Luteinizing hormone |
| LHRH | Luteinizing hormone releasing hormone |
| Lipoplexes | Nonviral nucleic acid NPs |
| LN | Lipid NPs |
| MCC | Mucociliary clearance |
| MCT | Medium-chain triglycerides |
| MI | Myocardial ischemia |
| miR-133 | MicroRNA-133 |
| MMP | Matrix metalloproteinase |
| $Mn^{2+}$ | Manganese ions |
| MnO | Manganese oxide |
| MNP | Melanin NPs |
| MOFs | Metal-organic frameworks |
| mPEG | MethoxyPEG |
| MPS | Mononuclear phagocyte system |
| MRI | Magnetic resonance imaging |
| MSNs | Mesoporous silica NPs |
| MTP | Metoprolol |
| MW | Molecular weight |
| NAC | N-acetyl cysteine |
| NALT | Nasal-associated lymphoid tissue |
| NIR | Near-infrared |
| NLC | Nanostructured lipid carrier |
| NMD | Nimodipine |
| NPs | Nanoparticles |
| nRGO | Reduced nano-GO |
| ORR | Objective response rate |
| OS | Overall survival |
| OVA | Ovalbumin |
| PB | Prussian blue |
| PBA | Phenylboronic acid |
| PBAE | Poly(β-amino ester) |
| PBAE-MPP | PBAE DNA NPs |
| PC | Protein corona |
| PCL | Poly(ε-caprolactone) |
| PCL-PEG | PEG-block-poly (-caprolactone) |
| PCP | Poly-methacrylic acid–chitosan-PEG |
| PDT | Photodynamic therapy |
| PE | Phosphatidylethanolamine |

| | |
|---|---|
| PEG | Polyethylene glycol |
| PEG-SWCNTs | PEG-modified single-walled carbon nanotubes |
| PEI | Polyethylenimine |
| PET | Positron emission tomography |
| PFC | Perfluorocarbon |
| PFS | Progression-free survival |
| PHBV | Poly(hydroxybutyrate-co-hydroxyvalerate) |
| PLA | Poly lactic acid |
| PLD | PEF-coated liposome |
| PLGA | Poly lactic-co-glycolic acid |
| PLGA-PEG-Fuc | Fucoxanthin inside PLGA-block-PEG |
| PLN | PEGylated lipid nano-carriers |
| PPa | Pyropheophorbide a |
| PRV | Herpesvirus |
| PS | Polystyrene |
| pTHPP | 5,10,15,20-Tetrakis(4-hydroxy-phenyl)-21H,23H-porphine |
| PUE | Puerarin |
| QDs | Quantum dots |
| RA | Rheumatoid arthritis |
| RES | Reticuloendothelial system |
| RGD | Arginyl-glycyl-aspartic acid |
| ROS | Reactive oxygen species |
| RT | Radiation therapy |
| RVL | Resveratrol |
| Sch B | Schisandrin B |
| SDT | Sonodynamic therapy |
| siRNA | Small interfering RNA |
| SLNs | Solid lipid NPs |
| SOD | Superoxide dismutase |
| SPIONs | Superparamagnetic iron oxide NPs |
| TAM | Tumor-associated macrophages |
| TGMS | Triglycerol monostearate |
| TLR | Toll-like receptor |
| UCNP/TPP | Upconverting nanophosphors and *meso*-Tetraphenyl porphine |
| WSONs | Water-soluble organic NPs |

## 18.1 Introduction

The ultra-small size and modifiable surface characteristics of nanostructures make them beneficial for targeted drug delivery applications (Mousa and Bharali 2011). The main obstacle to the NP-assisted delivery of therapeutic agents is the high tendency of individual particles to form irregular clusters or agglomerates under physiological conditions (Li et al. 2015a). The poor aqueous solubility of nanoparticles

(NPs) is another problem that limits their transfer to target organs, tissues, and cells. To avoid these problems, the surfaces of nano-carriers are usually coated with end-grafted biodegradable polymer moieties, such as poly lactic-co-glycolic acid (PLGA), polyethylene glycol (PEG), and surfactant copolymers (Li et al. 2015a; Gera et al. 2017).

In the fields of biomaterials and drug delivery, PEG is the polymer that is used the most. It is a Food and Drug Administration (FDA)-approved, nontoxic, neutral, and biologically compatible polymer that displays advantageous characteristics in aqueous solutions such as high stability, solubility, and preferential hydration (Harris 1992; Salim et al. 2007). When used as an outer shell of NPs, PEG coating can decrease protein opsonization on their surface and extend the circulation time (Immordino et al. 2006; Perrault et al. 2009). It is commonly incorporated into liposomes by the addition of PEGylated lipid derivatives into liposome constituents at the formulation stage (Heyes et al. 2006). Phosphatidylethanolamine (PE) is the hydrophobic anchor commonly used to form mixtures of PEG-phospholipid conjugates (Zalipsky 1995; Vernooij et al. 2002; Kibria et al. 2011).

To increase cellular uptake of PEGylated nanoconjugates, targeting ligands are conjugated to the free ends of tied chains covalently and can precisely target the receptors on the diseased cell surface (Chen et al. 2013a; Cho et al. 2010; Nativo et al. 2008; Oh et al. 2011). The physicochemical properties of NPs (e.g., particle size, morphology, and surface charge) directly affect their pharmacokinetic performance, biodistribution, and cellular absorption (Perrault et al. 2009; Chen et al. 2013a; Cho et al. 2010; Nativo et al. 2008; Oh et al. 2011; Tang et al. 2014). Accumulated evidence from in vitro experiments suggests that ultra-small NPs can be delivered to diseased cells more easily and efficiently, while the ones with spherical morphology are likely to have faster internalization rates (Li et al. 2015a; Cho et al. 2010).

PEGylation of drug surfaces was first introduced, in 1981, to develop new and improved versions of pharmaceuticals and then repeatedly applied to a wide range of biomaterial surfaces to render them more like the body (Hoffman 2016). The first PEGylated drug that got full FDA approval was called ADAGEN and was used for the treatment of immunodeficiencies in children (Holeab et al. 2017). The attachment of PEG to the surfaces of nanostructures modifies the conformation, immunogenicity, hydrophobicity, and electrostatic binding of the molecule, causing better epithelial permeation, slower absorption, decreased accumulation, and deprivation of nanoparticulate preparations by the reticuloendothelial system (RES), less frequent administrations, elongated steady-state concentrations, and minor dosages (Holeab et al. 2017). Surface engineering of NPs with PEG generates a steric barrier against macrophage uptake and increases opsonin protein adsorption (Howard et al. 2008; Salim et al. 2014). As well as facilitating drug delivery to target sites, improving their stability, and extending their retention time in the body, PEGylation is also used to enhance contrast and resolution in medical imaging (Fig. 18.1 and Table 18.1).

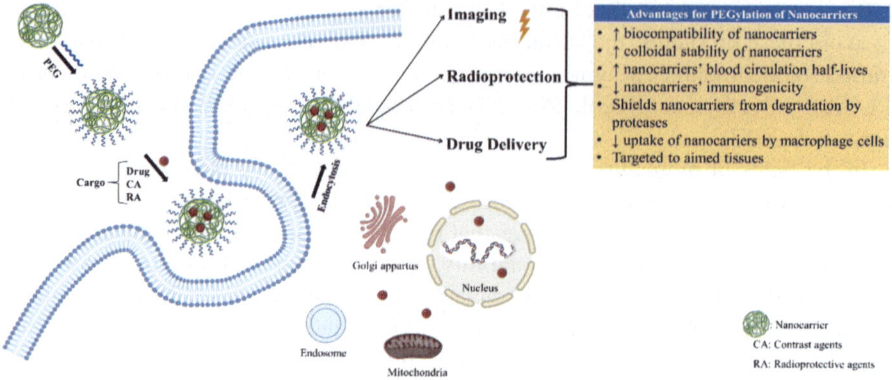


**Fig. 18.1** Therapeutic and pharmaceutical application of PEGylated nano-carriers. Cargo inside the PEGylated nano-carriers enters the cells via endocytosis. Cargo can be contrast agents (CAs), radioprotective agents, or pharmaceutical agents. In medical imaging, CAs are used to enhance the visibility of structures. Radioprotective agents help protect tissues from radiation damage and can be used in combination with radiotherapy for cancer treatments

**Table 18.1** PEGylated nano-carriers with medical and pharmaceutical application. Clinically approved nano-carriers have been highlighted in bold

| Nano-carrier | Application | References |
|---|---|---|
| PEG-b-PLA-UCNP/TPP | PDT | Shan et al. (2011) |
| WSON-PEG (3400) | PDT | Liu et al. (2021a) |
| PLGA-lipid-PEG hybrid NP containing pTHPP | PDT | Pramual et al. (2017) |
| PHBV-lipid-PEG hybrid NP containing pTHPP | PDT | Pramual et al. (2017) |
| PEG-Ce6-PEI@PB | PDT | Wang et al. (2022) |
| PLN-Cur | PDT | Abdel Fadeel et al. (2020) |
| Se-PEG-Cur NPs | Sonodynamic therapy | Ilbeigi et al. (2023) |
| DOXIL | Chemotherapy | Lao et al. (2013) |
| Onivyde | Chemotherapy | Papi et al. (2017) |
| PEGylated magnetic NPs functionalized with Tamoxifen-loaded FA | Chemotherapy | Majd et al. (2013) |
| PEG-PLGA NP containing docetaxel | Chemotherapy | Conte et al. (2018) |
| PEGylated Mn containing MOF NPs | Chemotherapy | Liu et al. (2021b) |
| PEGylated PLGA NPs incorporating TLR3/7 encapsulating OVA antigen | Immunotherapy | Cruz et al. (2014) |
| PEG-SWCNTs | Immunotherapy | Sacchetti et al. (2013) |

(continued)

**Table 18.1** (continued)

| Nano-carrier | Application | References |
|---|---|---|
| Mannose-modified PEG-sheddable NPs | Immunotherapy | Zhu et al. (2013) |
| PEGylated nano-vesicles (PEG-NPs) | Immunotherapy | Ochyl et al. (2018) |
| FNPs/rGO-PEG nanocomposites | Immunotherapy | Wang et al. (2019) |
| PEGylated AuNPs-based platform (AuRad™) | RT | Kumar et al. (2013) |
| AuNPs encapsulated in PEGylated nanogels | RT | Yasui et al. (2014) |
| PEG-Glu-AuNPs | RT | Geng et al. (2014) |
| PEG-coated Au@$MnO_2$ NPs | RT | Yi et al. (2016) |
| PEGylated MNP-PLGA-PEI NPs | Reducing nanoneurotoxicity of magnetic NPs<br>Enhancing transfection efficiency with magnetofection | Cui et al. (2019) |
| LDNPs | Diminishing the aggregation of NPs in the brain<br>Diminishing neuroinflammatory reactions | Huang et al. (2013) |
| Au-PEG nanoconjugate | Enhancing longevity of mice with ALS<br>Postponing the average age of ALS initiation<br>Ameliorating rotarod (a behavioral test) performance in mice with ALS | Vimal et al. (2022) |
| Trolox-functionalized Au@PEG | Reducing nanoneurotoxicity of Au@PEG | Zhang et al. (2020) |
| PLGA-PEG-Fuc containing fucoxanthin | Facilitating fucoxanthin's anti-Alzheimer's effects in vivo | Yang et al. (2021) |
| RGD-PEG-PLA NPs containing miR-133 | The facilitation of miR-133 delivery to infarcted tissue in rats | Sun et al. (2020) |
| $TiO_2$-PEG/Chitosan hydrogels | Enhancing cell adhesion of TiO2-PEG/CTS hydrogels on the cardiomyocytes cells | Liu et al. (2018b) |
| MMP-PEG-Sch B SLNs | Enhancing the Sch B concentration in cardiac tissue and blood circulation time of the drug | Shao et al. (2017) |
| RGD/PEG-PUE-SLN | Enhancing the half-life of puerarin in circulation<br>Strengthens protective effect of puerarin on acute MI rats | Dong et al. (2017) |

(continued)

**Table 18.1** (continued)

| Nano-carrier | Application | References |
|---|---|---|
| BN-PEG-NLC | Increasing the plasma half-life of the baicalin and its heart concentration | Zhang et al. (2016) |
| PCL–PEG containing DEXA | Enhance the half-life of DEXA in blood and targeted drug accumulation in joints with inflammation | Wang et al. (2016) |
| Ac2-26 peptide-loaded LDNPs | Enhance bioavailability and stability of Ac2-26 | Qin et al. (2021) |
| PEGylated GA NPs | Enhance bioavailability of GA | Nguyen et al. (2020) |
| DEXA-loaded MMP-responsive LDNPs | Lead to release of DEXA in areas where there are active MMPs | He et al. (2020) |
| PEGylated PLGA NPs | Diffuse more easily into the mucous layer of vaginal | Ensign et al. (2012a) |
| PEGylated silver-doped zinc oxide NPs | Photosensitizers for therapy of parasitic diseases | Nadhman et al. (2014) |
| Amphotericin B-loaded PLGA-PEG NPs | More effective than free amphotericin B against Leishmania donovani | Kumar et al. (2015) |
| PEGylated $CdS/Cd(OH)_2$ fluorescent nanocrystals | As biolabels of Trypanosoma cruzi | Chaves et al. (2008) |
| AMP bind to PLGA-PEG NPs | Speed up killing kinetics of *Pseudomonas aeruginosa* and *Staphylococcus aureus* | Ramôa et al. (2023) |
| Insulin-loaded PEGylated calcium phosphate NPs | Oral insulin delivery | Ramachandran et al. (2009) |
| PEGylated starch acetate NPs | Oral insulin delivery | Minimol et al. (2013) |
| Encapsulated insulin in silica NPs coated with PEG | Oral insulin delivery | Andreani et al. (2014) |
| pH-sensitive (PCP) NPs | Oral insulin delivery | Rostamizadeh et al. (2015) |
| B12-functionalized PEGylated liposomes | Oral insulin delivery | Sarhadi et al. (2022) |
| PEG-poly (Ser-Ketal) | The enzyme-based polymersome to deliver insulin in a glucose-dependent manner | Tai et al. (2014) |
| PEG-b-PAA-co-PAAPBA | Controlled insulin delivery | Wang et al. (2009) |
| PEG-polyboroxole block copolymer polymersome | Controlled insulin delivery | Kim et al. (2012) |

## 18.2 PEGylated Nano-Carriers for Bioimaging

Molecular imaging is a subfield of medical imaging that focuses on enlightening molecular processes involved in various diseases and conditions (Peng et al. 2013). Photo-based detection and treatment techniques are gaining importance due to their amplified spatial imaging resolution and negligibly invasive nature (Nguyen et al. 2013). Magnetic resonance imaging (MRI) and computerized tomography (CT) are both diagnostic techniques commonly used in clinics due to their advantages, including high-quality and reproducible output, a noninvasive nature with no tissue damage, and high 3D spatial resolution (Jin et al. 2015; Liu et al. 2013). Nanomedicine exploits nanotechnology to diagnose and cure debilitating diseases, leading to the prominence of photo-based nanomedicine in treating cancer and other diseases (Chow and Ho 2013). Commonly applicable ingredients for developing photo-based NP arrays range from metallics (gold, silica, and silver) to polymerics (dextran, chitosan, PLGA, and PEG). Among the polymers that are frequently used for light-responsive delivery systems, PEG is commonly applied for surface modification and developing nano-carrier systems (Nguyen et al. 2013). Single-modal molecular imaging has not had the ability to fulfill the advanced requirements of accuracy and efficacy in clinics (Liu et al. 2013). Nanostructured multimodal imaging probes (NP-based agents) take full advantage of different imaging modalities, e.g., MRI offers a great spatial resolution while optical imaging permits fast screening. However, optical MRI contrast media have inherent limitations that prevent their widespread use, such as low sensitivity and diagnosis efficiency. Gadolinium chelates and iodinated molecules have been used as modern CAs for MR and CT imaging, respectively, but face some restrictions related to low specificity, short imaging time, vascular penetration, and probable renal toxicity. New multifunctional probes are needed for contrast improvement in vivo, with major uses for specific imaging and angiography (Jin et al. 2015). Multimodal imaging by means of nanomaterials is hopeful for patients and affords exact information on the tumor size and location for doctors. Nanocomposites can be modified with PEG for biocompatibility improvement. Zhen Liu et al. developed a method of production of PEGylated europium ($Eu^{3+}$)-doped mesoporous gadolinium oxide (hybrid $Gd_2O_3$) nanorods for high-resolution multi-modality imaging studies (Liu et al. 2012).

### 18.2.1 MRI

MRI is a method of imaging that does not involve any risk of being exposed to ionizing radiation (Maddah et al. 2019; Wang et al. 2013). It produces diagnostic images on the basis of differences in the magnetic relaxation behavior of dispersed water molecules in tissues, but its low sensitivity disturbs the quality of its diagnosis (Mousa and Bharali 2011; Maddah et al. 2019; Grant et al. 2002; Liu et al. 2014a). MRI CAs agents can improve the imaging contrast between pathological and normal locations significantly (Zhang et al. 2013). They can shorten the relaxation time of water protons in order to improve diagnostic accuracy and sensitivity. Generally, commercial CAs are based on polydentate organic ligands comprising gadolinium ions ($Gd^{3+}$) and manganese ions ($Mn^{2+}$). Both CAs, based on metal ions and chelators, have restricted utilization for target tissues or cells in real-time monitoring due to a lack of target specificity and short blood half-lives (Maddah et al. 2019). Now, MRI is implemented with the aid of CAs in about one-third of clinical cases to reach an exact and reliable pathologic change diagnosis. In the last several decades, researchers have focused on the application of nanotechnology for the efficiency and sensitivity improvement of MRI for cancer diagnosis and imaging (Mousa and Bharali 2011). According to the fast improvement in nanotechnology investigation, researchers are exploring nano-carriers to overwhelm the MRI CAs problems. A suitable nano-carrier needs to be stable chemically, biocompatible, and exhibit little toxicity and great relaxation per volume (Maddah et al. 2019). Because of their low toxicity, very high specific surface area, and high water solubility, graphene-based materials, such as graphene oxide (GO), have recently been investigated for biomedical applications (Zhang et al. 2013). GO-based nanocomposites and functionalized GO, which demonstrate exciting magnetic and optical features, can be employed as CAs for different biological imaging techniques, such as MRI. Though GO is normally known to be safe, it may prompt blood platelets (Yang et al. 2013a), but PEGylated GO has not shown noticeable toxicity in a 3-month in vivo study (Yang et al. 2011). PEG enhances NPs' blood circulation half-lives and reduces their immunogenicity. Furthermore, the hydrated PEG moiety shields the NP from admission to peptidases and proteases, so decreases its degradation (Maddah et al. 2019). As PEG can protect their surface from the adsorption proteins, it can aid in the inhibition of NP phagocytosis by the mononuclear phagocyte system (MPS), such as macrophage cells. Recently, Robinson and colleagues have shown that GO PEGylation, acquired nGO–PEG reduction, and further PEGylated phospholipid coating can prepare ultra-small reduced nano-GO (nRGO) (Robinson et al. 2011). Mengxin Zhang and colleagues created a versatile theranostic nano-enabled platform for MRI and drug delivery based on GO. The cellular MRI quality of CA-based Gd has traditionally been limited because small Gd complex molecules, such as Magnevist, are likely to disperse in the extracellular space. One of the valuable strategies for elucidating this problem is the adaptation of nanosized carriers to bring numerous paramagnetic metal complexes to develop the MRI contrast of considered cells (Zhang et al. 2013).

The various other types of nano-carriers comprising liposomes, magnetic NPs, ultrasound nanobubbles, and quantum dots, which can progress the exact detection of diseases, have been developed recently (Lahooti et al. 2017). Despite developing various kinds of nano-carriers as MRI CAs, most of the nano-carriers require complicated surface modification and manufacturing steps for the binding process of paramagnetic ions prior to their use as MRI CAs (Maddah et al. 2019). Among the developed nano-carriers, superparamagnetic iron oxide NPs (SPIONs) coated with biocompatible materials such as PEG and dextran can be used as a CA for metastases diagnosis in the RES such as the spleen, liver, lymph nodes, and bone marrow (Lahooti et al. 2017). SPIONs have many medical applications as a negative CA in MRI, including cell tracking, cancer imaging, targeted delivery of genes and drugs, imaging of tumor-associated macrophages, theranostic applications, and critical biological procedures such as apoptosis. Lahooti et al. discovered a high accumulation of $^{68}$Ga-radiolabeled SPIONs in the spleen and liver and can be used in positron emission tomography (PET)/CT or PET/MR imaging for the diagnosis of spleen and liver malignancies (Lahooti et al. 2017). Through the study of Mahsa Maddah et al. the prepared melanin NPs (MNP)-Mn-PEG can be considered an MRI CA

(Maddah et al. 2019). Jingjing Li et al. have synthesized hydrophilic manganese oxide (MnO) NPs as T1 nano-CA for molecular MRI of renal cell carcinoma and synthesized the PEG-MnO NPs displaying good T1 relaxivity (Li et al. 2018). Wensheng Xie et al. have shown that the combination of doxorubicin-loaded PEG and 2-deoxy-D-glucose (2-DG)-modified $Fe_3O_4$@$MoS_2$ nanocubes can be used as an efficient theranostic platform for synergistic MRI-guided chemo-photothermal treatment of breast cancer (Xie et al. 2018a). Kai Yang et al. have developed PEGylated FeS nanoplates, for the first time, as a new agent with multifunctional theranostic capabilities for in vivo MRI-guided photothermal cancer therapy (Yang et al. 2015). Additionally, binding arginyl-glycyl-aspartic acid (RGD) peptide and PEG to the surface of ultrasmall iron oxide NPs enables targeted delivery of the NPs, which can be detected by MRI. This allows for the precise tracking of glioma cells both in vitro and in vivo (Luo et al. 2015).

MRI is a frequently used invasive imaging method in clinics for diagnosing diseases. Nano-carriers have been used to overcome the MRI CA's problems. They can be targeted to specific tissues by modifying them with PEG, which increases NP blood circulation half-lives while decreasing immunogenicity.

### 18.2.2 CT

CT is one of the most common molecular and diagnostic imaging methods in clinical settings due to its deep tissue penetration, great density and spatial resolution, rapid scan speed, cost-effectiveness, and facile post-image processing. CAs are generally required for effective, specific, and qualified CT imaging applications. Small molecular CAs based on iodine (for example, Omnipaque) have significant limitations, such as short imaging times due to quick elimination by the kidney, non-tissue specificity, and renal toxicity at high doses, which limit their CT imaging performances. Developing new CAs to overcome the mentioned drawbacks is important (Liu et al. 2014b; Peng et al. 2012; Zhou et al. 2014). Due to their unique quantum size-associated features, a wide variety of NPs as CT CAs have been evaluated for medicinal applications (Zhou et al. 2016a). Advantages of inorganic NPs include increased blood circulation duration, lower renal toxicity, and more selective targeting due to their nanoscale sizes, larger atomic numbers compared to iodine, and ease of surface functionalization. Among the established inorganic CT, CAs based on NP (e.g., nanosized gold, tantalum, ytterbium, platinum, and bismuth-based inorganics) have attracted enormous attention owing to their advantages. Compared to iodinated small molecular CT CAs, they have higher X-ray attenuation power due to the radiodense element's high atomic number. Other advantages include good biological compatibility after surface functionalization, ease of synthesis and manipulation, good stability, and an electron density permitting higher CT imaging sensitivity. On the other hand, they have been stated to be cytocompatible and non-toxic in a definite concentration after surface tailoring and tunable surface chemistry, providing easy functionalization by biological ligands (Peng et al. 2013; Connor et al. 2005). Individually, $Bi_2S_3$ NPs, Yb-based NPs, and gold NPs (AuNPs) are designed for different purposes in CT imaging (Liu et al. 2014b; Peng et al. 2012; Zhou et al. 2014). Dendrimers, among other polymer-based techniques for AuNP formation, can be used to template or stabilize AuNPs. This makes AuNPs ideal for use in CT imaging and producing high-quality images. The main advantage of using nanotechnology for dendrimers is that particle surface transformation can be facilitated with dendrimer modification prior to or following the formation of AuNPs (Zhou et al. 2014). Customizing AuNPs with targeting ligands is essential to increasing the performance of CT scans in cancer. The surface of AuNPs is frequently PEGylated to improve colloidal stability and cytocompatibility for use in the living body (Zhou et al. 2016b). Furthermore, modification of the surface with PEG can enlarge the dendrimer margin for more Au deposition within the reactive centers, permitting more sensitive CT imaging. Similarly, the dendrimers' modification with PEG provides the AuNPs formation with enhanced biological compatibility, reduced uptake of macrophage cells, and a prolonged blood circulation period, which is valuable for improving the blood pool and applications of CT imaging for tumors (Zhou et al. 2014). They can be accumulated in the phagocytic cells of the liver and spleen for hepatocellular carcinoma CT imaging (Zhou et al. 2016b). AuNP-incorporated PEG-attached dendrimers are a candidate for CT CA due to their ability to serve as nanoscale heat sources under light irradiation (Kojima et al. 2010). The characterization and synthesis of AuNPs trapped within PEGylated polyethylenimine (PEI) for blood pool and CT screening of cancer have been reported via the research of Zhou and coworkers. In addition to the improved penetrability and retention effect, the PEG-modified PEI-entrapped AuNPs (Au PENPs) could be utilized as a CA for effective tumor CT imaging (Zhou et al. 2014).

After particle formation, dendrimer terminal amines acetylation is frequently performed to provide the particles with their preferred biocompatibility. Prominently, the study of Hui Liu et al. has exhibited that the PEG modification of the 5th generation (G5) dendrimer surface allows a distended dendrimer margin for more Au element loading inside the dendrimer center. They have developed a robust method for PEGylated dendrimer-entrapped AuNPs (Au DENPs) fabrication for applying CT imaging with low-generation dendrimers. Their results have suggested that PEG modification of G2 polyamidoamine dendrimers can considerably enlarge the dendrimer periphery and geometry, which allows the effective trapping of AuNPs within the dendrimer centers. The designed PEG-Au DENPs with good cytocompatibility and stability in a certain concentration range have exhibited greatly enhanced X-ray attenuation capability compared to Omnipaque, the clinically used CT CA, at the same radiodense element (Au or iodine) concentration (Liu et al. 2014b). Chen Peng et al. have synthesized and characterized the folic acid (FA)-modified Au DENPs through a facile PEG linking approach for in vivo tumor-targeted CT imaging applications (Peng et al. 2013). Benqing Zhoua et al. demonstrated a simple method for producing folic acid (FA)-targeted multifunctional AuNPs via PEG modification via cost-effective branched PEI as a pattern for tumor CT imaging applications. PEI was sequentially modified by PEG monomethyl ether, FA-decorated PEG, and fluorescein isothiocyanate in their study, and then synthesized from a template for AuNPs, with the remaining PEI surface amines being transformed to acetamides (Zhou et al. 2016b).

Different NPs have been used for CT imaging applications, and their surfaces have frequently been modified with PEG to improve their colloidal stability and cytocompatibility for utilization in the living body.

### 18.2.3 Synergistic of CT and MR Imaging

Each imaging technique has some benefits and restrictions. Combining CT and MR imaging methods provides several advantages for diagnostic purposes, as it can cover multiple issues such as resolution, tissue penetration, and sensitivity (Zhou et al. 2016a). Up to now, multimode nanoparticulate CAs have been industrialized with the incidence of different nanoplatforms that are able to be simply combined with various imaging agents, especially radiodense agents for CT and MR imaging (Zhou et al. 2016a). Zhen Liu et al. created PEG-modified ytterbium NPs and investigated their use in vitro and in vivo as wide spectral range nanoprobes for CT imaging and MRI. The NPs produced were highly water-soluble and monodisperse (Liu et al. 2013). Dalong Ni et al. synthesized and explored a new type of $NaHoF_4$ NPs with different particle sizes as high-performance dual-modality CAs for MR and CT imaging. The PEGylated $NaHoF_4$ NPs improved imaging quality and could serve as CAs for CT and ultra-high field multimodal MR imaging, according to systematic in vitro and in vivo studies (Ni et al. 2016). Benqing Zhou et al. designed a simple method to produce cost-effective PEGylated branched PEI Au PENPs with gadolinium (Gd@Au PENPs) as a nanoscaffold for in vivo dual-mode CT/MR scanning of blood pools and main organs (Zhou et al. 2016a). Xiaoying Jin et al. designed and manufactured a new PEG-modified, Dy-doped $NaGdF_4$ nanoprobe that integrates $Gd^{3+}$ and $Dy^{3+}$ into a single nanoplatom. The nanoprobes have demonstrated low cytotoxicity, disregardable potential toxicity in vivo, and elimination from the injected mouse body. PEG-$NaGdF_4$-Dy NPs provided significantly greater contrast enhancement in CT and MR imaging in vitro and in vivo than Gd-chelated and iohexol. Recent improvements in nanotechnology have created a range of MR/CT multimodal imaging nanoprobes, including $Fe_3O_4$-Au NPs, $Gd^{3+}$/$Yb^{3+}$-doped ZnO NPs, and GO-$BaGdF_5$ NPs (Jin et al. 2015).

### 18.2.4 Radioprotection

Ionizing radiation (IR) protection studies have significant effects on human health. Radiation therapy (RT), comprising internal radioisotope therapy (IRT) and external beam radiotherapy (EBRT), is comprehensively used for numerous malignant diseases and cancers. According to the remarkable growth in the utilization of radiation strategies in medicine with the aim of radiotherapy and diagnosis, there is a danger of destructive radiation contact (Cao et al. 2020). The application of radioprotectors is the most common strategy to reduce the radiation's adverse effects

(Bagoria et al. 2022). The number of radioprotective drugs that can be employed in clinics is restricted as the radioprotectors' therapeutic efficacy is limited by drawbacks such as short circulation half-lives and adverse effects. As a result of the fast spread of drug-delivery system (DDS) nanotechnology, multifunctional nanostructures have been created to increase the efficiency of molecular radioprotectors and mitigate side effects (Cao et al. 2020). Until now, numerous radioprotectors have been developed and fabricated, most of which are organic molecular agents (Xie et al. 2018b).

Ceria NPs (CNPs) have recently been shown to protect cells and animals from radiation injury (Li et al. 2015b). Cerium oxide ($CeO_2$) NPs can scavenge free radicals by changing their surface charge state. Definitely, it has been reported that the $CeO_2$ NPs can catalyze the hydrogen peroxide decomposition in neutral media in healthy tissues, though in the acidic tumor microenvironment, the $CeO_2$ catalytic action will vanish. Thus, $CeO_2$ NPs can guard healthy tissues effectively against radiation-caused free radical injuries (Song et al. 2017). Most previous research on CNP radiation protection capability has used NPs without appropriate surface protection, which unavoidably leads to many impediments in vivo, particularly aggregation and elimination with RES, causing NP activity to decrease and even increase toxicity at the end. Numerous hydrophilic polymers, such as chitosan, acrylic acid, and dextran, have been employed for coating the CNPs surfaces to increase NP stability and modify the molecules' surface charges. Among all of the polymers, at present, PEG is the most general and was found to be the most effective one for increasing biocompatibility and modifying the inorganic NPs surface charge. PEGylated CNPs have been employed for the radioprotection of humanoid healthy liver cells (L-02). The PEG-modified CNPs have indicated greater radioprotective capacity in comparison to the pure ones. Therefore, the physicochemical factor optimization of NP based on cerium is important for additional improvement of their radioprotection efficacy (Li et al. 2015b). According to the research of Yang Z-Y and colleagues' research, alendronate (AL) has been discovered as a perfect anchor which can make the PEG conjugation to CNPs simple and result in developed biological properties, long-lasting blood circulation, enhanced stability, decreased cytotoxicity, improved superoxide dismutase (SOD) enzyme mimetic activity, and reduced liver and particularly spleen NPs accumulation (Yang et al. 2014). Hong Li et al. synthesized PEGylated CNPs (CNPs-AL-PEG) and studied their radioprotection application which showed less cytotoxicity and yielded more effective radiation protection to normal cells than free CNPs. It has been recognized that the elevated $Ce^{3+}/Ce^{4+}$ ratios, enhanced dispersion in cells, and the better CNPs-AL-PEG distribution in cells contribute to the ameliorated SOD expression, which caused improved healthy cells' protection from ROS and mitigated the oxidative DNA impairment (Li et al. 2015b). According to the perfluorocarbon (carbon-fluorine compounds) (PFC) ability for oxygen delivery and the Bi high-Z radiosensitization function, in the current work of Song et al. PEG-modified $Bi_2Se_3$ hollow NPs have been created and then loaded with PFC, providing PEG-$Bi_2Se_3$@PFC as a new radiosensitizer (Song et al. 2016). With PFC loading, PEG-$Bi_2Se_3$@PFC could perform as an oxygen reservoir (or oxygen carrier) and release oxygen gradually under motivation by a near-infrared (NIR) laser for improved radiotherapy, proposing a better radio-sensitization influence in vitro. Recently, Chao et al. loaded $WS_2$ nanoflakes with $^{188}Re$ in a chelator-free mode by only complexing the radioisotope with $WS_2$-PEG for tumor-directed IRT (Chao et al. 2016). The acquired $^{188}Re$-$WS_2$-PEG could act as a carrier to transport radioisotope $^{188}Re$ into the tumor efficiently after intravenous injection. Furthermore, similar to further high-Z agents as radiosensitizers, tungsten in $^{188}Re$-$WS_2$-PEG could absorb ionization radiation produced from $^{188}Re$, allowing "self-sensitization" to considerably increase the IRT efficiency. Organic biodegradable nano-carriers, such as albumins and liposomes, have previously been accepted for clinical trials among the various types of nano-carriers evaluated for IRT tumor-targeted delivery (Song et al. 2017). Jian Cao et al. have developed and produced radiation countermeasures based on metal-organic frameworks (MOFs) and investigated their radioprotective efficiencies. They have invented a novel nanoscale-based drug carrier MIL-101(Fe) with the aim of co-delivering two radioprotective agents by GSH and WR-1065, as amifostine active metabolite, encapsulation to prevent IR damage. Amifostine is an effective radioprotector that is often used in clinical settings to protect normal tissue from free radical damage caused by radiation. But amifostine has some common deficiencies, such as a fast metabolism, short distribution in the body, limited oral activity, insolubility in water, and adverse effects on normal organisms, which reduce its drug efficacy and restrict its employment (Cao et al. 2020). Yang et al. demonstrated that the PEGylated amifostine can prolong antioxidant effects and protect against cisplatin in vitro. PEG chains can be inserted into the MOFs to increase their mucus penetration and endow them with some significant properties, such as stability, solubility in water, biological compatibility, greater radioprotective efficiency, and better permeability for oral drug delivery. In their biomedical applications, PEG groups can protect the MOFs from long circulation time and increase cellular absorption (Yang et al. 2016). Obviously, the synthesis of qualified PEGylated MOF NPs is essential for large-scale drug delivery (Cao et al. 2021). According to the results of in vitro studies, the radioprotector based on MOF has shown lower cytotoxicity in comparison to naked WR-1065 and has improved cell viability via resisting DNA injury and scavenging reactive oxygen species (ROS) that have accumulated after irradiation. The MOF-based radiation countermeasures have resulted in important progress in the count of marginal blood cells, the clonogenic function of hematopoietic progenitor cells, and the frequency of progenitor and hematopoietic stem cells. The MOF-based radioprotectant can be applied as an encouraging therapeutic candidate to reduce total body irradiation-induced hematopoietic damage (Cao et al. 2020). The MIL-101 NP-based radioprotective agent has been produced on a large scale, with PEG modification, through a green continuous-flow approach to ameliorate its permeability and stability. These have confirmed that the MOF NPs can be sustainably fabricated at large scales and be applied to inhibit radiation damage (Cao et al. 2021).

Iqbal Ahmad et al. have successfully fabricated trans-resveratrol (RVL)-encapsulated solid lipid NPs (SLNs) by means of supercritical fluid technology with Gelucire®50/02 and Gelucire®50/13. The optimized formulation has been attained on the basis of particle size (nm), yield (%), entrapment efficacy (%), and polydispersity index and is PEGylated to provide covert properties. The pharmacokinetic profile has evidenced the improved residence time and long circulation of RVL after intravenous bolus injection of PEG-modified SLNs in Wistar rats in comparison to the free drug. Furthermore, an accelerated stability test predicted high stability and a minimum shelf life of two years at 8 °C. These findings are encouraging for ionizing radiation protection at a lower RVL dose over a longer period of time (Ahmad et al. 2016).

The use of nanomedicine approaches is advantageous to increasing tumor radiation responses as nanotechnology advances. As nanomaterials have high-Z elements to attract radiation rays, for example, X-rays, they can play as radio-sensitizers to trap radiation energy inside tumors and improve therapeutic efficiency. Nanoscale carriers have the ability to transport therapeutic radioisotopes into tumors for IRT or chemotherapy drugs for synergistic combination chemo-radiotherapy. Some nanocarriers with appropriate radioprotective drugs can decrease radiation-caused injury in healthy tissue. PEG modification of nano-carries/PEG can be generally performed to increase biocompatibility.

## 18.3 PEGylated Nano-carriers for Therapy

### 18.3.1 Targeted Cancer Therapy

#### 18.3.1.1 Photodynamic Therapy (PDT)

PDT is considered a minimally invasive means of treating tumors. It kills cancer cells by activating a photosensitizer using visible light. The excited photosensitizer molecule absorbs energy from the light, which causes it to become unstable. Because of this instability, it reacts with the surrounding environment, producing free radicals or ROS that can harm nearby tumor cells (Champeau et al. 2019). ROS generated from light exposure can only affect cells in their immediate vicinity. Therefore, the light can be targeted at the tumor while leaving healthy cells untouched in the distance (Zhou et al. 2016c). PDT works by targeting cancer cells with a light-sensitive drug triggered by a particular wavelength. This allows cancer cells to receive the drug directly, limiting damage to healthy tissues and minimizing the risk of drug resistance and systemic toxicity. Additionally, it has the potential to be more selective in targeting cancer cells, making it more effective than chemotherapy and radiotherapy (Agostinis et al. 2011). Limited light penetration in tissue to activate photosensitizers, the toxicity of soluble photosensitizers, such as Photofrin, and the difficult delivery of hydrophobic photosensitizers to the tumor are the main obstacles to PDT. PDT with PEGylated NPs is a procedure in which light-activated drugs are delivered to target tissues in combination with PEGylated nanomaterials, allowing more efficient and targeted delivery of the drugs. PEG-incorporated phospholipid coating helps stabilize hydrophobic photosensitizers in aqueous solutions (Li et al. 2007). The use of PEG-b-polylactic acid block copolymers has been shown to improve the stability of NPs containing upconverting nanophosphors and *meso*-Tetraphenyl porphine (UCNP/TPP). PEG-b-PLA-UCNP/TPP induces cell death in the HeLa cervical cancer cell line following 45 min of PDT with NIR light (978 nm). UCNPs act as transducers, converting as a NIR to Ultraviolet (UV)-visible light and TPP is a photosensitizer in this PEGylated NP (Shan et al. 2011). Alternatively, water-soluble organic NPs (WSONs) were constructed with a degradation product of chlorophyll-a, pyropheophorbide a (PPa), and PEG. In vitro and in vivo studies indicated that WSON-PEG (3400) with low toxicity in the absence of light has the highest efficiency in PDT (Liu et al. 2021a). 5,10,15,20-Tetrakis(4-hydroxy-phenyl)-21H,23H-porphine (pTHPP) serves as a photosensitizer during PDT. Encapsulating the pTHPP molecules within PLGA or PHBV NP cores with lipid-PEG as a stabilizer significantly improved their photophysical parameters, such as the production of singlet oxygen and the uptake of the molecules by the thyroid cancer cell line FTC-133 compared to nonencapsulated pTHPP molecules (Pramual et al. 2017). The capacity to self-supply oxygen and tumor acidity-induced PEI cytotoxicity were combined in a nano-drug platform for dual-enhanced PDT made from PEG-chlorine$_6$ (Ce$_6$)-PEI@prussian blue (PB). PEGylation diminishes the cytotoxicity of PEI and the phototoxicity of Ce$_6$ on non-target cells in blood circulation. The acidic condition of the extracellular medium of tumors induces degradation of PEG on the surface and exposes PEI-positive charges. The interaction of PEI with the negative charges of cancer cells increases cellular intake and increases PDT. PB as an artificial peroxidase enhances oxygen levels inside the tumor and enables Ce$_6$ to generate a more extensive number of cytotoxic ROS following laser irradiation to further enhance PDT (Wang et al. 2022). PEGylated lipid nano-carriers (PLN) containing curcumin (Cur) were fabricated to induce cell death in skin cancer cells by PDT. Treatment of skin cancer with PLN-Cur complexes was more successful and safer because they boosted cellular absorption of Cur by tumor cells and the photodynamic impact (Abdel Fadeel et al. 2020). Hence, photosensitizer-loaded PEGylated nano-carriers have the potential to transport large concentrations of medicines directly to the tumor tissue, thus reducing adverse effects and boosting the efficacy of the treatment. PEGylated nano-carriers are a promising delivery system because they are safe, economical, and effective and can be used to treat a variety of cancers.

Sonodynamic therapy (SDT) is a method to induce cell death in cancer cells using an agent known as sonosensitizer that is activated upon absorption of low-intensity ultrasound radiation. Selenium- and PEG-functionalized nano-Cur was introduced as a sonosensitizer with cytotoxicity activity against pancreatic ASPC1 cancer cell line (Ilbeigi et al. 2023).

#### 18.3.1.2 Chemotherapy

Chemotherapy is a chemical drug therapy that uses cytotoxic agents to kill fast-growing cancer cells. Chemotherapy drugs work by stopping the cancer cell's ability to grow and multiply, eventually leading to their death. Chemotherapy also affects healthy rapidly dividing cells, which can lead to a variety of side effects, including hair loss, anemia, nausea, and an increased risk of infection. Nano-carriers such as polymer-, lipid-, or metal-based nanomaterials can be used to deliver chemotherapy drugs more selectively to cancer cells and reduce the side effects of chemotherapy on healthy cells (Dastidar et al. 2022). The United States FDA has approved over 35 PEGylated NPs, and a greater number are currently in preclinical studies for both imaging and therapeutic purposes (Jokerst et al. 2011). Liposomes have attracted significant attention among the various PEGylated NP compositions for the targeted delivery of antitumor drugs. Due to their biocompatibility, stability, and capability to accommodate hydrophilic and lipophilic medicines, liposomes, which are formed of single or multiple concentric lipid bilayers, are an excellent alternative for drug administration (Liu et al. 2022). PEG incorporation into liposomes confers resistance to absorption by immune system cells, prolonging the serum half-life of the encapsulated drug. Additionally, PEG can form a protective layer on the liposomes, preventing the breakdown of their components and allowing extended circulation time in the blood (Allen et al. 1991). According to preclinical research with PEGylated liposomes, anticancer drugs encapsulated in PEGylated liposomes are more effectively concentrated in cancer tissues than in healthy cells (Newman et al. 1999). However, preclinical investigations of chemotherapeutic agents encapsulated in PEGylated liposomes in dogs and rodents revealed accelerated bloodstream elimination of the PEGylated drug carrier system because of enhanced anti-PEG-Immunoglobulin M (IgM) generation (Yang et al. 2013b; Suzuki et al. 2012). PEGylated liposomal doxorubicin (PLD), known as DOXIL, was developed to improve drug effectiveness in the chemotherapy of cancer. In this formulation, doxorubicin was encapsulated in unilamellar liposomes coated with PEG with an 80–90 nm size. This modification prolongs the drug's bioavailability, making it more effective in treating cancerous tumors (Lao et al. 2013). Almost all of the drug found in the blood after using PLD is PEGylated. Also, PLD is slowly cleared from circulation in comparison with the free form of the drug. With FDA clearance, PLD (DOXIL/Caelyx) is now being prescribed for recurrent ovarian cancer and Kaposi's sarcoma (Mishra et al. 2016; Gabizon 2006; Gabizon et al. 1994; Amantea et al. 1997). FDA-approved pancreatic cancer medication Onivyde is PEGylated liposomal. The presence of PEG on the surface of liposomes inhibits their quick elimination from circulation by the MPS. MPS recognizes opsonins and complement proteins attached to liposome surfaces. Liposomes are surrounded in biological media (such as blood or interstitial fluid) by a biomolecular layer referred to as a "protein corona" (PC). Liposome-PC plays a vital role in the transport of liposomes to their target sites, the interaction between liposomes and cells, and the response of cells to liposomes. It has been shown that PEG conjugation to liposome membranes decreases PC production in vivo, but does not entirely inhibit it, and PC significantly enhances Onivyde uptake by pancreatic duct adenocarcinoma cells (PANC-1) (Papi et al. 2017).

The use of magnetic NPs for targeted chemotherapeutic drug delivery is restricted due to disadvantages such as magnetic NPs self-assembly and nonspecific binding to plasma proteins following intravenous administration. A consequence of these problems is opsonization and immune system removal via the RES and greatly reduced magnetic NPs function against target cells. Surface modification of magnetic NPs with biocompatible polymers is one way to overcome these problems. In a study, PEGylated magnetic NPs conjugated with FA were used as carriers of Tamoxifen. FA facilitates the interaction between magnetic NPs and FA receptor-expressing cancer cells, preventing Tamoxifen from reaching those cells that lack these receptors. These magnetic NPs can also be used for imaging FR-positive cancer cells (Majd et al. 2013). In a study, NPs that are sensitive to oxidation-reduction conditions were developed by utilizing PEG-PLGA copolymers. Then, the NPs were employed to transport docetaxel to lung cancer cells. Hydrophobic PLGA served as the NPs' core, while PEG acted as the hydrophilic shell. The external hydrophilic shell of PEG was connected to the hydrophobic drug-carrying core (PLGA) by disulfide bonds. Since the NPs were functionalized with PEG on the outside, they were able to endure the harsh conditions of mucous for longer and pass through it more easily than uncoated NPs. When these NPs are exposed to the extracellular environment of cancer cells, thiols in the extracellular environment reduce the disulfide bonds between PEG and PLGA. This causes some PEG to be shed from the NPs. Reduced PEG shell thickness and density enhance the cellular absorption of NPs. In the cells, intracellular GSH reduces all the disulfide bonds of copolymers, which leads to the disintegration of the NPs and leakage of the drugs (Conte et al. 2018). Anticancer drugs can be delivered using nano-carriers of MOFs. According to in vitro and in vivo studies conducted, Mn-based MOF NPs bound to PEG have potent therapeutic benefits against fatal pancreatic cancers and activate innate antitumor immunity (Liu et al. 2021b).

#### 18.3.1.3 Immunotherapy

Recent advancements in immunotherapy have shown that the immune system has a significant function in cancer treatment. It operates as a kind of biotherapy by assisting the immune system in distinguishing and combating cancer cells. In recent years, an increasing number of immunotherapy medications have been licensed, and a number of therapies are in the early phases of clinical trials and preclinical investigations. Tumor cells are heterogenic, and the tumor microenvironment can suppress the immune response. For these reasons, immunotherapy is not successful in the majority of cancer therapies. Biomaterials such as NPs T-cell delivery therapy can help activate the immune system to target cancer cells and other pathogens while minimizing the risk of damaging healthy cells. These advanced biomaterials and drug delivery systems offer a way to improve the efficacy of immunotherapy treatments while reducing their side effects (Zhou et al. 2022). Physical features of pharmaceutical NPs, such as sizes, shapes, and surface charges, influence their therapeutic impact (Nel et al. 2009). NPs with almost neutral charges could be used to deliver tumor antigens because, when injected subcutaneously, they go to the lymph nodes that drain the tumor. The PEGylation of the NPs allowed them to reach the draining lymph nodes more efficiently and be more readily taken up by dendritic cells (DCs), which then presented the tumor antigens to the body's defense system. This resulted in an increased immune reaction, making the NPs more immune-stimulatory (Zhan et al. 2012). Accumulated scientific evidence has suggested that NP-mediated delivery of antigens to DCs can increase antigen presentation to T cells, thus inducing a stronger immune response than traditional delivery methods

(Bandyopadhyay et al. 2011; Bonifaz et al. 2004; Cruz et al. 2014). PEGylated PLGA NPs incorporating toll-like receptor (TLR) 3/7 and encapsulating ovalbumin (OVA) antigen were used to stimulate an effective cytotoxic T cell response by interaction with surface markers of DCs. According to in vitro cellular uptake results, NPs modified with TLR ligands were taken up more quickly and efficiently by DCs when compared to nontargeted NPs. Animal studies also indicated that ligand-targeted NP was more effective at stimulating CD8+ T cell responses than nontargeted NP (Cruz et al. 2014). PEG-modified single-walled carbon nanotubes (PEG-SWCNTs) were fabricated to increase the selectivity of the SWCNTs for targeting Tregs in the tumor microenvironment. PEG-modified and targeting ligand-attached SWCNTs were internalized predominantly by regulatory T cells that were located in the tumor microenvironment and activated by glucocorticoid-induced tumor necrosis factor (TNF) receptors (Sacchetti et al. 2013). Zhu and colleagues fabricated mannose-functionalized NPs with diffusible PEG (PEG-Mannose-NPs) to target tumor-associated macrophages (TAM). According to their findings, PEG-Mannose-NPs successfully targeted TAMs through the multilectin receptor proteins. A higher accumulation of NPs coated with caducous PEG was found in TAM compared to PEG which was difficult to detach. This may be attributed to the biochemical modification process, which diminishes NP opsonization and increases the duration of systemic circulation (Zhu et al. 2013).

αCTLA-4 and αPD-1 antibodies are among the checkpoint blockade inhibitors approved for clinical use. αPD-1 targeting is limited by the reliance on tumor-specific killer T cells (CTLs), which may not be present in all patients. Cancer vaccines may be able to solve this problem by inducing CTL responses, but first antigens specific to tumors must be identified and synthesized. Tumor cell lysate is a rich pool of tumor-associated and tumor-specific antigens that can be used to prepare vaccines. Tumor cell lysate-based vaccines have limited ability to elicit tumor-specific cytotoxic T lymphocytes. According to a study, NP vaccines made from tumor cell lysates coated with PEG are more effective at eliciting potent T-cell responses. The PEG coating improved serum stability and facilitated the delivery of vaccine NPs to the lymph nodes. Therefore, PEGylated NPs may be used as a potent platform for cancer vaccine development (Ochyl et al. 2018).

Recently, scientists explored the mechanisms of antitumor immunity, specifically checkpoint molecules in the immune checkpoint signaling route, which laid the ground for the development of immune checkpoint inhibitors (ICIs) drugs, a revolutionary in cancer therapy (Balar and Weber 2017; Li et al. 2016). Nonetheless, a number of studies have demonstrated that ICIs are not effective as monotherapy for all patients with cancer (Wang and Wu 2017; Debele et al. 2020; Sui et al. 2015). There have also been several scientific attempts to alleviate clinical challenges by combining NPs with existing therapeutics or treatment approaches (Nam et al. 2019). The first randomized, open-label phase trial, 3 JAVELIN Ovarian 200 (NCT02580058), examined the effectiveness of a human monoclonal antibody medication (Avelumab) as monotherapy or together with doxorubicin packaged in a PEG-coated liposome (PLD) in patients suffering from different forms of ovarian cancer. According to the results, the combination of PLD and avelumab significantly increased overall survival (OS) (15.7), a median progression-free survival (PFS) (3.7), and the objective response rate (ORR) (13.3), in comparison to both PLD (13.1, 3.5, and 4.2) and avelumab (11.8, 1.9, and 3.7) (Pujade-Lauraine et al. 2018). Additionally, RNA-formulated NPs, individually or together with immunotherapeutic drugs, including ICIs, have also been evaluated for clinical purposes (Debele et al. 2020; Shi 2020).

In recent years, there has been a lot of interest in using photothermal immunotherapy as a treatment for metastatic cancer. In a study, researchers prepared a PEGylated nanocomposite by integrating nanostructured iron oxide (FNPs) with GO after removing oxidized functional groups for photothermal immunotherapy. This PEGylated nanocomposite combined with laser irradiation (805 nm) effectively stimulated the host antitumor immune response, triggered immunogenic cell death, and stimulated DCs in tumor-draining lymph nodes. The strong magnetic properties of iron ions enable the use of designed nanocomposite for MRI-guided photothermal immunotherapy of cancer cells that spread from the original tumor (Wang et al. 2019).

#### 18.3.1.4 Radiotherapy

Radiotherapy is a pain management and cancer care procedure that employs high-energy X-ray photons to inhibit cell growth. Most cancer patients respond well to this treatment (Liu et al. 2018a; Baumann et al. 2016). In spite of this, when radiation therapy is used to treat a tumor, the radiation dose may not be evenly distributed throughout the tumor. This can lead to areas of the tumor receiving a lower or higher dose of radiation than other parts and can cause the tumor to become resistant to treatment. Additionally, since healthy tissues are also exposed to radiation, this increases the risk of long-term health complications (Liu et al. 2018a; Barazzuol et al. 2020). Thus, it is essential to find ways to minimize the amount of radiation delivered to healthy tissues while delivering a high enough dose to kill the tumor cells. There have been various approaches offered to find a balance between treatment efficacy and undesirable consequences (Chen and Kuo 2017). In recent years, radiosensitizers, defined as agents that increase the absorption of radiation by cancer cells, have become increasingly popular (Linam and Yang 2015). AuNPs have unique physicochemical characteristics that make them ideal candidates for use as tumor radiosensitizers (Chen et al. 2020). By coating AuNPs with specific molecules, the NPs are able to target receptors on the cancer cells. This allows for a more efficient drug delivery system and increases the effectiveness of the drug. The PEG molecules provide a protective layer that prevents the AuNPs from being recognized by the RES and also increases the circulation time of the AuNPs in the bloodstream (Zhang et al. 2012; Yasui et al. 2014; Alalaiwe et al. 2017). The use of a PEGylated AuNPs-based platform (AuRad™) in conjunction with X-ray irradiation enhanced cell death in HeLa cells when compared with radiation therapy alone (Kumar et al. 2013). AuNPs encapsulated in PEGylated nanogels sensitized murine squamous SCCVII cancer cells and V79 fibroblast cells isolated from the lung of a Chinese hamster to X-ray irradiation at 220 kVp. This increased apoptosis happens through the overexpression of proteins related to endoplasmic reticulum stress (Yasui et al.

2014). It has also been demonstrated in an in vitro study that PEGylated AuNPs increased cell uptake in B16F10 tumor cells and induced radiosensitivity when irradiated with 6 MeV (Mousavi et al. 2017). Pharmacokinetics and biodistribution play a significant role in intravenous drug efficacy. Researchers conjugated glucose and PEG ligands to optimize AuNP pharmacokinetics and biodistribution in order to employ the medicine for in vivo radiotherapy. As a result of the addition of PEG, the half-life period of PEG-Glu-AuNPs was significantly longer than that of Glu-AuNPs. Despite the fact that PEG-Glu-AuNPs are mostly found in the spleen, liver, lung, and kidneys, it was found that uterine and ovarian cancer cells contain more PEG-Glu-AuNPs than normal cells. Due to the fast consumption of glucose by cancer cells, PEG-Glu-AuNPs are more readily absorbed by cancer cells, and interactions between X-rays and AuNPs facilitate radiation therapy-induced cell death (Geng et al. 2014).

Oxygen is a crucial component in ionizing radiation-induced DNA damage. Treatment of cancer with radiation therapy often fails due to a lack of oxygen in the centers of solid tumors. It has been demonstrated through in vitro and in vivo experiments in mouse models that PEG-coated Au@$MnO_2$ NPs enhance the efficiency of radiation therapy by generating oxygen through the destruction of endogenous hydrogen peroxide (Yi et al. 2016).

### 18.3.2 Nervous System Diseases

There has been a lot of research on the possibility of using nano-carriers to transport drugs to the brain. The toxicity of NPs on neurons has limited their clinical application. Calcium channels are one of the mediators of the neurotoxicity of magnetic NPs. The positively charged polymer-coated magnetic NPs cause an increase in intracellular calcium levels in primary hippocampal neurons by disrupting the activity of calcium channels. PEG coatings on NPs have been shown to mitigate their harmful effects on cells by preventing them from interfering with calcium channel activity. PEGylated MNP-PLGA-PEI NPs may also enhance the gene transfection efficiency of green fluorescent protein (GFP) in primary hippocampal neurons in the presence of an external magnetic field (Cui et al. 2019). Moreover, coating lipid NPs (LN) with PEG not only diminish the aggregation of NPs in the brain parenchyma of mice, but also reduce the activity of caspase-1 and IL-1β as well as cerebrovascular lesions in P2X7−/− mice (Huang et al. 2013). AuNPs with PEG shielding (Au@PEG) are widely used in medicine. Investigating the curative capabilities of Au-PEG nanoconjugate in transgenic SOD1G93A mice model of amyotrophic lateral sclerosis (ALS), a rare neurological disease that is associated with gradual cell death of motor neurons, revealed that Au-PEG administration enhances longevity for about one week, postpones the average age of the disease initiation, and ameliorates rotarod (a behavioral test) performance (Vimal et al. 2022). It has been reported that 4.5 nm Au@PEG induces apoptosis in human neuroblastoma SHSY5Y cells and activation of astrocytes and neuronal apoptosis in mice treated with Au@PEG. Conjugating Trolox as an antioxidant on the surface of Au@PEG decreased the cytotoxicity of the NP in both cell culture and animal models (Zhang et al. 2020). The neuroprotective effects of fucoxanthin against Alzheimer's disease have been studied both in vitro and in vivo. Despite being safe in earliest-stage clinical trials with small size, the poor solubility and absorption of fucoxanthin in the central nervous system (CNS) have restricted its use in medical interventions. It was found that encapsulating fucoxanthin inside PLGA-block-PEG (PLGA-PEG-Fuc) NPs facilitates the penetration of the drug into the CNS and significantly inhibits the formation of Aβ fibrils and oligomers. Also, cognitive disorders in mice models treated with PLGA-PEG-Fuc NPs were less than in mice treated with free fucoxanthin (Yang et al. 2021). Thus, PEGylation not only facilitates the penetration of NPs into brain tissues but also improves their therapeutic properties for neurological diseases by reducing their toxic effects and preventing their aggregation in the brain.

### 18.3.3 Cardiovascular Diseases

#### 18.3.3.1 Myocardial Ischemia (MI)

Partial or complete closing of coronary arteries reduces blood flow through the heart and results in a condition known as MI. The investigation has revealed that microRNA-133 (miR-133) positively contributes to the recovery of acute myocardial infarction by helping cardiac muscle cells' growth and homeostasis. One of the major issues related to miRNA therapy is the administration of miRNA to a particular location. Sun and coworkers encapsulated miR-133 inside PEG-PLA NPs modified with the tripeptide RGD and injected them into the lateral tail veins of model rats with heart infarction lesions. Results revealed that RGD-PEG-PLA, by directing miR-123 to the site of heart damage, diminished the area of myocardial infarction and improved heart function (Sun et al. 2020). Liu and colleagues reported that loading spherical $TiO_2$ NPs into biologically active macromolecular (PEG/Chitosan) hydrogel matrixes improved $TiO_2$ NP properties and increased $TiO_2$-PEG/CTS hydrogel cell adhesion on cardiomyocyte cells (Liu et al. 2018b). Clinical trials have shown that Schisandrin B (Sch B) is effective in treating hepatitis and ischemia illness. Nevertheless, its therapeutic effectiveness is restricted because of its poor solubility in water and low bioavailability. Matrix metalloproteinase (MMP)-sensitive peptide-modified PEGylated SLNs for delivery of Sch B were prepared by Shao and colleagues and tested in a myocardial infarction model. The highest infarct size reduction was observed in the models administered with MMP-Sch B SLNs (Shao et al. 2017). Puerarin is commonly applied to treat heart patients. Short plasma half-life and low water solubility are the therapeutic limitations of puerarin. Dong and coworkers reported that encapsulating puerarin inside RGD-modified and PEGylated SLNs did not only dramatically enhance the half-life of puerarin in circulation, but also strengthened its protective effect on acute MI rats. So, RGD-modified and PEGylated SLN is a favorable carrier for medicine in the therapy of myocardial infarction (Dong et al. 2017). Baicalin's therapeutic limitations are its short plasma half-life and low water solubility. Zhang and coworkers investigated the bioavailability of PEG-functionalized lipid nano-carriers loaded with bioactive flavonoid glycosides in acute myocardial infarction rats. Encapsulating glycosyloxyflavone into lipid results in significantly reduced infarct size by increasing the drug's plasma half-life and heart concentration (Zhang et al. 2016).

#### 18.3.3.2 Rheumatoid Arthritis (RA)

Symptoms of RA include joint swelling and a dysfunctional immune system. Glucocorticoids (GCs), despite serious side effects in high doses, are the most important drugs used in RA therapy. Dexamethasone (DEXA) was encapsulated inside micelles constructed from the amphipathic PEG-block-poly (ε-caprolactone) (PCL-PEG) polymer and administered via intravenous infusion into rat models with arthritis in an attempt at low-dose GC therapy. Results revealed that the micelles enhanced the half-life of DEXA in the blood and targeted drug accumulation in joints with inflammation that was associated with a decline in the tenderness of joints, destructive bone lesions, and proinflammatory cytokine production in both joint tissue and serum (Wang et al. 2016). Ac2–26, a peptide with 25 amino acids, is a derivative of Annexin A that has good potential for anti-inflammatory treatments. The low bioavailability has restricted the application of Ac2–26 peptides in the clinic. It was found that encapsulating Ac2–26 peptides inside PEGylated lipid NPs (LDNPs) constructed from the co-assembly of a synthetic lipophilic ascorbic acid derivative and 1,2-distearoyl-sn-glycero-3-phospho-ethanolamine-PEG (DSPE-PEG2000) carboxylic acid not only enhance bioavailability and stability, but also decrease synovial inflammation and improve joint pathology upon intravenous infusion (Qin et al. 2021). Gambogic acid (GA) as an anti-inflammatory agent has a low solubility in water, which limits its efficacy in vivo. In a study, three NPs were made from the conjugation of GA with three different molecular weights (MWs) of methoxyPEG (mPEG) (550, 2000, and 5000). NP-2000, as the smallest particle and with the lowest critical micelle concentration (CMC), was the best candidate for biological investigations. The ability of NP-2000 to suppress TNF was comparable to that of free GA against in vitro cultured inflammatory macrophages. The ability of NP-2000 to suppress TNF-α was comparable to that of free GA against in vitro cultured inflammatory macrophages. Drug exposure in treated mice with NP-2000 was 185 times greater than in animals treated with free GA. Paw inflammation was notably diminished by about 27% in the treated murine model of RA (Nguyen et al. 2020). The increased expression of MMPs at the sites of inflammation in RA patients can be a guide for drug delivery. A study used the co-assembly of triglycerol monostearate (TGMS) and DSPE-PEG2000 to create MMP-sensitive PEGylated NPs. Upon exposure of the NPs to MMPs, the ester bond of TGMS is cleaved and the PEG coats are exposed. Intravenous administration of Dex-loaded MMP-responsive LDNPs to rat models with RT reduced joint swelling while suppressing TNF-α and IL-1β levels in joint tissues (He et al. 2020).

### 18.3.4 Infectious Diseases

Worldwide, infectious diseases contribute significantly to death and morbidity. A considerable mortality rate and the continued spread of malaria, tuberculosis, and hepatitis C indicate that treatment of these diseases is extremely complicated. A number of challenges in treating such diseases can be overcome by combining traditional and newly developed drugs in nanoscale carriers. Further, nano-carriers are useful for developing vaccines as a powerful tool for combating infectious diseases (Kirtane et al. 2021). Several viruses are capable of entering the body through the vaginal tract, including HIV and herpes simplex virus (HSV). By administering microbicides locally as a preventive measure, one can reduce the chances of contracting an infection. Multiple barriers prevent the delivery system from completely covering the vaginal tract and residing there for a prolonged period of time (das Neves et al. 2015). Mucus, as a dynamic protective barrier against drug particle penetration, is constantly being replaced, which means that particles that are trapped within it are eventually lost as new mucous is created. The additional barriers are the leakage of the formulated drug out of the vaginal tract and the folds or rugae on the surface of the vagina. The folds, or rugae, on the surface of the vagina can make it harder for drugs to reach their target, reducing drug accessibility. Also, the low pH and enzymes in the vaginal fluid can break down certain drugs, further reducing their effectiveness (das Neves et al. 2015; Leal et al. 2017). Various strategies have been employed to facilitate drug penetration across these barriers. Avidin and PEG are both molecules that can be used to modify the surface of NPs. Comparing the vaginal uptake of three types of dye-loaded PLGA NPs, including Avid-NP, PEG-NP, and unmodified NPs, showed that PEG-NPs had a higher uptake in the lower reproductive tract. This was likely due to increased diffusion of the PEGylated NPs through the mucous layer, as shown by multiple particle track analyses (Cu et al. 2011; Lai et al. 2007). PEGylated PLGA NPs in a hypotonic solution are diffused more easily into the folds of the vaginal wall in murine models (Ensign et al. 2012a).

Bacteria produce and secrete virulence-signaling molecules to regulate cooperative growth and form biofilms. Biofilms are collections of bacteria that communicate with each other through a process known as quorum sensing. Inhibitors of this quorum-sensing phenomenon can interfere with the ability of bacteria to communicate, which can disrupt the formation of the biofilm and, in turn, lead to its dissolution. Mucous barriers in the respiratory tract, particularly in individuals with cystic fibrosis, are a physical barrier for inhibitors of the quorum-sensing phenomenon. By encapsulating the quorum sensing inhibitor in lipid NPs and coating them with the surfactant (polysorbate80, a copolymer surfactant containing PEG), in vitro virulence control was improved (Nafee et al. 2014).

Longer circulation periods allow medications to be administered to target tissues with higher accuracy and with fewer adverse effects (Siegel et al. 2016). Macrophages are cells that recognize foreign substances in the body and ingest them. By altering the nano-carriers to be less recognizable to macrophages, they are less likely to be taken up and removed from circulation, resulting in longer circulation times.

Binding hydrophilic polymers such as PEG to nano-carriers can reduce the chance of the nano-carriers being detected and eliminated by the immune system, allowing them to reach their intended target (Gref et al. 1994). The accumulation of Gallium-67-labelled liposomes with varying half-lives at the area of infection was investigated in a rat model of unilateral Klebsiella pneumonia infection. Liposomes were able to flow more readily through the infected lung's lipid bilayer owing to enhanced permeability induced by inflammation and other factors. This was supported further by the increased concentration of liposomes in the infected lung vs the uninfected one. In rats with the most severe lung infection, PEG-coated liposomes (half-life = 27 h) increased concentrations about fourfold when compared to uncoated liposomes (half-life = 19 h). PEG coating was less beneficial at low infection levels (Bakker-Woudenberg et al. 1993; Bakker-Woudenberg et al. 1992).

Nano-carriers do not selectively distribute therapeutic agents into circulation. Therefore, there is an advantage in designing systems in which no medicine is delivered into the bloodstream until the entire substance is liberated following contact with the pathogens. Two nano-carriers were designed to release vancomycin following contact with the phosphatase and lipase enzymes in the bacteria. Phosphatase-activated nano-carriers were made of poly(phosphoester), PCL, and PEG, whereas lipase-activated nano-carriers were made of copolymers of poly(phosphoesters) and PEG (Xiong et al. 2012a; Xiong et al. 2012b). Activated nano-carriers by phosphatase significantly increased survival rates of zebrafish models of methicillin-resistant *S. aureus* infection compared to controls without nano-carriers (Kirtane et al. 2021; Xiong et al. 2012a).

Infectious agents can cause disease by altering the biological system. In addition to being life-threatening, many of these alterations result in disfiguring, disabling, and frequently stigmatizing conditions. As an example, leishmaniasis can cause disfigurement to the skin/face (cutaneous leishmaniasis) or have deadly effects (visceral leishmaniasis). The only treatments for leishmaniasis are chemotherapy and chromotherapy. There are several disadvantages associated with these treatment methods, including drug resistance and neurotoxicity. In addition, there is still no vaccine available to prevent this disease. Leishmania parasites are very sensitive to ROS. Leishmanial lesions can be exposed to ROS through PDT. PEGylated silver-doped zinc oxide NPs absorb visible light and transfer it to water and oxygen, resulting in ROS ($^1O_2$ and $OH^\bullet$) production in the biological system and ultimately the death of leishmania (Nadhman et al. 2014). Experimental evidence suggests that Amphotericin B-encapsulated PLGA-PEG NPs are more effective than free amphotericin B against Leishmania donovani (Kumar et al. 2015). PEGylated CdS/$Cd(OH)_2$ fluorescent nanocrystals were used in one study to label the live parasite *Trypanosoma cruzi* (*T. cruzi*) in order to evaluate endocytosis and cell differentiation (Chaves et al. 2008). Antimicrobial peptides (AMPs) failed in treating infected wounds due to loss of activity in vivo. In a research, AMP was bound to PLGA-PEG NPs, and their activity was evaluated on infectious agents. Based on the results, AMP grafting sped up the killing kinetics of *Pseudomonas aeruginosa* and *Staphylococcus aureus* (Ramôa et al. 2023).

### 18.3.5 Diabetes

High blood sugar levels are a hallmark of diabetes, a chronic condition. The pancreas secretes the hormone insulin, which is responsible for controlling the amount of sugar in the blood. Diabetics either do not create enough insulin or their bodies are unable to effectively use the insulin they do make. It is predicted that 642 million individuals will suffer from diabetes mellitus by the year 2040 (Jaacks et al. 2016). By using nanotechnology, researchers are able to create miniaturized devices that can measure glucose levels in the body more precisely and deliver insulin more quickly and accurately. This has the potential to make diabetes management easier and more effective (DiSanto et al. 2015). The injection of insulin is a painful and annoying procedure for diabetics. Other means of administering insulin, including oral and inhalation, have been the focus of extensive research and development efforts. As a protein, insulin is denatured by stomach acid when consumed orally. One strategy involves incorporating insulin into NPs that can withstand acidic environments. The passage of NPs through the intestine is size-dependent. Absorption occurs for particles smaller than 100 nm, but not for those larger than 300 nm (Jani et al. 1990; Abdel-Moneim and Ramadan 2022). It has been found that insulin delivery by NPs decreases glucose levels in the body for up to 10–20 h. Various polymers have been introduced to deliver insulin orally, but none has reached the clinical stage. In order to carry insulin, Ramachandaran and colleagues developed calcium phosphate NPs coated with PEG. It was demonstrated that these NPs, most of which were smaller than 100 nm, were well suited to transporting insulin and protecting it against acidic conditions. In acidic conditions, the insulin released from these NPs was very slow and sustained release for over eight hours at a pH similar to that of the intestine. These NPs have not been tested in this study for their ability to reduce blood sugar levels under physiological conditions (Abdel-Moneim and Ramadan 2022; Ramachandran et al. 2009). The PEGylated starch NPs have also been used to study insulin delivery via the oral route. PEG is a nontoxic, mucoadhesive, neutral, and water-soluble polymer that may block the enzymatic breakdown of PEGylated insulin and increase the drug's absorption in the gut (Abdel-Moneim and Ramadan 2022; Pisal et al. 2010; Minimol et al. 2013). There has been evidence of a hypoglycemic effect of PEGylated starch acetate NPs in vitro (Abdel-Moneim and Ramadan 2022; Minimol et al. 2013). For oral administration, Andreani and colleagues used the sol–gel technique to encapsulate insulin in silica NPs covered with PEG. Results revealed that the uncoated SiNP released insulin more slowly than the PEG-coated SiNP. Insulin secretion was not substantially altered by the high MW PEG coating ($p > 0.05$). Due to insulin's affinity for silanol groups on the silica surface, it is hypothesized that silica slows insulin release. The insulin release kinetics were then investigated, and the second-order kinetics model was shown to be the most appropriate for both SiNP and SiNPs coated with PEG (Abdel-Moneim and Ramadan 2022; Andreani et al. 2014). For insulin oral delivery, Rostamzadeh and coworkers synthesized the pH-sensitive poly-methacrylic acid–chitosan-PEG (PCP) NPs. In vitro release experiments showed that the insulin release was lower at pH 1.2 and higher at pH 7.4 (Abdel-Moneim and Ramadan

2022; Rostamizadeh et al. 2015). Sarhadi et al. fabricated a PEGylated insulin liposomal bound to B12. In vitro data demonstrated that B12 improves the absorption of PEGylated insulin liposomal in the Caco-2 cell line. Moreover, PEGylated insulin liposomal bound to B12 had hypoglycemic effects in diabetic rats (Sarhadi et al. 2022).

"Closed-loop" insulin systems are an intriguing alternative to regular insulin injections since they constantly monitor blood glucose levels and release insulin as required. These systems are attractive because they eliminate the need for manual monitoring and insulin injections, allowing for more precise, consistent, and timely adjustments to insulin levels. This can lead to better long-term glycemic control, reducing the risk of hypoglycemia and other health complications (DiSanto et al. 2015). The enzyme-based polymersomes are nanostructures that can be used to deliver insulin in a glucose-dependent manner. In a study, glucose oxidase, catalase, and insulin were included in polymersome nanovesicles composed of (PEG-poly (Ser-Ketal). The bilayer structure provides a protective environment for insulin and allows the diffusion of glucose and oxygen. High concentrations of glucose trigger the enzyme glucose oxidase, which converts glucose to gluconic acid, leading to polymer membrane disintegration and the subsequent release of insulin (Tai et al. 2014). Cross-linked polymer matrices that contain phenylboronic acid (PBA) are another strategy to slowly release insulin in response to changes in glucose levels. Upon glucose binding to the PBA, the matrices absorb water and swell which causes the release of insulin. Controlled release of insulin in response to glucose concentration was achieved by fabricating a copolymer micelle composed of PEG-block-poly-(acrylic acid-co-acryloamidophenylboronic acid)—(PEG-b-(PAA-co-PAAPBA). When glucose attaches to the PBA moiety in the micelle core, the moiety is transformed from hydrophobic to hydrophilic. Micelle swelling and insulin secretion are the results (Wang et al. 2009). PEG-polyboroxole block copolymer polymersome is an engineered nanostructure for insulin delivery in glucose-dependent manner (Kim et al. 2012).

So nanotechnology has enabled the development of glucose sensors and insulin delivery systems that are much more precise and accurate than previous systems. As described in this section, PEG has been used in the construction of some NPs for the purpose of measuring and controlling blood sugar levels.

## 18.4 Drug Administration Routes

Drugs can be administered in different ways, depending on the type of drug and what it is used for (Fig. 18.2). Each route has its own advantages and disadvantages in terms of effectiveness, safety, and cost. Drug delivery can be classified as systemic and nonsystemic based on the route of administration. Systemic drug delivery refers to drugs that are administered orally, intravenously, or intramuscularly and travel through the bloodstream to reach their target. Nonsystemic drug delivery refers to drugs that are applied topically, inhaled, or administered intranasally and do not enter the bloodstream (Suk et al. 2016).

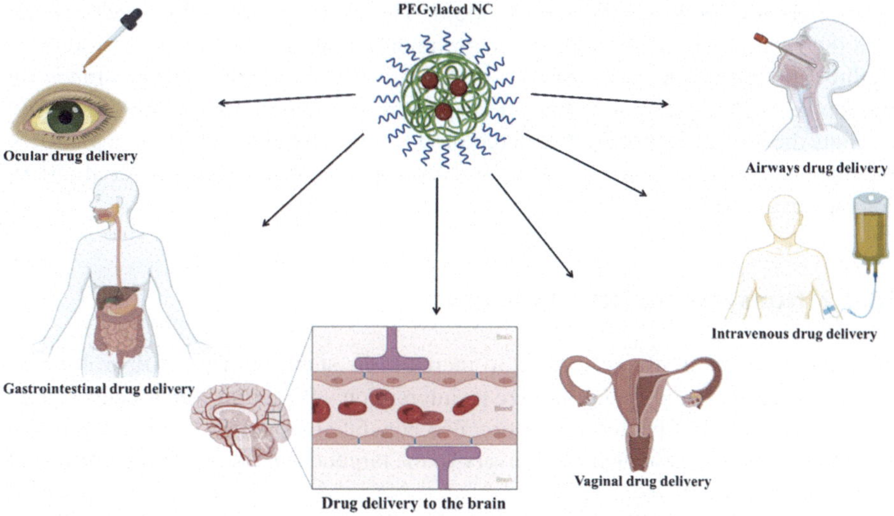


**Fig. 18.2** Drug delivery by PEGylated nano-carriers through various routes

### 18.4.1 Systemic Drug Delivery

To have an effect on the target tissue, systemically administrated drugs must circulate in the blood for a specific period of time. The use of NPs is one strategy for targeted drug delivery. NPs have a limited half-life in circulation due to aggregation, opsonization, and elimination by the MPS. As part of the immune system, the MPS is composed of monocytes in the blood, macrophages, and DCs, which are responsible for eliminating and degrading foreign substances. MPS recognizes NPs as foreign pathogens and rapidly removes them from the circulatory system. Moreover, the serum proteins known as opsonin binds to NP surfaces, facilitating the detection of NPs using MPS. Consequently, these factors not only reduce the half-life of NPs in the bloodstream, but also weaken their transfer to the target tissues (Moghimi et al. 2001; Vonarbourg et al. 2006). One approach to increasing the half-life of NPs in the blood is coating their surfaces with PEG polymer. PEG, one of the FDA-approved polymers, forms a hydrated cloud on the surface of the NP. This cloud not only increases the size of the NP, but also prevents the interaction of the NP with other NPs and blood components (Yang and Lai 2015). Doxil is the first PEGylated NP authorized by the FDA in 1995 to improve doxorubicin bioavailability for the therapy of certain malignancies (Stirland et al. 2013). Solid tumor vasculature has a very high permeability to large molecules. This unique characteristic is described as the enhanced permeability and retention effect (EPR) (Matsumura and Maeda 1986). Since PEGylated NPs have a large size, EPR is a means of directing systemically administered PEGylated NPs to target tissues. PEGylation decreases the systemic toxicity of NPs (Suk et al. 2016). One of the issues associated with systemically delivered NPs, particularly cationic NPs, is the lysis of erythrocytes and the release of hemoglobin (Dobrovolskaia et al. 2008). This is due to the NP's direct contact with the erythrocyte membrane and subsequent membrane breakdown. Moreover, erythrocyte cell lysates bind to the surface of NPs, facilitating NP clearance by macrophages. According to some studies, coating the surface of the NP with PEG prevents the attachment of the NP to red blood cells and hemolysis (Suk et al. 2016; Ziemba et al. 2012; Fischer et al. 2003; Eliyahu et al. 2002; Kurosaki et al. 2010; Petersen et al. 2002).

## 18.4.2 Nonsystemic Drug Delivery

PEGylation was initially performed to increase the stability of the NP, prolong its circulation in the blood, and minimize its interaction with serum components. Later, it was found that PEGylated NPs were also beneficial for nonsystemic drug delivery. PEG coatings facilitate drug delivery to the target tissue by removing biological barriers (Suk et al. 2016).

### 18.4.2.1 Vaginal NP Delivery

Local intravaginal delivery of drugs is a potential treatment route for several medical conditions related to the female reproductive system, ranging from bacterial and fungal infections to cervical cancer. Cervicovaginal mucus (CVM) layers due to viscoelasticity and renewal limit the uptake of NP-based drugs (Cone 2009; Lai et al. 2009; Ensign et al. 2012b). It has been demonstrated that modification of nanosurfaces with low MW PEG facilitates particle diffusion in the human CVM, as well as prevents particle core and PEG interactions with mucin. PEG increases the absorption of NPs into the cervical and vaginal tissues by reducing their adhesion to mucus (Suk et al. 2016). It was found that PEGylation increases the diffusion rate of large NPs, such as carboxylate-modified PS NPs with 200 and 500 nm in size, through human CVM (Lai et al. 2007). Wang and coworkers reported that PS-PEG NPs coated with high-density low-MW PEG diffuse rapidly in the human CVM in comparison with PS-PEG NPs coated with low-density high-MW PEG (Wang et al. 2008). It has been found that incorporating PEG on the surface of PLGA NPs enhances the diffusion of NPs through human cervical mucus in comparison with unPEGylated PLGA NPs (Cu and Saltzman 2009). The PEGylation process extended the retention time of PLGA NPs in the lower female genital tract of progestin-injected mice to 6 h (Cu et al. 2011). Yu and coworkers fabricated liposomes decorated with varying amounts of PEG and investigated the influence of the PEG modifiers on vaginal retention. They observed that liposomes containing 3 and 5 mol% PEG distribute more uniformly in the vagina in comparison with non-PEGylated liposomes. It was found that the further increase in PEG content to 12 mol% led to a decrease in vaginal tissue coverage compared to 7 mol% liposomes (Yu et al. 2015). Xu and coworkers mixed amphiphilic block copolymers in varied ratios to create PLGA NPs with variable PEG densities. They found that NPs with low PEG density had strong mucin binding in vitro and poor distribution throughout the mouse vagina in vivo, as well as being immobilized in human CVM ex vivo.

However, NPs with high PEG density had less mucin binding and were distributed uniformly in vivo and diffused in mucus ex vivo (Xu et al. 2015).

The PEG-coated NPs can be used to examine the structure of CVMs and figure out how the sizes of pores are spread out in mucus and tissues. Samuel and colleagues fabricated non-mucoadhesive PEG-coated NPs of different sizes and demonstrated their use for measuring the average mesh spacing (pore size) of cervico-vaginal mucus in healthy females. They reported that the majority of pore sizes were greater than infectious viruses (Lai et al. 2010).

#### 18.4.2.2 NP Delivery to the Airways

The respiratory tracts are coated with a viscous and thick colloid known as mucus that is renewed by mucociliary clearance (MCC) mechanisms. Mucus, an excellent insulator, protects the airways but is highly viscoelastic and resistant to penetration in the respiratory tracts of cystic fibrosis (CF) and chronic obstructive pulmonary disease (COPD) patients. It has been shown that 100 and 200 nm PS-PEG particles could pass through freshly expectorated CF sputum (Suk et al. 2009). Although PS-PEG particles (<200 nm) can penetrate CF sputum, the effectiveness of NPs transport can be affected by the severity of the illness (Forier et al. 2013).

The investigation has revealed that the compactness of the PEG layer of NPs has an influence on the permeation of CF sputum. Nevertheless, NPs with compacted PEG are not very efficient for gene transfection in the sputum of CF individuals (Boylan et al. 2012). To search for more effective NPs, PEG-dendron-based DNA NPs were fabricated and introduced as effective agents in gene transfer in vitro (Kim et al. 2013). Suk and colleagues prepared solid DNA NPs composed of PEGylated polymers for lung gene therapy. The densely PEGylated polymeric nanostructures promoted the efficiency of gene transfection due to good mucus penetration and uniform distribution through CF sputum and the airways of mice (Suk et al. 2014). PEI is an immunogenic and nondegradable polymer. In this regard, Mastorakos and colleagues fabricated compact DNA NPs using PEG and biodegradable poly(β-amino ester) (PBAE). PEGylated PBAE DNA NPs (PBAE-MPP) were able to enter the major airways and pulmonary parenchyma of the animal and were distributed uniformly. As a result, the gene expression pattern was correlated with the dispensation of the densely PEGylated NPs. Moreover, the level of gene expression did not decrease upon repeated treatments with the densely PEGylated PBAE-MPP, and no inflammation was observed in the mouse airways (Mastorakos et al. 2015a). PEGylation may protect viral vectors against neutralizing antibody binding in vitro and in vivo, resulting in persistent infectivity following repeated doses, as previously observed for adenovirus by O'Riordan et al. (1999). Although sputum obtained from CF patients is commonly used for ex vivo research, preparing undiluted sputum from healthy individuals presents a significant challenge in evaluating the barrier properties of NP formulations. A study conducted by Yang and colleagues evaluated the resistance characteristics of pig tracheal mucus against NPs and a contagious herpes virus (PRV). They observed that PEG-coated NPs quickly dispersed in pig tracheal mucus, whereas amine-modified and carboxylate-modified PS NPs as well as PRV were trapped (Yang et al. 2012). PEGylation leads to enhanced penetration of NPs that adhere to mucus. Schuster and colleagues studied the ability of NPs to pass through human respiratory mucus collected from endotracheal tubes of individuals who did not have a history of postoperative lung injury. Despite the presence of non-PEGylated NPs, highly PEGylated 200 nm NPs demonstrated the ability to pass through normal human respiratory mucus (Schuster et al. 2013).

Enhancing NP distribution to the nasal mucosa via PEGylation has been investigated. The mucus in the nose is an obstacle to the absorption of NP, especially in people with chronic rhinosinusitis (CRS), who accumulate highly viscoelastic mucus in the sinuses. PEG-coated NPs were capable of rapidly penetrating CRS mucus (CRSM) from patients with sinus inflammatory disease or tissue growth inside the nasal passages (Lai et al. 2011). Scientists examined how well PEGylated NPs were absorbed by the nasal mucosa of rats and how easily they entered the bloodstream. Lower NPs' entry into nasal mucus after PEGylation may be attributable to PEG-coated NPs' limited interactions with nasal-associated lymphoid tissue (NALT) (Brooking et al. 2001). On the other hand, one study found that PEG coating makes it easier for NPs to pass through nasal mucus. Tobio and his colleagues found that the nasal mucosa absorbs tetanus toxoid-loaded PLA-PEG NPs more easily than PLA NPs without coatings. The researchers hypothesized that the polymeric PLA-PEG NPs might get into NALT. They speculated that the NPs could then be carried through the lymphoid system and the bloodstream (Tobio et al. 1998). According to research by Vila and colleagues, increasing the PEG surface density resulted in increased nasal mucosal absorption of PLA-PEG NPs (Vila et al. 2004).

#### 18.4.2.3 Gastrointestinal NP Delivery

A convenient and favored form of drug and gene delivery is oral administration. Because of the ecosystem of the gastrointestinal (GI) tract, the continuous and rapid regeneration of mucus in the digestive tract, and the viscosity of the mucus, drug-carrying NPs are less stable and lose their effectiveness more quickly in the GI. There have been investigations into the use of PEG coatings for optimizing NP transport to the GI tract. The pH-sensitive multi-polymeric NPs were designed to be able to release their contents in a controlled manner in response to changes in the pH level of their environment, which would enable targeted drug delivery and improved drug efficacy. Metoprolol (MTP) tartrate was encapsulated in NPs as a model drug. The interaction of the NPs with mucin from the porcine stomach was found to be excellent, providing a promising platform for effective mucosal delivery (Pawar et al. 2012). Nimodipine could be beneficial in treating behavioral changes and impaired memory. This substance has a limited bioavailability, which makes oral administration challenging. PEGylated NPs significantly improve the oral delivery of nimodipine, allowing for more efficient absorption and distribution to the target tissues. NMD-NP/PEG was able to reduce the anxiety-like behavior of animals but not to the same extent as the control group. However, it was able to protect against the memory impairments caused by chronic corticosterone administration, suggesting that it could be a potential therapeutic intervention for cognitive impairments (Solas et al. 2018). In vitro, PLA-PEG NPs were shown to be more stable in digestive juices than PLA NPs, which improved the oral administration of tetanus toxoid to rats (Tobıo et al. 2000). The GI easily breaks down lipid-based formulations (LBFs) with medium-chain triglycerides (MCT), which makes drugs less soluble and makes them stick together. In order to increase the durability of MCT-LBFs, nonionic surfactants containing stealth PEG head groups with varying MWs were used. It was found that the increase in PEG MW to 2644 Da prevented the digestion of MCT-LBFs in the GI. Excessive hydrophilic PEG (PEG MW 4407) incorporation can cause MCT-LBF instability. Danazol, a drug administered to endometriosis patients, has a very low solubility in water. Adding PEGylated MCT-LBF carriers to the danazol and administering it orally to rats has been found to enhance absorption (Feeney et al. 2014). PEGylated adenoviral vectors were being investigated for oral gene delivery in colorectal cancer. As a result of PEGylation, adenoviral vectors were more efficiently transduced in Caco-2 cells and intestinal epithelium and were more stable in the presence of gastric juice and enzymes of the pancreas. The oral delivery of PEGylated adenoviral vectors activated by succinimidyl succinate to fasted Sprague-Dawley rats remarkably enhanced the expression of reporter proteins in the digestive tract compared to the delivery of the unmodified vectors (Cheng et al. 2003). In a study, the SLNs were PEGylated, and their capacity to penetrate mucus layers in the GI tract and deliver drugs across the mucus barrier was investigated. The results indicated that PEGylation not only helps to protect the NPs from the harsh digestive environment and increases their stability, but also facilitates their penetration through mucus barriers (Yuan et al. 2013). A study investigated the effectiveness of modifying cyclosporine A (CyA)-loaded liposomes with natural or synthetic polymers for oral administration. PEG-coated liposomes were more efficient than chitosan-coated liposomes for cyclosporine passing through the mucus layer of the small intestine (Chen et al. 2013b). PEG MW has an influence on the mucus permeation of NPs. PEG-coated NPs produced from the copolymer of methyl vinyl ether and maleic anhydride (Gantrez® AN) were more effective in interacting with the mucus and contacting the epithelium surface of the intestine upon oral delivery. A greater effect was observed for NPs with 2–6 kDa PEG than for those with higher MW (10 kDa) PEG (Inchaurraga et al. 2015). It has been shown that PEGylated mucoadhesive NPs with a size of 200 nm are uniformly distributed in the small intestine, while non-PEGylated mucoadhesive NPs are widely assembled distant from the epithelium. PEGylated mucoadhesive NPs are also diffused all over the rectum upon rectal delivery. The advantages of PEG coating on mucoadhesive NPs diffusion orally and rectally were also observed in animal models of inflammatory bowel disease (IBD). PEGylated mucoadhesive NPs were also able to penetrate inflamed tissue (Lautenschläger et al. 2013).

Intestinal diseases, such as colon cancer and inflammation, necessitate the use of targeted medications. Oral delivery is hampered by drug decomposition in the highly acidic environment of the stomach and poor penetration across mucus membranes of the intestine. It was found that adding polymeric surface grafts such as PEG and PEI to the mesoporous silica NPs (MSNs) increases their ability to penetrate the mucosal barrier and enhances their internalization into the cells. The different combinations of MSNs and targeting ligands allow for specific targeting of cells and tissues, as well as improved internalization and bioavailability of the drugs or agents encapsulated in the MSNs. It has been reported that the differentiation of intestinal goblet cells by functionalized MSNs loaded with Notch pathway inhibitors is enhanced when compared to free drugs. The Notch pathway is involved in intestinal progenitor development, inflammation, and colon cancer. The small intestine is where PEI coatings thrive, while the colon is where PEI-PEG coatings excel. MSN surface modification makes the drug target the desired area more specifically and effectively (Desai et al. 2016).

#### 18.4.2.4 NP Delivery to the Brain

NPs that have crossed the blood-brain barrier (BBB) or been injected directly into the brain face resistance from the extracellular matrix (ECM). Researchers determined that the typical pore size of the brain's ECM is between 38 and 64 nm in research employing PEGylated quantum dots (QDs) (Thorne and Nicholson 2006). Nevertheless, PS-PEG NPs with a high PEG density and 114 nm size are allowed to diffuse quickly through ex vivo human and rat brain cells. Twenty-five percent of all brain ECM holes were certainly within the size range of 100 nm, as determined by the injection of trace intravital particles into the mouse brain (Nance et al. 2012). Brain tumors have dense tissue that prevents the diffusion and distribution of NPs in the brain. Densely PEG-coated NPs, due to their high diffusion and distribution in the brain, can increase the efficiency of drug delivery in a rat glioma tumor model. It has been reported that PEG-coated PS (PS-PEG) with a size of 70 nm and PLA-PEG NPs with a size of 69 nm are well distributed in brain tumor tissue. In a study, paclitaxel was loaded into PLA-PEG NPs with high PEG density and injected directly into the tumor core within the skull. The results showed that paclitaxel spreads well throughout the brain and is able to inhibit tumor growth (Nance et al. 2014). Mastorakos and colleagues investigated the effect of PEGylation density on the transport of the NP with DNA cargo in the brain. Results confirmed that PEGylation enhanced the stability of the DNA NPs in artificial cerebrospinal fluid (ACSF) and attenuated the cytotoxicity in vitro and in vivo, with no remarkable effect on the in vitro cellular uptake. PEG-coated polymeric DNA NPs had uniform distribution and fast diffusion in comparison with low-density PEG NPs or uncoated NPs (Mastorakos et al. 2015b). Various publications have described how PEGylated NPs can be used to get drugs to certain parts of the brain. PEGylation makes it easier for nano-carriers to pass through the BBB. PEGylation also reduces the RES's non-specific uptake of the nano-carriers. This helps the nano-carriers stay in circulation longer in the body and reach their target site (Gajbhiye et al. 2020).

#### 18.4.2.5 Ocular NP Delivery

The most prevalent form of ocular drugs is topical drops, which have some disadvantages such as fast clearance and low permeation into ocular tissues. Various investigations have been done in order to enhance the half-life and absorption of topical ophthalmic drugs (Järvinen et al. 1995). Giannavola et al. investigated PEGylated PLA NPs for the ocular delivery of acyclovir (Giannavola et al. 2003). PEG facilitates the mucoadhesion of NPs by establishing hydrogen bonds and penetrating into the mucus (Wang et al. 2008). Based on the results, the aqueous humor of rabbits with acyclovir-loaded PLA-PEG injected into the conjunctival sac had a higher level of acyclovir than rabbits with acyclovir-loaded PLA NPs or free acyclovir. It must be noted that treatment of the eyes with N-acetyl cysteine (NAC) as a mucolytic agent before the use of acyclovir-loaded PEGylated NP decreases acyclovir delivery into the aqueous humor (Giannavola et al. 2003). Schopf and colleagues evaluated the corneal and retinal delivery of loteprednol etabonate (LE) incorporated into PEGylated NP following topical administration to the rabbit eyes. They reported that PEGylation not only ameliorated LE delivery by NPs, but also prevented vascular permeability in pigmented animal models (Schopf et al. 2015). Flurbiprofen, an anti-inflammatory drug, was encapsulated inside biodegradable PLGA NPs conjugated with PEG and the cell-penetrating peptide (PLGA-NPs-PEG-peptide) by Vasconcelus and colleagues. It was found that the positive charge on the surface of these NPs facilitated penetration into the corneal epithelium, resulting in more effective prevention of ocular inflammation (Vasconcelos et al. 2015). Licochalcone-A, a natural compound, has been found to reduce inflammation and oxidative stress, making it a powerful tool in the fight against inflammation-related diseases. However, its poor water solubility makes it challenging to use for the treatment of ocular inflammation. This is why the development of new delivery systems such as liposomes or nano-carriers to improve its solubility and bioavailability is being explored as a potential way to increase its effectiveness when treating ocular inflammation. Galindo and coworkers synthesized PLGA NPs with PEG and cell-penetrating peptides (Tet-1 and B6). This functionalization was designed to increase the solubility of licochalcone-A and reduce its clearance rate, which will allow it to remain in the ocular tissues for a longer period of time, improving its anti-inflammatory effects. Moreover, the addition of cell-penetrating peptides can enhance the cellular internalization of the NPs, allowing them to reach deeper layers of the ocular tissues and maximize their efficacy. Regardless, both Licochalcone-A Tet-1 and B6-targeted NPs were effective in treating ocular inflammation, but the B6 NPs displayed better therapeutic activity in vivo (Galindo et al. 2022).

Another significant challenge in ocular treatments is maintaining an appropriate medication level in the eye for an adequate amount of time. Considering administration convenience and practicability, eye drops are the most acceptable route of administration for treating local ophthalmic diseases. Because of their low bioavailability, they typically require frequent instillation to achieve therapeutic effects. Ofloxacin is a common antibiotic used to treat localized eye infections. It has been shown that coatings of chitosan and PEG on SLNs containing Ofloxacin increase transcorneal permeation and mucoadhesion strength of the drug (Eid et al. 2019). Antioxidants such as resveratrol and quercetin have the ability to reduce intraocular pressure efficiently. It has been shown that the concomitant administration of resveratrol and quercetin increases the bioavailability of resveratrol. By encapsulating resveratrol and quercetin into chitosan and PEG-modified chitosan NPs, it is possible to improve the bioavailability of resveratrol as well as increase the synergic effects of both resveratrol and quercetin on reducing intraocular pressure. This can help to be more effective to treat glaucoma (Natesan et al. 2017).

Some retinal diseases, such as pigmentation disorders, congenital retinal dystrophies, and rhegmatogenous retinal detachment, might be treatable by injecting DNA or small interfering RNA (siRNA) into the retina. In experimental animals with Leber congenital amaurosis, injecting viral gene carriers under the retina has been shown to repair the genetic problem. However, there are several disadvantages associated with viral gene carriers, as well as the risk of repeatedly injecting the subretinal space. Consequently, administering nonviral nucleic acid NPs intravitreally could be a more secure option for the treatment of serious eye illnesses. In vitreous, nonviral nucleic acid NPs (lipoplexes) dramatically assemble. It was found that shielding the lipoplex surface with PEG prevented them from clustering. Sanders and coworkers prepared PEGylated DNA NPs by attaching DNA onto positively charged lipids. PEG-coated DNA lipoplexes were examined for uptake by isolated vitreous from bovine eyes following injection, and it was concluded that PEG coatings enhanced dispersion and vitreous uptake, allowing the particles to reach the target cells more efficiently (Sanders et al. 2007).

The corneal epithelium acts as a main obstacle to the passage of drugs. Mu and colleagues evaluated the obstacle characteristics of the bovine eyes' cornea with silica NPs with 750 or 5000 Da PEG coating in comparison with thiolated silica NPs. They realized that thiolated silica NPs and silica NPs with a 750 Da PEG coating adhered to the cornea even in the absence of epithelium. Silica NPs with a 5000 Da PEG coating adhere to the surface of the stroma only in the absence of epithelium. It was concluded that the barrier characteristics of corneal epithelium and stroma are different. The cornea has greater resistance to drug permeation into the eye in comparison with the stroma (Mun et al. 2014). Martens and colleagues then thoroughly investigated the NP-obstruction features of the bovine vitreous. In order to prepare the eyes for microscopy, the cornea and lens were withdrawn, NPs were administered via the sclera into the vitreous, and the eyes were incubated for 24 h. Cationic PS NPs of different sizes clumped together in the vitreous and were not beyond the injection site. On the other hand, PS NPs that were negatively charged or PEGylated (2 kDa) could spread far beyond the injection site. NPs coated with anionic ligands have slower diffusion rates than NPs coated with PEG ligands of the same size. After proving that PEGylation makes NPs move more easily in the vitreous, they looked at the differences between PEGylated DNA polyplexes and those that were not. Most PEGylated polyplexes remained stable and dispersed in the vitreous, whereas a large number of noncationic polyplexes formed big, immobile clumps (Martens et al. 2013). Furthermore, PEG-coated NPs were used to examine the pore structure of bovine vitreous as well as length-scale-guided anelastic behavior. Data showed that the pore diameter of bovine vitreous was an average of 500 ± 50 nm and less than 2% of the pores had a diameter greater than 1000 nm. Moreover, highly condensed PEG-coated $CK_{30}PEG_{10k}$ DNA NPs diffused quickly through the vitreous gel in comparison with traditional PEI-based DNA NPs of highly positive surface charge (Xu et al. 2013).

## 18.5 Limitations of PEGylated Nano-Carriers

In the past few years, a lot of researches have been done on how PEG coating can change the biophysical and chemical properties of NPs. Even though there are some benefits to PEGylation, it also has some drawbacks. PEGylation renders the liposomes unable to effectively deliver the nucleic acids. This is because the PEG, as a hydrophilic shield on the surface of liposomes, limits cellular uptake and enhances the integrity of the lipid shell, but the procedure leads to a reduced endosomal escape through membrane fusion and cargo breakdown in lysosomes (Mishra et al. 2004; Remaut et al. 2007). Therefore, the application of PEG in the transportation of genes to tumor cells is described as the "PEG dilemma" (Hatakeyama et al. 2013). Upon systematically repeated administration of PEGylated NPs, their blood clearance rate increases with each subsequent dose, which is called accelerated blood clearance (ABC). ABC occurs as a result of the activation of the immune system. As a result of a single injection of PEGylated liposomes, the splenic B cells produce IgM antibodies against PEG, which attach to the liposomes, stimulating the complement system and increasing the uptake of the liposomes by Kupffer cells. ABC is influenced by a variety of factors, including the cargo inside the NPs, the physiochemical properties of the NPs, and preexisting anti-PEG antibodies. Doxorubicin-loaded liposomes produced less IgM against PEG than empty PEGylated liposomes. According to one hypothesis, doxorubicin's antiproliferative activity caused the death of spleen B cells, therefore inhibiting antibody production. The sequence of siRNA trapped inside PEGylated liposomes determines the production of IgM against PEG, and siRNA containing immune-stimulatory sequences increases this production. In addition, loading nucleic acids with methylated or cytosine-phosphate-guanine (CpG)-free sequences reduced the induction of inflammatory cytokines by NPs. In addition, there is an inverse correlation between the amount of PEGylated liposomes injected in the first administration and the production of anti-PEG antibodies. Increased exposure to PEGylated liposomes may suppress spleen B cells, resulting in immunological tolerance or anergy. Increasing liposome injection as part of the first dose could be an effective strategy for overcoming immunogenicity and accelerating elimination. In recent years, research has shown an increase from 0.2% to 22–25% in the level of anti-PEG antibodies in the bodies of individuals who have never been exposed to PEGylated NPs. This increase may be attributed to the development of PEG diagnostic methods with a high degree of sensitivity or the use of PEG in cosmetics, pharmaceuticals, and food products. Therefore, it is necessary to measure the amount of anti-PEG antibodies in patients prior to treatment with PEGylated NPs. The incidence of ABC and treatment resistance is lower in patients with low levels of anti-PEG antibodies (Suk et al. 2016). The physicochemical properties of NPs can also influence their immunogenicity. It has been shown that chemical bonds between PEG chains and core lipids and the physical and chemical characteristics of liposomes administered in the first and subsequent doses, such as PEG surface density, diameter, and charge distribution, are effective in producing anti-PEG antibodies (Suk et al. 2016).

Polymeric PEG-b-PLA NPs were designed for the transport of poorly water-soluble drugs to the brain bypassing the BBB and thus increasing the bioavailability of the drugs. Investigating the effect of PEG-b-PLA NPs and diethylstilbestrol (DES) on the reproductive system of neonatal female Wistar rats revealed that PEG-b-PLA NPs resulted in an increased luteinizing hormone (LH) release in response to luteinizing hormone releasing hormone (LHRH), indicating a potential disruption of the neuroendocrine system (Scsukova et al. 2015).

## 18.6 Conclusion

Nanomedicine is a field of research that is growing quickly. It looks at how nanotechnology can be used to diagnose and treat debilitating diseases, such as by controlling and directing the delivery of drugs. Nanomedicine has the potential to improve the accuracy of diagnoses and the efficacy of treatments, by allowing drugs to be specifically targeted to areas of the body that need them. This reduces the risk of adverse side effects and increases the effectiveness of the treatment. Naked NPs are poorly water-soluble and accumulate in physiological situations. As a result, they tend to accumulate in tissues and organs, leading to a variety of health problems. Opsonization is the process in which proteins bind to the surface of the NPs, leading to their clearance from the bloodstream. This is a very effective way to get rid of NPs because the proteins recognize them as foreign bodies and start the process of getting rid of them. As a result, the NPs do not have a chance to accumulate in the body. To avoid these problems, NP surfaces are frequently coated with end-grafted polymer moieties. The encapsulation of NPs with polymeric materials such as PLGA, PEG, and surfactant copolymers has been used by numerous researchers in the search for greater stability of the NP system in biological environments. PEG is used in numerous medical and consumer products due to its nontoxicity, neutrality, and biological compatibility. It is also approved by the FDA for use in medical and consumer products, making it a safe and reliable choice for many applications. PEGylation is the process of attaching a polymer known as PEG to a drug molecule. PEGylation helps to protect the drug-containing nano-carrier from being degraded and also increases its solubility, making it easier for it to be absorbed into the bloodstream and crossed by biological barriers such as mucus on the surface of the airways, vaginal, and digestive tracts. PEGylated nano-carriers are capable of delivering CAs directly to specific sites, leading to better contrast resolution and higher contrast-to-noise ratios. This helps to improve the accuracy and specificity of imaging results and can potentially reduce the dose of CA used in the imaging process. The PEGylated nano-carriers can enclose radioprotective agents, which can keep cells from being damaged by radiation. Generally, PEGylation facilitates the transfer of cargo to the target tissue by increasing nano-carriers' stability, improving their biocompatibility, increasing the half-life of nano-carriers in the blood circulation, reducing their immunogenicity, protecting the nano-carriers from degradation by proteases, and reducing their uptake by macrophages. NP PEGylation has been associated with many benefits; however, there are also reports describing its disadvantages. For example, PEGylation renders liposomes unable to deliver nucleic acids effectively. The production of anti-PEG antibodies is another disadvantage that reduces the effectiveness of PEG NPs in repeated doses.

## References

Abdel Fadeel DA, Kamel R, Fadel M (2020) PEGylated lipid nanocarrier for enhancing photodynamic therapy of skin carcinoma using curcumin: in vitro/in vivo studies and histopathological examination. Sci Rep 10(1):10435

Abdel-Moneim A, Ramadan H (2022) Novel strategies to oral delivery of insulin: current progress of nanocarriers for diabetes management. Drug Dev Res 83(2):301–316

Agostinis P et al (2011) Photodynamic therapy of cancer: an update. CA Cancer J Clin 61(4):250–281

Ahmad I et al (2016) Supercritical fluid technology-based trans-resveratrol sln for long circulation and improved radioprotection. J Pharm Innov 11(4):308–322

Alalaiwe A et al (2017) Influence of PEG coating on the oral bioavailability of gold nanoparticles in rats. Drug Deliv 24(1):591–598

Allen T et al (1991) Liposomes containing synthetic lipid derivatives of poly (ethylene glycol) show prolonged circulation half-lives in vivo. Biochim Biophys Acta 1066(1):29–36

Amantea MA et al (1997) Population pharmacokinetics and pharmacodynamics of pegylated-liposomal doxorubicin in patients with AIDS-related Kaposi's sarcoma. Clin Pharmacol Therap 61(3):301–311

Andreani T et al (2014) Preparation and characterization of PEG-coated silica nanoparticles for oral insulin delivery. Int J Pharm 473(1-2):627–635

Bagoria L, Yadav RK, Ratanpal S (2022) Role of nanoparticles as radioprotector. IJCRT 10:28–50

Bakker-Woudenberg IA et al (1992) Enhanced localization of liposomes with prolonged blood circulation time in infected lung tissue. Biochim Biophys Acta 1138(4):318–326

Bakker-Woudenberg I et al (1993) Liposomes with prolonged blood circulation and selective localization in Klebsiella pneumoniae-infected lung tissue. J Infect Dis 168(1):164–171

Balar AV, Weber JS (2017) PD-1 and PD-L1 antibodies in cancer: current status and future directions. Cancer Immunol Immunother 66:551–564

Bandyopadhyay A et al (2011) The impact of nanoparticle ligand density on dendritic-cell targeted vaccines. Biomaterials 32(11):3094–3105

Barazzuol L, Coppes RP, van Luijk P (2020) Prevention and treatment of radiotherapy-induced side effects. Mol Oncol 14(7):1538–1554

Baumann M et al (2016) Radiation oncology in the era of precision medicine. Nat Rev Cancer 16(4):234–249

Bonifaz LC et al (2004) In vivo targeting of antigens to maturing dendritic cells via the DEC-205 receptor improves T cell vaccination. J Exp Med 199(6):815–824

Boylan NJ et al (2012) Highly compacted DNA nanoparticles with low MW PEG coatings: in vitro, ex vivo and in vivo evaluation. J Control Release 157(1):72–79

Brooking J, Davis S, Illum L (2001) Transport of nanoparticles across the rat nasal mucosa. J Drug Target 9(4):267–279

Cao J et al (2020) Surface PEGylation of MIL-101 (Fe) nanoparticles for co-delivery of radioprotective agents. Chem Eng J 384:123363

Cao J et al (2021) Ultrasound-assisted continuous-flow synthesis of PEGylated MIL-101 (Cr) nanoparticles for hematopoietic radioprotection. Mater Sci Eng C 129:112369

Champeau M et al (2019) Photodynamic therapy for skin cancer: how to enhance drug penetration? J Photochem Photobiol B Biol 197:111544

Chao Y et al (2016) Rhenium-188 labeled tungsten disulfide nanoflakes for self-sensitized, near-infrared enhanced radioisotope therapy. Small 12(29):3967–3975

Chaves C et al (2008) Application of core–shell PEGylated CdS/Cd $(OH)_2$ quantum dots as biolabels of Trypanosoma cruzi parasites. Appl Surf Sci 255(3):728–730
Chen HH, Kuo MT (2017) Improving radiotherapy in cancer treatment: promises and challenges. Oncotarget 8(37):62742
Chen H et al (2013a) 'Living' PEGylation on gold nanoparticles to optimize cancer cell uptake by controlling targeting ligand and charge densities. Nanotechnology 24(35):355101
Chen D et al (2013b) Comparative study of Pluronic® F127-modified liposomes and chitosan-modified liposomes for mucus penetration and oral absorption of cyclosporine A in rats. Int J Pharm 449(1-2):1–9
Chen Y et al (2020) Gold nanoparticles as radiosensitizers in cancer radiotherapy. Int J Nanomed 15:9407–9430
Cheng X, Ming X, Croyle MA (2003) PEGylated adenoviruses for gene delivery to the intestinal epithelium by the oral route. Pharm Res 20(9):1444–1451
Cho EC et al (2010) The effects of size, shape, and surface functional group of gold nanostructures on their adsorption and internalization by cells. Small 6(4):517–522
Chow EK-H, Ho D (2013) Cancer nanomedicine: from drug delivery to imaging. Sci Transl Med 5:216
Cone RA (2009) Barrier properties of mucus. Adv Drug Deliv Rev 61(2):75–85
Connor EE et al (2005) Gold nanoparticles are taken up by human cells but do not cause acute cytotoxicity. Small 1(3):325–327
Conte C et al (2018) Enhanced uptake in 2D-and 3D-lung cancer cell models of redox responsive PEGylated nanoparticles with sensitivity to reducing extra-and intracellular environments. J Control Release 277:126–141
Cruz LJ et al (2014) Targeting nanoparticles to CD40, DEC-205 or CD11c molecules on dendritic cells for efficient CD8+ T cell response: a comparative study. J Control Release 192:209–218
Cu Y, Saltzman WM (2009) Controlled surface modification with poly (ethylene) glycol enhances diffusion of PLGA nanoparticles in human cervical mucus. Mol Pharm 6(1):173–181
Cu Y, Booth CJ, Saltzman WM (2011) In vivo distribution of surface-modified PLGA nanoparticles following intravaginal delivery. J Control Release 156(2):258–264
Cui Y et al (2019) Effect of PEGylated magnetic PLGA-PEI nanoparticles on primary hippocampal neurons: reduced nanoneurotoxicity and enhanced transfection efficiency with magnetofection. ACS Appl Mater Interfaces 11(41):38190–38204
das Neves J et al (2015) Polymer-based nanocarriers for vaginal drug delivery. Adv Drug Deliv Rev 92:53–70
Dastidar DG, Ghosh D, Das A (2022) Recent developments in nanocarriers for cancer chemotherapy. Open Nano 8:100080
Debele TA, Yeh C-F, Su W-P (2020) Cancer immunotherapy and application of nanoparticles in cancers immunotherapy as the delivery of immunotherapeutic agents and as the immunomodulators. Cancer 12(12):3773
Desai D et al (2016) Targeted modulation of cell differentiation in distinct regions of the gastrointestinal tract via oral administration of differently PEG-PEI functionalized mesoporous silica nanoparticles. Int J Nanomedicine 11:299
DiSanto RM, Subramanian V, Gu Z (2015) Recent advances in nanotechnology for diabetes treatment. Wiley Interdiscip Rev Nanomed Nanobiotechnol 7(4):548–564
Dobrovolskaia MA et al (2008) Preclinical studies to understand nanoparticle interaction with the immune system and its potential effects on nanoparticle biodistribution. Mol Pharm 5(4):487–495
Dong Z et al (2017) RGD modified and PEGylated lipid nanoparticles loaded with puerarin: formulation, characterization and protective effects on acute myocardial ischemia model. Biomed Pharmacother 89:297–304
Eid HM et al (2019) Development, optimization, and in vitro/in vivo characterization of enhanced lipid nanoparticles for ocular delivery of ofloxacin: the influence of pegylation and chitosan coating. AAPS PharmSciTech 20(5):1–14

Eliyahu H et al (2002) Lipoplex-induced hemagglutination: potential involvement in intravenous gene delivery. Gene Ther 9(13):850–858
Ensign LM et al (2012a) Mucus-penetrating nanoparticles for vaginal drug delivery protect against herpes simplex virus. Sci Transl Med 4(138):138
Ensign LM et al (2012b) Mucus penetrating nanoparticles: biophysical tool and method of drug and gene delivery. Adv Mater 24(28):3887–3894
Feeney OM et al (2014) 'Stealth' lipid-based formulations: poly (ethylene glycol)-mediated digestion inhibition improves oral bioavailability of a model poorly water soluble drug. J Control Release 192:219–227
Fischer D et al (2003) In vitro cytotoxicity testing of polycations: influence of polymer structure on cell viability and hemolysis. Biomaterials 24(7):1121–1131
Forier K et al (2013) Transport of nanoparticles in cystic fibrosis sputum and bacterial biofilms by single-particle tracking microscopy. Nanomedicine 8(6):935–949
Gabizon AA (2006) Applications of liposomal drug delivery systems to cancer therapy. In: Nanotechnology for cancer therapy. CRC Press, Boca Raton, pp 595–611
Gabizon A et al (1994) Prolonged circulation time and enhanced accumulation in malignant exudates of doxorubicin encapsulated in polyethylene-glycol coated liposomes. Cancer Res 54(4):987–992
Gajbhiye K et al (2020) PEGylated nanocarriers: a promising tool for targeted delivery to the brain. Colloids Surf B: Biointerfaces 187:110770
Galindo R et al (2022) Development of peptide targeted PLGA-PEGylated nanoparticles loading licochalcone-A for ocular inflammation. Pharmaceutics 14(2):285
Geng F et al (2014) Pegylated glucose gold nanoparticles for improved in-vivo bio-distribution and enhanced radiotherapy on cervical cancer. J Biomed Nanotechnol 10(7):1205–1216
Gera M et al (2017) Nanoformulations of curcumin: an emerging paradigm for improved remedial application. Oncotarget 8(39):66680
Giannavola C et al (2003) Influence of preparation conditions on acyclovir-loaded poly-d, l-lactic acid nanospheres and effect of PEG coating on ocular drug bioavailability. Pharm Res 20(4):584–590
Grant GA et al (2002) The utility of magnetic resonance imaging in evaluating peripheral nerve disorders. Muscle Nerve 25(3):314–331
Gref R et al (1994) Biodegradable long-circulating polymeric nanospheres. Science 263(5153):1600–1603
Harris JM (1992) Poly (ethylene glycol) chemistry: biotechnical and biomedical applications. Springer, Cham
Hatakeyama H, Akita H, Harashima H (2013) The polyethyleneglycol dilemma: advantage and disadvantage of PEGylation of liposomes for systemic genes and nucleic acids delivery to tumors. Biol Pharm Bull 36(6):892–899
He L et al (2020) Matrix metalloproteinase-responsive PEGylated lipid nanoparticles for controlled drug delivery in the treatment of rheumatoid arthritis. ACS Appl BioMater 3(5):3276–3284
Heyes J et al (2006) Synthesis and characterization of novel poly (ethylene glycol)-lipid conjugates suitable for use in drug delivery. J Control Release 112(2):280–290
Hoffman AS (2016) The early days of PEG and PEGylation (1970s–1990s). Acta Biomater 40:1–5
Holeab C, Păunică M, Curaj A (2017) A complex method of semantic bibliometrics for revealing conceptual profiles and trends in scientific literature. The case of future-oriented technology analysis (FTA) science. Econ Comput Econ Cybernet Stud Res 51(2):23–37
Howard MD et al (2008) PEGylation of nanocarrier drug delivery systems: state of the art. J Biomed Nanotechnol 4(2):133–148
Huang J-Y et al (2013) The effect of lipid nanoparticle PEGylation on neuroinflammatory response in mouse brain. Biomaterials 34(32):7960–7970
Ilbeigi S et al (2023) Sonodynamic therapy of pancreatic cancer cells based on synergistic chemotherapeutic effects of selenium-PEG-curcumin nanoparticles and gemcitabine. Appl Phys A 129(2):82

Immordino ML, Dosio F, Cattel L (2006) Stealth liposomes: review of the basic science, rationale, and clinical applications, existing and potential. Int J Nanomedicine 1(3):297
Inchaurraga L et al (2015) In vivo study of the mucus-permeating properties of PEG-coated nanoparticles following oral administration. Eur J Pharm Biopharm 97:280–289
Jaacks LM et al (2016) Type 2 diabetes: a 21st century epidemic. Best Pract Res Clin Endocrinol Metab 30(3):331–343
Jani P et al (1990) Nanoparticle uptake by the rat gastrointestinal mucosa: quantitation and particle size dependency. J Pharm Pharmacol 42(12):821–826
Järvinen K, Järvinen T, Urtti A (1995) Ocular absorption following topical delivery. Adv Drug Deliv Rev 16(1):3–19
Jin X et al (2015) An ultrasmall and metabolizable PEGylated NaGdF 4: Dy nanoprobe for high-performance T1/T2-weighted MR and CT multimodal imaging. Nanoscale 7(38):15680–15688
Jokerst JV et al (2011) Nanoparticle PEGylation for imaging and therapy. Nanomedicine 6(4):715–728
Kibria G et al (2011) Dual-ligand modification of PEGylated liposomes shows better cell selectivity and efficient gene delivery. J Control Release 153(2):141–148
Kim H et al (2012) Monosaccharide-responsive release of insulin from polymersomes of polyboroxole block copolymers at neutral pH. J Am Chem Soc 134(9):4030–4033
Kim AJ et al (2013) Use of single-site-functionalized PEG dendrons to prepare gene vectors that penetrate human mucus barriers. Angew Chem 125(14):4077–4080
Kirtane AR et al (2021) Nanotechnology approaches for global infectious diseases. Nat Nanotechnol 16(4):369–384
Kojima C et al (2010) X-ray computed tomography contrast agents prepared by seeded growth of gold nanoparticles in PEGylated dendrimer. Nanotechnology 21(24):245104
Kumar R et al (2013) Third generation gold nanoplatform optimized for radiation therapy. Transl Cancer Res 2(4):228–239
Kumar R et al (2015) Study the effects of PLGA-PEG encapsulated amphotericin B nanoparticle drug delivery system against Leishmania donovani. Drug Deliv 22(3):383–388
Kurosaki T et al (2010) Chondroitin sulfate capsule system for efficient and secure gene delivery. J Pharm Pharm Sci 13(3):351–361
Lahooti A et al (2017) PEGylated superparamagnetic iron oxide nanoparticles labeled with $^{68}$Ga as a PET/MRI contrast agent: a biodistribution study. J Radioanal Nucl Chem 311(1):769–774
Lai SK et al (2007) Rapid transport of large polymeric nanoparticles in fresh undiluted human mucus. Proc Natl Acad Sci 104(5):1482–1487
Lai SK, Wang Y-Y, Hanes J (2009) Mucus-penetrating nanoparticles for drug and gene delivery to mucosal tissues. Adv Drug Deliv Rev 61(2):158–171
Lai SK et al (2010) Nanoparticles reveal that human cervicovaginal mucus is riddled with pores larger than viruses. Proc Natl Acad Sci 107(2):598–603
Lai SK et al (2011) Drug carrier nanoparticles that penetrate human chronic rhinosinusitis mucus. Biomaterials 32(26):6285–6290
Lao J et al (2013) Liposomal doxorubicin in the treatment of breast cancer patients: a review. J Drug Deliv 2013:456409
Lautenschläger C et al (2013) PEG-functionalized microparticles selectively target inflamed mucosa in inflammatory bowel disease. Eur J Pharm Biopharm 85(3):578–586
Leal J, Smyth HD, Ghosh D (2017) Physicochemical properties of mucus and their impact on transmucosal drug delivery. Int J Pharm 532(1):555–572
Li L, Mu L, Torchilin V (2007) Mixed micelles made of pluronic and PEG-PE for solubilization of poorly soluble anticancer drug paclitaxel. J Drug Deliv Sci Technol 17(6):389–392
Li Y, Kröger M, Liu WK (2015a) Shape effect in cellular uptake of PEGylated nanoparticles: comparison between sphere, rod, cube and disk. Nanoscale 7(40):16631–16646
Li H et al (2015b) PEGylated ceria nanoparticles used for radioprotection on human liver cells under γ-ray irradiation. Free Radic Biol Med 87:26–35
Li Y et al (2016) A mini-review for cancer immunotherapy: molecular understanding of PD-1/PD-L1 pathway & translational blockade of immune checkpoints. Int J Mol Sci 17(7):1151

Li J et al (2018) One-pot preparation of hydrophilic manganese oxide nanoparticles as T1 nano-contrast agent for molecular magnetic resonance imaging of renal carcinoma in vitro and in vivo. Biosens Bioelectron 102:1–8

Linam J, Yang L-X (2015) Recent developments in radiosensitization. Anticancer Res 35(5):2479–2485

Liu Z et al (2012) Hybrid mesoporous gadolinium oxide nanorods: a platform for multimodal imaging and enhanced insoluble anticancer drug delivery with low systemic toxicity. J Mater Chem 22(30):14982–14990

Liu Z et al (2013) PEGylated hybrid ytterbia nanoparticles as high-performance diagnostic probes for in vivo magnetic resonance and X-ray computed tomography imaging with low systemic toxicity. Nanoscale 5(10):4252–4261

Liu B et al (2014a) Gastrointestinal imaging-practical magnetic resonance imaging approach. World J Radiol 6(8):544

Liu H et al (2014b) Synthesis of PEGylated low generation dendrimer-entrapped gold nanoparticles for CT imaging applications. Nanoscale 6(9):4521–4526

Liu Y et al (2018a) Metal-based nanoenhancers for future radiotherapy: radiosensitizing and synergistic effects on tumor cells. Theranostics 8(7):1824

Liu N et al (2018b) Fabrication of engineered nanoparticles on biological macromolecular (PEGylated chitosan) composite for bio-active hydrogel system in cardiac repair applications. Int J Biol Macromol 117:553–558

Liu Y et al (2021a) PEGylated water-soluble organic nanoparticles with chlorin derivative for biocompatible photodynamic therapy in vitro and in vivo. ChemNanoMat 7(2):165–173

Liu J et al (2021b) PEGylated Mn containing MOF nanoparticles for potential immunotherapy of pancreatic cancer via manganese induced activation of anti-tumor immunity. Colloid Interface Sci Commun 42:100409

Liu P, Chen G, Zhang J (2022) A review of liposomes as a drug delivery system: current status of approved products, regulatory environments, and future perspectives. Molecules 27(4):1372

Luo Y et al (2015) RGD-functionalized ultrasmall iron oxide nanoparticles for targeted T 1-weighted MR imaging of gliomas. Nanoscale 7(34):14538–14546

Maddah M, Delavari H, Mehravi B (2019) Preparation of bio-inspired melanin nanoplatforms chelated with manganese ions as a potential T1 MRI contrast agent. ChemistrySelect 4(19):5860–5865

Majd MH et al (2013) Tamoxifen loaded folic acid armed PEGylated magnetic nanoparticles for targeted imaging and therapy of cancer. Colloids Surf B: Biointerfaces 106:117–125

Martens TF et al (2013) Measuring the intravitreal mobility of nanomedicines with single-particle tracking microscopy. Nanomedicine 8(12):1955–1968

Mastorakos P et al (2015a) Highly compacted biodegradable DNA nanoparticles capable of overcoming the mucus barrier for inhaled lung gene therapy. Proc Natl Acad Sci 112(28):8720–8725

Mastorakos P et al (2015b) Highly PEGylated DNA nanoparticles provide uniform and widespread gene transfer in the brain. Adv Healthc Mater 4(7):1023–1033

Matsumura Y, Maeda H (1986) A new concept for macromolecular therapeutics in cancer chemotherapy: mechanism of tumoritropic accumulation of proteins and the antitumor agent smancs. Cancer Res 46(12):6387–6392

Minimol P, Paul W, Sharma CP (2013) PEGylated starch acetate nanoparticles and its potential use for oral insulin delivery. Carbohydr Polym 95(1):1–8

Mishra S, Webster P, Davis ME (2004) PEGylation significantly affects cellular uptake and intracellular trafficking of non-viral gene delivery particles. Eur J Cell Biol 83(3):97–111

Mishra P, Nayak B, Dey R (2016) PEGylation in anti-cancer therapy: an overview. Asian J Pharm Sci 11(3):337–348

Moghimi SM, Hunter AC, Murray JC (2001) Long-circulating and target-specific nanoparticles: theory to practice. Pharmacol Rev 53(2):283–318

Mousa SA, Bharali DJ (2011) Nanotechnology-based detection and targeted therapy in cancer: nano-bio paradigms and applications. Cancer 3(3):2888–2903

Mousavi M et al (2017) Enhancement of radiosensitivity of melanoma cells by pegylated gold nanoparticles under irradiation of megavoltage electrons. Int J Radiat Biol 93(2):214–221
Mun EA et al (2014) On the barrier properties of the cornea: a microscopy study of the penetration of fluorescently labeled nanoparticles, polymers, and sodium fluorescein. Mol Pharm 11(10):3556–3564
Nadhman A et al (2014) PEGylated silver doped zinc oxide nanoparticles as novel photosensitizers for photodynamic therapy against Leishmania. Free Radic Biol Med 77:230–238
Nafee N et al (2014) Antibiotic-free nanotherapeutics: ultra-small, mucus-penetrating solid lipid nanoparticles enhance the pulmonary delivery and anti-virulence efficacy of novel quorum sensing inhibitors. J Control Release 192:131–140
Nam J et al (2019) Cancer nanomedicine for combination cancer immunotherapy. Nat Rev Mater 4(6):398–414
Nance EA et al (2012) A dense poly (ethylene glycol) coating improves penetration of large polymeric nanoparticles within brain tissue. Sci Transl Med 4:149
Nance E et al (2014) Brain-penetrating nanoparticles improve paclitaxel efficacy in malignant glioma following local administration. ACS Nano 8(10):10655–10664
Natesan S et al (2017) Co-encapsulated resveratrol and quercetin in chitosan and peg modified chitosan nanoparticles: for efficient intra ocular pressure reduction. Int J Biol Macromol 104:1837–1845
Nativo P, Prior IA, Brust M (2008) Uptake and intracellular fate of surface-modified gold nanoparticles. ACS Nano 2(8):1639–1644
Nel AE et al (2009) Understanding biophysicochemical interactions at the nano–bio interface. Nat Mater 8(7):543–557
Newman MS et al (1999) Comparative pharmacokinetics, tissue distribution, and therapeutic effectiveness of cisplatin encapsulated in long-circulating, pegylated liposomes (SPI-077) in tumor-bearing mice. Cancer Chemother Pharmacol 43:1–7
Nguyen KT et al (2013) Nanomaterials for photo-based diagnostic and therapeutic applications. Theranostics 3(3):152–166
Nguyen A et al (2020) Utilization of click chemistry to study the effect of poly (ethylene) glycol molecular weight on the self-assembly of PEGylated gambogic acid nanoparticles for the treatment of rheumatoid arthritis. Biomater Sci 8(16):4626–4637
Ni D et al (2016) PEGylated NaHoF4 nanoparticles as contrast agents for both X-ray computed tomography and ultra-high field magnetic resonance imaging. Biomaterials 76:218–225
O'Riordan CR et al (1999) PEGylation of adenovirus with retention of infectivity and protection from neutralizing antibody in vitro and in vivo. Hum Gene Ther 10(8):1349–1358
Ochyl LJ et al (2018) PEGylated tumor cell membrane vesicles as a new vaccine platform for cancer immunotherapy. Biomaterials 182:157–166
Oh E et al (2011) Cellular uptake and fate of PEGylated gold nanoparticles is dependent on both cell-penetration peptides and particle size. ACS Nano 5(8):6434–6448
Papi M et al (2017) Clinically approved PEGylated nanoparticles are covered by a protein corona that boosts the uptake by cancer cells. Nanoscale 9(29):10327–10334
Pawar H, Douroumis D, Boateng JS (2012) Preparation and optimization of PMAA–chitosan–PEG nanoparticles for oral drug delivery. Colloids Surf B: Biointerfaces 90:102–108
Peng C et al (2012) PEGylated dendrimer-entrapped gold nanoparticles for in vivo blood pool and tumor imaging by computed tomography. Biomaterials 33(4):1107–1119
Peng C et al (2013) Targeted tumor CT imaging using folic acid-modified PEGylated dendrimer-entrapped gold nanoparticles. Polym Chem 4(16):4412–4424
Perrault SD et al (2009) Mediating tumor targeting efficiency of nanoparticles through design. Nano Lett 9(5):1909–1915
Petersen H et al (2002) Synthesis, characterization, and biocompatibility of polyethylenimine-g raft-poly (ethylene glycol) block copolymers. Macromolecules 35(18):6867–6874
Pisal DS, Kosloski MP, Balu-Iyer SV (2010) Delivery of therapeutic proteins. J Pharm Sci 99(6):2557–2575

Pramual S et al (2017) Polymer-lipid-PEG hybrid nanoparticles as photosensitizer carrier for photodynamic therapy. J Photochem Photobiol B Biol 173:12–22
Pujade-Lauraine E et al (2018) Avelumab (anti-PD-L1) in platinum-resistant/refractory ovarian cancer: JAVELIN Ovarian 200 Phase III study design. Future Oncol 14(21):2103–2113
Qin X et al (2021) Targeting the resolution pathway of inflammation using Ac2–26 peptide-loaded PEGylated lipid nanoparticles for the remission of rheumatoid arthritis. Asian J Pharm Sci 16(4):483–493
Ramachandran R, Paul W, Sharma CP (2009) Synthesis and characterization of PEGylated calcium phosphate nanoparticles for oral insulin delivery. J Biomed Mater Res B Appl Biomater 88(1):41–48
Ramôa AM et al (2023) Antimicrobial peptide-grafted PLGA-PEG nanoparticles to fight bacterial wound infections. Biomater Sci 11:499–508
Remaut K et al (2007) Pegylation of liposomes favours the endosomal degradation of the delivered phosphodiester oligonucleotides. J Control Release 117(2):256–266
Robinson JT et al (2011) Ultrasmall reduced graphene oxide with high near-infrared absorbance for photothermal therapy. J Am Chem Soc 133(17):6825–6831
Rostamizadeh K et al (2015) A hybrid modeling approach for optimization of PMAA–chitosan–PEG nanoparticles for oral insulin delivery. RSC Adv 5(85):69152–69160
Sacchetti C et al (2013) In vivo targeting of intratumor regulatory T cells using PEG-modified single-walled carbon nanotubes. Bioconjug Chem 24(6):852–858
Salim M et al (2007) Non-fouling microfluidic chip produced by radio frequency tetraglyme plasma deposition. Lab Chip 7(4):523–525
Salim M et al (2014) Amphiphilic designer nano-carriers for controlled release: from drug delivery to diagnostics. Med Chem Commun 5(11):1602–1618
Sanders NN et al (2007) Wanted and unwanted properties of surface PEGylated nucleic acid nanoparticles in ocular gene transfer. J Control Release 122(3):226–235
Sarhadi S et al (2022) B12-functionalized PEGylated liposomes for the oral delivery of insulin: in vitro and in vivo studies. J Drug Deliv Sci Technol 69:103141
Schopf LR et al (2015) Topical ocular drug delivery to the back of the eye by mucus-penetrating particles. Transl Vis Sci Technol 4(3):11–11
Schuster BS et al (2013) Nanoparticle diffusion in respiratory mucus from humans without lung disease. Biomaterials 34(13):3439–3446
Scsukova S et al (2015) Effect of polymeric nanoparticle poly (ethylene glycol)-block-poly (lactic acid)(PEG-b-PLA) on in vitro luteinizing hormone release from anterior pituitary cells of infantile and adult female rats. Neuro Endocrinol Lett 36:88–94
Shan J et al (2011) Pegylated composite nanoparticles containing upconverting phosphors and meso-tetraphenyl porphine (TPP) for photodynamic therapy. Adv Funct Mater 21(13):2488–2495
Shao M, Yang W, Han G (2017) Protective effects on myocardial infarction model: delivery of schisandrin B using matrix metalloproteinase-sensitive peptide-modified, PEGylated lipid nanoparticles. Int J Nanomedicine 12:7121
Shi Y (2020) Clinical translation of nanomedicine and biomaterials for cancer immunotherapy: progress and perspectives. Adv Therap 3(9):1900215
Siegel RA, Kirtane AR, Panyam J (2016) Assessing the benefits of drug delivery by nanocarriers: a partico/pharmacokinetic framework. IEEE Trans Biomed Eng 64(9):2176–2185
Solas M et al (2018) Pegylated nanoparticles for the oral delivery of nimodipine: pharmacokinetics and effect on the anxiety and cognition in mice. Int J Pharm 543(1-2):245–256
Song G et al (2016) Perfluorocarbon-loaded hollow $Bi_2Se_3$ nanoparticles for timely supply of oxygen under near-infrared light to enhance the radiotherapy of cancer. Adv Mater 28(14):2716–2723
Song G et al (2017) Emerging nanotechnology and advanced materials for cancer radiation therapy. Adv Mater 29(32):1700996
Stirland DL et al (2013) Mind the gap: a survey of how cancer drug carriers are susceptible to the gap between research and practice. J Control Release 172(3):1045–1064
Sui X et al (2015) The anticancer immune response of anti-PD-1/PD-L1 and the genetic determinants of response to anti-PD-1/PD-L1 antibodies in cancer patients. Oncotarget 6(23):19393

Suk JS et al (2009) The penetration of fresh undiluted sputum expectorated by cystic fibrosis patients by non-adhesive polymer nanoparticles. Biomaterials 30(13):2591–2597

Suk JS et al (2014) Lung gene therapy with highly compacted DNA nanoparticles that overcome the mucus barrier. J Control Release 178:8–17

Suk JS et al (2016) PEGylation as a strategy for improving nanoparticle-based drug and gene delivery. Adv Drug Deliv Rev 99:28–51

Sun B et al (2020) RGD-PEG-PLA delivers MiR-133 to infarct lesions of acute myocardial infarction model rats for cardiac protection. Pharmaceutics 12(6):575

Suzuki T et al (2012) Accelerated blood clearance of PEGylated liposomes containing doxorubicin upon repeated administration to dogs. Int J Pharm 436(1-2):636–643

Tai W et al (2014) Bio-inspired synthetic nanovesicles for glucose-responsive release of insulin. Biomacromolecules 15(10):3495–3502

Tang L et al (2014) Investigating the optimal size of anticancer nanomedicine. Proc Natl Acad Sci 111(43):15344–15349

Thorne RG, Nicholson C (2006) In vivo diffusion analysis with quantum dots and dextrans predicts the width of brain extracellular space. Proc Natl Acad Sci 103(14):5567–5572

Tobio M et al (1998) Stealth PLA-PEG nanoparticles as protein carriers for nasal administration. Pharm Res 15(2):270–275

Tobıo M et al (2000) The role of PEG on the stability in digestive fluids and in vivo fate of PEG-PLA nanoparticles following oral administration. Colloids Surf B: Biointerfaces 18(3-4):315–323

Vasconcelos A et al (2015) Conjugation of cell-penetrating peptides with poly (lactic-co-glycolic acid)-polyethylene glycol nanoparticles improves ocular drug delivery. Int J Nanomedicine 10:609

Vernooij E, Kettenes-van Den Bosch J, Crommelin D (2002) Fourier transform infrared spectroscopic determination of the hydrolysis of poly (ethylene glycol)–phosphatidylethanolamine-containing liposomes. Langmuir 18(9):3466–3470

Vila A et al (2004) Transport of PLA-PEG particles across the nasal mucosa: effect of particle size and PEG coating density. J Control Release 98(2):231–244

Vimal SK et al (2022) In vivo and in silico investigations of the pegylated gold nanoparticle treatment of amyotrophic lateral sclerosis in mice. New J Chem 46(25):12252–12264

Vonarbourg A et al (2006) Parameters influencing the stealthiness of colloidal drug delivery systems. Biomaterials 27(24):4356–4373

Wang Q, Wu X (2017) Primary and acquired resistance to PD-1/PD-L1 blockade in cancer treatment. Int Immunopharmacol 46:210–219

Wang YY et al (2008) Addressing the PEG mucoadhesivity paradox to engineer nanoparticles that "slip" through the human mucus barrier. Angew Chem Int Ed 47(50):9726–9729

Wang B et al (2009) Glucose-responsive micelles from self-assembly of poly (ethylene glycol)-b-poly (acrylic acid-co-acrylamidophenylboronic acid) and the controlled release of insulin. Langmuir 25(21):12522–12528

Wang G et al (2013) Multifunctional $Fe_3O_4$/graphene oxide nanocomposites for magnetic resonance imaging and drug delivery. Mater Chem Phys 141(2-3):997–1004

Wang Q et al (2016) Targeted delivery of low-dose dexamethasone using PCL–PEG micelles for effective treatment of rheumatoid arthritis. J Control Release 230:64–72

Wang L et al (2019) PEGylated reduced-graphene oxide hybridized with $Fe_3O_4$ nanoparticles for cancer photothermal-immunotherapy. J Mater Chem B 7(46):7406–7414

Wang H et al (2022) PEGylated Prussian blue nanoparticles for modulating polyethyleneimine cytotoxicity and attenuating tumor hypoxia for dual-enhanced photodynamic therapy. J Mater Chem B 10(28):5410–5421

Xie W et al (2018a) Doxorubicin-loaded $Fe_3O_4$@MoS 2-PEG-2DG nanocubes as a theranostic platform for magnetic resonance imaging-guided chemo-photothermal therapy of breast cancer. Nano Res 11:2470–2487

Xie J et al (2018b) Application of multifunctional nanomaterials in radioprotection of healthy tissues. Adv Healthc Mater 7(20):1800421

Xiong MH et al (2012a) Bacteria-responsive multifunctional nanogel for targeted antibiotic delivery. Adv Mater 24(46):6175–6180
Xiong M-H et al (2012b) Lipase-sensitive polymeric triple-layered nanogel for "on-demand" drug delivery. J Am Chem Soc 134(9):4355–4362
Xu Q et al (2013) Nanoparticle diffusion in, and microrheology of, the bovine vitreous ex vivo. J Control Release 167(1):76–84
Xu Q et al (2015) Impact of surface polyethylene glycol (PEG) density on biodegradable nanoparticle transport in mucus ex vivo and distribution in vivo. ACS Nano 9(9):9217–9227
Yang Q, Lai SK (2015) Anti-PEG immunity: emergence, characteristics, and unaddressed questions. Wiley Interdiscip Rev Nanomed Nanobiotechnol 7(5):655–677
Yang K et al (2011) In vivo pharmacokinetics, long-term biodistribution, and toxicology of PEGylated graphene in mice. ACS Nano 5(1):516–522
Yang X et al (2012) Immobilization of pseudorabies virus in porcine tracheal respiratory mucus revealed by single particle tracking. PLoS One 7(12):e51054
Yang K et al (2013a) Nano-graphene in biomedicine: theranostic applications. Chem Soc Rev 42(2):530–547
Yang Q et al (2013b) Accelerated drug release and clearance of PEGylated epirubicin liposomes following repeated injections: a new challenge for sequential low-dose chemotherapy. Int J Nanomedicine 8:1257–1268
Yang Z-Y et al (2014) Alendronate as a robust anchor for ceria nanoparticle surface coating: facile binding and improved biological properties. RSC Adv 4(104):59965–59969
Yang K et al (2015) FeS nanoplates as a multifunctional nano-theranostic for magnetic resonance imaging guided photothermal therapy. Biomaterials 38:1–9
Yang X et al (2016) Improvement of the in vitro safety profile and cytoprotective efficacy of amifostine against chemotherapy by PEGylation strategy. Biochem Pharmacol 108:11–21
Yang M et al (2021) PLGA-PEG nanoparticles facilitate in vivo anti-Alzheimer's effects of fucoxanthin, a marine carotenoid derived from edible brown algae. J Agric Food Chem 69(34):9764–9777
Yasui H et al (2014) Radiosensitization of tumor cells through endoplasmic reticulum stress induced by PEGylated nanogel containing gold nanoparticles. Cancer Lett 347(1):151–158
Yi X et al (2016) Core–shell Au@ MnO2 nanoparticles for enhanced radiotherapy via improving the tumor oxygenation. Nano Res 9(11):3267–3278
Yu T et al (2015) Liposome-based mucus-penetrating particles (MPP) for mucosal theranostics: demonstration of diamagnetic chemical exchange saturation transfer (diaCEST) magnetic resonance imaging (MRI). Nanomedicine 11(2):401–405
Yuan H et al (2013) Improved transport and absorption through gastrointestinal tract by PEGylated solid lipid nanoparticles. Mol Pharm 10(5):1865–1873
Zalipsky S (1995) Functionalized poly (ethylene glycols) for preparation of biologically relevant conjugates. Bioconjug Chem 6(2):150–165
Zhan X, Tran KK, Shen H (2012) Effect of the poly (ethylene glycol)(PEG) density on the access and uptake of particles by antigen-presenting cells (APCs) after subcutaneous administration. Mol Pharm 9(12):3442–3451
Zhang X-D et al (2012) Size-dependent radiosensitization of PEG-coated gold nanoparticles for cancer radiation therapy. Biomaterials 33(27):6408–6419
Zhang M et al (2013) Graphene oxide based theranostic platform for T 1-weighted magnetic resonance imaging and drug delivery. ACS Appl Mater Interfaces 5(24):13325–13332
Zhang S, Wang J, Pan J (2016) Baicalin-loaded PEGylated lipid nanoparticles: characterization, pharmacokinetics, and protective effects on acute myocardial ischemia in rats. Drug Deliv 23(9):3696–3703
Zhang X et al (2020) Surface functionalization of pegylated gold nanoparticles with antioxidants suppresses nanoparticle-induced oxidative stress and neurotoxicity. Chem Res Toxicol 33(5):1195–1205

Zhou B et al (2014) Synthesis and characterization of PEGylated polyethylenimine-entrapped gold nanoparticles for blood pool and tumor CT imaging. ACS Appl Mater Interfaces 6(19):17190–17199

Zhou B et al (2016a) PEGylated polyethylenimine-entrapped gold nanoparticles loaded with gadolinium for dual-mode CT/MR imaging applications. Nanomedicine 11(13):1639–1652

Zhou B et al (2016b) PEGylated polyethylenimine-entrapped gold nanoparticles modified with folic acid for targeted tumor CT imaging. Colloids Surf B: Biointerfaces 140:489–496

Zhou Z et al (2016c) Reactive oxygen species generating systems meeting challenges of photodynamic cancer therapy. Chem Soc Rev 45(23):6597–6626

Zhou L et al (2022) Nano drug delivery system for tumor immunotherapy: next-generation therapeutics. Front Oncol 12:864301

Zhu S et al (2013) Targeting of tumor-associated macrophages made possible by PEG-sheddable, mannose-modified nanoparticles. Mol Pharm 10(9):3525–3530

Ziemba B et al (2012) Influence of dendrimers on red blood cells. Cell Mol Biol Lett 17(1):21–35